Linear Algebra with Applications

third edition

Linear Algebra
with Applications

third edition

Gareth Williams
Stetson University

WCB **Wm. C. Brown Publishers**

Dubuque, IA Bogota Boston Buenos Aires Caracas Chicago
Guilford, CT London Madrid Mexico City Sydney Toronto

Book Team

Developmental Editor   *Daryl Bruflodt*
Cover Designer   *Kaye Farmer*

WCB Wm. C. Brown Publishers

President and Chief Executive Officer   *Beverly Kolz*
Vice President, Publisher   *Earl McPeek*
Vice President, Director of Sales and Marketing   *Virginia S. Moffat*
Vice President, Director of Production   *Colleen A. Yonda*
National Sales Manager   *Douglas J. DiNardo*
Marketing Manager   *Julie Joyce Keck*
Advertising Manager   *Janelle Keeffer*
Production Editorial Manager   *Renée Menne*
Publishing Services Manager   *Karen J. Slaght*
Royalty/Permissions Manager   *Connie Allendorf*

A Times Mirror Company

Interior design, production, composition, and illustrations by Jonathan Peck Typographers, Ltd.

Copyedited by *Patricia Steele*

Cover credit photo courtesy of Digital Stock Corporation, Space and Space Flight CD

Printed in the United States of America by Times Mirror Higher Education Group, Inc.,
2460 Kerper Boulevard, Dubuque, IA 52001

10 9 8 7 6 5 4 3 2 1

Contents

---

* Sections marked with an asterisk are optional. The instructor can use these sections to build around the core material to give the course the desired flavor.

MATLAB Discussions

Preface

CUPM, in its recommendations for *Reshaping College Mathematics*,[1] points out that linear algebra is at least as fundamental as calculus for a large part of modern applied mathematics. Linear algebra is a prerequisite for many other courses. The report suggests that while abstraction should remain a valuable purpose of the introductory linear algebra course, an important goal should be to emphasize applications and computational methods, opening up the course to the large group of students who need to use linear algebra.

This book is designed for an introductory course in linear algebra based on these recommendations. It is a flexible blend of theory, important computational techniques, and interesting applications. Instructors can select the topics that give the course the desired perspective and include other areas as general reading assignments to give students an all-around feel for the field.

The Aims

- To provide a solid foundation in the mathematics of linear algebra.
- To introduce some of the important computational aspects of the field. It is important, for example, that students be able to understand and analyze algorithms.
- To discuss interesting applications so that students may know when and how to apply linear algebra. Applications are taken from such areas as archaeology, coding theory, fractal geometry, and relativity.

The Arrangement   The book is arranged around a core of twenty-eight sections. Some of these sections contain discussions of numerical techniques and applications. A **theoretical** course can be based on these core sections. My hope is that even students in a course with a theoretical bias will be exposed to some computational considerations and applications. A **computational** course or **applications**-oriented course can touch lightly on parts of the theory in these core sections and include a variety of topics from the additional optional sections.

---

[1] *Reshaping College Mathematics*, a project of the Committee on the Undergraduate Program in Mathematics, MAA Notes Number 13, page 21, The Mathematical Association of America, 1989.

The Mathematics   Linear algebra is a central subject in undergraduate mathematics. Many important topics must be included in this course. Linear dependence, basis, inner product, and linear transformation must be covered carefully. Not only are such topics important in linear algebra; they are usually prerequisites for courses such as differential equations. This book gives a great deal of attention to presenting the "standard" linear algebra topics.

This course is often the first course in abstract mathematics for the student. The student should not be overwhelmed with proofs, but should nevertheless be taught how to prove theorems. The proofs of some theorems are given, when those proofs are instructive. Other proofs are given in outline form, while some proofs have been omitted. Students should be introduced slowly and carefully to the art of developing and writing proofs. This is at the heart of mathematics. The student should be trained to think "mathematically." For example, the idea of "if and only if" is extremely important in mathematics. It arises very naturally in linear algebra. We discuss this concept when it first arises, and the reader meets it again on a number of occasions.

One reason that linear algebra is an appropriate course to introduce abstract mathematical thinking is that much of the material has geometrical interpretation. The student can visualize results. Conversely, linear algebra helps develop the geometrical instinct of the student. Geometry and algebra go hand-in-hand in this course. The process of starting with known results in two-dimensional space and generalizing to n-dimensions arises naturally. I have attempted to teach the art of developing axioms.

Computation   While linear algebra has its elegant abstract side, it also has its numerical side. **Algorithm** is a word that the student should feel comfortable with by the end of the course. The algorithms of Gaussian elimination and Gauss-Jordan elimination, for example, are compared. The student participates in the process of determining exactly where one algorithm gains over the other. The Strassen fast matrix multiplication algorithm is discussed. The reader should be aware of ongoing research into finding faster algorithms for handling matrices.

The **computer** is often used when linear algebra is applied. The reader has the opportunity to think in terms of implementing the mathematics on the computer. The student is asked to predict the output if a random 3×3 matrix is entered into a Gauss-Jordan elimination program on a computer. Round-off error on the computer is introduced. The student is asked whether a computer is more likely to give a solution to a system of linear equations that has no solution, or vice versa. The computer is forcing us to rethink the way this subject should be taught.

While the course may be taught in the above "computer conscious" manner without actually using a computer, linear algebra software packages for both IBM and Macintosh computers, that go along with the text, are available for instructors who wish to integrate the computer into the class. The packages include not only computational programs such as Gauss-Jordan elimination with an all-steps option, but also applications such as digraphs, Markov chains, and a simulated space-time voyage.

Applications   Linear algebra is a subject of immense breadth. Its spectrum ranges from the abstract through numerical techniques to innumerable applications. In this book I have attempted to give the reader a glimpse of many interesting applications.

These applications range from such theoretical applications as the use of linear algebra in differential equations, difference equations, and least squares analyses to many practical applications in such fields as archaeology, demography, electrical engineering, traffic analysis, fractal geometry, relativity, and history. All such discussions are self-contained.

I have tried to stimulate the interest of the reader by selecting a wide variety of real-life applications. These range from looking at traffic flow through downtown Jacksonville to predicting the weather in Tel Aviv. A discussion of possible reasons for Moscow emerging as the capital of Russia is included. Linear algebra is used in relativity theory to predict that a person traveling at 0.8 the speed of light on a round-trip to Alpha Centauri would age 6 years, while a person on earth would age 10 years. There should be something here to interest most people! I have tried to involve the reader in the applications by exercises that extend the discussions given. Applying mathematics does not come easily to most people. They have to be trained in the art. Where better than in the linear algebra course with its wealth of applications?

New Features of This Edition   The book retains the philosophy of previous editions, namely, integrating mathematics, computation, and applications in a flexible manner. Two major changes involve rearrangement of topics.

*Eigenvalues and Eigenvectors   These are very important topics and should not be squeezed into the last part of the course. Their introduction has now been brought forward to the chapter on the Vector Space $\mathbf{R}^n$—to the earliest possible location. Discussions of bases and dimension lead naturally into talking about eigenspaces. Students now have ample time to master these topics and to apply them. They are used, for example, in discussing long-term trends of certain phenomena and in the diagonalization of matrices.

*Gaussian Elimination   The introduction to Gaussian elimination and the comparison with Gauss-Jordan elimination have been moved back from chapter 1 to chapter 9, the chapter on numerical techniques. Gauss-Jordan elimination remains the method of choice for solving small, uncomplicated systems of linear equations.

*Graphing Calculator   Discussions and exercises for the graphing calculator are included at the ends of appropriate sections in the first three chapters. Many students have experience working with graphing calculators prior to enrolling in linear algebra but are not experienced in performing matrix operations on them. The discussions are intended to introduce the necessary techniques. Upon completion of these materials in chapter 3, the student should be able to apply the techniques wherever desired throughout the remainder of the course.

*MATLAB Discussions   The use of the matrix software package MATLAB, with exercises, is discussed at the end of each chapter. Many schools are now using this software, and numerous summer workshops have been given on its integration into the introductory linear algebra course. The computer exercises should, however, be of general interest, as many can be implemented on any matrix algebra software package or on a graphing calculator.

The MATLAB programs that are included with this text were written by myself and Lisa Coulter. My co-workers in software development in linear algebra are Lisa and my wife Donna. I wish to thank both for their important contributions. I thank Cleve Moler of The MathWorks for the support that he gave in bringing out the IBM version and the National Science Foundation for a curriculum development grant that supported the development of the original MATLAB package.

The Flow of Material

This book contains mathematics with interesting applications integrated into the main body of the text. The student meets the application in its "natural surrounding." The approach that I have used is to develop the mathematics first and then provide applications, allowing the clearest text presentation. However, some instructors may prefer to look ahead with the class to an application and use it to motivate the mathematics. Historically, mathematics has developed through interplay with applications. For example, the analysis of the long-term behavior of a Markov chain model for analyzing population movement between U.S. cities and suburbs can be used to motivate eigenvalues and eigenvectors. This type of approach should not be overdone, but it can be very instructive.

Chapter 1   The reader is led step by step from solving systems of two linear equations to solving general systems. The method of Gauss-Jordan elimination is introduced as the standard method for the course because it is a "clean, uncomplicated" algorithm for the small systems encountered. It is important that the student master the method at this time since it will be used frequently throughout the course. The chapter closes with three optional applications. Fitting a polynomial of degree $n - 1$ to n data points leads to a system of linear equations that has a unique solution. In passing, we note that if the polynomial is of degree greater than $n - 1$, there are many solutions; while if the polynomial is of degree less than $n - 1$, there is in general no solution. (The reader will encounter the least squares problem later in the course.) The analyses of electrical networks and traffic flow give rise to systems that have unique solutions and many solutions. The model for traffic flow is similar to that of electrical networks, but has fewer restrictions, leading to more freedom and thus many solutions in place of a unique solution.

Chapter 2   In the first chapter matrices were used to handle systems of equations. This application motivates the algebraic development of the theory of matrices in this chapter. In addition to the standard development of the properties of matrix addition, scalar multiplication, matrix multiplication, and matrix inverse, interesting discussions of computation are included. The reader will see that the associative property of matrix multiplication enables the amount of computation for multiplying three matrices to be reduced by almost 50%. The fast matrix multiplication algorithm of Strassen is included. A brief but beautiful application of matrices in archaeology that illustrates the importance of matrix multiplication, transpose, and symmetric matrices is included.

The reader can anticipate, for physical reasons, why the product of a matrix and its transpose has to be symmetric, and then arrive at the result mathematically. This is mathematics at its best! The chapter closes with three optional sections on applications that should have broad appeal. The Leontief Input-Output Model in economics is used to analyze the interdependence of industries. Wassily Leontief received a Nobel prize in 1973 for his work in this area. A Markov chain model is used in demography and genetics, and digraphs are used in communication and sociology. Instructors who cannot fit these sections into their formal class schedule should encourage readers to browse through them. All discussions are self-contained.

Chapter 3 Determinants and their properties are introduced as quickly and painlessly as possible. Some proofs are included for the sake of completeness, but these can be skipped if the instructor so desires.

Chapter 4 The groundwork has been laid for a more theoretical turn in this chapter. The structure of the vector space $\mathbf{R}^n$ is developed. The concepts of subspace, linear dependence, basis, and dimension are introduced. These ideas are given geometrical interpretation whenever possible. The section on rank brings together many of the earlier concepts. The reader will see that matrix inverse, determinant, rank, and uniqueness of solutions are related. This chapter closes with an introduction to eigenvalues, eigenvectors, and eigenspaces. Interesting applications include an analysis of long-term behavior in demography and weather prediction.

Chapter 5 This chapter extends the familiar geometry of two-dimensional Euclidean space to n dimensions. The process of gradual generalization is extremely important in mathematics. The introduction of the dot product and then the use of the dot product to develop concepts of norm, angle, and distance that are compatible with the familiar definitions in two dimensions is beautiful and illustrates so well the way mathematics develops. The discussions of cross product and planes and lines in three-space are optional—these are often included in the calculus sequence.

Chapter 6 This chapter is a natural continuation of the thought process developed in the two previous chapters. The algebraic and geometrical structures of the vector space $\mathbf{R}^n$ are extended to spaces of matrices and functions. The axioms of vector spaces and inner product spaces are given. The theory is applied to the problem of approximating functions by polynomials, and the importance of such approximations to computer software is discussed. I could not resist including a discussion of the use of vector space theory to detect errors in codes. The Hamming code, whose elements are vectors over a finite field, is introduced. The reader is also introduced to non-Euclidean geometry, leading to a self-contained discussion of the special relativity model of space-time. Having developed the general inner product space, the reader finds that the framework is not appropriate for the mathematical description of space-time. The positive definite axiom is discarded, opening the door first for the pseudo inner product that is used in special relativity and later for one that describes gravity in general relativity. It is appropriate at this time to discuss the importance of first mastering standard mathematical structures such as inner product spaces, and then to indicate that mathematical research often involves changing the axioms of such standard structures.

Chapter 7   Linear transformations are the structure-preserving mappings of vector spaces. Core topics such as kernel, range, the rank/nullity theorem, and matrix representation are discussed. A great deal of attention has been given to notation in the discussion of matrix representation. Good notation here (as in all mathematics) can make all the difference in the world. It is made clear to the reader that mastering the notation is an important part of mathematical training. Linear transformations, kernel, and range are used to give the reader a geometrical picture of the sets of solutions to systems of linear equations, both homogeneous and nonhomogeneous. I felt that an introduction to affine transformations was most appropriate in this chapter even though they are not linear transformations. The reader is given an introduction to fractal geometry, where affine transformations are used to generate a fractal fern. This chapter contains possibly the most attractive application of the whole course. The concept of matrix pseudoinverse is used to arrive at the least squares solution to a system of over-determined equations. This tool is then used to determine the least squares curve for given data. In this discussion a number of theoretical results come together in a very elegant, powerful way to arrive at a very important computational tool.

Chapter 8   The reader will see that every finite dimensional vector space is isomorphic to $\mathbf{R}^n$, for some value of n, and that linear transformations can be represented by matrices. The central role of eigenvalues and eigenvectors in the diagonalization of matrices is discussed. The use of these diagonalization procedures in working with quadratic forms and difference equations is seen.

Chapter 9   This chapter on numerical techniques is important to the practitioner of linear algebra in today's computing environment. I have included Gaussian elimination, *LU* decomposition, and the Jacobi and Gauss-Seidel iterative methods. The merits of the various methods for solving linear systems are discussed. In addition to discussing the standard topics of round-off error, pivoting, and scaling, I felt it important, and well within the scope of the course, to introduce the concept of ill-conditioning. It is very interesting to return to some of the systems of equations that have arisen in the course and find out how dependable the solutions are! The matrix of coefficients of a least squares problem, for example, is very often a Vandermonde matrix, leading to an ill-conditioned system. The chapter concludes with the iterative method for finding dominant eigenvalues and eigenvectors. This discussion leads very naturally into a discussion of techniques used by geographers to measure the relative accessibility of nodes in a network.

Chapter 10   This final chapter gives the student a brief introduction to the ideas of linear programming. The field, developed by George Dantzig and his associates at the U.S. Department of the Air Force in 1947, and now widely used in industry, has its foundation in linear algebra. Problems are described by systems of linear inequalities. The reader sees how small systems can be solved in a geometrical manner but that large systems are solved by row operations on matrices using the simplex algorithm.

Chapter Features

- Each section begins with a motivating introduction that ties the material to previously learned topics.

- The pace of the book gradually increases. As the student matures mathematically, the explanations become gradually more sophisticated.

- Notation is carefully developed. It is important that notation at this level be standard; but there is some flexibility. Good notation helps understanding; poor notation clouds the picture.

- Boldface terms highlight important concepts and definitions. Much attention has been given to the layout of the text. Readability is vital.

- Many carefully explained examples illustrate the concepts.

- An abundance of exercises strengthen understanding. Initial exercises are usually of a computational nature; later ones become more theoretical in flavor.

- Many, but not all, exercises are based on examples given in the text. It is important that students have the maximum opportunity to develop their creative abilities.

- Review exercises at the end of each chapter have been carefully selected to give the student an overview of materials covered in that chapter.

Supplements

- *Instructor's Solutions Manual.* Contains detailed solutions to all exercises.

- *Student's Solutions Manual* with complete answers to selected exercises.

- Linear algebra stand-alone software packages for IBM and Macintosh computers, with manuals.

- The Linear Algebra with Applications Toolbox. MATLAB m-files.

- The complete textbook, workbook, and notebooks are available for Mathematica.

- Manual and subroutines for Maple.

- Printed Test Item File. This printed test bank contains about 100 questions per chapter. The questions are of various levels of difficulty and are in multiple-choice, free response, matching, and fill-in-the-blank formats.

- WCB Testwriter. This test generator allows you to select test questions from any of the test questions in the printed test bank. The questions may be selected manually or randomly by the computer subject to criteria you control. There are also algorithmically generated questions that provide a virtually limitless supply of questions. Graphics are incorporated and the test generator supports graphing and special math characters on a wide selection of printers. This is available in IBM-compatible and Macintosh versions. A key factor in choosing this system to accompany the text was its high degree of user friendliness.

Acknowledgments

It is a pleasure to acknowledge the help that made this book possible. My deepest thanks go to Wanda J. Mourant, who proofread and constructively criticized the manuscript and put together the instructor's manual. My thanks go to my colleagues Lisa Coulter, Dennis Kletzing, and Michael Branton of Stetson University for their suggestions that made this a better book. A special word of thanks to Lisa also for her work on software development. Many thanks to Giles Maloof of Boise State University, Gregory Davis of the University of Wisconsin–Green Bay, and Sudhir Kumar Goel of Valdosta State University for their very subtle proofreading and error checking. I thank the following reviewers for their constructive comments: David N. Adler, Dowling College; John Krajewski, University of Wisconsin–Eau Claire; Lala B. Krishna, The University of Akron; Steve Ligh, University of Southwestern Louisiana; Giles Wilson Maloof, Boise State University; James Osterburg, University of Cincinnati; Charles J. Searcy, New Mexico Highlands University; Jimmy L. Solomon, Mississippi State University.

I thank the National Science Foundation for grants under its Curriculum Development Program to develop part of the software.

Special thanks also goes to the participants of our Linear Algebra Focus Group at the January 1995 Joint Mathematics meeting in San Francisco: Robby Robson, Oregon State University; William Bauldry, Appalachian State University; Guershon Harel, Purdue University; and Brad Osgood, Stanford University.

A special word of thanks goes to Daryl Bruflodt of Wm. C. Brown Publishers for the manner in which he coordinated the production of this book. It was a pleasure to work with him.

I am as usual grateful to my wife Donna for all her mathematical and computing input and for her continued support in my writing.

Finally, a special mention of all those students who have gone through MS245 and MS345 at Stetson University, field testing this material. Their reaction and input over the years has been invaluable.

Reviewers

Joan E. Bell
Northeastern State University

David Buchthal
University of Akron

Philip E. Buechner
Cowley County Community College

Lisa Carrington
Belleville Area College

Dewey Furness
Ricks College

Sudhir Kumar Goel
Valdosta State University

Curtis Herink
Mercer University

Lala Krishna
University of Akron

Gary Neff
Ohio University–Eastern

Ronald D. Schwartz
Wilkes University

W. D. Wallis
Southern Illinois University

1

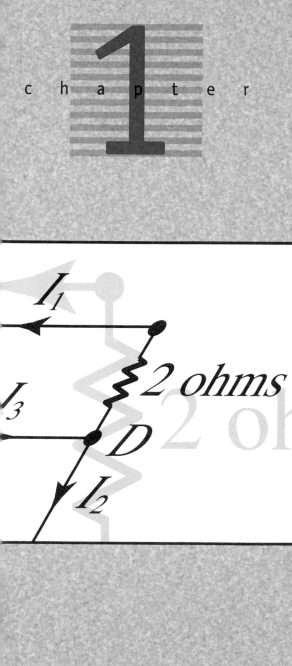

Systems of Linear Equations

Mathematics is a discipline in its own right. It is, however, more than that—it is a tool used in many other fields. Linear algebra, a branch of mathematics that plays a central role in modern mathematics, is also of increasing importance to engineers and physical, social, and behavioral scientists. In this course you will learn mathematics, and you will also be instructed in the art of applying mathematics; the course is a blend of theory, numerical techniques, and applications. You will, for example, extend two-dimensional Euclidean geometry to n dimensions and then generalize the results to arrive at non-Euclidean geometries. You will develop mathematical models—that is, mathematical "pictures" of problems—and see how such descriptions can lead to a deeper understanding of the problems. Applications of mathematics in the aerospace industry, electrical circuits, communication networks, archaeology, weather forecasting, population movement, relativity, and an analysis of trade routes in Russia will be among those discussed. The aim of the course is to impart a knowledge of core areas of linear algebra, to develop a skill in applying mathematics, and, at the same time, to teach the reader to "think mathematically."

When mathematics is used to solve a problem, it often becomes necessary to find a solution to a so-called system of linear equations. Historically linear algebra developed from studying methods for solving such equations. This chapter introduces methods for solving systems of linear equations and includes examples of problems that reduce to solving such equations. The techniques of this chapter will be used throughout the remainder of the text.

1.1   Systems of Two Linear Equations in Two Variables

In this section we discuss methods for solving systems of two equations in two variables. This discussion will give us insights into ways of solving larger systems of equations. Consider the following system of two equations in the two variables x and y.

$$x + 3y = 9$$
$$-2x + y = -4$$

Values of x and y that satisfy both equations form a **solution** to the system.

Finding solutions to such systems of equations is an algebraic problem. However, geometry can guide us in our understanding of such systems. Each of the above equations represents a straight line. Values of x and y that satisfy both equations correspond to a point (x, y) that lies on both lines. We sketch the lines in Figure 1.1. Their intersection represents the solution to the system of equations.

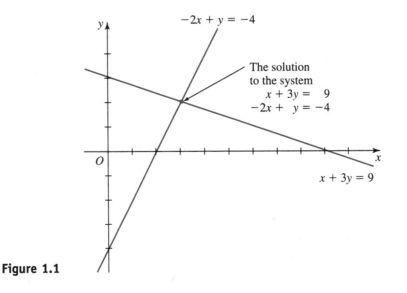

Figure 1.1

Let us now solve the system using algebra. Keep the first equation unchanged and eliminate x from the second equation by adding 2 times the first equation to it.

We get

$$x + 3y = 9$$
$$7y = 14$$

This system has the same solution as the original system, but is a simpler system. It leads to

$$x + 3y = 9$$
$$y = 2$$

Substitute this value of y into the first equation to get $x = 3$. The unique solution to this system of equations is

$$x = 3, \quad y = 2$$

Thus the point where the lines cross in Figure 1.1 is (3, 2).

An equation of the type $x + 3y = 9$, where the variables are of degree one, is called a **linear equation**. The graph of a linear equation in two variables is a line. The previous discussion raises the interesting questions of whether there will always be a solution to a system of two linear equations in two variables and whether there can be many solutions. Let us consider these questions.

Existence of a Solution

In the previous example the solution was a point that lay on both lines. If the equations represent parallel lines, there is no point that lies on both lines; thus there will be no solution. According to this reasoning we would expect that the following system has no solution.

$$-2x + y = 3$$
$$-4x + 2y = 2$$

Each equation represents a straight line of slope 2. See Figure 1.2. Let us see what happens when we try to solve this system with the algebraic approach used in the

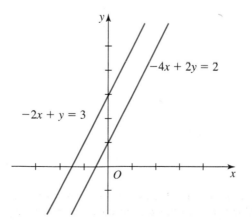

Figure 1.2

previous discussion. Eliminate x from the second equation by adding -2 times the first equation to it. We get the system

$$-2x + y = 3$$
$$0 = -4$$

There are no values of x and y that satisfy the second equation. Thus there is no solution.

Many Solutions

Since two lines can intersect in at most one point, it might appear that there can be at most one solution to a system of two linear equations in two variables. However, the case where both equations correspond to the same line should not be excluded. Any point on that line would satisfy both equations and would thus be a solution. Such a system of equations would have many solutions. Consider the following system of equations.

$$4x - 2y = 6$$
$$6x - 3y = 9$$

Each equation corresponds to a line of slope 2 having y intercept -3. See Figure 1.3.

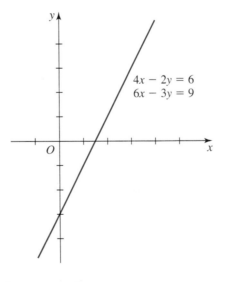

Figure 1.3

Any point on this line is a solution. Again, let us see what happens if we try to solve this system of equations algebraically. Eliminate x from the second equation by adding $-\frac{3}{2}$ times the first equation to it. The system becomes

$$4x - 2y = 6$$
$$0 = 0$$

The implication is that all values of x and y that satisfy the equation $4x - 2y = 6$ satisfy the system of equations. There are many solutions. At this time let us formulate a convenient way of representing such multiple solutions. Solve this first equation for y to get $y = 2x - 3$. Thus if x has the value r, the corresponding value of y in a solution is $2r - 3$. The solutions can be expressed

$$x = r, \quad y = 2r - 3, \qquad r \text{ a real number}$$

This form is said to be the **general solution** to the system of equations. We can generate any number of specific solutions to this system of equations by letting r have various values. For example, we get the following solutions.

$$\text{when } r = 2: \qquad x = 2, \quad y = 1$$
$$\text{when } r = -1: \qquad x = -1, \quad y = -5$$

r is called a **parameter**.

A system of equations that has solutions (one or many) is said to be **consistent**. A system that has no solutions is said to be **inconsistent**.

Example 1   Determine the values of k for which the following system of equations has (a) a unique solution, (b) many solutions.

$$4x + y = 8$$
$$3x + ky = 6$$

Solution   Eliminate x from the second equation by adding $-\frac{3}{4}$ times the first equation to it. The system becomes

$$4x + y = 8$$
$$\left(k - \tfrac{3}{4}\right)y = 0$$

(a)   When $k \neq \frac{3}{4}$, the second equation implies that $y = 0$. The first equation then gives $x = 2$. The system has the unique solution $x = 2$, $y = 0$.

(b)   When $k = \frac{3}{4}$, the second equation is satisfied for all values of y. The system has many solutions, corresponding to points that lie on the line $4x + y = 8$. The solutions are $x = r$, $y = -4r + 8$.

The solutions to the above system of equations depend on the value of the parameter k. There will be times in this course when we want such systems to have many solutions. There will be other times when we will want them to have a unique solution.

In this section we have discussed systems of two linear equations in two variables. Our aim is to analyze larger systems of linear equations.

Unique Solution

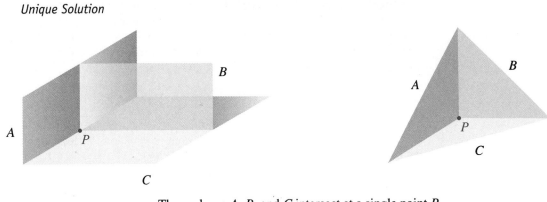

Three planes *A*, *B*, and *C* intersect at a single point *P*.
P corresponds to a unique solution.

No Solutions

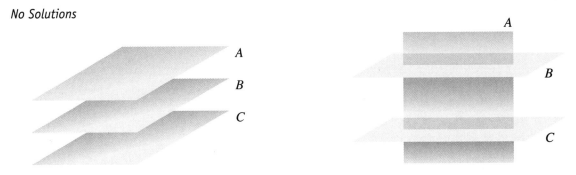

Planes *A*, *B*, and *C* have no points in common.
There are no solutions.

Many Solutions

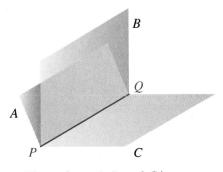
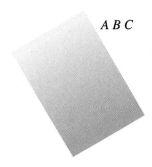

Three planes *A*, *B*, and *C* intersect
in a line *PQ*. Any point on a line
is a solution.

Three equations represent the same
plane. Any point on the plane
is a solution.

Figure 1.4

A **linear equation in n variables** $x_1, x_2, x_3, \ldots, x_n$ has the form

$$a_1 x_1 + a_2 x_2 + a_3 x_3 + \cdots + a_n x_n = b$$

where the coefficients $a_1, a_2, \ldots, a_n$ and b are real numbers. The following is an example of a system of three linear equations.

$$
\begin{aligned}
x_1 + 3x_2 - x_3 &= 4 \\
3x_1 + 2x_2 - 5x_3 &= -2 \\
2x_1 - x_2 + 3x_3 &= 8
\end{aligned}
$$

A linear equation in three variables corresponds to a plane in three-dimensional space. Solutions will be points that lie on all three planes. As for systems of two equations, there can be a unique solution, many solutions, or no solution. We illustrate some of the various possibilities in Figure 1.4.

As the number of variables increases, a geometrical interpretation of such a system of equations becomes increasingly complicated. Each equation will represent a space embedded in a larger space. Solutions will be points that lie on all the embedded spaces. The geometrical approach used in this section to visualize solutions becomes impractical. We have to rely solely on algebraic methods. We shall develop methods for solving larger systems of linear equations, based on the ideas that we have introduced in this section.

Exercise Set 1.1

1. Solve (if possible) each of the following systems of equations.

   *(a) $\begin{aligned} x + y &= 1 \\ x - y &= 2 \end{aligned}$

   (b) $\begin{aligned} 3x + 2y &= 0 \\ x - y &= 2 \end{aligned}$

   *(c) $\begin{aligned} 3x - 6y &= 1 \\ x - 2y &= 2 \end{aligned}$

   *(d) $\begin{aligned} x + 2y &= 4 \\ 2x + 4y &= 8 \end{aligned}$

   (e) $\begin{aligned} x + y &= 7 \\ x + 2y &= 3 \end{aligned}$

2. Solve (if possible) each of the following systems of equations.

   *(a) $\begin{aligned} 3x - 4y &= 2 \\ x + y &= 1 \end{aligned}$

   (b) $\begin{aligned} 2x - y &= 1 \\ 6x - 3y &= 2 \end{aligned}$

   *(c) $\begin{aligned} x + 2y &= 8 \\ 2x - y &= 6 \end{aligned}$

   (d) $\begin{aligned} 2x + 3y &= 3 \\ 3x - y &= 10 \end{aligned}$

   *(e) $\begin{aligned} -2x + 6y &= 4 \\ x - 3y &= -2 \end{aligned}$

3. Determine values of the constant c such that the following systems of equations have the indicated solutions.

   *(a) $\begin{aligned} x + 2y &= 5 \\ 2x + cy &= 8 \end{aligned}$
   Solution: $x = 1, y = 2$

   (b) $\begin{aligned} x - 3y &= 4 \\ 3x + cy &= 1 \end{aligned}$
   Solution: $x = 1, y = -1$

   *(c) $\begin{aligned} x - 2y &= 0 \\ cx + 3y &= 7 \end{aligned}$
   Solution: $x = 2, y = 1$

   (d) $\begin{aligned} 2x + 3y &= -1 \\ -cx - 2y &= 26 \end{aligned}$
   Solution: $x = 4, y = -3$

   *(e) $\begin{aligned} x - 3y &= -2 \\ 2x + 5y &= c \end{aligned}$
   Solution: $x = 1, y = 1$

   (f) $\begin{aligned} 3x + y &= -11 \\ 2x - 5y &= c \end{aligned}$
   Solution: $x = -4, y = 1$

---

* Answers to exercises marked with an asterisk are provided in the back of the book.

*4. Determine the values of k for which the following system of equations has **(a)** a unique solution, **(b)** many solutions.

$$x + 4y = 5$$
$$2x + ky = 10$$

*5. Determine the value of k for which the following system of equations has no solution.

$$x + 2y = 3$$
$$3x + ky = 4$$

6. Determine the value of c for which the following system of equations has many solutions.

$$x - 3y = 4$$
$$-2x + cy = -8$$

*7. Consider the following system of equations for various values of d.

$$x + 3y = 7$$
$$2x + dy = 12$$

Which values of d result in **(a)** a unique solution, **(b)** no solution? Prove that the system cannot have many solutions.

8. Consider the following system of equations for various values of c.

$$x + y = 4$$
$$cx + 2y = 2$$

Which values of c result in a system that has **(a)** a unique solution, **(b)** no solution? Prove that the system cannot have many solutions.

*9. Consider the following system of equations in x and y.

$$ax + by = e$$
$$cx + dy = f$$

(a) What are the conditions on a, b, c, d, e, and f for this system to have a unique solution? Find a formula for the unique solution and use it to solve the system when $a = 1$, $b = 3$, $c = 3$, $d = -2$, $e = 10$, and $f = -3$.

(b) What are the conditions on a, b, c, d, e, and f for this system to have no solutions?

(c) What are the conditions on a, b, c, d, e, and f for this system to have many solutions?

Graphing Calculator Introduction

Graphing calculator discussions are included for readers who want to use a graphing calculator to plot graphs and to perform matrix operations. Explanations are brief and to the point. It is assumed that the reader is comfortable with the basic operations of the graphing calculator, including entering and editing equations, graphing functions, zooming, and tracing. The discussions are meant to be flexible. These discussions are given at the ends of the sections in which the concepts are first introduced. Exercises give the reader practice at mastering each of the operations. Students can then use the graphing calculator whenever desirable in the course to graph or perform matrix computation.

Instructions are given here for the TI-82 since the keystrokes are straightforward on this calculator. However the primary aim is to convey concepts—to illustrate when and how to use a calculator. The reader should find out how to graph and perform each operation on his or her calculator.

C.1   Solving a System of Linear Equations in Two Variables Graphically

Graphing calculators can be used to find approximate solutions to systems of linear equations in two variables. Let us solve the system of Section 1.1.

$$x + 3y = 9$$
$$-2x + y = -4$$

Rewrite these equations in the form

$$y = -x/3 + 3$$
$$y = 2x - 4$$

We sketch the graphs of these equations in the same coordinate system. The point of intersection of the graphs will be the solution.

1.   *Enter the Equations*   **2.**   *Define the Viewing Window*
    ($\boxed{Y=}$ key on TI-82)   ($\boxed{WINDOW}$)

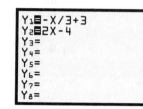

3.   *Display the Graph* ($\boxed{GRAPH}$)

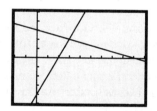

The point of intersection is the solution.

4.   *Trace/Zoom to find the Intersection* ($\boxed{TRACE}$ and $\boxed{ZOOM}$)

Trace/zoom to the desired accuracy. The solution to four decimal places is $x = 3$, $y = 2$. (This is seen to be the exact solution in Section 1.1)

Exercises C.1

In Exercises 1–5 solve (if possible) each of the following systems of equations on a calculator.

*1. $x + y = 1$
$x - y = 2$

2. $3x + 2y = 0$
$x - y = 2$

*3. $3x - 6y = 1$
$x - 2y = 2$

4. $x + 2y = 4$
$2x + 4y = 8$

*5. $x + y = 7$
$x + 2y = 3$

In Exercises 6–9 solve (if possible) each of the following systems of equations on a calculator.

*6. $3x - 4y = 2$
$x + y = 1$

7. $2x - y = 1$
$6x - 3y = 2$

*8. $x + 2y = 8$
$2x - y = 6$

9. $2x + 3y = 3$
$3x - y = 10$

*10. $-2x + 6y = 4$
$x - 3y = -2$

1.2 Systems of Linear Equations Having Unique Solutions

In the previous section we analyzed the behavior of systems of two linear equations in two variables. We now continue the discussion of systems of linear equations, looking at larger systems. We introduce the method of **Gauss-Jordan elimination*** as it applies to solving systems of n linear equations in n variables that have a unique solution. Such systems frequently arise in applications and have special properties; they deserve special attention. It is often known in advance that a system of equations has a unique solution because of the situation it describes. We shall see that currents in an electrical network are found by solving such a system of linear equations. The method of Gauss-Jordan elimination involves systematically eliminating variables from equations.

We shall use rectangular arrays of numbers called **matrices** to describe systems of linear equations. At this time we introduce the necessary terminology.

Definition   *A **matrix** is a rectangular array of numbers. The numbers in the array are called the **elements** of the matrix.*

---

* Carl Friedrich Gauss (1777–1855) was one of the greatest mathematical scientists of all time. He taught for forty-seven years at the University of Göttingen in Germany and made contributions to the fields of number theory, probability, statistics, and theory of functions. He discovered a way to calculate the orbits of asteroids. Gauss was described as "not really a physicist in the sense of searching for new phenomena, but rather a mathematician who attempted to formulate in exact mathematical terms the experimental results of others." His personal life was tragic, suffering from political turmoil and financial problems associated with the French Revolution and democratic revolutions in Germany.

Wilhelm Jordan (1842–1899) taught geodesy at the Technical College of Karlsruhe in Germany. His major work was a handbook on geodesy that contained his work on systems of equations. Jordan was considered a fine teacher and writer.

Matrices are usually denoted by capital letters. Examples of matrices in standard notation are

$$A = \begin{bmatrix} 2 & 3 & -4 \\ 7 & 5 & -1 \end{bmatrix}, \qquad B = \begin{bmatrix} 7 & 1 \\ 0 & 5 \\ -8 & 3 \end{bmatrix}, \qquad C = \begin{bmatrix} 3 & 5 & 6 \\ 0 & -2 & 5 \\ 8 & 9 & 12 \end{bmatrix}$$

Rows and Columns
Matrices consist of rows and columns. **Rows** are labeled from the top of the matrix, **columns** from the left. The following matrix has two rows and three columns.

$$\begin{bmatrix} 2 & 3 & -4 \\ 7 & 5 & -1 \end{bmatrix}$$

The rows are

$$[2 \quad 3 \quad -4], \qquad [7 \quad 5 \quad -1]$$
$$\text{row 1} \qquad\qquad\qquad \text{row 2}$$

The columns are

$$\begin{bmatrix} 2 \\ 7 \end{bmatrix}, \qquad \begin{bmatrix} 3 \\ 5 \end{bmatrix}, \qquad \begin{bmatrix} -4 \\ -1 \end{bmatrix}$$
$$\text{column 1} \qquad \text{column 2} \qquad \text{column 3}$$

Submatrix
A **submatrix** of a given matrix is an array obtained by deleting certain rows and columns of the matrix. For example, consider the following matrix A. The matrices P, Q, and R are submatrices of A.

$$A = \begin{bmatrix} 1 & 7 & 4 \\ 2 & 3 & 0 \\ 5 & 1 & -2 \end{bmatrix}, \qquad P = \begin{bmatrix} 1 & 7 \\ 2 & 3 \\ 5 & 1 \end{bmatrix}, \qquad Q = \begin{bmatrix} 7 \\ 3 \\ 1 \end{bmatrix}, \qquad R = \begin{bmatrix} 1 & 4 \\ 5 & -2 \end{bmatrix}$$
$$\text{matrix } A \qquad\qquad\qquad\qquad\qquad \text{submatrices of } A$$

Size and Type
The **size** of a matrix is described by specifying the number of rows and columns in the matrix. A matrix having two rows and three columns is said to be a 2×3 matrix; the first number indicates the number of rows, the second indicates the number of columns. When the number of rows is equal to the number of columns, the matrix is said to be a **square matrix**. A matrix consisting of one row is called a **row matrix**. A matrix consisting of one column is a **column matrix**. The following matrices are of the stated sizes and types.

$$\begin{bmatrix} 1 & 0 & 3 \\ -2 & 4 & 5 \end{bmatrix} \qquad \begin{bmatrix} 2 & 5 & 7 \\ -9 & 0 & 1 \\ -3 & 5 & 8 \end{bmatrix} \qquad [4 \quad -3 \quad 8 \quad 5] \qquad \begin{bmatrix} 8 \\ 3 \\ 2 \end{bmatrix}$$

2×3 matrix

3×3 matrix
a square matrix

1×4 matrix
a row matrix

3×1 matrix
a column matrix

Location   The **location** of an element in a matrix is described by giving the row and column in which the element lies. For example, consider the following matrix.

$$\begin{bmatrix} 2 & 3 & -4 \\ 7 & 5 & -1 \end{bmatrix}$$

The element 7 is in row 2, column 1. We say that it is in location (2, 1). The element in location (1, 3) is -4. Note that the convention is to give the row in which the element lies, followed by the column.

Identity Matrices   An **identity matrix** is a square matrix with 1s in the **diagonal** **locations**—(1, 1), (2, 2) (3, 3), etc.—and zeros elsewhere. We write I_n for the $n \times n$ identity matrix. The following matrices are identity matrices.

$$I_2 = \begin{bmatrix} 1 & 0 \\ 0 & 1 \end{bmatrix}, \qquad I_3 = \begin{bmatrix} 1 & 0 & 0 \\ 0 & 1 & 0 \\ 0 & 0 & 1 \end{bmatrix}, \qquad I_4 = \begin{bmatrix} 1 & 0 & 0 & 0 \\ 0 & 1 & 0 & 0 \\ 0 & 0 & 1 & 0 \\ 0 & 0 & 0 & 1 \end{bmatrix}$$

We are now ready to continue the discussion of systems of linear equations.

The following operations can be used to change a system of linear equations into another system of linear equations that has the same solution.

Elementary Transformations

1.   Interchange the positions of two equations.

2.   Multiply both sides of an equation by a nonzero constant.

3.   Add a multiple of one equation to another equation.

Elementary transformations preserve solutions since the order of the equations does not affect the solution, multiplying an equation throughout by a nonzero constant does not change the truth of the equality, and adding equal quantities to both sides of an equality maintains the equality.

Systems of equations that are related through elementary transformations, and thus have the same solutions, are called **equivalent systems**. The symbol $\approx$ is used to indicate equivalent systems of equations.

The method of Gauss-Jordan elimination uses elementary transformations to eliminate variables in a systematic manner, knowing that each new system of equations has the same solution as the original system. We illustrate the method with the following example.

Example 1   Solve the system of linear equations

$$\begin{aligned} x_1 + x_2 + x_3 &= 2 \\ 2x_1 + 3x_2 + x_3 &= 3 \\ x_1 - x_2 - 2x_3 &= -6 \end{aligned}$$

Solution We use the first equation to eliminate x_1 from the second and third equations.

$$
\begin{array}{rcr}
x_1 + x_2 + x_3 &=& 2 \\
2x_1 + 3x_2 + x_3 &=& 3 \\
x_1 - x_2 - 2x_3 &=& -6
\end{array}
\quad \approx \quad
\begin{array}{c}
\text{Eq 2} + (-2)\text{Eq 1} \\
\text{Eq 3} + (-1)\text{Eq 1}
\end{array}
\quad
\begin{array}{rcr}
x_1 + x_2 + x_3 &=& 2 \\
x_2 - x_3 &=& -1 \\
-2x_2 - 3x_3 &=& -8
\end{array}
$$

Next use the second equation to eliminate x_2 from the first and third equations.

$$
\approx \quad
\begin{array}{c}
\text{Eq 1} + (-1)\text{Eq 2} \\
\text{Eq 3} + 2\text{Eq 2}
\end{array}
\quad
\begin{array}{rcr}
x_1 + 2x_3 &=& 3 \\
x_2 - x_3 &=& -1 \\
-5x_3 &=& -10
\end{array}
$$

Multiply the last equation by $-\frac{1}{5}$ to get x_3.

$$
\approx \quad
-\tfrac{1}{5}\text{Eq 3}
\quad
\begin{array}{rcr}
x_1 + 2x_3 &=& 3 \\
x_2 - x_3 &=& -1 \\
x_3 &=& 2
\end{array}
$$

Finally, use the third equation to eliminate x_3 from the first and second equations.

$$
\approx \quad
\begin{array}{c}
\text{Eq 1} + (-2)\text{Eq 3} \\
\text{Eq 2} + \text{Eq 3}
\end{array}
\quad
\begin{array}{rcr}
x_1 &=& -1 \\
x_2 &=& 1 \\
x_3 &=& 2
\end{array}
$$

The solution to this system, and thus the solution also to the original system, is $x_1 = -1$, $x_2 = 1$, $x_3 = 2$.

Geometrically, each of the three original equations represents a plane in three-dimensional space. The fact that there is a unique solution means that these three planes intersect at a point. The solution $(-1, 1, 2)$ gives the coordinates of this point where the three planes intersect.

We now use matrices to describe systems of linear equations. There are two important matrices associated with every system of linear equations. The *coefficients of the variables* form a matrix called the **matrix of coefficients** of the system. The *coefficients, together with the constant terms*, form a matrix called the **augmented matrix** of the system. Consider the system

$$
\begin{array}{rcr}
x_1 + x_2 + x_3 &=& 2 \\
2x_1 + 3x_2 + x_3 &=& 3 \\
x_1 - x_2 - 2x_3 &=& -6
\end{array}
$$

The matrix of coefficients and augmented matrix for this system are as follows.

$$
\begin{bmatrix} 1 & 1 & 1 \\ 2 & 3 & 1 \\ 1 & -1 & -2 \end{bmatrix}
\qquad
\begin{bmatrix} 1 & 1 & 1 & 2 \\ 2 & 3 & 1 & 3 \\ 1 & -1 & -2 & -6 \end{bmatrix}
$$

matrix of coefficients   augmented matrix

Observe that the matrix of coefficients is a **submatrix** of the augmented matrix. The augmented matrix completely describes the system. Systems of linear equations are described on computers by their augmented matrices. In solving a system of linear equations, each equivalent system in the sequence can be represented by an augmented matrix. It becomes unnecessary to write down the variables x_1, x_2, x_3, ... Instead of performing elementary transformations on the systems of equations, we can perform analogous transformations called **elementary row operations** on the augmented matrices.

Elementary Row Operations on Matrices

1.   Interchange two rows.

2.   Multiply the elements of a row by a nonzero constant.

3.   Add a multiple of the elements of one row to the corresponding elements of another row.

Two matrices are said to be **row equivalent** if one can be obtained from the other by using a finite sequence of elementary row operations. We shall use the following notation to describe the row operations.

$$Ri \pm cRj$$

row i is the   by adding or subtracting a
row being modified   multiple of row j to it.

The first row mentioned is always the row being modified.

We now show how these row operations are used to solve a system of linear equations.

Example 2   Solve the following system of linear equations.

$$x_1 - 2x_2 + 4x_3 = 12$$
$$2x_1 - x_2 + 5x_3 = 18$$
$$-x_1 + 3x_2 - 3x_3 = -8$$

Solution   Start with the augmented matrix and use the first row to create zeros in the first column. (This corresponds to using the first equation to eliminate x_1 from the second and third equations.)

$$\begin{bmatrix} 1 & -2 & 4 & 12 \\ 2 & -1 & 5 & 18 \\ -1 & 3 & -3 & -8 \end{bmatrix} \approx \underset{\substack{R2 + (-2)R1 \\ R3 + R1}}{} \begin{bmatrix} 1 & -2 & 4 & 12 \\ 0 & 3 & -3 & -6 \\ 0 & 1 & 1 & 4 \end{bmatrix}$$

Now multiply row 2 by $\frac{1}{3}$. (This corresponds to making the coefficient of x_2 in the second equation 1.)

$$\approx \underset{\frac{1}{3}R2}{} \begin{bmatrix} 1 & -2 & 4 & 12 \\ 0 & 1 & -1 & -2 \\ 0 & 1 & 1 & 4 \end{bmatrix}$$

Next create zeros in the second column as follows. (This corresponds to using the second equation to eliminate x_2 from the first and third equations.)

$$\underset{\substack{R1 + 2R2 \\ R3 + (-1)R2}}{\approx} \begin{bmatrix} 1 & 0 & 2 & 8 \\ 0 & 1 & -1 & -2 \\ 0 & 0 & 2 & 6 \end{bmatrix}$$

Multiply row 3 by $\frac{1}{2}$. (This corresponds to making the coefficient of x_3 in the third equation 1.)

$$\underset{\frac{1}{2}R3}{\approx} \begin{bmatrix} 1 & 0 & 2 & 8 \\ 0 & 1 & -1 & -2 \\ 0 & 0 & 1 & 3 \end{bmatrix}$$

Finally, create zeros in the third column. (This corresponds to using the third equation to eliminate x_3 from the first and second equations.)

$$\underset{\substack{R1 + (-2)R3 \\ R2 + R3}}{\approx} \begin{bmatrix} 1 & 0 & 0 & 2 \\ 0 & 1 & 0 & 1 \\ 0 & 0 & 1 & 3 \end{bmatrix}$$

This matrix corresponds to the system

$$\begin{aligned} x_1 \qquad\quad &= 2 \\ x_2 \quad &= 1 \\ x_3 &= 3 \end{aligned}$$

The solution is $x_1 = 2$, $x_2 = 1$, $x_3 = 3$.

This method of reducing a system of linear equations to an equivalent simpler system, using matrices, involves creating 1s and 0s in certain locations of matrices. These numbers are created in a systematic manner, column by column. The following example illustrates that it may be necessary to interchange two rows at some stage to proceed in the above manner.

Example 3 Solve the system

$$\begin{aligned} 4x_1 + 8x_2 - 12x_3 &= 44 \\ 3x_1 + 6x_2 - 8x_3 &= 32 \\ -2x_1 - x_2 \qquad\quad &= -7 \end{aligned}$$

Solution We start with the augmented matrix and proceed as follows. (Note the use of zero in the augmented matrix as the coefficient of the missing variable x_3 in the third equation.)

$$\begin{bmatrix} 4 & 8 & -12 & 44 \\ 3 & 6 & -8 & 32 \\ -2 & -1 & 0 & -7 \end{bmatrix} \quad \underset{\frac{1}{4}R1}{\approx} \quad \begin{bmatrix} 1 & 2 & -3 & 11 \\ 3 & 6 & -8 & 32 \\ -2 & -1 & 0 & -7 \end{bmatrix}$$

$$\underset{\substack{R2 + (-3)R1 \\ R3 + 2R1}}{\approx} \quad \begin{bmatrix} 1 & 2 & -3 & 11 \\ 0 & 0 & 1 & -1 \\ 0 & 3 & -6 & 15 \end{bmatrix}$$

At this stage we need a nonzero element in the location (2, 2) to continue. To achieve this, we interchange the second row with the third row (a *later* row) and then proceed.

$$\underset{R2 \leftrightarrow R3}{\approx} \quad \begin{bmatrix} 1 & 2 & -3 & 11 \\ 0 & 3 & -6 & 15 \\ 0 & 0 & 1 & -1 \end{bmatrix} \quad \underset{\frac{1}{3}R2}{\approx} \quad \begin{bmatrix} 1 & 2 & -3 & 11 \\ 0 & 1 & -2 & 5 \\ 0 & 0 & 1 & -1 \end{bmatrix}$$

$$\underset{R1 + (-2)R2}{\approx} \quad \begin{bmatrix} 1 & 0 & 1 & 1 \\ 0 & 1 & -2 & 5 \\ 0 & 0 & 1 & -1 \end{bmatrix} \quad \underset{\substack{R1 + (-1)R3 \\ R2 + 2R3}}{\approx} \quad \begin{bmatrix} 1 & 0 & 0 & 2 \\ 0 & 1 & 0 & 3 \\ 0 & 0 & 1 & -1 \end{bmatrix}$$

The solution is $x_1 = 2$, $x_2 = 3$, $x_3 = -1$.

We now summarize this Gauss-Jordan method for solving a system of n linear equations in n variables that has a unique solution. The augmented matrix is made up of a matrix of coefficients A and a column matrix of constant terms B. Let us write $[A : B]$ for this matrix. Use row operations to gradually transform this matrix, column by column, into a matrix $[I_n : X]$, where I_n is the identity $n \times n$ matrix.

$$[A : B] \approx \cdots \approx [I_n : X]$$

This final matrix $[I_n : X]$ is called the **reduced echelon form** of the original augmented matrix. The matrix of coefficients of the final system of equations is I_n and X is the column matrix of constant terms. This implies that the elements of X are the unique solution. Observe that as $[A : B]$ is being transformed to $[I_n : X]$, A is being changed to I_n. Thus

If A is the matrix of coefficients of a system of n equations in n variables that has a unique solution, then it is row equivalent to I_n.

If $[A : B]$ cannot be transformed in this manner into a matrix of the form $[I_n : X]$, the system of equations does not have a unique solution. More will be said about such systems in the next section.

Many Systems

Certain applications involve solving a number of systems of linear equations, all having the same square matrix of coefficients A. Let the systems be

$$[A : B_1], \quad [A : B_2], \dots, \quad [A : B_k]$$

The constant terms $B_1, B_2, \dots, B_k$, might be test data, and one wants to know the solutions that would lead to these results. The situation often dictates that the solutions be unique. One could, of course, go through the method of Gauss-Jordan elimination for each system, solving each system independently. This procedure would lead to the reduced echelon forms

$$[I_n : X_1], \quad [I_n : X_2], \dots, \quad [I_n : X_k]$$

and the solutions would be $X_1, X_2, \dots, X_k$. However, the same reduction of A to I_n would be repeated for each system; this involves a great deal of unnecessary duplication. The systems can be represented by one large augmented matrix $[A : B_1 \quad B_2 \quad \cdots \quad B_k]$, and the Gauss-Jordan method can be applied to this one matrix. We would get

$$[A : B_1 \quad B_2 \quad \cdots \quad B_k] \approx \cdots \approx [I_n : X_1 \quad X_2 \quad \cdots \quad X_k]$$

leading to the solutions $X_1, X_2, \dots, X_k$.

Example 4   Solve the following three systems of linear equations, all of which have the same matrix of coefficients.

$$
\begin{aligned}
x_1 - x_2 + 3x_3 &= b_1 \\
2x_1 - x_2 + 4x_3 &= b_2 \\
-x_1 + 2x_2 - 4x_3 &= b_3
\end{aligned}
\quad \text{for} \quad
\begin{bmatrix} b_1 \\ b_2 \\ b_3 \end{bmatrix}
=
\begin{bmatrix} 8 \\ 11 \\ -11 \end{bmatrix},
\begin{bmatrix} 0 \\ 1 \\ 2 \end{bmatrix},
\begin{bmatrix} 3 \\ 3 \\ -4 \end{bmatrix}
\quad \text{in turn}
$$

Solution   Construct the large augmented matrix that describes all three systems and determine the reduced echelon form as follows.

$$
\begin{bmatrix}
1 & -1 & 3 & 8 & 0 & 3 \\
2 & -1 & 4 & 11 & 1 & 3 \\
-1 & 2 & -4 & -11 & 2 & -4
\end{bmatrix}
\underset{\substack{\text{R2} + (-2)\text{R1} \\ \text{R3} + \text{R1}}}{\approx}
\begin{bmatrix}
1 & -1 & 3 & 8 & 0 & 3 \\
0 & 1 & -2 & -5 & 1 & -3 \\
0 & 1 & -1 & 3 & 2 & 1
\end{bmatrix}
$$

$$
\underset{\substack{\text{R1} + \text{R2} \\ \text{R3} + (-1)\text{R2}}}{\approx}
\begin{bmatrix}
1 & 0 & 1 & 3 & 1 & 0 \\
0 & 1 & -2 & -5 & 1 & -3 \\
0 & 0 & 1 & 2 & 1 & 2
\end{bmatrix}
$$

$$
\underset{\substack{\text{R1} + (-1)\text{R3} \\ \text{R2} + 2\text{R3}}}{\approx}
\begin{bmatrix}
1 & 0 & 0 & 1 & 0 & -2 \\
0 & 1 & 0 & -1 & 3 & 1 \\
0 & 0 & 1 & 2 & 1 & 2
\end{bmatrix}
$$

The solutions to the three systems of equations are given by the last three columns of the reduced echelon form. They are

$$x_1 = 1, \quad x_2 = -1, \quad x_3 = 2$$
$$x_1 = 0, \quad x_2 = 3, \quad x_3 = 1$$
$$x_1 = -2, \quad x_2 = 1, \quad x_3 = 2$$

In this section we have limited our discussion to systems of n linear equations in n variables that have a unique solution. In the following section we shall extend the method of Gauss-Jordan elimination to accommodate other systems that have a unique solution, and also to include systems that have many solutions or no solutions.

Exercise Set 1.2

1. Give the sizes of the following matrices.

*(a) $\begin{bmatrix} 1 & 2 & 3 \\ 0 & 1 & 2 \\ 4 & 5 & 3 \end{bmatrix}$ (b) $\begin{bmatrix} 0 & 9 \\ -6 & 4 \\ -3 & 2 \end{bmatrix}$

*(c) $\begin{bmatrix} 1 & 2 & 3 & 0 \\ 1 & 2 & 4 & 5 \end{bmatrix}$ (d) $\begin{bmatrix} -7 \\ 4 \\ 3 \end{bmatrix}$

*(e) $\begin{bmatrix} 1 & 2 & 9 & -8 & 7 \\ 4 & 2 & 5 & 7 & 2 \\ 4 & -6 & 4 & 0 & 0 \end{bmatrix}$

(f) $\begin{bmatrix} 2 & -3 & 4 & 7 \end{bmatrix}$

*2. Give the (1, 1), (2, 2), (3, 3) (1, 5), (2, 4), (3, 2) elements of the following matrix.

$$\begin{bmatrix} 1 & 2 & 3 & 0 & -1 \\ -2 & 4 & -5 & 3 & 6 \\ 5 & 8 & 9 & 2 & 3 \end{bmatrix}$$

3. Give the (2, 3) (3, 2), (4, 1), (1, 3), (4, 4), (3, 1) elements of the following matrix.

$$\begin{bmatrix} 1 & 2 & 7 & 0 \\ -1 & 2 & 4 & 5 \\ 3 & 5 & 0 & -1 \\ 6 & 9 & 0 & 2 \end{bmatrix}$$

4. Write the identity matrix I_4.

5. Determine the matrix of coefficients and augmented matrix of each of the following systems of equations.

*(a) $\begin{aligned} x_1 + 3x_2 &= 7 \\ 2x_1 - 5x_2 &= -3 \end{aligned}$

(b) $\begin{aligned} 5x_1 + 2x_2 - 4x_3 &= 8 \\ x_1 + 3x_2 + 6x_3 &= 4 \\ 4x_1 + 6x_2 - 9x_3 &= 7 \end{aligned}$

*(c) $\begin{aligned} -x_1 + 3x_2 - 5x_3 &= -3 \\ 2x_1 - 2x_2 + 4x_3 &= 8 \\ x_1 + 3x_2 \qquad\;\; &= 6 \end{aligned}$

(d) $\begin{aligned} 5x_1 + 4x_2 &= 9 \\ 2x_1 - 8x_2 &= -4 \\ x_1 + 2x_2 &= 3 \end{aligned}$

*(e) $\begin{aligned} 5x_1 + 2x_2 - 4x_3 &= 8 \\ 4x_2 + 3x_3 &= 0 \\ x_1 \qquad\;\; - x_3 &= 7 \end{aligned}$

(f) $\begin{aligned} -x_1 + 3x_2 - 9x_3 &= -4 \\ x_1 \qquad\;\; - 4x_3 &= 11 \\ x_1 + 8x_2 \qquad\;\; &= 1 \end{aligned}$

(g) $\begin{aligned} x_1 \qquad\qquad &= -3 \\ x_2 \qquad &= 12 \\ x_3 &= 1 \end{aligned}$

(h) $\begin{aligned} -4x_1 + 2x_2 - 9x_3 + x_4 &= -1 \\ x_1 + 6x_2 - 8x_3 - 7x_4 &= 15 \\ -x_2 + 3x_3 - 5x_4 &= 0 \end{aligned}$

6. Interpret the following matrices as augmented matrices of systems of equations. Write each system of equations.

(a) $\begin{bmatrix} 1 & 2 & 3 \\ 4 & 5 & 6 \end{bmatrix}$ *(b) $\begin{bmatrix} 7 & 9 & 8 \\ 6 & 4 & -3 \end{bmatrix}$

(c) $\begin{bmatrix} 1 & 9 & -3 \\ 5 & 0 & 2 \end{bmatrix}$

*(d) $\begin{bmatrix} 8 & 7 & 5 & -1 \\ 4 & 6 & 2 & 4 \\ 9 & 3 & 7 & 6 \end{bmatrix}$

(e) $\begin{bmatrix} 2 & -3 & 6 & 4 \\ 7 & -5 & -2 & 3 \\ 0 & 2 & 4 & 0 \end{bmatrix}$

*(f) $\begin{bmatrix} 0 & -2 & 4 \\ 5 & 7 & -3 \\ 6 & 0 & 8 \end{bmatrix}$ (g) $\begin{bmatrix} 1 & 0 & 0 & 3 \\ 0 & 1 & 0 & 8 \\ 0 & 0 & 1 & 4 \end{bmatrix}$

*(h) $\begin{bmatrix} 1 & 2 & -1 & 6 \\ 0 & 1 & 4 & 5 \\ 0 & 0 & 1 & -2 \end{bmatrix}$

7. In the following exercises you are given a matrix followed by an elementary row operation. Determine each resulting matrix.

*(a) $\begin{bmatrix} 2 & 6 & -4 & 0 \\ 1 & 2 & -3 & 6 \\ 8 & 3 & 2 & 5 \end{bmatrix} \underset{\frac{1}{2}R1}{\approx}$

(b) $\begin{bmatrix} 0 & -8 & 4 & 3 \\ 2 & 7 & 5 & 1 \\ 3 & -5 & 8 & 9 \end{bmatrix} \underset{R1 \leftrightarrow R2}{\approx}$

*(c) $\begin{bmatrix} 1 & 2 & 3 & -1 \\ -1 & 1 & 7 & 1 \\ 2 & -4 & 5 & -3 \end{bmatrix} \underset{\substack{R2 + R1 \\ R3 + (-2)R1}}{\approx}$

(d) $\begin{bmatrix} 1 & 2 & 3 & -4 \\ 0 & 1 & 2 & 1 \\ 0 & -4 & 3 & -5 \end{bmatrix} \underset{\substack{R1 + (-2)R2 \\ R3 + 4R2}}{\approx}$

*(e) $\begin{bmatrix} 1 & 0 & 4 & -3 \\ 0 & 1 & -3 & 2 \\ 0 & 0 & 1 & 5 \end{bmatrix} \underset{\substack{R1 + (-4)R3 \\ R2 + 3R3}}{\approx}$

(f) $\begin{bmatrix} 1 & 0 & 2 & 7 \\ 0 & 1 & 5 & -3 \\ 0 & 0 & -2 & 8 \end{bmatrix} \underset{-\frac{1}{2}R3}{\approx}$

8. Interpret each of the following row operations as a stage in arriving at the reduced echelon form of a matrix. Why have the indicated operations been selected? What particular aims do they accomplish in terms of the systems of linear equations that are described by the matrices?

*(a) $\begin{bmatrix} 1 & -4 & 3 & 5 \\ -2 & 1 & 7 & 5 \\ 4 & 0 & -3 & 6 \end{bmatrix}$

$\underset{\substack{R2 + 2R1 \\ R3 + (-4)R1}}{\approx} \begin{bmatrix} 1 & -4 & 3 & 5 \\ 0 & -7 & 13 & 15 \\ 0 & 16 & -15 & -14 \end{bmatrix}$

(b) $\begin{bmatrix} 1 & 2 & -4 & 7 \\ 0 & 3 & 9 & -6 \\ 0 & 4 & 7 & -8 \end{bmatrix} \underset{\frac{1}{3}R2}{\approx} \begin{bmatrix} 1 & 2 & -4 & 7 \\ 0 & 1 & 3 & -2 \\ 0 & 4 & 7 & -8 \end{bmatrix}$

*(c) $\begin{bmatrix} 1 & 3 & -4 & 5 \\ 0 & 0 & -2 & 6 \\ 0 & 1 & 3 & -8 \end{bmatrix} \underset{R2 \leftrightarrow R3}{\approx} \begin{bmatrix} 1 & 3 & -4 & 5 \\ 0 & 1 & 3 & -8 \\ 0 & 0 & -2 & 6 \end{bmatrix}$

(d) $\begin{bmatrix} 1 & 2 & 5 & 0 \\ 0 & 1 & 2 & -3 \\ 0 & -3 & 1 & -2 \end{bmatrix}$

$\underset{\substack{R1 + (-2)R2 \\ R3 + 3R2}}{\approx} \begin{bmatrix} 1 & 0 & 1 & 6 \\ 0 & 1 & 2 & -3 \\ 0 & 0 & 7 & -11 \end{bmatrix}$

9. Interpret each of the following row operations as a stage in arriving at the reduced echelon form of a matrix. Why have these operations been selected?

*(a) $\begin{bmatrix} 1 & 0 & 2 & 6 \\ 0 & 1 & -1 & 3 \\ 0 & 0 & 1 & 2 \end{bmatrix} \underset{\substack{R1 + (-2)R3 \\ R2 + R3}}{\approx} \begin{bmatrix} 1 & 0 & 0 & 2 \\ 0 & 1 & 0 & 5 \\ 0 & 0 & 1 & 2 \end{bmatrix}$

(b) $\begin{bmatrix} 0 & 2 & 4 & -1 \\ 4 & 3 & 2 & -8 \\ 5 & -7 & 1 & 2 \end{bmatrix} \underset{R1 \leftrightarrow R2}{\approx} \begin{bmatrix} 4 & 3 & 2 & -8 \\ 0 & 2 & 4 & -1 \\ 5 & -7 & 1 & 2 \end{bmatrix}$

*(c) $\begin{bmatrix} 1 & 0 & 3 & 7 \\ 0 & 1 & 4 & 2 \\ 0 & 0 & 2 & 6 \end{bmatrix} \underset{\frac{1}{2}R3}{\approx} \begin{bmatrix} 1 & 0 & 3 & 7 \\ 0 & 1 & 4 & 2 \\ 0 & 0 & 1 & -3 \end{bmatrix}$

(d) $\begin{bmatrix} 1 & 0 & -2 & 4 \\ 0 & 1 & 3 & -4 \\ 0 & 0 & 1 & -3 \end{bmatrix} \underset{\substack{R1 + 2R3 \\ R2 + (-3)R3}}{\approx} \begin{bmatrix} 1 & 0 & 0 & -2 \\ 0 & 1 & 0 & 5 \\ 0 & 0 & 1 & -3 \end{bmatrix}$

10. The following systems of equations all have unique solutions. Solve these systems using the method of Gauss-Jordan elimination.

*(a) $x_1 - 2x_2 = -8$
$2x_1 - 3x_2 = -11$

(b) $2x_1 + 2x_2 = 4$
$3x_1 + 2x_2 = 3$

*(c) $x_1 \quad\quad + x_3 = 3$
$2x_2 - 2x_3 = -4$
$x_2 - 2x_3 = 5$

(d) $x_1 + x_2 + 3x_3 = 6$
$x_1 + 2x_2 + 4x_3 = 9$
$2x_1 + x_2 + 6x_3 = 11$

*(e) $x_1 - x_2 + 3x_3 = 3$
$2x_1 - x_2 + 2x_3 = 2$
$3x_1 + x_2 - 2x_3 = 2$

(f) $-x_1 + x_2 - x_3 = -2$
$3x_1 + x_2 + x_3 = 10$
$4x_1 + 2x_2 + 3x_3 = 14$

11. The following systems of equations all have unique solutions. Solve these systems using the methods of Gauss-Jordan elimination.

*(a) $2x_2 + 4x_3 = 8$
$2x_1 + 2x_2 \quad\quad = 6$
$x_1 + x_2 + x_3 = 5$

(b) $x_1 - 2x_2 - 4x_3 = -9$
$x_1 + 5x_2 + 10x_3 = 21$
$2x_1 - 3x_2 - 5x_3 = -13$

*(c) $x_1 + 2x_2 + 3x_3 = 14$
$2x_1 + 5x_2 + 8x_3 = 36$
$x_1 - x_2 \quad\quad = -4$

(d) $x_1 - x_2 - x_3 = -1$
$-2x_1 + 6x_2 + 10x_3 = 14$
$2x_1 + x_2 + 6x_3 = 9$

*(e) $2x_1 + 2x_2 - 4x_3 = 14$
$3x_1 + x_2 + x_3 = 8$
$2x_1 - x_2 + 2x_3 = -1$

12. The following systems of equations all have unique solutions. Solve these systems using the method of Gauss-Jordan elimination.

(a) $\frac{3}{2}x_1 \quad\quad + 3x_3 = 15$
$-x_1 + 7x_2 - 9x_3 = -45$
$2x_1 \quad\quad + 5x_3 = 22$

*(b) $-3x_1 - 6x_2 - 15x_3 = -3$
$x_1 + \frac{3}{2}x_2 + \frac{9}{2}x_3 = \frac{1}{2}$
$-2x_1 - \frac{7}{2}x_2 - \frac{17}{2}x_3 = -2$

(c) $3x_1 + 6x_2 \quad\quad - 3x_4 = 3$
$x_1 + 3x_2 - x_3 - 4x_4 = -12$
$x_1 - x_2 + x_3 + 2x_4 = 8$
$2x_1 + 3x_2 \quad\quad\quad = 8$

*(d) $x_1 + 2x_2 + 2x_3 + 5x_4 = 11$
$2x_1 + 4x_2 + 2x_3 + 8x_4 = 14$
$x_1 + 3x_2 + 4x_3 + 8x_4 = 19$
$x_1 - x_2 + x_3 \quad\quad = 2$

*(e) $x_1 + x_2 + 2x_3 + 6x_4 = 11$
$2x_1 + 3x_2 + 6x_3 + 19x_4 = 36$
$3x_2 + 4x_3 + 15x_4 = 28$
$x_1 - x_2 - x_3 - 6x_4 = -12$

13. The following exercises involve many systems of linear equations with unique solutions that have the same matrix of coefficients. Solve the systems by applying the method of Gauss-Jordan elimination to a large augmented matrix that describes many systems.

*(a) $x_1 + 2x_2 = b_1$
$3x_1 + 5x_2 = b_2$

for $\begin{bmatrix} b_1 \\ b_2 \end{bmatrix} = \begin{bmatrix} 3 \\ 8 \end{bmatrix}, \begin{bmatrix} 4 \\ 9 \end{bmatrix}, \begin{bmatrix} 3 \\ 7 \end{bmatrix}$ in turn

(b) $x_1 + x_2 = b_1$
$2x_1 + 3x_2 = b_2$

for $\begin{bmatrix} b_1 \\ b_2 \end{bmatrix} = \begin{bmatrix} 0 \\ 1 \end{bmatrix}, \begin{bmatrix} 5 \\ 13 \end{bmatrix}, \begin{bmatrix} 1 \\ 2 \end{bmatrix}$ in turn

*(c) $x_1 - 2x_2 + 3x_3 = b_1$
$x_1 - x_2 + 2x_3 = b_2$
$2x_1 - 3x_2 + 6x_3 = b_3$

for $\begin{bmatrix} b_1 \\ b_2 \\ b_3 \end{bmatrix} = \begin{bmatrix} 6 \\ 5 \\ 14 \end{bmatrix}, \begin{bmatrix} -5 \\ -3 \\ -8 \end{bmatrix}, \begin{bmatrix} 4 \\ 3 \\ 9 \end{bmatrix}$ in turn

(d) $x_1 + 2x_2 - x_3 = b_1$
$-x_1 - x_2 + x_3 = b_2$
$3x_1 + 7x_2 - x_3 = b_3$

for $\begin{bmatrix} b_1 \\ b_2 \\ b_3 \end{bmatrix} = \begin{bmatrix} -1 \\ 1 \\ -1 \end{bmatrix}, \begin{bmatrix} 6 \\ -4 \\ 18 \end{bmatrix}, \begin{bmatrix} 0 \\ -2 \\ -4 \end{bmatrix}$ in turn

C.2   Row Operations

Calculators are available that will perform elementary row operations on matrices. Let us solve the system of linear equations of Example 3, Section 1.2, using the TI-82 calculator. The system is

$$4x_1 + 8x_2 - 12x_3 = 44$$
$$3x_1 + 6x_2 - 8x_3 = 32$$
$$-2x_1 - x_2 \qquad = -7$$

with augmented matrix

$$A = \begin{bmatrix} 4 & 8 & -12 & 44 \\ 3 & 6 & -8 & 32 \\ -2 & -1 & 0 & -7 \end{bmatrix}$$

We shall need to enter the size of the matrix A and the elements of A. The matrix A has three rows and four columns. We say that A is a 3×4 matrix. The location of an element is given by the row and column in which it lies. Element $(1, 1) = 4$, element $(1, 2) = 8$, etc. Element $(2, 3)$, for example, would be -8.

Enter Matrix A

Press MATRX ◀ to get the Matrix Edit Menu

Press 1 to select Edit Screen for matrix A (2 for B, 3 for C, etc.)

Edit the matrix using ▶ ◀ ▼ ▲ keys and the ENTER key

   [A] 3×4
   $1, 1 = 4$
   $1, 2 = 8$, etc.

Press QUIT to return to Home Screen

Display Matrix A

Press MATRX to get the Matrix Names Menu

Press 1 ENTER to select and display matrix A (2 for B, 3 for C, etc.)

$$\begin{matrix} [& 4 & 8 & -12 & 44] \\ [& 3 & 6 & -8 & 32] \end{matrix}$$
Matrix A is displayed $[-2 \quad 1 \quad 0 \quad -7]$

(The display is not as clear as this—columns are not lined up. We shall show matrices with columns lined up for ease of reading.)

Perform (1/4) Row 1

Press MATRX ▶ to get the Matrix Math Menu

Select 0 to select *row(

Enter 1/4 ⎵ [A] ⎵ 1 ⎵ {Completed instruction is *row(1/4,[A],1)}

Press ENTER to perform the transformation

$$\begin{bmatrix} 1 & 2 & -3 & 11 \\ 3 & 6 & -8 & 32 \end{bmatrix}$$

Answer is displayed $[-2 \quad 1 \quad 0 \quad -7]$

Add (−3)Row(1) to Row 2 in the Answer Matrix

Press MATRX ▶ to get the Matrix Menu

Select A, perform the instruction *row+(−3,Ans,1,2)

$$\begin{bmatrix} 1 & 2 & -3 & 11 \\ 0 & 0 & 1 & -1 \end{bmatrix}$$

Answer is displayed $[-2 \quad 1 \quad 0 \quad -7]$

Add (2)Row(1) to Row 3

Press MATRX ▶ to get the Matrix Menu

Select A, perform the instruction *row+(2,Ans,1,3)

$$\begin{bmatrix} 1 & 2 & -3 & 11 \\ 0 & 0 & 1 & -1 \end{bmatrix}$$

Answer is displayed $[0 \quad 3 \quad -6 \quad 15]$

Swap Row 2 <−> Row 3

Press MATRX ▶ to get the Matrix Menu

Select 8, perform the instruction rowSwap(Ans,2,3)

$$\begin{bmatrix} 1 & 2 & -3 & 11 \\ 0 & 3 & -6 & 15 \end{bmatrix}$$

Answer is displayed $[0 \quad 0 \quad 1 \quad -1]$

$$\vdots$$

$$\vdots$$

Arrive at Reduced Form

$$\begin{bmatrix} 1 & 0 & 0 & 2 \\ 0 & 1 & 0 & 3 \\ 0 & 0 & 1 & -1 \end{bmatrix}$$

This matrix corresponds to the system

$$
\begin{aligned}
x_1 \quad &= \quad 2 \\
x_2 \quad &= \quad 3 \\
x_3 &= \ -1
\end{aligned}
$$

The solution is $x_1 = 2$, $x_2 = 3$, $x_3 = -1$.

Summary of Formats for Entering Row Operations

Press MATRX ▶ to get the Matrix Math Menu

Press 8, 9, 0, A to select the operation

8: Interchange two rows

rowSwap([A], 1st row, 2nd row)

9: Add a row to another row

row+([A], row to be added, row to be added to)

0: Multiply row by a number

*row(number, [A], row to be multiplied)

A: Add a multiple of a row to another row

*row(number, ([A], row to be multiplied and added, row to be added to)

Exercises C.2

1. Determine how to enter/display a matrix and how to perform elementary row operations on your calculator.

2. Solve the following system of equations

$$
\begin{aligned}
4x_1 + 8x_2 - 12x_3 &= \ 44 \\
3x_1 + 6x_2 - \ 8x_3 &= \ 32 \\
-2x_1 - \ x_2 \qquad\quad &= -7
\end{aligned}
$$

using the steps given above on your calculator.

In Exercises 3–8, you are given a matrix followed by an elementary row operation.
(a) Determine each resulting matrix by hand.
(b) Perform these transformations using a calculator.

*3. $\begin{bmatrix} 2 & 6 & -4 & 0 \\ 1 & 2 & -3 & 6 \\ 8 & 3 & 2 & 5 \end{bmatrix} \approx \left(\tfrac{1}{2}\right)R1$

4. $\begin{bmatrix} 0 & -8 & 4 & 3 \\ 2 & 7 & 5 & 1 \\ 3 & -5 & 8 & 9 \end{bmatrix} \underset{R1 <-> R2}{\sim}$

*5. $\begin{bmatrix} 1 & 2 & 3 & -1 \\ -1 & 1 & 7 & 1 \\ 2 & -4 & 5 & -3 \end{bmatrix} \approx$ Add R1 to R2 / Add (-2)R1 to R3

6. $\begin{bmatrix} 1 & 2 & 3 & -4 \\ 0 & 1 & 2 & 1 \\ 0 & -4 & 3 & -5 \end{bmatrix} \approx$ Add (-2)R2 to R1 / Add (4)R2 to R3

*7. $\begin{bmatrix} 1 & 0 & 4 & -3 \\ 0 & 1 & -3 & 2 \\ 0 & 0 & 1 & 5 \end{bmatrix}$ $\approx$ **8.** $\begin{bmatrix} 1 & 0 & 2 & 7 \\ 0 & 1 & 5 & -3 \\ 0 & 0 & -2 & 8 \end{bmatrix}$ $\approx$

Add (-4)R3 to R1 $\left(-\frac{1}{2}\right)$R3
Add (3)R3 to R2

In Exercises 9–14, the systems of equations all have unique solutions.
(a) Solve the systems by hand using the method of Gauss-Jordan elimination.
(b) Solve using the method of Gauss-Jordan elimination using a calculator.

*9.
$$2x_2 + 4x_3 = 8$$
$$2x_1 + 2x_2 \qquad = 6$$
$$x_1 + x_2 + x_3 = 5$$

10.
$$x_1 - 2x_2 - 4x_3 = -9$$
$$x_1 + 5x_2 + 10x_3 = 21$$
$$2x_1 - 3x_2 - 5x_3 = -13$$

*11.
$$x_1 + 2x_2 + 3x_3 = 14$$
$$2x_1 + 5x_2 + 8x_3 = 36$$
$$x_1 - x_2 \qquad = -4$$

12.
$$x_1 - x_2 - x_3 = -1$$
$$-2x_1 + 6x_2 + 10x_3 = 14$$
$$2x_1 + x_2 + 6x_3 = 9$$

*13.
$$2x_1 + 2x_2 - 4x_3 = 14$$
$$3x_1 + x_2 + x_3 = 8$$
$$2x_1 - x_2 + 2x_3 = -1$$

14.
$$x_1 + 2x_2 - x_3 = 3$$
$$x_1 + 3x_2 - 2x_3 = -6$$
$$-x_1 - x_2 + 3x_3 = 6$$

In Exercises 15 and 16, the systems of equations have unique solutions. Solve the systems using the method of Gauss-Jordan elimination using a calculator.

*15.
$$1.5x_1 \qquad + 3x_3 = 15$$
$$-x_1 + 7x_2 - 9x_3 = -45$$
$$2x_1 + \qquad 5x_3 = 22$$

16.
$$-3x_1 - 6x_2 - 15x_3 = -3$$
$$x_1 + 1.5x_2 + 4.5x_3 = 0.5$$
$$-2x_1 - 3.5x_2 - 8.5x_3 = -2$$

17. Construct a system of linear equations having solution $x_1 = 1$, $x_2 = 2$, and $x_3 = 3$, in which all three types of elementary row operations are used to arrive at this solution using Gauss-Jordan elimination.

18. Compute the reduced echelon forms of a number of 3×3 matrices entered at random. Why do you get these reduced echelon forms? How can you get another type of reduced echelon form? (*Hint:* Think about the geometry.)

1.3   Gauss-Jordan Elimination

In the previous section we used the method of Gauss-Jordan elimination to solve systems of n equations in n variables that had a unique solution. We shall now discuss the method in its more general setting, where the number of equations can differ from the number of variables and where there can be a unique solution, many solutions, or no solution. Our approach will again be to start from the augmented matrix of the given system and to perform a sequence of elementary row operations that will result in a simpler matrix (the reduced echelon form), which leads directly to the solution.

We now give the general definition of reduced echelon form. You will observe that the reduced echelon forms discussed in the previous section all conform to this definition.

Definition   *A matrix is in **reduced echelon form** if*

    1. *Any rows consisting entirely of zeros are grouped at the bottom of the matrix.*

    2. *The first nonzero element of each other row is 1. This element is called a **leading 1**.*

    3. *The leading 1 of each row after the first is positioned to the right of the leading 1 of the previous row.*

    4. *All other elements in a column that contains a leading 1 are zero.*

The following matrices are all in reduced echelon form.

$$\begin{bmatrix} 1 & 0 & 8 \\ 0 & 1 & 2 \\ 0 & 0 & 0 \end{bmatrix} \quad \begin{bmatrix} 1 & 0 & 0 & 7 \\ 0 & 1 & 0 & 3 \\ 0 & 0 & 1 & 9 \end{bmatrix} \quad \begin{bmatrix} 1 & 4 & 0 & 0 \\ 0 & 0 & 1 & 0 \\ 0 & 0 & 0 & 1 \end{bmatrix} \quad \begin{bmatrix} 1 & 2 & 3 & 0 \\ 0 & 0 & 0 & 1 \\ 0 & 0 & 0 & 0 \end{bmatrix}$$

$$\begin{bmatrix} 1 & 0 & 5 & 0 & 0 & 8 \\ 0 & 1 & 7 & 0 & 0 & 9 \\ 0 & 0 & 0 & 1 & 0 & 5 \\ 0 & 0 & 0 & 0 & 1 & 4 \end{bmatrix} \quad \begin{bmatrix} 1 & 2 & 0 & 3 & 0 & 4 \\ 0 & 0 & 1 & 2 & 0 & 7 \\ 0 & 0 & 0 & 0 & 1 & 6 \\ 0 & 0 & 0 & 0 & 0 & 0 \end{bmatrix}$$

The following matrices are not in reduced echelon form.

$$\begin{bmatrix} 1 & 2 & 0 & 4 \\ 0 & 0 & 0 & 0 \\ 0 & 0 & 1 & 3 \end{bmatrix} \quad \begin{bmatrix} 1 & 2 & 0 & 3 & 0 \\ 0 & 0 & 3 & 4 & 0 \\ 0 & 0 & 0 & 0 & 1 \end{bmatrix} \quad \begin{bmatrix} 1 & 0 & 0 & 2 \\ 0 & 0 & 1 & 4 \\ 0 & 1 & 0 & 3 \end{bmatrix} \quad \begin{bmatrix} 1 & 7 & 0 & 8 \\ 0 & 1 & 0 & 3 \\ 0 & 0 & 1 & 2 \\ 0 & 0 & 0 & 0 \end{bmatrix}$$

row of zeros   first nonzero   leading 1 in   nonzero
not at bottom   element in row 2 is   row 3 not to   element above
of matrix   not 1   the right of   leading 1 in
     leading 1 in   row 2
     row 2

Many sequences of row operations can usually be used to transform a given matrix to reduced echelon form; they all, however, lead to the same reduced echelon form. We say that *the reduced echelon form of a matrix is unique*. The method of Gauss-Jordan elimination is an important systematic way (called an algorithm) for arriving at the reduced echelon form. It can be programmed on the computer. We now summarize the method, then give examples of its implementation.

Gauss-Jordan Elimination

1. Write down the augmented matrix of the system of linear equations.

2. Derive the reduced echelon form of the augmented matrix using elementary row operations. This is done by creating leading 1s, then zeros above and below each leading 1, column by column starting with the first equation.

3. Write the system of equations corresponding to the reduced echelon form. This system gives the solution.

We stress the importance of mastering this algorithm. Not only is getting the correct solution important, the method of arriving at the solution is important. We shall be interested in the efficiency of this algorithm (the number of additions and multiplications used), and comparing it with other algorithms that can be used to solve systems of linear equations.

Example 1   Use the method of Gauss-Jordan elimination to find the reduced echelon form of the following matrix.

$$\begin{bmatrix} 0 & 0 & 2 & -2 & 2 \\ 3 & 3 & -3 & 9 & 12 \\ 4 & 4 & -2 & 11 & 12 \end{bmatrix}$$

Step 1   Interchange rows, if necessary, to bring a nonzero element to the top of the first nonzero column. This nonzero element is called a **pivot**.

$$\underset{R1 \leftrightarrow R2}{\approx} \begin{bmatrix} ③ & 3 & -3 & 9 & 12 \\ 0 & 0 & 2 & -2 & 2 \\ 4 & 4 & -2 & 11 & 12 \end{bmatrix} \quad \text{pivot}$$

Step 2   Create a 1 in the pivot location by multiplying the pivot row by $\dfrac{1}{\text{pivot}}$.

$$\underset{\frac{1}{3}R1}{\approx} \begin{bmatrix} 1 & 1 & -1 & 3 & 4 \\ 0 & 0 & 2 & -2 & 2 \\ 4 & 4 & -2 & 11 & 12 \end{bmatrix}$$

Step 3   Create zeros elsewhere in the pivot column by adding suitable multiples of the pivot row to all other rows of the matrix.

$$\underset{R3 + (-4)R1}{\approx} \begin{bmatrix} 1 & 1 & -1 & 3 & 4 \\ 0 & 0 & 2 & -2 & 2 \\ 0 & 0 & 2 & -1 & -4 \end{bmatrix}$$

Step 4   Cover the pivot row and all rows above it. Repeat steps 1 and 2 for the remaining submatrix. Repeat step 3 for the whole matrix. Continue thus until the reduced echelon form is reached.

$$\begin{bmatrix} 1 & 1 & -1 & 3 & 4 \\ 0 & 0 & 2 & -2 & 2 \\ 0 & 0 & 2 & -1 & -4 \end{bmatrix} = \begin{bmatrix} 1 & 1 & -1 & 3 & 4 \\ 0 & 0 & ② & -2 & 2 \\ 0 & 0 & 2 & -1 & -4 \end{bmatrix}$$

first nonzero column
of the submatrix

pivot

$$\underset{\frac{1}{2}R2}{\approx} \begin{bmatrix} 1 & 1 & -1 & 3 & 4 \\ 0 & 0 & 1 & -1 & 1 \\ 0 & 0 & 2 & -1 & -4 \end{bmatrix}$$

$$\approx \begin{array}{c} \\ \\ \text{R1 + R2} \\ \text{R3 + (-2)R2} \end{array} \begin{bmatrix} 1 & 1 & 0 & 2 & 5 \\ 0 & 0 & 1 & -1 & 1 \\ 0 & 0 & 0 & \textcircled{1} & -6 \end{bmatrix}$$

$$\underset{\uparrow}{\text{pivot}}$$

$$\approx \begin{array}{c} \\ \\ \text{R1 + (-2)R3} \\ \text{R2 + R3} \end{array} \begin{bmatrix} 1 & 1 & 0 & 0 & 17 \\ 0 & 0 & 1 & 0 & -5 \\ 0 & 0 & 0 & 1 & -6 \end{bmatrix}$$

This matrix is the reduced echelon form of the given matrix.

We now illustrate how this method is used to solve various systems of equations. The following example illustrates how to solve a system of linear equations that has many solutions. The reduced echelon form is derived. It then becomes necessary to interpret the reduced echelon form, expressing the many solutions in a clear manner.

Example 2   Solve, if possible, the system of equations

$$3x_1 - 3x_2 + 3x_3 = 9$$
$$2x_1 - x_2 + 4x_3 = 7$$
$$3x_1 - 5x_2 - x_3 = 7$$

Solution   Start with the augmented matrix and follow the Gauss-Jordan algorithm. Pivots and leading 1s are circled.

$$\begin{bmatrix} \textcircled{3} & -3 & 3 & 9 \\ 2 & -1 & 4 & 7 \\ 3 & -5 & -1 & 7 \end{bmatrix} \underset{\frac{1}{3}\text{R1}}{\approx} \begin{bmatrix} \textcircled{1} & -1 & 1 & 3 \\ 2 & -1 & 4 & 7 \\ 3 & -5 & -1 & 7 \end{bmatrix}$$

$$\underset{\begin{array}{c}\text{R2 + (-2)R1} \\ \text{R3 + (-3)R1}\end{array}}{\approx} \begin{bmatrix} 1 & -1 & 1 & 3 \\ 0 & \textcircled{1} & 2 & 1 \\ 0 & -2 & -4 & -2 \end{bmatrix} \underset{\begin{array}{c}\text{R1 + R2} \\ \text{R3 + 2R2}\end{array}}{\approx} \begin{bmatrix} 1 & 0 & 3 & 4 \\ 0 & 1 & 2 & 1 \\ 0 & 0 & 0 & 0 \end{bmatrix}$$

We have arrived at the reduced echelon form. The corresponding system of equations is

$$x_1 \quad + 3x_3 = 4$$
$$x_2 + 2x_3 = 1$$

Many values of x_1, x_2, and x_3 satisfy these equations. This is a system of equations that has many solutions. x_1 is called the **leading variable** of the first equation, and x_2 is the leading variable of the second equation. To express these many solutions, we write the leading variables in each equation in terms of the remaining variables. We get

$$x_1 = -3x_3 + 4$$
$$x_2 = -2x_3 + 1$$

Let us assign the arbitrary value r to x_3. The **general solution** to the system is

$$x_1 = -3r + 4, \quad x_2 = -2r + 1, \quad x_3 = r$$

As r ranges over the set of real numbers, we get many solutions; r is called a **parameter**. We can get specific solutions by giving r different values. For example,

$$r = 1 \quad \text{gives} \quad x_1 = 1, \quad x_2 = -1, \quad x_3 = 1$$
$$r = -2 \quad \text{gives} \quad x_1 = 10, \quad x_2 = 5, \quad x_3 = -2$$

Example 3   Solve the system of equations

$$2x_1 - 4x_2 + 12x_3 - 10x_4 = 58$$
$$-x_1 + 2x_2 - 3x_3 + 2x_4 = -14$$
$$2x_1 - 4x_2 + 9x_3 - 6x_4 = 44$$

Solution   Starting with the augmented matrix, we get

$$\begin{bmatrix} ② & -4 & 12 & -10 & 58 \\ -1 & 2 & -3 & 2 & -14 \\ 2 & -4 & 9 & -6 & 44 \end{bmatrix} \underset{\frac{1}{2}R1}{\approx} \begin{bmatrix} ① & -2 & 6 & -5 & 29 \\ -1 & 2 & -3 & 2 & -14 \\ 2 & -4 & 9 & -6 & 44 \end{bmatrix}$$

$$\underset{\substack{R2 + R1 \\ R3 + (-2)R1}}{\approx} \begin{bmatrix} 1 & -2 & 6 & -5 & 29 \\ 0 & 0 & ③ & -3 & 15 \\ 0 & 0 & -3 & 4 & -14 \end{bmatrix} \underset{\frac{1}{3}R2}{\approx} \begin{bmatrix} 1 & -2 & 6 & -5 & 29 \\ 0 & 0 & ① & -1 & 5 \\ 0 & 0 & -3 & 4 & -14 \end{bmatrix}$$

$$\underset{\substack{R1 + (-6)R2 \\ R3 + (3)R2}}{\approx} \begin{bmatrix} 1 & -2 & 0 & 1 & -1 \\ 0 & 0 & 1 & -1 & 5 \\ 0 & 0 & 0 & 1 & 1 \end{bmatrix} \underset{\substack{R1 + (-1)R3 \\ R2 + R3}}{\approx} \begin{bmatrix} 1 & -2 & 0 & 0 & -2 \\ 0 & 0 & 1 & 0 & 6 \\ 0 & 0 & 0 & 1 & 1 \end{bmatrix}$$

This is the reduced echelon form. The corresponding system of equations is

$$x_1 - 2x_2 = -2$$
$$x_3 = 6$$
$$x_4 = 1$$

There are many solutions. Once again write the leading variables in terms of the remaining variables. We get

$$x_1 = 2x_2 - 2, \quad x_3 = 6, \quad x_4 = 1$$

Assign the arbitrary value r to x_2. The general solution is

$$x_1 = 2r - 2, \quad x_2 = r, \quad x_3 = 6, \quad x_4 = 1$$

Specific solutions can be obtained by giving r various values.

Example 4   This example illustrates that it may be necessary to interchange rows at a certain stage. Solve the system of equations

$$
\begin{aligned}
x_1 + 2x_2 - x_3 + 3x_4 + x_5 &= 2 \\
2x_1 + 4x_2 - 2x_3 + 6x_4 + 3x_5 &= 6 \\
-x_1 - 2x_2 + x_3 - x_4 + 3x_5 &= 4
\end{aligned}
$$

Solution   On applying the Gauss-Jordan algorithm, we get

$$
\begin{bmatrix}
① & 2 & -1 & 3 & 1 & 2 \\
2 & 4 & -2 & 6 & 3 & 6 \\
-1 & -2 & 1 & -1 & 3 & 4
\end{bmatrix}
\underset{\substack{\text{R2} + (-2)\text{R1} \\ \text{R3} + \text{R1}}}{\approx}
\begin{bmatrix}
1 & 2 & -1 & 3 & 1 & 2 \\
0 & 0 & 0 & 0 & 1 & 2 \\
0 & 0 & 0 & ② & 4 & 6
\end{bmatrix}
$$

$$
\underset{\text{R2} \leftrightarrow \text{R3}}{\approx}
\begin{bmatrix}
1 & 2 & -1 & 3 & 1 & 2 \\
0 & 0 & 0 & ② & 4 & 6 \\
0 & 0 & 0 & 0 & 1 & 2
\end{bmatrix}
\underset{\frac{1}{2}\text{R2}}{\approx}
\begin{bmatrix}
1 & 2 & -1 & 3 & 1 & 2 \\
0 & 0 & 0 & ① & 2 & 3 \\
0 & 0 & 0 & 0 & 1 & 2
\end{bmatrix}
$$

$$
\underset{\text{R1} + (-3)\text{R2}}{\approx}
\begin{bmatrix}
1 & 2 & -1 & 0 & -5 & -7 \\
0 & 0 & 0 & 1 & 2 & 3 \\
0 & 0 & 0 & 0 & ① & 2
\end{bmatrix}
\underset{\substack{\text{R1} + 5\text{R3} \\ \text{R2} + (-2)\text{R3}}}{\approx}
\begin{bmatrix}
1 & 2 & -1 & 0 & 0 & 3 \\
0 & 0 & 0 & 1 & 0 & -1 \\
0 & 0 & 0 & 0 & 1 & 2
\end{bmatrix}
$$

We have arrived at the reduced echelon form. The corresponding system of equations is

$$
\begin{aligned}
x_1 + 2x_2 - x_3 &= 3 \\
x_4 &= -1 \\
x_5 &= 2
\end{aligned}
$$

Expressing the leading variables in terms of remaining variables, we get

$$x_1 = -2x_2 + x_3 + 3, \quad x_4 = -1, \quad x_5 = 2$$

Let us assign the arbitrary values r to x_2 and s to x_3. The general solution is

$$x_1 = -2r + s + 3, \quad x_2 = r, \quad x_3 = s, \quad x_4 = -1, \quad x_5 = 2$$

Example 5   The example illustrates a system that has no solution. Let us try to solve the system

$$
\begin{aligned}
x_1 - x_2 + 2x_3 &= 3 \\
2x_1 - 2x_2 + 5x_3 &= 4 \\
x_1 + 2x_2 - x_3 &= -3 \\
2x_2 + 2x_3 &= 1
\end{aligned}
$$

Solution   Starting with the augmented matrix, we get

$$
\begin{bmatrix} ① & -1 & 2 & 3 \\ 2 & -2 & 5 & 4 \\ 1 & 2 & -1 & -3 \\ 0 & 2 & 2 & 1 \end{bmatrix}
\underset{\substack{R2 + (-2)R1 \\ R3 + (-1)R1}}{\approx}
\begin{bmatrix} 1 & -1 & 2 & 3 \\ 0 & 0 & 1 & -2 \\ 0 & ③ & -3 & -6 \\ 0 & 2 & 2 & 1 \end{bmatrix}
$$

$$
\underset{R2 \leftrightarrow R3}{\approx}
\begin{bmatrix} 1 & -1 & 2 & 3 \\ 0 & ③ & -3 & -6 \\ 0 & 0 & 1 & -2 \\ 0 & 2 & 2 & 1 \end{bmatrix}
$$

$$
\underset{\frac{1}{3}R2}{\approx}
\begin{bmatrix} 1 & -1 & 2 & 3 \\ 0 & ① & -1 & -2 \\ 0 & 0 & 1 & -2 \\ 0 & 2 & 2 & 1 \end{bmatrix}
$$

$$
\underset{\substack{R1 + R2 \\ R4 + (-2)R2}}{\approx}
\begin{bmatrix} 1 & 0 & 1 & 1 \\ 0 & 1 & -1 & -2 \\ 0 & 0 & ① & -2 \\ 0 & 0 & 4 & 5 \end{bmatrix}
$$

$$
\underset{\substack{R1 + (-1)R3 \\ R2 + R3 \\ R4 + (-4)R3}}{\approx}
\begin{bmatrix} 1 & 0 & 0 & 3 \\ 0 & 1 & 0 & -4 \\ 0 & 0 & 1 & -2 \\ 0 & 0 & 0 & ⑬ \end{bmatrix}
$$

$$
\underset{\frac{1}{13}R4}{\approx}
\begin{bmatrix} 1 & 0 & 0 & 3 \\ 0 & 1 & 0 & -4 \\ 0 & 0 & 1 & -2 \\ 0 & 0 & 0 & ① \end{bmatrix}
$$

This matrix is still not in reduced echelon form; zeros still have to be created above the 1 in the last row. However, in such a situation, when the last nonzero row of a matrix is of the form $(0 \quad 0 \quad \cdots \quad 0 \quad 1)$, there is no need to proceed further. The system has no solution. To see this, let us write the equation that corresponds to the last row of the matrix.

$$0x_1 + 0x_2 + 0x_3 = 1$$

This equation cannot be satisfied by any values of x_1, x_2, and x_3. Thus the system of equations has no solution.

Homogeneous Systems of Linear Equations

A system of linear equations is said to be **homogeneous** if all the constant terms are zeros. As we proceed in the course, we shall find that homogeneous systems of linear equations have many interesting properties and play a key role in our discussions.

The following system is a homogeneous system of linear equations.

$$x_1 + 2x_2 - 5x_3 = 0$$
$$-2x_1 - 3x_2 + 6x_3 = 0$$

Observe that $x_1 = 0$, $x_2 = 0$, $x_3 = 0$ is a solution to this system. It is apparent that this result can be extended as follows to any homogeneous system of equations.

Theorem 1.1   A homogeneous system of linear equations in n variables always has the solution $x_1 = 0$, $x_2 = 0$, ..., $x_n = 0$. This solution is called the **trivial solution**.

Let us see if the above homogeneous system has any other solutions. We solve the above system using Gauss-Jordan elimination.

$$\begin{bmatrix} 1 & 2 & -5 & 0 \\ -2 & -3 & 6 & 0 \end{bmatrix} \underset{\text{R2 + 2R1}}{\approx} \begin{bmatrix} 1 & 2 & -5 & 0 \\ 0 & 1 & -4 & 0 \end{bmatrix}$$

$$\underset{\text{R1 + (−2)R2}}{\approx} \begin{bmatrix} 1 & 0 & 3 & 0 \\ 0 & 1 & -4 & 0 \end{bmatrix}$$

This reduced echelon form gives the system

$$x_1 \qquad + 3x_3 = 0$$
$$x_2 - 4x_3 = 0$$

Expressing the leading variables in terms of the remaining variable, we get

$$x_1 = -3x_3$$
$$x_2 = 4x_3$$

Letting $x_3 = r$, we see that the system has many solutions.

$$x_1 = -3r, \quad x_2 = 4r, \quad x_3 = r$$

Observe that the solution $x_1 = 0$, $x_2 = 0$, $x_3 = 0$ is obtained by letting $r = 0$.

In a similar manner the augmented matrix of any homogeneous system of linear equations that has more variables than equations will have a reduced echelon form that has more columns than rows, and has zeros in the last column. The corresponding system of equations, and thus the original system, will have many solutions, one of which is the trivial solution as in the last example. We summarize this important observation in the following theorem.

Theorem 1.2   A homogeneous system of linear equations that has more variables than equations has many solutions.

In this section we have completed the discussion of the method of Gauss-Jordan elimination for solving systems of linear equations. There is another popular elimination method for solving systems of linear equations called **Gaussian elimination**. We introduce that method and compare the merits of the two methods in Section 9.1.

Exercise Set 1.3

1. Determine whether the following matrices are in reduced echelon form. If a matrix is not in reduced echelon form, give a reason.

*(a) $\begin{bmatrix} 1 & 0 & 2 \\ 0 & 1 & 3 \end{bmatrix}$

(b) $\begin{bmatrix} 1 & 2 & 0 & 4 \\ 0 & 0 & 1 & 7 \end{bmatrix}$

*(c) $\begin{bmatrix} 1 & 2 & 5 & 6 \\ 0 & 1 & 3 & -7 \end{bmatrix}$

(d) $\begin{bmatrix} 1 & 4 & 0 & 5 \\ 0 & 0 & 2 & 9 \end{bmatrix}$

*(e) $\begin{bmatrix} 1 & 0 & 0 \\ 0 & 1 & 0 \\ 0 & 0 & 1 \end{bmatrix}$

(f) $\begin{bmatrix} 1 & 5 & 0 \\ 0 & 0 & 1 \\ 0 & 0 & 0 \end{bmatrix}$

(g) $\begin{bmatrix} 1 & 0 & 0 & 4 \\ 0 & 1 & 0 & 5 \\ 0 & 0 & 1 & 9 \end{bmatrix}$

*(h) $\begin{bmatrix} 1 & 0 & 0 & 3 & 2 \\ 0 & 2 & 0 & 6 & 1 \\ 0 & 0 & 1 & 2 & 3 \end{bmatrix}$

(i) $\begin{bmatrix} 1 & 0 & 3 & 0 \\ 0 & 1 & 6 & 0 \\ 0 & 0 & 0 & 1 \end{bmatrix}$

2. Determine whether the following matrices are in reduced echelon form. If a matrix is not in reduced echelon form, give a reason.

*(a) $\begin{bmatrix} 1 & 0 & 3 & -2 \\ 0 & 0 & 1 & 8 \\ 0 & 1 & 4 & 9 \end{bmatrix}$

(b) $\begin{bmatrix} 1 & 2 & 0 & 0 & 4 \\ 0 & 0 & 1 & 0 & 6 \\ 0 & 0 & 0 & 1 & 5 \end{bmatrix}$

*(c) $\begin{bmatrix} 1 & 5 & 0 & 2 & 0 \\ 0 & 0 & 1 & 9 & 0 \\ 0 & 0 & 0 & 0 & 1 \\ 0 & 0 & 0 & 0 & 0 \end{bmatrix}$

(d) $\begin{bmatrix} 1 & 0 & 4 & 2 & 6 \\ 0 & 1 & 2 & 3 & 4 \\ 0 & 0 & 0 & 1 & 2 \\ 0 & 0 & 0 & 0 & 1 \end{bmatrix}$

*(e) $\begin{bmatrix} 1 & 0 & 2 & 0 & 3 \\ 0 & 0 & 0 & 0 & 0 \\ 0 & 1 & 2 & 0 & 7 \\ 0 & 0 & 0 & 1 & 3 \end{bmatrix}$

(f) $\begin{bmatrix} 1 & 0 & 4 & 0 & 0 \\ 0 & 1 & 2 & 0 & 0 \\ 0 & 0 & 0 & 1 & 0 \\ 0 & 0 & 0 & 0 & 1 \end{bmatrix}$

*(g) $\begin{bmatrix} 1 & 0 & 0 & 5 & 3 \\ 0 & 0 & 1 & 0 & 3 \\ 0 & 1 & 2 & 3 & 7 \end{bmatrix}$

(h) $\begin{bmatrix} 0 & 0 & 1 & 0 & 4 \\ 0 & 0 & 0 & 1 & 5 \\ 0 & 1 & 0 & 0 & 3 \end{bmatrix}$

*(i) $\begin{bmatrix} 1 & 5 & -3 & 0 & 7 \\ 0 & 0 & 0 & 1 & 4 \\ 0 & 0 & 0 & 0 & 0 \end{bmatrix}$

3. Each of the following matrices is the reduced echelon form of the augmented matrix of a system of linear equations. Give the solution (if it exists) to each system of equations.

*(a) $\begin{bmatrix} 1 & 0 & 0 & 2 \\ 0 & 1 & 0 & 4 \\ 0 & 0 & 1 & -3 \end{bmatrix}$

(b) $\begin{bmatrix} 1 & 0 & -3 & 4 \\ 0 & 1 & 2 & 8 \\ 0 & 0 & 0 & 0 \end{bmatrix}$

*(c) $\begin{bmatrix} 1 & 3 & 0 & 6 \\ 0 & 0 & 1 & -2 \\ 0 & 0 & 0 & 0 \end{bmatrix}$

(d) $\begin{bmatrix} 1 & 0 & 5 & 0 \\ 0 & 1 & -7 & 0 \\ 0 & 0 & 0 & 1 \end{bmatrix}$

*(e) $\begin{bmatrix} 1 & 0 & 0 & 5 & 3 \\ 0 & 1 & 0 & 6 & -2 \\ 0 & 0 & 1 & 2 & -4 \end{bmatrix}$

(f) $\begin{bmatrix} 1 & 3 & 0 & 0 & 2 \\ 0 & 0 & 1 & 0 & 4 \\ 0 & 0 & 0 & 1 & 5 \end{bmatrix}$

4. Each of the following matrices is the reduced echelon form of the augmented matrix of a system of linear equations. Give the solution (if it exists) to each system of equations.

*(a) $\begin{bmatrix} 1 & 0 & 2 & 4 & 1 \\ 0 & 1 & -3 & 5 & -6 \\ 0 & 0 & 0 & 0 & 0 \end{bmatrix}$

(b) $\begin{bmatrix} 1 & -3 & 2 & 0 & 4 \\ 0 & 0 & 0 & 1 & -7 \\ 0 & 0 & 0 & 0 & 0 \end{bmatrix}$

*(c) $\begin{bmatrix} 1 & -2 & 0 & 3 & 0 & 4 \\ 0 & 0 & 1 & 2 & 0 & 9 \\ 0 & 0 & 0 & 0 & 1 & 8 \end{bmatrix}$

(d) $\begin{bmatrix} 1 & 0 & 2 & 0 & 3 & 6 \\ 0 & 1 & 5 & 0 & 4 & 7 \\ 0 & 0 & 0 & 1 & 9 & -3 \end{bmatrix}$

5. Solve (if possible) each of the following systems of three equations in three variables using the method of Gauss-Jordan elimination.

*(a) $\begin{aligned} x_1 + 4x_2 + 3x_3 &= 1 \\ 2x_1 + 8x_2 + 11x_3 &= 7 \\ x_1 + 6x_2 + 7x_3 &= 3 \end{aligned}$

(b) $\begin{aligned} x_1 + 2x_2 + 4x_3 &= 15 \\ 2x_1 + 4x_2 + 9x_3 &= 33 \\ x_1 + 3x_2 + 5x_3 &= 20 \end{aligned}$

*(c) $\begin{aligned} x_1 + x_2 + x_3 &= 7 \\ 2x_1 + 3x_2 + x_3 &= 18 \\ -x_1 + x_2 - 3x_3 &= 1 \end{aligned}$

(d) $\begin{aligned} x_1 + 4x_2 + x_3 &= 2 \\ x_1 + 2x_2 - x_3 &= 0 \\ 2x_1 + 6x_2 &= 3 \end{aligned}$

*(e) $\begin{aligned} x_1 - x_2 + x_3 &= 3 \\ 2x_1 - x_2 + 4x_3 &= 7 \\ 3x_1 - 5x_2 - x_3 &= 7 \end{aligned}$

(f) $\begin{aligned} 3x_1 - 3x_2 + 9x_3 &= 24 \\ 2x_1 - 2x_2 + 7x_3 &= 17 \\ -x_1 + 2x_2 - 4x_3 &= -11 \end{aligned}$

6. Solve (if possible) each of the following systems of three equations in three variables using the method of Gauss-Jordan elimination.

*(a) $\begin{aligned} 3x_1 + 6x_2 - 3x_3 &= 6 \\ -2x_1 - 4x_2 - 3x_3 &= -1 \\ 3x_1 + 6x_2 - 2x_3 &= 10 \end{aligned}$

(b) $\begin{aligned} x_1 + 2x_2 + x_3 &= 7 \\ x_1 + 2x_2 + 2x_3 &= 11 \\ 2x_1 + 4x_2 + 3x_3 &= 18 \end{aligned}$

*(c) $\begin{aligned} x_1 + 2x_2 - x_3 &= 3 \\ 2x_1 + 4x_2 - 2x_3 &= 6 \\ 3x_1 + 6x_2 + 2x_3 &= 1 \end{aligned}$

(d) $\begin{aligned} x_1 + 2x_2 + 3x_3 &= 8 \\ 3x_1 + 7x_2 + 9x_3 &= 26 \\ 2x_1 + 6x_3 &= 11 \end{aligned}$

*(e) $\begin{aligned} x_2 + 2x_3 &= 5 \\ x_1 + 2x_2 + 5x_3 &= 13 \\ x_1 + 2x_3 &= 4 \end{aligned}$

(f) $\begin{aligned} x_1 + 2x_2 + 8x_3 &= 7 \\ 2x_1 + 4x_2 + 16x_3 &= 14 \\ x_2 + 3x_3 &= 4 \end{aligned}$

7. Solve (if possible) each of the following systems of equations using the method of Gauss-Jordan elimination.

*(a) $\begin{aligned} x_1 + x_2 - 3x_3 &= 10 \\ -3x_1 - 2x_2 + 4x_3 &= -24 \end{aligned}$

(b) $\begin{aligned} 2x_1 - 6x_2 - 14x_3 &= 38 \\ -3x_1 + 7x_2 + 15x_3 &= -37 \end{aligned}$

*(c) $\begin{aligned} x_1 + 2x_2 - x_3 - x_4 &= 0 \\ x_1 + 2x_2 + x_4 &= 4 \\ -x_1 - 2x_2 + 2x_3 + 4x_4 &= 5 \end{aligned}$

(d) $\begin{aligned} x_1 + 2x_2 + 4x_4 &= 0 \\ -2x_1 - 4x_2 + 3x_3 - 2x_4 &= 0 \end{aligned}$
    (a homogeneous system)

*(e) $\begin{aligned} x_2 - 3x_3 + x_4 &= 0 \\ x_1 + x_2 - x_3 + 4x_4 &= 0 \\ -2x_1 - 2x_2 + 2x_3 - 8x_4 &= 0 \end{aligned}$
    (a homogeneous system)

8. Solve (if possible) each of the following systems of equations using the method of Gauss-Jordan elimination.

*(a) $\begin{aligned} x_1 + x_2 + x_3 - x_4 &= -3 \\ 2x_1 + 3x_2 + x_3 - 5x_4 &= -9 \\ x_1 + 3x_2 - x_3 - 6x_4 &= -7 \\ -x_1 - x_2 - x_3 &= 1 \end{aligned}$

(b) $\begin{aligned} x_2 + 2x_3 &= 7 \\ x_1 - 2x_2 - 6x_3 &= -18 \\ -x_1 - x_2 - 2x_3 &= -5 \\ 2x_1 - 5x_2 - 15x_3 &= -46 \end{aligned}$

*(c) $\begin{aligned} 2x_1 - 4x_2 + 16x_3 - 14x_4 &= 10 \\ -x_1 + 5x_2 - 17x_3 + 19x_4 &= -2 \\ x_1 - 3x_2 + 11x_3 - 11x_4 &= 4 \\ 3x_1 - 4x_2 + 18x_3 - 13x_4 &= 17 \end{aligned}$

*(d) $\begin{aligned} x_1 - x_2 + 2x_3 &= 7 \\ 2x_1 - 2x_2 + 2x_3 - 4x_4 &= 12 \\ -x_1 + x_2 - x_3 + 2x_4 &= -4 \\ -3x_1 + x_2 - 8x_3 - 10x_4 &= -29 \end{aligned}$

(e) $\begin{aligned} x_1 + 6x_2 - x_3 - 4x_4 &= 0 \\ -2x_1 - 12x_2 + 5x_3 + 17x_4 &= 0 \\ 3x_1 + 18x_2 - x_3 - 6x_4 &= 0 \end{aligned}$
    (a homogeneous system)

(f) $\begin{aligned} 4x_1 + 8x_2 - 12x_3 &= 28 \\ -x_1 - 2x_2 + 3x_3 &= -7 \\ 2x_1 + 4x_2 - 8x_3 &= 16 \\ -3x_1 - 6x_2 + 9x_3 &= -21 \end{aligned}$

*(g) $\begin{aligned} x_1 + x_2 &= 2 \\ 2x_1 + 3x_2 &= 3 \\ x_1 + 3x_2 &= 0 \\ x_1 + 2x_2 &= 1 \end{aligned}$

*9. Construct examples of the following:

    (a) A system of linear equations with more variables than equations, having no solution.

    (b) A system of linear equations with more equations than variables, having a unique solution.

10. The reduced echelon forms of the matrices of systems of two equations in two variables, and the types of solutions they represent, can be classified as follows (* corresponds to possible nonzero elements).

$$\begin{bmatrix} 1 & 0 & * \\ 0 & 1 & * \end{bmatrix} \quad \begin{bmatrix} 1 & * & 0 \\ 0 & 0 & 1 \end{bmatrix} \quad \begin{bmatrix} 1 & * & * \\ 0 & 0 & 0 \end{bmatrix}$$

      unique   no solutions   many
      solution   solutions

Classify, in a similar manner, the reduced echelon forms of the matrices, and the types of solutions they represent, of **(a)** systems of three equations in two variables, **(b)** systems of three equations in three variables.

*11. Determine the conditions on a, b, c, and d for the matrix $\begin{bmatrix} a & b \\ c & d \end{bmatrix}$ to have reduced echelon form **(a)** $\begin{bmatrix} 1 & 0 \\ 0 & 1 \end{bmatrix}$, **(b)** $\begin{bmatrix} 1 & 0 \\ 0 & 0 \end{bmatrix}$.

12. Consider the homogeneous system of linear equations

$$ax + by = 0$$
$$cx + dy = 0$$

    (a) Show that if $x = x_0$, $y = y_0$ is a solution, then $x = kx_0$, $y = ky_0$ is also a solution, for any value of the constant k.

    (b) Show that if $x = x_0$, $y = y_0$ and $x = x_1$, $y = y_1$ are any two solutions, then $x = x_0 + x_1$, $y = y_0 + y_1$ is also a solution.

13. Show that $x = 0$, $y = 0$ is a solution to the homogeneous system of linear equations

$$ax + by = 0$$
$$cx + dy = 0$$

Prove that this is the only solution if and only if $ad - bc \neq 0$.

14. Consider two systems of linear equations having augmented matrices $[A : B_1]$ and $[A : B_2]$, where the matrix of coefficients of both systems is the same 3×3 matrix A.

    *(a) Is it possible for $[A : B_1]$ to have a unique solution and $[A : B_2]$ to have many solutions?

    *(b) Is it possible for $[A : B_1]$ to have a unique solution and $[A : B_2]$ to have no solution?

    (c) Is it possible for $[A : B_1]$ to have many solutions and $[A : B_2]$ to have no solution?

15. Solve the following systems of linear equations by applying the method of Gauss-Jordan elimination to a large augmented matrix that represents two systems with the same matrix of coefficients.

    (a) $x_1 + x_2 + 5x_3 = b_1$
           $x_1 + 2x_2 + 8x_3 = b_2$
        $2x_1 + 4x_2 + 16x_3 = b_3$

    for $\begin{bmatrix} b_1 \\ b_2 \\ b_3 \end{bmatrix} = \begin{bmatrix} 2 \\ 5 \\ 10 \end{bmatrix}$, $\begin{bmatrix} 3 \\ 2 \\ 4 \end{bmatrix}$ in turn

    *(b) $x_1 + 2x_2 + 4x_3 = b_1$
          $x_1 + x_2 + 2x_3 = b_2$
       $2x_1 + 3x_2 + 6x_3 = b_3$

    for $\begin{bmatrix} b_1 \\ b_2 \\ b_3 \end{bmatrix} = \begin{bmatrix} 8 \\ 5 \\ 13 \end{bmatrix}$, $\begin{bmatrix} 5 \\ 3 \\ 11 \end{bmatrix}$ in turn

16. Write a 3×3 matrix at random. Find its reduced echelon form. The reduced echelon form is probably the identity matrix I_3! Explain this. (*Hint:* Think about the geometry.)

*17. If a 3×4 matrix is written at random, what type of reduced echelon form is it likely to have and why?

18. Computers can carry only a finite number of digits. This causes errors, called round-off errors, to occur when numbers are truncated. Because of this phenomenon, computers can give incorrect results. Much research goes into developing algorithms that minimize such round-off errors. (Readers who are interested in these algorithms should read Section 9.3.) A computer is used to determine the reduced echelon form of an augmented matrix of a system of linear equations. Which of the following is most likely to happen?

    (a) The computer gives a solution to the system, when in fact a solution does not exist.

    (b) The computer gives that a solution does not exist, when in fact a solution does exist.

*1.4 Curve Fitting, Electrical Networks, and Traffic Flow

Systems of linear equations are used in such diverse fields as electrical engineering, economics, and traffic analysis. We now discuss applications in some of these fields.

Curve Fitting

The following problem occurs in many different branches of science. A set of data points

$$(x_1, y_1), (x_2, y_2), \ldots, (x_n, y_n)$$

is given and it is necessary to find a polynomial whose graph passes through the points. The points are often measurements in an experiment. The x coordinates are called **base points**. It can be shown that if the base points are all distinct, then a unique polynomial of degree $n - 1$ (or less)

$$y = a_0 + a_1x + \cdots + a_{n-2}x^{n-2} + a_{n-1}x^{n-1}$$

can be **fitted** to the points. See Figure 1.5.

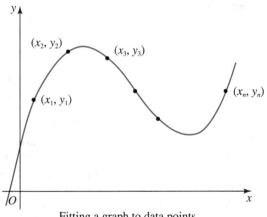

Figure 1.5   Fitting a graph to data points

The coefficients $a_0, a_1, \ldots, a_{n-2}, a_{n-1}$ of the appropriate polynomial can be found by substituting the points into the polynomial equation and then solving a system of linear equations. (It is usual to write the polynomial in terms of ascending powers of x for the purpose of finding these coefficients. The columns of the matrix of coefficients of the system of equations then often follow a pattern. More will be said about this later.)

---

* Sections and chapters marked with an asterisk are optional. The instructor can use these sections to build around the core material to give the course the desired flavor.

We now illustrate the procedure by fitting a polynomial of degree two, a parabola, to a set of three such data points.

Example 1   Determine the equation of the polynomial of degree two whose graph passes through the points (1, 6), (2, 3), (3, 2).

Solution   Observe that in this example we are given *three* points and we want to find a polynomial of degree *two* (one less than the number of data points). Let the polynomial be

$$y = a_0 + a_1x + a_2x^2$$

We are given three points and shall use these three sets of information to determine the three unknowns a_0, a_1 and a_2. Substituting

$$x = 1, y = 6; \; x = 2, y = 3; \; x = 3, y = 2$$

in turn into the polynomial leads to the following system of three linear equations in a_0, a_1, and a_2.

$$a_0 + a_1 + a_2 = 6$$
$$a_0 + 2a_1 + 4a_2 = 3$$
$$a_0 + 3a_1 + 9a_2 = 2$$

Solve this system for a_2, a_1, and a_0 using Gauss-Jordan elimination.

$$\begin{bmatrix} 1 & 1 & 1 & 6 \\ 1 & 2 & 4 & 3 \\ 1 & 3 & 9 & 2 \end{bmatrix} \underset{\substack{R2 + (-1)R1 \\ R3 + (-1)R1}}{\approx} \begin{bmatrix} 1 & 1 & 1 & 6 \\ 0 & 1 & 3 & -3 \\ 0 & 2 & 8 & -4 \end{bmatrix} \underset{\substack{R1 + (-1)R2 \\ R3 + (-2)R2}}{\approx} \begin{bmatrix} 1 & 0 & -2 & 9 \\ 0 & 1 & 3 & -3 \\ 0 & 0 & 2 & 2 \end{bmatrix}$$

$$\underset{\frac{1}{2}R3}{\approx} \begin{bmatrix} 1 & 0 & -2 & 9 \\ 0 & 1 & 3 & -3 \\ 0 & 0 & 1 & 1 \end{bmatrix} \underset{\substack{R1 + 2R3 \\ R2 + (-3)R3}}{\approx} \begin{bmatrix} 1 & 0 & 0 & 11 \\ 0 & 1 & 0 & -6 \\ 0 & 0 & 1 & 1 \end{bmatrix}$$

We get $a_0 = 11$, $a_1 = -6$, $a_2 = 1$. The parabola that passes through these points is $y = 11 - 6x + x^2$. See Figure 1.6.

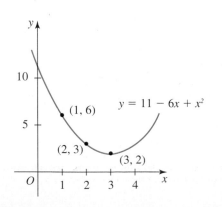

Figure 1.6

Electrical Network Analysis

Systems of linear equations are used to determine the currents through various branches of electrical networks. The following two laws, which are based on experimental verification in the laboratory, lead to the equations.

Kirchoff's Laws*

1. **Junctions:** All the current flowing into a junction must flow out of it.

2. **Paths:** The sum of the *IR* terms (*I* denotes current, *R* resistance) in any direction around a closed path is equal to the total voltage in the path in that direction.

Example 2 Consider the electrical network of Figure 1.7. Let us determine the currents through each branch of this network.

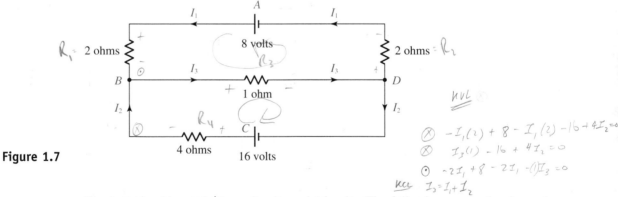

Figure 1.7

The batteries (denoted ⊩) are 8 volts and 16 volts. The following convention is used in electrical engineering to indicate the terminal of the battery out of which the current flows: ⬥. The resistances (denoted ⌇⌇) are one 1-ohm, one 4-ohm, and two 2-ohm. The current entering each battery will be the same as that leaving it.

Let the currents in the various branches of the above circuit be I_1, I_2, and I_3. Kirchhoff's laws refer to junctions and closed paths. There are two junctions in this circuit, namely the points B and D. There are three closed paths, namely $ABDA$, $CBDC$, and $ABCDA$. Apply the laws to the junctions and paths.

---

* Gustav Robert Kirchhoff (1824–1887) was educated at the University of Königsberg and did most of his teaching at the University of Heidelberg. His major contributions were in the experimental discovery and theoretical analysis of the laws of electromagnetic radiation. Kirchhoff was a master teacher whose texts set a standard for the teaching of theoretical physics in German universities. He was described as "not easily drawn out but of a cheerful and obliging disposition."

Junctions

$$\text{Junction } B: \quad I_1 + I_2 = I_3$$

$$\text{Junction } D: \quad I_3 = I_1 + I_2$$

These two equations result in a single linear equation:

$$I_1 + I_2 - I_3 = 0$$

Paths

$$\text{Path } ABDA: \quad 2I_1 + 1I_3 + 2I_1 = 8$$

$$\text{Path } CBDC: \quad 4I_2 + 1I_3 = 16$$

It is not necessary to look further at path *ABCDA*. We now have a system of three linear equations in three unknowns, I_1, I_2, and I_3. Path *ABCDA* leads to an equation that is a combination of the last two equations; there is no new information.

The problem thus reduces to solving the following system of three linear equations in three variables.

$$
\begin{aligned}
I_1 + \;\; I_2 - I_3 &= \;\; 0 \\
4I_1 \qquad\quad + I_3 &= \;\; 8 \\
4I_2 + I_3 &= 16
\end{aligned}
$$

Using the method of Gauss-Jordan elimination, we get

$$
\begin{bmatrix} 1 & 1 & -1 & 0 \\ 4 & 0 & 1 & 8 \\ 0 & 4 & 1 & 16 \end{bmatrix}
\underset{R2 + (-4)R1}{\approx}
\begin{bmatrix} 1 & 1 & -1 & 0 \\ 0 & -4 & 5 & 8 \\ 0 & 4 & 1 & 16 \end{bmatrix}
$$

$$
\underset{-\frac{1}{4}R2}{\approx}
\begin{bmatrix} 1 & 1 & -1 & 0 \\ 0 & 1 & -\frac{5}{4} & -2 \\ 0 & 4 & 1 & 16 \end{bmatrix}
\underset{\substack{R1 + (-1)R2 \\ R3 + (-4)R2}}{\approx}
\begin{bmatrix} 1 & 0 & \frac{1}{4} & 2 \\ 0 & 1 & -\frac{5}{4} & -2 \\ 0 & 0 & 6 & 24 \end{bmatrix}
$$

$$
\underset{\frac{1}{6}R3}{\approx}
\begin{bmatrix} 1 & 0 & \frac{1}{4} & 2 \\ 0 & 1 & -\frac{5}{4} & -2 \\ 0 & 0 & 1 & 4 \end{bmatrix}
\underset{\substack{R1 + -\frac{1}{4}R3 \\ R2 + \frac{5}{4}R3}}{\approx}
\begin{bmatrix} 1 & 0 & 0 & 1 \\ 0 & 1 & 0 & 3 \\ 0 & 0 & 1 & 4 \end{bmatrix}
$$

The currents are $I_1 = 1$, $I_2 = 3$, $I_3 = 4$. The units are amps. The solution is unique, as is to be expected in this physical situation.

Example 3   Determine the currents through the various branches of the electrical network in Figure 1.8. This example illustrates how one has to be conscious of direction in applying law 2 for closed paths.

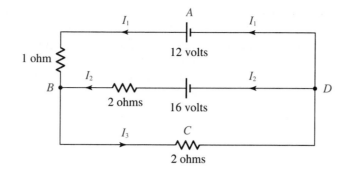

Figure 1.8

Solution   *Junctions*

$$\text{Junction } B: \quad I_1 + I_2 = I_3$$
$$\text{Junction } D: \quad I_3 = I_1 + I_2$$

giving $I_1 + I_2 - I_3 = 0$.

Paths

$$\text{Path } ABCDA: \quad 1I_1 + 2I_3 = 12$$
$$\text{Path } ABDA: \quad 1I_1 + 2(-I_2) = 12 + (-16)$$

Observe that we have selected the direction $ABDA$ around this last path. The current along the branch BD in this direction is $-I_2$, and the voltage is -16. We now have three equations in the three variables I_1, I_2, and I_3.

$$
\begin{aligned}
I_1 + I_2 - I_3 &= 0 \\
I_1 \phantom{{}+I_2} + 2I_3 &= 12 \\
I_1 - 2I_2 \phantom{{}+2I_3} &= -4
\end{aligned}
$$

Solving these equations, we get $I_1 = 2$, $I_2 = 3$, $I_3 = 5$ amps.

In practice, electrical networks can involve many resistances and circuits; determining currents through branches involves solving large systems of equations on the computer.

Traffic Flow

Network analysis, as we saw in the previous discussion, plays an important role in electrical engineering. In recent years the concepts and tools of network analysis have been found to be useful in many other fields, such as information theory and the study of transportation systems. The following analysis of traffic flow through a road network during peak period illustrates how systems of linear equations with many solutions can arise in practice.

Consider the typical road network of Figure 1.9. It represents an area of downtown Jacksonville, Florida. The flow of traffic in and out of the network is measured in vehicles per hour (vph). The figures given here are based on midweek peak traffic hours, 7 to 9 A.M. and 4 to 6 P.M. An increase of 2 percent in the overall flow should be allowed for during the Friday evening traffic flow. Let us construct a mathematical model that can be used to analyze this network.

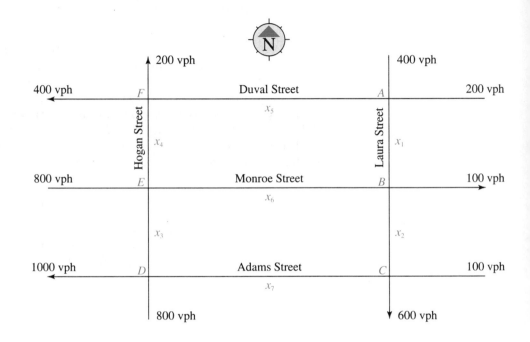

Figure 1.9   Downtown Jacksonville, FL

Let the traffic flows along the various branches be $x_1, \ldots, x_7$ as shown in Figure 1.9. Let us assume that the following traffic law applies.

All traffic entering a junction must leave that junction.

This conservation of flow constraint (compare it to the first of Kirchhoff's laws for electrical networks) leads to a system of linear equations.

Junction A: Traffic entering $= 400 + 200$. Traffic leaving $= x_1 + x_5$.
Thus $x_1 + x_5 = 600$.

Junction B: Traffic entering $= x_1 + x_6$. Traffic leaving $= x_2 + 100$.
Thus $x_1 + x_6 = x_2 + 100$.

Continuing thus for each junction and writing the resulting equations in convenient form with variables on the left and constants on the right, we get the following system of linear equations.

$$\begin{aligned}
\text{Junction } A \quad & x_1 && + x_5 && && = 600 \\
\text{Junction } B \quad & x_1 - x_2 && && + x_6 && = 100 \\
\text{Junction } C \quad & x_2 && && && - x_7 = 500 \\
\text{Junction } D \quad & -x_3 && && && + x_7 = 200 \\
\text{Junction } E \quad & - x_3 + x_4 && && + x_6 && = 800 \\
\text{Junction } F \quad & x_4 + x_5 && && && = 600
\end{aligned}$$

The method of Gauss-Jordan elimination is used to solve this system of equations. Observe that the augmented matrix contains many zeros. These zeros greatly reduce the amount of computation involved. We present the reduced echelon form, leaving its derivation as an exercise. In practice, networks are much larger than the one we have illustrated here, and the systems of linear equations that describe them are thus much larger. The systems are solved on a computer. However, the augmented matrices of all such systems contain many zeros. Much research has gone into the development of efficient algorithms for solving such *sparse* systems of equations efficiently. The augmented matrix and reduced echelon form of the above system are as follows:

$$
\begin{bmatrix}
1 & 0 & 0 & 0 & 1 & 0 & 0 & 600 \\
1 & -1 & 0 & 0 & 0 & 1 & 0 & 100 \\
0 & 1 & 0 & 0 & 0 & 0 & -1 & 500 \\
0 & 0 & -1 & 0 & 0 & 0 & 1 & 200 \\
0 & 0 & -1 & 1 & 0 & 1 & 0 & 800 \\
0 & 0 & 0 & 1 & 1 & 0 & 0 & 600
\end{bmatrix}
\approx \cdots \approx
\begin{bmatrix}
1 & 0 & 0 & 0 & 0 & 1 & -1 & 600 \\
0 & 1 & 0 & 0 & 0 & 0 & -1 & 500 \\
0 & 0 & 1 & 0 & 0 & 0 & -1 & -200 \\
0 & 0 & 0 & 1 & 0 & 1 & -1 & 600 \\
0 & 0 & 0 & 0 & 1 & -1 & 1 & 0 \\
0 & 0 & 0 & 0 & 0 & 0 & 0 & 0
\end{bmatrix}
$$

The system of equations that corresponds to this reduced echelon form is

$$\begin{aligned}
x_1 \quad && && + x_6 - x_7 &= 600 \\
x_2 \quad && && - x_7 &= 500 \\
x_3 \quad && && - x_7 &= -200 \\
x_4 \quad && + x_6 - x_7 &= 600 \\
x_5 && - x_6 + x_7 &= 0
\end{aligned}$$

Expressing each leading variable in terms of the remaining variables, we get

$$\begin{aligned}
x_1 &= -x_6 + x_7 + 600 \\
x_2 &= x_7 + 500 \\
x_3 &= x_7 - 200 \\
x_4 &= -x_6 + x_7 + 600 \\
x_5 &= x_6 - x_7
\end{aligned}$$

As was perhaps to be expected, the system of equations has many solutions—many traffic flows are possible. One does have a certain amount of choice at intersections.

Let us now use this mathematical model to arrive at information. Suppose it becomes necessary to perform road work on the stretch of Adams Street between Laura and Hogan. It is desirable to have as small a flow of traffic as possible along this stretch of road. The flows can be controlled along various branches by means of traffic lights at junctions. What is the minimum flow possible along Adams that would not lead to traffic congestion? What are the flows along the other branches when this is attained? Our model will enable us to answer these questions.

Minimizing the flow along Adams corresponds to minimizing x_7. Since all traffic flows must be greater than or equal to zero, the third equation implies that the minimum value of x_7 is 200, for otherwise x_3 could become negative. (A negative flow would be interpreted as traffic moving in the opposite direction to the one permitted on a one-way street.) Thus the road work must allow for a flow of at least 200 cars per hour on the branch CD in the peak period.

Let us now examine what the flows in the other branches will be when this minimum flow along Adams is attained; $x_7 = 200$ gives

$$
\begin{aligned}
x_1 &= -x_6 + 800 \\
x_2 &= 700 \\
x_3 &= 0 \\
x_4 &= -x_6 + 800 \\
x_5 &= x_6 - 200
\end{aligned}
$$

Since $x_7 = 200$ implies that $x_3 = 0$ and vice-versa, we see that the minimum flow in the branch x_7 can be attained by making $x_3 = 0$; that is, by closing DE to traffic.

Exercise Set 1.4

In Exercises 1–5, determine the equations of the polynomials of degree two whose graphs pass through the given points.

*1. (1, 2), (2, 2), (3, 4)

2. (1, 14), (2, 22), (3, 32)

*3. (1, 5), (2, 7), (3, 9)

*4. (1, 8), (3, 26), (5, 60). What is the value of y when $x = 2$?

5. (−1, −1), (0, 1), (1, −3). What is the value of y when $x = 3$?

*6. Find the equation of the polynomial of degree three whose graph passes through the points (1, −3), (2, −1), (3, 9), (4, 33).

In Exercises 7–14, determine the currents in the various branches of the electrical networks. The units of current are amps and the units of resistance are ohms. (Hint: In Exercise 14 it is difficult to decide the direction of the current along AB. Make a guess. A negative result for the current means that your guess was the wrong one—the current is in the opposite direction. However, the magnitude will be correct. There is no need to rework the problem.)

*7.

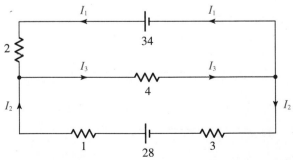

8.

*9.

10.

*11.

12.

*13.

14.

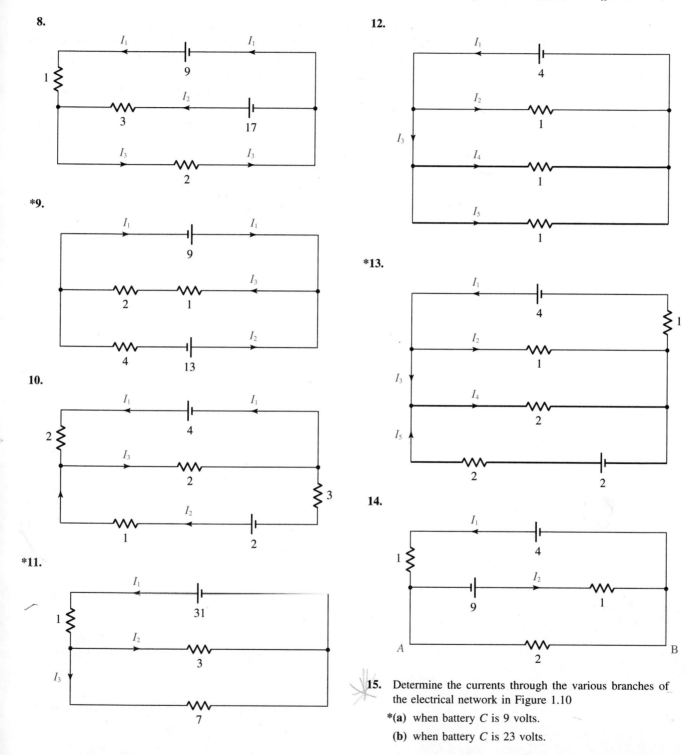

15. Determine the currents through the various branches of the electrical network in Figure 1.10

   *(a) when battery C is 9 volts.

   (b) when battery C is 23 volts.

Note how the current through the branch AB is reversed in (**b**). What would the voltage of C have to be for no current to pass through AB?

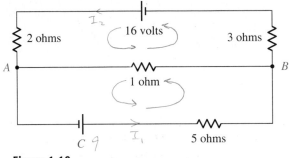

Figure 1.10

16. Construct a mathematical model that describes the traffic flow in the road network of Figure 1.11. All streets are one-way streets in the directions indicated. The units are in vehicles per hour. Give two distinct possible flows of traffic. What is the minimum possible flow that can be expected along branch AB?

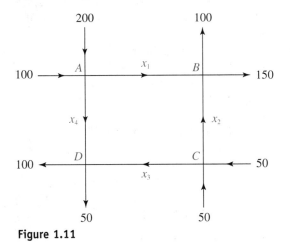

Figure 1.11

*17. Figure 1.12 represents the traffic entering and leaving a "roundabout" road junction. Such junctions are very common in Europe. Construct a mathematical model that describes the flow of traffic along the various branches. What is the minimum flow theoretically possible along the branch BC? Is this flow ever likely to be realized in practice?

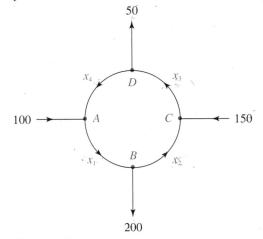

Figure 1.12

Review Exercises Chapter 1*

1. Solve (if possible) each of the following systems of equations.

   (a) $3x + 6y = 3$
   $2x + 4y = 2$

   (b) $-2x + y = 5$
   $2x + y = 1$

   (c) $4x - 8y = 7$
   $-3x + 6y = 9$

2. Determine values of the constant c such that the following systems of equations have the indicated solutions.

   (a) $2x - y = 1$
   $4x + cy = -1$
   Solution: $x = 2, y = 3$

   (b) $x + 2y = 8$
   $cx + 3y = 7$
   Solution: $x = -2, y = 5$

3. Consider the following system of equations for various values of d.

   $$x + y = 2$$
   $$2x + dy = 2d$$

   Which values of d result in a system with (a) a unique solution, (b) many solutions? Prove that the system cannot have no solution.

4. The following systems of equations have unique solutions. Solve these systems using the method of Gauss-Jordan elimination.

   (a) $x_1 - 2x_2 - 6x_3 = -17$
   $2x_1 - 6x_2 - 16x_3 = -46$
   $x_1 + 2x_2 - x_3 = -5$

   (b) $x_2 + 2x_3 + 6x_4 = 21$
   $x_1 - x_2 + x_3 + 5x_4 = 12$
   $x_1 - x_2 - x_3 - 4x_4 = -9$
   $3x_1 - 2x_2 - 6x_4 = -4$

5. Determine whether the following matrices are in reduced echelon form. If a matrix is not in reduced echelon form, give a reason.

   (a) $\begin{bmatrix} 1 & 0 & 4 \\ 0 & 1 & 7 \end{bmatrix}$

   (b) $\begin{bmatrix} 1 & 3 & 0 & 5 \\ 0 & 0 & 1 & 9 \\ 0 & 0 & 0 & 0 \end{bmatrix}$

   (c) $\begin{bmatrix} 1 & 2 & 0 & 6 \\ 0 & 1 & 0 & -7 \\ 0 & 0 & 1 & 9 \end{bmatrix}$

6. Solve (if possible) the following systems of equations using the method of Gauss-Jordan elimination.

   (a) $x_1 - x_2 + x_3 = 3$
   $-2x_1 + 3x_2 + x_3 = -8$
   $4x_1 - 2x_2 + 10x_3 = 10$

   (b) $x_1 + 3x_2 + 6x_3 - 2x_4 = -7$
   $-2x_1 - 5x_2 - 10x_3 + 3x_4 = 10$
   $x_1 + 2x_2 + 4x_3 = 0$
   $x_2 + 2x_3 - 3x_4 = -10$

7. Let A be an $n \times n$ matrix in reduced echelon form. Show that if $A \neq I_n$, then A has a row consisting entirely of zeros.

8. Let A and B be row equivalent matrices. Show that A and B have the same reduced echelon form.

9. Determine the equation of the polynomial of degree two whose graph passes through the points (1, 3), (2, 6), (3, 13).

10. Determine the currents through the branches of the following network.

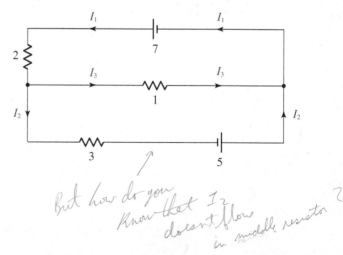

but how do you know that I_2 doesn't flow in middle resistor?

The Use of the Computer in Linear Algebra

A computer package can be used in the introductory linear algebra course to

■ relieve the student of the drudgery of doing arithmetic, so that he or she is free to concentrate on concepts.

■ provide the student with a tool for learning mathematics by investigating ideas and exploring patterns.

■ introduce the student to some of the benefits and pitfalls of working with computers.

■ provide a tool for implementing mathematical models.

MATLAB discussions are given at the end of each chapter. MATLAB is the most widely used software for working with matrices. It is available on a variety of computers, including Macintosh and IBM. MATLAB can be obtained from The Mathworks, Inc., Cochituate Place, 24 Prime Park Way, Natick, MA 01760-1520, Telephone (508) 653-1415. A student version is available from Prentice-Hall. Programs written in the MATLAB language to accompany this book are available from The Mathworks.

MATLAB discussions are written in computer manual style for flexibility of use. Explanations are brief and to the point. The topics covered are those in the chapter. A brief summary of each topic is included, and we show how to use MATLAB for that topic. The discussions include exercises. The exercises should be of general interest because many can be implemented on any matrix algebra software package or on a graphing calculator. Some exercises are open-ended and encourage exploration on the part of the student. Many of the assignments can be projects and students can work on them in groups.

MATLAB has built-in functions such as inv(A) for finding the inverse of a matrix. It is also possible to write programs called m-files in the MATLAB language to tap these built-in functions and make use of MATLAB matrix input/output.

The following additional software packages are available from the publisher to accompany this book.

Linear Algebra Computer Companion—Macintosh, DOS, Windows versions

Ideas for Linear Algebra with Maple

Linear Algebra with Applications—Mathematica Interactive Textbook

Linear Algebra Workbook—Mathematica Interactive Workbook

Linear Algebra Notebooks—Mathematica

Entering and Displaying a Matrix (Section 1.2)

The MATLAB prompt is »
To enter a matrix:

(a) start with [

(b) separate elements of the matrix with space (or comma)

(c) use ; to mark end of each row

(d) end the matrix with]

Example 1 Enter the matrix $A = \begin{bmatrix} 3 & 5 \\ 1 & 2 \end{bmatrix}$ into the workspace.

»A = [3   5;1   2]   {Type thus, followed by carriage return}
A =
   3   5   {Output.A ; at end of input can be used to suppress display}
   1   2

Example 2 A matrix that is in the workspace can be displayed at any time. For example, the above matrix A is in the workspace.

»A   {Type A, followed by carriage return to display the matrix}
A =
   3   5
   1   2

Exercises Enter and display the following matrices in MATLAB.

(a) $\begin{bmatrix} 5 & 4 \\ 2 & 9 \end{bmatrix}$ (b) $\begin{bmatrix} 2 & 4 & 1 \\ 5 & 0 & -3 \end{bmatrix}$ (c) $\begin{bmatrix} 9 \\ 3 \\ 2 \\ 0 \end{bmatrix}$ (d) $\begin{bmatrix} -2 & 3 & 7 & 1 \end{bmatrix}$

Solving Systems of Linear Equations (Sections 1.2, 1.3)

Gauss-Jordan Elimination

MATLAB contains functions for working with matrices. New functions can also be written in the MATLAB language. The function **gjelim** has been written for Gauss-Jordan elimination. It includes a "rational number" option (the default mode is the decimal number), a "count of arithmetic operations" option, and an "all steps" option that displays all the steps. gjelim and other MATLAB programs are found in The Linear Algebra with Applications Toolbox available from The Mathworks for use with this text.

Example 1 Solve the following system using Gauss-Jordan elimination in all steps mode.

$$x_1 - 2x_2 + 4x_3 = 12$$
$$2x_1 - x_2 + 5x_3 = 18$$
$$-x_1 + 3x_2 - 3x_3 = -8$$

»A = [−2 4 12;2 −1 5 18;   {enter the augmented matrix}
−1 3 −3 −8];
»gjelim(A)   {author-defined function for GJ elim for A}
Rational numbers? y/n: y   {option of rational arithmetic}
Count of operations? y/n: n   {option of keeping count of arith ops}
All steps? y/n: y   {option of displaying all the steps}

initial matrix   {start of output}

$$\begin{matrix} 1 & -2 & 4 & 12 \\ 2 & -1 & 5 & 18 \\ -1 & 3 & -3 & -8 \end{matrix}$$

[press return at each step to continue]

create zero   **−reduced echelon form−**

$$\begin{matrix} 1 & -2 & 4 & 12 \\ 0 & 3 & -3 & -6 \\ -1 & 3 & -3 & -8 \end{matrix} \quad \cdots \quad \begin{matrix} 1 & 0 & 0 & 2 \\ 0 & 1 & 0 & 1 \\ 0 & 0 & 1 & 3 \end{matrix} \quad \text{{end of output}}$$

We see that the system has the unique solution $x_1 = 2$, $x_2 = 1$, $x_3 = 3$.

Selecting Own Elementary Transformations

The function **transf** (for transform) can be used to perform your own sequence of elementary row transformations on matrices—the computer does the arithmetic for you.

Example 2   Solve the following system of equations using Gauss-Jordan elimination using **transf**—you have to select the transformations at each step.

$$\begin{aligned} x_1 - 2x_2 + 4x_3 &= 12 \\ 2x_1 - x_2 + 5x_3 &= 18 \\ -x_1 + 3x_2 - 3x_3 &= -8 \end{aligned}$$

»A = [1 −2 4 12;2 −1 5 18;−1 3 −3 −8]   {enter augmented matrix}
A =

$$\begin{matrix} 1 & -2 & 4 & 12 \\ 2 & -1 & 5 & 18 \\ -1 & 3 & -3 & -8 \end{matrix}$$

»**transf(A)**   {author-defined function}
Format: (r)ational numbers or (d)ecimal numbers? d   {choice of format}
Operation: (a)dd (m)ultiply (d)ivide (s)wap (e)xit a   {select operation, addition}
add the multiple: −2   {enter the operations}
　　　　　of row: 1   {   on   }
　　　　　to row: 2   {   these lines   }

$$\begin{matrix} 1 & -2 & 4 & 12 \\ 0 & 3 & -3 & -6 \\ -1 & 3 & -3 & -8 \end{matrix}$$ {the new matrix}

$$\vdots$$

Operation: (a)dd (m)ultiply (d)ivide (s)wap (e)xit e
ans =

1	0	0	2
0	1	0	1
0	0	1	3

{perform intermediate steps}

{exit at appropriate time}

The system has the unique solution $x_1 = 2$, $x_2 = 1$, $x_3 = 3$.

Help A help facility is available to get online information on MATLAB topics. To get information about a function, enter "help *function*." For example, to get information about the function **transf**, we would enter "help transf" as follows.

»help transf
Elementary row transformations for row reduction   {the information}
Add multiples of rows, multiply, divide, swap
Gives choice of format—rational numbers or decimal numbers
Calling format: transf(A)

rref MATLAB has a function **rref** to arrive directly at the reduced echelon form of a matrix. For example,

»A = [1 −2 4 12;2 −1 5 18;−1 3 −3 −8];
»rref(A)
ans =

1	0	0	2
0	1	0	1
0	0	1	3

{the reduced echelon form of A}

Exercises

1. Solve (if possible) each of the following systems of three equations in three variables using **gjelim** and the "all steps" option. Look closely at the purpose of each operation.

*(a) $x_1 + 4x_2 + 3x_3 = 1$
       $2x_1 + 8x_2 + 11x_3 = 7$
       $x_1 + 6x_2 + 7x_3 = 3$

*(b) $x_1 + 2x_2 - x_3 = 3$
       $2x_1 + 4x_2 - 2x_3 = 6$
       $3x_1 + 6x_2 + 2x_3 = -1$

(c) $x_1 + x_2 + 2x_3 = 2.4$
      $x_1 + 2x_2 + 2x_3 = 2.8$
      $2x_1 + 2x_2 + 3x_3 = 4.05$
      (use rational number option)

(d) $x_1 + 2x_2 - 4x_3 = -1.5$
      $x_1 + 3x_2 - 7x_3 = -2.95$
      $-2x_1 - 4x_2 + 9x_3 = 3.35$
      (use rational number option)

2. Use the function **transf** to solve the following systems of equations using Gauss-Jordan elimination. Use **rref** to check your answers. The function **frac**(A) can be used to express the elements of A as rational numbers ("frac" for fraction form). Express the solution to **(d)** in terms of rational numbers.

*(a) $x_1 - 2x_2 = -8$
       $3x_1 - 5x_2 = -19$

(c) $x_1 + 2x_2 = 4$
      $2x_1 + 4x_2 = 8$

*(b) $3x_1 - 6x_2 = 1$
       $x_1 - 2x_2 = 2$

(d) $3x_1 - 4x_2 = 2$
      $x_1 + x_2 = 1$

3. Use **transf** to solve the following systems using Gauss-Jordan elimination.

   *(a)
   $$x_1 \qquad + \; x_3 = \quad 3$$
   $$2x_2 - 2x_3 = -4$$
   $$- \; x_2 - 2x_3 = \quad 5$$

   (b)
   $$x_1 + \; x_2 + 3x_3 = \quad 6$$
   $$x_1 + 2x_2 + 4x_3 = \quad 9$$
   $$2x_1 + \; x_2 + 6x_3 = 11$$

   *(c)
   $$4x_1 + 8x_2 - 12x_3 = 44$$
   $$3x_1 + 6x_2 - \; 8x_3 = 32$$
   $$-2x_1 - \; x_2 \qquad = -7$$

4. Construct a system of linear equations having solutions $x_1 = 1$, $x_2 = 2$, $x_3 = 3$, in which all three types of elementary row operations are used to arrive at this solution using Gauss-Jordan elimination.

5. The function **gjpic** can be used to display the lines at each step in Gauss-Jordan elimination for a system of linear equations in two variables. Use **gjpic** to display the lines for the system

   $$x_1 + 3x_2 = \quad 9$$
   $$-2x_1 + \; x_2 = -4$$

   Note: Use "Enter" and "shg" (or "figure (gcf)") to alternate between command and graph windows on IBMs.

6. Compute the reduced echelon forms of a number of 3×3 matrices entered at random, using **rref**. Why do you get these reduced echelon forms? How can you get another type of reduced echelon form?
   (Use function **echtest** to test REFs of 5000 random 3×3 matrices—we got 97 not I_3, in 5000 matrices.)

7. Use the Gauss-Jordan elimination function **gjelim** to find the reduced echelon forms of

   $$\begin{bmatrix} 1 & 2 & 3 \\ 4 & 5 & 6 \\ 7 & 8 & 9 \end{bmatrix} \quad \text{and} \quad \begin{bmatrix} 7 & 8 & 9 \\ 4 & 5 & 6 \\ 1 & 2 & 3 \end{bmatrix}$$

   Do you expect these two reduced forms to differ? Why do they differ?

8. There are many ways of solving systems of equations. In this exercise we introduce the method of *Gaussian elimination*.

   The functions **gjalg** and **galg** have been written to give the reader practice at mastering the Gauss-Jordan and Gaussian algorithms by looking at patterns. Use the help facility to get information about these functions. Use these functions to see the difference in the algorithms in solving a system of three equations in three variables.

   Solve the following system of equations using the function **gelim** that performs Gaussian elimination with "all steps" option. You get the same REF as you would with **gjelim**, with a matrix called the echelon form of A appearing along the way. Describe the characteristics of a matrix in echelon form. Describe the algorithm used in Gaussian elimination.

   $$x_1 + \; x_2 + \; x_3 = \quad 3$$
   $$2x_1 + 3x_2 + \; x_3 = \quad 5$$
   $$x_1 - \; x_2 - 2x_3 = -5$$

*9. Determine the number of operations needed to solve the system of equations in Exercise 8,

   (a) using Gauss-Jordan elimination.

   (b) using Gaussian elimination.

   Explain, by investigating in "all steps" mode, how the arithmetic operations are counted. Where *exactly* does Gaussian elimination gain over Gauss-Jordan elimination?

10. Consider the algorithms for Gauss-Jordan elimination and Gaussian elimination for a system of n linear equations in n variables. Show, by analyzing these algorithms, that the counts of arithmetic operations (flop counts) for these methods are given by polynomials of degree three in n. Use the help facility to get information about **gjinfo** and **ginfo**. Use **gjinfo** and **ginfo** to determine the specific counts for such systems with $n = 2, 3, 4, 5$. Use this data in a system of linear equations to arrive at the specific polynomials that give the flop counts for these methods. Use the function **graph** to graph both these polynomials for the interval [2, 100] in steps of 10. Observe how the graphs diverge as n increases—what does this mean?

    Determine the total number of arithmetic operations needed to solve a system of linear equations having a 20 $\times$ 20 matrix of coefficients using each of the two methods.

11. Use Gauss-Jordan elimination to determine the equation of the unique polynomial of degree two that passes through the following points (Section 1.4).

    *(a) (1, 8), (3, 26), (5, 60)

    (b) (1, −3), (2, −1), (3, 9), (4, 33)

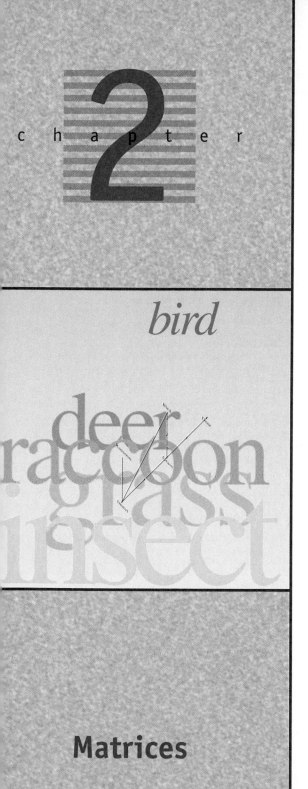

chapter

2

bird

deer
raccoon
grass
insect

Matrices

In this chapter we shall

develop a theory of matrices and see how the concepts are used in applications ranging from archaeology to economics. We shall introduce rules for adding and multiplying matrices and shall see how these operations are used in determining the chronological order of artifacts in archaeology and the interdependency of industries.

2.1 Addition, Scalar Multiplication, and Multiplication of Matrices

A convenient notation has been developed for working with matrices. The location of an element in a matrix is described by giving the row and column in which it lies. The element in row i, column j of matrix A is denoted a_{ij}.

$$a_{ij}$$

1st subscript ——⤒⤒—— 2nd subscript
indicates row   indicates column

We refer to a_{ij} as the (i,j)th element of the matrix A. We can visualize an arbitrary $m \times n$ matrix A as in Figure 2.1.

If the number of rows m is equal to the number of columns n, A is said to be a **square matrix**. The elements of a square matrix A where the subscripts are equal, namely $a_{11}, a_{22}, \ldots, a_{nn}$, form the **main diagonal**. See Figure 2.2.

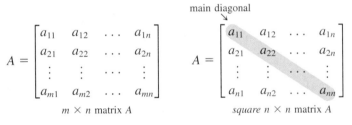

$$A = \begin{bmatrix} a_{11} & a_{12} & \cdots & a_{1n} \\ a_{21} & a_{22} & \cdots & a_{2n} \\ \vdots & \vdots & \cdots & \vdots \\ a_{m1} & a_{m2} & \cdots & a_{mn} \end{bmatrix}$$

$m \times n$ matrix A

Figure 2.1

main diagonal

$$A = \begin{bmatrix} a_{11} & a_{12} & \cdots & a_{1n} \\ a_{21} & a_{22} & \cdots & a_{2n} \\ \vdots & \vdots & \cdots & \vdots \\ a_{n1} & a_{n2} & \cdots & a_{nn} \end{bmatrix}$$

square $n \times n$ matrix A

Figure 2.2

Example 1   Determine the elements a_{12}, a_{23}, and a_{31} for the following matrix A. Give the main diagonal of A.

$$A = \begin{bmatrix} 1 & -2 & -1 \\ 3 & -3 & 4 \\ 2 & 7 & 5 \end{bmatrix}$$

Solution   Since a_{12} is the element in row 1, column 2, $a_{12} = -2$. We see that $a_{23} = 4$ and $a_{31} = 2$.

The main diagonal of A consists of the elements $a_{11} = 1$, $a_{22} = -3$, $a_{33} = 5$.

We commence the development of an algebraic theory of matrices by defining a concept of equality of matrices.

Definition   *Two matrices are **equal** if they are of the same size and if their corresponding elements are equal. Thus $A = B$ if $a_{ij} = b_{ij}$.*

This definition will enable us to introduce equations involving matrices. It immediately allows us to define an operation of addition for matrices.

Addition of Matrices

Definition *Let A and B be matrices of the same size. Their **sum** A + B is the matrix obtained by adding together the corresponding elements of A and B. The matrix A + B will be of the same size as A and B. If A and B are not of the same size, they cannot be added and we say that **the sum does not exist**.*

Thus if $C = A + B$, then $c_{ij} = a_{ij} + b_{ij}$.

Example 2 Let $A = \begin{bmatrix} 1 & 4 & 7 \\ 0 & -2 & 3 \end{bmatrix}$, $B = \begin{bmatrix} 2 & 5 & -6 \\ -3 & 1 & 8 \end{bmatrix}$, and $C = \begin{bmatrix} -5 & 4 \\ 2 & 7 \end{bmatrix}$. Determine $A + B$ and $A + C$, if the sums exist.

Solution A and B are both 2×3 matrices. They are the same size. The sum exists and will be a 2×3 matrix. On adding corresponding elements, we get

$$A + B = \begin{bmatrix} 1 & 4 & 7 \\ 0 & -2 & 3 \end{bmatrix} + \begin{bmatrix} 2 & 5 & -6 \\ -3 & 1 & 8 \end{bmatrix}$$

$$= \begin{bmatrix} 1 + 2 & 4 + 5 & 7 - 6 \\ 0 - 3 & -2 + 1 & 3 + 8 \end{bmatrix} = \begin{bmatrix} 3 & 9 & 1 \\ -3 & -1 & 11 \end{bmatrix}$$

A is a 2×3 matrix while C is a 2×2 matrix. A and C are not of the same size. $A + C$ does not exist.

Scalar Multiplication of Matrices

When working with matrices, it is customary to refer to numbers as **scalars**. We shall use *uppercase letters to denote matrices and lowercase letters for scalars*. The next step in the development of a theory of matrices is to introduce a rule for multiplying matrices by scalars.

Definition *Let A be a matrix and c be a scalar. The **scalar multiple** of A by c, denoted cA, is the matrix obtained by multiplying every element of A by c. The matrix cA will be the same size as A.*

Thus if $B = cA$, then $b_{ij} = ca_{ij}$.

Example 3   Let $A = \begin{bmatrix} 1 & -2 & 4 \\ 7 & -3 & 0 \end{bmatrix}$. Determine $3A$.

Solution   Multiply every element of A by 3 to get

$$3A = \begin{bmatrix} 3 & -6 & 12 \\ 21 & -9 & 0 \end{bmatrix}$$

Observe that A and $3A$ are both 2×3 matrices.

Negation and Subtraction

Let us define $-C$ to be the matrix $(-1)C$. This means that to **negate** a matrix we multiply every element in the matrix by -1. Thus if

$$C = \begin{bmatrix} 1 & 0 & -7 \\ -3 & 6 & 2 \end{bmatrix}, \quad \text{then} \quad -C = \begin{bmatrix} -1 & 0 & 7 \\ 3 & -6 & -2 \end{bmatrix}$$

Definition   *We now define the subtraction of matrices in such a way that it is compatible with addition, scalar multiplication, and negation. Let*

$$A - B = A + (-1)B$$

This definition implies that *subtraction is performed between matrices of the same size by subtracting corresponding elements.*

If $C = A - B$, then $c_{ij} = a_{ij} - b_{ij}$. Thus if

$$A = \begin{bmatrix} 5 & 0 & -2 \\ 3 & 6 & -5 \end{bmatrix} \quad \text{and} \quad B = \begin{bmatrix} 2 & 8 & -1 \\ 0 & 4 & 6 \end{bmatrix}$$

then

$$A - B = \begin{bmatrix} 5 - 2 & 0 - 8 & -2 - (-1) \\ 3 - 0 & 6 - 4 & -5 - 6 \end{bmatrix}$$

$$= \begin{bmatrix} 3 & -8 & -1 \\ 3 & 2 & -11 \end{bmatrix}$$

Multiplication of Matrices

We have introduced rules for adding matrices and for multiplying matrices by scalars. Let us now look at multiplication of matrices. The most natural way of multiplying two matrices A and B would seem to be to multiply corresponding elements of A and B. However, it has been found that this is not the most useful way of multiplying matrices. Mathematicians have introduced an alternative rule that involves multiplying the rows of the first matrix A by the columns of the second matrix B in a systematic manner. We state this rule by giving a method of arriving at an arbitrary element of the **product matrix** AB.

Definition   *Let the number of columns in a matrix A be the same as the number of rows in a matrix B. The product AB then exists. The element in row i and column j of AB is obtained by multiplying the corresponding elements of row i of A and column j of B and adding the products.*

If the number of columns in A does not equal the number of rows in B, *the product does not exist.*

The ith row of A is $[a_{i1} \quad a_{i2} \quad \ldots \quad a_{in}]$ and the jth column of B is $\begin{bmatrix} b_{1j} \\ b_{2j} \\ \vdots \\ b_{nj} \end{bmatrix}$. Thus

if $C = AB$, then $c_{ij} = a_{i1}b_{1j} + a_{i2}b_{2j} + \cdots + a_{in}b_{nj}$.

Example 4   Let $A = \begin{bmatrix} 1 & 3 \\ 2 & 0 \end{bmatrix}$, $B = \begin{bmatrix} 5 & 0 & 1 \\ 3 & -2 & 6 \end{bmatrix}$, and $C = [6 \quad -2 \quad 5]$.
Determine AB, BA, and AC, if the products exist.

Solution A has two columns and B has two rows; thus AB exists. Interpret A in terms of its rows and B in terms of its columns and multiply the rows by the columns in the following systematic manner.

$$AB = \begin{bmatrix} 1 & 3 \\ 2 & 0 \end{bmatrix}\begin{bmatrix} 5 & 0 & 1 \\ 3 & -2 & 6 \end{bmatrix}$$

$$= \begin{bmatrix} [1 \quad 3]\begin{bmatrix} 5 \\ 3 \end{bmatrix} & [1 \quad 3]\begin{bmatrix} 0 \\ -2 \end{bmatrix} & [1 \quad 3]\begin{bmatrix} 1 \\ 6 \end{bmatrix} \\[2ex] [2 \quad 0]\begin{bmatrix} 5 \\ 3 \end{bmatrix} & [2 \quad 0]\begin{bmatrix} 0 \\ -2 \end{bmatrix} & [2 \quad 0]\begin{bmatrix} 1 \\ 6 \end{bmatrix} \end{bmatrix}$$

$\leftarrow$ 1st row of A times each column of B in turn

$\leftarrow$ 2nd row of A times each column of B in turn

$$= \begin{bmatrix} (1 \times 5) + (3 \times 3) & (1 \times 0) + (3 \times (-2)) & (1 \times 1) + (3 \times 6) \\ (2 \times 5) + (0 \times 3) & (2 \times 0) + (0 \times (-2)) & (2 \times 1) + (0 \times 6) \end{bmatrix}$$

$$= \begin{bmatrix} 14 & -6 & 19 \\ 10 & 0 & 2 \end{bmatrix}$$

Let us now look at the two remaining products, named BA and AC.
The matrix B has three columns while A has two rows. Thus BA does not exist. The matrix A has two columns while C has one row. The product AC does not exist.

Observe that, in the previous example, AB exists while BA does not. We see that *the order in which two matrices are multiplied is important.* Unlike multiplication of real numbers, *matrix multiplication is not commutative.* In general, for two matrices A and B, $AB \neq BA$. Sometimes AB and BA will both exist, but only very rarely will they be equal.

Example 5   If $A = \begin{bmatrix} 2 & 1 \\ 7 & 0 \\ -3 & -2 \end{bmatrix}$ and $B = \begin{bmatrix} -1 & 0 \\ 3 & 5 \end{bmatrix}$, determine AB.

Solution   Since A has two columns and B has two rows, AB exists. Multiplying the rows of A by the columns of B in the appropriate manner, we get

$$AB = \begin{bmatrix} 2 & 1 \\ 7 & 0 \\ -3 & -2 \end{bmatrix} \begin{bmatrix} -1 & 0 \\ 3 & 5 \end{bmatrix}$$

$$= \begin{bmatrix} [2 \ 1]\begin{bmatrix} -1 \\ 3 \end{bmatrix} & [2 \ 1]\begin{bmatrix} 0 \\ 5 \end{bmatrix} \\ [7 \ 0]\begin{bmatrix} -1 \\ 3 \end{bmatrix} & [7 \ 0]\begin{bmatrix} 0 \\ 5 \end{bmatrix} \\ [-3 \ -2]\begin{bmatrix} -1 \\ 3 \end{bmatrix} & [-3 \ -2]\begin{bmatrix} 0 \\ 5 \end{bmatrix} \end{bmatrix}$$

← 1st row of A times each column of B in turn

← 2nd row of A times each column of B in turn

← 3rd row of A times each column of B in turn

$$= \begin{bmatrix} -2+3 & 0+5 \\ -7+0 & 0+0 \\ 3-6 & 0-10 \end{bmatrix} = \begin{bmatrix} 1 & 5 \\ -7 & 0 \\ -3 & -10 \end{bmatrix}$$

The following example illustrates that we can compute any desired element in a product matrix without calculating the whole product.

Example 6   Let $C = AB$ for the following matrices A and B. Determine the element c_{23} of C.

$$A = \begin{bmatrix} 2 & 1 \\ -3 & 4 \end{bmatrix} \quad \text{and} \quad B = \begin{bmatrix} -7 & 3 & 2 \\ 5 & 0 & 1 \end{bmatrix}$$

Solution c_{23} is the element in row 2, column 3 of C. It will be the product of row 2 of A and column 3 of B. We get

$$c_{23} = [-3 \ \ 4]\begin{bmatrix} 2 \\ 1 \end{bmatrix} = (-3 \times 2) + (4 \times 1) = -2$$

Size of a Product Matrix

Let us now discuss the size of a product matrix. Let A be an $m \times r$ matrix and B be an $r \times n$ matrix. A has r columns and B has r rows. AB thus exists. The first row of AB is obtained by multiplying the first row of A by each column of B in turn. Thus the number of columns in AB is equal to the number of columns in B. The first column of AB results from multiplying each row of A in turn with the first column of B. Thus the number of rows in AB is equal to the number of rows in A. AB will be an $m \times n$ matrix.

If *A* is an *m* × *r* matrix and *B* is an *r* × *n* matrix, then *AB* will be an *m* × *n* matrix.

We can picture this result as follows:

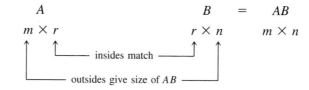

Example 7   If *A* is a 5 × 6 matrix and *B* is a 6 × 7 matrix, what is the size of the product matrix *AB*?

Solution   First we note that *A* has six columns while *B* has six rows. Thus *AB* exists. *AB* will be a 5 × 7 matrix.

$$
\begin{array}{ccccc}
A & & B & = & AB \\
5 \times 6 & & 6 \times 7 & & 5 \times 7
\end{array}
$$

match

5 × 7

Definition   *A **zero matrix** is a matrix all of whose elements are zeros. A **diagonal matrix** is a square matrix in which all the elements not on the main diagonal are zeros. An **identity matrix** is a diagonal matrix in which every diagonal element is 1. See Figure 2.3.*

$$
O_{mn} = \begin{bmatrix} 0 & 0 & \dots & 0 \\ 0 & 0 & \dots & 0 \\ \vdots & \vdots & \dots & \vdots \\ 0 & 0 & \dots & 0 \end{bmatrix} \qquad A = \begin{bmatrix} a_{11} & 0 & \dots & 0 \\ 0 & a_{22} & \dots & 0 \\ \vdots & \vdots & \dots & \vdots \\ 0 & 0 & \dots & a_{nn} \end{bmatrix} \qquad I_n = \begin{bmatrix} 1 & 0 & \dots & 0 \\ 0 & 1 & \dots & 0 \\ \vdots & \vdots & \dots & \vdots \\ 0 & 0 & \dots & 1 \end{bmatrix}
$$

Figure 2.3   zero matrix   diagonal matrix *A*   identity matrix

Zero matrices play a role in matrix theory similar to the role of the number 0 for real numbers, and identity matrices play a role similar to the number 1. These roles are described in the following theorem, which we illustrate by means of an example.

Theorem 2.1   Let A be an $m \times n$ matrix and 0_{mn} be the zero $m \times n$ matrix. Let B be an $n \times n$ square matrix, 0_n and I_n be the zero and identity $n \times n$ matrices. Then

$$A + 0_{mn} = 0_{mn} + A = A$$
$$B0_n = 0_n B = 0_n$$
$$BI_n = I_n B = B$$

Example 8   Let $A = \begin{bmatrix} 2 & 1 & -3 \\ 4 & 5 & 8 \end{bmatrix}$ and $B = \begin{bmatrix} 2 & 1 \\ -3 & 4 \end{bmatrix}$.

We see that

$$A + 0_{23} = \begin{bmatrix} 2 & 1 & -3 \\ 4 & 5 & 8 \end{bmatrix} + \begin{bmatrix} 0 & 0 & 0 \\ 0 & 0 & 0 \end{bmatrix} = \begin{bmatrix} 2 & 1 & -3 \\ 4 & 5 & 8 \end{bmatrix} = A$$

$$B0_2 = \begin{bmatrix} 2 & 1 \\ -3 & 3 \end{bmatrix}\begin{bmatrix} 0 & 0 \\ 0 & 0 \end{bmatrix} = \begin{bmatrix} 0 & 0 \\ 0 & 0 \end{bmatrix} = 0_2$$

$$BI_2 = \begin{bmatrix} 2 & 1 \\ -3 & 4 \end{bmatrix}\begin{bmatrix} 1 & 0 \\ 0 & 1 \end{bmatrix} = \begin{bmatrix} 2 & 1 \\ -3 & 4 \end{bmatrix} = B$$

Similarly $0_{23} + A = A$, $0_2 B = 0_2$, $I_2 B = B$.

Matrix Notation and Systems of Equations

This matrix notation is convenient for expressing arbitrary systems of linear equations. We can write a system of m linear equations in n variables as follows:

$$\begin{aligned} a_{11}x_1 + \cdots + a_{1n}x_n &= b_1 \\ \vdots \quad\quad \cdots \quad\quad \vdots \quad &\;\; \vdots \\ a_{m1}x_1 + \cdots + a_{mn}x_n &= b_m \end{aligned}$$

Let

$$A = \begin{bmatrix} a_{11} & \cdots & a_{1n} \\ \vdots & \cdots & \vdots \\ a_{m1} & \cdots & a_{mn} \end{bmatrix}, \qquad X = \begin{bmatrix} x_1 \\ \vdots \\ x_n \end{bmatrix}, \qquad \text{and} \qquad B = \begin{bmatrix} x_1 \\ \vdots \\ b_m \end{bmatrix}$$

matrix of coefficients

It can be shown by multiplying the matrices that we can write the system of equations in the matrix form

$$AX = B$$

Thus, for example,

$$\begin{array}{rcrcrcr} 3x_1 & + & 2x_2 & - & 5x_3 & = & 7 \\ x_1 & - & 8x_2 & + & 4x_3 & = & 9 \\ 2x_1 & + & 6x_2 & - & 7x_3 & = & -2 \end{array} \qquad \text{can be written} \qquad \begin{bmatrix} 3 & 2 & -5 \\ 1 & -8 & 4 \\ 2 & 6 & -7 \end{bmatrix} \begin{bmatrix} x_1 \\ x_2 \\ x_3 \end{bmatrix} = \begin{bmatrix} 7 \\ 9 \\ -2 \end{bmatrix}$$

This matrix way of expressing a system of linear equations will prove to be very useful later in the course.

*Fast Matrix Multiplication

Computations involving large matrices are done on computers. Much research goes into looking for fast ways of performing matrix computations on the computer. In 1969 V. Strassen caused quite a stir by coming up with a fast matrix multiplication algorithm for two square matrices of the same size. This is the algorithm that is now used to multiply such large matrices on computers. We present Strassen's ideas to illustrate this important numerical side of linear algebra.

Consider the number of arithmetic operations (multiplications and additions) required to multiply 2×2 matrices A and B.

$$A = \begin{bmatrix} a_{11} & a_{12} \\ a_{21} & a_{22} \end{bmatrix}, \qquad B = \begin{bmatrix} b_{11} & b_{12} \\ b_{21} & b_{22} \end{bmatrix},$$

$$AB = \begin{bmatrix} a_{11}b_{11} + a_{12}b_{21} & a_{11}b_{12} + a_{12}b_{22} \\ a_{21}b_{11} + a_{22}b_{21} & a_{21}b_{12} + a_{22}b_{22} \end{bmatrix}$$

Computing each element of AB involves 2 scalar multiplications and 1 addition. Since there are 4 elements in AB, the total number of arithmetic operations necessary to compute the product of two 2×2 matrices is 8 multiplications and 4 additions. Strassen came up with a way that requires fewer multiplications but more additions.

Consider the following identities:

$$\begin{array}{lll} m_1 = (a_{12} - a_{22})(b_{21} + b_{22}) & m_4 = (a_{11} + a_{12})b_{22} & m_7 = (a_{21} + a_{22})b_{11} \\ m_2 = (a_{11} + a_{22})(b_{11} + b_{22}) & m_5 = a_{11}(b_{12} - b_{22}) & \\ m_3 = (a_{11} - a_{21})(b_{11} + b_{12}) & m_6 = a_{22}(b_{21} - b_{11}) & \end{array}$$

The reader can verify that

$$AB = \begin{bmatrix} m_1 + m_2 - m_4 + m_6 & m_4 + m_5 \\ m_6 + m_7 & m_2 - m_3 + m_5 - m_7 \end{bmatrix}$$

It can be seen that there are 7 scalar multiplications required to compute AB this way, and 18 additions, The number of multiplications has been reduced by 1, but the number of additions has been increased by 14. The one saved multiplication turns out to be extremely important. We shall now convince you that this algorithm leads to a faster way of multiplying certain large matrices than the standard algorithm!

Let A and B be $n \times n$ matrices where n is a power of two. Write A and B in terms of $\left(\frac{n}{2}\right) \times \left(\frac{n}{2}\right)$ submatrices $A_{11}, A_{12}, A_{21}, A_{22}$ and $B_{11}, B_{12}, B_{21}, B_{22}$ as follows. Multiplying AB in the standard way is equivalent to multiplying these matrices as if they were 2×2 matrices with the submatrices being elements.

$$A = \begin{bmatrix} A_{11} & A_{12} \\ A_{21} & A_{22} \end{bmatrix}, \qquad B = \begin{bmatrix} B_{11} & B_{12} \\ B_{21} & B_{22} \end{bmatrix},$$

$$AB = \begin{bmatrix} A_{11}B_{11} + A_{12}B_{21} & A_{11}B_{12} + A_{12}B_{22} \\ A_{21}B_{11} + A_{22}B_{21} & A_{21}B_{12} + A_{22}B_{22} \end{bmatrix}$$

There are then 8 matrix multiplications of submatrices involved. However, we can reduce the number of matrix multiplications if we use Strassen's algorithm, as follows. Replace the scalars $a_{11}, \ldots, b_{22}$ in the Strassen algorithm by the matrices $A_{11}, \ldots,$ B_{22}. The algorithm can now be used to compute AB by carrying out 7 multiplications of $\left(\frac{n}{2}\right) \times \left(\frac{n}{2}\right)$ matrices. Each $\left(\frac{n}{2}\right) \times \left(\frac{n}{2}\right)$ matrix is then written in terms of $\left(\frac{n}{4}\right) \times \left(\frac{n}{4}\right)$ submatrices and the algorithm repeated, and so on, until the answer is found. There is one less matrix multiplication involved at each step of the method. The fewer matrix multiplications more than compensate for the additional matrix additions provided n is sufficiently large.

This method of repeatedly "calling" on the algorithm for the multiplication of smaller and smaller matrices lends itself to a computer programming technique called **recursion**. The method can still be used if the original matrices do not have a number of rows and columns that is a power of two, by padding the matrices with columns and rows of zeros to bring the number up to a power of two.

In this course we shall be developing standard methods and shall use the standard method of matrix multiplication. The reference to Strassen's original work is "Gaussian Elimination Is Not Optimal," *Numerische Mathematik,* **13**, 1969, pp. 354–356. Books that have discussions of Strassen's results and of other algorithms are *Algorithms and Complexity,* Herbert S. Wilf, Prentice-Hall, Inc., 1986, and *Algorithms, Theory and Practice,* Gilles Brassard and Paul Bratley, Prentice-Hall, 1988.

Exercise Set 2.1

1. Let $A = \begin{bmatrix} 5 & 4 \\ -1 & 7 \\ 9 & -3 \end{bmatrix}$, $B = \begin{bmatrix} -3 & 0 \\ 4 & 2 \\ 5 & -7 \end{bmatrix}$, $C = \begin{bmatrix} 1 & 2 \\ 3 & 4 \end{bmatrix}$,

$D = \begin{bmatrix} 9 & -5 \\ 3 & 0 \end{bmatrix}$

Compute the following (if they exist).

   *(a) $A + B$ **(b)** $2B$ *(c) $-D$

   (d) $C + D$ *(e) $A + D$ **(f)** $2A + B$

   *(g) $A - B$

2. Let $A = \begin{bmatrix} 9 \\ 2 \\ -1 \end{bmatrix}$, $B = \begin{bmatrix} 0 & -1 & 4 \\ 6 & -8 & 2 \\ -4 & 5 & 9 \end{bmatrix}$,

$C = \begin{bmatrix} 1 & 2 & -5 \\ -7 & 9 & 3 \\ 5 & -4 & 0 \end{bmatrix}$, $D = \begin{bmatrix} -3 \\ 0 \\ 2 \end{bmatrix}$

Compute the following (if they exist).

   *(a) $A + B$ **(b)** $4B$ **(c)** $-3D$

   *(d) $B - 3C$ **(e)** $-A$ *(f) $3A + 2D$

   (g) $A + D$

3. Let $A = \begin{bmatrix} 1 & 0 \\ 0 & 1 \end{bmatrix}$, $B = \begin{bmatrix} 0 & 1 \\ -2 & 5 \end{bmatrix}$, $C = \begin{bmatrix} 2 \\ 3 \end{bmatrix}$,

$D = \begin{bmatrix} -1 & 0 & 3 \\ 5 & 7 & 2 \end{bmatrix}$

Compute the following (if they exist).

   *(a) AB **(b)** BA **(c)** AC

   *(d) CA *(e) AD **(f)** DC

   *(g) BD **(h)** A^2 *(Hint: $A^2 = AA$.)*

4. Let $A = \begin{bmatrix} -1 \\ 2 \\ 5 \end{bmatrix}$, $B = \begin{bmatrix} 0 & 1 & 5 \\ 3 & -7 & 8 \\ 2 & 3 & 1 \end{bmatrix}$, $C = [-2 \ \ 0 \ \ 5]$,

$D = \begin{bmatrix} 9 & -5 \\ 3 & 0 \\ -4 & 2 \end{bmatrix}$

Compute the following (if they exist).

*(a) BA　　(b) AB　　(c) CB　　*(d) CA

(e) DA　　*(f) DB　　(g) AC　　*(h) B^2

5. Let $A = \begin{bmatrix} 0 & 1 \\ 0 & 3 \\ 5 & 6 \end{bmatrix}$, $B = \begin{bmatrix} -1 & 0 \\ 3 & 5 \\ 2 & 6 \end{bmatrix}$, $C = \begin{bmatrix} -4 & 0 \\ 3 & 2 \end{bmatrix}$,

$D = \begin{bmatrix} 5 & 0 \\ -2 & 1 \end{bmatrix}$

Compute the following (if they exist).

*(a) $2A - 3(BC)$　　　　　(b) AB

*(c) $AC - BD$　　　　　　(d) $CD - 2D$

*(e) BA　　　　　　　　　(f) $AD + 2(DC)$

*(g) $C^3 + 2(D^2)$ *(Hint: $C^3 = CCC$)*

6. Let A be a 3×5 matrix, B a 5×2 matrix, C a 3×4 matrix, D a 4×2 matrix, and E a 4×5 matrix. Determine which of the following matrix expressions exist, and give the sizes of the resulting matrices when they do exist.

*(a) AB　　　　　　　　　(b) EB

*(c) AC　　　　　　　　　(d) $AB + CD$

*(e) $3(EB) + 4D$　　　　　(f) $CD - 2(CE)B$

*(g) $2(EB) + DA$

7. Let A be a 2×2 matrix, B a 2×2 matrix, C a 2×3 matrix, D a 3×2 matrix, and E a 3×1 matrix. Determine which of the following matrix expressions exist, and give the sizes of the resulting matrices when they do exist.

(a) AB　　　　　　　　　*(b) $(A^2)C$

(c) $B^3 + 3(CD)$　　　　　*(d) $DC + BA$

*(e) $DA - 2(DB)$　　　　　(f) $C - 3D$

*(g) $3(BA)(CD) + (4A)(BC)D$

8. Let $A = \begin{bmatrix} 1 & -8 & 4 \\ 5 & -6 & 3 \\ 2 & 0 & -1 \end{bmatrix}$ and $B = \begin{bmatrix} 0 & 2 & -3 \\ 5 & 6 & 7 \\ -1 & 0 & 4 \end{bmatrix}$. Let O_3 and I_3 be the 3×3 zero and identity matrices. Show that

$$A + O_3 = O_3 + A = A, \qquad BO_3 = O_3B = O_3,$$

and

$$BI_3 = I_3B = B$$

9. Let $C = AB$ and $D = BA$ for the following matrices A and B.

$$A = \begin{bmatrix} 0 & 3 & -5 \\ 2 & 6 & 3 \\ 1 & 0 & -2 \end{bmatrix} \text{ and } B = \begin{bmatrix} -1 & 2 & -3 \\ 5 & 7 & 2 \\ 0 & 1 & 6 \end{bmatrix}$$

Determine the following elements of C and D without computing the complete matrices.

*(a) c_{31}　　(b) c_{23}　　*(c) d_{12}　　(d) d_{22}

10. Let $R = PQ$ and $S = QP$, where $P = \begin{bmatrix} 1 & -2 \\ 4 & 6 \\ -1 & 3 \end{bmatrix}$ and

$Q = \begin{bmatrix} 0 & 1 & 3 \\ 0 & -1 & 4 \end{bmatrix}$. Determine the following elements (if they exist) of R and S without computing the complete matrices.

(a) r_{21}　　*(b) r_{33}　　(c) s_{11}　　*(d) s_{23}

11. If $A = \begin{bmatrix} 1 & -3 \\ 0 & 4 \end{bmatrix}$, $B = \begin{bmatrix} 1 & 2 & -3 \\ 5 & 0 & -1 \end{bmatrix}$, and

$C = \begin{bmatrix} 2 & -4 & 5 \\ 7 & 1 & 0 \end{bmatrix}$, determine the following elements of $D = AB + 2C$ without computing the complete matrix.

*(a) d_{12}　　　　　(b) d_{23}

12. If $A = \begin{bmatrix} 1 & -3 & 0 \\ 4 & 5 & 1 \\ 3 & 8 & 0 \end{bmatrix}$, $B = \begin{bmatrix} 1 & 1 & -2 \\ 3 & 0 & 4 \\ -1 & 3 & 2 \end{bmatrix}$, and

$C = \begin{bmatrix} 2 & 0 & -2 \\ 4 & 7 & -5 \\ 1 & 0 & -1 \end{bmatrix}$, determine the following elements of $D = 2(AB) + C^2$ without computing the complete matrix.

*(a) d_{11}　　　　　(b) d_{21}　　　　　(c) d_{32}

13. Write each of the following systems of linear equations as a single matrix equation $AX = B$.

*(a) $\begin{aligned} 2x_1 + 3x_2 &= 4 \\ 3x_1 - 8x_2 &= -1 \end{aligned}$

(b) $\begin{aligned} 4x_1 + 7x_2 &= -2 \\ -2x_1 + 3x_2 &= -4 \end{aligned}$

(c) $\begin{aligned} -9x_1 - 3x_2 &= -4 \\ 6x_1 - 2x_2 &= 7 \end{aligned}$

14. Write each of the following systems of linear equations as a single matrix equation $AX = B$.

    (a) $\begin{aligned} x_1 + 8x_2 - 2x_3 &= 3 \\ 4x_1 - 7x_2 + x_3 &= -3 \\ -2x_1 - 5x_2 - 2x_3 &= 1 \end{aligned}$

    *(b) $\begin{aligned} 5x_1 + 2x_2 &= 6 \\ 4x_1 - 3x_2 &= -2 \\ 3x_1 + x_2 &= 9 \end{aligned}$ (c) $\begin{aligned} x_1 - 3x_2 + 6x_3 &= 2 \\ 7x_1 + 5x_2 + x_3 &= -9 \end{aligned}$

    *(d) $\begin{aligned} 2x_1 + 5x_2 - 3x_3 + 4x_4 &= 4 \\ x_1 \qquad\quad + 9x_3 + 5x_4 &= 12 \\ 3x_1 - 3x_2 - 8x_3 + 5x_4 &= -2 \end{aligned}$

*15. Let A be a matrix whose third row is all zeros. Let B be any matrix such that the product AB exists. Prove that the third row of AB is all zeros.

16. Let D be a matrix whose second column is all zeros. Let C be any matrix such that CD exists. Prove that the second column of CD is all zeros.

17. Let A be an $m \times r$ matrix, B be an $r \times n$ matrix, and $C = AB$. Let the column submatrices of B be $B_1, B_2, \ldots, B_n$ and of C be $C_1, C_2, \ldots, C_n$. We can write B in the form $[B_1 \ B_2 \ \ldots \ B_n]$ and C as $[C_1 \ C_2 \ \ldots \ C_n]$. Prove that $C_j = AB_j$.

18. Let A and B be the following matrices. Use the result of Exercise 17 to compute column 3 of matrix AB without computing the whole product.

$$A = \begin{bmatrix} 1 & 2 & 3 \\ 0 & 4 & 1 \\ 2 & 5 & 0 \end{bmatrix}, \quad B = \begin{bmatrix} 2 & 1 & 4 \\ 6 & 0 & 1 \\ 2 & 3 & 5 \end{bmatrix}$$

*19. Let A and B be the following matrices. Compute row 2 of the matrix AB without computing the whole product.

$$A = \begin{bmatrix} 2 & -3 & 1 \\ 4 & 0 & 3 \\ 5 & 1 & 0 \end{bmatrix}, \quad B = \begin{bmatrix} 8 & 1 & 3 \\ 2 & 1 & 0 \\ 4 & 6 & 3 \end{bmatrix}$$

20. Show, by listing all the multiplications and additions, that the Strassen algorithm for computing the product of two 2×2 matrices requires 7 multiplications and 18 additions.

21. Determine the product AB of the following 2×2 matrices using Strassen's algorithm.

    *(a) $A = \begin{bmatrix} 1 & 0 \\ 2 & 3 \end{bmatrix}, \quad B = \begin{bmatrix} 1 & 2 \\ 0 & 4 \end{bmatrix}$

    (b) $A = \begin{bmatrix} 2 & 1 \\ -1 & 3 \end{bmatrix}, \quad B = \begin{bmatrix} 0 & 5 \\ 1 & -1 \end{bmatrix}$

    *(c) $A = \begin{bmatrix} 2 & -1 \\ 3 & 0 \end{bmatrix}, \quad B = \begin{bmatrix} 5 & 1 \\ 2 & 1 \end{bmatrix}$

    (d) $A = \begin{bmatrix} 2 & 1 \\ 1 & -3 \end{bmatrix}, \quad B = \begin{bmatrix} 3 & 0 \\ 1 & -2 \end{bmatrix}$

22. Determine the product AB of the following 3×3 matrices using Strassen's algorithm.

    *(a) $A = \begin{bmatrix} 1 & -1 & 0 \\ 2 & 1 & 3 \\ 0 & 4 & 1 \end{bmatrix}, \quad B = \begin{bmatrix} 1 & 1 & 2 \\ 3 & 2 & -1 \\ 0 & 1 & 1 \end{bmatrix}$

    (b) $A = \begin{bmatrix} 4 & 1 & 2 \\ 0 & 1 & 3 \\ 5 & -2 & 1 \end{bmatrix}, \quad B = \begin{bmatrix} -3 & 0 & 1 \\ 2 & 5 & -1 \\ 2 & -2 & 3 \end{bmatrix}$

C.3   Matrix Addition and Scalar Multiplication

Let us compute $A + B$ and $3A$ for the following matrices A and B.

$$A = \begin{bmatrix} 2 & 4 \\ 1 & 3 \end{bmatrix}, \quad B = \begin{bmatrix} 5 & -2 \\ 0 & 3 \end{bmatrix}$$

Enter Matrix A

Press $\boxed{\text{MATRX}}$ $\boxed{\blacktriangleleft}$ to get the Matrix Edit Menu

Press 1 to select Edit Screen for matrix A (2 for B, 3 for C, etc.)

Edit the matrix using $\boxed{\blacktriangleright}$ $\boxed{\blacktriangleleft}$ $\boxed{\blacktriangledown}$ $\boxed{\blacktriangle}$ keys and the $\boxed{\text{ENTER}}$ key

Press $\boxed{\text{QUIT}}$ to return to Home Screen

Display Matrix *A*

Press ⌊MATRX⌋ to get the Matrix Names Menu

Press 1 ⌊ENTER⌋ to select and display matrix *A* (2 for *B*, 3 for *C*, etc.)

$$\begin{bmatrix} 2 & 4 \end{bmatrix}$$

Matrix *A* is displayed [1   3]

Add Matrices: *A* + *B*

Enter matrix *A* (as shown above)

Enter matrix *B*

Press ⌊QUIT⌋ to return to Home Screen

Press ⌊MATRX⌋ 1 ⌊+⌋ ⌊MATRX⌋ 2 ⌊ENTER⌋

$$\begin{bmatrix} 7 & 2 \end{bmatrix}$$

Matrix *A* + *B* is displayed [1   6]

Multiply a Matrix by a Scalar: 3*A*

Enter matrix *A*

Press ⌊QUIT⌋ to return to Home Screen

Press 3 ⌊MATRX⌋ 1 ⌊ENTER⌋

$$\begin{bmatrix} 6 & 12 \end{bmatrix}$$

Matrix 3*A* is displayed [3   9]

Exercises C.3

In Exercises 1–8, perform the indicated operations on a calculator using the following matrices A and B.

$$A = \begin{bmatrix} 3 & 5 \\ 0 & -2 \end{bmatrix}, \quad B = \begin{bmatrix} 1 & 7 \\ -3 & 9 \end{bmatrix}$$

*1. $A + B$ **2.** $A - B$ ***3.** $3A$

4. $-2B$ ***5.** $A + 2B$ **6.** $3A - 5B$

*7. $2.5A + 1.7B$ **8.** $-4.75A - 3.26B$

In Exercises 9–15, perform the indicated operations on a calculator, if possible, using the following matrices P, Q, and R. If the calculator gives an error message, use your manual to look up the error code and explain the error.

$$P = \begin{bmatrix} 0 & -2 \\ 4 & 7 \end{bmatrix}, \quad Q = \begin{bmatrix} 1 & 2 \\ -5 & 6 \end{bmatrix}, \quad R = \begin{bmatrix} 1 & 4 & -1 \\ 2 & 3 & -5 \end{bmatrix}$$

*9. $P + Q$ **10.** $P - Q$ ***11.** $2.5P$ **12.** $3.76Q$

*13. $5R$ **14.** $P - R$ ***15.** $Q + R$

C.4   Multiplication of Matrices and Powers of Matrices

Multiplication of Matrices: *AB*

Let us compute AB for the following matrices A and B:

$$A = \begin{bmatrix} 2 & 4 \\ 1 & 3 \end{bmatrix}, \quad B = \begin{bmatrix} 5 & -2 \\ 0 & 3 \end{bmatrix}$$

Enter matrix A (see Section C.2)

Enter matrix B

Press $\boxed{\text{QUIT}}$ to return to Home Screen

Press $\boxed{\text{MATRX}}$ 1 $\boxed{\times}$ $\boxed{\text{MATRX}}$ 2 $\boxed{\text{ENTER}}$

$$\begin{matrix} [10 & 8] \\ \end{matrix}$$
Matrix AB is displayed $\begin{bmatrix} 5 & 7 \end{bmatrix}$

Powers of a Matrix: *A⁴*

Let us compute the first four powers of $A = \begin{bmatrix} 1 & 2 \\ 3 & 4 \end{bmatrix}$.

Enter matrix A

$$\begin{matrix} [\ 7 & 10] \end{matrix}$$
Press $\boxed{\text{MATRX}}$ 1 $\boxed{\times}$ $\boxed{\text{MATRX}}$ 2 $\boxed{\text{ENTER}}$ $\boxed{\text{ENTER}}$ A^2 is displayed $\begin{bmatrix} 15 & 22 \end{bmatrix}$

$$\begin{matrix} [37 & 54] \end{matrix}$$
Press $\boxed{\times}$ $\boxed{\text{MATRX}}$ 1 $\boxed{\text{ENTER}}$
{screen shows Ans*[A]} A^3 is displayed $\begin{bmatrix} 81 & 118 \end{bmatrix}$

$$\begin{matrix} [199 & 290] \end{matrix}$$
Press $\boxed{\text{MATRX}}$ 1 $\boxed{\text{ENTER}}$ A^4 is displayed $\begin{bmatrix} 435 & 634 \end{bmatrix}$

Exercises C.4

In Exercises 1–8, perform the indicated operations on a calculator using the following matrices A and B.

$$A = \begin{bmatrix} 3 & 5 \\ 0 & -2 \end{bmatrix}, \quad B = \begin{bmatrix} 1 & 7 \\ -3 & 9 \end{bmatrix}$$

*1. AB 2. BA *3. $3AB$

 4. $AB + 2A$ *5. $AB - BA$ 6. A^4

*7. AB^5 8. $ABA - 2A^2B^3$

In Exercises 9–14, perform the indicated operations on a calculator, if possible, using the following matrices P, Q, and R. If the calculator gives an error message, use your manual to look up the error code and explain the error.

$$P = \begin{bmatrix} 0 & -2 \\ 4 & 7 \end{bmatrix}, \quad Q = \begin{bmatrix} 1 & 2 \\ -5 & 6 \end{bmatrix}, \quad R = \begin{bmatrix} 1 & 4 & -1 \\ 2 & 3 & -5 \end{bmatrix}$$

*9. PQ * 10. PQR 11. RP *12. $3PR + QR$

13. R^2 14. $P^2R - 3Q^3R$

2.2   Properties of Matrix Operations

We have defined operations of addition, scalar multiplication, and multiplication of matrices. In this section we discuss the algebraic properties of these operations. We have already seen that matrices have some, but not all, of the algebraic properties of real numbers. We have seen that identity matrices play a role similar to 1 for real numbers in that $AI_n = I_nA = A$. We have seen that multiplication of matrices is not commutative, whereas multiplication of real numbers is commutative.

Theorem 2.2   Let A, B, and C be matrices and a, b, and c be scalars. Assume that the sizes of the matrices are such that the operations can be performed.

Properties of Matrix Addition and Scalar Multiplication

1. $A + B = B + A$ *Commutative property of addition*
2. $A + (B + C) = (A + B) + C$ *Associative property of addition*
3. $A + 0 = 0 + A = A$ *Where 0 is the appropriate zero matrix*
4. $c(A + B) = cA + cB$ *Distributive property of addition*
5. $(a + b)C = aC + bC$ *Distributive property of addition*
6. $(ab)C = a(bC)$

Properties of Matrix Addition and Scalar Multiplication

1. $A(BC) = (AB)C$ *Associative property of multiplication*
2. $A(B + C) = AB + AC$ *Distributive property of multiplication*
3. $(A + B)C = AC + BC$ *Distributive property of multiplication*
4. $AI_n = I_nA = A$ *Where I_n is the appropriate identity matrix*
5. $c(AB) = (cA)B = A(cB)$

Note: $AB \neq BA$ in general.   *Multiplication of matrices is not commutative.*

Each one of these results asserts an equality between matrices. We know that two matrices are equal if they are of the same size and their corresponding elements are equal. Each result is verified by showing this to be the case. We illustrate the method for the commutative property of addition. You are asked to use the same approach to prove some of the other results in the exercises that follow.

$A + B = B + A$ By the rule of matrix addition we know that $A + B$ and $B + A$ are both matrices of the same size. It remains to show that their corresponding elements are equal. Consider the (i,j)th element of each matrix.

$$(i,j)\text{th element of } A + B = a_{ij} + b_{ij}$$
$$(i,j)\text{th element of } B + A = b_{ij} + a_{ij}$$
$$= a_{ij} + b_{ij} \quad \text{addition of real numbers is commutative}$$

The corresponding elements of $A + B$ and $B + A$ are equal. Thus $A + B = B + A$.

The associative properties of matrix addition and multiplication enable us to extend addition and multiplication to more than two matrices. We can write sums and products such as $A + B + C$ and ABC without parentheses, since the same results are obtained no matter how the matrices are grouped. The following examples illustrate these concepts.

Example 1   Determine $A + B + C$ for the following three matrices. Let $A = \begin{bmatrix} 1 & 3 \\ -4 & 5 \end{bmatrix}$, $B = \begin{bmatrix} 3 & -7 \\ 8 & 1 \end{bmatrix}$, and $C = \begin{bmatrix} 0 & -2 \\ 5 & -1 \end{bmatrix}$

Solution   We add A, B, and C by adding corresponding elements.

$$A + B + C = \begin{bmatrix} 1 & 3 \\ -4 & 5 \end{bmatrix} + \begin{bmatrix} 3 & -7 \\ 8 & 1 \end{bmatrix} + \begin{bmatrix} 0 & -2 \\ 5 & -1 \end{bmatrix}$$

$$= \begin{bmatrix} 1+3+0 & 3-7-2 \\ -4+8+5 & 5+1-1 \end{bmatrix} = \begin{bmatrix} 4 & -6 \\ 9 & 5 \end{bmatrix}$$

Certain products of matrices such as $ABCD$ will, of course, exist while other products will not. We can determine whether a product exists by comparing the numbers of rows and columns in adjacent matrices of the product, to see if they match. *If the product exists, the product matrix will have the same number of rows as the first matrix in the chain and the same number of columns as the last matrix.*

Example 2   Compute the product ABC of the following three matrices.

$$A = \begin{bmatrix} 1 & 2 \\ 3 & -1 \end{bmatrix}, B = \begin{bmatrix} 0 & 1 & 3 \\ -1 & 0 & -2 \end{bmatrix}, C = \begin{bmatrix} 4 \\ -1 \\ 0 \end{bmatrix}$$

Solution Let us check to see if the product ABC exists before we start spending time multiplying matrices. We get

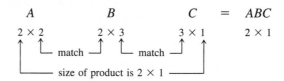

The product exists and will be a 2×1 matrix. Since matrix multiplication is associative, the matrices in the product ABC can be grouped together in any manner for multiplying as long as the order is maintained. Let us use the grouping $(AB)C$. This is probably the most natural. We get

$$AB = \begin{bmatrix} 1 & 2 \\ 3 & -1 \end{bmatrix}\begin{bmatrix} 0 & 1 & 3 \\ -1 & 0 & -2 \end{bmatrix}$$
$$= \begin{bmatrix} -2 & 1 & -1 \\ 1 & 3 & 11 \end{bmatrix}$$

and

$$(AB)C = \begin{bmatrix} -2 & 1 & -1 \\ 1 & 3 & 11 \end{bmatrix}\begin{bmatrix} 4 \\ -1 \\ 0 \end{bmatrix} = \begin{bmatrix} -9 \\ 1 \end{bmatrix}$$

Since matrix multiplication is associative, we have a choice in the way we group matrices together when multiplying a chain of matrices. With many matrices in a product, the sequence selected for multiplication can greatly influence the amount of computation involved. Research has gone into determining efficient algorithms for multiplying chains of matrices. The following example illustrates the concepts involved.

Example 3 Multiply the matrices A, B, and C of the previous example, using the sequence $A(BC)$ of products. Compare the number of multiplications involved in the two ways $(AB)C$ and $A(BC)$ of computing the product ABC.

Solution We get

$$BC = \begin{bmatrix} 0 & 1 & 3 \\ -1 & 0 & 2 \end{bmatrix}\begin{bmatrix} 4 \\ -1 \\ 0 \end{bmatrix} = \begin{bmatrix} -1 \\ -4 \end{bmatrix}$$

and

$$A(BC) = \begin{bmatrix} 1 & 2 \\ 3 & -1 \end{bmatrix}\begin{bmatrix} -1 \\ -4 \end{bmatrix} = \begin{bmatrix} -9 \\ 1 \end{bmatrix}$$

We get the same answer as previously, illustrating the associativity of matrix multiplication. However, if you do the computation both ways, you will sense that this second way is much quicker than the first! Computation such as this is, in practice, carried out on computers. Since multiplication is more time-consuming than addition on computers, analyses of efficiency are usually done in terms of multiplications. Let us count the number of multiplications involved both ways, ignoring additions.

The number of multiplications depends upon the sizes of the matrices, not on the specific elements. It can be shown (Exercise 8 following) that

> If A is an $m \times r$ matrix and B *is an* $r \times n$ matrix, the number of scalar multiplications involved in computing the product AB is *mrn*.

The sizes of our matrices are

$$\begin{matrix} A & B & C & = & D \\ 2 \times 2 & 2 \times 3 & 3 \times 1 & & 2 \times 1 \end{matrix}$$

We now look at the two distinct ways of computing the product, $(AB)C$ and $A(BC)$.

(AB)C   Compute $(AB)C$ in two steps, using an intermediate matrix F.

- Let $AB = F$. $FC = D$. F is a 2×3 matrix.
- The number of multiplications to compute F is $2 \times 2 \times 3 = 12$.
- The number of multiplications to compute D is $2 \times 3 \times 1 = 6$.
- Thus the total number of multiplications to compute $(AB)C$ is 18.

A(BC)   Compute $A(BC)$ in two steps. $BC = G$. $AG = D$. G is a 2×1 matrix.

- The number of multiplications to compute G is $2 \times 3 \times 1 = 6$.
- The number of multiplications to compute D is $2 \times 2 \times 1 = 4$.
- Thus the total number of multiplications to compute $A(BC)$ is 10.

We see that even for this small problem, the number of multiplications involved in computing ABC varies from 18, using the arrangement $(AB)C$, to 10, using the arrangement $A(BC)$. The second way of multiplying the matrices is almost twice as fast as the first.

We now illustrate the orders of magnitudes of numbers that can be involved in multiplying larger matrices. Suppose we wish to calculate the product $ABCD$ for matrices having the following sizes.

$$\begin{matrix} A & B & C & D & = & E \\ 5 \times 14 & 14 \times 87 & 87 \times 3 & 3 \times 42 & & 5 \times 42 \end{matrix}$$

There are five different ways of calculating the product. The number of multiplications required for each way are as follows:

$$((AB)C)D \qquad 8,025$$
$$(A(BC))D \qquad 4,494$$
$$(AB)(CD) \qquad 35,322$$
$$A((BC)D) \qquad 8,358$$
$$A(B(CD)) \qquad 65,058$$

The most efficient method is over 14 times faster than the slowest! (See exercises following for the derivation of these numbers.)

An algorithm has been developed for multiplying a given chain of matrices in the most efficient way. The algorithm uses a technique called dynamic programming. Readers who wish to look further into this topic should consult *Algorithms, Theory and Practice* by Gilles Brassard and Paul Bratley, Prentice-Hall, 1988, p. 147.

In algebra we know that the following cancellation laws apply.

- If $ab = ac$ and $a \neq 0$, then $b = c$.
- If $pq = 0$, then $p = 0$ or $q = 0$.

However, the corresponding results are not true for matrices.

- $AB = AC$ does not imply that $B = C$.
- $PQ = 0$ does not imply that $P = 0$ or $Q = 0$.

We demonstrate these possibilities by means of examples.

Consider the matrices $A = \begin{bmatrix} 1 & 2 \\ 2 & 4 \end{bmatrix}$, $B = \begin{bmatrix} -1 & 2 \\ 2 & 1 \end{bmatrix}$, $C = \begin{bmatrix} -3 & 8 \\ 3 & -2 \end{bmatrix}$. Observe that $AB = AC = \begin{bmatrix} 3 & 4 \\ 6 & 8 \end{bmatrix}$, but $B \neq C$.

Consider the matrices $P = \begin{bmatrix} 1 & -2 \\ -2 & 4 \end{bmatrix}$, $Q = \begin{bmatrix} 2 & -6 \\ 1 & -3 \end{bmatrix}$. Observe that $PQ = 0$, but $P \neq 0$ and $Q \neq 0$.

Powers of Matrices

The notation used for the powers of matrices is similar to that used for the powers of real numbers. If A is a square matrix, then A multiplied by itself k times is written A^k.

$$A^k = \underbrace{AA \ \ldots \ A}_{k \text{ times}}$$

Familiar laws of exponents of real numbers hold for matrices.

Theorem 2.3   If A is an $n \times n$ square matrix and r and s are nonnegative integers, then

1. $A^r A^s = A^{r+s}$ 2. $(A^r)^s = A^{rs}$ 3. $A^0 = I_n$ (by definition)

We verify the first property. The proof of the second property is similar.

$$A^r A^s = \underbrace{A \quad \ldots \quad A}_{r \text{ times}} \; \underbrace{A \quad \ldots \quad A}_{s \text{ times}} = \underbrace{A \quad \ldots \quad A}_{r + s \text{ times}} = A^{r+s}$$

Example 4 If $A = \begin{bmatrix} 1 & -2 \\ -1 & 0 \end{bmatrix}$, computer A^4.

Solution This example illustrates how the above rules can be used to reduce the amount of computation involved in multiplying matrices. We know that $A^4 = AAAA$. We could perform three matrix multiplications to arrive at A^4. However, we can apply rule 2 to write $A^4 = (A^2)^2$ and thus arrive at the result using two products. We get

$$A^2 = \begin{bmatrix} 1 & -2 \\ -1 & 0 \end{bmatrix}\begin{bmatrix} 1 & -2 \\ -1 & 0 \end{bmatrix} = \begin{bmatrix} 3 & -2 \\ -1 & 2 \end{bmatrix}$$

$$A^4 = \begin{bmatrix} 3 & -2 \\ -1 & 2 \end{bmatrix}\begin{bmatrix} 3 & -2 \\ -1 & 2 \end{bmatrix} = \begin{bmatrix} 11 & -10 \\ -5 & 6 \end{bmatrix}$$

The following example illustrates that the properties of matrix operations can be used to simplify matrix expressions in a way similar to the simplification of ordinary algebraic expressions.

Example 5 Simplify the following matrix expression.

$$A(A + 2B) + 3B(2A - B) - A^2 + 7B^2 - 5AB$$

Solution Using the properties of matrix operations, we get

$$A(A + 2B) + 3B(2A - B) - A^2 + 7B^2 - 5AB = A^2 + 2AB + 6BA - 3B^2 - A^2 + 7B^2 - 5AB$$
$$= -3AB + 6BA + 4B^2$$

Resist the temptation to simplify $-3AB + 6BA$; matrix multiplication is not commutative!

Exercise Set 2.2

Computation

1. Let $A = \begin{bmatrix} 1 & 2 \\ -1 & 0 \end{bmatrix}$, $B = \begin{bmatrix} 0 & 5 & 4 \\ 2 & 1 & 3 \end{bmatrix}$, $C = \begin{bmatrix} 2 & 3 \\ 6 & 1 \end{bmatrix}$, $D = \begin{bmatrix} 2 & -2 \\ 1 & 3 \end{bmatrix}$. Calculate, if possible, *(a) AB and BA,

(b) AC and CA, *(c) AD and DA. Observe that $AB \neq BA$ since BA does not exist, $AC \neq CA$ and $AD \neq DA$, illustrating all the different possibilities when order is reversed in matrix multiplication.

*2.** Compute $A(BC)$ and $(AB)C$ for the matrices

$$A = \begin{bmatrix} 1 & 2 \\ -1 & 0 \\ 1 & 1 \end{bmatrix}, \qquad B = \begin{bmatrix} 2 & 4 \\ -2 & 3 \end{bmatrix}, \qquad C = \begin{bmatrix} 1 \\ 2 \end{bmatrix}$$

Observe that these products are equal, illustrating the associative property of matrix multiplication.

3. Compute the product ABC for the following three matrices in two distinct ways.

$$A = \begin{bmatrix} 1 & 2 \\ -1 & 3 \end{bmatrix}, \qquad B = \begin{bmatrix} 3 & 1 & 2 \\ 4 & 3 & 1 \end{bmatrix}, \qquad C = \begin{bmatrix} 1 & 2 \\ 3 & 4 \\ 1 & 0 \end{bmatrix}$$

(2×2)　　(2×3)　　(3×2)

4. If $A = \begin{bmatrix} 1 & 2 \\ 3 & 4 \end{bmatrix}$, $B = \begin{bmatrix} 2 & -3 \\ 0 & 1 \end{bmatrix}$, and $C = \begin{bmatrix} -2 & 0 \\ 3 & 4 \end{bmatrix}$, compute each of the following.

 *(a) AC^2　　　　　　　(b) $A^2 - 3B$

 *(c) $BC^2 + B^3$　　　　(d) $3A^2 + 2A - I_2$

5. If $A = \begin{bmatrix} 2 & 0 \\ -1 & 5 \end{bmatrix}$, $B = \begin{bmatrix} -1 & 1 \\ 2 & 4 \end{bmatrix}$, and

 $C = \begin{bmatrix} 3 & 4 \\ 0 & 2 \end{bmatrix}$, compute each of the following.

 *(a) $(AB)^2$　　　　　　(b) $A - 3B^2$

 *(c) $A^2B + 2C^3$　　　(d) $2A^2 - 2A + 3I_2$

6. Given that A is a 4×2 matrix, B is 2×6, C is 3×4, and D is 6×3, determine the sizes of the following products, if they exist.

 *(a) ABC　　　(b) ABD　　　*(c) CAB

 (d) $DCAB$　　　(e) A^2BDC

7. If P is 3×2, Q is 2×1, R is 1×3, S is 3×1, and T is 3×3, determine the sizes of the following matrix expressions, if they exist.

 *(a) PQR　　　　　　(b) $PQ + TPQ$

 *(c) $5QR - 2TPR$　　(d) $4SPQ + 3PQ$

 *(e) $QRSR + QR$

*8. Prove that if A is an $m \times r$ matrix and B is an $r \times n$ matrix, then the number of scalar multiplications involved in computing the product AB is mrn. Let C be an $n \times s$ matrix. Derive formulas for the number of scalar multiplications involved in computing $(AB)C$ and $A(BC)$.

9. If A and B are matrices of the given sizes, determine the number of scalar multiplications necessary to compute AB.

 *(a) A 2×3, B 3×7　　(b) A 5×2, B 2×8

 *(c) A 1×9, B 9×27　(d) A 8×5, B 5×12

10. Let A be an $n \times n$ matrix. Determine the number of scalar multiplications necessary to compute A^2, A^3, and A^m.

11. Determine the number of multiplications needed to compute the products $(AB)C$ and $A(BC)$ when A, B, and C are the following sizes:

 *(a) A 2×4, B 4×3, C 3×1

 (b) A 3×7, B 7×5, C 5×2

 *(c) A 6×2, B 2×5, C 5×3

 (d) A 3×5, B 5×47, C 47×5

 *(e) A 7×97, B 97×2, C 2×3

12. Determine the number of multiplications needed to compute the products $((AB)C)D$, $(A(BC))D$, $(AB)(CD)$, $A((BC)D)$, $A(B(CD))$ when A, B, C, and D are the following sizes:

 $$A \ 5 \times 14, \ B \ 14 \times 87, \ C \ 87 \times 3, \ D \ 3 \times 42$$

 (the example given in this section)

*13. Let A be an $m \times r$ matrix and B be an $r \times n$ matrix. Derive a formula for the number of scalar additions involved in computing the product AB. Let C be an $n \times s$ matrix. Derive formulas for the number of scalar additions involved in computing $(AB)C$ and $A(BC)$. Test your formulas on a 2×2 matrix A, a 2×3 matrix B, and a 3×1 matrix C.

Addition, Scalar Multiplication, and Multiplication of Matrices

14. Let A be an $m \times n$ matrix. Prove that AB and BA will both exist only if B is an $n \times m$ matrix.

15. Verify the following properties of matrix operations given in this section.

 *(a) the associative property of matrix addition: $A + (B + C) = (A + B) + C$

 (b) the distributive property: $c(A + B) = cA + cB$

 (c) $AI_n = I_nA = A$ if A is an $n \times n$ matrix

16. Let A be an $m \times n$ matrix. Show that $AI_n = A$.

*17. Let A be an $m \times n$ matrix, 0_{mn} be the $m \times n$ zero matrix, and c be a scalar. Show that if $cA = 0_{mn}$, then either $c = 0$ or $A = 0_{mn}$.

18. Simplify the following matrix expressions.

 *(a) $A(A - 4B) + 2B(A + B) - A^2 + 7B^2 + 3AB$

 (b) $B(2I_n - BA) + B(4I_n + 5A)B - 3BAB + 7B^2A$

 *(c) $(A - B)(A + B) - (A + B)^2$

19. Simplify the following matrix expressions.

 (a) $A(A + B) - B(A + B)$

 *(b) $A(A - B)B + B2AB - 3A^2$

 *(c) $(A + B)^3 - 2A^3 - 3ABA - A3B^2 - B^3$

20. Find all the matrices that commute with the following matrices:

 *(a) $\begin{bmatrix} 1 & 0 \\ -1 & 0 \end{bmatrix}$　　(b) $\begin{bmatrix} 1 & 0 \\ 0 & 2 \end{bmatrix}$　　*(c) $\begin{bmatrix} 0 & 1 & 0 \\ 0 & 0 & 1 \\ 0 & 0 & 0 \end{bmatrix}$

*21. What is incorrect about the following proof?

Let $AX = B$ be a system of linear equations with solutions X_1 and X_2. Thus

$$AX_1 = B \quad \text{and} \quad AX_2 = B$$
$$AX_1 = AX_2$$
$$X_1 = X_2$$

Thus every system of linear equations has at most one solution.

Powers of Matrices

22. (a) Let A be an $n \times n$ matrix. Prove that A^2 is an $n \times n$ matrix.

(b) Let A be an $m \times n$ matrix, with $m \neq n$. Prove that A^2 does not exist.

Thus one can talk about powers of square matrices only.

*23. If A and B are square matrices of the same size, prove that, in general,

$$(A + B)^2 \neq A^2 + 2AB + B^2$$

Under what condition does equality hold?

24. If A and B are square matrices of the same size such that $AB = BA$, prove that $(AB)^2 = A^2B^2$. By constructing an example, show that this result does not hold for all square matrices of the same size.

25. If n is a nonnegative integer and A and B are square matrices of the same size such that $AB = BA$, prove that $(AB)^n = A^nB^n$. By constructing an example, show that this identity does not hold in general for all square matrices of the same size.

26. Show that nonnegative integer powers of the same matrix commute.

Diagonal Matrices

27. Let A and B be diagonal matrices of the same size and c a scalar. Prove that ***(a)** $A + B$ is diagonal, **(b)** cA is diagonal, **(c)** AB is diagonal.

28. If A and B are diagonal matrices of the same size, prove that $AB = BA$.

29. Prove that if a matrix A commutes with a diagonal matrix that has no two diagonal elements the same, then A is a diagonal matrix.

Idempotent and Nilpotent Matrices

A square matrix A is said to be **idempotent** if $A^2 = A$. A square matrix A is said to be **nilpotent** if there is a positive integer p such that $A^p = 0$. The least integer such that $A^p = 0$ is called the **degree of nilpotency** of the matrix.

30. Determine whether the following matrices are idempotent.

*(a) $\begin{bmatrix} 1 & 0 \\ 0 & 1 \end{bmatrix}$ **(b)** $\begin{bmatrix} 1 & 0 \\ 0 & 0 \end{bmatrix}$

*(c) $\begin{bmatrix} 0 & 1 \\ 1 & 0 \end{bmatrix}$ **(d)** $\begin{bmatrix} 3 & -6 \\ 1 & -2 \end{bmatrix}$

*(e) $\begin{bmatrix} 1 & 2 & 2 \\ 0 & 0 & -1 \\ 0 & 0 & 1 \end{bmatrix}$ **(f)** $\begin{bmatrix} 1 & 3 & 0 \\ 0 & 0 & 1 \\ 0 & 0 & 0 \end{bmatrix}$

*31. Determine b, c, and d such that $\begin{bmatrix} 1 & b \\ c & d \end{bmatrix}$ is idempotent.

32. Determine a, c, and d such that $\begin{bmatrix} a & 0 \\ c & d \end{bmatrix}$ is idempotent.

*33. Prove that if A and B are idempotent and $AB = BA$, then AB is idempotent.

34. Show that if A is idempotent and if n is a positive integer, then $A^n = A$.

35. Show that the following matrices are nilpotent with degree of nilpotency 2.

(a) $\begin{bmatrix} 1 & 1 \\ -1 & -1 \end{bmatrix}$ **(b)** $\begin{bmatrix} -4 & 8 \\ -2 & 4 \end{bmatrix}$ **(c)** $\begin{bmatrix} 3 & -9 \\ 1 & -3 \end{bmatrix}$

36. Show that the following matrix is nilpotent with degree of nilpotency 3.

$$\begin{bmatrix} 0 & 1 & 0 \\ 0 & 0 & 1 \\ 0 & 0 & 0 \end{bmatrix}$$

2.3 Symmetric Matrices and an Application in Archaeology

In this section we continue the algebraic development of matrices and we see how the theory developed is used by archaeologists to determine the chronological order of graves and artifacts.

Definition *The **transpose** of a matrix A, denoted A^t, is the matrix whose columns are the rows of the given matrix A.*

The first row of A becomes the first column of A^t, the second row of A becomes the second column of A^t, and so on. The (i,j)th element of A becomes the (j,i)th element of A^t. If A is an $m \times n$ matrix, then A^t is an $n \times m$ matrix.

Example 1 Determine the transpose of each of the following matrices.

$$A = \begin{bmatrix} 2 & 7 \\ -8 & 0 \end{bmatrix}, \qquad B = \begin{bmatrix} 1 & 2 & -7 \\ 4 & 5 & 6 \end{bmatrix}, \qquad C = [-1 \quad 3 \quad 4]$$

Solution Writing rows as columns, we get

$$A^t = \begin{bmatrix} 2 & -8 \\ 7 & 0 \end{bmatrix}, \qquad B^t = \begin{bmatrix} 1 & 4 \\ 2 & 5 \\ -7 & 6 \end{bmatrix}, \qquad C^t = \begin{bmatrix} -1 \\ 3 \\ 4 \end{bmatrix}$$

Observe that the $(1, 3)$ element of B, namely -7, becomes the $(3, 1)$ element of B^t. Note also that B is a 2×3 matrix while B^t is 3×2, C is a 1×3 matrix while C^t is 3×1.

We have defined three operations for matrices, namely addition, scalar multiplication, and multiplication. The following theorem tells us how transpose works in conjunction with these operations.

Theorem 2.4 **Properties of Transpose** Let A and B be matrices and c be a scalar. Assume that the sizes of the matrices are such that the operations can be performed.

1. $(A + B)^t = A^t + B^t$ *Transpose of a sum*
2. $(cA)^t = cA^t$ *Transpose of a scalar multiple*
3. $(AB)^t = B^t A^t$ *Transpose of a product*
4. $(A^t)^t = A$

We demonstrate the techniques that are used in proving results involving transposes by verifying the third property. You are asked to derive the other results in the exercises that follow.

$(AB)^t = B^t A^t$ The expressions $(AB)^t$ and $B^t A^t$ will be matrices, on carrying out all the products and transposes. We prove that these matrices are equal by showing that corresponding elements are equal.

$$(i,j)\text{th element of } (AB)^t = (j,i)\text{th element of } AB$$
$$= a_{j1}b_{1i} + \cdots + a_{jn}b_{ni}$$
$$(i,j)\text{th element of } B^t A^t = (\text{row } i \text{ of } B^t) \times (\text{column } j \text{ of } A^t)$$
$$= (\text{column } i \text{ of } B) \times (\text{row } j \text{ of } A)$$
$$= a_{j1}b_{1i} + \cdots + a_{jn}b_{ni}$$

The corresponding elements of $(AB)^t$ and $B^t A^t$ are equal, proving the result.

The results for the transpose of a sum and a product can be extended to any number of matrices. For example, for three matrices A, B, and C

$$(A + B + C)^t = A^t + B^t + C^t \qquad \text{and} \qquad (ABC)^t = C^t B^t A^t$$

Note the reversal of the order of the matrices in the transpose of a product. (See following exercises for the proofs.) We now introduce symmetric matrices. They are probably the most important class of matrices. They are used in areas of mathematics such as geometry, and in fields such as theoretical physics, mechanical and electrical engineering, and sociology.

Definition   *A **symmetric matrix** is a matrix that is equal to its transpose.*

The following are examples of symmetric matrices. Note the symmetry of these matrices about the main diagonal. All nondiagonal elements occur in pairs symmetrically located about the main diagonal.

$$\begin{bmatrix} 2 & 5 \\ 5 & -4 \end{bmatrix} \qquad \begin{bmatrix} 0 & 1 & -4 \\ 1 & 7 & 8 \\ -4 & 8 & -3 \end{bmatrix} \qquad \begin{bmatrix} 1 & 0 & -2 & 4 \\ 0 & 7 & 3 & 9 \\ -2 & 3 & 2 & -3 \\ 4 & 9 & -3 & 6 \end{bmatrix}$$

Example 2   Consider the following matrix, which represents distances between various U.S. cities.

	Chicago	LA	Miami	NY
Chicago	0	2092	1374	841
Los Angeles	2092	0	2733	2797
Miami	1374	2733	0	1336
New York	841	2797	1336	0

Observe that the matrix is symmetric. All elements occur in pairs, symmetrically located about the main diagonal. There is a reason for this, namely that the distance from city X to city Y is the same as the distance from city Y to city X. For example, the distance from Chicago to Miami is 1374 (row 1, column 3), which will be the same as the distance from Miami to Chicago (row, 3, column 1). All such mileage matrices will be symmetric.

The expressions *if and only if* and *necessary and sufficient* are frequently used in mathematics, and we shall use them periodically in this course. They mean the same thing. Let p and q be statements. Suppose that p implies q, written $p \Rightarrow q$, and also $q \Rightarrow p$. The second implication is called the **converse** of the first. We say that "p if and only if q" or "p is necessary and sufficient for q." The next example states a result using this language.

Example 3 Let A and B be symmetric matrices of the same size. Prove that the product AB is symmetric if and only if $AB = BA$.

Every "if and only if" problem such as this consists of two parts. We have to show that (a) if AB is symmetric, then $AB = BA$, and the converse (b) if $AB = BA$, then AB is symmetric.

Proof

(a) Let AB be symmetric. Then

$$AB = (AB)^t \quad \text{by definition of symmetric matrix}$$
$$= B^t A^t \quad \text{by the property of transpose}$$
$$= BA \quad \text{since } A \text{ and } B \text{ are symmetric matrices}$$

(b) Let $AB = BA$. Then

$$(AB)^t = (BA)^t$$
$$= A^t B^t \quad \text{by the property of transpose}$$
$$= AB \quad \text{since } A \text{ and } B \text{ are symmetric matrices}$$

Therefore AB is symmetric.

Thus, given two symmetric matrices of the same size, A and B, the product AB is symmetric if and only if $AB = BA$.

This result could also have been stated as follows: Given two symmetric matrices of the same size, A and B, then a necessary and sufficient condition for the product AB to be symmetric is that $AB = BA$.

We now introduce a number that is associated with every square matrix, called the trace of the matrix.

Definition   *Let A be a square matrix. The **trace** of A, denoted tr(A), is the sum of the diagonal elements of A. Thus if A is an n × n matrix,*

$$tr(A) = a_{11} + a_{22} + \cdots + a_{nn}$$

Example 4   Determine the trace of the matrix $A = \begin{bmatrix} 4 & 1 & -2 \\ 2 & -5 & 6 \\ 7 & 3 & 0 \end{bmatrix}$

Solution   We get

$$tr(A) = 4 + (-5) + 0 = -1$$

The trace of a matrix plays an important role in matrix theory and matrix applications because of its properties and the ease with which it can be evaluated. It is important in fields such as statistical mechanics, general relativity, and quantum mechanics where it has physical significance.

The following theorem tells us how the operation of trace interacts with the operations of matrix addition, scalar multiplication, multiplication, and transpose.

Theorem 2.5   **Properties of Trace**   Let A and B be matrices and c be a scalar. Assume that sizes of the matrices are such that the operations can be performed.

1. $tr(A + B) = tr(A) + tr(B)$
2. $tr(AB) = tr(BA)$
3. $tr(cA) = c\,tr(A)$
4. $tr(A^t) = tr(A)$

Proof   We prove the first property, leaving the proofs of the other properties for you to complete in the exercises. Since the diagonal elements of $A + B$ are $(a_{11} + b_{11}), (a_{22} + b_{22}), \ldots, (a_{nn} + b_{nn})$, we get

$$tr(A + B) = (a_{11} + b_{11}) + (a_{22} + b_{22}) + \cdots + (a_{nn} + b_{nn})$$
$$= (a_{11} + a_{22} + \cdots + a_{nn}) + (b_{11} + b_{22} + \cdots + b_{nn})$$
$$= tr(A) + tr(B)$$

*Matrices with Complex Elements

The elements of a matrix may be complex numbers. A **complex number** is of the form

$$z = a + bi$$

where a and b are real numbers and $i = \sqrt{-1}$. a is called the **real part** and b the **imaginary part** of z.

Let $z_1 = a + bi$, $z_2 = c + di$ be complex numbers. The rules of arithmetic for complex numbers are as follows.

Equality: $z_1 = z_2$ if $a = c$ and $b = d$

Addition: $z_1 + z_2 = (a + c) + (b + d)i$

Subtraction: $z_1 - z_2 = (a - c) + (b - d)i$

Multiplication: $z_1 z_2 = (a + bi)(c + di)$

$$= a(c + di) + bi(c + di)$$

$$= ac + adi + bci + bdi^2$$

$$= ac + bdi^2 + (ad + cb)i$$

$$= (ac - bd) + (ad + cb)i$$

The **conjugate** of a complex number $z = a + bi$ is defined and written $\bar{z} = a - bi$.

Example 5   Consider the complex numbers $z_1 = 2 + 3i$ and $z_2 = 1 - 2i$. Compute $z_1 + z_2$, $z_1 z_2$, and $\bar{z}_1$.

Solution   Using the above definitions, we get

$$z_1 + z_2 = (2 + 3i) + (1 - 2i) = (2 + 1) + (3 - 2)i = 3 + i$$

$$z_1 z_2 = (2 + 3i)(1 - 2i) = 2(1 - 2i) + 3i(1 - 2i) = 2 - 4i + 3i - 6i^2 = 8 - i$$

$$\bar{z}_1 = 2 - 3i$$

Matrices having complex elements are added, subtracted, and multiplied using the same rules as matrices having real elements.

Example 6   Let $A = \begin{bmatrix} 2 + i & 3 - 2i \\ 4 & 5i \end{bmatrix}$ and $B = \begin{bmatrix} 3 & 2i \\ 1 + i & 2 + 3i \end{bmatrix}$. Compute $A + B$, $2A$, and AB.

Solution   We get

$$A + B = \begin{bmatrix} 2 + i & 3 - 2i \\ 4 & 5i \end{bmatrix} + \begin{bmatrix} 3 & 2i \\ 1 + i & 2 + 3i \end{bmatrix} = \begin{bmatrix} 2 + i + 3 & 3 - 2i + 2i \\ 4 + 1 + i & 5i + 2 + 3i \end{bmatrix}$$

$$= \begin{bmatrix} 5 + i & 3 \\ 5 + i & 2 + 8i \end{bmatrix}$$

$$2A = 2 \begin{bmatrix} 2 + i & 3 - 2i \\ 4 & 5i \end{bmatrix} = \begin{bmatrix} 4 + 2i & 6 - 4i \\ 8 & 10i \end{bmatrix}$$

$$AB = \begin{bmatrix} 2 + i & 3 - 2i \\ 4 & 5i \end{bmatrix} \begin{bmatrix} 3 & 2i \\ 1 + i & 2 + 3i \end{bmatrix}$$

$$= \begin{bmatrix} (2 + i)3 + (3 - 2i)(1 + i) & (2 + i)(2i) + (3 - 2i)(2 + 3i) \\ (4)(3) + (5i)(1 + i) & 4(2i) + (5i)(2 + 3i) \end{bmatrix}$$

$$= \begin{bmatrix} 11 + 4i & 10 + 9i \\ 7 + 5i & -15 + 18i \end{bmatrix}$$

The **conjugate** of a matrix A is denoted $\overline{A}$ and is obtained by taking the conjugate of each element of the matrix. The **conjugate transpose** of A is written and defined by $A^* = \overline{A}^t$.

For example, if $A = \begin{bmatrix} 2 + 3i & 1 - 4i \\ 6 & 7i \end{bmatrix}$, then

$$\overline{A} = \begin{bmatrix} 2 - 3i & 1 + 4i \\ 6 & -7i \end{bmatrix} \quad \text{and} \quad A^* = \overline{A}^t = \begin{bmatrix} 2 - 3i & 6 \\ 1 + 4i & -7i \end{bmatrix}$$

A square matrix C is said to be **hermitian** if $C = C^*$.

Let us show that the matrix $C = \begin{bmatrix} 2 & 3 - 4i \\ 3 + 4i & 6 \end{bmatrix}$ is hermitian. We get

$$\overline{C} = \begin{bmatrix} 2 & 3 + 4i \\ 3 - 4i & 6 \end{bmatrix}; \quad C^* = \overline{C}^t = \begin{bmatrix} 2 & 3 - 4i \\ 3 + 4i & 6 \end{bmatrix} = C$$

Hermitian matrices are more important than symmetric matrices for matrices having complex elements.

The properties of conjugate transpose are similar to those of transpose. We list these properties in the following theorem, leaving the proofs for you to do in the exercises that follow.

Theorem 2.6   **Properties of Conjugate Transpose**   Let A amd B be matrices with complex elements, and let z be a complex number.

1. $(A + B)^* = A^* + B^*$ *Conjugate transpose of a sum*
2. $(zA)^* = \overline{z}A^*$ *Conjugate transpose of a scalar multiple*
3. $(AB)^* = B^*A^*$ *Conjugate transpose of a product*
4. $(A^*)^* = A$

The time is now ripe for a good application of matrix algebra!

Seriation in Archaeology

A problem confronting archaeologists is that of placing sites and artifacts in proper chronological order. This branch of archaeology, called **sequence dating** or **seriation** began with the work of Sir Flinders Petrie in the late nineteenth century. Petrie studied graves in the cemeteries of Nagada, Ballas, and Hu, all located in what was prehistoric Egypt. (Recent carbon dating shows that all the graves ranged from 6000 B.C. to 2500 B.C.) Petrie used the data from approximately 900 graves to order them and assign a time period to each type of pottery found.

Let us look at this general problem of seriation in terms of graves and varieties of pottery found in graves. An assumption usually made in archaeology is that two graves that have similar contents are more likely to lie close together in time than are two graves that have little in common. The mathematical model that we now construct leads to information concerning the common contents of graves and thus to the chrono-logical order of the graves.

We construct a matrix A, all of whose elements are either 1 or 0, that describes the pottery content of the graves. Label the graves 1, 2, . . . and the types of pottery 1, 2, ... Let the matrix A be defined by

$$a_{ij} = \begin{cases} 1 & \text{if grave } i \text{ contains pottery type } j \\ 0 & \text{if grave } i \text{ does not contain pottery type } j \end{cases}$$

The matrix A contains all the information about the pottery content of the various graves. The following result now tells us how information is extracted from A.

The element g_{ij} of the matrix $G = AA^t$ is equal to the number of types of pottery common to both grave i and grave j.

Thus the larger the element g_{ij}, the closer grave i and grave j are in time. By examining the elements of G, the archaeologist can arrive at the chronological order of the graves. Let us verify this result.

$$g_{ij} = \text{element in row } i, \text{ column } j \text{ of } G$$

$$= (\text{row } i \text{ of } A) \times (\text{column } j \text{ of } A^t)$$

$$= [a_{i1} \quad a_{i2} \quad \ldots \quad a_{in}] \begin{bmatrix} a_{j1} \\ a_{j2} \\ \vdots \\ a_{jn} \end{bmatrix}$$

$$= a_{i1}a_{j1} + a_{i2}a_{j2} + \cdots + a_{in}a_{jn}$$

Each term in this sum will be either 1 or 0. For example, the term $a_{i2}a_{j2}$ will be 1 if a_{i2} and a_{j2} are both 1—that is, if pottery type 2 is common to graves i and j. It will be 0 if pottery type 2 is not common to graves i and j. Thus the number of 1s in this expression for g_{ij} (the actual value of g_{ij}) is the number of types of pottery common to graves i and j.

The matrix $P = A^t A$ leads in an analogous manner to information about the sequence dating of the pottery. The assumption is made that the larger the number of graves in which two types of pottery appear, the closer they are chronologically. The element p_{ij} of the matrix $P = A^t A$ gives the number of graves in which the ith and jth types of pottery both appear. Thus the larger the element p_{ij}, the closer pottery types i and j are in time. By examining the elements of P, we can arrive at the chronological order of the pottery.

It can be shown mathematically that the matrices G ($=AA^t$) and P ($= A^t A$) are symmetric matrices (see Exercise 27 following). Furthermore it can be argued from the physical interpretation that G and P should be symmetric matrices. This illustrates the compatibility of the mathematics and the interpretation. The implication of this symmetry is that all the information is contained in the elements above the main diagonals of these matrices. The information is just duplicated in the elements below the main diagonals.

We now illustrate this method by means of an example. Let the following matrix A represent the three pottery contents of four graves.

$$A = \begin{bmatrix} 1 & 0 & 1 \\ 1 & 0 & 0 \\ 0 & 1 & 1 \\ 0 & 1 & 0 \end{bmatrix}$$

Thus, for example, $a_{13} = 1$ implies that grave 1 contains pottery type 3; $a_{23} = 0$ implies that grave 2 does not contain pottery type 3. G is calculated:

$$G = AA^t = \begin{bmatrix} 1 & 0 & 1 \\ 1 & 0 & 0 \\ 0 & 1 & 1 \\ 0 & 1 & 0 \end{bmatrix} \begin{bmatrix} 1 & 1 & 0 & 0 \\ 0 & 0 & 1 & 1 \\ 1 & 0 & 1 & 0 \end{bmatrix} = \begin{bmatrix} 2 & 1 & 1 & 0 \\ 1 & 1 & 0 & 0 \\ 1 & 0 & 2 & 1 \\ 0 & 0 & 1 & 1 \end{bmatrix}$$

Observe that G is indeed symmetric. The information contained in the elements above the main diagonal is duplicated in the elements below it. We systematically look at the elements above the main diagonal.

$g_{12} = 1$—graves 1 and 2 have one type of pottery in common.

$g_{13} = 1$—graves 1 and 3 have one type of pottery in common.

$g_{14} = 0$—graves 1 and 4 have no pottery in common.

$g_{23} = 0$—graves 2 and 3 have no pottery in common.

$g_{24} = 0$—graves 2 and 4 have no pottery in common.

$g_{34} = 1$—graves 3 and 4 have one type of pottery in common.

Graves 1 and 2 have pottery in common; they are close together in time. Let us start with graves 1 and 2 and construct a diagram.

$$1\text{--}2$$

Next add grave 3 to this diagram. $g_{13} = 1$ while $g_{23} = 0$. Thus grave 3 is close to grave 1 but not close to grave 2. We get

$$3\text{--}1\text{--}2$$

Finally, add grave 4 to the diagram. $g_{34} = 1$ while $g_{14} = 0$ and $g_{24} = 0$. Grave 4 is close to grave 3 but not close to grave 1 or to grave 2. We get

$$4\text{--}3\text{--}1\text{--}2$$

The mathematics does not tell us which way time flows in this diagram. There are two possibilities:

$$4 \rightarrow 3 \rightarrow 1 \rightarrow 2 \quad \text{and} \quad 4 \leftarrow 3 \leftarrow 1 \leftarrow 2$$

The archaeologist usually knows from other sources which of the two extreme graves (4 and 2 in our case) came first. Thus the chronological order of the graves is known.

The matrices G and P contain information about the chronological order of the graves and pottery, through the relative magnitudes of their elements. These matrices are, in practice, large, and the information cannot be sorted out as easily or give results that are as unambiguous as in the above illustration. For example, Petrie examined 900 graves; his matrix G would be a 900×900 matrix. Special mathematical techniques have been developed for extracting information from these matrices; these techniques are now being executed on computers. Readers who are interested in pursuing this topic further should consult "Some Problems and Methods in Statistical Archaeology," by David G. Kendall, *World Archaeology*, **1**, 1969, pp. 61–76.

In this section we have developed the algebraic theory of matrices. This theory is extremely important in applications of mathematics. We shall use matrices in mathematical models of communication and of population movements. Albert Einstein used a matrix equation to describe the relationship between geometry and matter in his general theory of relativity:

$$T = R - \tfrac{1}{2}rG$$

T is a matrix that represents geometry, R is a matrix, r is a scalar, and G is a matrix that describes gravity. All the matrices are symmetric 4×4 matrices. The theory involves solving this matrix equation to determine a gravitational field. The German physicist Werner Heisenberg made use of matrices in his development of quantum mechanics, for which he received a Nobel Prize. The theory of matrices is a cornerstone of two of the foremost physical theories of the twentieth century.

Exercise Set 2.3

Computation

1. Determine the transpose of each of the following matrices. Indicate whether or not the matrix is symmetric.

   *(a) $A = \begin{bmatrix} -1 & 2 \\ 2 & -3 \end{bmatrix}$

   (b) $B = \begin{bmatrix} 1 & 2 \\ 0 & 3 \end{bmatrix}$

   *(c) $C = \begin{bmatrix} 3 & -1 \\ 2 & 4 \end{bmatrix}$

   *(d) $D = \begin{bmatrix} 4 & 5 \\ -2 & 3 \\ 7 & 0 \end{bmatrix}$

   (e) $E = \begin{bmatrix} 4 & 5 & 6 \\ -1 & 2 & 3 \\ 0 & 1 & 2 \end{bmatrix}$

   *(f) $F = \begin{bmatrix} 1 & -1 & 3 \\ -1 & 2 & 0 \\ 3 & 0 & 4 \end{bmatrix}$

   (g) $G = \begin{bmatrix} -2 & 4 & 5 & 7 \\ 1 & 0 & 3 & -7 \end{bmatrix}$

   *(h) $H = \begin{bmatrix} 1 & -2 & 3 \\ 4 & 5 & 6 \\ -2 & 6 & 7 \end{bmatrix}$

   (i) $K = \begin{bmatrix} 7 & 0 & 0 \\ 0 & -3 & 0 \\ 0 & 0 & 9 \end{bmatrix}$

2. Each of the following matrices is to be symmetric. Determine the elements indicated with a *.

   *(a) $\begin{bmatrix} 1 & 2 & 4 \\ * & 6 & * \\ 4 & 5 & 2 \end{bmatrix}$

   (b) $\begin{bmatrix} 3 & 5 & * \\ * & 8 & 4 \\ -3 & * & 3 \end{bmatrix}$

   (c) $\begin{bmatrix} -3 & * & 8 & 9 \\ -4 & 7 & * & 7 \\ * & 2 & 6 & 4 \\ * & 7 & * & 9 \end{bmatrix}$

3. If A is 4×1, B is 2×3, C is 2×4, and D is 1×3, determine the sizes of the following matrices, if they exist.

*(a) ADB^t (b) $C^tB - 5AD$

*(c) $4CA - (CA)^2$ (d) $(ADB^tC)^2 + I_4$

*(e) $(B^tC)^t - AB$

Transpose

4. Prove the following properties of transpose given in Theorem 2.4.

(a) $(A + B)^t = A^t + B^t$ (b) $(cA)^t = cA^t$

(c) $(A^t)^t = A$

5. Prove the following properties of transpose using the results of Theorem 2.4.

*(a) $(A + B + C)^t = A^t + B^t + C^t$

(b) $(ABC)^t = C^tB^tA^t$

6. Let A be a diagonal matrix. Prove that $A = A^t$.

7. Let A be a square matrix. Prove that $(A^n)^t = (A^t)^n$.

Symmetric Matrices

8. Prove that a symmetric matrix is necessarily square.

*9. Let A be a symmetric matrix. Prove that A^t is symmetric.

10. Let A and B be symmetric matrices of the same size and c be a scalar. Prove that (a) $A + B$ is symmetric and that (b) cA is symmetric.

Antisymmetric Matrices

11. A matrix A is said to be **antisymmetric** if $A = -A^t$.

*(a) Give an example of an antisymmetric matrix.

(b) Prove that an antisymmetric matrix is a square matrix having diagonal elements zero.

*(c) Prove that the sum of two antisymmetric matrices of the same size is an antisymmetric matrix.

(d) Prove that the scalar multiple of an antisymmetric matrix is antisymmetric.

12. If A is a square matrix, prove that (a) $A + A^t$ is symmetric and that (b) $A - A^t$ is antisymmetric.

*13. Prove that any square matrix A can be decomposed into the sum of a symmetric matrix B and an antisymmetric matrix C: $A = B + C$.

14. (a) Prove that if A is idempotent, then A^t is also idempotent.

(b) Prove that if A^t is idempotent, then A is idempotent.

Trace of a Matrix

15. Determine the trace of each of the following matrices.

*(a) $\begin{bmatrix} 2 & 3 \\ -1 & -4 \end{bmatrix}$ (b) $\begin{bmatrix} 5 & 1 & 2 \\ 4 & -3 & 5 \\ -7 & 2 & 8 \end{bmatrix}$

*(c) $\begin{bmatrix} 0 & -1 & 2 & 3 \\ -4 & 5 & 3 & 2 \\ 1 & 6 & -7 & 2 \\ 3 & 9 & 2 & 1 \end{bmatrix}$

16. Prove the following properties of trace given in Theorem 2.5.

(a) $\text{tr}(cA) = c\text{tr}(A)$ (b) $\text{tr}(AB) = \text{tr}(BA)$

(c) $\text{tr}(A^t) = \text{tr}(A)$

*17. Prove the following property of trace using the results of Theorem 2.5: $\text{tr}(A + B + C) = \text{tr}(A) + \text{tr}(B) + \text{tr}(C)$.

If and Only If Condition

18. Consider two matrices A and B of the same size. Prove that $A = B$ if and only if $A^t = B^t$.

19. Prove that the matrix product AB exists if and only if the number of columns in A is equal to the number of rows in B.

20. Prove that $AB = 0_n$ for all $n \times n$ matrices B if and only if $A = 0_n$.

Complex Matrices

*21. Compute $A + B$, AB, and BA for the matrices

$$A = \begin{bmatrix} 5 & 3 - i \\ 2 + 3i & -5i \end{bmatrix}, \quad B = \begin{bmatrix} -2 + i & 5 + 2i \\ 3 - i & 4 + 3i \end{bmatrix}$$

22. Compute $A + B$, AB, and BA for the matrices

$$A = \begin{bmatrix} 4 + i & 2 - 3i \\ 6 + 2i & 1 - i \end{bmatrix}, \quad B = \begin{bmatrix} 2 + i & -3 \\ 2 & 4 - 5i \end{bmatrix}$$

*23. Find the conjugate and conjugate transpose of each of the following matrices. Determine which matrices are hermitian.

$$A = \begin{bmatrix} 2 - 3i & 5i \\ 2 & 5 - 4i \end{bmatrix}, \quad B = \begin{bmatrix} 4 & 5 - i \\ 5 + i & 6 \end{bmatrix}$$

$$C = \begin{bmatrix} 7i & 4 - 3i \\ 6 + 8i & -9 \end{bmatrix}, \quad D = \begin{bmatrix} -2 & 3 - 5i \\ 3 + 5i & 9 \end{bmatrix}$$

24. Find the conjugate and conjugate transpose of each of the following matrices. Determine which matrices are hermitian.

$$A = \begin{bmatrix} 3 & 7 + 2i \\ 7 - 2i & 5 \end{bmatrix}, \quad B = \begin{bmatrix} 3 + 5i & 1 - 2i \\ 1 + 2i & 5 + 6i \end{bmatrix},$$

$$C = \begin{bmatrix} 1 & 2 \\ 2 & 4 \end{bmatrix}, \quad D = \begin{bmatrix} 9 & -3i \\ 3i & 8 \end{bmatrix}$$

25. Prove the following four properties of conjugate transpose.

    *(a) $(A + B)^* = A^* + B^*$ **(b)** $(zA)^* = \bar{z}A^*$

    (c) $(AB)^* = B^*A^*$ **(d)** $(A^*)^* = A$

26. Show that a hermitian matrix is necessarily square. Prove that the diagonal elements have to be real numbers.

Applications

27. Let $G = AA^t$ and $P = A^tA$ for an arbitrary matrix A.

    *(a) Prove that G and P are both symmetric matrices.

    (b) G and P both have physical interpretation in the archaeological model. Use this physical interpretation to reason that G and P should be symmetric. The mathematical result and the physical interpretation are compatible.

28. The following matrices describe the pottery contents of various graves. For each situation determine possible chronological orderings of the graves and then the pottery types.

    *(a) $\begin{bmatrix} 1 & 0 \\ 0 & 1 \\ 1 & 1 \end{bmatrix}$ **(b)** $\begin{bmatrix} 0 & 0 & 1 \\ 1 & 1 & 0 \\ 1 & 0 & 1 \\ 0 & 1 & 0 \end{bmatrix}$

    *(c) $\begin{bmatrix} 1 & 0 & 1 & 0 \\ 0 & 1 & 1 & 1 \\ 1 & 1 & 1 & 1 \end{bmatrix}$ **(d)** $\begin{bmatrix} 0 & 0 & 0 & 1 \\ 1 & 1 & 0 & 0 \\ 0 & 0 & 1 & 1 \\ 1 & 0 & 1 & 0 \end{bmatrix}$

    *(e) $\begin{bmatrix} 1 & 0 & 1 & 0 \\ 1 & 0 & 0 & 0 \\ 0 & 1 & 0 & 1 \\ 0 & 1 & 1 & 0 \end{bmatrix}$ **(f)** $\begin{bmatrix} 0 & 1 & 0 & 0 \\ 1 & 0 & 1 & 1 \\ 0 & 0 & 1 & 0 \\ 0 & 1 & 0 & 0 \\ 1 & 1 & 0 & 1 \end{bmatrix}$

29. Derive the result for analyzing the pottery in graves: Let A describe the pottery contents of various graves. Let $P = A^tA$. Prove that

    The element p_{ij} of the matrix $P = A^tA$ gives the number of graves in which ith and jth types of pottery both appear.

Thus the larger the element p_{ij}, the closer pottery types i and j are in time. By examining the elements of P, the archaeologist can arrive at the chronological order of the pottery.

30. The model introduced here in archaeology is used in sociology to analyze relationships within a group of people. For example, consider the relationship of "friendship" within a group. Assume that all friendships are mutual. Label the people $1, \ldots, n$ and define a square matrix A as follows:

    $a_{ii} = 0$ for all i (diagonal elements of A are zero)

$$a_{ij} = \begin{cases} 1 & \text{if } i \text{ and } j \text{ are friends} \\ 0 & \text{if } i \text{ and } j \text{ are not friends} \end{cases}$$

    (a) Prove that if $F = AA^t$, then f_{ij} is the number of friends that i and j have in common.

    (b) Suppose that not all friendships are mutual. How does this affect the model?

C.5   Transpose of a Matrix

Let us compute the transpose of the following matrix A.

$$A = \begin{bmatrix} 1 & 2 & 3 \\ 4 & 5 & 6 \\ 7 & 8 & 9 \end{bmatrix}$$

Enter matrix A (see Section C.2)

Press $\boxed{\text{QUIT}}$ to return to Home Screen

Press MATRX 1 to select matrix A

Press MATRX ▶ to get the Matrix Math Menu

Press MATRX 2 to select T (transpose)

Press ENTER

Display is Ans$^\text{T}$
$$\begin{bmatrix} 1 & 4 & 7 \\ 2 & 5 & 8 \\ 3 & 6 & 9 \end{bmatrix}$$

The transpose of A is $A^t = \begin{bmatrix} 1 & 4 & 7 \\ 2 & 5 & 8 \\ 3 & 6 & 9 \end{bmatrix}$

Exercises C.5

In Exercises 1–6, compute the tranposes of the matrices using a calculator.

*1. $\begin{bmatrix} 1 & 0 \\ 2 & 1 \end{bmatrix}$ **2.** $\begin{bmatrix} 1 & 2 \\ 9 & 4 \end{bmatrix}$ ***3.** $\begin{bmatrix} 2 & 1 & 0 \\ 5 & 4 & 3 \end{bmatrix}$

4. $\begin{bmatrix} 0 & 1 & 1 & 3 \\ 7 & 9 & -1 & 2 \end{bmatrix}$ ***5.** $\begin{bmatrix} 1 & 2 \\ 3 & 6 \\ 0 & 8 \end{bmatrix}$ **6.** $\begin{bmatrix} 2 & -3 & 6 \\ -7 & 0 & 2 \\ 4 & -6 & -1 \end{bmatrix}$

In Exercises 7–10, compute the given matrix expressions (if possible) using the following matrices A, B, and C. If the calculator gives an error message, interpret the message.

$$A = \begin{bmatrix} 1 & -2 & 3 \\ 0 & 2 & -4 \end{bmatrix}, \quad B = \begin{bmatrix} 3 & 0 & -2 \\ 5 & 6 & -4 \end{bmatrix}, \quad C = \begin{bmatrix} 8 & 2 & 3 \\ -5 & 7 & 0 \\ -5 & 2 & -1 \end{bmatrix}$$

*7. AB^t ***8.** $(AB^t)^2$ **9.** B^tA **10.** ACB^t

11. Compute $G = AA^t$ for the following matrix A. (Archaeology example, Section 2.3)

$$A = \begin{bmatrix} 1 & 0 & 1 \\ 1 & 0 & 0 \\ 0 & 1 & 1 \\ 0 & 1 & 0 \end{bmatrix}$$

12. Compute $A + A^t$ for the following square matrix A. Examine the characteristics of the matrix $A + A^t$. Compute $A + A^t$ for other square matrices and examine their characteristics. Arrive at a result for $A + A^t$.

$$A = \begin{bmatrix} 2 & 5 & 4 \\ 6 & 3 & -3 \\ 0 & -8 & 9 \end{bmatrix}$$

13. Compute $A - A^t$ for the following square matrix A. Examine the characteristics of the matrix $A - A^t$. Compute $A - A^t$ for other square matrices and examine their characteristics. Arrive at a result for $A - A^t$. Such a matrix is called antisymmetric.

$$A = \begin{bmatrix} 0 & 5 & 4 \\ 6 & 0 & -3 \\ 0 & -8 & 0 \end{bmatrix}$$

2.4   The Inverse of a Matrix

In this section we introduce the concept of matrix inverse. We shall see how an inverse can be used to solve certain systems of linear equations, and we shall see an application of matrix inverse in cryptography, the study of codes.

We motivate the idea of the inverse of a matrix by looking at the multiplicative inverse of a real number. If number b is the inverse of a, then

$$ab = 1 \quad \text{and} \quad ba = 1$$

For example, $\frac{1}{4}$ is the inverse of 4 and we have

$$4\left(\tfrac{1}{4}\right) = \left(\tfrac{1}{4}\right)4 = 1$$

These are the ideas that we extend to matrices.

Definition   *Let A be an n × n matrix. If a matrix B can be found such that $AB = BA = I_n$, then A is said to be **invertible** and B is called the **inverse** of A. If such a matrix B does not exist, then A has no inverse.*

Example 1   Prove that the matrix $A = \begin{bmatrix} 1 & 2 \\ 3 & 4 \end{bmatrix}$ has inverse $B = \begin{bmatrix} -2 & 1 \\ \frac{3}{2} & -\frac{1}{2} \end{bmatrix}$

Solution   We have that

$$AB = \begin{bmatrix} 1 & 2 \\ 3 & 4 \end{bmatrix}\begin{bmatrix} -2 & 1 \\ \frac{3}{2} & -\frac{1}{2} \end{bmatrix} = \begin{bmatrix} 1 & 0 \\ 0 & 1 \end{bmatrix} = I_2$$

and

$$BA = \begin{bmatrix} -2 & 1 \\ \frac{3}{2} & -\frac{1}{2} \end{bmatrix}\begin{bmatrix} 1 & 2 \\ 3 & 4 \end{bmatrix} = \begin{bmatrix} 1 & 0 \\ 0 & 1 \end{bmatrix} = I_2$$

Thus $AB = BA = I_2$, proving that the matrix A has inverse B.

We know that a real number can have at most one inverse. We now see that this is also the case for a matrix.

Theorem 2.7   The inverse of an invertible matrix is unique.

Proof   Let B and C be inverses of A. Thus $AB = BA = I_n$, and $AC = CA = I_n$. Multiply both sides of the equation $AB = I_n$ by C and use the algebraic properties of matrices.

$$C(AB) = CI_n$$
$$(CA)B = C$$
$$I_nB = C$$
$$B = C$$

Thus an invertible matrix has only one inverse.

Notation   Let A be an invertible matrix. We denote its inverse A^{-1}.

$$\text{Thus } AA^{-1} = A^{-1}A = I_n$$
$$\text{Let } A^{-n} = (A^{-1})^n = \underbrace{A^{-1}A^{-1} \ \cdots \ A^{-1}}_{n \text{ times}}$$

We now derive a method, based on the Gauss-Jordan algorithm, for finding the inverse of a matrix.

Let A be an invertible matrix. Let the columns of A^{-1} be $X_1, X_2, \ldots, X_n$, and the columns of I_n be $C_1, C_2, \ldots, C_n$. Write A^{-1} and I_n in terms of their columns as follows:

$$A^{-1} = [X_1 \quad X_2 \quad \cdots \quad X_n] \quad \text{and} \quad I_n = [C_1 \quad C_2 \quad \cdots \quad C_n]$$

We shall find A^{-1} by finding $X_1, X_2, \ldots, X_n$. Since $AA^{-1} = I_n$, then

$$A[X_1 \quad X_2 \quad \cdots \quad X_n] = [C_1 \quad C_2 \quad \cdots \quad C_n]$$

Matrix multiplication, carried out in terms of columns, gives

$$[AX_1 \quad AX_2 \quad \cdots \quad AX_n] = [C_1 \quad C_2 \quad \cdots \quad C_n]$$

leading to

$$AX_1 = C_1, AX_2 = C_2, \ldots, AX_n = C_n$$

Thus $X_1, X_2, \ldots, X_n$ are solutions to the equations $AX = C_1$, $AX = C_2$, $\ldots$, $AX = C_n$, all having the same matrix of coefficients A. Solve these systems by using Gauss-Jordan elimination on the large augmented matrix $[A : C_1 \quad C_2 \quad \ldots \quad C_n]$.

$$[A : C_1 \quad C_2 \quad \cdots \quad C_n] \approx \cdots \approx [I_n : X_1 \quad X_2 \quad \cdots \quad X_n]$$

Thus

$$[A : I_n] \approx \cdots \approx [I_n : A^{-1}]$$

Therefore, if we compute the reduced echelon form of $[A : I_n]$ and get a matrix of the form $[I_n : B]$, then $B = A^{-1}$.

On the other hand, if we compute the reduced echelon form of $[A : I_n]$ and find that it is not of the form $[I_n : B]$, then A is not invertible.

We now summarize the results of this discussion.

Gauss-Jordan Elimination for Finding the Inverse of a Matrix

Let A be an $n \times n$ matrix.

1.   Adjoin the identity $n \times n$ matrix I_n to A to form the matrix $[A : I_n]$.

2.   Compute the reduced echelon form of $[A : I_n]$.
     If the reduced echelon form is of the type $[I_n : B]$, then B is the inverse of A.
     If the reduced echelon form is not of the type $[I_n : B]$, in that the first $n \times n$ submatrix is not I_n, then A has no inverse.

The following examples illustrate the method.

Example 2   Determine the inverse of the matrix

$$A = \begin{bmatrix} 1 & -1 & -2 \\ 2 & -3 & -5 \\ -1 & 3 & 5 \end{bmatrix}$$

Solution   Applying the method of Gauss-Jordan elimination, we get

$$[A : I_3] = \begin{bmatrix} 1 & -1 & -2 & 1 & 0 & 0 \\ 2 & -3 & -5 & 0 & 1 & 0 \\ -1 & 3 & 5 & 0 & 0 & 1 \end{bmatrix} \underset{\substack{R2 + (-2)R1 \\ R3 + R1}}{\approx} \begin{bmatrix} 1 & -1 & -2 & 1 & 0 & 0 \\ 0 & -1 & -1 & -2 & 1 & 0 \\ 0 & 2 & 3 & 1 & 0 & 1 \end{bmatrix}$$

$$\underset{(-1)R2}{\approx} \begin{bmatrix} 1 & -1 & -2 & 1 & 0 & 0 \\ 0 & 1 & 1 & 2 & -1 & 0 \\ 0 & 2 & 3 & 1 & 0 & 1 \end{bmatrix}$$

$$\underset{\substack{R1 + R2 \\ R3 + (-2)R2}}{\approx} \begin{bmatrix} 1 & 0 & -1 & 3 & -1 & 0 \\ 0 & 1 & 1 & 2 & -1 & 0 \\ 0 & 0 & 1 & -3 & 2 & 1 \end{bmatrix}$$

$$\underset{\substack{R1 + R3 \\ R2 + (-1)R3}}{\approx} \begin{bmatrix} 1 & 0 & 0 & 0 & 1 & 1 \\ 0 & 1 & 0 & 5 & -3 & -1 \\ 0 & 0 & 1 & -3 & 2 & 1 \end{bmatrix}$$

Thus

$$A^{-1} = \begin{bmatrix} 0 & 1 & 1 \\ 5 & -3 & -1 \\ -3 & 2 & 1 \end{bmatrix}$$

The following example illustrates the application of the method for a matrix that does not have an inverse.

Example 3   Determine the inverse of the following matrix, if it exists.

$$A = \begin{bmatrix} 1 & 1 & 5 \\ 1 & 2 & 7 \\ 2 & -1 & 4 \end{bmatrix}$$

Solution   Applying the method of Gauss-Jordan elimination, we get

$$[A : I_3] = \begin{bmatrix} 1 & 1 & 5 & 1 & 0 & 0 \\ 1 & 2 & 7 & 0 & 1 & 0 \\ 2 & -1 & 4 & 0 & 0 & 1 \end{bmatrix} \approx \begin{array}{c} \\ R2 + (-1)R1 \\ R3 + (-2)R1 \end{array} \begin{bmatrix} 1 & 1 & 5 & 1 & 0 & 0 \\ 0 & 1 & 2 & -1 & 1 & 0 \\ 0 & -3 & -6 & -2 & 0 & 1 \end{bmatrix}$$

$$\approx \begin{array}{c} \\ R1 + (-1)R2 \\ R3 + 3R2 \end{array} \begin{bmatrix} 1 & 0 & 3 & 2 & -1 & 0 \\ 0 & 1 & 2 & -1 & 1 & 0 \\ 0 & 0 & 0 & -5 & 3 & 1 \end{bmatrix}$$

There is no need to proceed further. The reduced echelon form cannot have a one in the (3, 3) location. The reduced echelon form cannot be of the form $[I_n : B]$. Thus A^{-1} does not exist.

The Gauss-Jordan method of computing the inverse of a matrix tells us that A is invertible if and only if the reduced echelon form of $[A : I_n]$ is $[I_n : B]$. As $[A : I_n]$ is transformed to $[I_n : B]$, A is transformed to I_n. This observation leads to the following result.

Theorem 2.8   An $n \times n$ matrix A is invertible if and only if it is row equivalent to I_n.

If the matrix of coefficients of a system of linear equations is invertible, then the inverse can be used to discuss the solutions. The following is a key result in such discussions.

Theorem 2.9   Let $AX = B$ be a system of n linear equations in n variables. If A^{-1} exists, the solution is unique and is given by $X = A^{-1}B$.

Proof   We first prove that $X = A^{-1}B$ is a solution.
Substitute $X = A^{-1}B$ into the matrix equation. Using the properties of matrices, we get

$$AX = A(A^{-1}B) = (AA^{-1})B = I_nB = B$$

$X = A^{-1}B$ satisfies the equation; thus it is a solution.
We now prove the uniqueness of the solution. Let X_1 be a solution. Thus $AX_1 = B$. Multiplying both sides of this equation by A^{-1} gives

$$A^{-1}AX_1 = A^{-1}B$$
$$I_nX_1 = A^{-1}B$$
$$X_1 = A^{-1}B$$

Thus there is a unique solution $X_1 = A^{-1}B$.

We now illustrate how this result can be used to solve a system of equations.

Example 4   Solve the system of equations

$$
\begin{aligned}
x_1 - x_2 - 2x_3 &= 1 \\
2x_1 - 3x_2 - 5x_3 &= 3 \\
-x_1 + 3x_2 + 5x_3 &= -2
\end{aligned}
$$

Solution   This system can be written in the following matrix form:

$$
\begin{bmatrix} 1 & -1 & -2 \\ 2 & -3 & -5 \\ -1 & 3 & 5 \end{bmatrix}
\begin{bmatrix} x_1 \\ x_2 \\ x_3 \end{bmatrix}
=
\begin{bmatrix} 1 \\ 3 \\ -2 \end{bmatrix}
$$

If the matrix of coefficients is invertible, the unique solution is

$$
\begin{bmatrix} x_1 \\ x_2 \\ x_3 \end{bmatrix}
=
\begin{bmatrix} 1 & -1 & -2 \\ 2 & -3 & -5 \\ -1 & 3 & 5 \end{bmatrix}^{-1}
\begin{bmatrix} 1 \\ 3 \\ -2 \end{bmatrix}
$$

This inverse has already been found in Example 2. Using that result, we get

$$
\begin{bmatrix} x_1 \\ x_2 \\ x_3 \end{bmatrix}
=
\begin{bmatrix} 0 & 1 & 1 \\ 5 & -3 & -1 \\ -3 & 2 & 1 \end{bmatrix}
\begin{bmatrix} 1 \\ 3 \\ -2 \end{bmatrix}
=
\begin{bmatrix} 1 \\ -2 \\ 1 \end{bmatrix}
$$

The unique solution is $x_1 = 1$, $x_2 = -2$, $x_3 = 1$.

The knowledge that the solution to the system of equations $AX = B$ is $X = A^{-1}B$, if A is invertible, is primarily of theoretical importance. It gives an algebraic expression for the solution. It is not usually used in practice to solve a specific system of equations. Gauss-Jordan elimination and Gaussian elimination, which we introduce in Section 9.1, are more efficient for solving systems of linear equations. Most systems of equations are solved on a computer. Two factors that are important when using a computer are *efficiency* and *accuracy*. To solve a system of n equations, the matrix inverse method requires $n^3 + n^2$ multiplications and $n^3 - n^2$ additions, while Gaussian elimination requires $\frac{1}{3}n^3 + n^2 - \frac{1}{3}n$ multiplications and $\frac{1}{3}n^3 + \frac{1}{2}n^2 - \frac{5}{6}n$ additions. Thus, for example, for a system of four equations, the matrix inverse method would involve 80 multiplications and 48 additions, while the more efficient Gaussian elimination would involve 36 multiplications and 26 additions. Furthermore, the more operations that are performed, the larger the possible round-off error. Thus Gaussian (and Gauss-Jordan) elimination is in general also more accurate than the matrix inverse method.

One may be tempted to assume that, given a number of linear systems $AX = B_1$, $AX = B_2, \ldots, AX = B_k$, all having the same invertible matrix of coefficients A, it would be efficient to calculate A^{-1} and then compute the solutions using $X_1 = A^{-1}B_1$, $X_2 = A^{-1}B_2, \ldots, X_n = A^{-1}B_k$. This approach would involve only one computation of A^{-1} and then a number of matrix multiplications. In general, however, it is more efficient and accurate to solve such systems using a large augmented matrix that represents all systems and an elimination method such as Gauss-Jordan elimination, as discussed earlier.

There are certain instances when the matrix method is used to arrive at specific solutions. We illustrate such a situation in a model for analyzing the interdependence of industries in the following section. The elements of the matrix of coefficients in that example lend themselves to an efficient algorithm for computing the inverse.

We now summarize some of the algebraic properties of matrix inverse.

Properties of Matrix Inverse

Let A and B be invertible matrices and c be a nonzero scalar. Then

1. $(A^{-1})^{-1} = A$

2. $(cA)^{-1} = \dfrac{1}{c}A^{-1}$

3. $(AB)^{-1} = B^{-1}A^{-1}$

4. $(A^n)^{-1} = (A^{-1})^n$

5. $(A^t)^{-1} = (A^{-1})^t$

We prove results 1 and 3 to illustrate the techniques involved, leaving the remaining results for you to prove in the exercises that follow.

$(A^{-1})^{-1} = A$ This result follows directly from the definition of inverse of a matrix. Since A^{-1} is the inverse of A, we have

$$AA^{-1} = A^{-1}A = I_n$$

This statement also tells us that A is the inverse of A^{-1}. Thus $(A^{-1})^{-1} = A$.

$(AB)^{-1} = B^{-1}A^{-1}$ We want to show that the matrix $B^{-1}A^{-1}$ is the inverse of the matrix AB. We get, using the properties of matrices,

$$(AB)(B^{-1}A^{-1}) = A(BB^{-1})A^{-1}$$
$$= AI_nA^{-1}$$
$$= AA^{-1}$$
$$= I_n$$

Similarly, it can be shown that $(B^{-1}A^{-1})(AB) = I_n$. Thus $B^{-1}A^{-1}$ is the inverse of the matrix AB.

Example 5   If $A = \begin{bmatrix} 4 & 1 \\ 3 & 1 \end{bmatrix}$, then it can be shown that $A^{-1} = \begin{bmatrix} 1 & -1 \\ -3 & 4 \end{bmatrix}$. Use this information to compute $(A^t)^{-1}$.

Solution   Result 5 above tells us that if know the inverse of a matrix, we also know the inverse of its transpose. We get

$$(A^t)^{-1} = (A^{-1})^t = \begin{bmatrix} 1 & -1 \\ -3 & 4 \end{bmatrix}^t = \begin{bmatrix} 1 & -3 \\ -1 & 4 \end{bmatrix}$$

Cryptography

Cryptography is the process of coding and decoding messages. The word comes from the Greek *kryptos*, meaning "hidden." The technique can be traced back to the ancient Greeks. Today governments use sophisticated methods of coding and decoding messages. One type of code, which is extremely difficult to break, makes use of a large matrix to encode a message. The receiver of the message decodes it using the inverse of the matrix. This first matrix is called the **encoding matrix** and its inverse is called the **decoding matrix**. We illustrate the method for a 3 × 3 matrix.

Let the message be

PREPARE TO ATTACK

and the encoding matrix be

$$\begin{bmatrix} -3 & -3 & -4 \\ 0 & 1 & 1 \\ 4 & 3 & 4 \end{bmatrix}$$

We assign a number to each letter of the alphabet. For convenience, let us associate each letter with its position in the alphabet: *A* is 1, *B* is 2, and so on. Let a space between words be denoted by the number 27. Thus the message becomes:

P	R	E	P	A	R	E	*	T	O	*	A	T	T	A	C	K
16	18	5	16	1	18	5	27	20	15	27	1	20	20	1	3	11

Since we are going to use a 3 × 3 matrix to encode the message, we break the enumerated message up into a sequence of 3 × 1 column matrices as follows:

$$\begin{bmatrix} 16 \\ 18 \\ 5 \end{bmatrix}, \quad \begin{bmatrix} 16 \\ 1 \\ 18 \end{bmatrix}, \quad \begin{bmatrix} 5 \\ 27 \\ 20 \end{bmatrix}, \quad \begin{bmatrix} 15 \\ 27 \\ 1 \end{bmatrix}, \quad \begin{bmatrix} 20 \\ 20 \\ 1 \end{bmatrix}, \quad \begin{bmatrix} 3 \\ 11 \\ 27 \end{bmatrix}$$

Observe that it was necessary to add a space at the end of the message to complete the last matrix. We now put the message into code by multiplying each of the above column matrices by the encoding matrix. This can be conveniently done by writing the above column matrices as columns of a matrix and premultiplying that matrix by the encoding matrix. We get

$$\begin{bmatrix} -3 & 3 & -4 \\ 0 & 1 & 1 \\ 4 & 3 & 4 \end{bmatrix} \begin{bmatrix} 16 & 16 & 5 & 15 & 20 & 3 \\ 18 & 1 & 27 & 27 & 20 & 11 \\ 5 & 18 & 20 & 1 & 1 & 27 \end{bmatrix}$$

$$= \begin{bmatrix} -122 & -123 & -176 & -130 & -124 & -150 \\ 23 & 19 & 47 & 28 & 21 & 38 \\ 138 & 139 & 181 & 145 & 144 & 153 \end{bmatrix}$$

The columns of this matrix give the encoded message. The message is transmitted in the following linear form.

$$-122,\ 23,\ 138,\ -123,\ 19,\ 139,\ -176,\ 47,\ 181,$$
$$-130,\ 28,\ 145,\ -124,\ 21,\ 144,\ -150,\ 38,\ 153$$

To decode the message, the receiver writes this string as a sequence of 3×1 column matrices and repeats the technique using the inverse of the encoding matrix. The inverse of this encoding matrix, the decoding matrix, is

$$\begin{bmatrix} 1 & 0 & 1 \\ 4 & 4 & 3 \\ -4 & -3 & -3 \end{bmatrix}$$

Thus, to decode the message,

$$\begin{bmatrix} 1 & 0 & 1 \\ 4 & 4 & 3 \\ -4 & -3 & -3 \end{bmatrix} \begin{bmatrix} -122 & -123 & -176 & -130 & -124 & -150 \\ 23 & 19 & 47 & 28 & 21 & 38 \\ 138 & 139 & 181 & 145 & 144 & 153 \end{bmatrix}$$

$$= \begin{bmatrix} 16 & 16 & 5 & 15 & 20 & 3 \\ 18 & 1 & 27 & 27 & 20 & 11 \\ 5 & 18 & 20 & 1 & 1 & 27 \end{bmatrix}$$

The columns of this matrix, written in linear form, give the original message.

16	18	5	16	1	18	5	27	20	15	27	1	20	20	1	3	11
P	R	E	P	A	R	E	*	T	O	*	A	T	T	A	C	K

Readers who are interested in an introduction to cryptography are referred to *Secret Codes and Ciphers* by Bernice Kohn, Prentice-Hall, Inc., 1968. For a more advanced book that gives an overview of recent developments and a guide to literature on cryptography, see *Codes and Cryptography* by Dominic Welsh, Oxford, 1988.

Exercise Set 2.4

1. Use the definition $AB = BA = I_2$ of inverse to check whether B is the inverse of A, for each of the following 2×2 matrices A and B.

*(a) $A = \begin{bmatrix} 7 & -3 \\ 5 & -2 \end{bmatrix} \quad B = \begin{bmatrix} -2 & 3 \\ -5 & 7 \end{bmatrix}$

(b) $A = \begin{bmatrix} 3 & 4 \\ 5 & 7 \end{bmatrix}, \quad B = \begin{bmatrix} 7 & -4 \\ -5 & 3 \end{bmatrix}$

*(c) $A = \begin{bmatrix} 2 & -4 \\ -5 & 3 \end{bmatrix}, \quad B = \begin{bmatrix} 3 & 1 \\ 5 & 2 \end{bmatrix}$

(d) $A = \begin{bmatrix} 7 & 6 \\ 8 & 7 \end{bmatrix}, \quad B = \begin{bmatrix} 7 & -6 \\ -8 & 7 \end{bmatrix}$

2. Use the definition $AB = BA = I_3$ of inverse to check whether B is the inverse of A, for each of the following 3×3 matrices A and B.

*(a) $A = \begin{bmatrix} 5 & 0 & 0 \\ 0 & \frac{1}{3} & 0 \\ 0 & 0 & -2 \end{bmatrix}, \quad B = \begin{bmatrix} \frac{1}{5} & 0 & 0 \\ 0 & 3 & 0 \\ 0 & 0 & -\frac{1}{2} \end{bmatrix}$

(b) $A = \begin{bmatrix} 1 & 1 & -1 \\ -3 & 2 & -1 \\ 3 & -3 & 2 \end{bmatrix}, \quad B = \begin{bmatrix} 1 & 1 & 1 \\ 3 & 5 & 4 \\ 3 & 6 & 5 \end{bmatrix}$

*(c) $A = \begin{bmatrix} 0 & 1 & -1 \\ 2 & -2 & -1 \\ -1 & 1 & 1 \end{bmatrix}, \quad B = \begin{bmatrix} 1 & 2 & 3 \\ 1 & 1 & 2 \\ 0 & 1 & 1 \end{bmatrix}$

3. Determine the inverse of each of the following 2×2 matrices, if it exists, using the method of Gauss-Jordan elimination.

*(a) $\begin{bmatrix} 1 & 0 \\ 2 & 1 \end{bmatrix}$

(b) $\begin{bmatrix} 1 & 2 \\ 9 & 4 \end{bmatrix}$

*(c) $\begin{bmatrix} 2 & 1 \\ 4 & 3 \end{bmatrix}$

(d) $\begin{bmatrix} 0 & 1 \\ 1 & 3 \end{bmatrix}$

*(e) $\begin{bmatrix} 1 & 2 \\ 3 & 6 \end{bmatrix}$

(f) $\begin{bmatrix} 2 & -3 \\ 6 & -7 \end{bmatrix}$

4. Determine the inverse of each of the following 3×3 matrices, if it exists, using the method of Gauss-Jordan elimination.

*(a) $\begin{bmatrix} 1 & 2 & 3 \\ 0 & 1 & 2 \\ 4 & 5 & 3 \end{bmatrix}$

(b) $\begin{bmatrix} 2 & 0 & 4 \\ -1 & 3 & 1 \\ 0 & 1 & 2 \end{bmatrix}$

*(c) $\begin{bmatrix} 1 & 2 & -3 \\ 1 & -2 & 1 \\ 5 & -2 & -3 \end{bmatrix}$

(d) $\begin{bmatrix} 1 & 2 & -1 \\ 3 & -1 & 0 \\ 2 & -3 & 1 \end{bmatrix}$

5. Determine the inverse of each of the following 3×3 matrices, if it exists, using the method of Gauss-Jordan elimination.

*(a) $\begin{bmatrix} 1 & 2 & 3 \\ 2 & -1 & 4 \\ 0 & -1 & 1 \end{bmatrix}$

(b) $\begin{bmatrix} 1 & 2 & -1 \\ 2 & 4 & -3 \\ 1 & -2 & 0 \end{bmatrix}$

*(c) $\begin{bmatrix} 1 & -2 & -1 \\ -2 & 4 & 6 \\ 0 & 0 & 5 \end{bmatrix}$

(d) $\begin{bmatrix} 7 & 0 & 0 \\ 0 & -3 & 0 \\ 0 & 0 & \frac{1}{5} \end{bmatrix}$

6. Determine the inverse of each of the following 4×4 matrices, if it exists, using the method of Gauss-Jordan elimination.

*(a) $\begin{bmatrix} -3 & -1 & 1 & -2 \\ -1 & 3 & 2 & 1 \\ 1 & 2 & 3 & -1 \\ -2 & 1 & -1 & -3 \end{bmatrix}$

(b) $\begin{bmatrix} 1 & 1 & 0 & 0 \\ 0 & 1 & 1 & 0 \\ 1 & 0 & 0 & 1 \\ 0 & 0 & 1 & 1 \end{bmatrix}$

*(c) $\begin{bmatrix} -1 & 0 & -1 & -1 \\ -3 & -1 & 0 & -1 \\ 5 & 0 & 4 & 3 \\ 3 & 0 & 3 & 2 \end{bmatrix}$

7. Solve the following systems of two equations in two variables by determining the inverse of the matrix of coefficients and then using matrix multiplication.

*(a) $\begin{aligned} x_1 + 2x_2 &= 2 \\ 3x_1 + 5x_2 &= 4 \end{aligned}$

(b) $\begin{aligned} x_1 + 5x_2 &= -1 \\ 2x_1 + 9x_2 &= 3 \end{aligned}$

*(c) $\begin{aligned} x_1 + 3x_2 &= 5 \\ 2x_1 + x_2 &= 10 \end{aligned}$

(d) $\begin{aligned} 2x_1 + x_2 &= 4 \\ 4x_1 + 3x_2 &= 6 \end{aligned}$

*(e) $\begin{aligned} 2x_1 + 4x_2 &= 6 \\ 3x_1 + 8x_2 &= 1 \end{aligned}$

(f) $\begin{aligned} 3x_1 + 9x_2 &= 9 \\ 2x_1 + 7x_2 &= 4 \end{aligned}$

8. Solve the following systems of three equations in three variables by determining the inverse of the matrix of coefficients and then using matrix multiplication.

*(a) $\begin{aligned} x_1 + 2x_2 - x_3 &= 2 \\ x_1 + x_2 + 2x_3 &= 0 \\ x_1 - x_2 - x_3 &= 1 \end{aligned}$

(b) $\begin{aligned} x_1 - x_2 &= 1 \\ x_1 + x_2 + 2x_3 &= 2 \\ x_1 + 2x_2 + x_3 &= 0 \end{aligned}$

*(c) $\begin{aligned} x_1 + 2x_2 + 3x_3 &= 1 \\ 2x_1 + 5x_2 + 3x_3 &= 3 \\ x_1 \quad\quad + 8x_3 &= 15 \end{aligned}$

(d) $\begin{aligned} x_1 - 2x_2 + 2x_3 &= 3 \\ -x_1 + x_2 + 3x_3 &= 2 \\ x_1 - x_2 - 4x_3 &= -1 \end{aligned}$

*(e) $\begin{aligned} -x_1 + x_2 &= 5 \\ -x_1 \quad\quad + x_3 &= -2 \\ 6x_1 - 2x_2 - 3x_3 &= 1 \end{aligned}$

*9. Solve the following system of four equations in four variables by determining the inverse of the matrix of coefficients and then using matrix multiplication.

$$\begin{aligned} x_1 + x_2 + 2x_3 + x_4 &= 5 \\ 2x_1 \quad\quad + 2x_3 + x_4 &= 6 \\ x_2 + 3x_3 - x_4 &= 1 \\ 3x_1 + 2x_2 \quad\quad + 2x_4 &= 7 \end{aligned}$$

10. Solve the following system of equations, all having the same matrix of coefficients, using the matrix inverse method,

$$\begin{aligned} x_1 + 2x_2 - x_3 &= b_1 \\ x_1 + x_2 + 2x_3 &= b_2 \\ x_1 - x_2 - x_3 &= b_3 \end{aligned} \quad \text{for}$$

$$\begin{bmatrix} b_1 \\ b_2 \\ b_3 \end{bmatrix} = \begin{bmatrix} 1 \\ 2 \\ 3 \end{bmatrix}, \quad \begin{bmatrix} 0 \\ 1 \\ 4 \end{bmatrix}, \quad \begin{bmatrix} 5 \\ 2 \\ -3 \end{bmatrix} \quad \text{in turn}$$

11. If $A = \begin{bmatrix} a & b \\ c & d \end{bmatrix}$, show that

$$A^{-1} = \frac{1}{(ad - bc)} \begin{bmatrix} d & -b \\ -c & a \end{bmatrix}.$$

This formula can be quicker than Gauss-Jordan elimination to compute the inverse of a 2×2 matrix. Compute the inverses of the following 2×2 matrices using both methods to see which you prefer.

*(a) $\begin{bmatrix} 3 & 8 \\ 1 & 3 \end{bmatrix}$ **(b)** $\begin{bmatrix} 3 & 5 \\ 2 & 4 \end{bmatrix}$

*(c) $\begin{bmatrix} 5 & 6 \\ 3 & 4 \end{bmatrix}$ **(d)** $\begin{bmatrix} 4 & -6 \\ 2 & -2 \end{bmatrix}$

(You will meet this formula again in the next chapter and at that time, will see where it fits into the general theory of matrix inverses.)

12. Prove the following properties of matrix inverse that were listed in this section.

*(a) $(cA)^{-1} = \frac{1}{c}A^{-1}$ **(b)** $(A^n)^{-1} = (A^{-1})^n$

(c) $(A^t)^{-1} = (A^{-1})^t$

13. If $A^{-1} = \begin{bmatrix} 2 & 1 \\ 4 & 3 \end{bmatrix}$, find A.

*14. If $A^{-1} = \frac{1}{2}\begin{bmatrix} -3 & 2 \\ -10 & 6 \end{bmatrix}$, find A.

15. Consider the matrix $A = \begin{bmatrix} 3 & 1 \\ 5 & 2 \end{bmatrix}$, having inverse $\begin{bmatrix} 2 & -1 \\ -5 & 3 \end{bmatrix}$. Determine

*(a) $(3A)^{-1}$ **(b)** $(A^2)^{-1}$ ***(c)** A^{-2} **(d)** $(A^t)^{-1}$

(*Hint:* Use the algebraic properties of matrix inverse.)

16. If $A = \begin{bmatrix} 5 & 1 \\ 9 & 2 \end{bmatrix}$, then $A^{-1} = \begin{bmatrix} 2 & -1 \\ -9 & 5 \end{bmatrix}$. Use this information to determine ***(a)** $(2A^t)^{-1}$, **(b)** A^{-3}, ***(c)** $(AA^t)^{-1}$.

*17. Find x such that $\begin{bmatrix} 2x & 7 \\ 1 & 2 \end{bmatrix}^{-1} = \begin{bmatrix} 2 & -7 \\ -1 & 4 \end{bmatrix}$.

18. Find x such that $2\begin{bmatrix} 2x & x \\ 5 & 3 \end{bmatrix}^{-1} = \begin{bmatrix} 3 & -2 \\ -5 & 4 \end{bmatrix}$.

*19. Find A such that $(4A^t)^{-1} = \begin{bmatrix} 2 & 3 \\ -4 & -4 \end{bmatrix}$.

20. Prove that $(ABC)^{-1} = C^{-1}B^{-1}A^{-1}$.

*21. Prove that $(A^tB^t)^{-1} = (A^{-1}B^{-1})^t$.

22. Prove that, in general $(A + B)^{-1} \neq A^{-1} + B^{-1}$.

23. Prove that if A has no inverse, then A^t also has no inverse.

24. Prove that if A is an invertible matrix such that ***(a)** $AB = AC$, then $B = C$; **(b)** $AB = 0$, then $B = 0$.

25. Prove that a matrix has no inverse if
   (a) two rows are equal.
   (b) two columns are equal. (*Hint:* Use the transpose.)

26. Prove that a diagonal matrix is invertible if and only if all its diagonal elements are nonzero. Can you find a quick way for determining the inverse of an invertible diagonal matrix?

*27. Let $AX = B$ be a system of five linear equations in five variables, where A is invertible. Find the number of multiplications and additions needed to solve this system using **(a)** Gaussian elimination, **(b)** the matrix inverse method.

28. Let A be an invertible 2×2 matrix. Show that it takes the same amount of computation to find the solution to the system of equations $AX = B$ using Gauss-Jordan elimination as it does to find the first column of the inverse of A. This exercise emphasizes the fact that it is more efficient to solve a system of equations using Gauss-Jordan elimination than using the inverse of the matrix of coefficients.

Cryptography Exercises

In Exercises 29–32, associate each letter with its position in the alphabet.

*29. Encode the message RETREAT using the matrix $\begin{bmatrix} 4 & -3 \\ 3 & -2 \end{bmatrix}$.

30. Encode the message THE BRITISH ARE COMING using the matrix

$$\begin{bmatrix} 1 & 2 & 1 \\ 2 & 3 & 1 \\ -2 & 0 & 1 \end{bmatrix}$$

*31. Decode the message 49, 38, −5, −3, −61, −39, which was encoded using the matrix of Exercise 29.

32. Decode the message 71, 100, −1, 28, 43, −5, 84, 122, −11, 63, 98, −27, 69, 102, −12, 88, 126, −3, which was encoded using the matrix of Exercise 30.

C.6 Inverse of a Matrix

Let us compute A^{-1} for the following matrix A.

$$A = \begin{bmatrix} 2 & 4 \\ 1 & 3 \end{bmatrix}$$

Enter matrix A (see Section C.2)

Press [QUIT] to return to Home Screen

Press [MATRX] 1 [x^{-1}] [ENTER]

Matrix A^{-1} is displayed $\begin{bmatrix} 1.5 & -2 \\ -.5 & 1 \end{bmatrix}$

Exercises C.6

In Exercises 1–6, compute the inverses of the matrices, if they exist, using a calculator.

*1. $\begin{bmatrix} 1 & 0 \\ 2 & 1 \end{bmatrix}$

2. $\begin{bmatrix} 1 & 2 \\ 9 & 4 \end{bmatrix}$

*3. $\begin{bmatrix} 2 & 1 \\ 4 & 3 \end{bmatrix}$

4. $\begin{bmatrix} 0 & 1 \\ 1 & 3 \end{bmatrix}$

*5. $\begin{bmatrix} 1 & 2 \\ 3 & 6 \end{bmatrix}$

6. $\begin{bmatrix} 2 & -3 \\ 6 & -7 \end{bmatrix}$

In Exercises 7–12, compute the inverses of the matrices, if they exist, using a calculator.

*7. $\begin{bmatrix} 1 & 2 & 3 \\ 0 & 1 & 2 \\ 4 & 5 & 3 \end{bmatrix}$

8. $\begin{bmatrix} 2 & 0 & 4 \\ -1 & 3 & 1 \\ 0 & 1 & 2 \end{bmatrix}$

*9. $\begin{bmatrix} 1 & 2 & -3 \\ 1 & -2 & 1 \\ 5 & -2 & -3 \end{bmatrix}$

10. $\begin{bmatrix} 1 & 2 & -1 \\ 3 & -1 & 0 \\ 2 & -3 & 1 \end{bmatrix}$

*11. $\begin{bmatrix} 1 & 2 & 3 \\ 2 & -1 & 4 \\ 0 & -1 & 1 \end{bmatrix}$

12. $\begin{bmatrix} 1 & 2 & -1 \\ 2 & 4 & -3 \\ 1 & -2 & 0 \end{bmatrix}$

In Exercises 13–18, solve the systems of two equations in two variables on a calculator by determining the inverse of the matrix of coefficients and then using matrix multiplication.

*13. $\begin{aligned} x_1 + 2x_2 &= 2 \\ 3x_1 + 5x_2 &= 4 \end{aligned}$

14. $\begin{aligned} x_1 + 5x_2 &= -1 \\ 2x_1 + 9x_2 &= 3 \end{aligned}$

*15. $\begin{aligned} x_1 + 3x_2 &= 5 \\ 2x_1 + x_2 &= 10 \end{aligned}$

16. $\begin{aligned} 2x_1 + x_2 &= 4 \\ 4x_1 + 3x_2 &= 6 \end{aligned}$

*17. $\begin{aligned} 2x_1 + 4x_2 &= 6 \\ 3x_1 + 8x_2 &= 1 \end{aligned}$

18. $\begin{aligned} 3x_1 + 9x_2 &= 9 \\ 2x_1 + 7x_2 &= 4 \end{aligned}$

In Exercises 19–21, solve the systems of three equations in three variables on a calculator by determining the inverse of the matrix of coefficients and then using matrix multiplication.

*19.
$$x_1 + 2x_2 - x_3 = 2$$
$$x_1 + x_2 + 2x_3 = 0$$
$$x_1 - x_2 - x_3 = 1$$

20.
$$x_1 - x_2 \quad\quad = 1$$
$$x_1 + x_2 + 2x_3 = 2$$
$$x_1 + 2x_2 + x_3 = 0$$

*21.
$$x_1 + 2x_2 + 3x_3 = 1$$
$$2x_1 + 5x_2 + 3x_3 = 3$$
$$x_1 + \quad\quad 8x_3 = 15$$

Similar Matrices

Two square matrices of the same size, A and B, are said to be similar *if there exists an invertible matrix C such that $B = C^{-1}AC$. The transformation of A into B is called a* similarity transformation. *In Exercises 22–25, use a calculator to perform the similarity transformation $B = C^{-1}AC$ for the given matrices A and C.*

*22. $A = \begin{bmatrix} 7 & -10 \\ 3 & -4 \end{bmatrix}$ and $C = \begin{bmatrix} 2 & 5 \\ 1 & 3 \end{bmatrix}$ **23.** $A = \begin{bmatrix} 1 & 2 \\ -1 & 3 \end{bmatrix}$ and $C = \begin{bmatrix} 2 & 5 \\ 1 & 3 \end{bmatrix}$

*24. $A = \begin{bmatrix} 1 & 2 & 3 \\ 0 & -1 & 2 \\ 1 & 1 & 0 \end{bmatrix}$ and $C = \begin{bmatrix} 3 & 5 & -1 \\ -2 & -3 & 1 \\ -1 & -2 & 1 \end{bmatrix}$

25. $A = \begin{bmatrix} 5 & 2 & 3 \\ 8 & 2 & 1 \\ -3 & 2 & 5 \end{bmatrix}$ and $C = \begin{bmatrix} 7 & -2 & 3 \\ 2 & 0 & 1 \\ 9 & -3 & 5 \end{bmatrix}$

*2.5   Leontief Input-Output Model in Economics

In this section we introduce the Leontief input-output model that is used to analyze the interdependence of economies. Wassily Leontief received a Nobel Prize in 1973 for his work in this field. The practical applications of this model have proliferated until it has now become a standard tool for investigating economic structures ranging from cities and corporations to states and countries.

Consider an economic situation that involves n interdependent industries. The output of any one industry is needed as input by other industries, and even possibly by the industry itself. We shall see how a mathematical model involving a system of linear equations can be constructed to analyze such a situation. Let us assume, for the sake of simplicity, that each industry produces one commodity. Let a_{ij} denote the amount of input of a certain commodity i to produce unit output of the commodity j. In our model let the amounts of input and output be measured in dollars. Thus, for example, $a_{34} = 0.45$ means that 45 cents' worth of commodity 3 is required to produce 1 dollar's worth of commodity 4.

$$a_{ij} = \text{amount of commodity } i \text{ in \$1 of commodity } j$$

The elements a_{ij}, called **input coefficients**, define a matrix A, called the **input-output matrix**, which describes the interdependence of the industries.

Example 1   National input-output matrices are used to describe interindustry relations that constitute the economic fabric of countries. We now display part of the matrix that describes the interdependency of the U.S. economy for 1972. The economic structure is actually described in terms of the flow among 79 producing sectors; the matrix is thus a 79 × 79 matrix. We cannot, of course, display the whole matrix. We list 10 sectors to give the reader a feel for the categories involved.

1. Livestock and livestock products

2. Agricultural crops

3. Forestry and fishery products

4. Agricultural, forestry, and fishery services

5. Iron and ferroalloy ores mining

6. Nonferrous metal ores mining

7. Coal mining

8. Crude petroleum and natural gas

9. Stone and clay mining and quarrying

10. Chemical and fertilizer mineral mining

The matrix A based on these sectors is

	1	2	3	4 ...
1	0.26110	0.02481	0	0.05278 ...
2	0.23277	0.03218	0	0.01444
3	0	0	0.00467	0.00294
4	0.02821	0.03673	0.02502	0.02959
5	0	0	0	0
6	0	0	0	0
7	0	0.00002	0	0
8	0	0	0	0
9	0.00001	0.00251	0	0.00034
10	0	0.00130	0	0
⋮	⋮			

Thus, for example, $a_{72} = 0.00002$ implies that \$0.00002 from the coal mining sector (sector 7) goes into producing each \$1 from the agricultural crops sector (sector 2).

We now extend the model to include an open sector. The products of industries may go, not only into other producing industries, but also into other nonproducing sectors of the economy such as consumers and governments. All such nonproducing sectors are grouped into what is called the **open sector**. The open sector in the above model of the 1972 U.S. economy included, for example, federal, state, and local government purchases. Let

d_i = demand of the open sector from industry i

x_i = total output of industry i necessary to meet demands of all n industries and the open sector

a_{ij} is the amount required from industry i to produce unit ouptput in industry j. Thus $a_{ij}x_j$ will be the amount required to produce x_j units of output in industry j. We get

$$x_i \quad = \quad a_{i1}x_1 \quad + \quad a_{i2}x_2 \quad + \cdots + \quad a_{in}x_n \quad + \quad d_i$$

total output	demand of	demand of		demand of	demand of
of industry i	industry 1	industry 2		industry n	open sector

The output levels required of the entire set of n industries to meet these demands are given by the following system of n linear equations:

$$x_1 = a_{11}x_1 + a_{12}x_2 + \cdots + a_{1n}x_n + d_1$$
$$x_2 = a_{21}x_1 + a_{22}x_2 + \cdots + a_{2n}x_n + d_2$$
$$\vdots$$
$$x_n = a_{n1}x_1 + a_{n2}x_2 + \cdots + a_{nn}x_n + d_n$$

This system of equations can be written in matrix form:

$$\begin{bmatrix} x_1 \\ x_2 \\ \vdots \\ x_n \end{bmatrix} = \begin{bmatrix} a_{11} & a_{12} & \cdots & a_{1n} \\ a_{21} & a_{22} & \cdots & a_{2n} \\ \vdots & & & \\ a_{n1} & a_{n2} & \cdots & a_{nn} \end{bmatrix} \begin{bmatrix} x_1 \\ x_2 \\ \vdots \\ x_n \end{bmatrix} + \begin{bmatrix} d_1 \\ d_2 \\ \vdots \\ d_n \end{bmatrix}$$

Let us introduce the following notation.

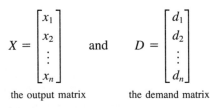

$$X = \begin{bmatrix} x_1 \\ x_2 \\ \vdots \\ x_n \end{bmatrix} \quad \text{and} \quad D = \begin{bmatrix} d_1 \\ d_2 \\ \vdots \\ d_n \end{bmatrix}$$

the output matrix   the demand matrix

The system of equations can now be written as a single matrix equation with the terms having the following significance:

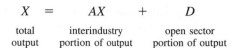

$$X = AX + D$$

total	interindustry	open sector
output	portion of output	portion of output

When the model is applied to the economy of a country, X represents the total output of each of the producing sectors of the economy, and AX describes the contributions made by the various sectors to fulfilling the intersectional input requirements of the economy. D is equal to $(X - AX)$, the difference between total output X and industry transaction AX.

D is thus the gross national products of the economy.

In practice, the equation $X = AX + D$ is applied in a variety of ways, depending on which variables are considered known and which are not known. For example, an analyst seeking to determine the implications of a change in government purchases or consumer demands on the economy described by A might assign values to D and solve the equation for X. The equation could be used in this manner to predict the amount of outputs from each sector needed to attain various gross national products (GNPs). (Example 2 below illustrates this application of the model.) On the other hand, an economist knowing the limited production capacity of an economic system described by A would consider X as known and solve the equation for D, to predict the maximum GNP the system can achieve. (Exercises 7, 8, and 9 illustrate this application of the model.)

Example 2 Consider an economy consisting of three industries having the following input-output matrix A. Determine the output levels required of the industries to meet the demands of the other industries and of the open sector in each case.

$$A = \begin{bmatrix} \frac{1}{5} & \frac{1}{5} & \frac{3}{10} \\ \frac{1}{2} & \frac{1}{2} & 0 \\ 0 & 0 & \frac{1}{5} \end{bmatrix}, \quad D = \begin{bmatrix} 9 \\ 12 \\ 16 \end{bmatrix}, \quad \begin{bmatrix} 6 \\ 9 \\ 8 \end{bmatrix}, \quad \begin{bmatrix} 12 \\ 18 \\ 32 \end{bmatrix}$$

The units of D are millions of dollars.

Solution We wish to compute the output levels X that correspond to the various open sector demands D. X is given by the equation $X = AX + D$. Rewrite as follows.

$$X - AX = D$$
$$(I - A)X = D$$

To solve this equation for X, we can use either Gauss-Jordan elimination or the matrix inverse method. In practice, the matrix inverse is used—a discussion of the merits of this approach is given below. We get

$$X = (I - A)^{-1}D$$

This is the equation that is used to determine X when A and D are known. For our matrix A, we get

$$I - A = \begin{bmatrix} 1 & 0 & 0 \\ 0 & 1 & 0 \\ 0 & 0 & 1 \end{bmatrix} - \begin{bmatrix} \frac{1}{5} & \frac{1}{5} & \frac{3}{10} \\ \frac{1}{2} & \frac{1}{2} & 0 \\ 0 & 0 & \frac{1}{5} \end{bmatrix} = \begin{bmatrix} \frac{4}{5} & -\frac{1}{5} & -\frac{3}{10} \\ -\frac{1}{2} & \frac{1}{2} & 0 \\ 0 & 0 & \frac{4}{5} \end{bmatrix}$$

$(I - A)^{-1}$ is computed using Gauss-Jordan elimination.

$$(I - A)^{-1} = \begin{bmatrix} \frac{5}{3} & \frac{2}{3} & \frac{5}{8} \\ \frac{5}{3} & \frac{8}{3} & \frac{5}{8} \\ 0 & 0 & \frac{5}{4} \end{bmatrix}$$

We can efficiently compute $X = (I - A)^{-1}D$ for each of the three values of D by forming a matrix having the various values of D as columns:

$$X = \begin{bmatrix} \frac{5}{3} & \frac{2}{3} & \frac{5}{8} \\ \frac{5}{3} & \frac{8}{3} & \frac{5}{8} \\ 0 & 0 & \frac{5}{4} \end{bmatrix} \begin{bmatrix} 9 & 6 & 12 \\ 12 & 9 & 18 \\ 16 & 8 & 32 \end{bmatrix} = \begin{bmatrix} 33 & 21 & 52 \\ 57 & 39 & 88 \\ 20 & 10 & 40 \end{bmatrix}$$

$\qquad\qquad\quad\uparrow \qquad\quad \uparrow\ \ \uparrow\ \ \uparrow \qquad\qquad \uparrow\ \ \uparrow\ \ \uparrow$

$\qquad\qquad (I - A)^{-1}$ various values   corresponding
$\qquad\qquad\qquad\qquad\quad$ of D $\qquad\qquad\qquad$ outputs

The output levels necessary to meet the demands

$$\begin{bmatrix} 9 \\ 12 \\ 16 \end{bmatrix}, \quad \begin{bmatrix} 6 \\ 9 \\ 8 \end{bmatrix}, \quad \begin{bmatrix} 12 \\ 18 \\ 32 \end{bmatrix} \quad \text{are} \quad \begin{bmatrix} 33 \\ 57 \\ 20 \end{bmatrix}, \quad \begin{bmatrix} 21 \\ 39 \\ 10 \end{bmatrix}, \quad \begin{bmatrix} 52 \\ 88 \\ 40 \end{bmatrix},$$

respectively. The units are millions of dollars.

Computational Note   In practice, analyses of this type usually involve many sectors (as we saw in the example of the U.S. economy), implying large input-output matrices. There is usually a great deal of computation involved in implementing the model, and an efficient algorithm is needed. The elements of an input-output matrix A are usually zero or very small. This characteristic of A has led to an appropriate numerical method for computing $(I - A)^{-1}$ that makes the matrix inverse method more efficient for solving the system of equations $(I - A)X = D$ than the Gauss-Jordan method. We now describe this method for computing $(I - A)^{-1}$. Consider the following matrix multiplication for any positive integer m:

$$(I - A)(I + A + A^2 + \cdots + A^m)$$
$$= I(I + A + A^2 + \cdots + A^m) - A(I + A + A^2 + \cdots + A^m)$$
$$= (I + A + A^2 + \cdots + A^m) - (A + A^2 + A^3 + \cdots + A^{m+1})$$
$$= I - A^{m+1}$$

The elements of successive powers of A become small rapidly and A^{m+1} approaches the zero matrix. Thus, for an appropriately large m,

$$(I - A)(I + A + A^2 + \cdots + A^m) = I$$

This implies that

$$(I - A)^{-1} = I + A + A^2 + \cdots + A^m$$

This expression is used on a computer to compute $(I - A)^{-1}$ in this model.

Readers who are interested in finding out more about applications of this model should read "The World Economy of the Year 2000," by Wassily W. Leontief, *Scientific American*, September 1980, p. 166. The article describes the application of this model to a world economy. The model was commissioned by the United Nations with special financial support from the Netherlands. In the model the world is divided into 15 distinct geographic regions, each one described by an individual input-output matrix. The regions are then linked by a larger matrix, which is used in an input-output model. Overall, more than 200 economic sectors are included in the model. By feeding in various values, economists use the model to create scenarios of future world economic conditions.

Exercise Set 2.5

1. Consider the following input-output matrix that defines the interdependency of five industries.

$$
\begin{array}{cc}
 & \begin{array}{ccccc} 1 & 2 & 3 & 4 & 5 \end{array} \\
\begin{array}{l} \text{1. Auto} \\ \text{2. Steel} \\ \text{3. Electricity} \\ \text{4. Coal} \\ \text{5. Chemical} \end{array} &
\left[\begin{array}{ccccc}
0.15 & 0.10 & 0.05 & 0.05 & 0.10 \\
0.40 & 0.20 & 0.10 & 0.10 & 0.10 \\
0.10 & 0.25 & 0.20 & 0.10 & 0.20 \\
0.10 & 0.20 & 0.30 & 0.15 & 0.10 \\
0.05 & 0.10 & 0.05 & 0.02 & 0.05
\end{array}\right]
\end{array}
$$

Determine

*(a) the amount of electricity consumed in producing $1 worth of steel.

(b) the amount of steel consumed in producing $1 worth in the auto industry.

*(c) the largest consumer of coal.

(d) the largest consumer of electricity.

*(e) the industry on which the auto industry is most dependent.

In Exercises 2–6, consider the economies consisting of either two or three industries. Determine the output levels required of each industry in each situation to meet the demands of the other industries and of the open sector.

*2. $A = \begin{bmatrix} 0.20 & 0.60 \\ 0.40 & 0.10 \end{bmatrix}$,

$D = \begin{bmatrix} 24 \\ 12 \end{bmatrix}$, $\begin{bmatrix} 8 \\ 6 \end{bmatrix}$, $\begin{bmatrix} 0 \\ 12 \end{bmatrix}$ in turn

3. $A = \begin{bmatrix} 0.10 & 0.40 \\ 0.30 & 0.20 \end{bmatrix}$,

$D = \begin{bmatrix} 6 \\ 12 \end{bmatrix}$, $\begin{bmatrix} 18 \\ 6 \end{bmatrix}$, $\begin{bmatrix} 24 \\ 12 \end{bmatrix}$ in turn

*4. $A = \begin{bmatrix} 0.30 & 0.60 \\ 0.35 & 0.10 \end{bmatrix}$,

$D = \begin{bmatrix} 42 \\ 84 \end{bmatrix}$, $\begin{bmatrix} 0 \\ 10 \end{bmatrix}$, $\begin{bmatrix} 14 \\ 7 \end{bmatrix}$, $\begin{bmatrix} 42 \\ 42 \end{bmatrix}$ in turn

*5. $A = \begin{bmatrix} 0.20 & 0.20 & 0.10 \\ 0 & 0.40 & 0.20 \\ 0 & 0.20 & 0.60 \end{bmatrix}$,

$D = \begin{bmatrix} 4 \\ 8 \\ 8 \end{bmatrix}$, $\begin{bmatrix} 0 \\ 8 \\ 16 \end{bmatrix}$, $\begin{bmatrix} 8 \\ 24 \\ 8 \end{bmatrix}$ in turn

6. $A = \begin{bmatrix} 0.20 & 0.20 & 0 \\ 0.40 & 0.40 & 0.60 \\ 0.40 & 0.10 & 0.40 \end{bmatrix}$,

$D = \begin{bmatrix} 36 \\ 72 \\ 36 \end{bmatrix}$, $\begin{bmatrix} 36 \\ 0 \\ 18 \end{bmatrix}$, $\begin{bmatrix} 36 \\ 0 \\ 0 \end{bmatrix}$, $\begin{bmatrix} 0 \\ 18 \\ 18 \end{bmatrix}$ in turn

In Exercises 7–9, consider the economies consisting of either two or three industries. The output levels of the industries are given. Determine the amounts available for the open sector from each industry.

*7. $A = \begin{bmatrix} 0.20 & 0.40 \\ 0.50 & 0.10 \end{bmatrix}$, $X = \begin{bmatrix} 8 \\ 10 \end{bmatrix}$

8. $A = \begin{bmatrix} 0.10 & 0.20 & 0.30 \\ 0 & 0.10 & 0.40 \\ 0.50 & 0.40 & 0.20 \end{bmatrix}$, $X = \begin{bmatrix} 10 \\ 10 \\ 20 \end{bmatrix}$

*9. $A = \begin{bmatrix} 0.10 & 0.10 & 0.20 \\ 0.20 & 0.10 & 0.30 \\ 0.40 & 0.30 & 0.15 \end{bmatrix}$, $X = \begin{bmatrix} 6 \\ 4 \\ 5 \end{bmatrix}$

10. Let a_{ij} be an arbitrary element of an input-output matrix. Why would you expect a_{ij} to satisfy the condition $0 \leq a_{ij} \leq 1$?

11. In an economically feasible situation the sum of the elements of each column of the input-output matrix is less than or equal to unity. Explain why this should be so.

*2.6 Stochastic Matrices: A Population Movement Model

Certain matrices, called **stochastic matrices**, are important in the study of random phenomena where the exact outcome is not known but probabilities can be determined. In this section we introduce stochastic matrices, derive some of their properties, and give examples of their application. One example is an analysis of population movement between cities and suburbs in the United States. The second example illustrates the use of stochastic matrices in genetics.

At this time we remind the reader of some basic ideas of probability. If the outcome of an event *is sure to occur*, we say that the probability of that outcome is 1. On the other hand, if it *will not occur*, we say that the probability is 0. Other probabilities are represented by the fractions between 0 and 1; *the larger the fraction, the greater the probability p of that outcome occurring*. Thus we have the restriction $0 \leq p \leq 1$ on a probability p.

If any one of n completely independent outcomes is equally likely to happen, and if m of these outcomes are of interest to us, then the probability p that one of these outcomes will occur is defined to be the fraction m/n.

As an example, consider the event of drawing a single card from a deck of 52 playing cards. What is the probability that the outcome will be an ace or a king? First we see that there are 52 possible outcomes. Since there are 4 aces and 4 kings in the pack, there are 8 outcomes of interest. Thus the probability of drawing an ace or a king is $\frac{8}{52}$, or $\frac{2}{13}$.

We now introduce matrices whose elements are probabilities.

Definition   *A **stochastic matrix** is a square matrix whose elements are probabilities and whose columns each sum to 1.*

The following matrices are stochastic matrices.

The following matrices are not stochastic.

the sum of the elements in
the first column is not 1

the 2 in the 1st row is not a probability
since it is greater than 1

A general 2 × 2 stochastic matrix can be written

$$\begin{bmatrix} x & y \\ 1-x & 1-y \end{bmatrix}$$

where $0 \leq x \leq 1$ and $0 \leq y \leq 1$.

Stochastic matrices have the following useful property. (You are asked to prove this result for 2 × 2 stochastic matrices in the exercises that follow.)

Theorem 2.10   If A and B are stochastic matrices of the same size, then AB is a stochastic matrix.

Thus if A is stochastic, then $A^2, A^3, A^4, \ldots$ are all stochastic.

Example 1   This example illustrates a stochastic matrix for an analysis of land use in center-city Toronto for the period 1952–1962.

The researchers collected data and wrote them in the form of the following stochastic matrix P. The rows and columns of P represent land uses. The element p_{ij} is the probability that land that was in use j in 1952 will be in use i in 1962.

Use in 1952										
1	2	3	4	5	6	7	8	9	10	Use in 1962
0.13	0.02	0.00	0.02	0.00	0.08	0.01	0.01	0.01	0.25	1. Low-density residential
0.34	0.41	0.07	0.01	0.00	0.05	0.03	0.02	0.18	0.08	2. High-density residential
0.10	0.05	0.43	0.09	0.11	0.14	0.02	0.02	0.14	0.03	3. Office
0.04	0.04	0.05	0.30	0.07	0.08	0.12	0.03	0.04	0.03	4. General commercial
0.04	0.00	0.01	0.09	0.70	0.12	0.03	0.03	0.10	0.05	5. Auto commercial
0.22	0.04	0.28	0.27	0.06	0.39	0.11	0.08	0.39	0.15	6. Parking
0.03	0.00	0.14	0.05	0.00	0.04	0.38	0.18	0.03	0.22	7. Warehousing
0.02	0.00	0.00	0.08	0.01	0.00	0.21	0.61	0.03	0.13	8. Industry
0.00	0.00	0.00	0.01	0.00	0.01	0.01	0.00	0.08	0.00	9. Transportation
0.08	0.44	0.02	0.08	0.05	0.09	0.08	0.02	0.00	0.06	10. Vacant

Let us interpret some of the information contained in this matrix. For example, $p_{63} = 0.28$. This tells us that land that was office space in 1952 had a probability of 0.28 of becoming parking area by 1962. The sixth row of P gives the probabilities that various areas of the city have become parking areas by 1962. These relatively large figures reveal the increasingly dominant role of parking in land use.

The diagonal elements give the probabilities that land use remained in the same category. For example, $p_{77} = 0.38$ is the probability that warehousing remained warehousing land. The relatively high figures of these diagonal elements reflect the tendency for land to remain in the same broad category of usage. The exceptions are transportation and vacant land.

It is interesting to note that $p_{10\ 2} = 0.44$. This is the probability that land that was high-density residential in 1952 had become vacant by 1962. Note that this is land usage for center-city Toronto.

Readers who are interested in continuing this analysis are encouraged to read the paper from which this matrix came: "Physical Adjustment Processes and Land Use Succession in Toronto," Larry S. Bourne, *Economic Geography*, **47**, No. 1, January 1971, pp. 1–15.

Example 2   In this example we develop a model of population movement between cities and surrounding suburbs in the United States. The numbers given are based on statistics in *Statistical Abstract of the United States*, 1987.

The number of people living in the United States during 1985 was 60 million. The number of people living in the surrounding suburbs was 125 million. Let us represent this information by the matrix $X_0 = \begin{bmatrix} 60 \\ 125 \end{bmatrix}$.

Consider the population flow from cities to suburbs. During 1985 the probability of a person staying in the city was 0.96. Thus the probability of moving to the suburbs was 0.04 (assuming that all those who moved went to the suburbs). Consider now the reverse population flow, from suburbia to the city. The probability of a person moving to the city was 0.01; the probability of remaining in suburbia was 0.99. These probabilities can be written as the elements of a stochastic matrix P:

$$P = \begin{bmatrix} 0.96 & 0.01 \\ 0.04 & 0.99 \end{bmatrix} \begin{matrix} \text{city} \\ \text{suburb} \end{matrix}$$

The probability of moving from location A to location B is given by the element in column A and row B. In this context the stochastic matrix is called a **matrix of transition probabilities**.

Consider the population distribution in 1986, one year later:

City population in 1986 = people who remained from 1985 + people who moved in from the suburbs
= $(0.96 \times 60) + (0.01 \times 125)$
= 58.85 million

$$\text{Suburban population in 1986} = \begin{array}{c}\text{people who moved} \\ \text{in from the city}\end{array} + \begin{array}{c}\text{people who stayed} \\ \text{from 1985}\end{array}$$

$$= (0.04 \times 60) + (0.99 \times 125)$$

$$= 126.15 \text{ million}$$

Note that we can arrive at these numbers using matrix multiplication:

$$\begin{bmatrix} 0.96 & 0.01 \\ 0.04 & 0.99 \end{bmatrix} \begin{bmatrix} 60 \\ 125 \end{bmatrix} = \begin{bmatrix} 58.85 \\ 126.15 \end{bmatrix}$$

Using 1985 as the base year, let X_1 be the population in 1986, one year later. We can write

$$X_1 = PX_0$$

Assume that the population flow represented by the matrix P is unchanged over the years. The population distribution X_2 after 2 years is given by

$$X_2 = PX_1$$

After 3 years the population distribution is given by

$$X_3 = PX_2$$

After n years we get

$$X_n = PX_{n-1}$$

The predictions of this model are

$$X_0 = \begin{bmatrix} 60 \\ 125 \end{bmatrix} \begin{array}{c}\text{city} \\ \text{suburb}\end{array}, \quad X_1 = \begin{bmatrix} 58.85 \\ 126.15 \end{bmatrix}, \quad X_2 = \begin{bmatrix} 57.7575 \\ 127.2425 \end{bmatrix},$$

$$X_3 = \begin{bmatrix} 56.7196 \\ 128.2804 \end{bmatrix}, \quad X_4 = \begin{bmatrix} 55.7336 \\ 129.2664 \end{bmatrix},$$

and so on.

Observe how the city population is decreasing annually, while that of the suburbs is increasing. We return to this model in Section 4.8. There we find that the sequence $X_0, X_1, X_2, \ldots$ approaches $\begin{bmatrix} 37 \\ 148 \end{bmatrix}$. If conditions do not change, the city population will gradually approach 37 million, while the population of suburbia will approach 148 million.

Further, note that the sequence $X_1, X_2, X_3, \ldots, X_n$ can be directly computed from X_0, as follows.

$$X_1 = PX_0, \quad X_2 = P^2X_0, \quad X_3 = P^3X_0, \ldots, X_n = P^nX_0$$

The matrix P^n is a stochastic matrix that takes X_0 into X_n, in n steps. This result can be generalized. That is, P^n can be used in this manner to predict the distribution n stages later, from any given distribution.

$$X_{i+n} = P^nX_i$$

P^n is called the ***n*-step transition matrix**. The (i, j)th element of P^n gives the probability of going from state j to state i in n steps. For example, it can be shown that (writing to 2 decimal places)

$$P^4 = \begin{matrix} & \text{(from)} & \text{(to)} \\ & \text{city} \quad \text{suburb} & \\ & \begin{bmatrix} 0.85 & 0.04 \\ 0.15 & 0.96 \end{bmatrix} & \begin{matrix} \text{city} \\ \text{suburb} \end{matrix} \end{matrix}$$

Thus, for instance, the probability of living in the city in 1985 and being in the suburbs 4 years later is 0.15.

The probabilities in this model depend only on the current state of a person—whether the person is living in the city or in suburbia. This type of a model, where the probability of going from one state to another depends only on the current state rather than on a more complete historical description, is called a **Markov chain**.*

A modification that allows for possible annual population growth or decrease would give improved estimates of future population distributions. You are asked to build such a factor into the model in the exercises that follow.

These concepts can be extended to Markov processes involving more than two states. The following example illustrates a Markov chain involving three states.

Example 3   Markov chains are useful tools for scientists in many fields. We now discuss the role of Markov chains in genetics.

Genetics is the branch of biology that deals with heredity. It is the study of units called **genes**, which determine the characteristics living things inherit from their parents. The inheritance of such traits as sex, height, eye color, and hair color of human beings, and such traits as petal color and leaf shape of plants, are governed by genes. Because many diseases are inherited, genetics is important in medicine. In agriculture, breeding methods based on genetic principles led to important advances in both plant and animal breeding. High-yield hybrid corn and disease-resistant rice rank among the most important contributions of genetics to increasing food production. We shall discuss a mathematical model developed for analyzing the behavior of traits involving a pair of genes. We illustrate the concepts involved in terms of crossing a pair of guinea pigs.

The traits that we shall study in guinea pigs are the traits of long hair and short hair. The length of hair is governed by a pair of genes, which we shall denote A and a. A guinea pig may have any one of the combinations AA, Aa, or aa. (aA is genetically the same as Aa.) Each of these classes is called a **genotype**. The AA type guinea pig is indistinguishable in appearance from the Aa type—both have long hair—while the

---

* Andrei Andreevich Markov (1856–1922) was educated and taught at the University of St. Petersburg, Russia. His areas of interest were number theory, probability, and function theory. It was said that "he gave distinguished lectures with irreproachable strictness of argument, and developed in his students that mathematical cast of mind that takes nothing for granted." He was faculty adviser to a student math circle. He developed chains to analyze literary texts where the states were vowels and consonants. Markov was a man of firm opinions who participated in the liberal movement in Russia at the beginning of the twentieth century. When "pompous officials" celebrated the 300th anniversary of the House of Romanov, Markov organized a celebration of the 200th anniversary of the law of large numbers!

aa type has short hair. The *A* gene is said to **dominate** the *a* gene. An animal is called **dominant** if it has *AA* genes, **hybrid** with *Aa* genes, and **recessive** with *aa* genes.

When two guinea pigs are crossed, the offspring inherits one gene from each parent in a random manner. Given the genotypes of the parents, we can determine the probabilities of the genotype of the offspring. Consider a given population of guinea pigs. Let us perform a series of experiments in which we *keep crossing offspring with dominant animals only.* Thus we keep crossing *AA*, *Aa*, and *aa* with *AA*. What are the probabilities of the offspring being *AA*, *Aa*, or *aa* in each of these cases?

Consider the crossing of *AA* with *AA*. The offspring will have one gene from each parent, so it will be of type *AA*. Thus the probabilities of *AA*, *Aa*, and *aa* resulting are 1, 0, and 0, respectively. All offspring have long hair.

Next consider the crossing of *Aa* with *AA*. Taking one gene from each parent, we have the possibilities of *AA*, *AA* (taking *A* from the first parent and each *A* in turn from the second parent), *aA*, and *aA* (taking the *a* from the first parent and each *A* in turn from the second parent). Thus the probabilities of *AA*, *Aa*, and *aa*, respectively, are $\frac{1}{2}$, $\frac{1}{2}$, and 0. All offspring again have long hair.

Finally, on crossing *aa* with *AA* there is only one possibility, namely *aA*. Thus the probabilities of *AA*, *Aa*, and *aa* are 0, 1, and 0, respectively.

All offspring resulting from these experiments have long hair. This series of experiments is a Markov chain having transition matrix

$$P = \begin{array}{c} \\ \\ \end{array} \begin{array}{ccc} AA & Aa & aa \\ \end{array}$$
$$P = \begin{bmatrix} 1 & \frac{1}{2} & 0 \\ 0 & \frac{1}{2} & 1 \\ 0 & 0 & 0 \end{bmatrix} \begin{array}{c} AA \\ Aa \\ aa \end{array}$$

Consider an initial population of guinea pigs made up of an equal number of each genotype. Let the initial distribution be $X_0 = \begin{bmatrix} \frac{1}{3} \\ \frac{1}{3} \\ \frac{1}{3} \end{bmatrix}$, representing the fraction of guinea pigs of each type initially. The components of X_1, X_2, X_3, ... will give the fractions of the following generations that are of types *AA*, *Aa*, and *aa*, respectively. We get

$$X_0 = \begin{bmatrix} \frac{1}{3} \\ \frac{1}{3} \\ \frac{1}{3} \end{bmatrix} \begin{array}{c} AA \\ Aa, \\ aa \end{array} \quad X_1 = \begin{bmatrix} \frac{1}{2} \\ \frac{1}{2} \\ 0 \end{bmatrix}, \quad X_2 = \begin{bmatrix} \frac{3}{4} \\ \frac{1}{4} \\ 0 \end{bmatrix}, \quad X_3 = \begin{bmatrix} \frac{7}{8} \\ \frac{1}{8} \\ 0 \end{bmatrix}, \quad X_4 = \begin{bmatrix} \frac{15}{16} \\ \frac{1}{16} \\ 0 \end{bmatrix},$$

and so on.

Observe that the *aa* type disappears after the initial generation and that the *Aa* type becomes a smaller and smaller fraction of each successive generation. The sequence, in fact, approaches the matrix

$$X = \begin{bmatrix} 1 \\ 0 \\ 0 \end{bmatrix} \begin{array}{c} AA \\ Aa \\ aa \end{array}$$

The genotype *AA* in this model is called an **absorbing state**.

Here we have considered the case of crossing offspring with a dominant animal. You are asked to construct a similar model that describes the crossing of offspring with a hybrid in the exercises that follow. Some of the offspring will have long hair and some short hair in that series of experiments.

Exercise Set 2.6

1. State which of the following matrices are stochastic and which are not. Explain why a matrix is not stochastic.

*(a) $\begin{bmatrix} \frac{1}{4} & 0 \\ \frac{3}{4} & 1 \end{bmatrix}$

(b) $\begin{bmatrix} \frac{1}{2} & -1 \\ \frac{1}{2} & 12 \end{bmatrix}$

*(c) $\begin{bmatrix} \frac{1}{3} & \frac{1}{7} \\ \frac{2}{3} & \frac{5}{7} \end{bmatrix}$

(d) $\begin{bmatrix} 1 & 0 & 0 \\ 0 & 1 & 0 \\ 0 & 0 & 1 \end{bmatrix}$

*(e) $\begin{bmatrix} 0 & \frac{3}{8} & 0 \\ \frac{1}{2} & \frac{1}{8} & 1 \\ \frac{1}{2} & \frac{1}{2} & 0 \end{bmatrix}$

(f) $\begin{bmatrix} 0 & \frac{1}{5} & \frac{3}{4} \\ \frac{5}{6} & \frac{2}{5} & \frac{3}{4} \\ \frac{1}{6} & \frac{2}{5} & -\frac{1}{2} \end{bmatrix}$

2. Prove that the product of two 2×2 stochastic matrices is a stochastic matrix.

*3. A stochastic matrix, the sum of whose rows is 1, is called a **doubly stochastic matrix**. Give examples of 2×2 and 3×3 doubly stochastic matrices. Is the product of two doubly stochastic matrices doubly stochastic?

4. Use the stochastic matrix of Example 1 of this section to answer the following questions:

*(a) What is the probability that land used for industry in 1952 was used for offices in 1962?

(b) What is the probability that land used for parking in 1952 was in a high-density residential area in 1962?

*(c) Vacant land in 1952 had the highest probability of becoming what kind of land in 1962?

(d) Which was the most stable usage of land over the period 1952–1962?

5. In the model of Example 2, determine

*(a) the probability of moving from city to suburb in 2 years.

(b) the probability of moving from the suburb to city in 3 years.

6. Construct a model of population flow between metropolitan and nonmetropolitan areas of the United States, given that their respective populations in 1985 were 185 and 55 million. The probabilities are given by the matrix

	(from)	(to)
	metro nonmetro	

$$\begin{bmatrix} 0.99 & 0.02 \\ 0.01 & 0.98 \end{bmatrix} \begin{matrix} \text{metro} \\ \text{nonmetro} \end{matrix}$$

Predict the population distributions of metropolitan and nonmetropolitan areas for the years 1986 through 1990. If a person was living in a metropolitan area in 1985, what is the probability that the person was still living in a metropolitan area in 1987?

*7. Construct a model of population flows between cities, suburbs, and nonmetropolitan areas of the United States. Their respective populations in 1985 were 60, 125, and 55 million. The stochastic matrix giving the probabilities of the moves is

	(from)		(to)
	city suburb nonmetro		

$$\begin{bmatrix} 0.96 & 0.01 & 0.015 \\ 0.03 & 0.98 & 0.005 \\ 0.01 & 0.01 & 0.98 \end{bmatrix} \begin{matrix} \text{city} \\ \text{suburb} \\ \text{nonmetro} \end{matrix}$$

This model is a refinement on the model of the previous exercise in that the metropolitan population is broken down into city and suburb. It is also a more inclusive model than that of Example 2 of this section, which did not allow for any population outside cities and suburbs.

Predict the populations of city, suburban, and nonmetropolitan areas for 1986 and 1987. If a person is living in the city in 1985, what is the probability that the person was living in a nonmetropolitan area in 1987?

*8. In the period 1980 to 1985, the total population of the United States increased by 0.2 % per annum. Assume that the population increases annually by 0.2% during the years immediately following. Build this factor into the model of Example 2 and predict the populations of city and suburbia in 1990.

9. In the period 1980 to 1985, the total population of the United States increased annually by 0.2%, with 0.25% of this growth taking place in the cities and 0.1% of the growth taking place in suburbia. Assume that the same growth took place in the years immediately following. Build this factor into the model of Example 2 and predict the populations of city and suburbia in 1990. This model is a refinement on that of the previous exercise in that it allows for unequal population growths in city and suburbia.

10. Consider the population movement model for flow between cities and suburbs of Example 2 of this section. Determine the population distributions for 1980 to 1984—prior to 1985. Is the chain going from 1985 into the past a Markov chain? What are the characteristics of the matrix that takes one from distribution to distribution into the past?

*11. The following stochastic matrix gives occupational transition probabilities.

(initial generation)

$$\begin{matrix} \text{white-collar} & \text{manual} \\ \begin{bmatrix} 1 & 0.2 \\ 0 & 0.8 \end{bmatrix} & \begin{matrix} \text{white-collar} \\ \text{manual} \end{matrix} \end{matrix} \text{(next generation)}$$

(a) If the father is a manual worker, what is the probability that the son will be a white-collar worker?

(b) If there are 10,000 in the white-collar category and 20,000 in the manual category, what will the distribution be one generation later?

12. The following matrix gives occupational transition probabilities.

(initial generation)

$$\begin{matrix} \text{nonfarming} & \text{farming} \\ \begin{bmatrix} 1 & 0.4 \\ 0 & 0.6 \end{bmatrix} & \begin{matrix} \text{nonfarming} \\ \text{farming} \end{matrix} \end{matrix} \text{(next generation)}$$

(a) If the father is a farmer, what is the probability that the son will be a farmer?

(b) If there are 10,000 in the nonfarming category and 1,000 in the farming category at a certain time, what will the distribution be one generation later? Four generations later?

(c) If the father is a farmer, what is the probability that the grandson will be a farmer?

*13. A market analysis of car purchasing trends in a certain region has concluded that a family purchases a new car once every 3 years on an average. The buying patterns are described by the matrix

$$P = \begin{matrix} & \text{small} & \text{large} \\ & \begin{bmatrix} 80\% & 40\% \\ 20\% & 60\% \end{bmatrix} & \begin{matrix} \text{small} \\ \text{large} \end{matrix} \end{matrix}$$

The elements of P are to be interpeted as follows. The first column indicates that, of the current small cars, 80% will be replaced with small cars, 20% with a large car. The second column implies that 40% of the current large cars will be replaced with small cars while 60% will be replaced with large cars. Write the elements of P as follows to get a stochastic matrix that defines a Markov chain.

$$P = \begin{bmatrix} \frac{80}{100} & \frac{40}{100} \\ \frac{20}{100} & \frac{60}{100} \end{bmatrix} = \begin{bmatrix} 0.8 & 0.4 \\ 0.2 & 0.6 \end{bmatrix}$$

If there are currently 40,000 small cars and 50,000 large cars in the region, what is your prediction of the distribution in 12 years' time?

14. The conclusion of an analysis of voting trends in a certain state is that the voting patterns of successive generations are described by the following matrix P.

$$P = \begin{matrix} & \text{Dem.} & \text{Rep.} & \text{Ind.} \\ & \begin{bmatrix} 80\% & 20\% & 60\% \\ 15\% & 70\% & 30\% \\ 5\% & 10\% & 10\% \end{bmatrix} & \begin{matrix} \text{Democrat} \\ \text{Republican} \\ \text{Independent} \end{matrix} \end{matrix}$$

Among the Democrats of one generation, 80% of the next generation are Democrats, 15% are Republican, and 5% are Independents, etc. Express P as a stochastic matrix that defines a Markov chain model of the voting patterns. If there are 2.5 million registered Démocrats, 1.5 million registered Republicans, and 0.25 million registered Independents at a certain time, what is the distribution likely to be in the next generation?

*15. Determine the transition matrix for a Markov chain that describes the crossing of offspring of guinea pigs with hybrids only. There is no absorbing state in this model.

Let the initial matrix $X_0 = \begin{bmatrix} \frac{1}{3} \\ \frac{1}{3} \\ \frac{1}{3} \end{bmatrix}$ be the fraction of guinea pigs of each type initially. Determine the distributions for the next three generations.

Communication Model and Group Relationships

Many branches of the physical sciences, social sciences, and business use models from **graph theory** to analyze relationships. We introduce the reader to this important area of mathematics, which uses linear algebra, with an example from the field of communication.

Consider a communication network involving five stations, labeled $P_1, \ldots, P_5$. The communication links could be roads, phone lines, etc. Certain stations are linked by two-way communication, others by one-way links. Still others may have only indirect communication by way of intermediate stations. Suppose the network of interest is described in Figure 2.4. Lines joining stations represent direct communication links; the arrows give the directions of those links. For example, stations P_1 and P_2 have two-way direct communication. Station P_4 can send a message to P_1 by way of stations P_3 and P_2 or by way of P_5 and P_2. This communication network is an example of a digraph.

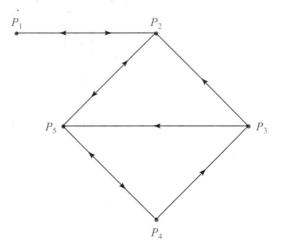

Figure 2.4

Definition   *A **diagraph** is a finite collection of **vertices** $P_1, P_2, \ldots, P_n$, together with **directed arcs** joining certain pairs of vertices. A **path** between vertices is a sequence of arcs that allows one to proceed in a continuous manner from one vertex to another. The **length** of a path is its number of arcs. A path of length n is called an **n-path**.*

In the above communication network there are five vertices, namely $P_1, \ldots, P_5$. Suppose that we are interested in sending a message from P_3 to P_1. From the figure we see that various paths can be taken. The path $P_3 \rightarrow P_2 \rightarrow P_1$ is of length 2 (a 2-path), while the path $P_3 \rightarrow P_5 \rightarrow P_2 \rightarrow P_1$ is of length 3 (a 3-path). The path $P_3 \rightarrow P_5 \rightarrow P_4 \rightarrow P_3 \rightarrow P_2 \rightarrow P_1$, a 5-path, takes in the vertex P_3 twice. Such paths could be of interest if, for example, P_3 wanted to consult with P_5 and P_4 before sending a message to P_1. In this section, however, we shall be interested in finding the shortest route to send a message from one station to another.

Communication networks can be vast, involving many stations. It is impractical to get information about large networks from diagrams, as we have done above. The mathematical theory that we now present, from graph theory, can be used to give information about large networks. The mathematics can be implemented on the computer.

A digraph can be described by a matrix A, called its adjacency matrix. This matrix consists of zeros and ones and is defined as follows.

Definition *Consider a diagraph with vertices $P_1, \ldots, P_n$. The **adjacency matrix** A of the digraph is such that*

$$a_{ij} = \begin{cases} 1 & \text{if there is an arc from vertex } P_i \text{ to } P_j \\ 0 & \text{otherwise} \end{cases}$$

The adjacency matrix of the communication network is

$$A = \begin{bmatrix} 0 & 1 & 0 & 0 & 0 \\ 1 & 0 & 0 & 0 & 1 \\ 0 & 1 & 0 & 0 & 1 \\ 0 & 0 & 1 & 0 & 1 \\ 0 & 1 & 0 & 1 & 0 \end{bmatrix}$$

For example, $a_{12} = 1$ since there is an arc from P_1 to P_2; $a_{13} = 0$ since there is no arc from P_1 to P_3.

The network is completely described by the adjacency matrix. This matrix is a mathematical "picture" of the network that can be given to a computer. We could look at the sketch of the network given in Figure 2.4 and decide how best to send a message from P_3 to P_1. How can we extract such information from the adjacency matrix? The following theorem from graph theory gives information about paths with digraphs.

Theorem 2.11 If A is the adjacency matrix of a digraph, let $a_{ij}^{(m)}$ be the element in row i and column j of A^m.

The number of m-paths from P_i to $P_j = a_{ij}^{(m)}$

Proof Consider a digraph with n vertices. a_{i1} is the number of arcs from P_i to P_1, and a_{1j} if the number of arcs from P_1 to P_j. Thus $a_{i1}a_{1j}$ is the number of 2-paths from P_i to P_j, passing through P_1. Summing up over all such possible intermediate stations, we see that the total number of 2-paths from P_i to P_j is

$$a_{i1}a_{1j} + a_{i2}a_{2j} + \cdots + a_{in}a_{nj}$$

This is the element in row i, column j of A^2. Thus $a_{ij}^{(2)}$ is the number of 2-paths from P_i to P_j.

(continues)

Let us now look at 3-paths. Interpret a 3-path as a 2-path followed by an arc. The number of 2-paths from P_i to P_1 followed by an arc from P_1 to P_j is $a_{i1}^{(2)}a_{1j}$. Summing over all such possible intermediate stations, we see that the total number of 3-paths from P_i to P_j is

$$a_{i1}^{(2)}a_{1j} + a_{i2}^{(2)}a_{2j} + \cdots + a_{in}^{(2)}a_{nj}$$

This is the element in row i, column j, of the matrix product A^2A—that is, of A^3. Thus $a_{ij}^{(3)}$ is the number of 3-paths from P_i to P_j.

Continuing this, we can interpret a 4-path as a 3-path followed by an arc, and so on, arriving at the result that $a_{ij}^{(m)}$ is the number of m-paths from P_i to P_j.

We now illustrate the application of this theorem to our communication network. Successive powers of the adjacency matrix are determined.

$$A = \begin{bmatrix} 0 & 1 & 0 & 0 & 0 \\ 1 & 0 & 0 & 0 & 1 \\ 0 & 1 & 0 & 0 & 1 \\ 0 & 0 & 1 & 0 & 1 \\ 0 & 1 & 0 & 1 & 0 \end{bmatrix}, \quad A^2 = \begin{bmatrix} 1 & 0 & 0 & 0 & 1 \\ 0 & 2 & 0 & 1 & 0 \\ 1 & 1 & 0 & 1 & 1 \\ 0 & 2 & 0 & 1 & 1 \\ 1 & 0 & 1 & 0 & 2 \end{bmatrix}$$

$$A^3 = \begin{bmatrix} 0 & 2 & 0 & 1 & 0 \\ 2 & 0 & 1 & 0 & 3 \\ 1 & 2 & 1 & 1 & 2 \\ 2 & 1 & 1 & 1 & 3 \\ 0 & 4 & 0 & 2 & 1 \end{bmatrix}, \quad A^4 = \begin{bmatrix} 2 & 0 & 1 & 0 & 3 \\ 0 & 6 & 0 & 3 & 1 \\ 2 & 4 & 1 & 2 & 4 \\ 1 & 6 & 1 & 3 & 3 \\ 4 & 1 & 2 & 1 & 6 \end{bmatrix}, \quad A^5 = \cdots$$

Let us use these matrices to discuss paths from P_4 to P_1.

A gives $a_{41} = 0$. There is no direct communication.

A^2 gives $a_{41}^{(2)} = 0$. There are no 2-paths from P_4 to P_1.

A^3 gives $a_{41}^{(3)} = 2$. There are two distinct 3-paths from P_4 to P_1.

These are the shortest paths from P_4 to P_1. If we check with Figure 2.4, we see that this is the case. The two 3-paths are

$$P_4 \to P_3 \to P_2 \to P_1 \quad \text{and} \quad P_4 \to P_5 \to P_2 \to P_1$$

As a second example, let us determine the length of the shortest path from P_1 to P_3.

A gives $a_{13} = 0$. There is no direct communication.

A^2 gives $a_{13}^{(2)} = 0$. There are no 2-paths from P_1 to P_3.

A^3 gives $a_{13}^{(3)} = 0$. There are no 3-paths from P_1 to P_3.

A^4 gives $a_{13}^{(4)} = 1$. There is a single 4-path from P_1 to P_3.

This result is confirmed when we examine the digraph. The quickest way to send a message from P_1 to P_3 is the 4-path

$$P_1 \rightarrow P_2 \rightarrow P_5 \rightarrow P_4 \rightarrow P_3$$

This model that we have discussed gives the lengths of the shortest paths of a digraph; it does not give the intermediate stations that make up that path. Mathematicians have not, as yet, been able to derive this information from the adjacency matrix. An algorithm for finding the shortest paths for a specific digraph, using a search procedure, has been developed by a Dutch computer scientist named Dijkstra. See, for example, *Algorithms, Practice and Theory* by Gilles Brassard and Paul Bratley, Prentice-Hall, 1988, p. 87, for a discussion of this algorithm. The following discussion leads to an application where the *lengths* of the shortest paths, not the actual paths, are important.

Distance in a Digraph

The **distance** from one vertex of a digraph to another is the length of the shortest path from that vertex to the other. If there is no path from the one vertex to the other, we say that the distance is **undefined**. The distances between the various vertices of a digraph form the elements of a matrix. The **distance matrix** D is defined as follows:

$$d_{ij} = \begin{cases} \text{number of arcs in shortest path from vertex } P_i \text{ to vertex } P_j \\ 0 \quad \text{if } i = j \\ x \quad \text{if there is no path from } P_i \text{ to } P_j \end{cases}$$

If the digraph is small, the distance matrix can be constructed by observation. Powers of the adjacency matrix are used to construct the distance matrix of a large digraph. The distance matrix of the above communication network is

$$D = \begin{bmatrix} 0 & 1 & 4 & 3 & 2 \\ 1 & 0 & 3 & 2 & 1 \\ 2 & 1 & 0 & 2 & 1 \\ 3 & 2 & 1 & 0 & 1 \\ 2 & 1 & 2 & 1 & 0 \end{bmatrix}$$

Note that the distance from P_i to P_j is not necessarily equal to the distance from P_j to P_i in a digraph, implying that a distance matrix in graph theory is not necessarily symmetric.

We now illustrate how a digraph and a distance matrix can be used to analyze group relations in sociology.

Group Relationships in Sociology

Consider a group of five people. A sociologist is interested in finding out which one of the five has most influence over, or dominates, the other members. The group is asked to fill out the following questionnaire:

- Your name   _____
- Person whose opinion you value most   _____

These answers are then tabulated. Let us for convenience label the group members $M_1, M_2, \ldots, M_5$. Suppose the results are as given in Table 2.1.

Table 2.1

Group member	Person whose opinion valued
M_1	M_4
M_2	M_1
M_3	M_2
M_4	M_2
M_5	M_4

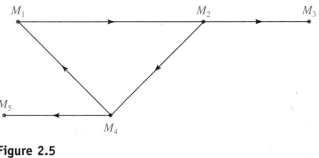

Figure 2.5

The sociologist makes the assumption that the person whose opinion a member values most is the person who influences that member most. Thus influence goes from the right column to the left column in the above table. We can represent these results by a digraph. The group members are represented by vertices, direct influence by an arc, the direction of influence being the direction of the arc. See Figure 2.5. Construct the distance matrix of this digraph, and sum all the elements in each row.

$$
D = \begin{bmatrix} 0 & 1 & 2 & 2 & 3 \\ 2 & 0 & 1 & 1 & 2 \\ x & x & 0 & x & x \\ 1 & 2 & 3 & 0 & 1 \\ x & x & x & x & 0 \end{bmatrix} \quad \begin{array}{c} \text{row sums} \\ 8 \\ 6 \\ 4x \\ 7 \\ 4x \end{array}
$$

In this graph arcs correspond to direct influence; 2-paths, 3-paths, etc., correspond to direct influence. Thus, presumably, the smaller the distance from M_i to M_j, the greater the influence M_i has on M_j. The sum of the elements in row i gives the total distance of M_i to the other vertices. This leads to the following interpretation of row sums.

The smaller the row sum i, the greater the influence of person M_i on the group.

We see that the smallest row sum is 6, for row 2. Thus M_2 is the most influential person in the group, followed by M_4 and then M_1.

Readers who are interested in learning more about graph theory are referred to the following two books: *Introduction to Graph Theory* by Robin J. Wilson, John Wiley and Sons, 1987, and *Discrete Mathematical Structures* by Fred S. Roberts, Prentice-Hall, 1976. The former book is a beautiful introduction to the mathematics of graph theory; the latter has a splendid collection of applications. "Predicting Chemistry by Topology," by Dennis H. Rouvray, *Scientific American*, **40**, September 1986, contains a fascinating account of how graph theoretical methods are being used to predict chemical properties of molecules that have not yet been synthesized.

Exercise Set 2.7

1. Determine the adjacency matrix and the distance matrix of each of the digraphs in Figure 2.6.

2. The **diameter** of a digraph is the largest of the distances between the vertices. Determine the diameters of the digraphs of Figure 2.6.

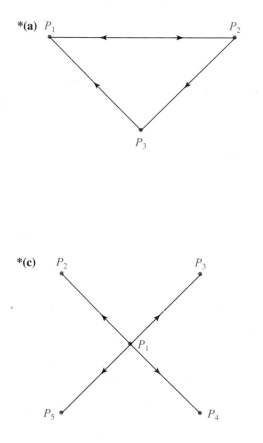

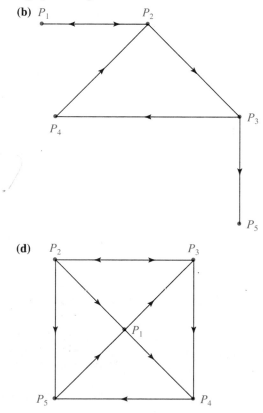

Figure 2.6 (a–d)

3. Sketch the digraphs that have the following adjacency matrices.

*(a)
$$\begin{bmatrix} 0 & 1 & 1 & 1 \\ 1 & 0 & 0 & 0 \\ 1 & 0 & 0 & 0 \\ 1 & 0 & 0 & 0 \end{bmatrix}$$

(b)
$$\begin{bmatrix} 0 & 1 & 1 & 0 \\ 0 & 0 & 1 & 1 \\ 0 & 0 & 0 & 1 \\ 1 & 0 & 0 & 0 \end{bmatrix}$$

(c)
$$\begin{bmatrix} 0 & 1 & 1 & 0 \\ 1 & 0 & 1 & 0 \\ 0 & 0 & 0 & 1 \\ 1 & 1 & 0 & 0 \end{bmatrix}$$

*(d)
$$\begin{bmatrix} 0 & 1 & 1 & 0 & 0 \\ 0 & 0 & 1 & 0 & 1 \\ 0 & 0 & 0 & 1 & 0 \\ 0 & 0 & 0 & 0 & 1 \\ 1 & 1 & 0 & 0 & 0 \end{bmatrix}$$

(e)
$$\begin{bmatrix} 0 & 1 & 0 & 1 & 0 \\ 1 & 0 & 1 & 0 & 0 \\ 0 & 1 & 0 & 1 & 0 \\ 1 & 0 & 0 & 0 & 0 \\ 1 & 1 & 1 & 1 & 0 \end{bmatrix}$$

4. The network given in Figure 2.7 describes a system of streets in a city downtown area. Many of the streets are one-way. Interpret the network as a digraph. Give its adjacency matrix and distance matrix.

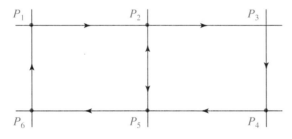

Figure 2.7

*5. Graph theory is being used in mathematical models to better understand the delicate balance of nature. Figure 2.8 gives the digraph that describes the food web of an ecological community in the Ocala National Forest, central Florida. Determine the adjacency matrix.

6. When all arcs in a network are two-way, it is customary not to include arrows, since they are not necessary. The term **graph** is then used. Scientists are using graphs to predict the chemical properties of molecules. The graphs in Figure 2.9 describe the molecular structures of butane and isobutane. Both have the same chemical formula, C_4H_{10}. Determine the adjacency matrices of these graphs. Note that the matrices are symmetric. Would you expect the adjacency matrix of any graph to be symmetric?

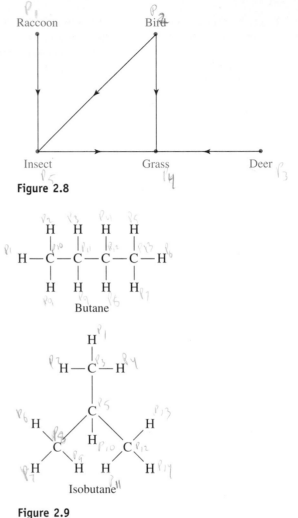

Figure 2.8

Figure 2.9

*7. The following matrix defines a communication network. Sketch the network. Determine the shortest path for sending a message from

(a) P_2 to P_5 (b) P_3 to P_2

Find the distance matrix of the digraph.

$$\begin{bmatrix} 0 & 1 & 1 & 0 & 0 \\ 1 & 0 & 1 & 0 & 0 \\ 0 & 0 & 0 & 1 & 0 \\ 0 & 0 & 0 & 0 & 1 \\ 1 & 0 & 0 & 0 & 0 \end{bmatrix}$$

8. In each of the following exercises the matrix A is the adjacency matrix for a communication network. Sketch the networks. Powers of the adjacency matrices are given. Interpret the elements that have been circled.

*(a) $A = \begin{bmatrix} 0 & 1 & 0 & 0 \\ 1 & 0 & 0 & 0 \\ 0 & 1 & 0 & 0 \\ 0 & 0 & 1 & 0 \end{bmatrix}$,

$A^2 = \begin{bmatrix} 1 & 0 & 0 & 0 \\ 0 & ① & 0 & ⓪ \\ ① & 0 & 0 & 0 \\ 0 & ① & 0 & 0 \end{bmatrix}$,

$A^3 = \begin{bmatrix} 0 & ① & 0 & 0 \\ 1 & 0 & 0 & ⓪ \\ 0 & ① & 0 & 0 \\ ① & 0 & 0 & 0 \end{bmatrix}$

(b) $A = \begin{bmatrix} 0 & 0 & 1 & 0 \\ 0 & 0 & 1 & 0 \\ 0 & 1 & 0 & 0 \\ 0 & 0 & 1 & 0 \end{bmatrix}$,

$A^2 = \begin{bmatrix} 0 & 1 & 0 & 0 \\ 0 & 1 & 0 & ⓪ \\ 0 & 0 & ① & 0 \\ 0 & ① & 0 & 0 \end{bmatrix}$,

$A^3 = \begin{bmatrix} 0 & 0 & ① & 0 \\ 0 & 0 & 1 & 0 \\ 0 & ① & 0 & ⓪ \\ 0 & 0 & 1 & 0 \end{bmatrix}$

(c) $A = \begin{bmatrix} 0 & 1 & 0 & 1 \\ 0 & 0 & 1 & 0 \\ 1 & 0 & 0 & 0 \\ 0 & 0 & 1 & 0 \end{bmatrix}$,

$A^2 = \begin{bmatrix} 0 & 0 & ② & 0 \\ ① & 0 & 0 & 0 \\ 0 & 1 & 0 & ① \\ 1 & 0 & 0 & 0 \end{bmatrix}$,

$A^3 = \begin{bmatrix} ② & 0 & 0 & 0 \\ 0 & 1 & 0 & ① \\ 0 & 0 & ② & 0 \\ 0 & ① & 0 & 1 \end{bmatrix}$

(d) $A = \begin{bmatrix} 0 & 1 & 0 & 0 & 1 \\ 0 & 0 & 1 & 1 & 0 \\ 0 & 0 & 0 & 0 & 0 \\ 0 & 0 & 1 & 0 & 0 \\ 0 & 0 & 0 & 1 & 0 \end{bmatrix}$,

$A^2 = \begin{bmatrix} 0 & 0 & 1 & ② & 0 \\ 0 & 0 & ① & 0 & 0 \\ 0 & 0 & 0 & 0 & 0 \\ 0 & 0 & 0 & 0 & 0 \\ 0 & 0 & ① & 0 & 0 \end{bmatrix}$,

$A^3 = \begin{bmatrix} 0 & 0 & ② & 0 & 0 \\ 0 & 0 & 0 & 0 & 0 \\ ⓪ & 0 & 0 & 0 & 0 \\ 0 & 0 & 0 & 0 & 0 \\ 0 & 0 & 0 & 0 & 0 \end{bmatrix}$

(e) $A = \begin{bmatrix} 0 & 0 & 0 & 0 & 0 \\ 1 & 0 & 0 & 1 & 0 \\ 0 & 1 & 0 & 1 & 0 \\ 0 & 0 & 1 & 0 & 1 \\ 1 & 1 & 0 & 0 & 0 \end{bmatrix}$,

$A^2 = \begin{bmatrix} 0 & 0 & 0 & 0 & 0 \\ 0 & 0 & 1 & 0 & ① \\ ① & 0 & 1 & 1 & 1 \\ 1 & ② & 0 & 1 & 0 \\ ① & 0 & ⓪ & ① & 0 \end{bmatrix}$,

$A^3 = \begin{bmatrix} 0 & 0 & 0 & 0 & 0 \\ ① & ② & 0 & 1 & 0 \\ 1 & ② & 1 & ① & ① \\ ② & 0 & 1 & 2 & 1 \\ 0 & 0 & ① & 0 & 1 \end{bmatrix}$

9. Let A be the adjacency matrix of a digraph. What do you know about the digraph in each of the following cases?

*(a) The third row of A is all zeros.

(b) The fourth column of A is all zeros.

*(c) The sum of the elements in the fifth row of A is 3.

(d) The sum of the elements in the second column of A is 2.

(e) The second row of A^3 is all zeros.

*(f) The third column of A^4 is all zeros.

10. Consider digraphs with adjacency matrices having the following characteristics. What can you tell about each digraph?

   *(a) The second row is all zeros.

   (b) The third column is all zeros.

   *(c) The fourth row has all ones except for zero in the diagonal location.

   (d) The fifth column has all ones except for zero in the diagonal location.

   *(e) The sum of the elements in the third row is 5.

   (f) The sum of the elements in the second column is 4.

   *(g) The number of ones in the matrix is 7.

   (h) The sum of the elements in row 2 of the fourth power is 3.

   *(i) The sum of the elements in column 3 of the fifth power is 4.

   (j) The fourth row of the square of the matrix is all zeros.

   (k) The third column of the fourth power is all zeros.

*11. Let A be the adjacency matrix of a digraph. Sketch the digraph if A^2 is as follows. Use the digraph to find A^3.

$$A^2 = \begin{bmatrix} 0 & 1 & 0 & 0 \\ 0 & 0 & 1 & 0 \\ 0 & 0 & 0 & 1 \\ 0 & 1 & 0 & 0 \end{bmatrix}$$

(*Hint:* You are given all the 2-paths in the digraph.)

12. Let A be the adjacency matrix of a digraph. Sketch all the possible digraphs described by A if

$$A^2 = \begin{bmatrix} 0 & 0 & 1 & 1 \\ 0 & 0 & 0 & 0 \\ 0 & 0 & 0 & 0 \\ 0 & 0 & 0 & 0 \end{bmatrix}$$

13. The following tables represent information obtained from questionnaires given to groups of people. In each case construct the digraph that describes the leadership structure within the group. Rank the members according to their influence on the group.

*(a)
Group member	Person whose opinion valued
M_1	M_4
M_2	M_1
M_3	M_2
M_4	M_2

(b)
Group member	Person whose opinion valued
M_1	M_5
M_2	M_1
M_3	M_2
M_4	M_3
M_5	M_4

(c)
Group member	Person whose opinion valued
M_1	M_4
M_2	M_1 and M_5
M_3	M_2
M_4	M_3
M_5	M_1

(d)
Group member	Person whose opinion valued
M_1	M_5
M_2	M_1
M_3	M_1 and M_4
M_4	M_5
M_5	M_3

14. The following matrices describe the relationship "friendship" between groups of people. $a_{ij} = 1$ if M_i is a friend of M_j; $a_{ij} = 0$ otherwise. Draw the digraphs that describe these relationships. Note that all the matrices are symmetric. What is the significance of this symmetry? Can such a relationship be described by a matrix that is not symmetric?

(a) $A = \begin{bmatrix} 0 & 1 & 0 & 0 & 0 \\ 1 & 0 & 0 & 0 & 1 \\ 0 & 0 & 0 & 1 & 0 \\ 0 & 0 & 1 & 0 & 0 \\ 0 & 1 & 0 & 0 & 0 \end{bmatrix}$

(b) $A = \begin{bmatrix} 0 & 0 & 0 & 0 & 0 & 1 \\ 0 & 0 & 1 & 0 & 0 & 0 \\ 0 & 1 & 0 & 0 & 1 & 0 \\ 0 & 0 & 0 & 0 & 0 & 1 \\ 0 & 0 & 1 & 0 & 0 & 0 \\ 1 & 0 & 0 & 1 & 0 & 0 \end{bmatrix}$

*15. A structure in a digraph that is of interest to social scientists is a clique. A **clique** is the largest subset of a digraph consisting of three or more vertices, each pair of which is mutually related. The application of this concept to the relationship "friendship" is immediate: three or more people form a clique if they are all friends and if they do not all have any mutual friendships with any single person outside that set. Give an example of a digraph that contains a clique.

16. Prove that the adjacency matrix of a digraph is necessarily square.

*17. Let A be the adjacency matrix of a digraph. The matrix A^t is a square matrix consisting of zeros and ones. It is also the adjacency matrix of a digraph. How are the digraphs of A and A^t related?

18. If the adjacency matrix of a digraph is symmetric, what does this tell you about the digraph?

*19. Prove that the shortest path from one vertex of a digraph to another vertex cannot contain any repeated vertices.

20. In a graph with n vertices what is the greatest possible distance between two vertices?

*21. Let A be the adjacency matrix of a communication digraph. Let $C = AA^t$. Show that c_{ij} = number of stations that can receive a message directly from both stations i and j.

*22. The **reachability matrix** R of the digraph is defined as follows:

$$r_{ij} = \begin{cases} 1 & \text{if there is a path from vertex } P_i \text{ to } P_j \\ 1 & \text{if } i = j \\ 0 & \text{if there is no path from } P_i \text{ to } P_j \end{cases}$$

Determine the reachability matrices of the digraphs of Exercise 8.

*23. The reachability matrix of a digraph can be constructed using information from the adjacency matrix and its various powers. How many powers of the adjacency matrix of a digraph having n vertices would have to be calculated to get all the information needed?

24. (a) If the adjacency matrix of a digraph is symmetric, does this mean that the reachability matrix is symmetric?

   *(b) If the reachability matrix of a digraph is symmetric, does this mean that the adjacency matrix is symmetric?

25. Let R be the reachability matrix of a communication digraph. Let $r(i)$ be the sum of the elements of row i of R, and let $c(j)$ be the sum of the elements of column j of R. What information about the digraph do $r(i)$ and $c(j)$ give?

26. Let R be the reachability matrix of a digraph. What information about the digraph does the element in row i, column j of R^2 give?

27. Let R be the reachability matrix of a digraph. What information about the digraph does R^t give?

28. The adjacency matrix A and reachability matrix R of a digraph are both made up of elements that are either zero or one. Can

(a) A and R have the same number of ones?

(b) A have more ones than R?

*(c) A have fewer ones than R?

29. The ILLIAC IV supercomputer at NASA Ames Research Center had 64 processors under the control of one control unit. The processors are networked together. If one imagines the processors as elements of an 8×8 matrix, then each processor at location (i, j) can communicate with its immediate neighbors, the processors in locations $(i, j - 1)$, $(i - 1, j)$, $(i, j + 1)$, and $(i + 1, j)$. The processors in the first row communicate with those in the last row, in wraparound fashion. Those in the first column communicate with those in the last column in wraparound fashion. See Figure 2.10. Determine the maximum distance between processors in this network.

Routing network in the
ILLIAC 1V supercomputer
(64 microprocessors)

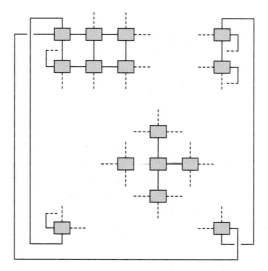

Figure 2.10

Review Exercises Chapter 2

1. Let $A = \begin{bmatrix} 2 & 0 \\ 7 & -1 \end{bmatrix}$, $B = \begin{bmatrix} 7 & 0 \\ -1 & 3 \end{bmatrix}$,

   $C = \begin{bmatrix} 6 & -1 & 3 \\ 5 & 0 & -2 \end{bmatrix}$, $D = \begin{bmatrix} 6 \\ 4 \end{bmatrix}$. Compute the following (if they exist).

   (a) $2AB$

   (b) $AB + C$

   (c) $BA + AB$

   (d) $AD - 3D$

   (e) $AC + BC$

   (f) $2DA + B$

2. Let A be a 2×2 matrix, B a 2×2 matrix, C a 2×3 matrix, D a 3×2 matrix, and E a 3×1 matrix. Determine which of the following matrix expressions exist, and give the size of the resulting matrices when they do exist.

   (a) AB

   (b) $(A^2)C$

   (c) $B^3 + 3(CD)$

   (d) $DC + BA$

   (e) $DA - 2(DB)$

   (f) $C - 3D$

   (g) $3(BA)(CD) + (4A)(BC)D$

3. If $A = \begin{bmatrix} 1 & -3 \\ 0 & 4 \end{bmatrix}$, $B = \begin{bmatrix} 1 & 2 & -3 \\ 5 & 0 & -1 \end{bmatrix}$, and

   $C = \begin{bmatrix} 2 & -4 & 5 \\ 7 & 1 & 0 \end{bmatrix}$, determine the following elements of $D = 2AB - 3C$, without computing the complete matrix.

   (a) d_{12} (b) d_{23}

4. Determine the product AB of the following 2×2 matrices using Strassen's algorithm.

   $$A = \begin{bmatrix} 2 & 3 \\ 1 & -5 \end{bmatrix}, \quad B = \begin{bmatrix} 1 & 3 \\ 2 & -4 \end{bmatrix}$$

5. If $A = \begin{bmatrix} 3 & 1 \\ 0 & 2 \end{bmatrix}$, $B = \begin{bmatrix} -2 & 1 \\ 3 & 1 \end{bmatrix}$, and $C = \begin{bmatrix} 0 & 1 \\ 3 & 2 \end{bmatrix}$, compute each of the following.

   (a) $(A')^2$

   (b) $A' - B^2$

   (c) $AB^3 + 2C^2$

   (d) $A^2 - 3A + 4I_2$

6. If A and B are matrices of the given sizes, determine the number of scalar multiplications necessary to compute AB.

   (a) $A\ 2 \times 2,\ B\ 2 \times 6$

   (b) $A\ 4 \times 2,\ B\ 2 \times 3$

   (c) $A\ 1 \times 7,\ B\ 7 \times 25$

   (d) $A\ 9 \times 5,\ B\ 5 \times 11$

7. Determine the inverse of each of the following matrices, if it exists, using the method of Gauss-Jordan elimination.

   (a) $\begin{bmatrix} 1 & 4 \\ 2 & -1 \end{bmatrix}$

   (b) $\begin{bmatrix} 0 & 3 & 3 \\ 1 & 2 & 3 \\ 1 & 4 & 6 \end{bmatrix}$

   (c) $\begin{bmatrix} 1 & 2 & 3 \\ 2 & 5 & 3 \\ 1 & 0 & 8 \end{bmatrix}$

8. Use the matrix inverse method to solve the following system of equations.

   $$\begin{aligned} x_1 + 3x_2 - 2x_3 &= 1 \\ 2x_1 + 5x_2 - 3x_3 &= 5 \\ -3x_1 + 2x_2 - 4x_3 &= 7 \end{aligned}$$

9. Find A such that $3A^{-1} = \begin{bmatrix} 5 & -6 \\ -2 & 3 \end{bmatrix}$.

10. Verify the associative property of multiplication, $A(BC) = (AB)C$.

11. Simplify $A(AB + A^2) - B(A^2 + AB) + 3ABA - 4AB^2$.

12. If n is a nonnegative integer and c is a scalar, prove that $(cA)^n = c^n A^n$.

13. Let A be a matrix such at $AA' = 0$. Show that $A = 0$.

14. A matrix is said to be **normal** if $AA' = A'A$. Prove that all symmetric matrices are normal.

15. A matrix A is **idempotent** if $A^2 = A$. Prove that if A is idempotent, then A' is also idempotent.

16. A matrix A is **nilpotent** if $A^p = 0$ for some positive integer p. The least such integer p is called the **degree of nilpotency**. Prove that if A is nilpotent, then A' is also nilpotent with the same degree of nilpotency.

17. Prove that if A is symmetric and invertible, then A^{-1} is also symmetric.

18. Prove that a matrix with a row of zeros or a column of zeros has no inverse.

19. Compute $A + B$ and AB for the following matrices, and show that A is hermitian.

    $$A = \begin{bmatrix} 2 & 4 - 3i \\ 4 + 3i & -1 \end{bmatrix}, \quad B = \begin{bmatrix} 3 + i & 1 - 2i \\ 2 + 7i & -2 + i \end{bmatrix}$$

20. Prove that every real symmetric matrix is hermitian.

21. The following matrix A describes the pottery contents of various graves. Determine possible chronological orderings of the graves and then the pottery types.

$$A = \begin{bmatrix} 1 & 0 & 0 & 0 \\ 1 & 1 & 0 & 0 \\ 0 & 0 & 1 & 1 \\ 0 & 0 & 0 & 1 \\ 0 & 1 & 1 & 0 \end{bmatrix}$$

22. The following stochastic matrix P gives the probabilities for a certain region of college- and noncollege-educated households having at least one college-educated child. By college-educated we understand that at least one parent is college educated, while by noncollege-educated we mean that neither parent is college educated.

household

    coll ed.   noncoll ed.

$$P = \begin{bmatrix} 0.9 & 0.25 \\ 0.1 & 0.75 \end{bmatrix} \begin{array}{l} \text{college educated} \\ \text{noncollege educated} \end{array} \text{child}$$

If there are currently 300,000 college-educated households and 750,000 noncollege-educated households, what is the predicted distribution for two generations hence? What is the probability that a couple that has no college education will have at least one grandchild with college education?

23. Let A be the adjacency matrix of a digraph. What do you know about the digraph in each of the following cases?

(a) All the elements in the fourth column of A are zero.

(b) The sum of the elements in the third row of A is 2.

(c) The sum of the elements in the second row of A^3 is 4.

(d) The third column of A^2 is all zeros.

(e) The element in the (4, 4) location of A^3 is 2.

(f) The number of nonzero elements in A^4 is 3.

MATLAB Discussion

M.3   Matrix Operations (Sections 2.1, 2.2, 2.3)

MATLAB can be used to perform matrix algebra. The symbols for standard operations are

Addition: +   Subtraction: −   Multiplication: ∗   Power: ^   Transpose: '

Let us enter the matrices $A = \begin{bmatrix} 3 & 5 \\ 1 & 2 \end{bmatrix}$, $B = \begin{bmatrix} 1 & 0 \\ 2 & -1 \end{bmatrix}$, and $C = \begin{bmatrix} 3 & -2 \\ 0 & -1 \end{bmatrix}$ into the workspace. We shall work with these matrices.

»A = [3   5;1   2]; B = [1   0;2   −1]; C = [3   −2;0   −1];

Example 1   Compute AB.

»A∗B
ans =
     13 −5
      5 −2

Example 2   Compute A^3.

»A^3
ans =
     67 120
     24   43

Exercise 1

(a) Compute $A + B$.

(b) Compute B^{64}. (Explore various powers of matrix B.)

(c) Use your answer to (b) to predict the value of B^{129}. Use MATLAB to check your answer.

(d) Compute $(B∗A)^t$.

(e) Compute $(P + P^t)$ for various matrices. Make a conjecture and prove it.

(f) Enter various square matrices P and compute PP^t for each matrix. Examine the outputs and make a conjecture. Prove this conjecture.

Algebraic Expressions

Algebraic expressions involving matrices can be computed. The hierarchy of operations is ^, ∗, +, and −.

Example 3   Compute $2A - B^3 + 4C$ for the matrices A, B, and C (page 122) in the workspace.

```
»2*A−B^3+4*C
ans =
       17   2
        0   1
```

Exercise 2   Compute the following expressions, using the matrices A, B, and C on page 122.

(a) $A^2 + 3B - 2AB$ **(b)** $4A^3 + BC^t$

$$A = \begin{bmatrix} 3 & 5 \\ 1 & 2 \end{bmatrix}, \quad B = \begin{bmatrix} 1 & 0 \\ 2 & -1 \end{bmatrix}, \quad C = \begin{bmatrix} 3 & -2 \\ 0 & -1 \end{bmatrix}$$

Matrix Elements

Individual matrix elements can be listed and manipulated.

Example 4   List the element in row 2, column 1 of B. Change this element to 3.

```
»B(2,1)
ans =
        2

»B(2,1)=3
B=
    1    0
    3   -1
```

Exercise 3   Enter the matrix $P = \begin{bmatrix} 1 & 2 & -4 \\ 5 & 7 & 0 \end{bmatrix}$. Change the element p_{23} to -3.

This technique can be used to correct input in the elements of a matrix, without having to reenter the whole matrix.

*Exercise 4   Let $P = \begin{bmatrix} 0 & -2 & 3 \\ 1 & 4 & 5 \end{bmatrix}$ and $Q = \begin{bmatrix} 5 & 2 & -1 \\ 0 & 6 & 3 \end{bmatrix}$.

Compute $4p_{13} - 5q_{23} + 2(p_{22})^2$.

Submatrices, Rows, and Columns of a Matrix

In work involving matrices it is often important to be able to manipulate rows and columns of matrices, and also to work with submatrices. MATLAB has a special function for selecting a submatrix of a given matrix. The functions for selecting a row or column become special cases of this function.

A(i : j, p : q)—**select the submatrix of** *A* **lying from row** *i* **to row** *j,* **and column**
p **to column** *q.*
A(i, :)—**select row** *i* **of** *A*
A(:, p)—**select column** *p* **of** *A*

Example 5   Consider the following matrix *A*.

(a)   Let *X* be the second row of *A*, and *Y* be the third column. Compute *XAY.*

(b)   Let *B* be the matrix lying from row 2 to row 4, column 1 to column 3 of *A*.
Compute B^5.

(c)   Let *C* be the matrix consisting of columns 1, 2, and 4 of *A*. Construct *C*.

$$A = \begin{bmatrix} 1 & 3 & 2 & 5 \\ 0 & 3 & 5 & 1 \\ -6 & 3 & 5 & 9 \\ -8 & -2 & 3 & 4 \end{bmatrix}$$

(a)   »A = [1   3   2   5;0   3   5   1;−6   3   5   9;−8   −2   3   4];
     »X = A(2, :); Y = A(:, 3);
     »X*A*Y
     ans =
         405

(b)   »B = A(2 : 4, 1 : 3);
     »B ^ 5
     ans =
        29778   12117   1385
        11406   18135   19205
       −23600   13702   31203

(c)   »C = [A(:, 1)   A(:, 2)   A(:, 4)]
     C=
        1   3   5
        0   3   1
       −6   3   9
       −8   −2   4

Exercise 5   Consider the following matrix *A*.

(a)   Let *X* be the third row of *A*, and *Y* be the fourth column. Compute $XA^tY.$

(b)   Let *B* be the matrix lying from row 1 to row 3, column 2 to column 4 of *A*.
Compute B^7.

(c) Let C be the matrix consisting of rows 1, 3, and 4 of A. Compute CA.

$$A = \begin{bmatrix} 4 & -2 & 5 & 8 \\ 0 & 1 & 2 & 7 \\ -3 & 2 & 5 & 4 \\ 1 & 0 & -2 & 3 \end{bmatrix}$$

M.4 Computational Considerations (Section 2.2)

Most matrix computation is now done on computers. Efficient algorithms are important—they mean time and accuracy. In computing a product such as ABC, we can use $(AB)C$ or $A(BC)$ since multiplication is *associative*. This property is useful since the number of scalar multiplications involved one way is usually less than the number involved the other way. (Scalar multiplication is more time-consuming than addition; thus analyses of efficiency are done in terms of multiplications.)

Example 1 A is 2×2, B is 2×3, and C is 3×1. Compute the number of multiplications involved in performing $(AB)C$ and $A(BC)$.

»ops   {author-defined function for computing # multiplications}
Give # matrices (3 or 4): 3
rows in A: 2 ... cols in C: 1
ways =
　　　　AB×C   A×BC
numbers =
　　　　　18   10

Thus the number of scalar multiplications in $(AB)C$ is 18, while in $A(BC)$ it is 10. When products of many large matrices are involved the savings can be substantial.

Exercise 1   Verify by hand that the above multiplications are indeed 18 and 10.

Exercise 2   Let A be an $m \times r$ matrix, B $r \times n$, and C $n \times s$. Determine a formula that gives the number of scalar multiplications in computing AB. Use this result to derive a formula that gives the number of scalar multiplications in computing $(AB)C$ and $A(BC)$. Check your formula with your MATLAB results in the following exercise.

*Exercise 3   Use MATLAB to find the most and least efficient ways of performing the following products. A is 5×14, B is 14×87, C is 87×3, D is 3×42.

(a) ABC (b) $ABCD$

Exercise 4   Use MATLAB to find the most and least efficient ways of performing the following products. A is 2×45, B is 45×45, C is 45×3.

(a) ABC (b) AB^2 (c) AB^3 (d) AB^2C

Exercise 5   Let A be an $m \times r$ matrix, B $r \times n$, and C $n \times s$. Determine a formula that gives the number of scalar additions in computing AB. Use this result to derive a formula for the number of scalar additions in computing $(AB)C$ and $A(BC)$. Copy the function **ops** onto your own diskette. Modify it to take into account additions as well as multiplications. Test your formula and function on a 2×2 matrix A, a 2×3 matrix B, and a 3×1 matrix C.

M.5   Inverse of a Matrix (Section 2.4)

MATLAB has a matrix inverse function **inv**(A).

Example 1   Find the inverse function of the matrix $A = \begin{bmatrix} 3 & 5 \\ 1 & 2 \end{bmatrix}$.

```
»A = [3   5;1   2];
»inv(A)
ans =
        2.0000   -5.0000
       -1.0000    3.0000
```

(MATLAB uses a default "short" format of 5 decimal digits. Computations use IEEE arithmetic, and are to 2^{-52} [approx. 2.22×10^{-16}].) Try some matrix outputs using other formats available in the FORMAT menu bar.

Exercise 1   Use MATLAB to prove that $C = \begin{bmatrix} 2 & -5 \\ -1 & 3 \end{bmatrix}$ is the inverse of $A = \begin{bmatrix} 3 & 5 \\ 1 & 2 \end{bmatrix}$ by showing that $AC = CA = I_2$.

Exercise 2   Enter the matrix $B = \begin{bmatrix} 1 & 0 \\ 2 & -1 \end{bmatrix}$. Compute the first four powers of B.

(a)   By looking at these results, why do you expect $(B^2)^{-1} = B^2$? Compute $(B^2)^{-1}$ using MATLAB to verify that this is indeed the case.

(b)   Predict the value of $(B^{194})^{-1}$. Use MATLAB to check your answer.

Example 2   It is often preferable to use rational numbers rather than decimal numbers. The function **frac**(x) returns the rational approximation to x. For example,

$$\text{frac}(3.375) = 27/8$$

Let us compute the inverse of the matrix $A = \begin{bmatrix} 1/2 & 3 \\ -3/5 & 6 \end{bmatrix}$ and express the answer with rational elements.

A = [1/2   3;−3/5   6];   {the matrix can be entered
»inv(A)   with rational elements}
ans =
   1.25000000000000   −0.62500000000001   {using "long" format}
   0.12500000000000   0.10416666666667
»frac(inv(A))   {frac for "fraction"}
ans =
   5/4   −5/8   {inverse in terms of
   1/8   5/48   rational numbers}

Exercise 3   Compute the inverse of each of the following matrices. Express your answers in terms of both decimal and rational numbers.

(a) $A = \begin{bmatrix} 1 & 4 \\ 3 & 5 \end{bmatrix}$ **(b)** $B = \begin{bmatrix} 1 & 9 \\ 11 & 13 \end{bmatrix}$

Similar Matrices

Two square matrices of the same size, A and B, are said to be **similar** if there exists an invertible matrix C such that $B = C^{-1}AC$. The transformation of A into B is called a **similarity transformation**.

Exercise 4   Use MATLAB to perform the similarity transformation $B = C^{-1}AC$, where

(a) $A = \begin{bmatrix} 7 & -10 \\ 3 & -4 \end{bmatrix}$ and $C = \begin{bmatrix} 2 & 5 \\ 1 & 3 \end{bmatrix}$ **(b)** $A = \begin{bmatrix} 1 & 2 \\ -1 & 3 \end{bmatrix}$ and $C = \begin{bmatrix} 2 & 5 \\ 1 & 3 \end{bmatrix}$

(c) $A = \begin{bmatrix} 1 & 2 & 3 \\ 0 & -1 & 2 \\ 1 & 1 & 0 \end{bmatrix}$ and $C = \begin{bmatrix} 3 & 5 & -1 \\ -2 & -3 & 1 \\ -1 & -2 & 1 \end{bmatrix}$

(d) $A = \begin{bmatrix} 5 & 2 & 3 \\ 8 & 2 & 1 \\ -3 & 2 & 5 \end{bmatrix}$ and $C = \begin{bmatrix} 7 & -2 & 3 \\ 2 & 0 & 1 \\ 9 & -3 & 5 \end{bmatrix}$

Exercise 5   Write your own function **sim**(A,C) for performing a similarity transformation $B = C^{-1}AC$. The function should be on your own diskette, call matrices A and C from the matrix command window and return B to the command window. Check your function on the similarity transformations of the above exercise.

Solving Systems of Equations Using Matrix Inverse (Section 2.4)

A system of linear equations can be written in matrix form $AX = B$. If A is invertible, there is a unique solution $X = A^{-1}B$.

Example 1   Solve the system

$$3x_1 + 5x_2 = 1$$
$$x_1 + 2x_2 = 2$$

We can write this system and solution in the following matrix forms:

$$\overset{A}{\begin{bmatrix} 3 & 5 \\ 1 & 2 \end{bmatrix}} \overset{X}{\begin{bmatrix} x_1 \\ x_2 \end{bmatrix}} = \overset{B}{\begin{bmatrix} 1 \\ 2 \end{bmatrix}}, \qquad \overset{X}{\begin{bmatrix} x_1 \\ x_2 \end{bmatrix}} = \overset{A^{-1}}{\begin{bmatrix} 3 & 5 \\ 1 & 2 \end{bmatrix}^{-1}} \overset{B}{\begin{bmatrix} 1 \\ 2 \end{bmatrix}}$$

Enter A and B and compute the solution X.

```
»A= [3   5;1   2];   B= [1;2];
»X= inv(A)*B
X =
   -8.0000
    5.0000      {The solution is x_1 = -8, x_2 = 5}
```

Exercises   Solve the following systems of equations.

*1. $x_1 + 2x_2 = 6$
$x_1 + 4x_2 = 2$

2. $x_1 + 2x_2 = 4$
$2x_1 + x_2 - x_3 = 2$
$3x_1 + x_2 + x_3 = -2$

Left Division in MATLAB

Again consider the system of equations $AX = B$, where A is nonsingular. The unique solution can be conveniently found in MATLAB using left division, $X = A\backslash B$.

Example 2   Solve the system of equations

$$x_1 + x_2 + x_3 = 2$$
$$2x_1 + 3x_2 + x_3 = 3$$
$$x_1 - x_2 - 2x_3 = -6$$

We can write this system in the matrix form

$$\overset{A}{\begin{bmatrix} 1 & 1 & 1 \\ 2 & 3 & 1 \\ 1 & -1 & -2 \end{bmatrix}} \overset{X}{\begin{bmatrix} x_1 \\ x_2 \\ x_3 \end{bmatrix}} = \overset{B}{\begin{bmatrix} 2 \\ 3 \\ -6 \end{bmatrix}}$$

Enter A and B and compute the solution X.
```
»A = [1   1   1;2   3   1;1   -1   -2];   B = [2;3; -6];
»X = A\B
X =
   -1.0000
    1.0000
    2.0000      {The solution is x_1 = -1, x_2 = 1, x_3 = 2}
```

Exercises   Solve the following systems using left division.

*3. $2x_1 + x_2 = 4$
$4x_1 + 3x_2 = -6$

4. $x_1 + 2x_2 - x_3 = 2$
$x_1 + x_2 + 2x_3 = 0$
$x_1 - x_2 - x_3 = 1$

Left division can be conveniently used to solve many systems, all having the same invertible matrix of coefficients:

Example 3   Solve the systems

$$x_1 - x_2 + 3x_3 = b_1$$
$$2x_1 - x_2 + 4x_3 = b_2 \quad \text{where} \quad \begin{bmatrix} b_1 \\ b_2 \\ b_3 \end{bmatrix} = \begin{bmatrix} 8 \\ 11 \\ -11 \end{bmatrix}, \begin{bmatrix} 0 \\ 1 \\ 2 \end{bmatrix}, \begin{bmatrix} 3 \\ 3 \\ -4 \end{bmatrix} \text{ in turn}$$
$$-x_1 + 2x_2 - 4x_3 = b_3$$

Let $A = \begin{bmatrix} 1 & -1 & 3 \\ 2 & -1 & 4 \\ -1 & 2 & -4 \end{bmatrix}$ and $B = \begin{bmatrix} 8 & 0 & 3 \\ 11 & 1 & 3 \\ -11 & 2 & -4 \end{bmatrix}$.

Letting $X = A \backslash B$ in MATLAB, the columns of X give the three solutions.

```
»A = [1  -1  3;2  -1  4;-1  2  -4];
»B = [8  0  3;11  1  3;-11  2  -4];
»X = A\B
X =
     1.0000    0.0000   -2.0000
    -1.0000    3.0000    1.0000
     2.0000    1.0000    2.0000
```

The solutions to the three systems are

$$x_1 = 1, \quad x_2 = -1, \quad x_3 = 2$$
$$x_1 = 0, \quad x_2 = 3, \quad x_3 = 1$$
$$x_1 = -2, \quad x_2 = 1, \quad x_3 = 2$$

Exercise 5   Solve the systems

$$x_1 - 2x_2 + 3x_3 = b_1$$
$$x_1 - x_2 + 2x_3 = b_2 \quad \text{where} \quad \begin{bmatrix} b_1 \\ b_2 \\ b_3 \end{bmatrix} = \begin{bmatrix} 6 \\ 5 \\ 14 \end{bmatrix}, \begin{bmatrix} -5 \\ -3 \\ -8 \end{bmatrix}, \begin{bmatrix} 4 \\ 3 \\ 9 \end{bmatrix} \text{ in turn}$$
$$2x_1 - 3x_2 + 6x_3 = b_3$$

Cryptography (Section 2.4)

Matrix algebra is used in cryptography to code and decode messages. Let A be an invertible $n \times n$ matrix—the secret, encoding matrix. Assign a number to each letter in the alphabet and space between words. Write the message in numerical form. Break the enumerated message into a sequence of $n \times 1$ columns and write them as columns of a matrix X, padding any remaining locations with the "space number." The multiplication AX encodes the message. The receiver uses A^{-1}, the decoding matrix, to decode the message.

Example 1   Let us send the message "Prepare to attack" using the encoding matrix

$$A = \begin{bmatrix} -3 & -3 & -4 \\ 0 & 1 & 1 \\ 4 & 3 & 4 \end{bmatrix}$$

Let us associate each letter with its position in the alphabet. A is 1, B is 2, and so on. Let a space between words be denoted by the number 27. Thus the message becomes

P	R	E	P	A	R	E	*	T	O	*	A	T	T	A	C	K
16	18	5	16	1	18	5	27	20	15	27	1	20	20	1	3	11

Enter this form of the message as columns of a matrix X, which has three rows, as follows.

$$X = \begin{bmatrix} 16 & 16 & 5 & 15 & 20 & 3 \\ 18 & 1 & 27 & 27 & 20 & 11 \\ 5 & 18 & 20 & 1 & 1 & 27 \end{bmatrix}$$

Multiply by A to get the message in encoded matrix form Y.

»A = [−3 −3 −4;0 1 1;4 3 4];
»X = [16 16 5 15 20 3;18 1 27 27 20 11;5 18 20 1 1 27];
»Y = A*X
Y =
    −122   −123   −176   −130   −124   −150
      23   19   47   28   21   38
     138   139   181   145   144   153

The receiver gets this encoded message and uses **inv**(A) as follows to reproduce X and thus the original message.

»A = [−3 −3 −4;0 1 1;4 3 4];
»Y = [−122 −123 −176 −130 −124 −150;23 19 47 28 21 38;
        138 139 181 145 144 153];
»X = inv(A)*Y
X =
    16   16   5   15   20   3
    18   1 27   27   20 11
     5   18 20   1   1 27

The columns of this matrix, written in linear form, give the original message.

16	18	5	16	1	18	5	27	20	15	27	1	20	20	1	3	11
P	R	E	P	A	R	E	*	T	O	*	A	T	T	A	C	K

Readers who are interested in an introduction to cryptography are referred to *Secret Codes and Ciphers* by Bernice Kohn, Prentice-Hall, Inc., 1968. For a more advanced book that gives an overview of recent developments and a guide to literature in cryptography, see *Codes and Cryptography* by Dominic Welsh, Oxford, 1988.

Exercise 1   Decode the following messages, which were sent using the above encoding matrix *A*.

(a) {398, 25, −104, 359, 411, 12, −124, 409, 280, 41, −67, 243, 122, −9, −50, 152, 346, 19, −89, 309}.

(b) {392, 17, −108, 367, 106, 9, −25, 90, 181, −27, −69, 210, 238, −11, −86, 268, 83, −43, −57, 152, 191, 75, −11, 92}

Leontief I/O Model (Section 2.5)

Let *A* be the I/O matrix that describes the interdependence of industries in an economy. Let *X* describe industrial production and *D* the demand of the industries and the open sector. Then

$$X \quad = \quad AX \quad + \quad D$$

total output   interindustry portion   open sector portion

The output level required to meet the demands is

$$X = (I - A)^{-1} D$$

Example 1   An economy of three industries has the following I/O matrix *A* and demand matrix *D*. Find the output levels necessary to meet these demands. Units are millions of dollars.

$$A = \begin{bmatrix} 0.2 & 0.2 & 0.3 \\ 0.5 & 0.5 & 0 \\ 0 & 0 & 0.2 \end{bmatrix}, D = \begin{bmatrix} 9 \\ 12 \\ 16 \end{bmatrix}$$

The identity *n* × *n* matrix in MATLAB is **eye**(n). Thus

$$X = (I - A)^{-1} D \text{ is written } X = \text{inv}(\text{eye}(3) - A)*D$$

Enter the following sequence:

»A = [.2   .2   .3;.5   .5   0;0   0   .2];   D = [9;12;16];
»X = inv(eye(3)−A)*D
X =
    33.0000
    57.0000
    20.0000

The necessary output levels in dollars are 33, 57, and 20 million.

*Exercise 1   An economy of three industries has the following I/O matrix A and demand matrix D. Find the output levels necessary to meet these demands. Units are millions of dollars.

$$A = \begin{bmatrix} 0.2 & 0.2 & 0 \\ 0.4 & 0.4 & 0.6 \\ 0.4 & 0.1 & 0.4 \end{bmatrix}, D = \begin{bmatrix} 36 \\ 72 \\ 36 \end{bmatrix}$$

Exercise 2   An economy of two industries has the following I/O matrix A and demand matrix D. Find the output levels necessary to meet these demands. Units are millions of dollars (*Hint:* Combine D_1, D_2, and D_3 into one matrix D.)

$$A = \begin{bmatrix} 0.2 & 0.6 \\ 0.4 & 0.1 \end{bmatrix}, D_1 = \begin{bmatrix} 24 \\ 12 \end{bmatrix}, D_2 = \begin{bmatrix} 8 \\ 6 \end{bmatrix}, D_3 = \begin{bmatrix} 0 \\ 12 \end{bmatrix}$$

*Exercise 3   Consider the following economy consisting of three industries. The output levels of the industries are given. Determine the amounts available for the open sector from each industry.

$$A = \begin{bmatrix} 0.10 & 0.10 & 0.20 \\ 0.20 & 0.10 & 0.30 \\ 0.40 & 0.30 & 0.15 \end{bmatrix}, X = \begin{bmatrix} 6 \\ 4 \\ 5 \end{bmatrix}$$

Exercise 4   An economy of two industries has the following I/O matrix A and demand matrix D. Apply the Leontief model to this situation. Why does it not work? Explain the situation from both a mathematical and practical viewpoint. Extend your results to the general model

$$A = \begin{bmatrix} 0.8 & 0.4 \\ 0.2 & 0.6 \end{bmatrix}, D = \begin{bmatrix} 20 \\ 32 \end{bmatrix}$$

M.9   Markov Chains (Sections 2.6, 4.8)

Let P be the matrix of transition probabilities for a Markov chain. Let X be the initial distribution. Then

$$\text{distribution after } n \text{ steps } X_n = P^n X$$

probability of starting in state i and being in state j after n steps is $p_{ji}^{(n)}$

If P is **regular** (some power of P has all positive elements), then long-term trends are

$$X_n \to X \quad \text{and} \quad P^n \to P$$

Example 1   Annual population movement between U.S. cities and suburbs in 1985 is described by the following stochastic matrix P. The population distribution in 1985 is given by the matrix X, in units of 1 million.

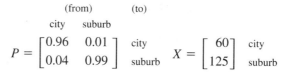

$$P = \begin{bmatrix} 0.96 & 0.01 \\ 0.04 & 0.99 \end{bmatrix} \begin{matrix} \text{city} \\ \text{suburb} \end{matrix} \qquad X = \begin{bmatrix} 60 \\ 125 \end{bmatrix} \begin{matrix} \text{city} \\ \text{suburb} \end{matrix}$$

(a) Predict the population distribution for the next 5 years.

(b) What is the probability that a person who was in the city in 1985 is living in the suburbs in 1990?

(a) »**P = [.96   .01;.04   .99];   X = [60;125];**
»**Markov(P,X)**   {author-defined function to compute distributions}

Give number of steps: 5
Display all steps? y/n: y   {option is available for displaying all steps or for going immediately to the final answer. Latter is useful when long-term predictions are desired.}

   step =   . . .   **step =**
      1   **5**
distribution =   **distribution =**
      58.85 {City in 1986}   **54.80** {City in 1990}
     126.15 {Suburb in 1986}   **130.20** {Suburb in 1990}

Bar graphs? y/n: y   {option of drawing bar graphs
wait   is available for two states}

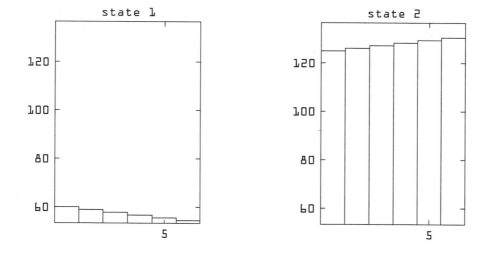

Note: Use "Enter" and "shg" (or "figure (gcf)") to alternate between command and graph windows on IBMs.

The predicted population distributions (in millions) for 1990 are $\begin{bmatrix} 54.8 \\ 130.20 \end{bmatrix} \begin{matrix} \text{city} \\ \text{suburb} \end{matrix}$.

City populations are steadily decreasing while suburban populations are rising.

(b)　1990 is 5 years after 1985. We compute P^5.

```
»P = [.96  .01;.04  .99];
»P^5
ans =
   0.8190   0.0452
   0.1810   0.9548
```

The probability that a person who was in the city (state 1) in 1985 is living in the suburbs (state 2) in 1990 (5 years later) is $p_{21}^{(5)} = 0.1810$.

*Exercise 1　Construct a model of population flow between metropolitan and non-metropolitan areas of the United States, given that their respective populations in 1985 were 185 and 55 million. The probabilities are given by the matrix

$$\begin{array}{cc} & \text{(from)} \\ \begin{array}{cc} \text{metro} & \text{nonmetro} \end{array} & \text{(to)} \\ \begin{bmatrix} 0.99 & 0.02 \\ 0.01 & 0.98 \end{bmatrix} & \begin{array}{l} \text{metro} \\ \text{nonmetro} \end{array} \end{array}$$

(a)　Predict the population distributions of metropolitan and nonmetropolitan areas for the years 1986 through 1990.

(b)　If a person was living in a metropolitan area in 1985, what is the probability that the person was still living in a metropolitan area in 1989?

Exercise 2　Consider the city/suburb model of this section.

(a)　By computing $P^n X$ using MATLAB, for various increasing values of n (e.g., $n = 50, 100, 150$, etc.) show that the population distributions of the cities and suburbs will approach 37 and 148 million, unless conditions change. Arrive at these same numbers using mathematics.

(b)　Let $P^n \rightarrow P$. Determine P and use it to show that long-term probabilities of living in city and suburbia are 0.2 and 0.8.

Exercise 3　Consider the city/suburb model of this section.

(a)　The population of the United States increased by 1% per annum during the period 1985 to 1990. Allow for this increase in your model, and use MATLAB to predict the populations for the years 1986 to 1990.

(b)　Assume that the populations of cities due to births, deaths, and immigration increased by 1.2% during the period 1985 to 1990 and that the population of the suburbs increased by 0.8% due to these factors. Allow for these increases in your model and predict the populations for the years 1986 to 1990.

Exercise 4 Consider the population movement model of this section.

(a) Interchange the columns of P to get a matrix Q. Look at the bar graphs for the behavior of city and suburban populations for 5 years, and compare them with those obtained with transition matrix P. Which model is most realistic, the P or Q model?

(b) Consider other such matrices P and corresponding matrices Q. Examine the bar graphs and make a hypothesis. Prove or disprove your hypothesis.

Exercise 5 Consider the above population movement model for flow between cities and suburbs. Determine the population distributions for 1980 to 1984—prior to 1985. Is the chain going from 1985 into the past a Markov chain? What are the characteristics of the matrix that takes one from distribution to distribution?

*Exercise 6 Construct a model of population flows between cities, suburbs, and nonmetropolitan areas of the United States. Their respective populations in 1985 were 60, 125, and 55 million. The stochastic matrix giving the probabilities of the moves is

$$\begin{array}{cccc} & \multicolumn{3}{c}{\text{(from)}} \\ \text{city} & \text{suburb} & \text{nonmetro} & \text{(to)} \\ \begin{bmatrix} 0.96 & 0.01 & 0.015 \\ 0.03 & 0.98 & 0.005 \\ 0.01 & 0.01 & 0.98 \end{bmatrix} & & & \begin{array}{c} \text{city} \\ \text{suburb} \\ \text{nonmetro} \end{array} \end{array}$$

(a) Predict the population distributions for the years 1986 to 1990.

(b) Determine the long-term population distributions.

(c) Determine the long-term probabilities of living in the city, suburbs, and nonmetropolitan areas.

M.10 Digraphs (Section 2.7)

Let A be the adjacency matrix of a digraph. Let $a_{ij}^{(n)}$ be the element in row i, column j of A^n. Then

$$a_{ij}^{(n)} = \text{number of paths of length } n \text{ from vertex } i \text{ to vertex } j$$

The *distance* from one vertex to another is the length of the shortest path from that vertex to the other. The function **digraph**(A) sketches a digraph and uses powers of the adjacency matrix A to give the distance from one vertex to the other. It also gives the number of distinct shortest paths.

Example 1 Consider the digraph defined by the following adjacency matrix A. Sketch the digraph. Determine the distance from vertex 4 to vertex 1. Find the number of paths from vertex 4 to vertex 1 that give this distance.

$$A = \begin{bmatrix} 0 & 1 & 0 & 0 & 0 \\ 1 & 0 & 0 & 0 & 1 \\ 0 & 1 & 0 & 0 & 1 \\ 0 & 0 & 1 & 0 & 1 \\ 0 & 1 & 0 & 1 & 0 \end{bmatrix}$$

A = [0 1 0 0 0;1 0 0 0 1;0 1 0 0 1;0 0 1 0 1;0 1 0 1 0];
»digraph(A)　　{author-defined function to sketch digraph and find distance}
Want distances between vertices? y/n: y
give 1st vertex: 4
give 2nd vertex: 1

distance =
**　　　　3**

numberpaths =
**　　　　2**

Again? y/n: n

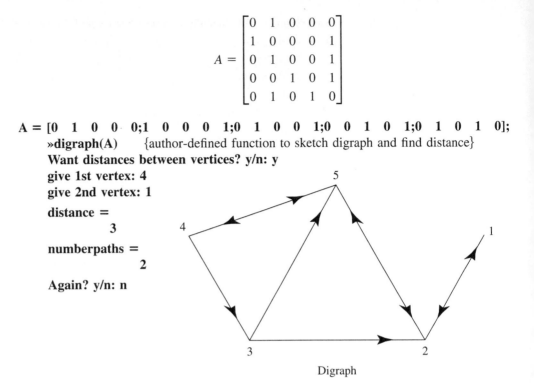

Digraph

Note: Use "Enter" and "shg" (or "figure (gcf)") to alternate between command and graph windows on IBMs.

The distance from vertex 4 to vertex 1 is 3. There are two distinct paths that give this distance. From the sketch we see that they are

$$4 \to 3 \to 2 \to 1 \quad \text{and} \quad 4 \to 5 \to 2 \to 1$$

The user has the opportunity to repeat for other vertices, thus finding the distance matrix.

*Exercise 1　　Consider the digraph defined by the following adjacency matrix A. Sketch the digraph. Determine the distance from vertex 2 to vertex 1. Find the number of paths from vertex 2 to vertex 1 that give this distance.

$$A = \begin{bmatrix} 0 & 1 & 0 & 0 & 0 \\ 0 & 0 & 0 & 1 & 0 \\ 1 & 0 & 0 & 0 & 0 \\ 0 & 0 & 1 & 0 & 1 \\ 1 & 0 & 0 & 1 & 0 \end{bmatrix}$$

Exercise 2   The following tables represent information obtained from questionnaires given to groups of people. Construct the digraphs that describe the leadership structures within the groups. Compute the distance matrix and rank the members according to their influence on each group. Who should M_3 strive to influence directly to become the most influential person in the group?

(a)	Group member	Person whose opinion valued		(b)	Group member	Person whose opinion valued
	M_1	M_4			M_1	M_5
	M_2	M_1 and M_5			M_2	M_1
	M_3	M_2			M_3	M_1 and M_4
	M_4	M_3			M_4	M_5
	M_5	M_1			M_5	M_3

chapter 3

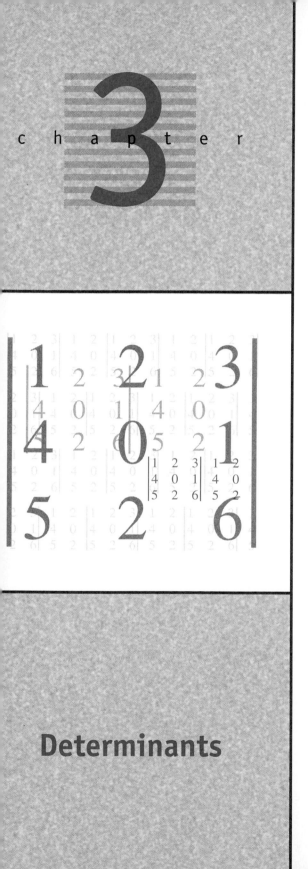

Associated with every square matrix is a number called its **determinant**. The determinant of a matrix is a tool used in many branches of mathematics, science, and engineering. In this chapter the determinant is defined and its properties are developed. We shall see that it can give us information about the solutions of systems of linear equations and that it can also be used in a formula for the inverse of a matrix.

Determinants

3.1   Introduction to Determinants

We commence our discussion of determinants by defining the determinant of a 2 × 2 matrix.

Definition

The determinant of a 2 × 2 matrix A is denoted $|A|$ and is given by

$$\begin{vmatrix} a_{11} & a_{12} \\ a_{21} & a_{22} \end{vmatrix} = a_{11}a_{22} - a_{12}a_{21}$$

*Observe that the determinant of a 2 × 2 matrix is given by **the difference of the products of the two diagonals** of the matrix.*
The notation det(A) is also used for the determinant of A.

Example 1

Find the determinant of the matrix.

$$A = \begin{bmatrix} 2 & 4 \\ -3 & 1 \end{bmatrix}$$

Solution

Applying the above definition, we get

$$\begin{vmatrix} 2 & 4 \\ -3 & 1 \end{vmatrix} = (2 \times 1) - (4 \times (-3)) = 2 + 12 = 14$$

The determinant of a 3 × 3 matrix is defined in terms of determinants of 2 × 2 matrices. The determinant of a 4 × 4 matrix is defined in terms of determinants of 3 × 3 matrices, and so on. For these definitions we need the following concepts of minor and cofactor.

Definition

Let A be a square matrix.
*The **minor** of the element a_{ij} is denoted M_{ij} and is the determinant of the matrix that remains after deleting row i and column j of A.*
*The **cofactor** of a_{ij} is denoted C_{ij} and is given by*

$$C_{ij} = (-1)^{i+j}M_{ij}$$

Note that the minor and cofactor differ only in sign: $C_{ij} = \pm M_{ij}$.

Example 2

Determine the minors and cofactors of the elements a_{11} and a_{32} of the following matrix A.

$$A = \begin{bmatrix} 1 & 0 & 3 \\ 4 & -1 & 2 \\ 0 & -2 & 1 \end{bmatrix}$$

Solution

Applying the above definitions, we get the following.

Minor of a_{11}: $M_{11} = \begin{vmatrix} 1 & 0 & 3 \\ 4 & -1 & 2 \\ 0 & -2 & 1 \end{vmatrix} = \begin{vmatrix} -1 & 2 \\ -2 & 1 \end{vmatrix} = (-1 \times 1) - (2 \times (-2)) = 3$

Cofactor of a_{11}: $C_{11} = (-1)^{1+1}M_{11} = (-1)^2(3) = 3$

Minor of a_{32}: $M_{32} = \begin{vmatrix} 1 & 0 & 3 \\ 4 & -1 & 2 \\ 0 & -2 & 1 \end{vmatrix} = \begin{vmatrix} 1 & 3 \\ 4 & 2 \end{vmatrix} = (1 \times 2) - (3 \times 4) = -10$

Cofactor of a_{32}: $C_{32} = (-1)^{3+2}M_{32} = (-1)^5(-10) = 10$

We now define determinants of larger matrices.

Definition *The determinant of a square matrix is the sum of the products of the elements of the first row and their cofactors.*

$$\text{If } A \text{ is } 3 \times 3, \ |A| = a_{11}C_{11} + a_{12}C_{12} + a_{13}C_{13}$$
$$\text{If } A \text{ is } 4 \times 4, \ |A| = a_{11}C_{11} + a_{12}C_{12} + a_{13}C_{13} + a_{14}C_{14}$$
$$\vdots$$
$$\text{If } A \text{ is } n \times n, \ |A| = a_{11}C_{11} + a_{12}C_{12} + a_{13}C_{13} + \cdots + a_{1n}C_{1n}$$

*These equations are called **cofactor expansions** of* $|A|$.

Example 3 Evaluate the determinant of the following matrix A.

$$A = \begin{bmatrix} 1 & 2 & -1 \\ 3 & 0 & 1 \\ 4 & 2 & 1 \end{bmatrix}$$

Solution Using the elements of the first row and their corresponding cofactors, we get

$$|A| = a_{11}C_{11} + a_{12}C_{12} + a_{13}C_{13}$$
$$= 1(-1)^2 \begin{vmatrix} 0 & 1 \\ 2 & 1 \end{vmatrix} + 2(-1)^3 \begin{vmatrix} 3 & 1 \\ 4 & 1 \end{vmatrix} + (-1)(-1)^4 \begin{vmatrix} 3 & 0 \\ 4 & 2 \end{vmatrix}$$
$$= [(0 \times 1) - (1 \times 2)] - 2[(3 \times 1) - (1 \times 4)] - [(3 \times 2) - (0 \times 4)]$$
$$= -2 + 2 - 6 = -6$$

We have defined the determinant of a matrix in terms of its first row. It can be shown that the determinant can be found, according to the following rules, using any row or column.

Theorem 3.1 The determinant of a square matrix is the sum of the products of the elements of any row or column and their cofactors.

ith row expansion: $|A| = a_{i1}C_{i1} + a_{i2}C_{i2} + \cdots + a_{in}C_{in}$

jth column expansion: $|A| = a_{1j}C_{1j} + a_{2j}C_{2j} + \cdots + a_{nj}C_{nj}$

There is a useful rule that can be used to give the sign part, $(-1)^{i+j}$, of the cofactors in these expansions. The rule is summarized in the following array.

$$\begin{bmatrix} + & - & + & - & \cdots \\ - & + & - & + & \cdots \\ + & - & + & - & \cdots \\ \vdots & & & & \end{bmatrix}$$

If, for example, one expands in terms of the second row, the signs will be $- + -$ etc. The signs alternate as one goes along any row or column.

Example 4   Find the determinant of the following matrix using the second row.

$$A = \begin{bmatrix} 1 & 2 & -1 \\ 3 & 0 & 1 \\ 4 & 2 & 1 \end{bmatrix}$$

Solution   Expanding the determinant in terms of the second row, we get

$$|A| = a_{21}C_{21} + a_{22}C_{22} + a_{23}C_{23}$$
$$= -3\begin{vmatrix} 2 & -1 \\ 2 & 1 \end{vmatrix} + 0\begin{vmatrix} 1 & -1 \\ 4 & 1 \end{vmatrix} - 1\begin{vmatrix} 1 & 2 \\ 4 & 2 \end{vmatrix}$$
$$= -3[(2 \times 1) - (-1 \times 2)] + 0[(1 \times 1) - (-1 \times 4)] - 1[(1 \times 2) - (2 \times 4)]$$
$$= -12 + 0 + 6 = -6$$

Note that we have already evaluated this determinant in terms of the first row in Example 3. As is to be expected, we obtained the same value.

The computation in evaluating a determinant can be minimized by expanding in terms of the row or column that contains the most 0s. The following example illustrates this.

Example 5   Evaluate the determinant of the following 4×4 matrix.

$$\begin{bmatrix} 2 & 1 & 0 & 4 \\ 0 & -1 & 0 & 2 \\ 7 & -2 & 3 & 5 \\ 0 & 1 & 0 & -3 \end{bmatrix}$$

Solution   The third column of this matrix contains the most 0s. Expanding the determinant in terms of the third column, we get

$$|A| = a_{13}C_{13} + a_{23}C_{23} + a_{33}C_{33} + a_{43}C_{43}$$
$$= 0(C_{13}) + 0(C_{23}) + 3(C_{33}) + 0(C_{43})$$
$$= +3 \begin{vmatrix} 2 & 1 & 4 \\ 0 & -1 & 2 \\ 0 & 1 & -3 \end{vmatrix}$$

The first column of this determinant contains the most 0s. Expand in terms of the first column.

$$|A| = 3(2) \begin{vmatrix} -1 & 2 \\ 1 & -3 \end{vmatrix} = 6(3 - 2) = 6$$

The elements of determinants can be variables. Equations can involve determinants. The following example illustrates such a **determinantal equation**.

Example 6 Solve the following equation for the variable x.

$$\begin{vmatrix} x & x + 1 \\ -1 & x - 2 \end{vmatrix} = 7$$

Solution Expand the determinant to get the equation

$$x(x - 2) - (x + 1)(-1) = 7$$

Proceed to simplify this equation and solve for x.

$$x^2 - 2x + x + 1 = 7$$
$$x^2 - x - 6 = 0$$
$$(x + 2)(x - 3) = 0$$
$$x = -2 \text{ or } 3$$

There are two solutions to this equation, namely $x = -2$ and $x = 3$.

Computing Determinants of 2 × 2 and 3 × 3 Matrices

The determinants of 2 × 2 and 3 × 3 matrices can be found quickly using diagonals. For a 2 × 2 matrix the actual diagonals are used, while in the case of a 3 × 3 matrix the diagonals of an array consisting of the matrix with the two first columns added to the right are used. A determinant is equal to the sum of the diagonal products that go from left to right minus the sum of the diagonal products that go from right to left, as follows.

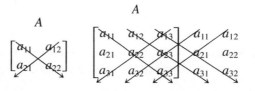

$$|A| = a_{11}a_{22} - a_{12}a_{21}$$

$$|A| = a_{11}a_{22}a_{33} + a_{12}a_{23}a_{31} + a_{13}a_{21}a_{32} - a_{11}a_{22}a_{31} - a_{11}a_{23}a_{32} - a_{12}a_{21}a_{33}$$

(diagonal products from left to right)   (diagonal products from right to left)

(We leave it to you to show that the expansion of a 3×3 arbitrary determinant can be written thus.)

For example,

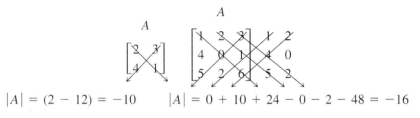

$$|A| = (2 - 12) = -10 \qquad |A| = 0 + 10 + 24 - 0 - 2 - 48 = -16$$

There are no such shortcuts for computing determinants of larger matrices.

*Alternative Definition of Determinant

We now introduce an alternative definition of determinant that involves permutations. This definition is used mainly for theoretical purposes. We first give the necessary background about permutations.

Let $S = \{1, 2, \ldots, n\}$, be the set of integers from 1 to n. A rearrangement $i_1 i_2 \ldots i_n$ of the elements of S is called a **permutation** of S. Thus, for example, 2314 is a permutation of the set $\{1, 2, 3, 4\}$. The arrangement $i_1 i_2 \ldots i_n$ is called an **even permutation** if it can be rearranged in the form $12 \ldots n$ through an even number of interchanges, and $i_1 i_2 \ldots i_n$ is an **odd permutation** if it can be rearranged in the form $12 \ldots n$ through an odd number of interchanges. (Zero is considered an even number.)

Example 7   Show that (a) 13425 is an even permutation and that (b) 13452 is an odd permutation.

Solution   (a) 13425 can be rearranged as follows.

$$13425 \rightarrow 13245 \rightarrow 12345$$

There are two interchanges—an even number. Thus 13425 is an even permutation.
(b) 13452 can be rearranged as follows.

$$13452 \rightarrow 13425 \rightarrow 13245 \rightarrow 12345$$

There are three interchanges—an odd number. 13452 is an odd permutation.

Example 8   Give all the permutations of $\{1, 2, 3\}$ and indicate whether the permutations are even or odd.

Solution Systematically consider the permutations that have 1 as first element, then those that have 2 as first element, and finally those that have 3 as first element. We get

$$123 \quad \text{(even)}$$
$$132 \to 123 \quad \text{(odd)}$$
$$213 \to 123 \quad \text{(odd)}$$
$$231 \to 213 \to 123 \quad \text{(even)}$$
$$312 \to 132 \to 123 \quad \text{(even)}$$
$$321 \to 231 \to 213 \to 123 \quad \text{(odd)}$$

These are all the permutations of the set $\{1, 2, 3\}$.

Definition *Let A be an $n \times n$ matrix. Then*

$$|A| = \sum \pm a_{1j_1} a_{2j_2} \ldots a^{nj}{}_n$$

$\sum$ means that the terms are to be summed over all permutations $j_1 j_2 \ldots j_n$ of the set $\{1, 2, \ldots, n\}$. The sign $+$ or $-$ is selected for each term according to whether the permutation is even or odd.

Example 9 We have given two definitions of determinant, one in terms of cofactors, the other in terms of permutations. Show that these definitions are equivalent when applied to a 2×2 matrix.

Solution Let A be a 2×2 matrix.
Cofactor definition: This definition gives

$$|A| = \begin{vmatrix} a_{11} & a_{12} \\ a_{21} & a_{22} \end{vmatrix} = a_{11}a_{22} - a_{12}a_{21}$$

Permutation definition: The permutations of $\{1, 2\}$ are 12 (even) and 21 (odd). Thus this definition gives

$$|A| = \sum \pm a_{1j_1} a_{2j_2} = a_{11}a_{22} - a_{12}a_{21}$$

The two definitions agree for 2×2 matrices.

Exercise Set 3.1

1. Evaluate the determinants of the following 2×2 matrices.

*(a) $\begin{bmatrix} 2 & 1 \\ 3 & 5 \end{bmatrix}$

(b) $\begin{bmatrix} 3 & -2 \\ 1 & 2 \end{bmatrix}$

*(c) $\begin{bmatrix} 4 & 1 \\ -2 & 3 \end{bmatrix}$

(d) $\begin{bmatrix} 5 & -2 \\ -3 & -4 \end{bmatrix}$

2. Evaluate the determinants of the following 2×2 matrices.

*(a) $\begin{bmatrix} 1 & -5 \\ 0 & 3 \end{bmatrix}$

(b) $\begin{bmatrix} 1 & 2 \\ 3 & 4 \end{bmatrix}$

*(c) $\begin{bmatrix} -3 & 1 \\ 2 & -5 \end{bmatrix}$

(d) $\begin{bmatrix} 3 & -2 \\ -1 & 0 \end{bmatrix}$

3. Let

$$A = \begin{bmatrix} 1 & 2 & -3 \\ 5 & 0 & 6 \\ 7 & 1 & -4 \end{bmatrix}$$

Find the following minors and cofactors of A.

*(a) M_{11} and C_{11} (b) M_{21} and C_{21}

*(c) M_{23} and C_{23} (d) M_{33} and C_{33}

4. Let

$$A = \begin{bmatrix} 5 & 0 & 1 \\ -2 & 3 & 7 \\ 0 & -6 & 2 \end{bmatrix}$$

Find the following minors and cofactors of A.

*(a) M_{13} and C_{13} (b) M_{22} and C_{22}

*(c) M_{31} and C_{31} (d) M_{33} and C_{33}

5. Let

$$A = \begin{bmatrix} 2 & 0 & 1 & -5 \\ 8 & -1 & 2 & 1 \\ 4 & -3 & -5 & 0 \\ 1 & 4 & 8 & 2 \end{bmatrix}$$

Find the following minors and cofactors of A.

*(a) M_{12} and C_{12} (b) M_{24} and C_{24}

*(c) M_{33} and C_{33} (d) M_{43} and C_{43}

6. Evaluate the determinants of the following matrices (1) using the first row and (2) using the "diagonals" method.

*(a) $\begin{bmatrix} 1 & 2 & 4 \\ 4 & -1 & 5 \\ -2 & 2 & 1 \end{bmatrix}$ (b) $\begin{bmatrix} 3 & 1 & 4 \\ -7 & -2 & 1 \\ 9 & 1 & -1 \end{bmatrix}$

*(c) $\begin{bmatrix} 4 & 1 & -2 \\ 5 & 3 & -1 \\ 2 & 4 & 1 \end{bmatrix}$

7. Evaluate the determinants of the following matrices (1) using the first row and (2) using the "diagonals" method.

*(a) $\begin{bmatrix} 2 & 0 & 7 \\ 8 & -1 & -2 \\ 5 & 6 & 1 \end{bmatrix}$ (b) $\begin{bmatrix} 2 & -1 & 3 \\ 4 & 0 & 2 \\ 1 & 1 & 1 \end{bmatrix}$

*(c) $\begin{bmatrix} 0 & 0 & 5 \\ 1 & 1 & 1 \\ 2 & 2 & 2 \end{bmatrix}$

8. Evaluate the determinants of each of the following matrices in two ways, using the indicated rows and columns. Observe that you get the same answers both ways.

*(a) $\begin{bmatrix} 0 & 3 & 2 \\ 1 & 5 & 7 \\ -2 & -6 & -1 \end{bmatrix}$ (b) $\begin{bmatrix} 4 & 2 & 1 \\ -6 & 3 & -2 \\ 7 & 1 & -1 \end{bmatrix}$

2nd row and 1st column   3rd row and 2nd column

*(c) $\begin{bmatrix} 5 & -1 & 2 \\ 3 & 0 & 6 \\ -4 & 3 & 1 \end{bmatrix}$ (d) $\begin{bmatrix} 6 & 3 & 0 \\ -2 & -1 & 5 \\ 4 & 6 & -2 \end{bmatrix}$

1st row and 3rd row   2nd column and 3rd column

9. Evaluate the determinants of each of the following matrices in two ways, using the indicated rows and columns. Observe that you get the same answers both ways.

*(a) $\begin{bmatrix} 1 & 3 & -1 \\ 2 & 0 & 5 \\ 1 & 4 & 3 \end{bmatrix}$ (b) $\begin{bmatrix} -2 & -1 & 1 \\ 9 & 3 & 2 \\ 4 & 0 & 0 \end{bmatrix}$

2nd row and 1st column   1st row and 3rd column

*(c) $\begin{bmatrix} 1 & 0 & 2 \\ 3 & -2 & 1 \\ 4 & 0 & 2 \end{bmatrix}$

1st column and 2nd column

(d) $\begin{bmatrix} 1 & 2 & 3 \\ -1 & -4 & 0 \\ 0 & 0 & 4 \end{bmatrix}$

3rd row and 1st column

10. Evaluate the determinants of the following matrices using the row or column that involves the least computation.

*(a) $\begin{bmatrix} 1 & -2 & 3 \\ 1 & 4 & 0 \\ 2 & -1 & 0 \end{bmatrix}$ (b) $\begin{bmatrix} 3 & -1 & 2 \\ 0 & 4 & 0 \\ -5 & 1 & 9 \end{bmatrix}$

*(c) $\begin{bmatrix} 9 & 2 & 1 \\ -3 & 2 & 6 \\ 0 & 0 & -3 \end{bmatrix}$

11. Evaluate the determinants of the following matrices using as little computation as possible.

*(a) $\begin{bmatrix} 1 & -2 & 3 & 0 \\ 4 & 0 & 5 & 0 \\ 7 & -3 & 8 & 4 \\ -3 & 0 & 4 & 0 \end{bmatrix}$

(b) $\begin{bmatrix} 1 & 4 & 5 & 9 \\ 2 & 3 & -7 & 1 \\ 0 & 0 & 0 & -3 \\ 0 & 1 & 0 & 8 \end{bmatrix}$

*(c) $\begin{bmatrix} 9 & 3 & 7 & -8 \\ 1 & 0 & 4 & 2 \\ 1 & 0 & 0 & -1 \\ -2 & 0 & -1 & 3 \end{bmatrix}$

*12. Solve the following equation for x.

$$\begin{vmatrix} x+1 & x \\ 3 & x-2 \end{vmatrix} = 3$$

13. Solve the following equation for x.

$$\begin{vmatrix} 2x & -3 \\ x-1 & x+2 \end{vmatrix} = 1$$

*14. Find all the values of x that make the following determinant zero.

$$\begin{vmatrix} x-1 & -2 \\ x-2 & x-1 \end{vmatrix}$$

15. Find all the values of x that make the following determinant zero.

$$\begin{vmatrix} x & 0 & 2 \\ 2x & x-1 & 4 \\ -x & x-1 & x+1 \end{vmatrix}$$

*16. Why would you expect the following two determinants to have the same value?

$$\begin{vmatrix} 4 & -1 & 0 \\ 5 & 6 & 3 \\ 2 & 1 & 0 \end{vmatrix} \quad \text{and} \quad \begin{vmatrix} 4 & -1 & 0 \\ 9 & 2 & 3 \\ 2 & 1 & 0 \end{vmatrix}$$

17. Why would you expect the following determinant to have the same value whatever values are given to a and b?

$$\begin{vmatrix} 3 & 5 & a \\ 9 & 1 & b \\ 0 & 0 & 2 \end{vmatrix}$$

18. Determine whether the following permutations are even or odd.

*(a) 4213   (b) 3142   *(c) 3214

(d) 2413   *(e) 4321

19. Determine whether the following permutations are even or odd.

*(a) 35241   (b) 43152   *(c) 54312

(d) 25143   *(e) 32514

20. Give all the permutations of $\{1, 2, 3, 4\}$ and indicate whether they are even or odd.

21. We have given two definitions of determinant in this section. Show that these definitions are equivalent when applied to a 3×3 matrix.

C.7   Determinant of a Matrix

Let us compute $|A|$ for the following matrix A.

$$A = \begin{bmatrix} 5 & 2 & 4 \\ 1 & 3 & -7 \\ 0 & 1 & 3 \end{bmatrix}$$

Enter matrix A (See Section C.2)

Press [QUIT] to return to Home Screen

Press [MATRX] [▶] to get the Matrix Math Menu

Press 1 to select det

Press MATRX 1 to select matrix A

Press ENTER

det[A]   78   is displayed

    Thus $|A| = 78$

Exercises C.7

In Exercises 1–8, compute the determinants of the matrices using a calculator.

*1. $\begin{bmatrix} 1 & 2 \\ 3 & 4 \end{bmatrix}$ **2.** $\begin{bmatrix} 5 & -3 \\ 0 & 1 \end{bmatrix}$ ***3.** $\begin{bmatrix} 7.2 & 9.3 \\ -6.5 & -8.1 \end{bmatrix}$

4. $\begin{bmatrix} 1.23 & 5.32 \\ -9.78 & 4.32 \end{bmatrix}$ ***5.** $\begin{bmatrix} 8 & 4 & 3 \\ -2 & -7 & 1 \\ 0 & 2 & 9 \end{bmatrix}$ **6.** $\begin{bmatrix} 2 & 1 & 3 \\ 8 & -7 & -2 \\ 0 & -4 & 3 \end{bmatrix}$

*7. $\begin{bmatrix} -8 & 4 & 3 \\ 2 & 4 & 6 \\ -7 & 3 & -9 \end{bmatrix}$ **8.** $\begin{bmatrix} 6 & 3 & 2 & 6 \\ -2 & 5 & -8 & 5 \\ 1 & 3 & 5 & 6 \\ -6 & -8 & 0 & 1 \end{bmatrix}$

In Exercises 9–12, compute the determinants of the matrices and explain the results.

*9. $\begin{bmatrix} 1 & -2 & 3 \\ 0 & 4 & 1 \\ 3 & -6 & 9 \end{bmatrix}$ **10.** $\begin{bmatrix} 2 & 3 & -3 \\ -4 & 1 & 6 \\ 6 & 2 & -9 \end{bmatrix}$

*11. $\begin{bmatrix} 1 & 5 & 5 \\ 0 & -2 & -2 \\ 3 & 1 & 1 \end{bmatrix}$ **12.** $\begin{bmatrix} 1 & -2 & 4 \\ 0 & -1 & 3 \\ -3 & 6 & -12 \end{bmatrix}$

Properties of Determinants

In this section we discuss the algebraic properties of determinants. Many properties of determinants hold for both rows and columns. When this is the case, we write "row (column)" to indicate that the word *column* may be substituted for *row* in the statement. Proofs are either given or suggested as exercises only when we feel that they illustrate useful techniques.

The following theorem tells us how elementary row operations affect determinants. It also tells us that these operations can be extended to columns.

Theorem 3.2 Let A be an $n \times n$ matrix and c be a nonzero scalar.

(a) If a matrix B is obtained from A by multiplying the elements of a row (column) by c, then $|B| = c|A|$.

(b) If a matrix B is obtained from A by interchanging two rows (columns), then $|B| = -|A|$.

(c) If a matrix B is obtained from A by adding a multiple of one row (column) to another row (column), then $|B| = |A|$.

Proof (a) Let the matrix B be obtained by multiplying the kth row of A by c. The kth row of B is thus

$$ca_{k1} \quad ca_{k2} \quad ca_{k3} \quad \ldots \quad ca_{kn}$$

Expand $|B|$ in terms of this row.

$$\begin{aligned} |B| &= ca_{k1}C_{k1} + ca_{k2}C_{k2} + \cdots + ca_{kn}C_{kn} \\ &= c(a_{k1}C_{k1} + a_{k2}C_{k2} + \cdots + a_{kn}C_{kn}) \\ &= c|A| \end{aligned}$$

The corresponding result for columns is obtained by expanding B in terms of the kth column.

The following example illustrates how row and column operations can be used to create zeros in a determinant, making it easier to compute the determinant.

Example 1 Evaluate the determinant

$$\begin{vmatrix} 3 & 4 & -2 \\ -1 & -6 & 3 \\ 2 & 9 & -3 \end{vmatrix}$$

Solution We examine the rows and columns of the determinant to see if we can create zeros in a row or column using the above operations. Note that we can create zeros in the second column by adding twice the third column to it:

$$\begin{vmatrix} 3 & 4 & -2 \\ -1 & -6 & 3 \\ 2 & 9 & -3 \end{vmatrix} \underset{\text{C2 + 2C3}}{=} \begin{vmatrix} 3 & 0 & -2 \\ -1 & 0 & 3 \\ 2 & 3 & -3 \end{vmatrix}$$

Expand this determinant in terms of the second column to take advantage of the zeros.

$$= (-3)\begin{vmatrix} 3 & -2 \\ -1 & 3 \end{vmatrix} = (-3)(9 - 2) = -21$$

Example 2   If $A = \begin{bmatrix} 1 & 4 & 3 \\ 0 & 2 & 5 \\ -2 & -4 & 10 \end{bmatrix}$, then it can be shown, by expanding in terms of cofactors,

that $|A| = 12$. Use this information, together with the above row and column properties of determinants, to evaluate the determinants of the following matrices.

(a) $B_1 = \begin{bmatrix} 1 & 12 & 3 \\ 0 & 6 & 5 \\ -2 & -12 & 10 \end{bmatrix}$ (b) $B_2 = \begin{bmatrix} 1 & 4 & 3 \\ -2 & -4 & 10 \\ 0 & 2 & 5 \end{bmatrix}$

(c) $B_3 = \begin{bmatrix} 1 & 4 & 3 \\ 0 & 2 & 5 \\ 0 & 4 & 16 \end{bmatrix}$

Solution   (a) Observe that B_1 can be obtained from A by multiplying column 2 by 3. Thus $|B_1| = 3|A| = 36$.

(b) Observe that B_2 can be obtained from A by interchanging row 2 and row 3. Thus $|B_2| = -|A| = -12$.

(c) Observe that B_3 can be obtained from A by adding 2 times row 1 to row 3. Thus $|B_3| = |A| = 12$.

We shall find that matrices that have zero determinant play a significant role in the theory of matrices.

Definition   *A square matrix A is said to be **singular** if $|A| = 0$.*

The following theorem gives information about some of the circumstances under which we can expect a matrix to be singular.

Theorem 3.3   Let A be a square matrix. A is singular if

(a) all the elements of a row (column) are zero.

(b) two rows (columns) are equal.

(c) two rows (columns) are proportional.

Note: (b) is a special case of (c), but we list it to give it special emphasis.

Proof   (a) Let all the elements of the kth row of A be zero. Expand $|A|$ in terms of the kth row.

$$|A| = a_{k1}C_{k1} + a_{k2}C_{k2} + \cdots + a_{kn}C_{kn}$$
$$= 0C_{k1} + 0C_{k2} + \cdots + 0C_{kn}$$
$$= 0$$

The corresponding result can be seen to hold for columns by expanding the determinant in terms of the column of zeros.

(b) Interchange the equal rows of A to get a matrix B that is equal to A. Thus $|B| = |A|$.

We know that interchanging two rows of a matrix negates the determinant. Thus $|B| = -|A|$.

The two results $|B| = |A|$ and $|B| = -|A|$ combine to give $|A| = -|A|$. Thus $2|A| = 0$, implying that $|A| = 0$.

The proof for columns is similar.

Example 3   Show that the following matrices are singular.

$$\text{(a)}\quad A = \begin{bmatrix} 2 & 0 & -7 \\ 3 & 0 & 1 \\ -4 & 0 & 9 \end{bmatrix} \qquad \text{(b)}\quad B = \begin{bmatrix} 2 & -1 & 3 \\ 1 & 2 & 4 \\ 2 & 4 & 8 \end{bmatrix}$$

Solution   (a)   All the elements in column 2 of A are zero. Thus $|A| = 0$.

(b)   Observe that every element in row 3 of B is twice the corresponding element in row 2. We write

$$(\text{row } 3) = 2(\text{row } 2)$$

Row 2 and row 3 are proportional. Thus $|B| = 0$.

The following theorem tells us how determinants interact with various matrix operations.

Theorem 3.4   Let A and B be $n \times n$ matrices and c be a nonzero scalar.

(a)   Determinant of a scalar multiple: $|cA| = c^n|A|$

(b)   Determinant of a product: $|AB| = |A||B|$

(c)   Determinant of a transpose: $|A'| = |A|$

(d)   Determinant of an inverse: $|A^{-1}| = \dfrac{1}{|A|}$ (assuming A^{-1} exists)

Example 4   If A is a 2×2 matrix with $|A| = 4$, use the properties of determinants to compute the following determinants.

(a) $|3A|$ (b) $|A^2|$

(c) $|5A'A^{-1}|$, assuming A^{-1} exists

Solution   (a) $|3A| = (3^2)|A|$, by Theorem 3.4(a)
               $= 9 \times 4 = 36$

(b) $|A^2| = |AA|$
$= |A||A|$, by Theorem 3.4(b)
$= 4 \times 4 = 16$

(c) $|5A^tA^{-1}| = (5^2)|A^tA^{-1}|$, by Theorem 3.4(a), since A^tA^{-1} is a 2×2 matrix
$= 25|A^t||A^{-1}|$, by Theorem 3.4(b)
$= 25|A|\dfrac{1}{|A|}$, by Theorem 3.4(c) and (d)
$= 25$

Example 5   Prove that $|A^{-1}A^tA| = |A|$.

Solution   By the properties of matrices, determinants, and real numbers, we get

$$|A^{-1}A^tA| = |(A^{-1}A^t)A| = |A^{-1}A^t||A| = |A^{-1}||A^t||A| = \frac{1}{|A|}|A||A| = |A|$$

We ask you to justify each step in this proof.

Example 6   Prove that if A and B are square matrices of the same size, with A being singular, then AB is also singular. Is the converse true?

Solution   The matrix A is singular. Thus $|A| = 0$. Applying the properties of determinants, we get

$$|AB| = |A||B| = 0$$

Thus the matrix AB is singular.

We now investigate the converse: does AB being singular mean that A is singular? We get

$$|AB| = 0 \Rightarrow |A||B| = 0 \Rightarrow |A| = 0 \text{ or } |B| = 0$$

Thus AB being singular implies that either A or B is singular. (We do not exclude the possibility of both being singular.) The converse is not true.

Exercise Set 3.2

1. Simplify the determinants of the following matrices by creating many zeros in a single row or column, then expand the determinant in terms of that row or column.

   *(a) $\begin{bmatrix} 1 & 2 & 3 \\ 2 & 4 & 1 \\ 1 & 1 & 1 \end{bmatrix}$

   (b) $\begin{bmatrix} 0 & 1 & 5 \\ 1 & 1 & 6 \\ 2 & 2 & 7 \end{bmatrix}$

   *(c) $\begin{bmatrix} 2 & 1 & -1 \\ 3 & -1 & 1 \\ 1 & 4 & -4 \end{bmatrix}$

   (d) $\begin{bmatrix} 3 & -1 & 0 \\ 4 & 2 & 1 \\ 1 & 1 & 2 \end{bmatrix}$

2. Simplify the determinants of the following matrices by creating many zeros in a single row or column, then expand the determinant in terms of that row or column.

   *(a) $\begin{bmatrix} 2 & -1 & 2 \\ 1 & 2 & -4 \\ 3 & 1 & 2 \end{bmatrix}$

   (b) $\begin{bmatrix} 5 & 1 & 3 \\ 1 & 2 & 4 \\ -1 & 1 & -4 \end{bmatrix}$

   *(c) $\begin{bmatrix} 1 & -2 & 3 \\ -3 & 6 & -9 \\ 4 & 5 & 7 \end{bmatrix}$

   (d) $\begin{bmatrix} -1 & 3 & 2 \\ 2 & 5 & -4 \\ 4 & 1 & -8 \end{bmatrix}$

3. If $A = \begin{bmatrix} 1 & -1 & -3 \\ 2 & 0 & -4 \\ -1 & 1 & 2 \end{bmatrix}$, then $|A|$ can be expanded using

cofactors to get $|A| = -2$. Use this information, together with the properties of determinants, to compute the determinants of the following matrices.

*(a) $\begin{bmatrix} 1 & -1 & -3 \\ 2 & 0 & -4 \\ -2 & 2 & 4 \end{bmatrix}$ (b) $\begin{bmatrix} 2 & 0 & -4 \\ 1 & -1 & -3 \\ -1 & 1 & 2 \end{bmatrix}$

*(c) $\begin{bmatrix} 1 & -1 & -3 \\ 4 & -2 & -10 \\ -1 & 1 & 2 \end{bmatrix}$

4. If $A = \begin{bmatrix} 8 & 5 & 2 \\ 1 & 3 & 2 \\ -1 & -2 & -1 \end{bmatrix}$, then $|A| = 5$. Use this infor-

mation, together with the row and column properties of determinants, to compute determinants of the following matrices.

*(a) $\begin{bmatrix} 8 & 2 & 5 \\ 1 & 2 & 3 \\ -1 & -1 & -2 \end{bmatrix}$ (b) $\begin{bmatrix} 8 & 1 & 2 \\ 1 & -1 & 2 \\ -1 & 0 & -1 \end{bmatrix}$

*(c) $\begin{bmatrix} 8 & 1 & -1 \\ 5 & 3 & -2 \\ 2 & 2 & -1 \end{bmatrix}$

*5. We have used elementary row operations to compute a determinant in two different ways below and obtained two different answers. One approach is correct, while the other is incorrect. Which is the correct answer? (The mistake illustrated by this exercise is a very common one. Beware!)

$\begin{vmatrix} 1 & 2 & 3 \\ -1 & 2 & 4 \\ 2 & 4 & 7 \end{vmatrix} \underset{(2 \times R_1) - R_3}{\overset{2R1 - R3}{=}} \begin{vmatrix} 0 & 0 & -1 \\ -1 & 2 & 4 \\ 2 & 4 & 7 \end{vmatrix}$

$= (-1) \begin{vmatrix} -1 & 2 \\ 2 & 4 \end{vmatrix}$

$= (-1)(-4 - 4) = 8$

$\begin{vmatrix} 1 & 2 & 3 \\ -1 & 2 & 4 \\ 2 & 4 & 7 \end{vmatrix} \overset{R3 - 2R1}{=} \begin{vmatrix} 1 & 2 & 3 \\ -1 & 2 & 4 \\ 0 & 0 & 1 \end{vmatrix}$

$\underset{+(-2R_1)}{=} (1) \begin{vmatrix} 1 & 2 \\ -1 & 2 \end{vmatrix}$

$= (1)(2 + 2) = 4$

6. The following matrices are singular because of certain column or row properties. Give the reason.

*(a) $\begin{bmatrix} 1 & -2 & 3 \\ 7 & 5 & 4 \\ 0 & 0 & 0 \end{bmatrix}$ (b) $\begin{bmatrix} 1 & 5 & 5 \\ 0 & -2 & -2 \\ 3 & 1 & 1 \end{bmatrix}$

*(c) $\begin{bmatrix} 1 & -2 & 4 \\ 0 & -1 & 3 \\ -3 & 6 & -12 \end{bmatrix}$ (d) $\begin{bmatrix} 1 & 2 & 0 \\ 0 & 0 & 3 \\ 0 & 0 & 0 \end{bmatrix}$

7. The following matrices are singular because of certain column or row properties. Give the reason.

*(a) $\begin{bmatrix} 7 & 9 & 0 \\ -2 & 3 & 0 \\ 4 & 5 & 0 \end{bmatrix}$ (b) $\begin{bmatrix} 1 & 0 & 4 \\ 0 & 1 & 9 \\ 0 & 0 & 0 \end{bmatrix}$

*(c) $\begin{bmatrix} 1 & -2 & 3 \\ 0 & 4 & 1 \\ 3 & -6 & 9 \end{bmatrix}$ (d) $\begin{bmatrix} 2 & 3 & -3 \\ -4 & 1 & 6 \\ 6 & 2 & -9 \end{bmatrix}$

8. If A is a 2×2 matrix with $|A| = 3$, use the properties of determinants to compute the following determinants.

*(a) $|2A|$ (b) $|3A^t|$ *(c) $|A^2|$
(d) $|(A^t)^2|$ *(e) $|(A^2)^t|$ (f) $|4A^{-1}|$

9. If A and B are 3×3 matrices and $|A| = -3$, $|B| = 2$, compute the following determinants.

*(a) $|AB|$ (b) $|AA^t|$ *(c) $|A^t B|$
(d) $|3A^2 B|$ *(e) $|2AB^{-1}|$ (f) $|(A^2 B^{-1})^t|$

10. If $A = \begin{bmatrix} a & b & c \\ d & e & f \\ g & h & i \end{bmatrix}$ and $|A| = 3$, compute the following determinants.

*(a) $\begin{bmatrix} d & e & f \\ g & h & i \\ a & b & c \end{bmatrix}$ (b) $\begin{bmatrix} d & a & g \\ e & b & h \\ f & c & i \end{bmatrix}$

*(c) $\begin{bmatrix} d & f & e \\ a & c & b \\ g & i & h \end{bmatrix}$ (d) $\begin{bmatrix} d & e & f \\ a & b & c \\ 2g & 2h & 2i \end{bmatrix}$

11. Prove the following identity without evaluating the determinants.

$\begin{vmatrix} a+b & c+d & e+f \\ p & q & r \\ u & v & w \end{vmatrix} = \begin{vmatrix} a & c & e \\ p & q & r \\ u & v & w \end{vmatrix} + \begin{vmatrix} b & d & f \\ p & q & r \\ u & v & w \end{vmatrix}$

*12. Prove that the determinant of a diagonal matrix is the product of the diagonal elements.

13. Prove that if a square matrix B is obtained from a matrix A by adding a multiple of one row (column) to another row (column), then $|B| = |A|$.

14. Prove that if two rows (columns) of a square matrix A are proportional, then $|A| = 0$.

*15. Prove that if the sum of the elements of each row (column) of a square matrix A is zero, then $|A| = 0$.

16. Let A be a square matrix. Prove that if there exists a positive integer n such that $A^n = 0$, then $|A| = 0$.

*17. Prove that if A and B are square matrices of the same size, then $|AB| = |BA|$.

18. Prove that if A, B, and C are square matrices of the same size, then $|ABC| = |A||B||C|$.

19. Let A and B be square matrices of the same size. Is the result $|A + B| = |A| + |B|$ true in general? (*Hint:* Construct an example.)

20. Let A be a square matrix. Let B be a matrix obtained from A using an elementary row operation. Prove that $|B| \neq 0$ if and only if $|A| \neq 0$.

21. (a) Let A be a 2×2 matrix. Show that the only reduced echelon form that has no zero rows is I_2.

   (b) Let A be a 3×3 matrix. Show that the only reduced echelon form that has no zero rows is I_3.

   (c) Let A be an $n \times n$ matrix. Show that the only reduced echelon form that has no zero rows is I_n.

   (d) Let A be an $n \times n$ matrix with reduced echelon form E. Show that $|A| \neq 0$ if and only if $E = I_n$.

3.3   Numerical Evaluation of a Determinant

In the previous sections we discussed a method for evaluating a determinant that used cofactors. That method, based on the definition, involves much computation. In this section we introduce a more efficient numerical method that is suitable for programming on the computer.

In the method of Gauss-Jordan elimination for solving equations, we used elementary row operations to change the initial system of equations into another simpler system that gave the solution. We use a similar approach to compute a determinant. In the previous section we discussed the effect of elementary row operations on determinants. We shall use elementary row transformations to transform a determinant into a so-called upper triangular form. This form immediately leads to the value of the determinant.

Definition   *A square matrix is called an **upper triangular matrix** if all the elements below the main diagonal are zero. It is called a **lower triangular matrix** if all the elements above the main diagonal are zero.*

The following are examples of triangular matrices.

$$\begin{bmatrix} 3 & 8 & 2 \\ 0 & 1 & 5 \\ 0 & 0 & 9 \end{bmatrix}, \begin{bmatrix} 1 & 4 & 0 & 7 \\ 0 & 2 & 3 & 5 \\ 0 & 0 & 0 & 9 \\ 0 & 0 & 0 & 1 \end{bmatrix} \qquad \begin{bmatrix} 7 & 0 & 0 \\ 2 & 1 & 0 \\ 3 & 9 & 8 \end{bmatrix}, \begin{bmatrix} 8 & 0 & 0 & 0 \\ 1 & 4 & 0 & 0 \\ 7 & 0 & 2 & 0 \\ 4 & 5 & 8 & 1 \end{bmatrix}$$

upper triangular matrices   lower triangular matrices

The following theorem gives information about determinants of triangular matrices.

Theorem 3.5   The determinant of a triangular matrix is the product of its diagonal elements.

Proof   Expand the determinant of an upper triangular matrix A, and each following determinant, in terms of the first column as follows

$$
\begin{vmatrix}
a_{11} & a_{12} & \cdots & a_{1n} \\
 & a_{22} & \cdots & a_{2n} \\
 & & \ddots & \\
0 & & & a_{nn}
\end{vmatrix}
= a_{11}
\begin{vmatrix}
a_{22} & a_{23} & \cdots & a_{2n} \\
 & a_{33} & \cdots & a_{3n} \\
 & & \ddots & \\
0 & & & a_{nn}
\end{vmatrix}
$$

$$
a_{11}a_{22}
\begin{vmatrix}
a_{33} & a_{34} & \cdots & a_{3n} \\
 & a_{44} & \cdots & a_{4n} \\
 & & \ddots & \\
0 & & & a_{nn}
\end{vmatrix}
= \cdots = a_{11}a_{22} \cdots a_{nn}
$$

This proves the theorem for an upper triangular matrix. The proof for a lower triangular matrix is similar.

Example 1   Evaluate the determinant of the matrix

$$
A = \begin{bmatrix} 2 & -1 & 9 \\ 0 & 3 & -4 \\ 0 & 0 & -5 \end{bmatrix}
$$

This is an upper triangular matrix. Thus the determinant is the product of the diagonal elements. We get

$$
|A| = 2 \times 3 \times (-5) = -30
$$

We now summarize the numerical method for computing a determinant.

Numerical Evaluation of a Determinant

Transform the given determinant into upper triangular form using two types of elementary row operations:

1. Adding a multiple of one row to another row. This transformation leaves the determinant unchanged.

2. Interchanging two rows of the determinant. This transformation multiplies the determinant by -1.

The zeros *below* the main diagonal are created systematically in the columns, from left to right, according to the Gauss-Jordan pattern.

The final determinant, in upper triangular form, is the product of the diagonal elements. The given determinant will be either this number or its negative, depending on the number of row interchanges performed. The following examples illustrate the method.

Example 2   Evaluate the determinant.

$$\begin{vmatrix} 2 & 4 & 1 \\ -2 & -5 & 4 \\ 4 & 9 & 10 \end{vmatrix}$$

Solution   Applying the algorithm, we get

$$\begin{vmatrix} 2 & 4 & 1 \\ -2 & -5 & 4 \\ 4 & 9 & 10 \end{vmatrix} \underset{\substack{R2 + R1 \\ R3 + (-2)R1}}{=} \begin{vmatrix} 2 & 4 & 1 \\ 0 & -1 & 5 \\ 0 & 1 & 8 \end{vmatrix} \underset{R3 + R2}{=} \begin{vmatrix} 2 & 4 & 1 \\ 0 & -1 & 5 \\ 0 & 0 & 13 \end{vmatrix}$$

$$= 2 \times (-1) \times 13 = -26$$

(handwritten) if you wrote
(2) $R_1 - R_3 \rightarrow R_3$

IT IS LIKE
MULTIPLYING $R_1 \times 2$
and $R_2 \times (-1)$ and
ADDING!!! WRONG.
And you would obtain
(in R_3) $\rightarrow$ $\begin{vmatrix} 0 & -1 & -8 \end{vmatrix}$
WRONG !!!

(handwritten arrow) ← BECAUSE ← MUSTADD

Example 3   Evaluate the determinant

$$\begin{vmatrix} 1 & 0 & 2 & 1 \\ 2 & -1 & 1 & 0 \\ 1 & 0 & 0 & 3 \\ -1 & 0 & 2 & 1 \end{vmatrix}$$

Solution   Applying the algorithm, we get

$$\begin{vmatrix} 1 & 0 & 2 & 1 \\ 2 & -1 & 1 & 0 \\ 1 & 0 & 0 & 3 \\ -1 & 0 & 2 & 1 \end{vmatrix} \underset{\substack{R2 + (-2)R1 \\ R3 + (-1)R1 \\ R4 + R1}}{=} \begin{vmatrix} 1 & 0 & 2 & 1 \\ 0 & -1 & -3 & -2 \\ 0 & 0 & -2 & 2 \\ 0 & 0 & 4 & 2 \end{vmatrix}$$

$$\underset{R4 + 2R3}{=} \begin{vmatrix} 1 & 0 & 2 & 1 \\ 0 & -1 & -3 & -2 \\ 0 & 0 & -2 & 2 \\ 0 & 0 & 0 & 6 \end{vmatrix}$$

$$= 1 \times (-1) \times (-2) \times 6 = 12$$

(handwritten) ∘∘ place in R4
∘ result in R4
∘∘ 2R3 + R4 → R4

(handwritten) you can't put $R_1 - R_3 \rightarrow R_3$
because here
you are multiplying
R_3 by (-1) and
placing result in
R_3 !!!

(handwritten) Note
first
∘∘ look at
where you will place
result, then multiply
the other row/column by the
∘ notice how they multiply
by a NEGATIVE
then ADD
adding rows/columns
NEGATIVE

It sometimes becomes necessary to interchange rows to obtain the upper triangular matrix, as the following example illustrates. Whenever rows are interchanged, remember to negate thc determinant.

Example 4   Evaluate the determinant

$$\begin{vmatrix} 1 & -2 & 4 \\ -1 & 2 & -5 \\ 2 & -2 & 11 \end{vmatrix}$$

Solution   We get

$$
\begin{vmatrix} 1 & -2 & 4 \\ -1 & 2 & -5 \\ 2 & -2 & 11 \end{vmatrix}
\quad \underset{\substack{R2 + R1 \\ R3 + (-2)R1}}{=} \quad
\begin{vmatrix} 1 & -2 & 4 \\ 0 & 0 & -1 \\ 0 & 2 & 3 \end{vmatrix}
$$

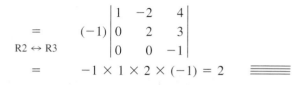

∴ place result in R_3 ∴ multiply R_1 by (-2) (not R_2 then subtract) (and add)

At this time we need a nonzero element in the (2, 2) location to create a zero in the (3, 2) location. Interchange row 2 and row 3 and then proceed.

$$
\underset{R2 \leftrightarrow R3}{=} \quad (-1)\begin{vmatrix} 1 & -2 & 4 \\ 0 & 2 & 3 \\ 0 & 0 & -1 \end{vmatrix}
$$

$$
= \quad -1 \times 1 \times 2 \times (-1) = 2
$$

If at any stage the diagonal element is zero, and all elements below it in that column are also zero, it is not possible to interchange rows to obtain a nonzero diagonal element at that stage. The final upper triangular matrix will have a zero diagonal element. It is not necessary to continue beyond this stage; the determinant is zero. The following example illustrates this situation.

Example 5   Evaluate the determinant

$$
\begin{vmatrix} 1 & -1 & 0 & 2 \\ -1 & 1 & 2 & 3 \\ 2 & -2 & 3 & 4 \\ 6 & -6 & 5 & 1 \end{vmatrix}
$$

Solution   We get

$$
\begin{vmatrix} 1 & -1 & 0 & 2 \\ -1 & 1 & 2 & 3 \\ 2 & -2 & 3 & 4 \\ 6 & -6 & 5 & 1 \end{vmatrix}
\quad \underset{\substack{R2 + R1 \\ R3 + (-2)R1 \\ R4 + (-6)R1}}{=} \quad
\begin{vmatrix} 1 & -1 & 0 & 2 \\ 0 & 0 & 2 & 5 \\ 0 & 0 & 3 & 0 \\ 0 & 0 & 5 & -11 \end{vmatrix} = 0
$$

Look

diagonal element is zero, and all elements below diagonal elements are zero

Exercise Set 3.3

1. Find the determinants of the following triangular matrices.

   *(a) $\begin{bmatrix} 3 & 9 & 6 \\ 0 & -1 & 2 \\ 0 & 0 & 4 \end{bmatrix}$ (b) $\begin{bmatrix} 3 & 4 & -7 \\ 0 & 1 & 2 \\ 0 & 0 & 5 \end{bmatrix}$

   *(c) $\begin{bmatrix} 9 & 0 & 0 \\ 2 & 3 & 0 \\ 5 & 2 & 1 \end{bmatrix}$

2. Find the determinants of the following triangular matrices.

   *(a) $\begin{bmatrix} 2 & 5 & 1 & 0 \\ 0 & 3 & 2 & -7 \\ 0 & 0 & 5 & 1 \\ 0 & 0 & 0 & -2 \end{bmatrix}$

   (b) $\begin{bmatrix} 3 & 0 & 0 & 0 \\ 1 & -2 & 0 & 0 \\ -7 & 9 & 5 & 0 \\ 6 & 4 & 2 & 1 \end{bmatrix}$

   *(c) $\begin{bmatrix} 7 & -2 & 4 & 3 \\ 0 & 6 & 2 & 1 \\ 0 & 0 & 0 & -4 \\ 0 & 0 & 0 & 8 \end{bmatrix}$

3. Evaluate the following 3×3 determinants using elementary row operations.

   *(a) $\begin{vmatrix} 1 & 0 & -1 \\ 2 & 1 & 2 \\ -1 & 1 & 1 \end{vmatrix}$ (b) $\begin{vmatrix} 2 & 1 & 1 \\ 4 & 0 & 1 \\ -1 & 2 & 0 \end{vmatrix}$

   *(c) $\begin{vmatrix} 1 & -2 & 3 \\ -1 & 2 & 1 \\ 2 & 1 & 3 \end{vmatrix}$

4. Evaluate the following 3×3 determinants using elementary row operations.

   *(a) $\begin{vmatrix} 2 & 3 & 8 \\ -2 & -3 & 4 \\ 4 & 6 & -2 \end{vmatrix}$ (b) $\begin{vmatrix} 1 & -1 & 2 \\ 3 & -1 & 7 \\ -2 & 2 & 0 \end{vmatrix}$

   *(c) $\begin{vmatrix} 2 & -1 & 4 \\ -2 & 1 & -3 \\ 0 & 3 & -2 \end{vmatrix}$

5. Evaluate the following 3×3 determinants using elementary row operations.

   *(a) $\begin{vmatrix} 2 & 4 & 1 \\ 0 & 0 & 3 \\ 0 & 1 & 2 \end{vmatrix}$ *(b) $\begin{vmatrix} 0 & 4 & 1 \\ 1 & 2 & 3 \\ 4 & 1 & 5 \end{vmatrix}$

   (c) $\begin{vmatrix} 2 & -1 & 6 \\ 0 & 0 & 1 \\ 0 & 0 & 2 \end{vmatrix}$

6. Evaluate the following 4×4 determinants using elementary row operations.

   *(a) $\begin{vmatrix} -1 & 2 & 0 & 1 \\ 1 & 1 & -1 & 0 \\ 2 & 1 & 1 & 0 \\ -1 & -1 & 0 & 1 \end{vmatrix}$

   (b) $\begin{vmatrix} 1 & 0 & -2 & 1 \\ 2 & 1 & 0 & 2 \\ -1 & 1 & -2 & 1 \\ 3 & 1 & -1 & 0 \end{vmatrix}$

   *(c) $\begin{vmatrix} -1 & 1 & 2 & 1 \\ 1 & -1 & 3 & -1 \\ 2 & -2 & 3 & 1 \\ 1 & -1 & 0 & 1 \end{vmatrix}$

7. Evaluate the following 4×4 determinants using elementary row operations.

   *(a) $\begin{vmatrix} 1 & -1 & 0 & 2 \\ -1 & 1 & 0 & 0 \\ 2 & -2 & 0 & 1 \\ 3 & 1 & 5 & -1 \end{vmatrix}$

   (b) $\begin{vmatrix} 1 & 2 & -1 & 0 \\ 1 & 4 & 2 & 1 \\ -1 & 2 & 6 & 6 \\ 2 & 2 & -4 & -2 \end{vmatrix}$

   *(c) $\begin{vmatrix} 2 & 1 & 3 & 1 \\ -2 & 3 & -1 & 2 \\ 2 & 1 & 2 & 3 \\ -4 & -2 & 0 & -1 \end{vmatrix}$

8. I_n is the identity $n \times n$ matrix. Prove that $|I_n| = 1$.

9. Prove that if A is invertible, then $|A||A^{-1}| = 1$.

10. Prove that the determinant of a diagonal matrix is the product of its diagonal elements.

11. Prove that the determinant of a lower triangular matrix is the product of the diagonal elements.

12. *(a) Show that if $A = A^{-1}$, then $|A| = \pm 1$.

    (b) Show that if $A^t = A^{-1}$, then $|A| = \pm 1$.

13. Prove that $|A| \neq 0$ if and only if $|A^{-1}| \neq 0$.

*14. Show that if $AB = I_n$, then $|A| \neq 0$ and $|B| \neq 0$.

15. Prove that every symmetric triangular matrix is a diagonal matrix.

3.4 Determinants, Matrix Inverses, and Systems of Linear Equations

We have introduced the concept of the determinant, discussed various ways of computing determinants, and looked at the algebraic properties of determinants. We shall now see how a determinant can give information about the inverse of a matrix and how a determinant can also give information about solutions to systems of equations.

The determinant of a matrix can tell us whether the inverse of a matrix exists or not.

Theorem 3.6 A square matrix A is invertible if and only if $|A| \neq 0$.

Proof Assume that A is invertible. Thus $AA^{-1} = I_n$. This implies that $|AA^{-1}| = |I_n|$. Properties of determinants give $|A||A^{-1}| = 1$. Thus $|A| \neq 0$.

Conversely, assume that $|A| \neq 0$. Consider the sequence of matrices in the Gauss-Jordan elimination process, as A is transformed into its reduced echelon form. Theorem 3.2 implies that since $|A| \neq 0$, then the determinant of each consecutive matrix in this sequence is nonzero. In particular, the determinant of the reduced echelon form is nonzero. However, the only $n \times n$ reduced echelon form that has nonzero determinant is I_n. Thus the reduced echelon form of A is I_n. The Gauss-Jordan method of finding the inverse of a matrix would therefore lead to an inverse for A. A is invertible.

The result of the previous theorem, relating the existence of the inverse of a matrix to its determinant, is extremely important. It is often stated

$$A^{-1} \text{ exists if and only if } A \text{ is nonsingular}$$

Example 1 Determine which of the following matrices are invertible.

$$A = \begin{bmatrix} 1 & -1 \\ 3 & 2 \end{bmatrix}, \qquad B = \begin{bmatrix} 4 & 2 \\ 2 & 1 \end{bmatrix},$$

$$A = \begin{bmatrix} 2 & 4 & -3 \\ 4 & 12 & -7 \\ -1 & 0 & 1 \end{bmatrix}, \qquad D = \begin{bmatrix} 1 & 2 & -1 \\ -1 & 1 & 2 \\ 2 & 8 & 0 \end{bmatrix}$$

Solution　Compute the determinant of each matrix and apply the previous theorem. We get

$$|A| = 5 \neq 0. \; A \text{ is invertible.}$$
$$|B| = 0. \; B \text{ is singular. The inverse does not exist.}$$
$$|C| = 0. \; C \text{ is singular. The inverse does not exist.}$$
$$|D| = 2 \neq 0. \; D \text{ is invertible.}$$

Having derived a criterion for the existence of a matrix inverse, we now introduce the tools necessary for developing a formula for the inverse.

Definition　*Let A be an n × n matrix and C_{ij} be the cofactor of a_{ij}. The matrix whose (i,j)th element is C_{ij} is called the **matrix of cofactors** of A. The transpose of this matrix is called the **adjoint** of A and is denoted adj(A).*

$$\begin{bmatrix} C_{11} & C_{12} & \cdots & C_{1n} \\ C_{21} & C_{22} & \cdots & C_{2n} \\ \vdots & \vdots & & \vdots \\ C_{n1} & C_{n2} & \cdots & C_{nn} \end{bmatrix} \qquad \begin{bmatrix} C_{11} & C_{12} & \cdots & C_{1n} \\ C_{21} & C_{22} & \cdots & C_{2n} \\ \vdots & \vdots & & \vdots \\ C_{n1} & C_{n2} & \cdots & C_{nn} \end{bmatrix}^t$$

matrix of cofactors　　　　　adjoint matrix

Example 2　Give the matrix of cofactors and the adjoint matrix of the following matrix A.

$$A = \begin{bmatrix} 2 & 0 & 3 \\ -1 & 4 & -2 \\ 1 & -3 & 5 \end{bmatrix}$$

Solution　The cofactors of A are as follows.

$$C_{11} = \begin{vmatrix} 4 & -2 \\ -3 & 5 \end{vmatrix} = 14 \qquad C_{12} = -\begin{vmatrix} -1 & -2 \\ 1 & 5 \end{vmatrix} = 3 \qquad C_{13} = \begin{vmatrix} -1 & 4 \\ 1 & -3 \end{vmatrix} = -1$$

$$C_{21} = -\begin{vmatrix} 0 & 3 \\ -3 & 5 \end{vmatrix} = -9 \qquad C_{22} = \begin{vmatrix} 2 & 3 \\ 1 & 5 \end{vmatrix} = 7 \qquad C_{23} = -\begin{vmatrix} 2 & 0 \\ 1 & -3 \end{vmatrix} = 6$$

$$C_{31} = \begin{vmatrix} 0 & 3 \\ 4 & -2 \end{vmatrix} = -12 \qquad C_{32} = -\begin{vmatrix} 2 & 3 \\ -1 & -2 \end{vmatrix} = 1 \qquad C_{33} = \begin{vmatrix} 2 & 0 \\ -1 & 4 \end{vmatrix} = 8$$

The matrix of cofactors of A is

$$\begin{bmatrix} 14 & 3 & -1 \\ -9 & 7 & 6 \\ -12 & 1 & 8 \end{bmatrix}$$

The adjoint of A is the transpose of this matrix.

$$\text{adj}(A) = \begin{bmatrix} 14 & -9 & -12 \\ 3 & 7 & 1 \\ -1 & 6 & 8 \end{bmatrix}$$

The following theorem gives a formula for the inverse of a matrix.

Theorem 3.7   Let A be a square matrix with $|A| \neq 0$. A is invertible with

$$A^{-1} = \frac{1}{|A|} \, \text{adj}(A)$$

Proof   Consider the matrix product $A \, \text{adj}(A)$. The (i,j)th element of this product is

$$(i,j)\text{th element} = (\text{row } i \text{ of } A) \times (\text{column } j \text{ of adj}(A))$$

$$= [a_{i1} \quad a_{i2} \quad \dots \quad a_{in}] \begin{bmatrix} C_{j1} \\ C_{j2} \\ \vdots \\ C_{jn} \end{bmatrix}$$

$$= a_{i1}C_{j1} + a_{i2}C_{j2} + \cdots + a_{in}C_{jn}$$

If $i = j$, this is the expansion of $|A|$ in terms of the ith row. If $i \neq j$, then it is the expansion of the determinant of a matrix in which the jth row of A has been replaced by the ith row of A, a matrix having two identical rows. Therefore

$$(i,j)\text{th element} = \begin{cases} |A| & \text{if } i = j \\ 0 & \text{if } i \neq j \end{cases}$$

The product $A \, \text{adj}(A)$ is thus a diagonal matrix with the diagonal elements all being $|A|$. Factor out all the diagonal elements to get $|A|I_n$. Thus

$$A \, \text{adj}(A) = |A|I_n$$

Since $|A| \neq 0$, we can rewrite this equation

$$A\left(\frac{1}{|A|} \, \text{adj}(A) \right) = I_n$$

Similarly, it can be shown that

$$\left(\frac{1}{|A|} \, \text{adj}(A) \right) A = I_n$$

Thus

$$A^{-1} = \frac{1}{|A|} \, \text{adj}(A)$$

proving the theorem.

The importance of this result lies in that it gives a *formula* for the inverse of an arbitrary matrix that can be used in theoretical work. The Gauss-Jordan algorithm presented earlier is much more efficient than this formula for computing the inverse of a specific matrix. However, the Gauss-Jordan method cannot be used to describe the inverse of an arbitrary matrix.

Having said that, we now reinforce our understanding of this result by using it to compute the inverse of a matrix.

Example 3   Use the formula for the inverse of a matrix to compute the inverse of the matrix

$$
A = \begin{bmatrix} 2 & 0 & 3 \\ -1 & 4 & -2 \\ 1 & -3 & 5 \end{bmatrix}
$$

Solution $|A|$ is computed and found to be 25. Thus the inverse of A exists. This matrix was discussed in Example 2. There we found that

$$
\text{adj}(A) = \begin{bmatrix} 14 & -9 & -12 \\ 3 & 7 & 1 \\ -1 & 6 & 8 \end{bmatrix}
$$

The formula for the inverse of a matrix gives

$$
A^{-1} = \frac{1}{25} \text{adj}(A) = \begin{bmatrix} \frac{14}{25} & -\frac{9}{25} & -\frac{12}{25} \\ \frac{3}{25} & \frac{7}{25} & \frac{1}{25} \\ -\frac{1}{25} & \frac{6}{25} & \frac{8}{25} \end{bmatrix}
$$

We now discuss the relationship between the existence and uniqueness of the solutions to a system of n linear equations in n variables and the determinant of the matrix of coefficients of the system.

Theorem 3.8   Let $AX = B$ be a system of n linear equations in n variables. If $|A| \neq 0$, there is a unique solution. If $|A| = 0$, there will be many or no solutions.

Proof   If $|A| \neq 0$, we know that A^{-1} exists (Theorem 3.6) and that there is then a unique solution given by $X = A^{-1}B$ (Theorem 2.9).

If $|A| = 0$, the determinant of every matrix leading up to, and including, the reduced echelon form of A is zero (Theorem 3.2). Thus the reduced echelon form of A is not I_n. The solution to the system $AX = B$ is not unique. The following two systems of linear equations, each of which has a singular matrix of coefficients, illustrate that there may be many or no solutions.

$$
\begin{aligned}
x_1 - 2x_2 + 3x_3 &= 1 \\
3x_1 - 4x_2 + 5x_3 &= 3 \\
2x_1 - 3x_2 + 4x_3 &= 2
\end{aligned}
\qquad
\begin{aligned}
x_1 + 2x_2 + 3x_3 &= 3 \\
2x_1 + x_2 + 3x_3 &= 3 \\
x_1 + x_2 + 2x_3 &= 0
\end{aligned}
$$

many solutions: no solutions
$x_1 = r + 1, x_2 = 2r, x_3 = r$

Example 4   Determine whether or not the following system of equations has a unique solution

$$3x_1 + 3x_2 - 2x_3 = 2$$
$$4x_1 + x_2 + 3x_3 = -5$$
$$7x_1 + 4x_2 + x_3 = 9$$

Solution   We compute the determinant of the matrix of coefficients. We get

$$\begin{vmatrix} 3 & 3 & -2 \\ 4 & 1 & 3 \\ 7 & 4 & 1 \end{vmatrix} = 0$$

Thus the system does not have a unique solution.

We now introduce a result called **Cramer's rule**, for solving a system of n linear equations in n variables that has a unique solution. This rule is of theoretical importance in that it gives us a formula for the solution of a system of equations.

Theorem 3.9   **Cramer's Rule***   Let $AX = B$ be a system of n linear equations in n variables such that $|A| \neq 0$. The system has a unique solution given by

$$x_1 = \frac{|A_1|}{|A|}, \qquad x_2 = \frac{|A_2|}{|A|}, \cdots, \qquad x_n = \frac{|A_n|}{|A|}$$

where A_i is the matrix obtained by replacing column i of A with B.

Proof   Since $|A| \neq 0$, the solution to the system $AX = B$ is unique and is given by

$$X = A^{-1}B$$

$$= \frac{1}{|A|} \operatorname{adj}(A)B$$

(continues)

---

* Gabriel Cramer (1704–1752) was educated and lived in Geneva, Switzerland, and traveled widely in Europe. He made contributions to the fields of geometry and probability and was described as "a great poser of stimulating problems." He is best known for Cramer's rule, which was, however, not his original work. He explained it and encouraged its use. His greatest contribution was probably in the thankless (but important) task of editing and distributing the works of others. Cramer's interests were broad, being involved in civic life, in the restoration of cathedrals, and in excavations. He was said to be "friendly, good-humored, pleasant in voice and appearance, possessing a good memory, judgment, and health."

x_i, the ith element of X, is given by

$$x_i = \frac{1}{|A|}[\text{row } i \text{ of adj}(A)] \times B$$

$$= \frac{1}{|A|}[C_{1i} \quad C_{2i} \quad \cdots \quad C_{ni}]\begin{bmatrix} b_1 \\ b_2 \\ \vdots \\ b_n \end{bmatrix}$$

$$= \frac{1}{|A|}(b_1C_{1i} + b_2C_{2i} + \cdots + b_nC_{ni})$$

The expression in parentheses is the cofactor expansion of $|A_i|$ in terms of the ith column. Thus

$$x_i = \frac{|A_i|}{|A|}$$

proving the theorem.

Example 5   Solve the following system of equations using Cramer's rule.

$$x_1 + 3x_2 + x_3 = -2$$
$$2x_1 + 5x_2 + x_3 = -5$$
$$x_1 + 2x_2 + 3x_3 = 6$$

Solution   The matrix of coefficients A and column matrix of constants B are

$$A = \begin{bmatrix} 1 & 3 & 1 \\ 2 & 5 & 1 \\ 1 & 2 & 3 \end{bmatrix} \quad \text{and} \quad B = \begin{bmatrix} -2 \\ -5 \\ 6 \end{bmatrix}$$

It is found that $|A| = -3 \neq 0$. Thus Cramer's rule can be applied. We get

$$A_1 = \begin{bmatrix} -2 & 3 & 1 \\ -5 & 5 & 1 \\ 6 & 2 & 3 \end{bmatrix}, \quad A_2 = \begin{bmatrix} 1 & -2 & 1 \\ 2 & -5 & 1 \\ 1 & 6 & 3 \end{bmatrix}, \quad A_3 = \begin{bmatrix} 1 & 3 & -2 \\ 2 & 5 & -5 \\ 1 & 2 & 6 \end{bmatrix}$$

giving $|A_1| = -3, |A_2| = 6, |A_3| = -9$.

Cramer's rule now gives

$$x_1 = \frac{|A_1|}{|A|} = \frac{-3}{-3} = 1, \quad x_2 = \frac{|A_2|}{|A|} = \frac{6}{-3} = -2, \quad x_3 = \frac{|A_3|}{|A|} = \frac{-9}{-3} = 3$$

The unique solution is $x_1 = 1, x_2 = -2, x_3 = 3$.

It is important that the reader get to understand Cramer's rule by solving systems of equations like this. However, we stress again that the importance of Cramer's rule lies in that it gives a *formula* for the solution of a system of equations that has a unique solution. It is not used in practice to solve specific systems; the methods of Gauss-Jordan and Gaussian elimination (Section 9.1) are both more efficient for doing that. Gaussian elimination involves $\frac{1}{3}n^3 + n^2 - \frac{1}{3}n$ multiplications and $\frac{1}{3}n^3 + \frac{1}{2}n^2 - \frac{5}{6}n$ additions, while it can be shown that Cramer's rule uses $\frac{1}{3}n^4 + \frac{1}{3}n^3 + \frac{2}{3}n^2 + \frac{1}{3}n - 1$ multiplications and $\frac{1}{3}n^4 - \frac{1}{6}n^3 - \frac{1}{3}n^2 + \frac{1}{6}n$ additions. This means, for example, that a system of four equations in four variables can be solved with 36 multiplications and 26 additions using Gaussian elimination in contrast to 115 multiplications and 70 additions using Cramer's rule.

Theorem 3.8 implies that the solution to a linear system of n equations in n variables $AX = B$ will be unique if and only if $|A| \neq 0$. Consider a homogeneous system $AX = 0$. Observe that $X = 0$ is a solution, called the **trivial solution**. If we are going to have nontrivial solutions to such a system, as is often the case in applications, we must have $|A| = 0$. The following example illustrates this situation.

Example 6   Determine values of λ for which the following system of equations has nontrivial solutions. Find the solutions for each value of λ.

$$(\lambda + 2)x_1 + (\lambda + 4)x_2 = 0$$
$$2x_1 + (\lambda + 1)x_2 = 0$$

Solution   This system is a homogeneous system of linear equations. It thus has the trivial solution $x_1 = 0$, $x_2 = 0$. Theorem 3.8 tells us that there is the possibility of other solutions only if the determinant of the matrix of coefficients is zero. Equating this determinant to zero, we get

$$\begin{vmatrix} \lambda + 2 & \lambda + 4 \\ 2 & \lambda + 1 \end{vmatrix} = 0$$
$$(\lambda + 2)(\lambda + 1) - 2(\lambda + 4) = 0$$
$$\lambda^2 + \lambda - 6 = 0$$
$$(\lambda - 2)(\lambda + 3) = 0$$

The determinant is zero if $\lambda = -3$ or $\lambda = 2$.
   $\lambda = -3$ results in the system

$$-x_1 + x_2 = 0$$
$$2x_1 - 2x_2 = 0$$

This system has many solutions, $x_1 = r$, $x_2 = r$.

$\lambda = 2$ results in the system

$$4x_1 + 6x_2 = 0$$
$$2x_1 + 3x_2 = 0$$

This system has many solutions, $x_1 = -3r/2$, $x_2 = r$.

Exercise Set 3.4

1. Use determinants to find out whether the following matrices have inverses. You need not compute inverses.

   *(a) $\begin{bmatrix} 4 & 7 \\ 1 & 3 \end{bmatrix}$

   (b) $\begin{bmatrix} -3 & 1 \\ 6 & -2 \end{bmatrix}$

   *(c) $\begin{bmatrix} 6 & 4 \\ 3 & 2 \end{bmatrix}$

   (d) $\begin{bmatrix} 7 & 2 \\ 3 & 1 \end{bmatrix}$

2. Determine whether the following matrices have inverses. You need not compute inverses.

   *(a) $\begin{bmatrix} 0 & 2 \\ 3 & 4 \end{bmatrix}$

   (b) $\begin{bmatrix} 10 & 5 \\ 4 & 2 \end{bmatrix}$

   *(c) $\begin{bmatrix} 5 & 3 \\ 1 & 2 \end{bmatrix}$

   (d) $\begin{bmatrix} 4 & -1 \\ -8 & 2 \end{bmatrix}$

3. Determine whether the following matrices have inverses. You need not compute inverses.

   *(a) $\begin{bmatrix} 2 & 4 & -7 \\ 0 & 1 & 3 \\ 0 & 0 & 9 \end{bmatrix}$

   (b) $\begin{bmatrix} 4 & -2 & 9 \\ 0 & 0 & 3 \\ 0 & 0 & 6 \end{bmatrix}$

   *(c) $\begin{bmatrix} -3 & 0 & 0 \\ 1 & 7 & 0 \\ -2 & 8 & 5 \end{bmatrix}$

   (d) $\begin{bmatrix} 1 & 2 & 3 \\ 2 & 4 & 6 \\ 7 & 3 & -1 \end{bmatrix}$

4. Determine whether the following matrices have inverses. You need not compute inverses.

   *(a) $\begin{bmatrix} 1 & -2 & 3 \\ 2 & 7 & 6 \\ -3 & 5 & -9 \end{bmatrix}$

   (b) $\begin{bmatrix} 4 & 0 & 5 \\ 1 & 3 & 7 \\ 2 & 0 & 6 \end{bmatrix}$

   *(c) $\begin{bmatrix} 7 & 1 & 3 \\ -1 & 2 & 0 \\ 5 & 4 & 1 \end{bmatrix}$

   (d) $\begin{bmatrix} 1 & -2 & 3 \\ 4 & -3 & 2 \\ 1 & -1 & 1 \end{bmatrix}$

5. Determine whether the following matrices have inverses. If a matrix has an inverse, find the inverse using the formula for the inverse of a matrix.

   *(a) $\begin{bmatrix} 1 & 4 \\ 3 & 2 \end{bmatrix}$

   (b) $\begin{bmatrix} -2 & -1 \\ 7 & 3 \end{bmatrix}$

   *(c) $\begin{bmatrix} 1 & 2 \\ 2 & 4 \end{bmatrix}$

   (d) $\begin{bmatrix} 2 & 1 \\ 4 & 3 \end{bmatrix}$

6. Determine whether the following matrices have inverses. If a matrix has an inverse, find the inverse using the formula for the inverse of a matrix.

   *(a) $\begin{bmatrix} 1 & 2 & 3 \\ 0 & 1 & 2 \\ 4 & 5 & 3 \end{bmatrix}$

   (b) $\begin{bmatrix} 0 & 3 & 3 \\ 1 & 2 & 3 \\ 1 & 4 & 6 \end{bmatrix}$

   *(c) $\begin{bmatrix} 1 & 2 & -1 \\ 2 & 4 & -3 \\ 1 & -2 & 0 \end{bmatrix}$

   (d) $\begin{bmatrix} 5 & 1 & 3 \\ 6 & 1 & 4 \\ 1 & 0 & 1 \end{bmatrix}$

7. Determine whether the following matrices have inverses. If a matrix has an inverse, find the inverse using the formula for the inverse of a matrix.

   *(a) $\begin{bmatrix} 5 & 2 & 4 \\ 2 & 1 & 2 \\ 4 & 2 & 3 \end{bmatrix}$

   (b) $\begin{bmatrix} -3 & -2 & -5 \\ 3 & 4 & 3 \\ 1 & 1 & 1 \end{bmatrix}$

   *(c) $\begin{bmatrix} 7 & 8 & 3 \\ 3 & 6 & 3 \\ 1 & 2 & 1 \end{bmatrix}$

   (d) $\begin{bmatrix} 2 & 2 & 1 \\ 4 & -1 & 4 \\ 7 & 4 & 5 \end{bmatrix}$

8. Solve the following sytems of equations using Cramer's rule.

   *(a) $x_1 + 2x_2 = 8$
   $2x_1 + 5x_2 = 19$

   (b) $x_1 + 2x_2 = 3$
   $3x_1 + x_2 = -1$

   *(c) $x_1 + 3x_2 = 11$
   $-2x_1 + x_2 = -1$

9. Solve the following systems of equations using Cramer's rule.

   *(a) $3x_1 + x_2 = -1$
   $x_1 + x_2 = 3$

   *(b) $3x_1 + 2x_2 = 11$
   $2x_1 + 3x_2 = 14$

   (c) $2x_1 + x_2 = -1$
   $-2x_1 + 2x_2 = 10$

10. Solve the following systems of equations using Cramer's rule.

*(a) $x_1 + 3x_2 + 4x_3 = 3$
$2x_1 + 6x_2 + 9x_3 = 5$
$3x_1 + x_2 - 2x_3 = 7$

(b) $x_1 + 2x_2 + x_3 = 9$
$x_1 + 3x_2 - x_3 = 4$
$x_1 + 4x_2 - x_3 = 7$

*(c) $2x_1 + x_2 + 3x_3 = 2$
$3x_1 - 2x_2 + 4x_3 = 2$
$x_1 + 4x_2 - 2x_3 = 1$

11. Solve the following systems of equations using Cramer's rule.

*(a) $x_1 + 4x_2 + 2x_3 = 5$
$x_1 + 4x_2 - x_3 = 2$
$2x_1 + 6x_2 + x_3 = 7$

(b) $2x_1 - x_2 + 3x_3 = 7$
$x_1 + 4x_2 + 2x_3 = 10$
$3x_1 + 2x_2 + x_3 = 0$

*(c) $8x_1 - 2x_2 + x_3 = 1$
$2x_1 - x_2 + 6x_3 = 3$
$6x_1 + x_2 + 4x_3 = 3$

12. Solve the following systems of equations using Cramer's rule (if possible).

*(a) $2x_1 + 7x_2 + 3x_3 = 7$
$x_1 + 2x_2 + x_3 = 2$
$x_1 + 5x_2 + 2x_3 = 5$

(b) $3x_1 + x_2 - x_3 = 7$
$x_1 + 2x_2 + x_3 = 3$
$2x_1 + 6x_2 = -4$

*(c) $3x_1 + 6x_2 - x_3 = 3$
$x_1 - 2x_2 + 3x_3 = 2$
$4x_1 - 2x_2 + 5x_3 = 5$

13. By examining the determinant of the matrix of coefficients, decide whether or not the following systems of equations have a unique solution.

*(a) $2x_1 - 3x_2 = -1$
$-4x_1 + 6x_2 = 3$

(b) $3x_1 + x_2 = 11$
$4x_1 - 2x_2 = 14$

*(c) $6x_1 + 9x_2 = -1$
$-2x_1 - 3x_2 = 10$

14. By examining the determinant of the matrix of coefficients, decide whether or not the following systems of equations have a unique solution.

*(a) $3x_1 + 2x_2 + x_3 = 4$
$x_1 - 3x_2 + 2x_3 = -11$
$2x_1 + x_2 - 3x_3 = -5$

(b) $2x_1 + 7x_2 + x_3 = 8$
$-x_1 - 5x_2 = 3$
$x_1 + 2x_2 + x_3 = -4$

*(c) $5x_1 + 6x_2 + 4x_3 = 3$
$7x_1 + 8x_2 + 6x_3 = 1$
$6x_1 + 7x_2 + 5x_3 = 0$

*15. Determine the values of λ for which the following system of equations has nontrivial solutions. Find the solutions for each value of λ.

$$(1 - \lambda)x_1 + 6x_2 = 0$$
$$5x_1 + (2 - \lambda)x_2 = 0$$

16. Determine the values of λ for which the following system of equations has nontrivial solutions. Find the solutions for each value of λ.

$$(\lambda + 4)x_1 + (\lambda - 2)x_2 = 0$$
$$4x_1 + (\lambda - 3)x_2 = 0$$

*17. Determine the values of λ for which the following system of equations has nontrivial solutions. Find the solutions for each value of λ.

$$(5 - \lambda)x_1 + 4x_2 + 2x_3 = 0$$
$$4x_1 + (5 - \lambda)x_2 + 2x_3 = 0$$
$$2x_1 + 2x_2 + (2 - \lambda)x_3 = 0$$

*18. Let $AX = \lambda X$ be a system of n linear equations in n variables. Prove that this system has a nontrivial solution if and only if $|A - \lambda I_n| = 0$.

19. Prove that if A is a symmetric matrix, then adj(A) is also symmetric.

20. Prove that if the square matrix A is not invertible, then adj(A) is not invertible.

21. Prove that if a matrix A is invertible, then adj(A) is also invertible, with inverse given by

$$[\text{adj}(A)]^{-1} = \frac{1}{|A|}A$$

*22. Prove that $[\text{adj}(A)]^{-1} = \text{adj}(A^{-1})$.

23. Prove that if A and B are square matrices of the same size, with AB being invertible, then A and B are invertible. Is the converse true?

24. Prove that if A is an invertible upper triangular matrix, then A^{-1} is also an upper triangular matrix.

*25. Let A be a matrix all of whose elements are integers. Show that if $|A| = \pm 1$, then all the elements of A^{-1} are integers.

26. Let $AX = 0$ be a homogeneous system of n linear equations in n variables that has only the trivial solution. If k is a positive integer, show that the system $A^k X = 0$ also has only the trivial solution.

*27. Let $AX = B_1$ and $AX = B_2$ be two systems of n linear equations in n variables, having the same matrix of coefficients A. Prove that $AX = B_2$ has a unique solution if and only if $AX = B_1$ has a unique solution.

Review Exercises Chapter 3

1. Evaluate the determinants of the following 2×2 matrices.

(a) $\begin{bmatrix} 3 & 2 \\ 5 & 1 \end{bmatrix}$ **(b)** $\begin{bmatrix} -3 & 0 \\ 1 & 6 \end{bmatrix}$ **(c)** $\begin{bmatrix} 9 & 7 \\ 1 & 4 \end{bmatrix}$

2. Let $A = \begin{bmatrix} 2 & 1 & 0 \\ -3 & 4 & 1 \\ 7 & 9 & 2 \end{bmatrix}$. Find the following minors and cofactors.

(a) M_{12} and C_{12} **(b)** M_{31} and C_{31}

(c) M_{22} and C_{22}

3. Evaluate the determinants of each of the following matrices in two ways, using the indicated rows and columns. Observe that you get the same answers both ways.

(a) $\begin{bmatrix} 1 & 2 & -3 \\ 0 & 2 & 5 \\ 4 & 1 & 2 \end{bmatrix}$ **(b)** $\begin{bmatrix} 0 & 5 & 3 \\ 2 & -3 & 1 \\ 2 & 7 & 3 \end{bmatrix}$

1st row and 1st column   3rd row and 2nd column

4. Solve the following equation for x.

$$\begin{vmatrix} x & x \\ 2 & x-3 \end{vmatrix} = -6$$

5. Simplify the determinants of the following matrices by creating many zeros in a single row or column, then expand the determinant in terms of that row or column.

(a) $\begin{bmatrix} 1 & 2 & -1 \\ 3 & 1 & 1 \\ 2 & 4 & 1 \end{bmatrix}$ **(b)** $\begin{bmatrix} 5 & 3 & 4 \\ 4 & 6 & 1 \\ 2 & -3 & 7 \end{bmatrix}$

(c) $\begin{bmatrix} 1 & 4 & -2 \\ 2 & 3 & 1 \\ -1 & 5 & 6 \end{bmatrix}$

6. If $A = \begin{bmatrix} 2 & -1 & 3 \\ 4 & -1 & 6 \\ 2 & -3 & 4 \end{bmatrix}$, then $|A| = 2$. Use this information, together with the row and column properties of determinants, to compute determinants of the following matrices.

(a) $\begin{bmatrix} 2 & -1 & 3 \\ 12 & -3 & 18 \\ 2 & -3 & 4 \end{bmatrix}$ **(b)** $\begin{bmatrix} 2 & -1 & 3 \\ 0 & 1 & 0 \\ 2 & -3 & 4 \end{bmatrix}$

(c) $\begin{bmatrix} 4 & -2 & 6 \\ -4 & 1 & -6 \\ 6 & -9 & 12 \end{bmatrix}$

7. Evaluate the following 3×3 determinants using elementary row operations.

(a) $\begin{vmatrix} 1 & 2 & 4 \\ -1 & 4 & 3 \\ 2 & 0 & 5 \end{vmatrix}$ **(b)** $\begin{vmatrix} -1 & 3 & 2 \\ 0 & 5 & 2 \\ 1 & 7 & 6 \end{vmatrix}$

(c) $\begin{vmatrix} 2 & -3 & 5 \\ 4 & 0 & 6 \\ 1 & 2 & 7 \end{vmatrix}$

8. If A is a 3×3 matrix with $|A| = -2$, compute the following determinants.

(a) $|3A|$ **(b)** $|2AA^t|$ **(c)** $|A^3|$

(d) $|(A^t A)^2|$ **(e)** $|(A^t)^3|$ **(f)** $|2A^t(A^{-1})^2|$

9. Prove that if A is a square matrix and c a nonzero scalar, then $|cA| = c^n |A|$.

10. Prove that if A and B are square matrices of the same size, with $|B| \neq 0$, then there exists a matrix C such that $CB = A$.

11. Let A and C be matrices of the same size with C being invertible. Prove that $|A| = |C^{-1}AC|$.

12. If A is a triangular matrix, prove that A^2 is also triangular.

13. Show that if A is an invertible matrix such that $A^2 = A$, then $|A| = 1$.

14. Determine whether the following matrices have inverses. If a matrix has an inverse, find the inverse using the formula for the inverse of a matrix.

(a) $\begin{bmatrix} 3 & 5 \\ 1 & 2 \end{bmatrix}$

(b) $\begin{bmatrix} 3 & 2 \\ -1 & 5 \end{bmatrix}$

(c) $\begin{bmatrix} 1 & 4 & -1 \\ 0 & 2 & 0 \\ 1 & 6 & -1 \end{bmatrix}$

(d) $\begin{bmatrix} 2 & 1 & 3 \\ 0 & 2 & 9 \\ 4 & 2 & 11 \end{bmatrix}$

15. Solve the following systems of equations using Cramer's rule.

(a) $2x_1 + x_2 = -1$
$3x_1 - 5x_2 = 18$

(b) $x_1 + x_2 + x_3 = 1$
$2x_1 - x_2 + 3x_3 = 5$
$4x_1 + 5x_2 + x_2 = 3$

16. Prove that if the square matrix A is not invertible, then $A[\text{adj}(A)]$ is the zero matrix.

17. Prove that if A is an invertible triangular matrix, then all the diagonal elements must be nonzero.

18. Let $AX = B$ be a system of n linear equations in n variables with the elements of both A and B being integers. Prove that if $|A| = \pm 1$, the solution X has all integer components.

MATLAB Discussion

M.11   Determinants (Sections 3.1, 3.2, 3.3)

MATLAB has a determinant function **det**(A).

Example 1   Find the determinant of the matrix $A = \begin{bmatrix} 1 & 2 & -1 \\ 3 & 0 & 1 \\ 4 & 2 & 1 \end{bmatrix}$.

»A = [1 2 −1;3 0 1;4 2 1]
»det(A)
ans =
    −6

Example 2   Determine the minor of the element a_{32} of the matrix $A = \begin{bmatrix} 1 & 0 & 3 \\ 4 & -1 & 2 \\ 0 & -2 & 1 \end{bmatrix}$.

»A = [1 0 3;4 −1 2;
    0 −2 1];
»minor(A, 3, 2)   {author-defined function for finding minors}
The submatrix is   {The submatrix obtained on deleting
  1   3   row 3 and column 2}
  4   2
The minor is   {The minor is the determinant
  −10   of the submatrix}

Exercise 1   Use the determinant function to compute the determinants of the following matrices. Use the minor function to determine the minors of the indicated elements.

*(a) $A = \begin{bmatrix} -1 & 2 & 3 \\ 0 & 1 & 4 \\ 1 & 1 & 2 \end{bmatrix}$

Find minors of a_{22} and a_{31}.

(b) $B = \begin{bmatrix} 1 & -7 & 6 \\ 2 & 1 & 0 \\ 3 & 1 & -5 \end{bmatrix}$

Find minors of b_{21} and b_{33}.

Exercise 2   Compute the determinants of the following matrices and explain the results.

*(a) $\begin{bmatrix} 1 & -2 & 3 \\ 0 & 4 & 1 \\ 3 & -6 & 9 \end{bmatrix}$

(b) $\begin{bmatrix} 2 & 3 & -3 \\ -4 & 1 & 6 \\ 6 & 2 & -9 \end{bmatrix}$

*(c) $\begin{bmatrix} 1 & 5 & 5 \\ 0 & -2 & -2 \\ 3 & 1 & 1 \end{bmatrix}$ (d) $\begin{bmatrix} 1 & -2 & 4 \\ 0 & -1 & 3 \\ -3 & 6 & -12 \end{bmatrix}$

Exercise 3   Section 3.3 gives an elimination method for computing the determinant of a matrix. Use the function **transf** to perform the row operations to compute the determinants of the following matrices. Check your answers with **det**(A).

*(a) $\begin{vmatrix} 1 & -1 & 0 & 2 \\ -1 & 1 & 0 & 0 \\ 2 & -2 & 0 & 1 \\ 3 & 1 & 5 & -1 \end{vmatrix}$ (b) $\begin{vmatrix} 2 & 1 & 3 & 1 \\ -2 & 3 & -1 & 2 \\ 2 & 1 & 2 & 3 \\ -4 & -2 & 0 & -1 \end{vmatrix}$

M.12   Matrix Inverses and Systems of Linear Equations (Section 3.4)

Inverse of a Matrix Using Its Adjoint

If A is a square matrix with $|A| \neq 0$, then A is invertible with

$$A^{-1} = \frac{1}{|A|} \text{adj}(A)$$

There is an author-defined function **adj**(A) for computing the adjoint of a matrix.

Exercise 1   Use MATLAB to compute the inverses of the following matrices using both **inv**(A) and **adj**(A)/**det**(A). Compare the results using the two methods.

*(a) $\begin{bmatrix} 1 & 2 & 3 \\ 0 & 1 & 2 \\ 4 & 5 & 3 \end{bmatrix}$ (b) $\begin{bmatrix} 0 & 3 & 3 \\ 1 & 2 & 3 \\ 1 & 4 & 6 \end{bmatrix}$ *(c) $\begin{bmatrix} 1 & 2 & -1 \\ 2 & 4 & -3 \\ 1 & -2 & 0 \end{bmatrix}$

Cramer's Rule

Let $AX = B$ be a system of n linear equations in n variables such that $|A| \neq 0$. The system has a unique solution

$$x_1 = \frac{|A_1|}{|A|}, \ x_2 = \frac{|A_2|}{|A|}, \ \dots, \ x_n = \frac{|A_n|}{|A|}$$

where A_i is the matrix obtained by replacing column i of A with B.

Example 1   Solve the following system of equations using Cramer's rule.

$$x_1 + 3x_2 + x_3 = -2$$
$$2x_1 + 5x_2 + x_3 = -5$$
$$x_1 + 2x_2 + 3x_3 = 6$$

The matrix of coefficients A and column matrix of constants B, are

$$A = \begin{bmatrix} 1 & 3 & 1 \\ 2 & 5 & 1 \\ 1 & 2 & 3 \end{bmatrix} \text{ and } B = \begin{bmatrix} -2 \\ -5 \\ 6 \end{bmatrix}$$

Using MATLAB, we get

»A = [1   3   1;2   5   1;1   2   3];   B = [−2;−5;6];
»Cramer(A, B)   {author-defined function for Cramer's rule}
All information? y/n: y   {option of getting all determinants, or just the solution}
det−matrx−coeffs =
  −3
variable =   {x_1 variable for this system}
  1

Matrix =	**determinant =**	... **determinant =**	**ans =**
−2 3 1	**−3**	**−9**	**1 −2 3**
−5 5 1	{$|A_1|$}	{$|A_3|$}	{unique solution,
6 2 3			$x_1 = 1, x_2 = -2, x_3 = 3$}
{A_1}			

Exercise 2   Solve the following systems of equations using Cramer's rule.

*(a) $\quad x_1 + 3x_2 + 4x_3 = 3$
$\quad\quad 2x_1 + 6x_2 + 9x_3 = 5$
$\quad\quad 3x_1 + \quad x_2 - 2x_3 = 7$

(b) $x_1 + 2x_2 + x_3 = 9$
$\quad\quad x_1 + 3x_2 - x_3 = 4$
$\quad\quad x_1 + 4x_2 - x_3 = 7$

*(c) $2x_1 + \quad x_2 + 3x_3 = 2$
$\quad\quad 3x_1 - 2x_2 + 4x_3 = 2$
$\quad\quad\quad x_1 + 4x_2 - 2x_3 = 1$

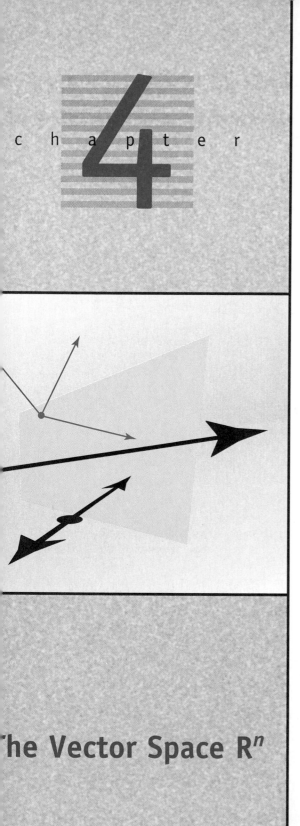

The Vector Space R^n

At this time the course

becomes geometrical in nature. We shall discuss two-dimensional and three-dimensional spaces, leading up to *n*-dimensional spaces. Matrices and systems of linear equations will be valuable tools for this discussion.

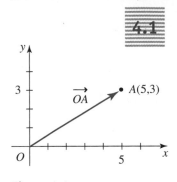

Figure 4.1

4.1   Introduction to Vectors

The locations of points in a plane are usually discussed in terms of a coordinate system. For example, in Figure 4.1 the location of each point in the plane can be described using a **rectangular coordinate system**. The point A is the point $(5, 3)$.

Furthermore, A is a certain distance in a certain direction from the **origin** $(0, 0)$. The distance and direction are characterized by the length and direction of the line segment from the origin O, to A. We call such a directed line segment a **position vector** and denote it $\overrightarrow{OA}$. O is called the **initial point** of $\overrightarrow{OA}$ and A is called the **terminal point**. There are thus two ways of interpreting $(5, 3)$; it defines the location of a point in a plane, and it also defines the position vector $\overrightarrow{OA}$.

Example 1   Sketch the position vectors $\overrightarrow{OA} = (4, 1)$, $\overrightarrow{OC} = (-3, 4)$, and $\overrightarrow{OB} = (-5, -2)$. See Figure 4.2.

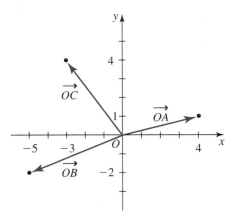

Figure 4.2

Denote the collection of all **ordered pairs** of real numbers **R**^{2}. Note the significance of "ordered" here; for example, $(5, 3)$ is not the same vector as $(3, 5)$. The order is significant.

These concepts can be extended to arrays consisting of three real numbers, such as $(2, 4, 3)$, which can be interpreted in two ways—as the location of a point in three-space relative to an *xyz* coordinate system, or as a position vector. These interpretations are illustrated in Figure 4.3. We shall denote the set of all **ordered triples** of real numbers by **R**^{3}.

Figure 4.3

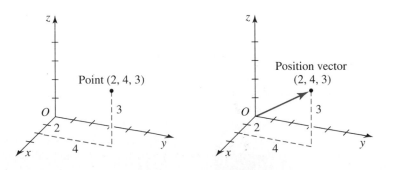

We now generalize these concepts with the following definition.

Definition
Let $(u_1, u_2, \ldots, u_n)$ be a sequence of n real numbers. The set of all such sequences is called n-space and is denoted $\mathbf{R}^n$.

*u_1 is the **first component** of $(u_1, u_2, \ldots, u_n)$. u_2 is the **second component** and so on.*

Many of the results and techniques that we develop for $\mathbf{R}^n$ with $n > 3$ will be useful mathematical tools, without direct geometrical significance. The elements of $\mathbf{R}^n$ can, however, be interpreted as points in n-space or as position vectors in n-space. It is difficult to visualize an n-space for $n > 3$, but the reader is encouraged to try to form an intuitive picture. A geometrical "feel" for what is taking place often makes an algebraic discussion easier to follow. The mathematics that we shall develop on $\mathbf{R}^n$ will be motivated by the geometry that we are familiar with on $\mathbf{R}^2$ and $\mathbf{R}^3$.

In keeping with this scheme, let $\mathbf{R}$ denote the **set of real numbers**.

Example 2
$\mathbf{R}^4$ is the collection of all sets of four ordered real numbers. For example, $(1, 2, 3, 4)$ and $\left(-1, \frac{3}{4}, 0, 5\right)$ are elements of $\mathbf{R}^4$.

$\mathbf{R}^5$ is the collection of all sets of five ordered real numbers. For example, $\left(-1, 2, 0, \frac{7}{8}, 9\right)$ is in this collection.

Definition
*Let $\mathbf{u} = (u_1, \ldots, u_n)$ and $\mathbf{v} = (v_1, \ldots, v_n)$ be two elements of $\mathbf{R}^n$. We say that $\mathbf{u}$ and $\mathbf{v}$ are **equal** if $u_1 = v_1, \ldots, u_n = v_n$. Thus two elements of $\mathbf{R}^n$ are equal if their **corresponding components** are equal.*

Let us now develop the algebraic structure for $\mathbf{R}^n$. The following definitions of addition and scalar multiplication are similar to those for matrices. We shall also find that the algebraic properties are similar.

Definition
Let $\mathbf{u} = (u_1, \ldots, u_n)$ and $\mathbf{v} = (v_1, \ldots, v_n)$ be elements of $\mathbf{R}^n$ and let c be a scalar. Addition and scalar multiplication are performed as follows.

$$\text{Addition:} \qquad \mathbf{u} + \mathbf{v} = (u_1 + v_1, \ldots, u_n + v_n)$$

$$\text{Scalar multiplication:} \quad c\mathbf{u} = (cu_1, \ldots, cu_n)$$

To add two elements of $\mathbf{R}^n$, we add corresponding components. To multiply an element of $\mathbf{R}^n$ by a scalar, we multiply every component by that scalar. Observe that the resulting elements are in $\mathbf{R}^n$. We say that $\mathbf{R}^n$ is **closed** under addition and under scalar multiplication.

$\mathbf{R}^n$ with operations of componentwise addition and scalar multiplication is an example of a **vector space**, and its elements are called **vectors**.

We shall henceforth in this course interpret $\mathbf{R}^n$ to be a vector space.

Example 3   Let $\mathbf{u} = (-1, 4, 3, 7)$ and $\mathbf{v} = (-2, -3, 1, 0)$ be vectors in **R**^{4}. Find $\mathbf{u} + \mathbf{v}$ and $3\mathbf{u}$.

Solution   We get

$$\mathbf{u} + \mathbf{v} = (-1, 4, 3, 7) + (-2, -3, 1, 0) = (-3, 1, 4, 7)$$
$$3\mathbf{u} = 3(-1, 4, 3, 7) = (-3, 12, 9, 21)$$

Note that the resulting vector under each operation is in the original vector space **R**^{4}.

We now give examples to illustrate geometrical interpretations of these vectors and their operations.

Example 4   This example gives a geometrical interpretation of vector addition. Consider the sum of the vectors (4, 1) and (2, 3). We get

$$(4, 1) + (2, 3) = (6, 4)$$

In Figure 4.4 we interpret these vectors as position vectors. Construct the parallelogram having the vectors (4, 1) and (2, 3) as adjacent sides. The vector (6, 4), the sum, will be the diagonal of the parallelogram.

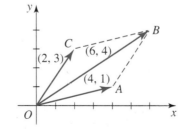

Figure 4.4

In general, if $\mathbf{u}$ and $\mathbf{v}$ are vectors in the same vector space, then $\mathbf{u} + \mathbf{v}$ is the diagonal of the parallelogram defined by $\mathbf{u}$ and $\mathbf{v}$. See Figure 4.5. (We shall use boldface for vectors and italic text for scalars.) This way of visualizing vector addition is useful in all vector spaces.

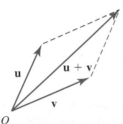

Figure 4.5 O

Example 5

This example gives a geometrical interpretation of scalar multiplication. Consider the scalar multiple of the vector (3, 2) by 2. We get

$$2(3, 2) = (6, 4)$$

Observe in Figure 4.6 that (6, 4) is a vector in the same direction as (3, 2), and 2 times it in length.

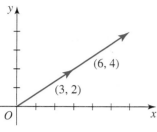

Figure 4.6

The direction will depend upon the sign of the scalar. The general result is as follows. Let **u** be a vector and c a scalar. The direction of c**u** will be the same as the direction of **u** if $c > 0$, the opposite direction to **u** if $c < 0$. The length of c**u** is $|c|$ times the length of **u**. See Figure 4.7.

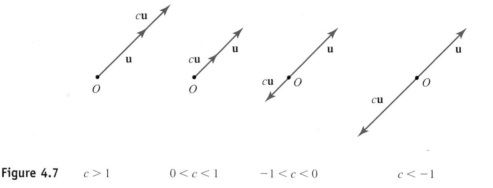

Figure 4.7 $\quad c > 1 \qquad\qquad 0 < c < 1 \qquad -1 < c < 0 \qquad\qquad c < -1$

Zero Vector The vector $(0, 0, \dots, 0)$, having n zero components, is called the **zero vector** of $\mathbf{R}^n$ and is denoted **0**. For example, $(0, 0, 0)$ is the zero vector of $\mathbf{R}^3$. We shall find that zero vectors play a central role in the development of vector spaces.

Negative Vector The vector (-1)**u** is written $-$**u** and is called the **negative** of **u**. It is a vector having the same magnitude as **u**, but lies in the opposite direction to **u**.

Subtraction Subtraction is defined on elements of $\mathbf{R}^n$ by subtracting corresponding components. For example, in $\mathbf{R}^3$,

$$(5, 3, -6) - (2, 1, 3) = (3, 2, -9)$$

Observe that this is equivalent to

$$(5, 3, -6) + (-1)(2, 1, 3) = (3, 2, -9)$$

Thus subtraction is not a new operation on $\mathbf{R}^n$, but a combination of addition and scalar multiplication by -1. We have only two independent operations on $\mathbf{R}^n$, namely addition and scalar multiplication.

We now discuss some of the properties of vector addition and scalar multiplication. The properties are similar to those of matrices.

Theorem 4.1   Let **u**, **v**, and **w** be vectors in **R**^{n} and let c and d be scalars.

(a) $\mathbf{u} + \mathbf{v} = \mathbf{v} + \mathbf{u}$ (e) $c(d\mathbf{u}) = (cd)\mathbf{u}$

(b) $\mathbf{u} + (\mathbf{v} + \mathbf{w}) = (\mathbf{u} + \mathbf{v}) + \mathbf{w}$ (f) $c(\mathbf{u} + \mathbf{v}) = c\mathbf{u} + c\mathbf{v}$

(c) $\mathbf{u} + \mathbf{0} = \mathbf{0} + \mathbf{u} = \mathbf{u}$ (g) $(c + d)\mathbf{u} = c\mathbf{u} + d\mathbf{u}$

(d) $\mathbf{u} + (-\mathbf{u}) = \mathbf{0}$ (h) $1\mathbf{u} = \mathbf{u}$

Proof   These results are proved by writing the vectors in terms of components and using the definitions of vector addition and scalar multiplication, and the properties of real numbers. We give the proofs of (a) and (f).

Let $\mathbf{u} = (u_1, \ldots, u_n)$ and $\mathbf{v} = (v_1, \ldots, v_n)$. Then

$$\mathbf{u} + \mathbf{v} = \mathbf{v} + \mathbf{u}: \quad \mathbf{u} + \mathbf{v} = (u_1, \ldots, u_n) + (v_1, \ldots, v_n)$$
$$= (u_1 + v_1, \ldots, u_n + v_n)$$
$$= (v_1 + u_1, \ldots, v_n + u_n)$$
$$= \mathbf{v} + \mathbf{u}$$

$$c(\mathbf{u} + \mathbf{v}) = c\mathbf{u} + c\mathbf{v}: \quad c(\mathbf{u} + \mathbf{v}) = c((u_1, \ldots, u_n) + (v_1, \ldots, v_n))$$
$$= c((u_1 + v_1, \ldots, u_n + v_n))$$
$$= (c(u_1 + v_1), \ldots, c(u_n + v_n))$$
$$= (cu_1 + cv_1, \ldots, cu_n + cv_n)$$
$$= (cu_1, \ldots, cu_n) + (cv_1, \ldots, cv_n)$$
$$= c(u_1, \ldots, u_n) + c(v_1, \ldots, v_n)$$
$$= c\mathbf{u} + c\mathbf{v}$$

Some of the above theorems can be illustrated geometrically. The commutative property of vector addition is illustrated in Figure 4.8. Note that we get the same diagonal to the parallelogram whether we add the vectors in the order $\mathbf{u} + \mathbf{v}$ or in the order $\mathbf{v} + \mathbf{u}$. One implication of the above results is that we can write certain algebraic expressions involving vectors, without parentheses, as was the case for matrices.

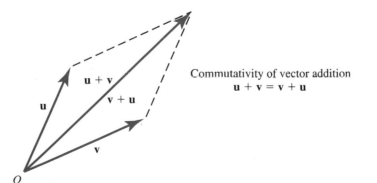

Commutativity of vector addition
$\mathbf{u} + \mathbf{v} = \mathbf{v} + \mathbf{u}$

Figure 4.8 O

Example 6   Let $\mathbf{u} = (2, 5, -3)$, $\mathbf{v} = (-4, 1, 9)$, and $\mathbf{w} = (4, 0, 2)$. Determine the vector $2\mathbf{u} - 3\mathbf{v} + \mathbf{w}$.

Solution   We get

$$2\mathbf{u} - 3\mathbf{v} + \mathbf{w} = 2(2, 5, -3) - 3(-4, 1, 9) + (4, 0, 2)$$
$$= (4, 10, -6) - (-12, 3, 27) + (4, 0, 2)$$
$$= (4 + 12 + 4, 10 - 3 + 0, -6 - 27 + 2)$$
$$= (20, 7, -31)$$

Physical quantities, such as forces and velocities, that have both direction and magnitude are called **vectors**. They can be represented mathematically by elements of $\mathbf{R}^2$ and $\mathbf{R}^3$, as the following example illustrates.

Example 7   If forces acting on a body all lie in a plane, they can be represented by elements of $\mathbf{R}^2$; if they lie in three dimensions, they can be represented by elements of $\mathbf{R}^3$. Two forces acting simultaneously on a body are equivalent to a single force called the **resultant force**. Experiments show that if two forces are represented by elements of $\mathbf{R}^2$ or $\mathbf{R}^3$, the resultant force is represented by the sum of these vectors. This law is called the **parallelogram of forces**. We have seen that such a vector sum is the diagonal of the parallelogram having the given vectors as adjacent sides—hence the term *parallelogram of forces*.

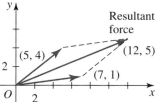

Figure 4.9

Let a body be located at the origin of a rectangular coordinate system. Two forces, represented by the vectors $(5, 4)$ and $(7, 1)$ act on the body. See Figure 4.9. Let us find the resultant force and its magnitude. On adding $(5, 4)$ and $(7, 1)$, we get

$$(5, 4) + (7, 1) = (12, 5)$$

The resultant force is $(12, 5)$.

The magnitude of a force is given by the length of the vector. This length can be found using the Theorem of Pythagoras. The length of the vector $(12, 5)$ is computed as follows.

$$\text{Length of } (12, 5) = \sqrt{12^2 + 5^2} = \sqrt{144 + 25} = \sqrt{169} = 13$$

The magnitude of the resultant force is 13.

Suppose we wish to double the magnitude of the resultant force by changing the force $(7, 1)$ but keeping the force $(5, 4)$ unchanged. How can this be done? Scalar multiplication is used. The new resultant force will be $2(12, 5) = (24, 10)$. This force is in the same direction as $(12, 5)$, but twice it in length. See Figure 4.10. We see that

$$(24, 10) = (5, 4) + (19, 6)$$

The force $(7, 1)$ would have to be replaced by $(19, 6)$ to double the resultant force on the body.

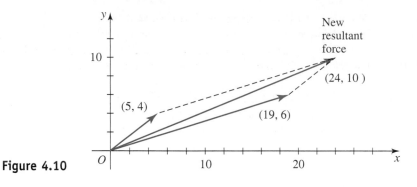

Figure 4.10

We now give an example of addition and scalar multiplication of elements of $\mathbf{R}^4$.

Example 8   Consider the following homogeneous system of linear equations.

$$\begin{aligned}
x_1 + 2x_2 - x_3 - 7x_4 &= 0 \\
2x_1 + 5x_2 \qquad\quad - 9x_4 &= 0 \\
x_1 + 3x_2 + x_3 - 2x_4 &= 0 \\
x_1 + 4x_2 + 4x_3 + 6x_4 &= 0
\end{aligned}$$

It can be shown that the system has many solutions:

$$x_1 = 2r, \qquad x_2 = r, \qquad x_3 = -3r, \qquad x_4 = r$$

We can write these solutions as elements of $\mathbf{R}^4$ as follows:

$$(2r, r, -3r, r)$$

Let us consider two specific solutions:

$$r = 1 \text{ gives the solution } (2, 1, -3, 1)$$
$$r = 2 \text{ gives the solution } (4, 2, -6, 2)$$

Add these two solutions to get the vector $(6, 3, -9, 3)$. Observe that this is a solution corresponding to $r = 3$.

Multiply the solution $(2, 1, -3, 1)$ by a scalar, say $\frac{3}{2}$. We get the vector $\left(3, \frac{3}{2}, -\frac{9}{2}, \frac{3}{2}\right)$. Observe that this is a solution corresponding to $r = \frac{3}{2}$.

These results prompt us to investigate whether the vector sum of any two solutions is a solution and whether the scalar product of any solution is a solution.

Let $(2r_1, r_1, -3r_1, r_1)$ and $(2r_2, r_2, -3r_2, r_2)$ be two solutions. If we add these solutions, we get the vector

$$(2(r_1 + r_2), r_1 + r_2, -3(r_1 + r_2), r_1 + r_2)$$

This is a solution corresponding to $r = r_1 + r_2$.

Consider the scalar multiple of $(2r_1, r_1, -3r_1, r_1)$ by an arbitrary scalar c. We get the vector

$$(2cr_1, cr_1, -3cr_1, cr_1)$$

This is a solution corresponding to $r = cr_1$.

Thus the vector sum of any two solutions is a solution, and a scalar multiple of any solution is a solution. New solutions can be generated from known solutions using the operations of vector addition and scalar multiplication.

This example suggests a future role of vector space theory in the discussion of systems of linear equations. You may have anticipated a result that we shall later derive, namely that the vector sum of any two solutions to a homogeneous system of linear equations is a solution. The scalar multiple of a solution is also a solution.

Column Vectors

Up to this point we have defined only **row vectors**; that is, the components of a vector were written in row form. We shall find that it is more suitable at times to use **column vectors,** with the components written in column form. We again define addition and scalar multiplication of column vectors in $\mathbf{R}^n$ in a componentwise manner:

$$\begin{bmatrix} u_1 \\ \vdots \\ u_n \end{bmatrix} + \begin{bmatrix} v_1 \\ \vdots \\ v_n \end{bmatrix} = \begin{bmatrix} u_1 + v_1 \\ \vdots \\ u_n + v_n \end{bmatrix} \quad \text{and} \quad c \begin{bmatrix} u_1 \\ \vdots \\ u_n \end{bmatrix} = \begin{bmatrix} cu_1 \\ \vdots \\ cu_n \end{bmatrix}$$

For example, in $\mathbf{R}^2$,

$$\begin{bmatrix} 1 \\ 2 \end{bmatrix} + \begin{bmatrix} 4 \\ 7 \end{bmatrix} = \begin{bmatrix} 5 \\ 9 \end{bmatrix} \quad \text{and} \quad 4 \begin{bmatrix} 2 \\ 1 \end{bmatrix} = \begin{bmatrix} 8 \\ 4 \end{bmatrix}$$

Exercise Set 4.1

*1. Sketch the position vectors $(1, 0)$ and $(0, 1)$ in $\mathbf{R}^2$. The notation **i** and **j** is often used in science for these vectors.

*2. Sketch the position vectors $(1, 0, 0)$, $(0, 1, 0)$, $(0, 0, 1)$ in $\mathbf{R}^3$. The notation **i, j,** and **k** is often used in science for these vectors.

3. Sketch the following position vectors in $\mathbf{R}^2$.
   *(a) $\overrightarrow{OA} = (5, 6)$, $\overrightarrow{OB} = (-3, 2)$, $\overrightarrow{OC} = (1, -3)$
   (b) $\overrightarrow{OP} = (2, 4)$, $\overrightarrow{OQ} = (-4, 5)$, $\overrightarrow{OR} = (3, -3)$
   *(c) $\mathbf{u} = (1, 1)$, $\mathbf{v} = (-1, -4)$, $\mathbf{w} = (5, 3)$

4. Sketch the following position vectors in $\mathbf{R}^3$.
   *(a) $\overrightarrow{OA} = (2, 3, 1)$, $\overrightarrow{OB} = (0, 5, -1)$, $\overrightarrow{OC} = (-1, 3, 4)$
   (b) $\mathbf{u} = (1, 1, 1)$, $\mathbf{v} = (-1, -2, -4)$, $\mathbf{w} = (0, 0, -3)$

5. Multiply the following vectors by the given scalars.
   *(a) $(1, 4)$ by 3
   (b) $(-1, 3)$ by -2
   *(c) $(2, 6)$ by $\frac{1}{2}$
   *(d) $(2, 4, 2)$ by $-\frac{1}{2}$

   (e) $(-1, 2, 3)$ by 3
   (f) $(-1, 2, 3, -2)$ by 4
   *(g) $(1, -4, 3, -2, 5)$ by -5
   (h) $(3, 0, 4, 2, -1)$ by 3

6. Compute the following vector expressions for $\mathbf{u} = (1, 2)$, $\mathbf{v} = (4, -1)$, and $\mathbf{w} = (-3, 5)$.
   (a) $\mathbf{u} + \mathbf{w}$
   *(b) $\mathbf{u} + 3\mathbf{v}$
   (c) $\mathbf{v} + \mathbf{w}$
   *(d) $2\mathbf{u} + 3\mathbf{v} - \mathbf{w}$
   *(e) $-3\mathbf{u} + 4\mathbf{v} - 2\mathbf{w}$

7. Compute the following vector expressions for $\mathbf{u} = (2, 1, 3)$, $\mathbf{v} = (-1, 3, 2)$, and $\mathbf{w} = (2, 4, -2)$.
   *(a) $\mathbf{u} + \mathbf{w}$
   *(b) $2\mathbf{u} + \mathbf{v}$
   (c) $\mathbf{u} + 3\mathbf{w}$
   *(d) $5\mathbf{u} - 2\mathbf{v} + 6\mathbf{w}$
   (e) $2\mathbf{u} - 3\mathbf{v} - 4\mathbf{w}$

8. Prove the following properties of vector addition and scalar multiplication that were introduced in this section.

(a) $\mathbf{u} + (\mathbf{v} + \mathbf{w}) = (\mathbf{u} + \mathbf{v}) + \mathbf{w}$

(b) $\mathbf{u} + (-\mathbf{u}) = \mathbf{0}$

(c) $(c + d)\mathbf{u} = c\mathbf{u} + d\mathbf{u}$ (d) $1\mathbf{u} = \mathbf{u}$

9. If $\mathbf{u}$, $\mathbf{v}$, and $\mathbf{w}$ are the following column vectors in $\mathbf{R}^2$, determine

(a) $\mathbf{u} + \mathbf{v}$ *(b) $2\mathbf{v} - 3\mathbf{w}$

(c) $2\mathbf{u} + 4\mathbf{v} - \mathbf{w}$ *(d) $-3\mathbf{u} - 2\mathbf{v} + 4\mathbf{w}$

$$\mathbf{u} = \begin{bmatrix} 2 \\ 3 \end{bmatrix}, \quad \mathbf{v} = \begin{bmatrix} -1 \\ -4 \end{bmatrix}, \quad \mathbf{w} = \begin{bmatrix} 4 \\ -6 \end{bmatrix}$$

10. If $\mathbf{u}$, $\mathbf{v}$, and $\mathbf{w}$ are the following column vectors in $\mathbf{R}^3$, determine

*(a) $\mathbf{u} + 2\mathbf{v}$ (b) $-4\mathbf{v} + 3\mathbf{w}$

*(c) $3\mathbf{u} - 2\mathbf{v} + 4\mathbf{w}$ (d) $2\mathbf{u} + 3\mathbf{v} - 8\mathbf{w}$

$$\mathbf{u} = \begin{bmatrix} 1 \\ 2 \\ -1 \end{bmatrix}, \quad \mathbf{v} = \begin{bmatrix} 3 \\ 0 \\ 1 \end{bmatrix}, \quad \mathbf{w} = \begin{bmatrix} -1 \\ 0 \\ 5 \end{bmatrix}$$

11. Consider the following homogeneous system of linear equations:

$$\begin{aligned} x_1 + 2x_2 - \ x_3 - 2x_4 &= 0 \\ 2x_1 + 5x_2 \qquad - 2x_4 &= 0 \\ 4x_1 + 9x_2 - 2x_3 - 6x_4 &= 0 \\ x_1 + 3x_2 + \ x_3 \qquad &= 0 \end{aligned}$$

You are given the following two solutions: $(5, -2, 1, 0)$ and $(11, -4, 1, 1)$. Using the operations of $\mathbf{R}^4$, generate five other solutions.

12. You are given that $(0, -1, 1, -1)$ and $(3, 7, 2, 1)$ are solutions of the following homogeneous system of linear equations. Use the operations of $\mathbf{R}^4$ to generate five further solutions.

$$\begin{aligned} x_1 - \ x_2 + \ x_3 + 2x_4 &= 0 \\ 3x_1 - 2x_2 + \ x_3 + 3x_4 &= 0 \\ 5x_1 - 4x_2 + 3x_3 + 7x_4 &= 0 \\ 2x_1 - \ x_2 \qquad + \ x_4 &= 0 \end{aligned}$$

13. Determine the resultants of the following forces.

*(a) $(3, 2)$ and $(5, -5)$ (b) $(-1, 3)$ and $(-2, 3)$

*(c) $(9, -1)$ and $(-3, 2)$

14. Determine the resultants of the following forces.

*(a) $(1, 4, 6)$ and $(2, 5, 1)$

(b) $(3, -1, 7)$ and $(-2, 5, -1)$

*15. Find the resultant of the forces $(0, 3)$ and $(3, 1)$. How can the resultant be doubled in magnitude by changing $(0, 3)$ but keeping $(3, 1)$ unchanged?

*16. Find the resultant of the forces $(5, 1)$ and $(3, 4)$. How can the resultant be changed to $(9, 7)$ by changing $(3, 4)$ but keeping $(5, 1)$ unchanged?

17. Find the resultant of the forces $(2, 8)$ and $(-1, 3)$. How can the resultant be changed to $(2, 5)$ by changing $(2, 8)$ but keeping $(-1, 3)$ unchanged?

*18. Which of the following physical quantities are scalars, and which are vectors?

(a) temperature   (b) acceleration

(c) pressure   (d) frequency

(e) gravity   (f) position

(g) time   (h) sound

(i) cost

4.2   Subspaces of $\mathbf{R}^n$

The vector space $\mathbf{R}^n$ is a set of elements called vectors, on which operations of addition and scalar multiplication have been defined. Observe that if we add two elements of $\mathbf{R}^n$, we get an element of $\mathbf{R}^n$. Furthermore, if we multiply an element of $\mathbf{R}^n$ by a scalar, we get an element of $\mathbf{R}^n$. For example, in $\mathbf{R}^3$

$$(1, 2, 5) + (3, 1, 7) = (4, 3, 12) \quad \text{and} \quad 3(1, -2, 5) = (3, -6, 15)$$

$\uparrow$ $\uparrow$ $\uparrow$ $\uparrow$ $\uparrow$
elements of $\mathbf{R}^3$ element of $\mathbf{R}^3$ element of $\mathbf{R}^3$ element of $\mathbf{R}^3$

$\mathbf{R}^n$ is said to be **closed under addition and scalar multiplication**.

These closure properties give the vector space **R**^{n} a certain completeness. We now look at certain subsets of **R**^{n} that have these same closure characteristics.

Consider the subset V of **R**^{3} consisting of vectors of the form (a, a, b). V consists of all the elements of **R**^{3} that have the first two components the same. For example, $(2, 2, 3)$ and $(-1, -1, 5)$ would be in V; $(1, 2, 3)$ would not be in V. Observe that if we add two elements of V, we get an element of V and that if we multiply an element of V by a scalar, we get an element of V. Let (a, a, b) and (c, c, d) be elements of V and let k be a scalar. Then

$$(a, a, b) + (c, c, d) = (a + c, a + c, b + d) \in V$$

$$k(a, a, b) = (ka, ka, kb) \in V$$

V has operations of addition and scalar multiplication defined on it. It is also closed under these operations. It has the algebraic characteristics of the vector space **R**^{3}. We thus define it to be a vector space, **embedded** in **R**^{3}. We call such an embedded vector space a **subspace** of the larger space.

Let us look at the geometrical interpretation of V. **R**^{3} is the set of all points in 3-space. V will be the subset of all points that have equal x and y components. These make up a plane perpendicular to the xy plane, through the line $y = x$, $z = 0$. See Figure 4.11. The sum of any two position vectors that lie in this plane will lie in the plane. The scalar multiple of any vector that lies in the plane also lies in the plane. This discussion leads to the following definition of a subspace of a vector space.

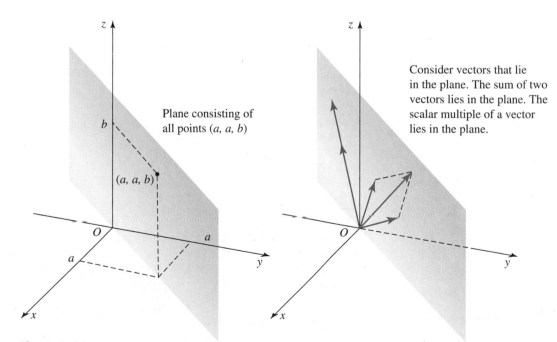

Plane consisting of
all points (a, a, b)

Consider vectors that lie
in the plane. The sum of two
vectors lies in the plane. The
scalar multiple of a vector
lies in the plane.

Figure 4.11

Definition   *A nonempty subset of the vector space* **R**^{n} *forms a* ***subspace*** *of* **R**^{n} *if it is closed under addition and under scalar multiplication.*

        A subspace is a vector space in its own right.

Example 1   Let U be the subset of **R**^{3} consisting of all vectors of the form $(a, 0, 0)$ (with zeros as second and third components). Show that U is a subspace of **R**^{3}.

Solution   Let $(a, 0, 0)$ and $(b, 0, 0)$ be two elements of U and let k be a scalar. We get

$$(a, 0, 0) + (b, 0, 0) = (a + b, 0, 0) \in U$$

$$k(a, 0, 0) = (ka, 0, 0) \in U$$

The sum and scalar product are in U. Thus U is a subspace of **R**^{3}.

    Geometrically, U is the set of vectors that lie on the x axis. Note that the sum of two such vectors lies on the x axis and so does the scalar multiple of any such vector.

Example 2   Let W be the set of vectors of the form (a, a^2, b). Show that W is not a subspace of **R**^{3}.

Solution W consists of all elements of **R**^{3} for which the second component is the square of the first. Thus, for example, the vector $(2, 4, 3)$ is in W, whereas the vector $(2, 5, 3)$ is not.

    Let (a, a^2, b) and (c, c^2, d) be elements of W. We get

$$(a, a^2, b) + (c, c^2, d) = (a + c, a^2 + c^2, b + d)$$
$$\neq (a + c, (a + c)^2, b + d)$$

Thus $(a, a^2, b) + (c, c^2, d)$ is not an element of W. W is not closed under addition. W is not a subspace.

    The following theorem gives an important characteristic of all subspaces.

Theorem 4.2   Let V be a subspace of a vector space **R**^{n}. V contains the zero vector.

Proof   Let $\mathbf{v} = (v_1, v_2, \ldots, v_n)$ be an element of V. Let 0 be the zero scalar and $\mathbf{0}$ the zero vector.

    Since V is a subspace, it is closed under scalar multiplication. Thus $0\mathbf{v}$ is an element of V. We get

$$0\mathbf{v} = 0(v_1, v_2, \ldots, v_n)$$
$$= (0v_1, 0v_2, \ldots, 0v_n)$$
$$= (0, 0, \ldots, 0)$$
$$= \mathbf{0}$$

Therefore $\mathbf{0} \in V$, proving the theorem.

This theorem tells us, for example, that all subspaces of **R**^{3} contain (0, 0, 0). This means that all subspaces of 3-space pass through the origin. This theorem can sometimes be used as a quick check to show that certain subsets cannot be subspaces. If a given subset does not contain the zero vector, it cannot be a subspace. The following example illustrates this.

Example 3   Let W be the set of vectors of the form $(a, a, a + 2)$. Show that W is not a subspace of **R**^{3}.

Solution   We check to see if (0, 0, 0) is in W. Is there a value of a for which $(a, a, a + 2)$ is (0, 0, 0)? On equating $(a, a, a + 2)$ to (0, 0, 0), we get

$$(a, a, a + 2) = (0, 0, 0)$$

Equating corresponding components, we get

$$a = 0$$
$$a + 2 = 0$$

This system of equations has no solution. Thus (0, 0, 0) is not an element of W. W is not a subspace.

In the following sections we develop tools that will enable us to understand subspaces further. We shall find that all the significant subspaces of **R**^{3} are either lines or planes through the origin, and that those of **R**^{2} are lines through the origin.

Exercise Set 4.2

1. Consider the sets of vectors of the following form. Prove that the sets are subspaces of **R**^{2}. Give the geometrical interpretation of each subspace.

    (a) $(a, 0)$ ****(b)** (a, a)

    (c) $(a, 2a)$ ****(d)** $(a, a + 3b)$

2. Consider the sets of vectors of the following form. Prove that the sets are subspaces of **R**^{3}. Give the geometrical interpretation of cach subspace.

    ****(a)** $(a, b, 0)$ **(b)** $(0, a, 0)$

    ****(c)** $(a, 2a, b)$ **(d)** $(a, b, a + b)$

3. Consider the sets of vectors of the following form. Prove that the sets are subspaces of **R**^{3}. Give the geometrical interpretation of each subspace.

    ****(a)** (a, a, a) **(b)** $(0, a, 2a)$

    ****(c)** $(a, a + b, 3a)$ **(d)** $(a, 2a, 3a + 5b)$

4. Consider the sets of vectors of the following form. Prove that the sets are subspaces of **R**^{4}.

    (a) $(a, 2a, b, 3b)$ **(b)** $(a, 2a, 3a, 4a)$

    (c) $(0, a, b, a + 2b)$ **(d)** $(a, b, c, a + 2b + 3c)$

*5.   Let A be the set of vectors of the form $(a, 2a)$, B be the set of vectors of the form (a, a^2), and C be the set of vectors of the form $(a, a^2 + 3)$. Determine whether A, B, and C are subspaces of **R**^{2}.

6. Consider the sets of vectors of the following form. Determine which of the sets are subspaces of **R**^{3} by checking for closure under addition and scalar multiplication.

    ****(a)** $(a, b, a + 3)$ **(b)** $(a, 4a, -3a)$

    ****(c)** $(a, b, 2)$ **(d)** $(a, b, 4a - 1)$

7. Consider the sets of vectors of the following form. Determine which of the sets are subspaces of **R**^{3} by checking for closure under addition and scalar multiplication.

(a) $(a, b, a - 4b)$ *(b) $(a, a^2, 5a)$

(c) $(a, 1, 1)$ *(d) $(a, b, 2a + 3b + 6)$

8. Are the following sets subspaces of **R**^{3}? The set of all vectors of the form (a, b, c) where *(a) $a + b + c = 0$, (b) $a + b + c = 1$, *(c) $ab = 0$, (d) $ab = 5$, *(e) $ab = ac$, (f) $a = b + c$.

9. Which of the following subsets of **R**^{3} are subspaces? The set of all vectors of the form (a, b, c) where a, b, and c are *(a) integers, (b) nonnegative real numbers, *(c) rational numbers.

10. Are the following sets subspaces of **R**^{2}? The set of all vectors of the form (a) (a, b^2), *(b) (a, b^3), *(c) (a, b) where $a > 0$, (d) (a, b) where $ab < 0$, *(e) (a, b) where a is nonpositive and b is nonnegative.

*11. Give an example of a subset of **R**^{3} that is

(a) closed under addition, but not closed under scalar multiplication.

(b) closed under scalar multiplication, but not closed under addition.

Such examples illustrate the **independence** of these two conditions.

12. Prove that the following sets are not subspaces of **R**^{3} by showing that they do not contain the zero vector. The set of all vectors of the form *(a) $(a, a + 1, b)$,

(b) $(a, 3, 2a)$, *(c) $(a, b, a + b - 4)$, (d) (a, b, c) where $a > 0$.

13. *(a) Give an example of a subset of **R**^{2} that contains the zero vector but is not a subspace.

(b) Give an example of a subset of **R**^{3} that contains the zero vector but is not a subspace.

These examples illustrate that the property of containing the zero vector is a *necessary but not sufficient* condition for a subset to be a subspace.

*14. Let U be the set of all vectors of the form $(a, 2a, b)$ and V be the set of all vectors of the form $(p, 2p, q)$. Show that U and V are the same set. Is this set a subspace of **R**^{3}?

15. Let U be the set of all vectors of the form (a, b, c) and V be the set of all vectors of the form $(a, a + b, c)$. Show that U and V are the same set. Is this set a subspace of **R**^{3}?

*16. Let U be the set of all vectors of the form $(a, 3a, b + 2)$ and V be the set of all vectors of the form $(p, 3p, r^3)$. Show that U and V are the same set. Is this set a subspace of **R**^{3}?

17. Let U be the set of all vectors of the form $(a, b, 2a)$ and V be the set of all vectors of the form $(a, a + b, 2a)$. Show that U and V are the same set. Is this set a subspace of **R**^{3}?

18. Let U be a subset of **R**^{3}, let u_1 and u_2 be vectors in U, and let a and b be scalars. Prove that U is a subspace of **R**^{3} if and only if $au_1 + bu_2$ is a vector in U for all values of a and b.

4.3   Linear Combinations of Vectors

In the previous section we discussed the subspace of **R**^{3} consisting of all vectors of the form (a, a, b). Observe that an arbitrary vector in this space can be written

$$(a, a, b) = a(1, 1, 0) + b(0, 0, 1)$$

The implication is that every vector in the subspace can be expressed in terms of $(1, 1, 0)$ and $(0, 0, 1)$. For example,

$$(2, 2, 3) = 2(1, 1, 0) + 3(0, 0, 1)$$

and $$(-1, -1, 7) = -1(1, 1, 0) + 7(0, 0, 1)$$

The vectors $(1, 1, 0)$ and $(0, 0, 1)$ in some sense characterize the subspace. In this section and the following ones we pursue this approach to understanding vector spaces in terms of certain vectors that represent the whole space.

Definition *Let $v_1, v_2, \ldots, v_m$ be vectors in a vector space V. We say that v, a vector in V, is a **linear combination** of $v_1, v_2, \ldots, v_m$ if there exist scalars $c_1, c_2, \ldots, c_m$ such that v can be written*

$$v = c_1 v_1 + c_2 v_2 + \cdots + c_m v_m$$

Example 1 The vector $(5, 4, 2)$ is a linear combination of the vectors $(1, 2, 0)$, $(3, 1, 4)$, and $(1, 0, 3)$ since it can be written

$$(5, 4, 2) = (1, 2, 0) + 2(3, 1, 4) - 2(1, 0, 3)$$

The problem of determining whether or not a vector is a linear combination of other vectors becomes that of solving a system of linear equations.

Example 2 Determine whether or not the vector $(-1, 1, 5)$ is a linear combination of the vectors $(1, 2, 3)$, $(0, 1, 4)$, and $(2, 3, 6)$.

Solution We examine the identity

$$c_1(1, 2, 3) + c_2(0, 1, 4) + c_3(2, 3, 6) = (-1, 1, 5)$$

Can we find scalars c_1, c_2, and c_3 such that this identity holds?

Using the operations of addition and scalar multiplication, we get

$$(c_1, 2c_1 \; 3c_1) + (0, c_2, 4c_2) + (2c_3, 3c_3, 6c_3) = (-1, 1, 5)$$
$$(c_1 + 2c_3, 2c_1 + c_2 + 3c_3, 3c_1 + 4c_2 + 6c_3) = (-1, 1, 5)$$

Equating components leads to the following system of linear equations:

$$
\begin{aligned}
c_1 \qquad\quad + 2c_3 &= -1 \\
2c_1 + c_2 + 3c_3 &= \;\;\; 1 \\
3c_1 + 4c_2 + 6c_3 &= \;\;\; 5
\end{aligned}
$$

It can be shown that this system of equations has the unique solution

$$c_1 = 1, \qquad c_2 = 2, \qquad c_3 = -1$$

Thus the vector $(-1, 1, 5)$ is the following linear combination of the vectors $(1, 2, 3)$, $(0, 1, 4)$, and $(2, 3, 6)$:

$$(-1, 1, 5) = (1, 2, 3) + 2(0, 1, 4) - 1(2, 3, 6)$$

The following example illustrates that it may be possible to express a vector as a linear combination of other vectors in more than one way.

Example 3 Express the vector $(4, 5, 5)$ as a linear combination of the vectors $(1, 2, 3)$, $(-1, 1, 4)$, and $(3, 3, 2)$.

Solution   Examine the following identity for values of c_1, c_2, and c_3.

$$c_1(1, 2, 3) + c_2(-1, 1, 4) + c_3(3, 3, 2) = (4, 5, 5)$$

We get

$$(c_1, 2c_1, 3c_1) + (-c_2, c_2, 4c_2) + (3c_3, 3c_3, 2c_3) = (4, 5, 5)$$
$$(c_1 - c_2 + 3c_3, 2c_1 + c_2 + 3c_3, 3c_1 + 4c_2 + 2c_3) = (4, 5, 5)$$

Equating components leads to the following system of linear equations:

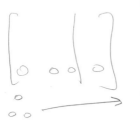

$$c_1 - c_2 + 3c_3 = 4$$
$$2c_1 + c_2 + 3c_3 = 5$$
$$3c_1 + 4c_2 + 2c_3 = 5$$

This system of equations has many solutions,

$$c_1 = -2r + 3, \qquad c_2 = r + 1, \qquad c_3 = r$$

Thus the vector $(4, 5, 5)$ can be expressed in many ways as a linear combination of the vectors $(1, 2, 3)$, $(-1, 1, 4)$, and $(3, 3, 2)$:

$$(4, 5, 5) = (-2r + 3)(1, 2, 3) + (r - 1)(-1, 1, 4) + r(3, 3, 2)$$

For example,

$$r = 3 \text{ gives } (4, 5, 5) = -3(1, 2, 3) + 2(-1, 1, 4) + 3(3, 3, 2)$$
$$r = -1 \text{ gives } (4, 5, 5) = 5(1, 2, 3) - 2(-1, 1, 4) - (3, 3, 2)$$

The following example illustrates that it may not be possible to express a vector as a linear combination of other vectors.

Example 4   Show that the vector $(3, -4, -6)$ cannot be expressed as a linear combination of the vectors $(1, 2, 3)$, $(-1, -1, -2)$, and $(1, 4, 5)$.

Solution   Consider the identity

$$c_1(1, 2, 3) + c_2(-1, -1, -2) + c_3(1, 4, 5) = (3, -4, -6)$$

This identity leads to the following system of linear equations.

$$c_1 - c_2 + c_3 = 3$$
$$2c_1 - c_2 + 4c_3 = -4$$
$$3c_1 - 2c_2 + 5c_3 = -6$$

This system has no solution. Thus $(3, -4, -6)$ is not a linear combination of vectors $(1, 2, 3)$, $(-1, -1, -2)$, and $(1, 4, 5)$.

Definition   *The vectors* $\mathbf{v}_1, \mathbf{v}_2, \ldots, \mathbf{v}_m$ *are said to span a vector space if every vector in the space can be expressed as a linear combination of these vectors.*

A spanning set of vectors in a sense defines the vector space, since every vector in the space can be obtained from this set.

Example 5   Show that the vectors $(1, 2, 0)$, $(0, 1, -1)$, and $(1, 1, 2)$ span $\mathbf{R}^3$.

Solution   Let (x, y, z) be an arbitrary element of $\mathbf{R}^3$. We have to determine whether we can write

$$(x, y, z) = c_1(1, 2, 0) + c_2(0, 1, -1) + c_3(1, 1, 2)$$

Multiply and add vectors to get

$$(x, y, z) = (c_1 + c_3, 2c_1 + c_2 + c_3, -c_2 + 2c_3)$$

Thus

$$
\begin{aligned}
c_1 \quad\quad + \; c_3 &= x \\
2c_1 + c_2 + \; c_3 &= y \\
- c_2 + 2c_3 &= z
\end{aligned}
$$

This system of equations in the variables c_1, c_2, and c_3 is solved by the method of Gauss-Jordan elimination. It is found to have the solution

$$c_1 = 3x - y - z, \quad\quad c_2 = -4x + 2y + z, \quad\quad c_3 = -2x + y + z$$

The vectors $(1, 2, 0)$, $(0, 1, -1)$, and $(1, 1, 2)$ thus span $\mathbf{R}^3$. We can write an arbitrary vector of $\mathbf{R}^3$ as a linear combination of these vectors as follows.

$$(x, y, z) = (3x - y - z)(1, 2, 0) + (-4x + 2y + z)(0, 1, -1) + (-2x + y + z)(1, 1, 2)$$

This vector formula enables us to quickly express any vector in $\mathbf{R}^3$ as a linear combination of $(1, 2, 0)$, $(0, 1, -1)$, and $(1, 1, 2)$. For example, if we want to know how $(2, 4, -1)$ looks in terms of these vectors, we let $x = 2$, $y = 4$, and $z = -1$ in this formula. We get

$$(2, 4, -1) = 3(1, 2, 0) - (0, 1, -1) - (1, 1, 2)$$

We have developed the mathematics for looking at a vector space in terms of a set of vectors that spans the space. It is also useful to be able to do the converse, namely to use a set of vectors to **generate** a vector space.

Theorem 4.3   Let $\mathbf{v}_1, \ldots, \mathbf{v}_m$ be vectors in a vector space V. Let U be the set consisting of all linear combinations of $\mathbf{v}_1, \ldots, \mathbf{v}_m$. Then U is a subspace of V spanned by the vectors $\mathbf{v}_1, \ldots, \mathbf{v}_m$.

U is said to be the vector space *generated* by $\mathbf{v}_1, \ldots, \mathbf{v}_m$.

Proof   Let

$$\mathbf{u}_1 = a_1\mathbf{v}_1 + \cdots + a_m\mathbf{v}_m \quad \text{and} \quad \mathbf{u}_2 = b_1\mathbf{v}_1 + \cdots + b_m\mathbf{v}_m$$

be arbitrary elements of U. Then

$$\mathbf{u}_1 + \mathbf{u}_2 = (a_1\mathbf{v}_1 + \cdots + a_m\mathbf{v}_m) + (b_1\mathbf{v}_1 + \cdots + b_m\mathbf{v}_m)$$
$$= (a_1 + b_1)\mathbf{v}_1 + \cdots + (a_m + b_m)\mathbf{v}_m$$

$\mathbf{u}_1 + \mathbf{u}_2$ is a linear combination of $\mathbf{v}_1, \ldots, \mathbf{v}_m$. Thus $\mathbf{u}_1 + \mathbf{u}_2$ is in U. U is closed under vector addition.

Let c be an arbitrary scalar. Then

$$c\mathbf{u}_1 = c(a_1\mathbf{v}_1 + \cdots + a_m\mathbf{v}_m)$$
$$= ca_1\mathbf{v}_1 + \cdots + ca_m\mathbf{v}_m$$

$c\mathbf{u}_1$ is a linear combination of $\mathbf{v}_1, \ldots, \mathbf{v}_m$. Therefore $c\mathbf{u}_1$ is in U. U is closed under scalar multiplication. Thus U is a subspace of V.

By the definition of U, every vector in U can be written as a linear combination of $\mathbf{v}_1, \ldots, \mathbf{v}_m$. Thus $\mathbf{v}_1, \ldots, \mathbf{v}_m$ span U.

Example 6   Consider the vector space $\mathbf{R}^3$. The vectors $(-1, 5, 3)$ and $(2, -3, 4)$ are in $\mathbf{R}^3$. Let U be the subset of $\mathbf{R}^3$ consisting of all vectors of the form

$$c_1(-1, 5, 3) + c_2(2, -3, 4)$$

Then U is a subspace of $\mathbf{R}^3$.

The following are examples of some of the vectors in U, obtained by giving c_1 and c_2 various values.

$$c_1 = 1, c_2 = 0; \text{ vector } (-1, 5, 3)$$
$$c_1 = 0, c_2 = 1; \text{ vector } (2, -3, 4)$$
$$c_1 = 0, c_2 = 0; \text{ vector } (0, 0, 0)$$
$$c_1 = 2, c_2 = 3; \text{ vector } (4, 1, 18)$$

We can visualize U. U is made up of all vectors in the plane defined by the vectors $(-1, 5, 3)$ and $(2, -3, 4)$. See Figure 4.12.

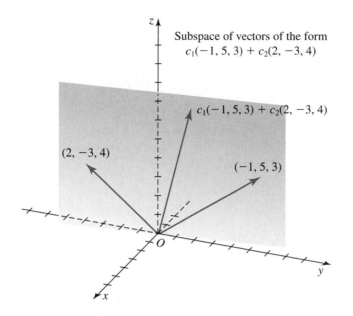

Figure 4.12

We can generalize this result. Let $\mathbf{v}_1$ and $\mathbf{v}_2$ be vectors in a vector space V. The subspace U generated by $\mathbf{v}_1$ and $\mathbf{v}_2$ is the set of all vectors of the form $c_1\mathbf{v}_1 + c_2\mathbf{v}_2$. If $\mathbf{v}_1$ and $\mathbf{v}_2$ are not collinear, then U is the plane defined by $\mathbf{v}_1$ and $\mathbf{v}_2$. See Figure 4.13.

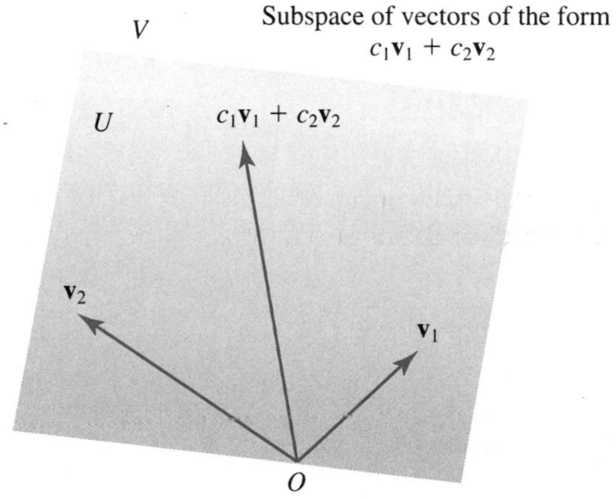

Figure 4.13

Example 7   Let $\mathbf{v}_1$ and $\mathbf{v}_2$ span a subspace U of a vector space V. Let k_1 and k_2 be nonzero scalars. Show that $k_1\mathbf{v}_1$ and $k_2\mathbf{v}_2$ also span U.

Solution   Let $\mathbf{v}$ be a vector in U. Since $\mathbf{v}_1$ and $\mathbf{v}_2$ span U, there exist scalars a and b such that

$$\mathbf{v} = a\mathbf{v}_1 + b\mathbf{v}_2$$

We can write

$$\mathbf{v} = \frac{a}{k_1}(k_1\mathbf{v}_1) + \frac{b}{k_2}(k_2\mathbf{v}_1)$$

Thus the vectors $k_1\mathbf{v}_1$ and $k_2\mathbf{v}_2$ span U.

If $\mathbf{v}_1$ and $\mathbf{v}_2$ are vectors in **R**^{3} that are not collinear, then we can visualize U as a plane in three dimensions. $k_1\mathbf{v}_1$ and $k_2\mathbf{v}_2$ will be vectors on the same lines as $\mathbf{v}_1$ and $\mathbf{v}_2$. See Figure 4.14.

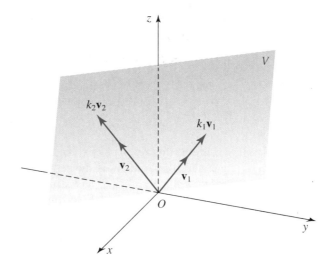

Figure 4.14

Two sets are equal if they have the same elements. A standard method of showing that two sets A and B are equal is to show that every element of A is in B, and that every element of B is in A. We use this method in the following example to show that two subspaces are equal.

Example 8   Let U be the subspace of **R**^{3} generated by the vectors $(1\ 2,\ 0)$ and $(-3,\ 1,\ 2)$. Let V be the subspace of **R**^{3} generated by the vectors $(-1,\ 5,\ 2)$ and $(4,\ 1,\ -2)$. Show that $U = V$.

Solution   Let $\mathbf{u}$ be a vector in U. Let us show that $\mathbf{u}$ is in V. Since $\mathbf{u}$ is in U, there exist scalars a and b such that

$$\mathbf{u} = a(1,\ 2,\ 0) + b(-3,\ 1,\ 2) = (a - 3b,\ 2a + b,\ 2b)$$

Let us see if we can write $\mathbf{u}$ as a linear combination of $(-1,\ 5,\ 2)$ and $(4,\ 1,\ -2)$.

$$\mathbf{u} = p(-1,\ 5,\ 2) + q(4,\ 1,\ -2) = (-p + 4q,\ 5p + q,\ 2p - 2q)$$

Such p and q would have to satisfy

$$-p + 4q = a - 3b$$
$$5p + q = 2a + b$$
$$2p - 2q = 2b$$

This system of equations has the unique solution $p = (a + b)/3$, $q = (a - 2b)/3$. Thus **u** can be written

$$\mathbf{u} = \frac{a + b}{3}(-1, 5, 2) + \frac{a - 2b}{3}(4, 1, -2)$$

Therefore **u** is a vector in V.

Conversely, let $\mathbf{v} = c(-1, 5, 2) + d(4, 1, -2)$ be a vector in V. We leave it to you to show that **v** can be written

$$\mathbf{v} = (2c + d)(1, 2, 0) + (c + d)(-3, 1, 2)$$

Thus **v** is a vector in U.

Therefore $U = V$. This subspace will be the plane through the origin defined by the vectors $(1, 2, 0)$, $(-3, 1, 2)$. See Figure 4.15.

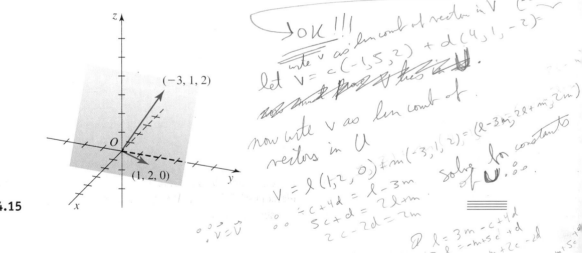

Figure 4.15

Exercise Set 4.3

1. In the following sets of vectors, determine whether the first vector is a linear combination of the other vectors.

*(a) $(-1, 7)$; $(1, -1)$, $(2, 4)$

(b) $(8, 13)$; $(1, 2)$, $(2, 3)$

*(c) $(-1, 15)$; $(-1, 4)$, $(2, -8)$

(d) $(13, 6)$; $(1, 3)$, $(4, 1)$

2. In the following sets of vectors, determine whether the first vector is a linear combination of the other vectors.

*(a) $(7, 5)$; $(1, -1)$, $(5, 1)$

(b) $(6, 22)$; $(2, 3)$, $(-1, 5)$

(c) $(2, 1)$; $(3, -1)$, $(9, -2)$

*(d) $(4, 0)$; $(-1, 2)$, $(3, 2)$, $(6, 4)$

(handwritten: $(x, y, z) = u_1(1, 2, 3)$ let u = anything)

3. In the following sets of vectors, determine whether the first vector is a linear combination of the other vectors.

(a) $(-3, 3, 7)$; $(1, -1, 2)$, $(2, 1, 0)$, $(-1, 2, 1)$

(b) $(-2, 11, 7)$; $(1, -1, 0)$, $(2, 1, 4)$, $(-2, 4, 1)$

(c) $(2, 7, 13)$; $(1, 2, 3)$, $(-1, 2, 4)$, $(1, 6, 10)$

(d) $(0, 10, 8)$; $(-1, 2, 3)$, $(1, 3, 1)$, $(1, 8, 5)$

4. In the following sets of vectors, determine whether the first vector is a linear combination of the other vectors.

(a) $(4, -4, 6)$; $(1, 2, -3)$, $(2, -4, 6)$, $(-1, 2, -3)$

(b) $(4, 3, 8)$; $(-1, 0, 1)$, $(2, 1, 3)$, $(0, 1, 5)$

(c) $(-1, 4, -9)$; $(-1, 3, 1)$, $(1, 1, 1)$, $(0, 1, 4)$

(d) $(1, 5, 3)$; $(1, 2, 1)$, $(-1, 3, 2)$, $(-1, 8, 5)$

5. Show that the following sets of vectors span $\mathbf{R}^2$. Express the vector $(3, 5)$ in terms of each spanning set.

(a) $(1, 1)$, $(1, -1)$ (b) $(1, 4)$, $(-2, 0)$

(c) $(1, 3)$, $(3, 10)$ (d) $(1, 0)$, $(0, 1)$

6. Show that the following sets of vectors span $\mathbf{R}^3$. Express the vector $(1, 3, -2)$ in terms of each spanning set.

(a) $(1, 2, 3)$, $(-1, -1, 0)$, $(2, 5, 4)$

(b) $(1, 3, 1)$, $(-1, 1, 0)$, $(4, 1, 1)$

(c) $(5, 1, 3)$, $(2, 0, 1)$, $(-2, -3, -1)$

(d) $(1, 0, 0)$, $(0, 1, 0)$, $(0, 0, 1)$

7. Determine whether the following vectors span $\mathbf{R}^2$.

(a) $(1, -3)$, $(2, -5)$ (b) $(1, 1)$, $(-2, 1)$

(c) $(-3, 1)$, $(3, -1)$ *(d)* $(3, 2)$, $(1, 1)$, $(1, 0)$

(e) $(4, -1)$, $(2, 3)$, $(6, 5)$

8. Determine whether the following vectors span $\mathbf{R}^3$.

(a) $(2, 1, 0)$, $(-1, 3, 1)$, $(4, 5, 0)$

(b) $(4, 0, 1)$, $(0, 1, 0)$, $(0, 0, 1)$

(c) $(1, 2, 1)$, $(-1, 3, 0)$, $(0, 5, 1)$

(d) $(1, -1, -1)$, $(0, 1, 2)$, $(1, 2, 1)$

*9. Give three other vectors in the subspace of $\mathbf{R}^3$ generated by the vectors $(1, 2, 3)$ and $(1, 2, 0)$.

10. Give three other vectors in the subspace of $\mathbf{R}^3$ generated by the vectors $(1, 2, 1)$ and $(2, 1, 4)$.

*11. Give three other vectors in the subspace of $\mathbf{R}^3$ generated by the vector $(1, 2, 3)$. Sketch the subspace.

12. Give three other vectors in the subspace of $\mathbf{R}^3$ generated by the vector $(4, -1, 3)$.

13. Give three other vectors in the subspace of $\mathbf{R}^2$ generated by the vector $(1, 2)$. Sketch the subspace.

*14. Give three other vectors in the subspace of $\mathbf{R}^4$ generated by the vector $(1, 2, -1, 3)$. *let u = any no.* *(handwritten: $(x, y, z) = u_1(1, 2, -1$)*

*15. Give three other vectors in the subspace of $\mathbf{R}^4$ generated by the vectors $(2, 1, -3, 4)$, $(-3, 0, 1, 5)$, and $(4, 1, 2, 0)$.

*16. Let U be the subspace of $\mathbf{R}^2$ generated by the vector $(-1, 3)$. Let V be the subspace of $\mathbf{R}^2$ generated by the vector $(-2, 6)$. Show that $U = V$.

17. Let U be the subspace of $\mathbf{R}^3$ generated by the vectors $(1, 2, 3)$ and $(-1, 2, 5)$. Let V be the subspace of $\mathbf{R}^3$ generated by the vectors $(1, 6, 11)$ and $(2, 0, -2)$. Show that $U = V$.

18. Let U be the subspace of $\mathbf{R}^3$ generated by the vectors $(-2, 1, 1)$ and $(0, 1, 3)$. Let V be the subspace of $\mathbf{R}^3$ generated by the vectors $(-2, 2, 4)$ and $(-2, 3, 7)$. Show that $U = V$.

19. Let U be the subspace of $\mathbf{R}^3$ generated by the vectors $(3, -1, 2)$ and $(1, 0, 4)$. Let V be the subspace of $\mathbf{R}^3$ generated by the vectors $(4, -1, 6)$ and $(1, -1, -6)$. Show that $U = V$.

*20. Let $\mathbf{u}$ be a nonzero vector in $\mathbf{R}^2$. Show that the subspace generated by $\mathbf{u}$ consists of vectors that lie on a line through the origin.

*21. Let $\mathbf{v}$ be a linear combination of $\mathbf{v}_1$ and $\mathbf{v}_2$. Let c_1 and c_2 be nonzero scalars. Show that $\mathbf{v}$ is also a linear combination of $c_1\mathbf{v}_1$ and $c_2\mathbf{v}_2$.

22. Let c_1 and c_2 be nonzero scalars. Show that if $\mathbf{v}$ is not a linear combination of $\mathbf{v}_1$ and $\mathbf{v}_2$, then neither is $\mathbf{v}$ a linear combination of $c_1\mathbf{v}_1$ and $c_2\mathbf{v}_2$.

23. Let c_1 and c_2 be nonzero scalars. Show that if $\mathbf{v}_1$ and $\mathbf{v}_2$ do not span V, then neither do $c_1\mathbf{v}_1$ and $c_2\mathbf{v}_2$.

*24. Let $\mathbf{v}_1$ and $\mathbf{v}_2$ span a vector space V. Let $\mathbf{v}_3$ be any other vector in V. Show that $\mathbf{v}_1$, $\mathbf{v}_2$, and $\mathbf{v}_3$ also span V.

4.4   Linear Dependence and Independence

In this section we continue the development of vector space structure. We introduce concepts of dependence and independence of vectors. These will be useful tools in constructing "efficient" spanning sets for vector spaces—sets where there are no redundant vectors.

Let us motivate the idea of dependence of vectors. Observe that the vector $(4, -1, 0)$ is a linear combination of the vectors $(2, 1, 3)$ and $(0, 1, 2)$ since it can be written

$$(4, -1, 0) = 2(2, 1, 3) - 3(0, 1, 2)$$

The above equation can be rewritten in a number of ways. Each vector can be expressed in terms of the other vectors:

$$(2, 1, 3) = \left(\tfrac{1}{2}\right)(4, -1, 0) + \left(\tfrac{3}{2}\right)(0, 1, 2)$$
$$(0, 1, 2) = \left(\tfrac{2}{3}\right)(2, 1, 3) - \left(\tfrac{1}{3}\right)(4, -1, 0)$$

Each of the three vectors is, in fact, dependent on the other two vectors. We express this by writing

$$(4, -1, 0) + 2(2, 1, 3) + 3(0, 1, 2) = (0, 0, 0)$$

This concept of dependence of vectors is made precise with the following definition.

Definition *(a)   The set of vectors $\{\mathbf{v}_1, \ldots, \mathbf{v}_m\}$ in a vector space V is said to be **linearly dependent** if there exists scalars $c_1, \ldots, c_m$, not all zero, such that*

$$c_1\mathbf{v}_1 + \cdots + c_m\mathbf{v}_m = \mathbf{0}$$

*(b)   The set of vectors $\{\mathbf{v}_1, \ldots, \mathbf{v}_m\}$ is **linearly independent** if*

$$c_1\mathbf{v}_1 + \cdots + c_m\mathbf{v}_m = \mathbf{0} \text{ can be satisfied only when } c_1 = 0, \ldots, c_m = 0.$$

Example 1 Show that the set $\{(1, 2, 3), (-2, 1, 1), (8, 6, 10)\}$ is linearly dependent in $\mathbf{R}^3$.

Solution Let us examine the identity

$$c_1(1, 2, 3) + c_2(-2, 1, 1) + c_3(8, 6, 10) = \mathbf{0}$$

We want to show that at least one of the cs can be nonzero. We get

$$(c_1, 2c_1, 3c_1) + (-2c_2, c_2, c_2) + (8c_3, 6c_3, 10c_3) = \mathbf{0}$$
$$(c_1 - 2c_2 + 8c_3, 2c_1 + c_2 + 6c_3, 3c_1 + c_2 + 10c_3) = \mathbf{0}$$

Equating each component of this vector to zero gives the system of equations

$$c_1 - 2c_2 + 8c_3 = 0$$
$$2c_1 + c_2 + 6c_3 = 0$$
$$3c_1 + c_2 + 10c_3 = 0$$

This system has the solution $c_1 = 4$, $c_2 = -2$, and $c_3 = -1$. Since at least one of the cs is nonzero, the set of vectors is linearly dependent.

The linear dependence is expressed by the equation

$$4(1, 2, 3) - 2(-2, 1, 1) - (8, 6, 10) = \mathbf{0}$$

Example 2   Show that the set $\{(3, -2, 2), (3, -1, 4), (1, 0, 5)\}$ is linearly independent in **R**^{3}.

Solution   We examine the identity

$$c_1(3, -2, 2) + c_2(3, -1, 4) + c_3(1, 0, 5) = \mathbf{0}$$

We want to show that this identity can only hold if c_1, c_2, and c_3 are all zero. We get

$$(3c_1, -2c_1, 2c_1) + (3c_2, -c_2, 4c_2) + (c_3, 0, 5c_3) = \mathbf{0}$$
$$(3c_1 + 3c_2 + c_3, -2c_1 - c_2, 2c_1 + 4c_2 + 5c_3) = \mathbf{0}$$

Equating the components to zero gives

$$\begin{aligned}
3c_1 + 3c_2 + c_3 &= 0 \\
-2c_1 - c_2 &= 0 \\
2c_1 + 4c_2 + 5c_3 &= 0
\end{aligned}$$

This system has the unique solution $c_1 = 0$, $c_2 = 0$, $c_3 = 0$. Thus the set is linearly independent.

We now present an important result that relates the concepts of linear dependence and linear combination.

Theorem 4.4   A set consisting of two or more vectors in a vector space is linearly dependent if and only if it is possible to express one of the vectors as a linear combination of the other vectors.

Proof   Let the set $\{\mathbf{v}_1, \mathbf{v}_2, \ldots, \mathbf{v}_m\}$ be linearly dependent. Therefore there exist scalars $c_1, c_2, \ldots, c_m$, not all zero, such that

$$c_1\mathbf{v}_1 + c_2\mathbf{v}_2 + \cdots + c_m\mathbf{v}_m = \mathbf{0}$$

Assume that $c_1 \neq 0$. The above identity can be rewritten

$$\mathbf{v}_1 = \left(\frac{-c_2}{c_1}\right)\mathbf{v}_2 + \cdots + \left(\frac{-c_m}{c_1}\right)\mathbf{v}_m$$

Thus $\mathbf{v}_1$ is a linear combination of $\mathbf{v}_2, \ldots, \mathbf{v}_m$.

Conversely, assume that $\mathbf{v}_1$ is a linear combination of $\mathbf{v}_2, \ldots, \mathbf{v}_n$. Therefore there exist scalars $d_2, \ldots, d_m$ such that

$$\mathbf{v}_1 = d_2\mathbf{v}_2 + \cdots d_m\mathbf{v}_m$$

Rewrite this equation as

$$1\mathbf{v}_1 + (-d_2)\mathbf{v}_2 + \cdots + (-d_m)\mathbf{v}_m = \mathbf{0}$$

Thus the set $\{\mathbf{v}_1, \mathbf{v}_2, \ldots, \mathbf{v}_m\}$ is linearly dependent, completing the proof.

This theorem leads to the following ways of visualizing what linear dependence means for sets of two and three vectors.

Linear Dependence of $\{v_1, v_2\}$

The set $\{\mathbf{v}_1, \mathbf{v}_2\}$ is linearly dependent if and only if it is possible to write one vector as a scalar multiple of the other vector. Let $\mathbf{v}_2 = c\mathbf{v}_1$. This means that $\mathbf{v}_1$ and $\mathbf{v}_2$ are collinear. See Figure 4.16.

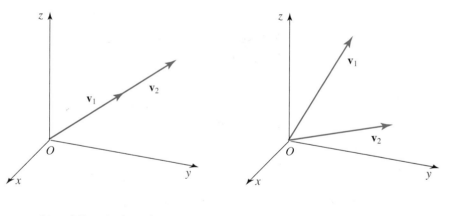

$\{\mathbf{v}_1, \mathbf{v}_2\}$ linearly dependent;
vectors lie on a line

$\{\mathbf{v}_1, \mathbf{v}_2\}$ linearly independent;
vectors do not lie on a line

Figure 4.16　　　　Linear dependence and independence of $\{\mathbf{v}_1, \mathbf{v}_2\}$ in $\mathbf{R}^3$

Linear Dependence of $\{v_1, v_2, v_3\}$

The set $\{\mathbf{v}_1, \mathbf{v}_2, \mathbf{v}_3\}$ is linearly dependent if and only if it is possible to write one of the vectors, say $\mathbf{v}_3$, as a linear combination of the other two vectors $\mathbf{v}_1$ and $\mathbf{v}_2$. Let $\mathbf{v}_3 = c_1\mathbf{v}_1 + c_2\mathbf{v}_2$. In general, when $\mathbf{v}_1$ and $\mathbf{v}_2$ are linearly independent, this means that $\mathbf{v}_3$ lies in the plane generated by $\mathbf{v}_1$ and $\mathbf{v}_2$. See Figure 4.17. If $\mathbf{v}_1$ and $\mathbf{v}_2$ happen to be linearly dependent, then $\mathbf{v}_3$ will lie on the line containing $\mathbf{v}_1$ and $\mathbf{v}_2$.

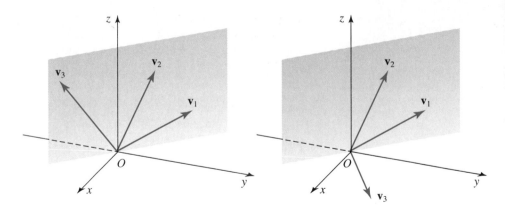

{$\mathbf{v}_1, \mathbf{v}_2, \mathbf{v}_3$} linearly dependent;
vectors lie in a plane

{$\mathbf{v}_1, \mathbf{v}_2, \mathbf{v}_3$} linearly independent;
vectors do not lie in a plane

Figure 4.17   Linear dependence and independence of {$\mathbf{v}_1, \mathbf{v}_2, \mathbf{v}_3$} in **R**^{3}

We complete this section with two further results about vector spaces.

Theorem 4.5   Let V be a vector space. Any set of vectors in V that contains the zero vector is linearly dependent.

Proof   Consider the set {$\mathbf{0}, \mathbf{v}_2, \ldots, \mathbf{v}_m$}, which contains the zero vector. Let us examine the identity

$$c_1\mathbf{0} + c_2\mathbf{v}_2 + \cdots + c_m\mathbf{v}_m = \mathbf{0}$$

We see that the identity is true for $c_1 = 1, c_2 = 0, \ldots, c_m = 0$ (not all zero). Thus the set of vectors is linearly dependent, proving the theorem.

Theorem 4.6   Let the set {$\mathbf{v}_1, \ldots, \mathbf{v}_m$} be linearly dependent in a vector space V. Any set of vectors in V that contains these vectors will also be linearly dependent.

Proof   Since the set {$\mathbf{v}_1, \ldots, \mathbf{v}_m$} is linearly dependent, there exist scalars $c_1, \ldots, c_m$, not all zero, such that

$$c_1\mathbf{v}_1 + \cdots + c_m\mathbf{v}_m = \mathbf{0}$$

Consider the set of vectors {$\mathbf{v}_1, \ldots, \mathbf{v}_m, \mathbf{v}_{m+1}, \ldots, \mathbf{v}_n$}, which contains the given vectors. There are scalars, not all zero, namely $c_1, \ldots, c_m, 0, \ldots, 0$, such that

$$c_1\mathbf{v}_1 + \cdots + c_m\mathbf{v}_m + 0\mathbf{v}_{m+1} + \cdots + 0\mathbf{v}_n = \mathbf{0}$$

Thus the set {$\mathbf{v}_1, \ldots, \mathbf{v}_m, \mathbf{v}_{m+1}, \ldots, \mathbf{v}_n$} is linearly dependent.

Example 3   Let the set $\{v_1, v_2\}$ be linearly independent. Prove that $\{v_1 + v_2, v_1 - v_2\}$ is also linearly independent.

Solution   Let us examine the identity

$$a(v_1 + v_2) + b(v_1 - v_2) = 0 \qquad (1)$$

If we can show that this identity implies $a = 0$ and $b = 0$, then $\{v_1 + v_2, v_1 - v_2\}$ will be linearly independent. We get

$$av_1 + av_2 + bv_1 - bv_2 = 0$$
$$(a + b)v_1 + (a - b)v_2 = 0$$

Since $\{v_1, v_2\}$ is linearly independent,

$$a + b = 0$$
$$a - b = 0$$

This system has the unique solution $a = 0$, $b = 0$. Returning to Identity (1), we get that $\{v_1 + v_2, v_1 - v_2\}$ is linearly independent.

Exercise Set 4.4

1. Prove that the following sets of vectors are linearly dependent in $\mathbf{R}^2$.

*(a) $\{(-1, 2), (2, -4)\}$ (b) $\{(3, 1), (9, 3)\}$

*(c) $\{(-2, 3), (6, -9)\}$ (d) $\{(1, 5), (0, 0)\}$

2. Prove that the following sets of vectors are linearly independent in $\mathbf{R}^2$.

*(a) $\{(1, 0), (0, 1)\}$ (b) $\{(1, 2), (3, 2)\}$

*(c) $\{(-1, 3), (2, 5)\}$ (d) $\{(2, -4), (5, 3)\}$

3. Prove that the following sets of vectors are linearly dependent in $\mathbf{R}^3$. Express one vector in each set as a linear combination of the other vectors.

*(a) $\{(1, -2, 3), (-2, 4, 1), (-4, 8, 9)\}$

(b) $\{(1, 0, 2), (2, 6, 4), (1, 12, 2)\}$

*(c) $\{(3, 4, 1), (2, 1, 0), (9, 7, 1)\}$

(d) $\{(1, 2, -3), (2, 1, -1), (0, 0, 0)\}$

4. Prove that the following sets of vectors are linearly independent in $\mathbf{R}^3$.

*(a) $\{(1, 2, 5), (1, -2, 1), (2, 1, 4)\}$

(b) $\{(1, 1, 1), (-4, 3, 2), (4, 1, 2)\}$

*(c) $\{(1, 3, -4), (3, -1, 4), (1, 0, -2)\}$

(d) $\{(3, 4, 7), (2, -1, 1), (4, 1, 3)\}$

(e) $\{(1, 0, 0), (0, 1, 0), (0, 0, 1)\}$

5. Use Theorem 4.6 to show that the following sets of vectors are linearly dependent in $\mathbf{R}^3$.

*(a) $\{(2, -1, 3), (-4, 2, -6), (8, 0, 1)\}$

(b) $\{(1, -2, 3), (7, 4, -2), (3, -6, 9)\}$

*(c) $\{(5, 2, -3), (3, 0, 4), (-3, 0, -4)\}$

(d) $\{(1, 1, 1), (2, 2, 2), (0, 1, 5), (3, -1, 4)\}$

6. Determine whether the following sets of vectors are linearly dependent or independent.

*(a) $\{(1, 2), (-1, 4), (-1, 16)\}$

(b) $\{(1, 3), (-2, 1)\}$

(c) $\{(1, -1, 3), (0, 2, 3), (1, -1, 2), (-2, 6, 3)\}$

*(d) $\{(1, 2, 8), (1, -1, -1), (1, 0, 3)\}$

*(e) $\{(1, 0, 0, 0), (0, 1, 0, 0), (0, 0, 1, 0), (0, 0, 0, 1)\}$

7. Find values of t for which the following sets are linearly dependent.

*(a) $\{(-1, 2), (t, -4)\}$ (b) $\{(3, t), (6, t - 1)\}$

*(c) $\{(2, -t), (2t + 6, 4t)\}$

8. (a) Prove that the set $\{(1, 1), (0, 2)\}$ is linearly independent in **R**^{2}.

(b) Prove that the set $\{(1, 1, 2), (0, -1, 3), (0, 0, 5)\}$ is linearly independent in **R**^{3}.

(c) Prove that the set $\{(3, -2, 4, 5), (0, 2, 3, -4), (0, 0, 2, 7), (0, 0, 0, 4)\}$ is linearly independent in **R**^{4}.

(d) Discuss the pattern of the zero components in the vectors of (a), (b), (c) above. How can you use this pattern to construct a set of five linearly independent vectors in **R**^{5}?

9. Consider the following matrix, which is in reduced echelon form.

$$\begin{bmatrix} 1 & 0 & 0 & 7 \\ 0 & 1 & 0 & 4 \\ 0 & 0 & 1 & 3 \end{bmatrix}$$

Show that the row vectors form a linearly independent set. Is the set of nonzero row vectors of any matrix in reduced echelon form linearly independent?

*10. Let the set $\{\mathbf{v}_1, \mathbf{v}_2\}$ be linearly dependent. Prove that $\{\mathbf{v}_1 + \mathbf{v}_2, \mathbf{v}_1 - \mathbf{v}_2\}$ is also linearly dependent.

11. Let the set $\{\mathbf{v}_1, \mathbf{v}_2, \mathbf{v}_3\}$ be linearly dependent. Let c be a nonzero scalar. Prove that the following sets are also linearly dependent.

*(a) $\{\mathbf{v}_1, \mathbf{v}_1 + \mathbf{v}_2, \mathbf{v}_3\}$ **(b)** $\{\mathbf{v}_1 \ c\mathbf{v}_2, \mathbf{v}_3\}$

(c) $\{\mathbf{v}_1, \mathbf{v}_1 + c\mathbf{v}_2, \mathbf{v}_3\}$

12. Let the set $\{\mathbf{v}_1, \mathbf{v}_2, \mathbf{v}_3\}$ be linearly independent in **R**^{3}. Let c be a nonzero scalar. Prove that the following sets are also linearly independent.

*(a) $\{\mathbf{v}_1, \mathbf{v}_1 + \mathbf{v}_2, \mathbf{v}_3\}$ **(b)** $\{\mathbf{v}_1, c\mathbf{v}_2, \mathbf{v}_3\}$

(c) $\{\mathbf{v}_1, \mathbf{v}_1, + c\mathbf{v}_2, \mathbf{v}_3\}$

13. Let a set S be linearly independent. Prove that every subset of S is also linearly independent. Let P be linearly dependent. Is every subset of P linearly dependent?

14. Let $\{\mathbf{v}_1, \mathbf{v}_2\}$ be linearly independent. Show that if a vector $\mathbf{v}_3$ is not of the form $a\mathbf{v}_1 + b\mathbf{v}_2$, then the set $\{\mathbf{v}_1, \mathbf{v}_2, \mathbf{v}_3\}$ is linearly independent.

15. Prove that a set of two or more vectors in a vector space is linearly independent if and only if no vector in the set can be expressed as a linear combination of the other vectors.

*16. Write two vectors from **R**^{3} at random. Which is the more likely—that these vectors are linearly dependent or linearly independent?

17. Write three vectors from **R**^{3} at random. Which is the more likely—that these vectors are linearly dependent or linearly independent?

*18. A computer program accepts a number of vectors in **R**^{3} as input and gives the information whether the vectors are linearly dependent or independent as output. Which is the more likely to happen due to round-off error—that the computer states that a given set of linearly independent vectors is linearly dependent, or vice versa?

4.5 Bases and Dimension

We talk about a line as being one-dimensional and a plane as being two-dimensional. In this section we introduce the mathematical definition of dimension. It will be compatible with our intuitive ideas. We first bring together the concepts of spanning set and linear independence.

Definition   *A finite set of vectors* $\{\mathbf{v}_1, \ldots, \mathbf{v}_m\}$ *is called a **basis** for a vector space V if the set spans V and is linearly independent.*

Intuitively, a basis is an efficient set for characterizing a vector space, in that any vector can be expressed as a linear combination of the basis vectors, and the basis vectors are independent of one another.

Standard Basis   *The set of n vectors*

$$\{(1, 0, \ldots, 0), (0, 1, \ldots, 0), \ldots, (0, \ldots, 1)\}$$

*is a basis for $\mathbf{R}^n$. This basis is called the **standard basis** for $\mathbf{R}^n$.*

Let us verify this result. We have to show that this set spans $\mathbf{R}^n$ and that it is linearly independent. Let $(x_1, x_2, \ldots, x_n)$ be an arbitrary element of $\mathbf{R}^n$. We can write

$$(x_1, x_2, \ldots, x_n) = x_1(1, 0, \ldots, 0) + x_2(0, 1, \ldots, 0) + \cdots + x_n(0, \ldots, 1)$$

Thus the set spans $\mathbf{R}^n$.

Furthermore, when we examine this set for linear independence, we get

$$c_1(1, 0, \ldots, 0) + c_2(0, 1, \ldots, 0) + \cdots + c_n(0, \ldots, 1) = (0, 0, \ldots, 0)$$
$$(c_1, 0, \ldots, 0) + (0, c_2, \ldots, 0) + \cdots + (0, \ldots, c_n) = (0, 0, \ldots, 0)$$
$$(c_1, c_2 \ldots, c_n) = (0, 0, \ldots, 0)$$

Thus $c_1 = 0, c_2 = 0, \ldots, c_n = 0$. The set is linearly independent.

Therefore the set $\{(1, 0, \ldots, 0), (0, 1, \ldots, 0), \ldots, (0, \ldots, 1)\}$ is a basis for $\mathbf{R}^n$. We shall find that there are many bases for $\mathbf{R}^n$. However, the standard basis is the most important basis.

Example 1   Show that the set $\{(1, 0, -1), (1, 1, 1), (1, 2, 4)\}$ is a basis for $\mathbf{R}^3$.

Solution   Let us first show that the set spans $\mathbf{R}^3$. Let (x_1, x_2, x_3) be an arbitrary element of $\mathbf{R}^3$. We try to find scalars a, a_2, a_3 such that

$$(x_1, x_2, x_3) = a_1(1, 0, -1) + a_2(1, 1, 1) + a_3(1, 2, 4)$$

This identity leads to the system of equations

$$a_1 + a_2 + a_3 = x_1$$
$$a_2 + 2a_3 = x_2$$
$$-a_1 + a_2 + 4a_3 = x_3$$

This system of equations has the solution

$$a_1 = 2x_1 - 3x_2 + x_3, \ a_2 = -2x_1 + 5x_2 - 2x_3, \ a_3 = x_1 - 2x_2 + x_3$$

Thus the set spans the space.

We now show that the set is linearly independent. Consider the identity

$$b_1(1, 0, -1) + b_2(1, 1, 1) + b_3(1, 2, 4) = (0, 0, 0)$$

This identity leads to the system of equations

$$b_1 + b_2 + b_3 = 0$$
$$b_2 + 2b_3 = 0$$
$$-b_1 + b_2 + 4b_3 = 0$$

This system has the unique solution $b_1 = 0$, $b_2 = 0$, and $b_3 = 0$. Thus the set is linearly independent.

We have shown that the set $\{(1, 0, -1), (1, 1, 1), (1, 2, 4)\}$ spans $\mathbf{R}^3$ and is linearly independent. It thus forms a basis for $\mathbf{R}^3$. ≡≡≡

The following theorem leads to the key result that all bases consist of the same number of vectors.

Theorem 4.7   Let $\{\mathbf{v}_1, \ldots, \mathbf{v}_n\}$ be a basis for a vector space V. If $\{\mathbf{w}_1, \ldots, \mathbf{w}_m\}$ is a set of more than n vectors in V, then this set is linearly dependent.

Proof   We examine the identity

$$c_1\mathbf{w}_1 + \cdots + c_m\mathbf{w}_m = 0 \tag{1}$$

We shall show that values of $c_1, \ldots, c_m$, not all zero, exist, satisfying this identity and thus proving that the vectors are linearly dependent.

The set $\{\mathbf{v}_1, \ldots, \mathbf{v}_n\}$ is a basis for V. Thus each of the vectors $\mathbf{w}_1, \ldots, \mathbf{w}_m$ can be expressed as a linear combination of $\mathbf{v}_1, \ldots, \mathbf{v}_n$. Let

$$\mathbf{w}_1 = a_{11}\mathbf{v}_1 + a_{12}\mathbf{v}_2 + \cdots + a_{1n}\mathbf{v}_n$$
$$\vdots$$
$$\mathbf{w}_m = a_{m1}\mathbf{v}_1 + a_{m2}\mathbf{v}_2 + \cdots + a_{mn}\mathbf{v}_n$$

Substituting for $\mathbf{w}_1 \ldots, \mathbf{w}_m$ into Equation (1), we get

$$c_1(a_{11}\mathbf{v}_1 + a_{12}\mathbf{v}_2 + \cdots + a_{1n}\mathbf{v}_n) + \cdots + c_m(a_{m1}\mathbf{v}_1 + a_{m2}\mathbf{v}_2 + \cdots + a_{mn}\mathbf{v}_n)$$
$$= \mathbf{0}$$

Rearranging, we get

$$(c_1a_{11} + c_2a_{21} + \cdots + c_ma_{m1})\mathbf{v}_1 + \cdots + (c_1a_{1n} + c_2a_{2n} + \cdots + c_ma_{mn})\mathbf{v}_n$$
$$= \mathbf{0}$$

Since $\mathbf{v}_1, \ldots, \mathbf{v}_n$ are linearly independent, this identity can be satisfied only if the coefficients are all zero. Thus

$$a_{11}c_1 + a_{21}c_2 + \cdots + a_{m1}c_m = 0$$
$$\vdots$$
$$a_{1n}c_1 + a_{2n}c_2 + \cdots + a_{mn}c_n = 0$$

Thus finding cs that satisfy Equation (1) reduces to finding solutions to this system of n equations in m variables. Since $m > n$, the number of variables is greater than the number of equations. We know that such a system of homogeneous equations has many solutions. There are therefore nonzero values of cs that satisfy Equation (1). Thus the set $\{\mathbf{w}_1, \ldots, \mathbf{w}_m\}$ is linearly dependent.

Theorem 4.8   Any two bases for a vector space V consist of the same number of vectors.

Proof   Let $\{\mathbf{v}_1, \ldots, \mathbf{v}_n\}$ and $\{\mathbf{w}_1, \ldots, \mathbf{w}_m\}$ be two bases for V.

If we interpret $\{\mathbf{v}_1, \ldots, \mathbf{v}_n\}$ as a basis for V and $\{\mathbf{w}_1, \ldots, \mathbf{w}_m\}$ as a set of linearly independent vectors in V, then the previous theorem tells us that $m \leq n$. Conversely, if we interpret $\{\mathbf{w}_1, \ldots, \mathbf{w}_m\}$ as a basis for V and $\{\mathbf{v}_1, \ldots, \mathbf{v}_n\}$ as a set of linearly independent vectors in V, then $n \leq m$. Thus $n = m$, proving that both bases consist of the same number of vectors.

We are now in a position to give the promised definition of dimension of a vector space.

Definition   *If a vector space V has a basis consisting of n vectors, then the **dimension** of V is said to be n. We write $\dim(V)$ for the dimension of V.*

The set of n vectors $\{(1, 0, \ldots, 0), \ldots, (0, \ldots, 0, 1)\}$ forms a basis for $\mathbf{R}^n$. Thus the dimension of $\mathbf{R}^n$ is n.

Note that we have defined the basis for a vector space to be a *finite* set of vectors that spans the space and is linearly independent. Such a set does not exist for all vector spaces. When such a finite set exists, we say that the vector space is **finite dimensional**. If such a finite set does not exist, we say that the vector space is **infinite dimensional**. We shall meet some function vector spaces that are infinite dimensional in the next chapter. We are primarily interested in finite dimensional vector spaces in this course.

Example 2   Consider the set $\{(1, 2, 3), (-2, 4, 1)\}$ of vectors in $\mathbf{R}^3$. These vectors generate a subspace V of $\mathbf{R}^3$ consisting of all vectors of the form

$$\mathbf{v} = c_1(1, 2, 3) + c_2(-2, 4, 1)$$

The vectors $(1, 2, 3)$ and $(-2, 4, 1)$ span this subspace.

Furthermore, since the second vector is not a scalar multiple of the first vector, the vectors are linearly independent.

Therefore $\{(1, 2, 3), (-2, 4, 1)\}$ is a basis for V. Thus $\dim(V) = 2$. We know that V is, in fact, a plane through the origin.

In this last example we saw that a certain plane through the origin was a two-dimensional subspace of $\mathbf{R}^3$. The following theorem gives us more information about the subspaces of $\mathbf{R}^3$ and their dimensions.

Theorem 4.9   (a) The origin is a subspace of $\mathbf{R}^3$. The dimension of this subspace is defined to be zero.

(b) The one-dimensional subspaces of $\mathbf{R}^3$ are lines through the origin.

(c) The two-dimensional subspaces of $\mathbf{R}^3$ are planes through the origin. See Figure 4.18.

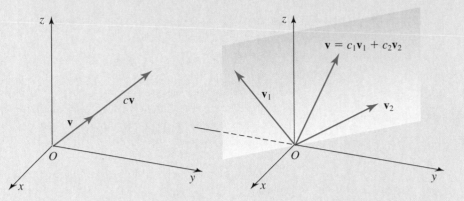

One-dimensional subspace of $\mathbf{R}^3$ with basis $\{\mathbf{v}\}$ is a line through the origin

Two-dimensional subspace of $\mathbf{R}^3$ with basis $\{\mathbf{v}_1, \mathbf{v}_2\}$ is a plane through the origin

Figure 4.18   One and two-dimensional subspaces of $\mathbf{R}^3$

Proof

(a) Let V be the set $\{(0, 0, 0)\}$, consisting of a single element, the zero vector of $\mathbf{R}^3$. Let c be an arbitrary scalar. Since

$$(0, 0, 0) + (0, 0, 0) = (0, 0, 0) \quad \text{and} \quad c(0, 0, 0) = (0, 0, 0)$$

V is closed under addition and scalar multiplication. It is thus a subspace of $\mathbf{R}^3$. The dimension of this subspace is defined to be zero.

(b) Let $\mathbf{v}$ be a basis for a one-dimensional subspace V of $\mathbf{R}^3$. Every vector in V is thus of the form $c\mathbf{v}$, for some scalar c. We know that these vectors form a line through the origin.

(c) Let $\{\mathbf{v}_1, \mathbf{v}_2\}$ be a basis for a two-dimensional subspace V of $\mathbf{R}^3$. Every vector in V is of the form $c_1\mathbf{v}_1 + c_2\mathbf{v}_2$. V is thus a plane through the origin.

We have discussed the fact that a set of vectors that spans a vector space characterizes the space in that every vector in the space can be expressed as a linear combination of the spanning set. In general, there may be more than one such linear combination.

Consider the space spanned by the set $\{(1, 2, 3), (-1, 1, 4), (3, 3, 2)\}$. The vector $(4, 5, 5)$ can be expressed

$$(4, 5, 5) = -(1, 2, 3) + (-1, 1, 4) + 2(3, 3, 2)$$

and

$$(4, 5, 5) = 5(1, 2, 3) - 2(-1, 1, 4) - (3, 3, 2)$$

The following theorem tells us that if the spanning set is a basis, each such combination is unique. A basis thus characterizes a vector space more specifically than does a general spanning set.

Theorem 4.10   Let $\{\mathbf{v}_1, \dots, \mathbf{v}_n\}$ be a basis for a vector space V. Then each vector in V can be expressed uniquely as a linear combination of these vectors.

Proof   Let $\mathbf{v}$ be a vector in V. Since $\{\mathbf{v}_1, \dots, \mathbf{v}_n\}$ is a basis, we can express $\mathbf{v}$ as a linear combination of these vectors. Suppose we can write

$$\mathbf{v} = a_1\mathbf{v}_1 + \cdots + a_n\mathbf{v}_n \qquad \text{and} \qquad \mathbf{v} = b_1\mathbf{v}_1 + \cdots + b_n\mathbf{v}_n$$

Then

$$a_1\mathbf{v}_1 + \cdots + a_n\mathbf{v}_n = b_1\mathbf{v}_1 + \cdots + b_n\mathbf{v}_n$$

giving

$$(a_1 - b_1)\mathbf{v}_1 + \cdots + (a_n - b_n)\mathbf{v}_n = \mathbf{0}$$

Since $\{\mathbf{v}_1, \dots, \mathbf{v}_n\}$ is a basis, the vectors $\mathbf{v}_1, \dots, \mathbf{v}_n$ are linearly independent. Thus $(a_1 - b_1) = 0, \dots, (a_n - b_n) = 0$, implying that $a_1 = b_1, \dots, a_n = b_n$. There is thus only one way of expressing $\mathbf{v}$ as a linear combination of the basis.

Suppose that a vector space is known to be of dimension n. The following theorem (which we state without proof) tells us that we do not have to check both the linear dependence and spanning conditions to see if a given set is a basis.

Theorem 4.11   Let V be a vector space of dimension n.

(a) If $S = \{\mathbf{v}_1, \dots, \mathbf{v}_n\}$ is a set of n linearly independent vectors in V, then S is a basis for V.

(b) If $S = \{\mathbf{v}_1, \dots, \mathbf{v}_n\}$ is a set of n vectors that spans V, then S is a basis for V.

Example 3   Prove that the set $\{(1, 3, -1), (2, 1, 0), (4, 2, 1)\}$ is a basis for **R**^{3}.

Solution   The dimension of **R**^{3} is three. Thus a basis for **R**^{3} consists of three vectors. We have the correct number of vectors for a basis. Normally we would have to show that this set is linearly independent and that it spans **R**^{3}. Theorem 4.11 tells us that we need check only one of these two conditions. Let us check for linear independence. We get

$$c_1(1, 3, -1) + c_2(2, 1, 0) + c_3(4, 2, 1) = (0, 0, 0)$$

This identity leads to the system of equations

$$c_1 + 2c_2 + 4c_3 = 0$$
$$3c_1 + c_2 + 2c_3 = 0$$
$$-c_1 + c_3 = 0$$

This system has the unique solution $c_1 = 0$, $c_2 = 0$, $c_3 = 0$. Thus the vectors are linearly independent. The set $\{(1, 3, -1), (2, 1, 0), (4, 2, 1)\}$ is therefore a basis for **R**^{3}.

The preceding four sections form a unit in which we have developed the algebraic structure of the vector space **R**^{n}. We have introduced the concepts of subspace, linear dependence/independence, spanning set, basis, and dimension. The following example brings all these concepts together. You should strive for an intuitive "feel" for each of the situations discussed.

Example 4   State (with a brief explanation) whether the following statements are true or false.

(a)   The vectors $(1, 2)$, $(-1, 3)$, $(5, 2)$ are linearly dependent in **R**^{2}.

(b)   The vectors $(1, 0, 0)$, $(0, 2, 0)$, $(1, 2, 0)$ span **R**^{3}.

(c) $\{(1, 0, 2), (0, 1, -3)\}$ is a basis for the subspace of **R**^{3} consisting of vectors of the form $(a, b, 2a - 3b)$.

(d)   Any set of two vectors can be used to generate a two-dimensional subspace of **R**^{3}.

Solution   (a)   True: The dimension of **R**^{2} is two. Thus any three vectors are linearly dependent.

(b)   False: The three vectors are linearly dependent. Thus they cannot span a three-dimensional space.

(c)   True: The vectors span the subspace since

$$(a, b, 2a - 3b) = a(1, 0, 2) + b(0, 1, -3)$$

The vectors are also linearly independent since they are not collinear.

(d)   False: The two vectors must be linearly independent.

Exercise Set 4.5

1. Prove that the following sets are bases for $\mathbf{R}^2$ by showing that they span the space and are linearly independent.

   *(a) $\{1, 2), (3, 1)\}$ (b) $\{(-1, 4), (2, 5)\}$

   *(c) $\{(1, 1), (-1, 1)\}$ (d) $\{(1, 0), (1, 1)\}$

2. Prove that the following sets are bases for $\mathbf{R}^2$ by applying Theorem 4.11.

   *(a) $\{(1, 3), (-1, 2)\}$ (b) $\{(2, 6), (4, 1)\}$

   *(c) $\{(-1, 2), (3, 4)\}$ (d) $\{(0, 1), (1, 1)\}$

3. Which of the following sets of vectors are bases for $\mathbf{R}^2$?

   *(a) $\{(3, 1), (2, 1)\}$ *(b) $\{(1, -3), (-2, 6)\}$

   (c) $\{(4, 5), (3, 2)\}$ (d) $\{(-1, 2), (3, -6)\}$

4. Prove that the following sets are bases for $\mathbf{R}^3$.

   *(a) $\{(1, 1, 1), (0, 1, 2), (3, 0, 1)\}$

   (b) $\{(1, 2, 3), (2, 4, 1), (3, 0, 0)\}$

   (c) $\{(0, 0, 1), (2, 3, 1), (4, 1, 2)\}$

   (d) $\{(1, 1, 4), (2, 1, 3), (0, 1, 6)\}$

5. Which of the following sets are bases for $\mathbf{R}^3$?

   *(a) $\{(1, -1, 2), (2, 0, 1), (3, 0, 0)\}$

   *(b) $\{(2, 1, 0), (-1, 1, 1), (3, 3, 1)\}$

   (c) $\{(3, 1, -1), (-1, -1, 0), (4, 0, -2)\}$

   (d) $\{(1, 2, 2), (-1, 0, 1), (-3, 1, -1)\}$

6. Explain, without performing any computation, why the following sets cannot be bases for the indicated vector spaces.

   *(a) $\{(3, -2), (6, -4)\}$ for $\mathbf{R}^2$

   *(b) $\{(1, 3), (4, 1), (1, 1)\}$ for $\mathbf{R}^2$

   (c) $\{(0, 0), (1, 3)\}$ for $\mathbf{R}^2$

   *(d) $\{(1, 0, 1), (2, -1, 3), (-4, 2, -6)\}$ for $\mathbf{R}^3$

   (e) $\{(1, 1, 1), (2, 1, 3), (5, 0, 0), (-1, -2, 4)\}$ for $\mathbf{R}^3$

   *(f) $\{(1, 4), (3, 1, 2), (2, 4, 5)\}$ for $\mathbf{R}^3$

   (g) $\{(4, 3, 2), (-1, 0, 5), (2, 7, 1)\}$ for $\mathbf{R}^4$

*7. Prove that the subspace of $\mathbf{R}^3$ generated by the vectors $(-1, 2, 1), (2, -1, 0), (1, 4, 3)$ is a two-dimensional subspace of $\mathbf{R}^3$. Give a basis for this subspace.

*8. Prove that the vector $(1, 2, -1)$ lies in the two-dimensional subspace of $\mathbf{R}^3$ generated by the vectors $(1, 3, 1)$ and $(1, 4, 3)$.

9. Prove that the vector $(2, 1, 4)$ lies in the two-dimensional subspace of $\mathbf{R}^3$ generated by the vectors $(1, 0, 2)$ and $(1, 1, 2)$.

*10. Prove that the vector $(-3, 3, -6)$ lies in the one-dimensional subspace of $\mathbf{R}^3$ generated by the vector $(2, -2, 4)$.

11. Does the vector $(1, 2, -1)$ lie in the subspace of $\mathbf{R}^3$ generated by the vectors $(1, -1, 0)$ and $(3, -1, 2)$?

*12. Find a basis for $\mathbf{R}^2$ that includes the vector $(1, 2)$.

*13. Find a basis for $\mathbf{R}^3$ that includes the vectors $(1, 1, 1)$ and $(1, 0, -2)$.

14. Find a basis for $\mathbf{R}^3$ that includes the vectors $(-1, 0, 2)$ and $(0, 1, 1)$.

15. Determine a basis for each of the following subspaces of $\mathbf{R}^3$. Give the dimension of each subspace.

   *(a) the set of vectors of the form (a, a, b)

   (b) the set of vectors of the form $(a, a, 2a)$

   *(c) the set of vectors of the form $(a, b, a + b)$

   (d) the set of vectors of the form $(a, 2b, a + 3b)$

   *(e) the set of vectors of the form (a, b, c), where $a + b + c = 0$

16. Determine a basis for each of the following subspaces of $\mathbf{R}^4$. Give the dimension of each subspace.

   *(a) the set of vectors of the form $(a, b, a + b, a - b)$

   (b) the set of vectors of the form $(a, 2a, b, 0)$

   *(c) the set of vectors of the form $(2a, b, a + 3b, c)$

   (d) the set of vectors of the form (a, a, a, a)

*17. Let $\{\mathbf{v}_1, \mathbf{v}_2\}$ be a basis for a vector space V. Show that the set of vectors $\{\mathbf{u}_1, \mathbf{u}_2\}$, where $\mathbf{u}_1 = \mathbf{v}_1 + \mathbf{v}_2$, and $\mathbf{u}_2 = \mathbf{v}_1 - \mathbf{v}_2$, is also a basis for V.

18. Let $\{\mathbf{v}_1, \mathbf{v}_2, \mathbf{v}_3\}$ be a basis for a vector space V. Show that the set of vectors $\{\mathbf{u}_1, \mathbf{u}_2, \mathbf{u}_3\}$, where $\mathbf{u}_1 = \mathbf{v}_1$, and $\mathbf{u}_2 = \mathbf{v}_1 + \mathbf{v}_2$, $\mathbf{u}_3 = \mathbf{v}_1 + \mathbf{v}_2 + \mathbf{v}_3$, is also a basis for V.

*19. Let $\{\mathbf{v}_1, \mathbf{v}_2, \ldots, \mathbf{v}_3\}$ be a basis for a vector space V. Let c be a nonzero scalar. Show that the set $\{c\mathbf{v}_1, c\mathbf{v}_2, \ldots, c\mathbf{v}_n\}$ is also a basis for V.

20. Let V be a vector space of dimension greater than 1. Determine whether the following statements are true or false.

   *(a) A basis for V can include the zero vector.

   (b) There is more than one basis for V.

   *(c) There exists a set that spans V but which is not linearly independent.

   (d) There exists a set that is linearly independent but does not span V.

21. Let V be a vector space of dimension n. Prove that no set of $n - 1$ vectors can span V.

*22. Let V be a vector space, and let W be a subspace of V. If $\dim(V) = n$ and $\dim(W) = m$, prove that $m \leq n$.

23. State (with a brief explanation) whether the following statements are true or false.

   *(a) The set $\{(-1, 2)(3, -6)\}$ is a basis for **R**^{2}.

(b) The vectors $(1, 2, 3), (-1, 4, 6)$ span **R**^{3}.

   *(c) The subspace of **R**^{3} generated by the vectors $(1, 2, 3), (0, 1, 4), (1, 3, 7)$ is of dimension two.

   (d) Any set of two linearly independent vectors that spans **R**^{2} forms a basis for **R**^{2}.

24. State (with a brief explanation) whether the following statements are true or false.

   (a) The set $\{(2, 0, 0), (3, 4, 0), (200, 567, 0)\}$ is linearly independent.

   *(b) A single vector can be added to any two vectors in **R**^{3} to get a basis for **R**^{3}.

   (c) The maximum number of linearly dependent vectors in **R**^{2} is two.

   *(d) If there are three linearly independent vectors in a vector space, the dimension must be greater than or equal to three.

4.6 Rank of a Matrix

In this section the reader is introduced to the concept of the rank of a matrix. Rank enables one to relate matrices to vectors, and vice versa. Rank is a unifying tool that enables us to bring together many of the concepts discussed in the course. Solutions to certain systems of linear equations, singularity of a matrix, and invertibility of a matrix all come together under the umbrella of rank.

Definition   *Let A be an $m \times n$ matrix. The rows of A may be viewed as row vectors $\mathbf{r}_1, \ldots, \mathbf{r}_m$, and the columns as column vectors $\mathbf{c}_1, \ldots, \mathbf{c}_n$. Each row vector will have n components, and each column vector will have m components. The row vectors will span a subspace of $\mathbf{R}^n$ called the **row space** of A, and the column vectors will span a subspace of $\mathbf{R}^m$ called the **column space** of A.*

Example 1   Consider the matrix

$$\begin{bmatrix} 1 & 2 & -1 & 2 \\ 3 & 4 & 1 & 6 \\ 5 & 4 & 1 & 0 \end{bmatrix}$$

The row vectors of A are

$$\mathbf{r}_1 = (1, 2, -1, 2), \quad \mathbf{r}_2 = (3, 4, 1, 6), \quad \mathbf{r}_3 = (5, 4, 1, 0)$$

These vectors span a subspace of $\mathbf{R}^4$ called the row space of A.

The column vectors of A are

$$\mathbf{c}_1 = \begin{bmatrix} 1 \\ 3 \\ 5 \end{bmatrix}, \qquad \mathbf{c}_2 = \begin{bmatrix} 2 \\ 4 \\ 4 \end{bmatrix}, \qquad \mathbf{c}_3 = \begin{bmatrix} -1 \\ 1 \\ 1 \end{bmatrix}, \qquad \mathbf{c}_4 = \begin{bmatrix} 2 \\ 6 \\ 0 \end{bmatrix}$$

These vectors span a subspace of $\mathbf{R}^3$ called the column space of A.

Theorem 4.12   The row space and the column space of a matrix A have the same dimension.

Proof   Let $\mathbf{u}_1, \ldots, \mathbf{u}_m$ be the row vectors of A. The ith vector is

$$\mathbf{u}_i = (a_{i1}, a_{i2}, \ldots, a_{in})$$

Let the dimension of the row space be s. Let the vectors $\mathbf{v}_1, \ldots, \mathbf{v}_s$ form a basis for the row space. Let the jth vector of this set be

$$\mathbf{v}_j = (b_{j1}, b_{j2}, \ldots b_{jn})$$

Each of the row vectors of A is a linear combination of $\mathbf{v}_1, \ldots, \mathbf{v}_s$. Let

$$\mathbf{u}_1 = c_{11}\mathbf{v}_1 + c_{12}\mathbf{v}_2 + \cdots + c_{1s}\mathbf{v}_s$$
$$\vdots$$
$$\mathbf{u}_m = c_{m1}\mathbf{v}_1 + c_{m2}\mathbf{v}_2 + \cdots + c_{ms}\mathbf{v}_s$$

Equating the ith components of the vectors on the left and right, we get

$$a_{1i} = c_{11}b_{1i} + c_{12}b_{2i} + \cdots + c_{1s}b_{si}$$
$$\vdots$$
$$a_{mi} = c_{m1}b_{1i} + c_{m2}b_{2i} + \cdots + c_{ms}b_{si}$$

This may be written

$$\begin{bmatrix} a_{1i} \\ \vdots \\ a_{mi} \end{bmatrix} = b_{1i}\begin{bmatrix} c_{11} \\ \vdots \\ c_{m1} \end{bmatrix} + b_{2i}\begin{bmatrix} c_{12} \\ \vdots \\ c_{m2} \end{bmatrix} + \cdots + b_{si}\begin{bmatrix} c_{1s} \\ \vdots \\ c_{ms} \end{bmatrix}$$

This implies that each column vector of A lies in a space spanned by a single set of s vectors. Since s is the dimension of the row space of A, we get

$$\dim(\text{column space of } A) \leq \dim(\text{row space of } A)$$

By similar reasoning, we can show that

$$\dim(\text{row space of } A) \leq \dim(\text{column space of } A)$$

Combining these two results, we see that

$$\dim(\text{row space of } A) = \dim(\text{column space of } A)$$

proving the theorem.

Definition   *The dimension of the row space and the column space of a matrix A is called the **rank** of A. The rank of A is denoted* rank(A).

As we proceed, we shall find that rank of a matrix will be a useful computational and geometrical tool.

Example 2   Determine the rank of the matrix

$$A = \begin{bmatrix} 1 & 2 & 3 \\ 0 & 1 & 2 \\ 2 & 5 & 8 \end{bmatrix}$$

Solution   We see by inspection that the third row of A is a linear combination of the first two rows:

$$(2, 5, 8) = 2(1, 2, 3) + (0, 1, 2)$$

Hence the three rows of A are linearly dependent. The rank of A must be less than 3. Since $(1, 2, 3)$ is not a scalar multiple of $(0, 1, 2)$, these two vectors are linearly independent. These vectors form a basis for the row space of A. Thus rank$(A) = 2$.

This method, based on the definition, is not practical for determining the ranks of larger matrices. We shall give a more systematic method for finding the rank of a matrix. The following theorem, which paves the way for the method, tells us that the rank of a matrix that is in reduced echelon form is immediately known.

Theorem 4.13   The nonzero row vectors of a matrix A that is in reduced echelon form are a basis for the row space of A. The rank of A is the number of nonzero row vectors.

Proof   Let A be an $m \times n$ matrix with nonzero row vectors of $\mathbf{r}_1, \ldots, \mathbf{r}_t$. Consider the identity

$$k_1\mathbf{r}_1 + k_2\mathbf{r}_2 + \cdots + k_t\mathbf{r}_t = \mathbf{0}$$

where $k_1, \ldots, k_t$ are scalars.

The first nonzero element of $\mathbf{r}_1$ is 1. $\mathbf{r}_1$ is the only one of the row vectors to have a nonzero number in this component. Thus, on adding the vectors $k_1\mathbf{r}_1$, $k_2\mathbf{r}_2, \ldots, k_t\mathbf{r}_t$, we get a vector whose first component is k_1. On equating this vector to zero, we get $k_1 = 0$. The identity then reduces to

$$k_2\mathbf{r}_2 + \cdots + k_t\mathbf{r}_t = \mathbf{0}$$

The first nonzero element of $\mathbf{r}_2$ is 1, and it is the only one of these remaining row vectors with a nonzero number in this component. Thus $k_2 = 0$. Similarly, $k_3, \ldots, k_t$ are all zero. The vectors $\mathbf{r}_1, \ldots, \mathbf{r}_t$ are therefore linearly independent. These vectors span the row space of A. They thus form a basis for the row space of A. The dimension of the row space is t. The rank of A is t, the number of nonzero row vectors in A.

Example 3   Find the rank of the matrix

$$A = \begin{bmatrix} 1 & 2 & 0 & 0 \\ 0 & 0 & 1 & 0 \\ 0 & 0 & 0 & 1 \\ 0 & 0 & 0 & 0 \end{bmatrix}$$

This matrix is in reduced echelon form. There are three nonzero row vectors, namely (1, 2, 0, 0), (0, 0, 1, 0), and (0, 0, 0, 1). According to the previous theorem, these three vectors form a basis for the row space of A. Rank(A) = 3.

The following theorem relates the row spaces of row equivalent matrices.

Theorem 4.14   Let A and B be row equivalent matrices. Then A and B have the same row space. Rank(A) = rank(B).

Proof   Since A and B are row equivalent, the rows of B can be obtained from the rows of A through a sequence of elementary row operations. Therefore each row of B is a linear combination of the rows of A. Thus the row space of B is contained in the row space of A.

In the same way, the rows of A can be obtained from the rows of B through a sequence of elementary row operations, implying that the row space of A is contained in the row space of B.

It follows that the row spaces of A and B are equal. Since their row spaces are equal, their ranks must be equal.

The next result brings the last two results together to give a method for finding a basis for the row space of a matrix, and the rank of the matrix.

Theorem 4.15   Let E be a reduced echelon form of a matrix A. The nonzero row vectors of E form a basis for the row space of A. The rank of A is the number of nonzero vectors in E.

Example 4   Find a basis for the row space of the following matrix A, and determine its rank.

$$A = \begin{bmatrix} 1 & 2 & 3 \\ 2 & 5 & 4 \\ 1 & 1 & 5 \end{bmatrix}$$

Solution   Use elementary row operations to find a reduced echelon form of the matrix A. We get

$$\begin{bmatrix} 1 & 2 & 3 \\ 2 & 5 & 4 \\ 1 & 1 & 5 \end{bmatrix} \approx \begin{bmatrix} 1 & 2 & 3 \\ 0 & 1 & -2 \\ 0 & -1 & 2 \end{bmatrix} \approx \begin{bmatrix} 1 & 0 & 7 \\ 0 & 1 & -2 \\ 0 & 0 & 0 \end{bmatrix}$$

The two vectors $(1, 0, 7)$ and $(0, 1, -2)$ form a basis for the row space of A. Rank$(A) = 2$.

The column vectors of a matrix A become the row vectors of the matrix A^t. Thus the column space of A is the row space of A^t. We can find a basis for the column space of a matrix A by using the above method to find a basis for the row space of A^t. The following example illustrates the technique.

Example 5   Find a basis for the column space of the following matrix A.

$$A = \begin{bmatrix} 1 & 1 & 0 \\ 2 & 3 & -2 \\ -1 & -4 & 6 \end{bmatrix}$$

Solution   The transpose of A is

$$A^t = \begin{bmatrix} 1 & 2 & -1 \\ 1 & 3 & -4 \\ 0 & -2 & 6 \end{bmatrix}$$

The column space of A becomes the row space of A^t. Let us find a basis for the row space of A^t. Compute a reduced echelon form of A^t.

$$\begin{bmatrix} 1 & 2 & -1 \\ 1 & 3 & -4 \\ 0 & -2 & 6 \end{bmatrix} \approx \begin{bmatrix} 1 & 2 & -1 \\ 0 & 1 & -3 \\ 0 & -2 & 6 \end{bmatrix} \approx \begin{bmatrix} 1 & 0 & 5 \\ 0 & 1 & -3 \\ 0 & 0 & 0 \end{bmatrix}$$

The nonzero row vectors of this reduced echelon form, namely $(1, 0, 5)$ and $(0, 1, -3)$, are a basis for the row space of A^t. Write these vectors in column form to get a basis for the column space of A. The following vectors are a basis for the column space of A.

$$\begin{bmatrix} 1 \\ 0 \\ 5 \end{bmatrix}, \quad \begin{bmatrix} 0 \\ 1 \\ -3 \end{bmatrix}.$$

Theorem 4.15 can also be used to determine a basis for the subspace V spanned by a given set of vectors. The vectors are written as the row vectors of a matrix, and a reduced echelon form of that matrix is computed. The nonzero row vectors of this reduced echelon form give a basis for V. The following example illustrates the method.

Example 6   Find a basis for the subspace V of $\mathbf{R}^4$ spanned by the vectors

$$(1, 2, 3, 4), (-1, -1, -4, -2), (3, 4, 11, 8)$$

Solution   We construct a matrix A having these vectors as row vectors.

$$A = \begin{bmatrix} 1 & 2 & 3 & 4 \\ -1 & -1 & -4 & -2 \\ 3 & 4 & 11 & 8 \end{bmatrix}$$

Determine a reduced echelon form of A. We get

$$\begin{bmatrix} 1 & 2 & 3 & 4 \\ -1 & -1 & -4 & -2 \\ 3 & 4 & 11 & 8 \end{bmatrix} \approx \begin{bmatrix} 1 & 2 & 3 & 4 \\ 0 & 1 & -1 & 2 \\ 0 & -2 & 2 & -4 \end{bmatrix} \approx \begin{bmatrix} 1 & 0 & 5 & 0 \\ 0 & 1 & -1 & 2 \\ 0 & 0 & 0 & 0 \end{bmatrix}$$

The nonzero vectors of this reduced echelon form, namely $(1, 0, 5, 0)$ and $(0, 1, -1, 2)$, are a basis for the subspace of V.

Our next theorem brings together in a convenient manner a number of results and concepts that have appeared in this course.

Theorem 4.16   Let A be an $n \times n$ matrix. The following statements are equivalent.

(a) A is invertible.

(b) $|A| \neq 0$ (A is nonsingular.)

(c) The system of equations $AX = B$ has a unique solution.

(d) A is row equivalent to I_n.

(e) Rank$(A) = n$.

Proof   We have previously seen that (a), (b), (c), and (d) are equivalent. You are asked to prove that (d) and (e) are equivalent in the exercises that follow. This proves that all four statements are equivalent.

The previous theorem told us how rank gives information about the uniqueness of the solution to a system of n linear equations in n variables. The concept of rank plays an important role in understanding the behavior of systems of linear equations of all sizes. We have seen how systems of linear equations can have a unique solution, many solutions, or no solution at all. These situations can be conveniently categorized in terms of the ranks of the augmented matrix and the matrix of coefficients.

Theorem 4.17 Consider a system of m equations in n variables.

(a) If the augmented matrix and the matrix of coefficients have the same rank r and $r = n$, the solution is unique.

(b) If the augmented matrix and the matrix of coefficients have the same rank r and $r < n$, there are many solutions.

(c) If the augmented matrix and the matrix of coefficients do not have the same rank, a solution does not exist.

Proof Let the system of equations be

$$a_{11}x_1 + \cdots + a_{1n}x_n = b_1$$
$$\vdots$$
$$a_{m1}x_1 + \cdots + a_{mn}x_n = b_m$$

This system can be written

$$x_1 \begin{bmatrix} a_{11} \\ \vdots \\ a_{m1} \end{bmatrix} + \cdots + x_n \begin{bmatrix} a_{1n} \\ \vdots \\ a_{mn} \end{bmatrix} = \begin{bmatrix} b_1 \\ \vdots \\ b_m \end{bmatrix}$$

That is,

$$x_1\mathbf{a}_1 + \cdots + x_n\mathbf{a}_n = \mathbf{b} \tag{1}$$

Let us now look at the three possibilities.

(a) Since the ranks of the matrix of coefficients and augmented matrix are the same, $\mathbf{b}$ must be linearly dependent on $\mathbf{a}_1, \ldots, \mathbf{a}_n$. Furthermore, since the rank is n, the vectors $\mathbf{a}_1, \ldots, \mathbf{a}_n$ are linearly independent and thus form a basis for the column space of the augmented matrix. Therefore Equation (1) has a unique solution; the solution to the system is unique.

(b) Since the ranks of the matrix of coefficients and augmented matrix are the same, $\mathbf{b}$ must be linearly dependent on $\mathbf{a}_1, \ldots, \mathbf{a}_n$. However, since rank $< n$, the vectors $\mathbf{a}_1, \ldots, \mathbf{a}_n$ are linearly dependent. $\mathbf{b}$ can therefore be expressed in more than one way as a linear combination of $\mathbf{a}_1, \ldots, \mathbf{a}_n$. Thus Equation (1) has many solutions; the solution to the system exists but is not unique.

(c) Since the rank of the augmented matrix is not equal to the rank of the matrix of coefficients, $\mathbf{b}$ is linearly independent of $\mathbf{a}_1, \ldots, \mathbf{a}_n$. Thus Equation (1) has no solution; a solution to the system does not exist.

Exercise Set 4.6

1. Determine the ranks of the following matrices using the definition of rank.

   *(a) $\begin{bmatrix} 1 & 2 \\ 3 & 4 \end{bmatrix}$ (b) $\begin{bmatrix} 1 & -3 \\ -2 & 6 \end{bmatrix}$

   *(c) $\begin{bmatrix} 4 & 0 \\ 1 & 5 \end{bmatrix}$ (d) $\begin{bmatrix} 2 & 4 \\ 3 & 6 \end{bmatrix}$

2. Determine the ranks of the following matrices using the definition of rank.

   *(a) $\begin{bmatrix} 1 & 2 & 1 \\ 2 & 4 & 2 \\ 1 & 2 & 3 \end{bmatrix}$ (b) $\begin{bmatrix} 1 & 0 & 0 \\ 0 & 1 & 0 \\ 0 & 0 & 1 \end{bmatrix}$

   *(c) $\begin{bmatrix} 2 & 1 & 3 \\ 4 & 2 & 6 \\ 2 & 1 & 3 \end{bmatrix}$ (d) $\begin{bmatrix} 1 & 4 & 2 \\ 0 & 1 & 5 \\ 0 & 0 & 1 \end{bmatrix}$

   (e) $\begin{bmatrix} 1 & 3 & 4 \\ -1 & 3 & 1 \\ 0 & 6 & 5 \end{bmatrix}$ (f) $\begin{bmatrix} 1 & 2 & 3 \\ 4 & 5 & 6 \\ 7 & 8 & 9 \end{bmatrix}$

3. Find the reduced echelon form for each of the following matrices. Use the reduced echelon form to determine a basis for the row space, and the rank of each matrix.

   *(a) $\begin{bmatrix} 1 & 2 & -1 \\ 2 & 5 & 2 \\ 0 & 2 & 9 \end{bmatrix}$ (b) $\begin{bmatrix} 1 & 1 & 8 \\ 0 & 1 & 3 \\ -1 & 1 & -2 \end{bmatrix}$

   *(c) $\begin{bmatrix} 1 & -3 & 2 \\ -2 & 6 & -4 \\ -1 & 3 & -2 \end{bmatrix}$

4. Find the reduced echelon form for each of the following matrices. Use the reduced echelon form to determine a basis for the row space, and the rank of each matrix.

   *(a) $\begin{bmatrix} 1 & 4 & 0 \\ -1 & -3 & 3 \\ 2 & 9 & 5 \end{bmatrix}$ (h) $\begin{bmatrix} 1 & 2 & 0 \\ 0 & 1 & 1 \\ -1 & 2 & 3 \end{bmatrix}$

   *(c) $\begin{bmatrix} 1 & 2 & 3 \\ 0 & -1 & -1 \\ 3 & 4 & 7 \end{bmatrix}$

5. Find the reduced echelon form for each of the following matrices. Use the reduced echelon form to determine a basis for the row space, and the rank of each matrix.

   *(a) $\begin{bmatrix} 1 & 2 & 3 & 4 \\ -1 & 2 & 0 & 1 \\ 0 & 1 & 0 & 2 \end{bmatrix}$

   *(b) $\begin{bmatrix} 1 & 2 & -1 & 4 \\ 0 & 1 & -2 & 3 \\ -1 & 0 & -3 & 2 \end{bmatrix}$

   (c) $\begin{bmatrix} 1 & 1 & 0 & -1 \\ 2 & 1 & 0 & 0 \\ 3 & 2 & 0 & -1 \\ -1 & 0 & 1 & 1 \end{bmatrix}$

6. Find bases for the subspaces of $\mathbf{R}^3$ spanned by the following vectors.

   *(a) $(1, 3, 2), (0, 1, 4), (1, 4, 9)$

   (b) $(1, -2, 5), (1, -1, 4), (2, -5, 14)$

   *(c) $(1, -1, 3), (1, 0, 1), (-2, 1, -4)$

   (d) $(2, -6, 4), (-1, 3, -2), (3, -9, 6)$

7. Find bases for the subspaces of $\mathbf{R}^4$ spanned by the following vectors.

   *(a) $(1, 3, -1, 4), (1, 3, 0, 6), (-1, -3, 0, -8)$

   (b) $(1, 2, 0, 1), (-1, -1, 3, 1), (2, 3, -2, 4)$

   *(c) $(1, 2, 3, 4), (0, -1, 2, 3), (2, 3, 8, 11), (2, 3, 6, 8)$

   (d) $(1, -3, -1, 2), (0, 1, -4, 1), (1, -4, 5, 1),$ $(2, -5, -6, 5)$

*8. Find bases for both the row and column spaces of the following matrix A. Show that the dimensions of both row space and column space are the same.

$$A = \begin{bmatrix} 1 & 2 & -1 \\ 0 & 1 & 3 \\ 1 & 4 & 6 \end{bmatrix}$$

9. Find bases for both the row and column spaces of the following matrix A. Show that the dimensions of both the row space and the column space are the same.

$$A = \begin{bmatrix} 1 & 3 & 2 \\ 1 & 4 & 1 \\ 2 & 5 & 5 \end{bmatrix}$$

10. (a) If A is a 2×5 matrix, what is the largest possible rank of A?

*(b) If A is an $m \times n$ matrix, what is the largest possible rank of A?

*11. Let A be an $m \times n$ matrix where $m < n$. Prove that

$$\dim(\text{column space of } A) < n$$

12. *(a) Let A be a 3×4 matrix. Prove that the column vectors of A are linearly dependent.

(b) Let A be a 7×3 matrix. Prove that the row vectors of A are linearly dependent.

(c) Let A be an $m \times n$ matrix with $m < n$. Prove that the column vectors of A are linearly dependent.

13. (a) Let A be an $n \times n$ matrix with $|A| \neq 0$. Prove that the row vectors of A form a basis for **R**^{n}.

(b) Let A be an $n \times n$ invertible matrix. Prove that the row vectors of A form a basis for **R**^{n}.

*14. Let A be an $n \times n$ matrix. Prove that the columns of A are linearly independent if and only if rank$(A) = n$.

15. Let A be an $n \times n$ matrix. Prove that the columns of A span **R**^{n} if and only if the rows of A are linearly independent.

16. If A and B are matrices of the same size, prove that

$$\text{rank}(A + B) \leq \text{rank}(A) + \text{rank}(B)$$

4.7   Eigenvalues and Eigenvectors

In this section we define eigenvalues and eigenvectors for the vector space **R**^{n}, and introduce the basic concepts.

Definition   *Let A be an $n \times n$ matrix. A scalar λ is called an **eigenvalue** of A if there exists a nonzero vector **x** in* **R**^{n} *such that*

$$A\mathbf{x} = \lambda\mathbf{x}$$

*The vector **x** is called an **eigenvector** corresponding to λ.*

Let us look at the geometrical significance of an eigenvector that corresponds to a nonzero eigenvalue. The vector $A\mathbf{x}$ is in the same direction as **x**, or the opposite direction, depending on the sign of λ. See Figure 4.19. An eigenvector of A is thus a vector whose direction is unchanged or reversed when multiplied by A.

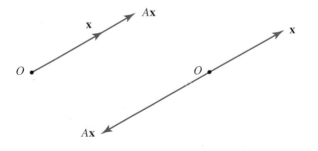

Figure 4.19   **x** is an eigenvector of A
$A\mathbf{x}$ is in same or opposite direction as **x**, if $\lambda \neq 0$

Computation of Eigenvalues and Eigenvectors

Let A be an $n \times n$ matrix with eigenvalue λ and corresponding eigenvector $\mathbf{x}$. Thus $A\mathbf{x} = \lambda\mathbf{x}$. This equation may be rewritten

$$A\mathbf{x} - \lambda\mathbf{x} = \mathbf{0}$$

giving

$$(A - \lambda I_n)\mathbf{x} = \mathbf{0}$$

This matrix equation represents a system of homogeneous linear equations having matrix of coefficients $(A - \lambda I_n)$. $\mathbf{x} = \mathbf{0}$ is a solution to this system. However, eigenvectors have been defined to be nonzero vectors. Further, nonzero solutions to this system of equations can exist only if the matrix of coefficients is singular, $|A - \lambda I_n| = 0$. Hence, solving the equation $|A - \lambda I_n| = 0$ for λ leads to all the eigenvalues of A.

On expanding the determinant $|A - \lambda I_n|$, we get a polynomial in λ. This polynomial is called the **characteristic polynomial** of A. The equation $|A - \lambda I_n| = 0$ is called the **characteristic equation** of A.

The eigenvalues are then substituted back into the equation $(A - \lambda I_n)\mathbf{x} = \mathbf{0}$ to find the corresponding eigenvectors.

We now give a number of examples to illustrate this method.

Example 1   Find the eigenvalues and eigenvectors of the matrix

$$A = \begin{bmatrix} -4 & -6 \\ 3 & 5 \end{bmatrix}$$

Solution   Let us first derive the characteristic polynomial of A. We get

$$A - \lambda I_2 = \begin{bmatrix} -4 & -6 \\ 3 & 5 \end{bmatrix} - \lambda \begin{bmatrix} 1 & 0 \\ 0 & 1 \end{bmatrix} = \begin{bmatrix} -4 - \lambda & -6 \\ 3 & 5 - \lambda \end{bmatrix}$$

Note that the matrix $A - \lambda I_2$ is obtained by subtracting λ from the diagonal elements of A. The characteristic polynomial of A is

$$|A - \lambda I_2| = (-4 - \lambda)(5 - \lambda) + 18 = \lambda^2 - \lambda - 2$$

We now solve the characteristic equation of A.

$$\lambda^2 - \lambda - 2 = 0$$
$$(\lambda - 2)(\lambda + 1) = 0$$
$$\lambda = 2 \quad \text{or} \quad -1$$

The eigenvalues of A are 2 and -1.

The corresponding eigenvectors are found by using these values of λ in the equation $(A - \lambda I_2)\mathbf{x} = \mathbf{0}$. There are many eigenvectors corresponding to each eigenvalue.

$\boldsymbol{\lambda = 2}$ We solve the equation $(A - 2I_2)\mathbf{x} = \mathbf{0}$ for $\mathbf{x}$. The matrix $(A - 2I_2)$ is obtained by subtracting 2 from the diagonal elements of A. We get

$$\begin{bmatrix} -6 & -6 \\ 3 & 3 \end{bmatrix} \begin{bmatrix} x_1 \\ x_2 \end{bmatrix} = \mathbf{0}$$

This leads to the system of equations

$$-6x_1 - 6x_2 = 0$$
$$3x_1 + 3x_2 = 0$$

giving $x_1 = -x_2$. The solutions to this system of equations are $x_1 = -r$ and $x_2 = r$, where r is a scalar. Thus the eigenvectors of A corresponding to $\lambda = 2$ are nonzero vectors of the form

$$r \begin{bmatrix} -1 \\ 1 \end{bmatrix}$$

$\boldsymbol{\lambda = -1}$ We solve the equation $(A + 1I_2)\mathbf{x} = \mathbf{0}$ for $\mathbf{x}$. The matrix $(A + 1I_2)$ is obtained by adding 1 to the diagonal elements of A. We get

$$\begin{bmatrix} -3 & -6 \\ 3 & 6 \end{bmatrix} \begin{bmatrix} x_1 \\ x_2 \end{bmatrix} = \mathbf{0}$$

This leads to the equations

$$-3x_1 - 6x_2 = 0$$
$$3x_1 + 6x_2 = 0$$

Thus $x_1 = -2x_2$. The solutions to these equations are $x_1 = -2s$ and $x_2 = s$, where s is a scalar. Thus the eigenvectors of A corresponding to $\lambda = -1$ are nonzero vectors of the form

$$s \begin{bmatrix} -2 \\ 1 \end{bmatrix}$$

We are always interested in knowing whether sets of vectors form subspaces! Observe that the set of eigenvectors of $\lambda = 2$, together with the zero vector, is a one-dimensional subspace of **R**^{2} with basis $\left\{ \begin{bmatrix} -1 \\ 1 \end{bmatrix} \right\}$. Further, the set of eigenvectors of $\lambda = -1$, together with the zero vector, is a one-dimensional subspace of **R**^{2} with basis $\left\{ \begin{bmatrix} -2 \\ 1 \end{bmatrix} \right\}$. These observations lead to the following result.

Theorem 4.18   Let A be an $n \times n$ matrix and λ an eigenvalue of A. The set of all eigenvectors corresponding to λ, together with the zero vector, is a subspace of **R**^{n}. This subspace is called the **eigenspace** of λ.

Proof   To show that the eigenspace is a subspace, we have to show that it is closed under vector addition and scalar multiplication.

Let $\mathbf{x}_1$ and $\mathbf{x}_2$ be two vectors in the eigenspace of λ and c be a scalar. Then $A\mathbf{x}_1 = \lambda\mathbf{x}_1$ and $A\mathbf{x}_2 = \lambda\mathbf{x}_2$. Hence,

$$A\mathbf{x}_1 + A\mathbf{x}_2 = \lambda\mathbf{x}_1 + \lambda\mathbf{x}_2$$
$$A(\mathbf{x}_1 + \mathbf{x}_2) = \lambda(\mathbf{x}_1 + \mathbf{x}_2)$$

Thus $\mathbf{x}_1 + \mathbf{x}_2$ is an element of the eigenspace of λ. The eigenspace is closed under addition.

Further, since $A\mathbf{x}_1 = \lambda\mathbf{x}_1$,

$$cA\mathbf{x}_1 = c\lambda\mathbf{x}_1$$
$$A(c\mathbf{x}_1) = \lambda(c\mathbf{x}_1)$$

Therefore $c\mathbf{x}_1$ is an element of the eigenspace of λ. The eigenspace is closed under scalar multiplication.

Thus the eigenspace is a subspace.

Example 2 Find the eigenvalues and eigenvectors of the matrix

$$A = \begin{bmatrix} 5 & 4 & 2 \\ 4 & 5 & 2 \\ 2 & 2 & 2 \end{bmatrix}$$

Solution The matrix $A - \lambda I_3$ is obtained by subtracting λ from the diagonal elements of A. Thus

$$A - \lambda I_3 = \begin{bmatrix} 5 - \lambda & 4 & 2 \\ 4 & 5 - \lambda & 2 \\ 2 & 2 & 2 - \lambda \end{bmatrix}$$

The characteristic polynomial of A is $|A - \lambda I_3|$. Using row and column operations to simplify determinants, we get

$$|A - \lambda I_3| = \begin{vmatrix} 5 - \lambda & 4 & 2 \\ 4 & 5 - \lambda & 2 \\ 2 & 2 & 2 - \lambda \end{vmatrix} = \begin{vmatrix} 1 - \lambda & -1 + \lambda & 0 \\ 4 & 5 - \lambda & 2 \\ 2 & 2 & 2 - \lambda \end{vmatrix}$$

$$= \begin{vmatrix} 1 - \lambda & 0 & 0 \\ 4 & 9 - \lambda & 2 \\ 2 & 4 & 2 - \lambda \end{vmatrix}$$

$$= (1 - \lambda)[(9 - \lambda)(2 - \lambda) - 8] = (1 - \lambda)[\lambda^2 - 11\lambda + 10]$$
$$= (1 - \lambda)(\lambda - 10)(\lambda - 1) = -(\lambda - 10)(\lambda - 1)^2$$

We now solve the characteristic equation of A:

$$-(\lambda - 10)(\lambda - 1)^2 = 0$$
$$\lambda = 10 \quad \text{or} \quad 1$$

The eigenvalues of A are 10 and 1.

The corresponding eigenvectors are found by using these values of λ in the equation $(A - \lambda I_3)\mathbf{x} = \mathbf{0}$.

$\lambda = 10$ We get

$$(A - 10I_3)\mathbf{x} = \mathbf{0}$$

$$\begin{bmatrix} -5 & 4 & 2 \\ 4 & -5 & 2 \\ 2 & 2 & -8 \end{bmatrix} \begin{bmatrix} x_1 \\ x_2 \\ x_3 \end{bmatrix} = \mathbf{0}$$

The solutions to this system of equations are $x_1 = 2r$, $x_2 = 2r$, and $x_3 = r$, where r is a scalar. Thus the eigenspace of $\lambda = 10$ is the one-dimensional space of vectors of the form

$$r\begin{bmatrix} 2 \\ 2 \\ 1 \end{bmatrix}$$

$\lambda = 1$ Let $\lambda = 1$ in $(A = \lambda I_3)\mathbf{x} = \mathbf{0}$. We get

$$(A - 1I_3)\mathbf{x} = 0$$

$$\begin{bmatrix} 4 & 4 & 2 \\ 4 & 4 & 2 \\ 2 & 2 & 1 \end{bmatrix} \begin{bmatrix} x_1 \\ x_2 \\ x_3 \end{bmatrix} = \mathbf{0}$$

The solutions to this system of equations can be shown to be $x_1 = -s - t$, $x_2 = s$, and $x_3 = 2t$, where s and t are scalars. Thus the eigenspace of $\lambda = 1$ is the space of vectors of the form

$$\begin{bmatrix} -s - t \\ s \\ 2t \end{bmatrix}$$

Separating the parameters s and t, we can write

$$\begin{bmatrix} -s - t \\ s \\ 2t \end{bmatrix} = s\begin{bmatrix} -1 \\ 1 \\ 0 \end{bmatrix} + t\begin{bmatrix} -1 \\ 0 \\ 2 \end{bmatrix}$$

Thus the eigenspace of λ is a two-dimensional subspace of **R**^{2} with basis

$$\left\{ \begin{bmatrix} -1 \\ 1 \\ 0 \end{bmatrix}, \begin{bmatrix} -1 \\ 0 \\ 2 \end{bmatrix} \right\}$$

If an eigenvalue occurs as a k times repeated root of the characteristic equation, we say that it is of **multiplicity** k. Thus $\lambda = 10$ has multiplicity 1, while $\lambda = 1$ has multiplicity 2 in the preceding example.

Example 3   Let A be an $n \times n$ matrix A with eigenvalues $\lambda_1, \ldots, \lambda_n$ and corresponding eigenvectors $X_1, \ldots, X_n$. Prove that if $c \neq 0$, then the eigenvalues of cA are $c\lambda_1, \ldots, c\lambda_n$ with corresponding eigenvectors $X_1, \ldots, X_n$.

Solution   Let λ_1 be one of the eigenvalues of A with corresponding eigenvector X_i. Then $AX_i = \lambda_i X_i$. Multiply both sides of this equation by c to get

$$cAX_i = c\lambda_i X_i$$

Thus $c\lambda_i$ is an eigenvalue of cA with corresponding eigenvector X_i.

Further, since cA is an $n \times n$ matrix, the characteristic polynomial of A is of degree n. The characteristic equation has n roots, implying that cA has n eigenvalues. The eigenvalues of cA are therefore $c\lambda_1, \ldots, c\lambda_n$ with corresponding eigenvectors $X_1, \ldots, X_n$.

Exercise Set 4.7

In Exercises 1–8, determine the characteristic polynomials, eigenvalues, and corresponding eigenspaces of the given 2×2 matrices.

*1. $\begin{bmatrix} 5 & 4 \\ 1 & 2 \end{bmatrix}$
  2. $\begin{bmatrix} 1 & -2 \\ 1 & 4 \end{bmatrix}$
  3. $\begin{bmatrix} 5 & 6 \\ -2 & -2 \end{bmatrix}$

*4. $\begin{bmatrix} 5 & 2 \\ -8 & -3 \end{bmatrix}$
  *5. $\begin{bmatrix} 1 & 2 \\ 2 & 1 \end{bmatrix}$
  6. $\begin{bmatrix} 2 & 1 \\ -1 & 4 \end{bmatrix}$

7. $\begin{bmatrix} 3 & -1 \\ 2 & 0 \end{bmatrix}$
  *8. $\begin{bmatrix} 2 & -4 \\ -1 & 2 \end{bmatrix}$

In Exercises 9–14, determine the characteristic polynomials, eigenvalues, and corresponding eigenspaces of the given 3×3 matrices.

*9. $\begin{bmatrix} 3 & 2 & -2 \\ -3 & -1 & 3 \\ 1 & 2 & 0 \end{bmatrix}$
  *10. $\begin{bmatrix} 1 & -2 & 2 \\ -2 & 1 & 2 \\ -2 & 0 & 3 \end{bmatrix}$

11. $\begin{bmatrix} 1 & 0 & 0 \\ -2 & 1 & 2 \\ -2 & 0 & 3 \end{bmatrix}$
  12. $\begin{bmatrix} 1 & 0 & 0 \\ -2 & 5 & -2 \\ -2 & 4 & -1 \end{bmatrix}$

*13. $\begin{bmatrix} 15 & 7 & -7 \\ -1 & 1 & 1 \\ 13 & 7 & -5 \end{bmatrix}$
  14. $\begin{bmatrix} 5 & -2 & 2 \\ 4 & -3 & 4 \\ 4 & -6 & 7 \end{bmatrix}$

In Exercises 15 and 16, determine the characteristic polynomials, eigenvalues, and corresponding eigenspaces of the given 4×4 matrices.

*15. $\begin{bmatrix} 4 & 2 & -2 & 2 \\ 1 & 3 & 1 & -1 \\ 0 & 0 & 2 & 0 \\ 1 & 1 & -3 & 5 \end{bmatrix}$
  16. $\begin{bmatrix} 3 & 5 & -5 & 5 \\ 3 & 1 & 3 & -3 \\ -2 & 2 & 0 & 2 \\ 0 & 4 & -6 & 8 \end{bmatrix}$

In Exercises 17–19, determine the characteristic polynomials, eigenvalues, and corresponding eigenvectors of the given matrices. Interpret your results geometrically.

*17. $\begin{bmatrix} 1 & 0 \\ 0 & 1 \end{bmatrix}$
  18. $\begin{bmatrix} 3 & 0 \\ 0 & 3 \end{bmatrix}$
  *19. $\begin{bmatrix} -2 & 0 \\ 0 & -2 \end{bmatrix}$

*20. Show that the following matrix has no real eigenvalues. Readers who have a knowledge of complex numbers will observe that it does have complex eigenvalues. All matrices have eigenvalues in the complex number system. The eigenvectors are vectors in $\mathbf{C}^n$.

$$\begin{bmatrix} 1 & 1 \\ -2 & -1 \end{bmatrix}$$

21. Show that the following matrix has no real eigenvalues and thus no eigenvectors. Interpret your result geometrically.

$$\begin{bmatrix} 0 & -1 \\ 1 & 0 \end{bmatrix}$$

*22. Prove that if A is a diagonal matrix, then its eigenvalues are the diagonal elements.

23. Prove that if A is an upper triangular matrix, then its eigenvalues are the diagonal elements.

*24. Prove that A and A^t have the same eigenvalues.

25. Prove that $\lambda = 0$ is an eigenvalue of a matrix A if and only if A is singular.

26. Prove that if the eigenvalues of a matrix A are $\lambda_1, \ldots, \lambda_n$, with corresponding eigenvectors $X_1, \ldots, X_n$, then $\lambda_1^m, \ldots, \lambda_n^m$ are eigenvalues of A^m with corresponding eigenvectors $X_1, \ldots, X_n$.

*27. Show that 0 is an eigenvalue of matrix A if and only if $|A| = 0$.

28. A matrix A is said to be **nilpotent** if $A^k = 0$ for some integer k. Prove that if A is nilpotent, then 0 is the only eigenvalue of A.

*29. Prove that the constant term of the characteristic polynomial of a matrix A is $|A|$.

30. Determine the eigenvalues and corresponding eigenvectors of the matrices

$$\begin{bmatrix} 1 & 2 \\ 1 & 0 \end{bmatrix}, \begin{bmatrix} 2 & 3 \\ 1 & 0 \end{bmatrix}, \begin{bmatrix} 3 & 4 \\ 1 & 0 \end{bmatrix}, \begin{bmatrix} 4 & 5 \\ 1 & 0 \end{bmatrix}$$

Use these results to make a conjecture about the eigenvalues and eigenvectors of the general matrix, $\begin{bmatrix} a & a+1 \\ 1 & 0 \end{bmatrix}$, where a is any real number. Verify your conjecture.

31. The **Cayley-Hamilton theorem** states that *a square matrix satisfies its characteristic equation*. That is, if the characteristic equation of a square matrix A is

$$\lambda^n + c_{n-1}\lambda^{n-1} + \cdots + c_0 = 0$$

then

$$A^n + c_{n-1}A^{n-1} + \cdots + c_0 I = 0$$

Show that the following matrices satisfy their characteristic equations.

*(a) $\begin{bmatrix} 0 & 2 \\ -1 & 3 \end{bmatrix}$ **(b)** $\begin{bmatrix} 8 & -10 \\ 5 & -7 \end{bmatrix}$

*(c) $\begin{bmatrix} 6 & -8 \\ 4 & -6 \end{bmatrix}$ **(d)** $\begin{bmatrix} -1 & 5 \\ -10 & 14 \end{bmatrix}$

*4.8   Demography and Weather Prediction

We now discuss how eigenvalues and eigenvectors can be used to predict the long-term behavior of certain Markov chains. Applications in demography and weather prediction are given.

Let us return to the population movement model of Section 2.6. There we found that annual population distributions could be described by a sequence of vectors $\mathbf{x}_0$, $\mathbf{x}_1(= P\mathbf{x}_0)$, $\mathbf{x}_2 (= P\mathbf{x}_1)$, $\mathbf{x}_3 (= P\mathbf{x}_2)$, ... P is a matrix of transition probabilities that takes us from one vector in the sequence to the following vector. Such a sequence (or chain) of vectors is called a **Markov chain**. Of special interest are Markov chains where the sequence $\mathbf{x}_0, \mathbf{x}_1, \mathbf{x}_2, \ldots$ converges to some fixed vector $\mathbf{x}$, where $P\mathbf{x} = \mathbf{x}$. The population movement would then be in a "steady state" with the total city population and total suburban population remaining constant thereafter. We then write

$$\mathbf{x}_0, \mathbf{x}_1, \mathbf{x}_2, \ldots \rightarrow \mathbf{x}$$

Since such a vector $\mathbf{x}$ satisfies $P\mathbf{x} = \mathbf{x}$, it would be an eigenvector of P corresponding to the eigenvalue 1. Knowledge of the existence and value of such a vector would give us information about the long-term behavior of the population distribution.

A special class of Markov chains has these desired properties. We now define this class, discuss their properties, and apply the results to our population movement model to determine long-term predictions.

Definition   *The transition matrix P of a Markov chain is said to be **regular** if for some power of P all the components are positive. The chain is then called a **regular Markov chain**.*

Example 1   Determine whether the following transition matrices are regular.

(a) $A = \begin{bmatrix} 0.3 & 0.6 \\ 0.7 & 0.4 \end{bmatrix}$ (b) $B = \begin{bmatrix} 0.7 & 1 \\ 0.3 & 0 \end{bmatrix}$ (c) $C = \begin{bmatrix} 0.4 & 0 \\ 0.6 & 1 \end{bmatrix}$

Solution   (a)   The elements of A are all positive, thus A is regular.

(b) $B^2 = \begin{bmatrix} 0.79 & 0.7 \\ 0.21 & 0.3 \end{bmatrix}$. Thus B is regular.

(c)   It can be shown that $C^2 = \begin{bmatrix} 0.16 & 0 \\ 0.84 & 1 \end{bmatrix}$, $C^3 = \begin{bmatrix} 0.064 & 0 \\ 0.936 & 1 \end{bmatrix}, \ldots$

The element in row 1, column 2 will always be zero. Thus C is not regular.

The following theorem, which we do not prove, leads to information about the long-term behavior of regular Markov chains.

Theorem 4.19   Consider a regular Markov chain having initial vector $\mathbf{x}_0$ and transition matrix P. Then

(1) $\mathbf{x}_0, \mathbf{x}_1, \mathbf{x}_2, \ldots \to \mathbf{x}$, where $\mathbf{x}$ satisfies $P\mathbf{x} = \mathbf{x}$.

(2) $P, P^2, P^3, \ldots \to Q$, where Q is a stochastic matrix. The columns of Q are all identical, each being an eigenvector of P corresponding to $\lambda = 1$.

Let us now apply this theorem to population movements.

Example 2   Determine the long-term trends in population movements between U.S. cities and suburbs.

Solution   We remind the reader of the model that was developed earlier. The populations of U.S. cities and suburbs in 1985 are described by the following vector $\mathbf{x}_0$ (in units of one million), and the populations in the following years are given by a Markov chain with the transition matrix P.

$$\mathbf{x}_0 = \begin{bmatrix} 60 \\ 125 \end{bmatrix} \begin{matrix} \text{city} \\ \text{suburb} \end{matrix} \qquad P = \begin{bmatrix} 0.96 & 0.01 \\ 0.04 & 0.99 \end{bmatrix} \begin{matrix} \text{city} \\ \text{suburb} \end{matrix}$$

$$\begin{matrix} & \text{initial} \\ & \text{populations} \end{matrix} \qquad\qquad \begin{matrix} & \text{(from)} \\ \text{city} & \text{suburb} & \text{(to)} \end{matrix}$$

Observe that all the elements of P are positive. The chain is therefore regular, and the results of the above theorem can be applied to give the long-term trends. The theorem tells us that P will have an eigenvalue of 1 and the steady-state vector $\mathbf{x}$ is a corresponding eigenvector. Thus

$$P\mathbf{x} = \mathbf{x}$$
$$(P - I)\mathbf{x} = \mathbf{0}$$

$$\begin{bmatrix} 0.96 - 1 & 0.01 \\ 0.04 & 0.99 - 1 \end{bmatrix} \begin{bmatrix} x_1 \\ x_2 \end{bmatrix} = \mathbf{0}$$

This leads to the system of equations

$$-0.04x_1 + 0.01x_2 = 0$$
$$0.04x_1 - 0.01x_2 = 0$$

giving $x_2 = 4x_1$. The solutions to this system of equations are $x_1 = r$ and $x_2 = 4r$, where r is a scalar. Thus the eigenvectors of P corresponding to $\lambda = 1$ are nonzero vectors of the form

$$r\begin{bmatrix} 1 \\ 4 \end{bmatrix}$$

The steady-state vector $\mathbf{x}$ will be a vector of this type. Let us assume that there is no total annual population change over the years. Therefore the sums of the elements of $\mathbf{x}$ and $\mathbf{x}_0$ are equal.

$$r + 4r = 60 + 125$$
$$r = 37$$

The steady-state vector is thus

$$\mathbf{x} = \begin{bmatrix} 37 \\ 148 \end{bmatrix}$$

This implies the following long-term predictions:

$$\text{U.S. city populations} \rightarrow 37 \text{ million}$$
$$\text{U.S. suburban populations} \rightarrow 148 \text{ million}$$

The above theorem gives further information about long-term population trends. Each column of the matrix Q is an eigenvector corresponding to the eigenvalue 1. Let

$$Q = \begin{bmatrix} s & s \\ 4s & 4s \end{bmatrix}$$

Since Q is a stochastic matrix, the sum of the elements in each column is 1. Thus

$$s + 4s = 1$$
$$s = 0.2$$

We get (exhibiting the elements to two decimal places for ease of reading)

$$
\begin{array}{cccc}
P & P^2 & P^3 & Q \\
\begin{bmatrix} 0.96 & 0.01 \\ 0.04 & 0.99 \end{bmatrix}, &
\begin{bmatrix} 0.92 & 0.02 \\ 0.08 & 0.98 \end{bmatrix}, &
\begin{bmatrix} 0.89 & 0.03 \\ 0.11 & 0.97 \end{bmatrix}, \dots \rightarrow &
\begin{bmatrix} 0.2 & 0.2 \\ 0.8 & 0.8 \end{bmatrix}
\end{array}
$$

Let us interpret this result. We focus on the (2, 1) element in each matrix—a similar interpretation will apply to the other elements. We get the sequence

$$0.04, 0.08, 0.11, \dots \rightarrow 0.8$$

These are the probabilities of moving from city to suburbia in 1 year, 2 years, 3 years, and so on. The probability gradually increases, approaching 0.8. Q is thus the **long-term transition matrix** of the model. It gives the long-term probabilities of living in city or suburbia.

$$
Q = \begin{array}{c} \\ \begin{bmatrix} 0.2 & 0.2 \\ 0.8 & 0.8 \end{bmatrix} \end{array}
\begin{array}{l} \text{city} \\ \text{suburb} \end{array}
$$

(from) city suburb (to)

Observe that the long-term probability of living in the city is 0.2, while the long-term probability of living in suburbia is 0.8. These probabilities are independent of initial location. The long-term probabilities being independent of initial state is a characteristic of regular Markov chains.

We now discuss an interesting application of Markov chains in a model that describes rainfall in Tel Aviv.

Example 3 K. R. Gabriel and J. Nemann have developed "A Markov Chain Model for Daily Rainfall Occurrence at Tel Aviv," *Quart. J. R. Met. Soc.*, **88**(1962), 90–95. The probabilities used were based on data of daily rainfall in Tel Aviv (Nahami Street) for the 27 years 1923 to 1950. Days were classified as wet or dry according to whether or not there had been recorded at least 0.1 mm of precipitation in the 24-hour period from 8 A.M. to 8 A.M. the following day. A Markov chain was constructed for each of the months November through April, these months constituting the rainy season. We discuss the chain developed for November. The model assumes that the probability of rainfall on any day depends only on whether the previous day was wet or dry. The statistics accumulated over the years for November were

A Given Day	Following Day Wet
Wet	117 out of 195
Dry	80 out of 615

Thus the probability of a wet day following a wet day is $\frac{117}{195} = 0.6$. The probability of a wet day following a dry day is $\frac{80}{615} = 0.13$. These probabilities lead to the following transition matrix for the weather pattern in November.

$$
\begin{array}{cc}
 & \text{(a given day)} \\
 & \begin{array}{cc} \text{wet} & \text{dry} \end{array} \\
P = \begin{bmatrix} 0.6 & 0.13 \\ 0.4 & 0.87 \end{bmatrix} & \begin{array}{l} \text{wet} \\ \text{dry} \end{array}
\end{array} \quad \text{(following day)}
$$

On any given day in November, one can use P to predict the weather on a future day in November. For example, if today, a Wednesday, is wet, let us compute the probability that next Saturday will be dry. Saturday is three days hence, thus the various probabilities for the weather on Saturday will be given by the elements P^3. It can be shown that (exhibiting the elements to two decimal places for ease of reading)

$$
\begin{array}{cc}
 & \text{(today, Wednesday)} \\
 & \begin{array}{cc} \text{wet} & \text{dry} \end{array} \\
P^3 = \begin{bmatrix} 0.32 & 0.22 \\ 0.68 & 0.78 \end{bmatrix} & \begin{array}{l} \text{wet} \\ \text{dry} \end{array}
\end{array} \quad \text{(Saturday)}
$$

If today is wet, the probability of Saturday being dry is 0.68.

Observe that P, the matrix of transition probabilities, is regular. Eigenvectors can thus be used to obtain long-term weather predictions. The eigenvectors of P corresponding to the eigenvalue 1 are found to be nonzero vectors of the form

$$
r \begin{bmatrix} 0.325 \\ 1 \end{bmatrix}
$$

The column vectors of the long-term transition matrix Q will be eigenvectors of this type, whose components sum to 1 since Q is stochastic. Therefore $0.325r + r = 1$, giving $r = 0.75$ (to 2 decimal places). Thus

$$
Q = \begin{bmatrix} 0.25 & 0.25 \\ 0.75 & 0.75 \end{bmatrix}
$$

We can interpret this matrix as follows:

$$
\begin{array}{cc}
 & \text{(today)} \\
 & \begin{array}{cc} \text{wet} & \text{dry} \end{array} \\
\begin{bmatrix} 0.25 & 0.25 \\ 0.75 & 0.75 \end{bmatrix} & \begin{array}{l} \text{wet} \\ \text{dry} \end{array}
\end{array} \quad \text{(day in the distant future)}
$$

This implies the following weather forecast for Tel Aviv in November!

Long-range Forecast for Tel Aviv in November

0.25 probability wet
0.75 probability dry

Exercise Set 4.8

*1. The populations of U.S. metropolitan and nonmetropolitan areas in 1985 are described by the following vector $\mathbf{x}_0$ (in units of one million), and the populations in the following years are given by a Markov chain with the transition matrix P. (See Exercise 6, Section 2.6.) Determine the long-term predictions for metro and nonmetro populations, assuming no change in their total population.

initial
populations

$$\mathbf{x}_0 = \begin{bmatrix} 185 \\ 55 \end{bmatrix} \begin{array}{l} \text{metro} \\ \text{nonmetro} \end{array}$$

(from)
metro   nonmetro   (to)

$$P = \begin{bmatrix} 0.99 & 0.02 \\ 0.01 & 0.98 \end{bmatrix} \begin{array}{l} \text{metro} \\ \text{nonmetro} \end{array}$$

2. The populations of U.S. cities, suburbs, and nonmetro areas in 1985 are described by the following vector $\mathbf{x}_0$ (in units of one million), and the populations in the following years are given by a Markov chain with the transition matrix P. (See Exercise 7, Section 2.6.) Determine the long-term predictions for the populations of these regions, assuming no change in their total population. (This model is a refinement of the model of the previous exercise in that the metro areas are subdivided into city and suburbia.)

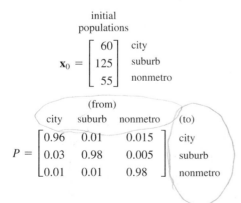

initial
populations

$$\mathbf{x}_0 = \begin{bmatrix} 60 \\ 125 \\ 55 \end{bmatrix} \begin{array}{l} \text{city} \\ \text{suburb} \\ \text{nonmetro} \end{array}$$

(from)
city   suburb   nonmetro   (to)

$$P = \begin{bmatrix} 0.96 & 0.01 & 0.015 \\ 0.03 & 0.98 & 0.005 \\ 0.01 & 0.01 & 0.98 \end{bmatrix} \begin{array}{l} \text{city} \\ \text{suburb} \\ \text{nonmetro} \end{array}$$

*3. Consider a genetics model in which the offspring of guinea pigs are crossed with hybrids only. (See Exercise 15, Section 2.6.) The transition matrix P for that model is as follows. Prove that P is regular. Let P, P^2, P^3, ... $\rightarrow Q$. Determine Q. What information about the long-term distribution of guinea pigs does it give?

(from)
AA   Aa   aa   (to)

$$P = \begin{bmatrix} 0.5 & 0.25 & 0 \\ 0.5 & 0.5 & 0.5 \\ 0 & 0.25 & 0.5 \end{bmatrix} \begin{array}{l} AA \\ Aa \\ aa \end{array}$$

*4. The statistics for rainfall for the month of December in Tel Aviv are as follows:

A Given Day	Following Day Wet
Wet	213 out of 326
Dry	117 out of 511

Use these figures to construct a Markov chain for predicting weather in December in Tel Aviv. (See Example 3.)

(a) If today is a dry Thursday in December in Tel Aviv, what is the probability that next Saturday will also be dry?

(b) Determine the long-range forecast for December in Tel Aviv—what are the probabilities that a distant day will be wet/dry?

5. A psychologist conducts an experiment in which 20 rats are placed at random in a compartment that has been divided into rooms labeled 1, 2, and 3, as shown in Figure 4.20.

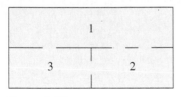

Figure 4.20

Observe that there are four doors in the arrangement. There are three possible states for the rats: they can be in room 1, 2, or 3. Let us assume that the rats move from room to room. A rat in room 1 has the probabilities $p_{11} = 0$, $p_{12} = \frac{2}{3}$, and $p_{13} = \frac{1}{3}$ of moving to the various rooms, based on the distribution of doors. This approach leads to a Markov chain with the following transition matrix P. Predict the long-term distribution of rats. What is the long-term probability that a given marked rat is in room 2?

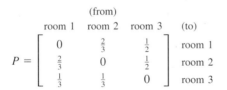

*6. Forty rats are placed at random in a compartment having four rooms labeled 1, 2, 3, and 4 as shown in Figure 4.21. Construct a Markov chain to describe the movement of the rats between the rooms. Predict the long-term distribution of rats. What is the long-term probability that a given marked rat is in room 4?

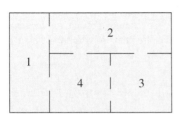

Figure 4.21

*7. Two car rental companies, A and B, are competing for customers at certain airports. A study has been made of customer satisfaction with the various companies. The results are expressed by the following matrix R. The first column of R implies that 75% of those currently using rental company A are satisfied and intend to use A next time, while 25% of those currently using rental company A are dissatisfied and plan to use B next time. There is a similar interpretation to the second column of R.

$$\text{(from)}$$
$$R = \begin{matrix} & A & B & \text{(to)} \\ & \begin{bmatrix} 75\% & 20\% \\ 25\% & 80\% \end{bmatrix} & & \begin{matrix} A \\ B \end{matrix} \end{matrix}$$

Modify the matrix R to obtain a transition matrix P for a Markov chain that describes the rental patterns. If the current trends continue, how will the rental distribution eventually settle? Express the distribution in percentages that use companies A and B.

8. A market research group has been studying the buying patterns for competing products I, II, and III. The results of the analysis are described by the following matrix A.

$$\text{(from)}$$
$$R = \begin{matrix} & \text{I} & \text{II} & \text{III} & \text{(to)} \\ & \begin{bmatrix} 80\% & 20\% & 5\% \\ 5\% & 75\% & 5\% \\ 15\% & 5\% & 90\% \end{bmatrix} & & & \begin{matrix} \text{I} \\ \text{II} \\ \text{III} \end{matrix} \end{matrix}$$

Column 1 implies that of those people currently using product I, 80% plan to continue using it, while 5% plan to switch to product II and 15% to product III. Columns 2 and 3 are to be interpreted similarly. Use the matrix A to construct a Markov chain that describes buying trends. If the current buying patterns continue, determine the likely eventual distribution of sales, in terms of percentages.

*9. Prove that 1 is an eigenvalue of every stochastic matrix. (*Hint:* Prove that $|A - 1I| = 0$, using the properties of determinants.)

Review Exercises Chapter 4

1. Compute the following vector expressions for
   $\mathbf{u} = (3, -1, 5)$, $\mathbf{v} = (2, 3, 7)$, and $\mathbf{w} = (0, 1, -3)$.

   (a) $\mathbf{u} + \mathbf{w}$ **(b)** $3\mathbf{u} + \mathbf{v}$

   (c) $\mathbf{u} - 2\mathbf{w}$ **(d)** $4\mathbf{u} - 2\mathbf{v} + 3\mathbf{w}$

   (e) $2\mathbf{u} - 5\mathbf{v} - \mathbf{w}$

2. Consider the sets of vectors of the following form. Determine which of the sets are subspaces of $\mathbf{R}^3$.

   (a) $(a, b, a - 2)$ **(b)** $(a, -2a, 3a)$

   (c) $(a, b, 2a - 3b)$ **(d)** $(a, 2, b)$

3. Which of the following subsets of $\mathbf{R}^3$ are subspaces? The set of all vectors of the form $p(1, 2, 3)$ where p is **(a)** a real number, **(b)** an integer, **(c)** a nonnegative real number, **(d)** a rational number.

4. In the following sets of vectors, determine whether the first vector is a linear combination of the other vectors.

   (a) $(3, 15, -4), (1, 2, -1), (2, 4, 0), (5, 1, 2)$

   (b) $(-3, -4, 7), (5, -1, 3), (2, 0, 3), (4, 1, 0)$

5. Determine whether the following vectors span $\mathbf{R}^3$.

   (a) $(1, 2, 3), (-2, 4, 1), (0, 6, 4)$

   (b) $(-2, 1, 0), (1, 2, -1), (-1, 8, -3)$

6. Determine whether the following sets are linearly dependent or independent.

   (a) $\{(1, -2, 0), (0, 1, 3), (2, 0, 12)\}$

   (b) $\{(-1, 18, 7), (-1, 4, 1), (1, 3, 2)\}$

   (c) $\{(5, -1, 3), (2, 1, 0), (3, -2, 2)\}$

7. Prove that the following sets are bases for $\mathbf{R}^2$ or $\mathbf{R}^3$.

   (a) $\{(-1, 2), (3, 1)\}$

   (b) $\{(4, 2), (1, -1)\}$

   (c) $\{(1, 2, 3), (-1, 0, 5), (0, 2, 7)\}$

   (d) $\{(8, 1, 0), (4, -1, 3), (5, 2, -3)\}$

8. Show that the vector $(10, 9, 8)$ lies in the vector space spanned by $\{(-1, 3, 1), (4, 1, 2)\}$.

9. Find a basis for $\mathbf{R}^3$ that includes the vectors $(1, -2, 3)$ and $(4, 1, -1)$.

10. Determine a basis for the subspace of $\mathbf{R}^4$ that consists of vectors of the form $(a, b, c, a - 2b + 3c)$.

11. Determine the ranks of the following matrices.

$$\text{(a)} \begin{bmatrix} 1 & 2 & -1 \\ -1 & 3 & 4 \\ 0 & 5 & 3 \end{bmatrix} \quad \text{(b)} \begin{bmatrix} 2 & 1 & 4 \\ -2 & 0 & -1 \\ 3 & 2 & 7 \end{bmatrix}$$

$$\text{(c)} \begin{bmatrix} -2 & 4 & 8 \\ 1 & -2 & 4 \\ 4 & -8 & 16 \end{bmatrix}$$

12. Find a basis for the subspace of $\mathbf{R}^4$ spanned by the following set: $\{(1, -2, 3, 4), (-1, 3, 1, -2), (2, -3, 10, 10)\}$.

13. Let $\mathbf{v}$ be a linear combination of $\mathbf{v}_1$ and $\mathbf{v}_2$. Let $\mathbf{v}_3$ be any other vector in the same vector space as $\mathbf{v}$, $\mathbf{v}_1$, and $\mathbf{v}_2$. Show that $\mathbf{v}$ is also a linear combination of $\mathbf{v}_1$, $\mathbf{v}_2$, and $\mathbf{v}_3$.

14. Let the set $\{\mathbf{v}_1, \mathbf{v}_2, \mathbf{v}_3\}$ be linearly independent in $\mathbf{R}^3$. Prove that each of the following subsets is also linearly independent: $\{\mathbf{v}_1, \mathbf{v}_2\}, \{\mathbf{v}_1, \mathbf{v}_3\}, \{\mathbf{v}_2, \mathbf{v}_3\}, \{\mathbf{v}_1\}, \{\mathbf{v}_2\}, \{\mathbf{v}_3\}$.

15. Let the set $\{\mathbf{v}_1, \mathbf{v}_2\}$ be linearly dependent. Prove that $\{\mathbf{v}_1 + 2\mathbf{v}_2, 3\mathbf{v}_1 - \mathbf{v}_2\}$ is also linearly dependent.

16. State (with a brief explanation) whether the following statements are true or false.

   (a) $\{(1, 0, 1), (2, 1, 5)\}$ is a basis for the subspace of vectors in $\mathbf{R}^3$ of the form $(a, b, a + 3b)$.

   (b) There are three linearly independent vectors in $\mathbf{R}^2$.

   (c) Any set of three vectors in $\mathbf{R}^3$ that are linearly independent span the space.

   (d) Any two nonzero vectors in $\mathbf{R}^2$ that do not form a basis are collinear.

   (e) Let $\mathbf{v}_1$, $\mathbf{v}_2$, and $\mathbf{v}_3$ be vectors in $\mathbf{R}^3$ written at random. They are probably linearly dependent.

17. Let *V* be a vector space of dimension *n*. Determine whether the following statements are true or false.

(a) *n* is the largest number of linearly dependent vectors in *V*.

(b) *n* is the largest number of linearly independent vectors in *V*.

(c) No set of fewer than *n* vectors can span *V*.

(d) Any set of more than *n* vectors in *V* must be linearly dependent.

(e) A set of more than *n* vectors can span *V* but cannot be linearly dependent.

18. Prove that an $n \times n$ matrix *A* is row equivalent of I_n if and only if rank(*A*) = *n*.

19. Determine the characteristic polynomial, eigenvalues, and corresponding eigenspaces of the matrix

$$\begin{bmatrix} 5 & -7 & 7 \\ 4 & -3 & 4 \\ 4 & -1 & 2 \end{bmatrix}$$

20. Let *A* be an invertible matrix. Show that the eigenvalues of A^{-1} are the inverses of the eigenvalues of *A*. Prove that *A* and A^{-1} have the same eigenvectors.

21. Let λ be an eigenvalue of a matrix *A* with corresponding eigenvector **x**. If *k* is a scalar, show that $\lambda - k$ is an eigenvalue of $A - kI$ and that **x** is a corresponding eigenvector.

MATLAB Discussion

M.13   Linear Combinations, Linear Dependence, Basis (Sections 4.2–4.6)

Discussions of linear combination, dependence, and basis often reduce to solving systems of linear equations.

Example 1   Determine whether or not $(-1, 1, 5)$ is a linear combination of $(1, 2, 3)$, $(0, 1, 4)$, and $(2, 3, 6)$.

Examine the identity

$$a(1, 2, 3) + b(0, 1, 4) + c(2, 3, 6) = (-1, 1, 5)$$

This reduces to the system

$$
\begin{aligned}
a + \quad\; + 2c &= -1 \\
2a + b + 3c &= 1 \\
3a + 4b + 6c &= 5
\end{aligned}
$$

Observe that the given vectors become the *columns* of the augmented matrix A. Using MATLAB, we get

```
»A = [1   0   2   −1;2   1   3   1;3   4   6   5]
»rref(A)
ans =
   1   0   0    1
   0   1   0    2
   0   0   1   −1      {the reduced echelon form}
```

Thus $a = 1$, $b = 2$, and $c = -1$. The linear combination is unique.

$$(-1, 1, 5) = (1, 2, 3) + 2(0, 1, 4) - (2, 3, 6)$$

Example 2   Prove that the vectors $(1, 2, 3)$, $(-2, 1, 1)$, and $(8, 6, 10)$ are linearly dependent. We examine the identity

$$a(1, 2, 3) + b(-2, 1, 1) + c(8, 6, 10) = 0$$

This reduces to the system

$$
\begin{aligned}
a - 2b + 8c &= 0 \\
2a + b + 6c &= 0 \\
3a + b + 10c &= 0
\end{aligned}
$$

Observe that again the given vectors become the *columns* of the augmented matrix A. Using MATLAB, we get

»A = [1 −2 8 0;2 1 6 0;3 1 10 0]
»rref(A)
ans =

1	0	4	0
0	1	−2	0
0	0	0	0

{the reduced echelon form}

Thus $a = -4r$, $b = 2r$, $c = r$. The vectors are linearly dependent

$$-4r(1, 2, 3) + 2r(-2, 1, 1) + r(8, 6, 10) = 0$$

Example 3   Determine a basis for the subspace of **R**^{4} spanned by the vectors $(1, 2, 3, 4)$, $(0, -1, 2, 3)$, $(2, 3, 8, 11)$, $(2, 3, 6, 8)$. Write these vectors as the row vectors of a matrix A and compute the reduced echelon form of A.

»A = [1 2 3 4;0 −1 2 3;2 3 8 11;2 3 6 8];
»rref(A)
ans =

1	0	0	−.5
0	1	0	0
0	0	1	1.5
0	0	0	0

{the reduced echelon form}

The vectors form a three-dimensional subspace of **R**^{4} having basis $\{(1, 0, 0, -.5)$, $(0, 1, 0, 0)$, $(0, 0, 1, 1.5)\}$.

Exercise 1   Determine whether the first vector is a linear combination of the other vectors.

*(a) $(-3, 3, 7)$; $(1, -1, 2)$, $(2, 1, 0)$, $(-1, 2, 1)$

*(b) $(2, 7, 13)$; $(1, 2, 3)$, $(-1, 2, 4)$, $(1, 6, 10)$

(c) $(0, 10, 8)$; $(-1, 2, 3)$, $(1, 3, 1)$, $(1, 8, 5)$

Exercise 2   Determine whether the following sets of vectors are linearly dependent.

*(a) $\{(-1, 3, 2), (1, -1, -3), (-5, 9, 13)\}$

(b) $\{(1, -1, 2, 1), (4, -1, 6, 2), (-2, -1, 1, -2)\}$

Exercise 3   Use the determinant function **det**(A) and rank function **rank**(A) to determine whether or not the following vectors are linearly dependent.

*(a) $(2, 4, 5, -3)$, $(-3, 2, 7, 0)$, $(9, 2, 4, 6)$, $(3, 2, -6, 5)$

(b) $(1, -1, 3, 2)$, $(-1, 2, 4, -5)$, $(3, -4, 2, 9)$, $(7, 2, 1, 3)$

Exercise 4   Use the computer to determine the rank of the matrix

$$\begin{bmatrix} 1 & 2 & 3 & 4 & 5 \\ 6 & 7 & 8 & 9 & 10 \\ 11 & 12 & 13 & 14 & 15 \\ 16 & 17 & 18 & 19 & 20 \\ 21 & 22 & 23 & 24 & 25 \end{bmatrix}$$

(a) Explain the result.   **(b)** Generalize the result.

Exercise 5   Determine the bases for the subspaces spanned by the following sets of vectors.

*__(a)__ $\{(2, 4, 6), (3, -2, 1), (-7, 14, 7)\}$

(b) $\{(2, 1, 3, 0, 4), (-1, 2, 3, 1, 0), (3, -1, 0, -1, 4)\}$

(c) $\{(1, 2, -1, 0, 4, 1), (3, 4, -2, 0, 8, 2), (2, -1, 3, 2, 1, 4), (3, 1, 2, 2, 5, 5),$
$(2, 2, -1, 0, 4, 1)\}$

M.14 Eigenvalues and Eigenvectors and Applications
(Sections 4.7 and 4.8)

Let A be a square matrix. Let X be a nonzero vector such that

$$A\mathbf{X} = \lambda\mathbf{X}$$

λ is said to be an **eigenvalue** of A with corresponding **eigenvector X**.

Example 1   Find the eigenvalues and corresponding eigenvectors of $A = \begin{bmatrix} 1 & 6 \\ 5 & 2 \end{bmatrix}$.

```
»A = [1   6;5   2]
»[X, D] — eig(A)        {Gives a diagonal matrix D of eigenvalues of A.}
                        {Columns of X will be corresponding normalized
                          eigenvectors.}
```

X =   D =
  -0.7682   -0.7071   -4   0
   0.6402   -0.7071   0   7

Thus $\lambda_1 = -4$, $\mathbf{v}_1 = \begin{bmatrix} -0.7682 \\ 0.6402 \end{bmatrix}$, and $\lambda_2 = 7$, $\mathbf{v}_2 = \begin{bmatrix} -0.7071 \\ -0.7071 \end{bmatrix}$. Note that MATLAB scales each eigenvector so that the norm of each is 1.

*****Exercise 1**　Compute the eigenvalues and corresponding eigenvectors of

$$A = \begin{bmatrix} 5 & 4 & 2 \\ 4 & 5 & 2 \\ 2 & 2 & 2 \end{bmatrix}$$

Exercise 2　Use MATLAB to compute the eigenvalues and eigenvectors of the matrices $\begin{bmatrix} 1 & 2 \\ 1 & 0 \end{bmatrix}$, $\begin{bmatrix} 2 & 3 \\ 1 & 0 \end{bmatrix}$, and $\begin{bmatrix} 3 & 4 \\ 1 & 0 \end{bmatrix}$. Use these results to make a conjecture about the eigenvalues and eigenvectors of $\begin{bmatrix} a & a+1 \\ 1 & 0 \end{bmatrix}$. Verify your conjecture.

Exercise 3　Construct a symmetric matrix A. Use $[X, D] = $ **eig**(A) to find eigenvalues and eigenvectors of A. Use MATLAB to compute $B = X^{-1}AX$. Investigate such similarity transformations on other symmetric matrices. Make some conjectures.

Long-Term Behavior

Eigenvalues and eigenvectors are used in understanding the long-term behavior of phenomena. Consider a regular Markov chain having initial vector $\mathbf{X}_0$ and transition matrix P. Suppose P is regular (some power of P has all positive elements). Then the following sequences are convergent.

$$\mathbf{X}_1 \ (= P\mathbf{X}_0), \ \mathbf{X}_2 \ (= P\mathbf{X}_1), \ \ldots, \ \mathbf{X}_n \ (= P\mathbf{X}_{n-1}), \ \ldots \to X$$

and

$$P, P^2, P^3 \ldots \to P \quad \text{(a stochastic matrix)}$$

Further,

(a)　X is an eigenvector of P corresponding to eigenvalue 1.

(b)　the column vectors of P are all identical, each being an eigenvector of P corresponding to the eigenvalue 1.

Example 2　Let us return to the population movement model of Section M.9 in Chapter 2. Annual population movement between U.S. cities and suburbs in 1985 is described by the following stochastic matrix P. Observe that P is regular. The population distribution in 1985 is given by the matrix X, in units of one million.

$$\begin{array}{cc} & \text{(from)} \\ & \text{city} \quad \text{suburb} \quad \text{(to)} \end{array}$$

$$P = \begin{bmatrix} 0.96 & 0.01 \\ 0.04 & 0.99 \end{bmatrix} \begin{array}{l} \text{city} \\ \text{suburb} \end{array} \qquad \mathbf{X}_0 = \begin{bmatrix} 60 \\ 125 \end{bmatrix} \begin{array}{l} \text{city} \\ \text{suburb} \end{array}$$

Let us use eigenconcepts to find X and P and interpret these results.

The eigenvalues and eigenvectors of P are computed.

»**P = [0.96   0.01;0.04   0.99];**
»**[V, D] = eig(P)**

V =   **D =**
   −0.7071   −0.2425   **0.9500   0**
   0.7071   −0.9701   **0   1.0000**

Thus $\lambda_1 = 0.95$, $\mathbf{v}_1 = \begin{bmatrix} -0.7071 \\ 0.7071 \end{bmatrix}$, and $\lambda_2 = 1$, $\mathbf{v}_2 = \begin{bmatrix} -0.2425 \\ -0.9701 \end{bmatrix}$.

We use the eigenvector corresponding to the eigenvalue 1. Each step in the Markov process redistributes the initial total of 185, eventually reaching an eigenvector, a vector whose components are in the same ratio as $\mathbf{v}_2$ above. Thus

$$X = \begin{bmatrix} 0.2425 \times 185/1.2126 \\ 0.9701 \times 185/1.2126 \end{bmatrix} = \begin{bmatrix} 37 \\ 148 \end{bmatrix} \begin{matrix} \text{city} \\ \text{suburb} \end{matrix}$$

City and suburban populations will gradually approach 37 and 148 million, respectively.

The columns of P are an eigenvector. Since P is stochastic, the components of each column will be in the same ratio as those of $\mathbf{v}_2$, but now adding up to 1. We get

	(from)		
	city	suburb	(to)

$$P = \begin{bmatrix} 0.2425/1.2126 & 0.2425/1.2126 \\ 0.9701/1.2126 & 0.9701/1.2126 \end{bmatrix} = \begin{bmatrix} 0.2 & 0.2 \\ 0.8 & 0.8 \end{bmatrix} \begin{matrix} \text{city} \\ \text{suburb} \end{matrix}$$

The long-term probability of living in the city is 0.2, while the long-term probability of living in suburbia is 0.8—independent of initial location.

*Exercise 4   Construct a model of population flow between metropolitan and non-metropolitan areas of the United States, given that their respective populations in 1985 were 185 million and 55 million. The probabilities are given by the matrix

	(from)		
	metro	nonmetro	(to)

$$\begin{bmatrix} 0.99 & 0.02 \\ 0.01 & 0.98 \end{bmatrix} \begin{matrix} \text{metro} \\ \text{nonmetro} \end{matrix}$$

Determine the long-term predictions for metropolitan and nonmetropolitan populations, assuming no change in their total populations.

Exercise 5   Forty rats are placed at random in a compartment having four rooms labeled 1, 2, 3, and 4 as shown in the figure below. Construct a Markov chain to describe the movement of the rats between the rooms. Predict the long-term distribution of rats. What is the long-term probability that a given marked rat is in room 4?

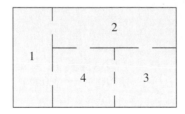

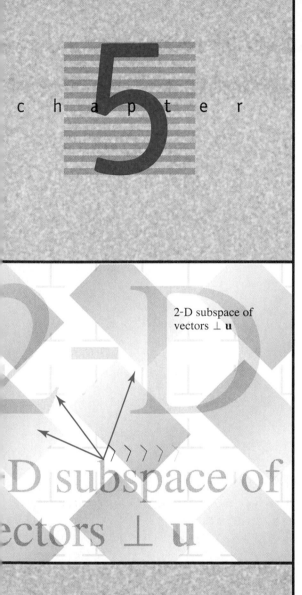

2-D subspace of
vectors ⊥ **u**

n-Dimensional Euclidean Space

In the previous chapter we
developed algebraic structure for the vector
space **R**^{n}. We defined subspaces, spanning
sets, and linear independence, leading up to
dimension. In this chapter we continue the
development of mathematical structure on
R^n. We introduce angle, length of a vector,
and distance between points. We shall
extend the familiar mathematical structures
of two-dimensional Euclidean space to *n*
dimensions.

5.1   Dot Product, Norm, Angle, and Distance

As we develop the geometry of the vector space $\mathbf{R}^n$, you should pay close attention to the approach we use. While the results are, of course, important, the way we arrive at the results is also very important. Magnitude, angle, and distance are defined on $\mathbf{R}^n$ by generalizing expressions for magnitude, angle, and distance from $\mathbf{R}^2$. We extend the Pythagorean theorem to $\mathbf{R}^n$. This process of gradually extending familiar concepts to more general surroundings is fundamental to mathematics.

We start the discussion with the definition of the dot product of two vectors. The dot product is a tool that is used to control the geometry of $\mathbf{R}^n$.

Definition   *Let $\mathbf{u} = (u_1, \ldots, u_n)$ and $\mathbf{v} = (v_1, \ldots, v_n)$ be two vectors in $\mathbf{R}^n$. The **dot product** of $\mathbf{u}$ and $\mathbf{v}$ is denoted $\mathbf{u} \cdot \mathbf{v}$ and is defined by*

$$\mathbf{u} \cdot \mathbf{v} = u_1v_1 + \cdots + u_nv_n$$

The dot product assigns a real number to each pair of vectors.

Example 1   Find the dot product of

$$\mathbf{u} = (1, -2, 4) \quad \text{and} \quad \mathbf{v} = (3, 0, 2)$$

Solution   We get

$$\mathbf{u} \cdot \mathbf{v} = (1 \times 3) + (-2 \times 0) + (4 \times 2) = 3 + 0 + 8 = 11$$

Properties of the Dot Product

Let $\mathbf{u}$, $\mathbf{v}$, and $\mathbf{w}$ be vectors in $\mathbf{R}^n$ and let c be a scalar. Then

1. $\mathbf{u} \cdot \mathbf{v} = \mathbf{v} \cdot \mathbf{u}$
2. $(\mathbf{u} + \mathbf{v}) \cdot \mathbf{w} = \mathbf{u} \cdot \mathbf{w} + \mathbf{v} \cdot \mathbf{w}$
3. $c\mathbf{u} \cdot \mathbf{v} = c(\mathbf{u} \cdot \mathbf{v}) = \mathbf{u} \cdot c\mathbf{v}$
4. $\mathbf{u} \cdot \mathbf{u} \geq 0$ and $\mathbf{u} \cdot \mathbf{u} = 0$ if and only if $\mathbf{u} = \mathbf{0}$

Proof   We prove properties 1 and 4, leaving the other properties for you to prove in the exercises that follow.

1. Let $\mathbf{u} = (u_1, \ldots, u_n)$ and $\mathbf{v} = (v_1, \ldots, v_n)$. We get

$$\mathbf{u} \cdot \mathbf{v} = u_1v_1 + \cdots + u_nv_n$$
$$= v_1u_1 + \cdots + v_nu_n, \text{ by the commutative property}$$
$$\text{of real numbers}$$
$$= \mathbf{v} \cdot \mathbf{u}$$

4. $\mathbf{u} \cdot \mathbf{u} = u_1u_1 + \cdots + u_nu_n = (u_1)^2 + \cdots + (u_n)^2$
   $(u_1)^2 + \cdots + (u_n)^2 \geq 0$, since it is the sum of squares. Thus $\mathbf{u} \cdot \mathbf{u} \geq 0$. Further, $(u_1)^2 + \cdots + (u_n)^2 = 0$ if and only if $u_1 = 0, \ldots, u_n = 0$. Thus $\mathbf{u} \cdot \mathbf{u} = 0$ if and only if $\mathbf{u} = \mathbf{0}$.

We shall use these properties to simplify expressions involving dot products. In this section, for example, we shall use the properties in arriving at a suitable expression for the angle between two vectors in $\mathbf{R}^n$. In the next chapter we shall use the properties to generalize the concept of a dot product to spaces of matrices and functions.

Norm of a Vector in $\mathbf{R}^n$

Let $\mathbf{u} = (u_1, u_2)$ be a vector in $\mathbf{R}^2$. We know that the length of this vector is $\sqrt{(u_1)^2 + (u_2)^2}$. See Figure 5.1. We generalize this result to define the length (the technical name is norm) of a vector in $\mathbf{R}^n$.

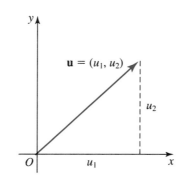

Figure 5.1   Norm of vector $(u_1, u_2) = \sqrt{(u_1)^2 + (u_2)^2}$

Definition   *The* **norm** *(**length** or **magnitude**) of a vector* $\mathbf{u} = (u_1, \ldots, u_n)$ *in* $\mathbf{R}^n$ *is denoted* $\|\mathbf{u}\|$ *and defined by*

$$\|\mathbf{u}\| = \sqrt{(u_1)^2 + \cdots + (u_n)^2}$$

Note: The norm of a vector can also be written in terms of the dot product:

$$\|\mathbf{u}\| = \sqrt{\mathbf{u} \cdot \mathbf{u}}$$

Example 2   Find the norms of the vectors $\mathbf{u} = (1, 3, 5)$ of $\mathbf{R}^3$ and $\mathbf{v} = (3, 0, 1, 4)$ of $\mathbf{R}^4$.

Solution   Using the above definition, we get

$$\|\mathbf{u}\| = \sqrt{(1)^2 + (3)^2 + (5)^2} = \sqrt{1 + 9 + 25} = \sqrt{35}$$
$$\|\mathbf{v}\| = \sqrt{(3)^2 + (0)^2 + (1)^2 + (4)^2} = \sqrt{9 + 0 + 1 + 16} = \sqrt{26}$$

Definition   *A **unit vector** is a vector whose norm is one. If* **v** *is a nonzero vector, then the vector*

$$\mathbf{u} = \frac{1}{\|\mathbf{v}\|}\mathbf{v}$$

is a unit vector in the direction of **v**. *(See following exercises.) This procedure of constructing a unit vector in the same direction as a given vector is called **normalizing** the vector.*

Example 3   (a)   Show that the vector $(1, 0)$ is a unit vector.

(b)   Find the norm of the vector $(2, -1, 3)$. Normalize this vector.

Solution   (a) $\|(1, 0)\| = \sqrt{1^2 + 0^2} = 1$. Thus $(1, 0)$ is a unit vector. It can be similarly shown that $(0, 1)$ is a unit vector in $\mathbf{R}^2$.

(b) $\|(2, -1, 3)\| = \sqrt{2^2 + (-1)^2 + 3^2} = \sqrt{14}$. The norm of $(2, -1, 3)$ is $\sqrt{14}$. The normalized vector is

$$\frac{1}{\sqrt{14}}(2, -1, 3)$$

This vector may also be written $\left(\dfrac{2}{\sqrt{14}}, \dfrac{-1}{\sqrt{14}}, \dfrac{3}{\sqrt{14}}\right)$. This vector is a unit in the direction of $(2, -1, 3)$.

The following are important unit vectors.

$(1, 0)$, $(0, 1)$ are unit vectors in $\mathbf{R}^2$

$(1, 0, 0)$, $(0, 1, 0)$, $(0, 0, 1)$ are unit vectors in $\mathbf{R}^3$

$(1, 0, \ldots, 0)$, $\ldots$, $(0, 0, \ldots, 1)$ are unit vectors in $\mathbf{R}^n$

We now introduce a useful mathematical result called the **Cauchy-Schwartz Inequality**. The inequality will, for example, enable us to extend the definition of angle that we have in $\mathbf{R}^2$ to $\mathbf{R}^n$.

Theorem 5.1   **The Cauchy-Schwartz Inequality.*** If **u** and **v** are vectors in $\mathbf{R}^n$, then

$$|\mathbf{u} \cdot \mathbf{v}| \leq \|\mathbf{u}\| \|\mathbf{v}\|$$

Here $|\mathbf{u} \cdot \mathbf{v}|$ denotes the absolute value of the number $\mathbf{u} \cdot \mathbf{v}$.

Proof   We first look at the special case when $\mathbf{u} = \mathbf{0}$. If $\mathbf{u} = \mathbf{0}$, then $|\mathbf{u} \cdot \mathbf{v}| = 0$ and $\|\mathbf{u}\| \|\mathbf{v}\| = 0$. Equality holds.

Let us now consider $\mathbf{u} \neq \mathbf{0}$. The proof of the inequality for this case involves an interesting application of the quadratic formula. Consider the vector $r\mathbf{u} + \mathbf{v}$, where r is any real number. Using the properties of the dot product, we get

$$(r\mathbf{u} + \mathbf{v}) \cdot (r\mathbf{u} + \mathbf{v}) = r^2(\mathbf{u} \cdot \mathbf{u}) + 2r(\mathbf{u} \cdot \mathbf{v}) + \mathbf{v} \cdot \mathbf{v}$$

The properties of the dot product also tell us that

$$(r\mathbf{u} + \mathbf{v}) \cdot (r\mathbf{u} + \mathbf{v}) \geq 0$$

Combining these results, we get

$$r^2(\mathbf{u} \cdot \mathbf{u}) + 2r(\mathbf{u} \cdot \mathbf{v}) + \mathbf{v} \cdot \mathbf{v} \geq 0$$

Let $a = \mathbf{u} \cdot \mathbf{u}$, $b = 2\mathbf{u} \cdot \mathbf{v}$, $c = \mathbf{v} \cdot \mathbf{v}$. Thus

$$ar^2 + br + c \geq 0$$

This implies that the **quadratic function** $f(r) = ar^2 + br + c$ is never negative. Therefore $f(r)$, whose graph is a parabola, must have either one zero or no zeros. The zeros of $f(r)$ are the roots of the equation

$$ar^2 + br + c = 0$$

The **discriminant** gives information about these zeros. There is one zero if $b^2 - 4ac = 0$ and no zeros if $b^2 - 4ac < 0$. Thus

$$b^2 - 4ac \leq 0$$

$$b^2 \leq 4ac$$

$$(2\mathbf{u} \cdot \mathbf{v})^2 \leq 4(\mathbf{u} \cdot \mathbf{u})(\mathbf{v} \cdot \mathbf{v})$$

$$(\mathbf{u} \cdot \mathbf{v})^2 \leq (\mathbf{u} \cdot \mathbf{u})(\mathbf{v} \cdot \mathbf{v})$$

$$(\mathbf{u} \cdot \mathbf{v})^2 \leq \|\mathbf{u}\|^2 \|\mathbf{v}\|^2$$

Taking the square root of both sides gives the Cauchy-Schwartz Inequality.

$$|\mathbf{u} \cdot \mathbf{v}| \leq \|\mathbf{u}\| \|\mathbf{v}\|$$

---

* Augustin Louis Cauchy (1789–1857) was educated in Paris and then worked for some time as an engineer before returning to teach in Paris. He was one of the greatest mathematicians ever and one of the most prolific. He wrote seven books and over eight hundred papers. More concepts and theorems have been named after him than any other mathematician. He had a turbulent life. After the revolution of 1830 he refused to take an oath of allegiance to Louis-Philippe and went into exile in Turin but was later able to return to Paris. Cauchy was a devout Catholic and was involved in social work. He was involved in charities for unwed mothers, aid for the starving in Ireland, and the rehabilitation of criminals. On giving one year's salary to the mayor of Sceaux for charity he commented, "Do not worry, it is only my salary; it is not my money, it is the emperor's."

Herman Amandus Schwartz (1843–1921) was educated in Berlin and taught in Zurich, Göttingen, and Berlin. His great strength was his geometric intuition. He worked on narrowly defined problems. The methods he developed transcended the significance of the actual problems. His teaching and his concern for students did not leave him much time for publishing. Contributing elements may have been his propensity for handling the important and the trivial with the same thoroughness.

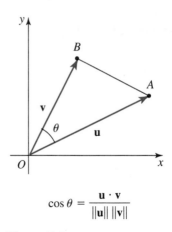

$$\cos \theta = \frac{\mathbf{u} \cdot \mathbf{v}}{\|\mathbf{u}\| \, \|\mathbf{v}\|}$$

Figure 5.2

Angle between Vectors

Let $\mathbf{u} = (a, b)$ and $\mathbf{v} = (c, d)$ be position vectors in $\mathbf{R}^2$. See Figure 5.2. Let us arrive at an expression for the cosine of the angle θ formed by these vectors.

The **law of cosines** gives

$$AB^2 = OA^2 + OB^2 - 2(OA)(OB) \cos \theta$$

Solve for $\cos \theta$.

$$\cos \theta = \frac{OA^2 + OB^2 - AB^2}{2(OA)(OB)}$$

We get that

$$OA^2 + OB^2 - AB^2 = \|\mathbf{u}\|^2 + \|\mathbf{v}\|^2 - \|\mathbf{v} - \mathbf{u}\|^2$$
$$= a^2 + b^2 + c^2 + d^2 - [(c - a)^2 + (d - b)^2]$$
$$= 2ac + 2db = 2\mathbf{u} \cdot \mathbf{v}$$

Further,

$$2(OA)(OB) = 2\|\mathbf{u}\| \, \|\mathbf{v}\|$$

Thus the angle θ between two vectors in $\mathbf{R}^2$ is given by

$$\cos \theta = \frac{\mathbf{u} \cdot \mathbf{v}}{\|\mathbf{u}\| \, \|\mathbf{v}\|}$$

We know that the cosine of any angle such as θ has to satisfy the condition $|\cos \theta| \leq 1$. The Cauchy-Schwartz Inequality assures us that

$$\frac{\mathbf{u} \cdot \mathbf{v}}{\|\mathbf{u}\| \, \|\mathbf{v}\|} \leq 1$$

We can thus extend this concept of angle between two vectors in $\mathbf{R}^2$ to $\mathbf{R}^{2n}$ as follows.

Definition   *Let $\mathbf{u}$ and $\mathbf{v}$ be two nonzero vectors in $\mathbf{R}^n$. The cosine of the angle θ between these vectors is*

$$\cos \theta = \frac{\mathbf{u} \cdot \mathbf{v}}{\|\mathbf{u}\| \, \|\mathbf{v}\|}, \qquad 0 \leq \theta \leq \pi$$

Example 4   Determine the angle between the vectors $\mathbf{u} = (1, 0, 0)$ and $\mathbf{v} = (1, 0, 1)$ in $\mathbf{R}^3$.

Solution   We get

$$\mathbf{u} \cdot \mathbf{v} = (1, 0, 0) \cdot (1, 0, 1) = 1, \qquad \|\mathbf{u}\| = \sqrt{1^2 + 0^2 + 0^2} = 1,$$
$$\|\mathbf{v}\| = \sqrt{1^2 + 0^2 + 1^2} = \sqrt{2}$$

Thus

$$\cos \theta = \frac{\mathbf{u} \cdot \mathbf{v}}{\|\mathbf{u}\| \, \|\mathbf{v}\|} = \frac{1}{\sqrt{2}}$$

The angle between $\mathbf{u}$ and $\mathbf{v}$ is $45°$.

We say that two nonzero vectors are **orthogonal** if the angle between them is a right angle. The following theorem, which you are asked to prove in the exercises that follow, gives us a condition for orthogonality.

Theorem 5.2 Two nonzero vectors **u** and **v** are orthogonal if and only if $\mathbf{u} \cdot \mathbf{v} = \mathbf{0}$.

Example 5 Show that the following pairs of vectors are orthogonal.

(a) $(1, 0)$ and $(0, 1)$

(b) $(2, -3, 1)$ and $(1, 2, 4)$

Solution On taking the dot products, we get

(a) $(1, 0) \cdot (0, 1) = (1 \times 0) + (0 \times 1) = 0$. The vectors are orthogonal.

(b) $(2, -3, 1) \cdot (1, 2, 4) = (2 \times 1) + (-3 \times 2) + (1 \times 4) = 2 - 6 + 4 = 0$. The vectors are orthogonal.

The following are important pairwise orthogonal vectors (taken in pairs).

$(1, 0), (0, 1)$ are orthogonal vectors in $\mathbf{R}^2$

$(1, 0, 0), (0, 1, 0) \, (0, 0, 1)$ are orthogonal vectors in $\mathbf{R}^3$

$(1, 0, \ldots, 0), \ldots, (0, 0, \ldots, 1)$ are orthogonal vectors in $\mathbf{R}^n$

Example 6 Determine a vector in $\mathbf{R}^2$ that is orthogonal to $(3, -1)$. Show that there are many such vectors and prove that these vectors form a subspace of $\mathbf{R}^2$.

Solution Let the vector (a, b) be orthogonal to $(3, -1)$. We get

$$(a, b) \cdot (3, -1) = 0$$
$$(a \times 3) + (b \times (-1)) = 0$$
$$3a - b = 0$$
$$b = 3a$$

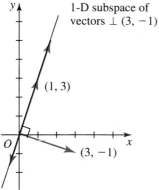

Thus any vector of the form $(a, 3a)$ is orthogonal to the vectors $(3, -1)$.

Any vector of this form can be written

$$a(1, 3)$$

The set of all such vectors thus forms a one-dimensional subspace of $\mathbf{R}^2$ with basis $\{(1, 3)\}$. See Figure 5.3.

In this section we are extending concepts and results from $\mathbf{R}^2$ and $\mathbf{R}^3$ to $\mathbf{R}^n$. The above example leads us to the conjecture that the set of vectors orthogonal to a given vector in $\mathbf{R}^n$ forms a subspace.

Figure 5.3

Example 7 Let **u** be a vector in $\mathbf{R}^n$. Let V be the set of vectors in $\mathbf{R}^n$ that are orthogonal to **u**. Is V a subspace of $\mathbf{R}^n$?

Solution   Let $\mathbf{v}$ and $\mathbf{w}$ be elements of V and let c be a scalar. Since $\mathbf{v}$ and $\mathbf{w}$ are orthogonal to $\mathbf{u}$,

$$\mathbf{v} \cdot \mathbf{u} = 0 \quad \text{and} \quad \mathbf{w} \cdot \mathbf{u} = 0$$

Adding these equations and using a property of dot products, we get

$$\mathbf{v} \cdot \mathbf{u} + \mathbf{w} \cdot \mathbf{u} = 0$$
$$(\mathbf{v} + \mathbf{w}) \cdot \mathbf{u} = 0$$

Therefore $\mathbf{v} + \mathbf{w}$ is orthogonal to $\mathbf{u}$; $\mathbf{v} + \mathbf{w}$ is an element of V. Thus V is closed under vector addition.

Further, since $\mathbf{v} \cdot \mathbf{u} = 0$, multiplication by c and a property of dot products gives

$$c(\mathbf{v} \cdot \mathbf{u}) = 0$$
$$(c\mathbf{v}) \cdot \mathbf{u} = 0$$

Therefore $c\mathbf{v}$ is orthogonal to $\mathbf{u}$; $c\mathbf{v}$ is an element of V. Thus V is closed under scalar multiplication. Since V is closed under addition and scalar multiplication, it is a subspace.

If the vector space is $\mathbf{R}^3$, then the subspace of vectors orthogonal to any vector $\mathbf{u}$ is a plane. See Figure 5.4. You are asked to prove this result in the following exercises.

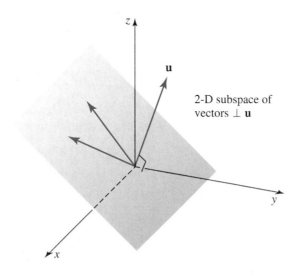

Figure 5.4

We now extend two familiar geometrical results from $\mathbf{R}^2$ and $\mathbf{R}^3$ to $\mathbf{R}^n$.

Theorem 5.3   Let **u** and **v** be vectors in $\mathbf{R}^n$.

(a) **Triangle Inequality:** $\|\mathbf{u} + \mathbf{v}\| \le \|\mathbf{u}\| + \|\mathbf{v}\|$.

This inequality tells us that the length of one side of a triangle cannot exceed the sum of the lengths of the other two sides. See Figure 5.5(a).

(b) **Pythagorean Theorem:** If $\mathbf{u} \cdot \mathbf{v} = 0$, then $\|\mathbf{u} + \mathbf{v}\|^2 = \|\mathbf{u}\|^2 + \|\mathbf{v}\|^2$.

The square of the hypotenuse of a right triangle is equal to the sum of the squares of the other two sides. See Figure 5.5(b).

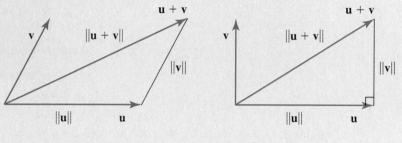

$$\|\mathbf{u} + \mathbf{v}\| \le \|\mathbf{u}\| + \|\mathbf{v}\|$$
Triangle Inequality

$$\|\mathbf{u} + \mathbf{v}\|^2 = \|\mathbf{u}\|^2 + \|\mathbf{v}\|^2$$
Pythagorean Theorem

Figure 5.5   (a)   (b)

Proof

(a) By the properties of norm,

$$\begin{aligned}
\|\mathbf{u} + \mathbf{v}\|^2 &= (\mathbf{u} + \mathbf{v}) \cdot (\mathbf{u} + \mathbf{v}) \\
&= \mathbf{u} \cdot \mathbf{u} + 2\mathbf{u} \cdot \mathbf{v} + \mathbf{v} \cdot \mathbf{v} \\
&= \|\mathbf{u}\|^2 + 2\mathbf{u} \cdot \mathbf{v} + \|\mathbf{v}\|^2 \\
&\le \|\mathbf{u}\|^2 + 2|\mathbf{u} \cdot \mathbf{v}| + \|\mathbf{v}\|^2
\end{aligned}$$

The Cauchy-Schwartz Inequality implies that

$$\begin{aligned}
\|\mathbf{u} + \mathbf{v}\|^2 &\le \|\mathbf{u}\|^2 + 2\|\mathbf{u}\|\,\|\mathbf{v}\| + \|\mathbf{v}\|^2 \\
&= (\|\mathbf{u}\| + \|\mathbf{v}\|)^2
\end{aligned}$$

Taking the square root of each side gives the triangle inequality.

(b) By the properties of norm, and the fact that $\mathbf{u} \cdot \mathbf{v} = 0$, we get

$$\begin{aligned}
\|\mathbf{u} + \mathbf{v}\|^2 &= (\mathbf{u} + \mathbf{v}) \cdot (\mathbf{u} + \mathbf{v}) \\
&= \mathbf{u} \cdot \mathbf{u} + 2\mathbf{u} \cdot \mathbf{v} + \mathbf{v} \cdot \mathbf{v} \\
&= \|\mathbf{u}\|^2 + \|\mathbf{v}\|^2
\end{aligned}$$

proving the Pythagorean Theorem.

Distance between Points

The distance between the points $\mathbf{x} = (x_1, x_2)$ and $\mathbf{y} = (y_1, y_2)$ in $\mathbf{R}^2$ is $d(\mathbf{x}, \mathbf{y}) = \sqrt{(x_1 - y_1)^2 + (x_2 - y_2)^2}$. We define distance in $\mathbf{R}^n$ by generalizing this expression as follows.

Definition   *Let $\mathbf{x} = (x_1, \ldots, x_n)$ and $\mathbf{y} = (y_1, \ldots, y_n)$ be two points in $\mathbf{R}^n$. The distance between $\mathbf{x}$ and $\mathbf{y}$ is denoted $d(\mathbf{x}, \mathbf{y})$ and is defined by*

$$d(\mathbf{x}, \mathbf{y}) = \sqrt{(x_1 - y_1)^2 + \cdots + (x_n - y_n)^2}$$

Note: We can also write this distance as follows.

$$d(\mathbf{x}, \mathbf{y}) = \|\mathbf{x} - \mathbf{y}\|$$

Example 8   Determine the distance between the points $\mathbf{x} = (1, -2, 3, 0)$ and $\mathbf{y} = (4, 0, -3, 5)$ in $\mathbf{R}^4$.

Solution   Applying the above formula for distance, we get

$$\begin{aligned} d(\mathbf{x}, \mathbf{y}) &= \sqrt{(1 - 4)^2 + (-2 - 0)^2 + (3 + 3)^2 + (0 - 5)^2} \\ &= \sqrt{9 + 4 + 36 + 25} \\ &= \sqrt{74} \end{aligned}$$

Example 9   Prove that the distance function on $\mathbf{R}^n$ has the following **symmetric property**: $d(\mathbf{x}, \mathbf{y}) = d(\mathbf{y}, \mathbf{x})$.

Solution   Let $\mathbf{x} = (x_1, \ldots, x_n)$ and $\mathbf{y} = (y_1, \ldots, y_n)$. We get

$$\begin{aligned} d(\mathbf{x}, \mathbf{y}) &= \sqrt{(x_1 - y_1)^2 + \cdots + (x_n - y_n)^2} \\ &= \sqrt{(y_1 - x_1)^2 + \cdots + (y_n - x_n)^2} = d(\mathbf{y}, \mathbf{x}) \end{aligned}$$

This result tells us that the distance from $\mathbf{x}$ to $\mathbf{y}$ is the same as the distance from $\mathbf{y}$ to $\mathbf{x}$. This property is one we would naturally want a distance function to have. We summarize in the following table the most important properties of norm and distance. In the exercises that follow, you are asked to derive those properties not already discussed. The norms and distances introduced in this section were all derived from the dot product. In numerical work it is often convenient to define other norms and distances, using these properties as **axioms**—basic properties that any norm or distance function should have. You will meet some of these ideas in later sections.

Properties of Norm	Properties of Distance
1. $\|\mathbf{u}\| \geq 0$ (the length of a vector cannot be negative)	1. $d(\mathbf{x}, \mathbf{y}) \geq 0$ (the distance between two points cannot be negative)
2. $\|\mathbf{u}\| = 0$ if and only if $\mathbf{u} = 0$ (the length of a vector is zero if and only if the vector is the zero vector)	2. $d(\mathbf{x}, \mathbf{y}) = 0$ if and only if $\mathbf{x} = \mathbf{y}$ (the distance between two points is zero if and only if the points are coincident)
3. $\|c\mathbf{u}\| = |c| \, \|\mathbf{u}\|$ (the length of $c\mathbf{u}$ is $|c|$ times the length of $\mathbf{u}$)	3. $d(\mathbf{x}, \mathbf{y}) = d(\mathbf{y}, \mathbf{x})$ (symmetry property) (the distance between $\mathbf{x}$ and $\mathbf{y}$ is the same as the distance between $\mathbf{y}$ and $\mathbf{x}$)
4. $\|\mathbf{u} + \mathbf{v}\| \leq \|\mathbf{u}\| + \|\mathbf{v}\|$ (triangle inequality)	4. $d(\mathbf{x}, \mathbf{z}) \leq d(\mathbf{x}, \mathbf{y}) + d(\mathbf{y}, \mathbf{z})$ (triangle inequality)

(the sum of the lengths of two sides of a
triangle is at least as large as the third side)

We now bring the results of this section together with the following definition.

Definition   *The vector space $\mathbf{R}^n$ with the following additional mathematical structure of dot product, norm of vector, angle between vectors, and distance between points is called a **Euclidean n-space**.*

Dot product of vectors $\mathbf{u}$ *and* $\mathbf{v}$:	$\mathbf{u} \cdot \mathbf{v} = u_1 v_1 + \cdots + u_n v_n$
Norm of a vector $\mathbf{u}$:	$\|\mathbf{u}\| = \sqrt{(u_1)^2 + \cdots + (u_n)^2}$
Angle between vectors $\mathbf{u}$ *and* $\mathbf{v}$:	$\cos \theta = \dfrac{\mathbf{u} \cdot \mathbf{v}}{\|\mathbf{u}\| \, \|\mathbf{v}\|}$
Distance between points $\mathbf{x}$ *and* $\mathbf{y}$:	$d(\mathbf{x}, \mathbf{y}) = \sqrt{(x_1 - y_1)^2 + \cdots (x_n - y_n)^2}$

The geometry of Euclidean space is, of course, called Euclidean geometry. In this section we have extended the familiar Euclidean two-dimensional geometry to n dimensions. Unless we state otherwise, *we shall assume henceforth in this course that $\mathbf{R}^n$ is a Euclidean space.*

Exercise Set 5.1

Dot Product

1. Determine the dot products of the following pairs of vectors.

   *(a) (2, 1), (3, 4)   (b) (1, −4), (3, 0)

   *(c) (2, 0), (0, −1)   (d) (5, −2), (−3, −4)

2. Determine the dot products of the following pairs of vectors.

   *(a) (1, 2, 3), (4, 1, 0)   (b) (3, 4, −2), (5, 1, −1)

   *(c) (7, 1, −2), (3, −5, 8)

3. Determine the dot products of the following pairs of vectors.

   *(a) (5, 1), (2, −3)   (b) (−3, 1, 5), (2, 0, 4)

   *(c) (7, 1, 2, −4), (3, 0, −1, 5)

   (d) (2, 3, −4, 1, 6), (−3, 1, −4, 5, −1)

   *(e) (1, 2, 3, 0, 0, 0), (0, 0, 0, −2, −4, 9)

Norms of Vectors

4. Find the norms of the following vectors.

   *(a) (1, 2)   (b) (3, −4)   *(c) (4, 0)

   *(d) (−3, 1)   (e) (0, 27)

5. Find the norms of the following vectors.

   *(a) (1, 3, −1)   (b) (3, 0, 4)

   *(c) (5, 1, 1)   (d) (0, 5, 0)

   *(e) (7, −2, −3)

6. Find the norms of the following vectors.

   *(a) (5, 2)   *(b) (−4, 2, 3)

   (c) (1, 2, 3, 4)   (d) (4, −2, 1, 3)

   *(e) (−3, 0, 1, 4, 2)   (f) (0, 0, 0, 7, 0, 0)

7. Normalize the following vectors.

   *(a) (1, 3)   (b) (2, −4)   *(c) (1, 2, 3)

   *(d) (−2, 4, 0)   (e) (0, 5, 0)

8. Normalize the following vectors.

   *(a) (4, 2)   (b) (4, 1, 1)

   *(c) (7, 2, 0, 1)   (d) (3, −1, 1, 2)

   *(e) (0, 0, 0, 7, 0, 0)

Angles between Vectors

9. Determine the angles between the following pairs of vectors.

   *(a) (−1, 1), (0, 1)   *(b) (2, 0), (1, √3)

   (c) (2, 3), (3, −2)   (d) (5, 2), (−5, −2)

10. Determine the cosine of the angle between the following pairs of vectors.

    *(a) (4, −1), (2, 3)   *(b) (3, −1, 2), (4, 1, 1)

    (c) (2, −1, 0), (5, 3, 1)

    *(d) (7, 1, 0, 0), (3, 2, 1, 0)

    (e) (1, 2, −1, 3, 1), (2, 0, 1, 0, 4)

11. Show that the following pairs of vectors are orthogonal.

    *(a) (1, 3), (3, −1)   (b) (−2, 4), (4, 2)

    (c) (3, 0), (0, −2)   (d) (7, −1), (1, 7)

12. Show that the following pairs of vectors are orthogonal.

    (a) (3, −5), (5, 3)   (b) (1, 2, −3), (4, 1, 2)

    *(c) (7, 1, 0), (2, −14, 3)

    (d) (5, 1, 0, 2), (−3, 7, 9, 4)

    (e) (1, −1, 2, −5, 9), (4, 7, 4, 1, 0)

13. Determine nonzero vectors that are orthogonal to the following vectors.

    *(a) (1, 3)   (b) (7, −1)   *(c) (−4, −1)

    (d) (−3, 0)

14. Determine nonzero vectors that are orthogonal to the following vectors.

    (a) (5, −1)   *(b) (1, −2, 3)

    (c) (5, 1, −1)   *(d) (5, 0, 1, 1)

    (e) (6, −1, 2, 3)   *(f) (0, −2, 3, 1, 5)

*15. Determine a nonzero vector that is orthogonal to both (1, 2, −1) and (3, 1, 0).

Distances between Points

16. Find the distances between the following pairs of points.

   *(a) (6, 5), (2, 2)   **(b)** (3, 1), (−4, 0)

   *(c) (7, −3), (2, 2)   **(d)** (1, −3), (5, 1)

17. Find the distances between the following pairs of points.

   (a) (4, 1), (2, −3)   *(b) (1, 2, 3), (2, 1, 0)

   (c) (−3, 1, 2), (4, −1, 1)

   *(d) (5, 1, 0, 0), (2, 0, 1, 3)

   *(e) (−3, 1, 1, 0, 2), (2, 1, 4, 1, −1)

Miscellaneous Exercises

18. Prove the following two properties of the dot product.

   (a) $(\mathbf{u} + \mathbf{v}) \cdot \mathbf{w} = \mathbf{u} \cdot \mathbf{w} + \mathbf{v} \cdot \mathbf{w}$

   (b) $c\mathbf{u} \cdot \mathbf{v} = c(\mathbf{u} \cdot \mathbf{v}) = \mathbf{u} \cdot c\mathbf{v}$

*19. Prove that if $\mathbf{v}$ is a nonzero vector, then the following vector $\mathbf{u}$ is a unit vector in the direction of $\mathbf{v}$.

$$\mathbf{u} = \frac{1}{\|\mathbf{v}\|}\mathbf{v}$$

20. Prove that two nonzero vectors $\mathbf{u}$ and $\mathbf{v}$ are orthogonal if and only if $\mathbf{u} \cdot \mathbf{v} = 0$.

*21. Show that if $\mathbf{v}$ and $\mathbf{w}$ are two vectors in a space U, and $\mathbf{u} \cdot \mathbf{v} = \mathbf{u} \cdot \mathbf{w}$ for all vectors $\mathbf{u}$ in U, then $\mathbf{v} = \mathbf{w}$.

22. Let $\mathbf{u}, \mathbf{v}_1, \ldots, \mathbf{v}_n$ be vectors in a given Euclidean space. Let $a_1, \ldots, a_n$ be scalars. Prove that

$$\mathbf{u} \cdot (a_1\mathbf{v}_1 + \cdots + a_n\mathbf{v}_n) = a_1\mathbf{u} \cdot \mathbf{v}_1 + \cdots + a_n\mathbf{u} \cdot \mathbf{v}_n$$

23. Let $\mathbf{u}, \mathbf{v},$ and $\mathbf{w}$ be vectors in a given Euclidean space and let c and d be nonzero scalars. Tell whether each of the following expressions is a scalar, a vector, or makes no sense.

   *(a) $(\mathbf{u} \cdot \mathbf{v})\mathbf{w}$ **(b)** $(\mathbf{u} \cdot \mathbf{v}) \cdot \mathbf{w}$

   *(c) $\mathbf{u} \cdot \mathbf{v} + c\mathbf{w}$ **(d)** $\mathbf{u} \cdot \mathbf{v} + c$

   (e) $c(\mathbf{u} \cdot \mathbf{v}) + d\mathbf{w}$ *(f) $\|\mathbf{u} + c\mathbf{v}\| + d$

   (g) $c\mathbf{u} \cdot d\mathbf{v} + \|\mathbf{w}\|\mathbf{v}$ *(h) $\|\mathbf{u} \cdot \mathbf{v}\|$

*24. Find all the values of c such that $\|c(3, 0, 4)\| = 15$.

25. Prove that $\mathbf{u}$ and $\mathbf{v}$ are orthogonal vectors if and only if

$$\|\mathbf{u} + \mathbf{v}\|^2 = \|\mathbf{u}\|^2 + \|\mathbf{v}\|^2$$

*26. Let (a, b) be a vector in $\mathbf{R}^2$. Prove that the vector $(-b, a)$ is orthogonal to (a, b).

*27. Let $\mathbf{u} = (u_1, u_2)$ be a nonzero vector in $\mathbf{R}^2$. Prove that the set of vectors orthogonal to $\mathbf{u}$ forms a one-dimensional subspace of $\mathbf{R}^2$.

28. Let $\mathbf{u} = (u_1, u_2, u_3)$ be a nonzero vector in $\mathbf{R}^3$. Prove that the set of vectors orthogonal to $\mathbf{u}$ forms a two-dimensional subspace of $\mathbf{R}^3$.

29. Let $\mathbf{u} = (u_1, u_2, u_3)$ and $\mathbf{v} = (v_1, v_2, v_3)$ be two nonzero linearly independent vectors in $\mathbf{R}^3$. Prove that the set of vectors orthogonal to both $\mathbf{u}$ and $\mathbf{v}$ forms a one-dimensional subspace of $\mathbf{R}^3$.

*30. Let $\mathbf{u}$ and $\mathbf{v}$ be vectors in $\mathbf{R}^n$. Prove that $\|\mathbf{u}\| = \|\mathbf{v}\|$ if and only if $\mathbf{u} + \mathbf{v}$ and $\mathbf{u} - \mathbf{v}$ are orthogonal.

*31. Let U and V be two subspaces of $\mathbf{R}^n$. U is said to be orthogonal to V if and only if every vector in U is orthogonal to every vector in V. Give an example of two orthogonal subspaces of $\mathbf{R}^3$.

32. Let $\{\mathbf{u}_1, \ldots, \mathbf{u}_n\}$ be a set of n mutually orthogonal vectors in $\mathbf{R}^n$. (Any two of the vectors are orthogonal.) Prove that the set $\{\mathbf{u}_1, \ldots, \mathbf{u}_n\}$ is linearly independent. (*Hint:* Use the result of Exercise 22.)

33. Let $\mathbf{u}$ be a vector in $\mathbf{R}^n$ and c a scalar. Prove that the norm of a vector has the following properties.

   (a) $\|\mathbf{u}\| \geq 0$

   (b) $\|\mathbf{u}\| = 0$ if and only if $\mathbf{u} = 0$

   (c) $\|c\mathbf{u}\| = |c| \|\mathbf{u}\|$

34. Consider the vector space $\mathbf{R}^n$. Let $\mathbf{u} = (u_1, \ldots, u_n)$ be a vector in $\mathbf{R}^n$. Prove that both the following functions satisfy the properties of norm mentioned in Exercise 33. These functions, even though they do not lead to Euclidean geometry, have all the algebraic properties we expect a norm to have and are used in numerical mathematics.

   (a) $\|\mathbf{u}\| = |u_1| + \cdots + |u_n|$ **sum of magnitudes norm**

   *(b) $\|\mathbf{u}\| = \max_{i=1\ldots n} |u_n|$ **maximum magnitude norm**

Compute these two norms for the vectors (1, 2), (−3, 4), (1, 2, −5), (0, −2, 7).

35. Let $\mathbf{x}, \mathbf{y},$ and $\mathbf{z}$ be points in $\mathbf{R}^n$. Prove that the distance function has the following properties.

   (a) $d(\mathbf{x}, \mathbf{y}) \geq 0$

   (b) $d(\mathbf{x}, \mathbf{y}) = 0$ if and only if $\mathbf{x} = \mathbf{y}$

   *(c) $d(\mathbf{x}, \mathbf{z}) \leq d(\mathbf{x}, \mathbf{y}) + d(\mathbf{y}, \mathbf{z})$

These properties are used as axioms to generalize the concept of distance for certain spaces.

5.2   Orthonormal Vectors, Projections

We now focus our attention on sets of unit, orthogonal vectors. We shall introduce a method for constructing such sets. Special properties of these sets will be discussed.

Definition   *A set of vectors is said to be an **orthogonal set** if every pair of vectors in the set is orthogonal. The set is said to be an **orthonormal set** if it is orthogonal and each vector is a unit vector.*

Example 1   Show that the set $\left\{(1, 0, 0), \left(0, \frac{3}{5}, \frac{4}{5}\right), \left(0, \frac{4}{5}, -\frac{3}{5}\right)\right\}$ is an orthonormal set.

Solution   We first show that each pair of vectors in this set is orthogonal.

$$(1, 0, 0) \cdot \left(0, \tfrac{3}{5}, \tfrac{4}{5}\right) = 0; \quad (1, 0, 0) \cdot \left(0, \tfrac{4}{5}, -\tfrac{3}{5}\right) = 0; \quad \left(0, \tfrac{3}{5}, \tfrac{4}{5}\right) \cdot \left(0, \tfrac{4}{5}, -\tfrac{3}{5}\right) = 0$$

Thus the vectors are **mutually orthogonal**. It remains to show that each vector is a unit vector. We get

$$\|(1, 0, 0)\| = \sqrt{1^2 + 0^2 + 0^2} = 1$$

$$\left\|\left(0, \tfrac{3}{5}, \tfrac{4}{5}\right)\right\| = \sqrt{0^2 + \left(\tfrac{3}{5}\right)^2 + \left(\tfrac{4}{5}\right)^2} = 1$$

$$\left\|\left(0, \tfrac{4}{5}, -\tfrac{3}{5}\right)\right\| = \sqrt{0^2 + \left(\tfrac{4}{5}\right)^2 + \left(-\tfrac{3}{5}\right)^2} = 1$$

The vectors are unit, orthogonal vectors. The set is thus an orthonormal set.

The next theorem tells us that any orthogonal set of nonzero vectors is linearly independent. This result, of course, also implies that an orthonormal set of nonzero vectors is linearly independent.

Theorem 5.4   An orthogonal set of nonzero vectors in a vector space is linearly independent.

Proof   Let $\{\mathbf{v}_1, \ldots, \mathbf{v}_m\}$ be an orthogonal set of nonzero vectors in a vector space V. Let us examine the identity

$$c_1\mathbf{v}_1 + c_2\mathbf{v}_2 + \cdots + c_m\mathbf{v}_m = \mathbf{0}$$

We shall show that $c_1 = 0, \ldots, c_m = 0$, proving that the vectors are linearly independent. Let $\mathbf{v}_i$ be the ith vector of the orthogonal set. Take the dot product of each side of this equation with $\mathbf{v}_i$ and use the properties of the dot product. We get

$$(c_1\mathbf{v}_1 + c_2\mathbf{v}_2 + \cdots + c_m\mathbf{v}_m) \cdot \mathbf{v}_i = \mathbf{0} \cdot \mathbf{v}_i$$

$$c_1\mathbf{v}_1 \cdot \mathbf{v}_i + c_2\mathbf{v}_2 \cdot \mathbf{v}_i + \cdots + c_m\mathbf{v}_m \cdot \mathbf{v}_i = 0$$

Since the vectors $\mathbf{v}_1, \ldots, \mathbf{v}_m$ are mutually orthogonal, $\mathbf{v}_j \cdot \mathbf{v}_i = 0$ unless $j = i$. Thus

$$c_i\mathbf{v}_i \cdot \mathbf{v}_i = 0$$

Since $\mathbf{v}_i$ is a nonzero vector, then $\mathbf{v}_i \cdot \mathbf{v}_i \neq 0$. Thus $c_i = 0$. Letting $i = 1, \ldots, m$, we get $c_1 = 0, \ldots, c_m = 0$, proving that the vectors are linearly independent.

There are many bases for a vector space. The basis that one uses depends upon the problem under consideration—one selects the basis that best fits the situation. Very often the most suitable basis is either an orthogonal set or an orthonormal set.

Definition   *A basis that is an orthogonal set is said to be an **orthogonal basis**. A basis that is an orthonormal set is said to be an **orthonormal basis**.*

Note that the standard bases are orthonormal bases.

Standard Bases

$\mathbf{R}^2$: $\{(1, 0), (0, 1)\}$
$\mathbf{R}^3$: $\{(1, 0, 0), (0, 1, 0), (0, 0, 1)\}$ } orthonormal bases
$\mathbf{R}^n$: $\{(1, \ldots, 0), \ldots, (0, \ldots, 1)\}$

We know that any vector in a vector space can be expressed as a unique linear combination of basis vectors. The following theorem tells us that an orthonormal basis is particularly useful for discussing a vector space because the linear combination can be easily found.

Theorem 5.5   Let $\{\mathbf{u}_1, \ldots, \mathbf{u}_n\}$ be an orthonormal basis for a vector space V. Let $\mathbf{v}$ be a vector in V. $\mathbf{v}$ can be written as a linear combination of these basis vectors as follows:

$$\mathbf{v} = (\mathbf{v} \cdot \mathbf{u}_1)\mathbf{u}_1 + (\mathbf{v} \cdot \mathbf{u}_2)\mathbf{u}_2 + \cdots + (\mathbf{v} \cdot \mathbf{u}_n)\mathbf{u}_n$$

Proof   Since $\{\mathbf{u}_1, \ldots, \mathbf{u}_n\}$ is a basis, there exist scalars $c_1, c_2, \ldots, c_n$ such that

$$\mathbf{v} = c_1\mathbf{u}_1 + c_2\mathbf{u}_2 + \cdots + c_n\mathbf{u}_n$$

We shall show that $c_1 = \mathbf{v} \cdot \mathbf{u}_1, \ldots, c_n = \mathbf{v} \cdot \mathbf{u}_n$.

Let $\mathbf{u}_i$ be the ith basis vector. Take the dot product of each side of this equation with $\mathbf{u}_i$ and use the properties of the dot product. We get

$$\mathbf{v} \cdot \mathbf{u}_i = (c_1\mathbf{u}_1 + c_2\mathbf{u}_2 + \cdots + c_n\mathbf{u}_n) \cdot \mathbf{u}_i$$
$$= c_1\mathbf{u}_1 \cdot \mathbf{u}_i + c_2\mathbf{u}_2 \cdot \mathbf{u}_i + \cdots + c_n\mathbf{u}_n \cdot \mathbf{u}_i$$

Since the vectors $\mathbf{u}_1, \ldots, \mathbf{u}_n$ are mutually orthogonal, $\mathbf{u}_j \cdot \mathbf{u}_i = 0$ unless $j = i$. Thus

$$\mathbf{v} \cdot \mathbf{u}_i = c_i\mathbf{u}_i \cdot \mathbf{u}_i$$

Furthermore, since the basis is orthonormal, $\mathbf{u}_i \cdot \mathbf{u}_i = 1$. Therefore

$$\mathbf{v} \cdot \mathbf{u}_i = c_i$$

Thus, letting $i = 1, \ldots, n$, we get $c_1 = \mathbf{v} \cdot \mathbf{u}_1, \ldots, c_n = \mathbf{v} \cdot \mathbf{u}_n$. We can write

$$\mathbf{v} = (\mathbf{v} \cdot \mathbf{u}_1)\mathbf{u}_1 + (\mathbf{v} \cdot \mathbf{u}_2)\mathbf{u}_2 + \cdots + (\mathbf{v} \cdot \mathbf{u}_n)\mathbf{u}_n$$

Example 2   The following vectors $\mathbf{u}_1$, $\mathbf{u}_2$, and $\mathbf{u}_3$ form an orthonormal basis for $\mathbf{R}^3$. Express the vector $\mathbf{v} = (7, -5, 10)$ as a linear combination of these vectors.

$$\mathbf{u}_1 = (1, 0, 0), \ \mathbf{u}_2 = \left(0, \tfrac{3}{5}, \tfrac{4}{5}\right), \ \mathbf{u}_3 = \left(0, \tfrac{4}{5}, -\tfrac{3}{5}\right)$$

Solution   We get

$$\mathbf{v} \cdot \mathbf{u}_1 = (7, -5, 10) \cdot (1, 0, 0) = 7$$
$$\mathbf{v} \cdot \mathbf{u}_2 = (7, -5, 10) \cdot \left(0, \tfrac{3}{5}, \tfrac{4}{5}\right) = 5$$
$$\mathbf{v} \cdot \mathbf{u}_3 = (7, -5, 10) \cdot \left(0, \tfrac{4}{5}, -\tfrac{3}{5}\right) = -10$$

Thus

$$(7, -5, 10) = 7(1, 0, 0) + 5\left(0, \tfrac{3}{5}, \tfrac{4}{5}\right) - 10\left(0, \tfrac{4}{5}, -\tfrac{3}{5}\right)$$

We now introduce the concept of projection of one vector onto another. Projection intuitively tells us "how much" of one vector is pointing in the direction of another. It is particularly useful in analyzing forces, where one wants to know the physical effect of a force in a certain direction. We shall use projections to construct sets of orthonormal vectors.

Projection of One Vector onto Another Vector

Let $\mathbf{v}$ and $\mathbf{u}$ be vectors in $\mathbf{R}^n$ with angle α between them. See Figure 5.6(a). The vector $\overrightarrow{OA}$ tells us "how much" of $\mathbf{v}$ is pointing in the direction of $\mathbf{u}$. We call $\overrightarrow{OA}$ the **projection of $\mathbf{v}$ onto $\mathbf{u}$**. Let us find an expression for $\overrightarrow{OA}$. We see that

$$OA = OB \cos \alpha$$
$$= \|\mathbf{v}\| \cos \alpha$$
$$= \|\mathbf{v}\| \frac{\mathbf{v}}{\|\mathbf{v}\|} \cdot \frac{\mathbf{u}}{\|\mathbf{u}\|}$$
$$= \mathbf{v} \cdot \frac{\mathbf{u}}{\|\mathbf{u}\|}$$

The direction of the vector $\overrightarrow{OA}$ is defined by the unit vector $\mathbf{u}/\|\mathbf{u}\|$. Thus

$$\overrightarrow{OA} = \left(\mathbf{v} \cdot \frac{\mathbf{u}}{\|\mathbf{u}\|}\right) \frac{\mathbf{u}}{\|\mathbf{u}\|} = \frac{\mathbf{v} \cdot \mathbf{u}}{\mathbf{u} \cdot \mathbf{u}}\mathbf{u}$$

This expression for projection also holds if $\alpha > 90°$. See Figure 5.6(b). In that case the projection is in the opposite direction to $\mathbf{u}$, and the sign of $(\mathbf{v} \cdot \mathbf{u})/(\mathbf{u} \cdot \mathbf{u})$ is negative.

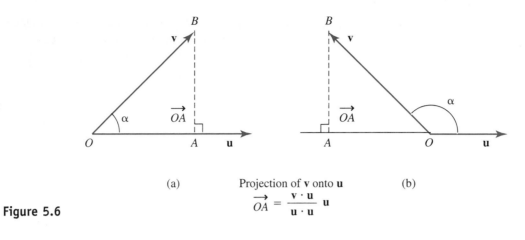

(a)   Projection of **v** onto **u**   (b)

$$\overrightarrow{OA} = \frac{\mathbf{v} \cdot \mathbf{u}}{\mathbf{u} \cdot \mathbf{u}} \mathbf{u}$$

Figure 5.6

This discussion leads to the following definition of projection.

Definition   *The **projection** of a vector* **v** *onto a nonzero vector* **u** *in* **R**^{n} *is denoted* proj$_\mathbf{u}$ **v** *and is defined by*

$$\text{proj}_\mathbf{u} \, \mathbf{v} = \frac{\mathbf{v} \cdot \mathbf{u}}{\mathbf{u} \cdot \mathbf{u}} \mathbf{u}$$

See Figure 5.7. The term **component** of **v** in the direction of **u** is also used for proj$_\mathbf{u}$ **v**.

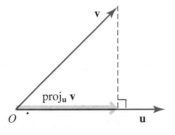

Figure 5.7 $\text{proj}_\mathbf{u} \, \mathbf{v} = \dfrac{\mathbf{v} \cdot \mathbf{u}}{\mathbf{u} \cdot \mathbf{u}} \, \mathbf{u}$

Example 3   Determine the projection of the vector **v** = (6, 7) onto the vector **u** = (1, 4).

Solution   For these vectors, we get

$$\mathbf{v} \cdot \mathbf{u} = (6, 7) \cdot (1, 4) = 6 + 28 = 34$$
$$\mathbf{u} \cdot \mathbf{u} = (1, 4) \cdot (1, 4) = 1 + 16 = 17$$

Thus

$$\text{proj}_\mathbf{u} \, \mathbf{v} = \frac{\mathbf{v} \cdot \mathbf{u}}{\mathbf{u} \cdot \mathbf{u}} \mathbf{u} = \frac{34}{17}(1, 4) = (2, 8)$$

The projection of **v** onto **u** is (2, 8).

Suppose the vector $\mathbf{v} = (6, 7)$ represents a force acting on a body located at the origin. Then $\text{proj}_{\mathbf{u}}\, \mathbf{v} = (2, 8)$ is the component of the force in the direction of the vector $\mathbf{u} = (1, 4)$. Physically, $(2, 8)$ is the effect of the force in that direction.

We now give a method for constructing an orthonormal basis from a given basis. The method uses vector projections.

Theorem 5.6   **The Gram-Schmidt Orthogonalization Process***   Let $\{\mathbf{v}_1, \ldots, \mathbf{v}_n\}$ be a basis for a vector space V. The set of vectors $\{\mathbf{u}_1, \ldots, \mathbf{u}_n\}$ defined as follows is orthogonal. To obtain an orthonormal basis for V, normalize each of the vectors $\mathbf{u}_1, \ldots, \mathbf{u}_n$.

$$\mathbf{u}_1 = \mathbf{v}_1$$
$$\mathbf{u}_2 = \mathbf{v}_2 - \text{proj}_{u_1}\, \mathbf{v}_2$$
$$\mathbf{u}_3 = \mathbf{v}_3 - \text{proj}_{u_1}\, \mathbf{v}_3 - \text{proj}_{u_2}\, \mathbf{v}_3$$
$$\cdot \quad \cdot \quad \cdot \quad \cdot \quad \cdot \quad \cdot \quad \cdot$$
$$\mathbf{u}_n = \mathbf{v}_n - \text{proj}_{u_1}\, \mathbf{v}_n - \text{proj}_{u_2}\, \mathbf{v}_n - \cdots - \text{proj}_{u_{n-1}}\, \mathbf{v}_n$$

Proof   If $\mathbf{v}$ and $\mathbf{u}$ are vectors in $\mathbf{R}^n$, then vector addition tells us that $(\mathbf{v} - \text{proj}_{\mathbf{u}}\, \mathbf{v})$ is orthogonal to $\mathbf{u}$. See Figure 5.8. We apply this result in proving the theorem. Consider the construction of the ith vector $\mathbf{u}_i$. By subtracting $\text{proj}_{u_1}\, \mathbf{v}_i$ from $\mathbf{v}_i$ we get a vector that is orthogonal to $\mathbf{u}_1$. Subtracting $\text{proj}_{u_2}\, \mathbf{v}_i$ from the resulting vector gives a vector that is orthogonal to $\mathbf{u}_2$, while still preserving the orthogonality to $\mathbf{u}_1$, and so on. This leads to a vector $\mathbf{u}_i$ that is orthogonal to each of the previous vectors $\mathbf{u}_1, \ldots, \mathbf{u}_{i-1}$. See the following exercises.

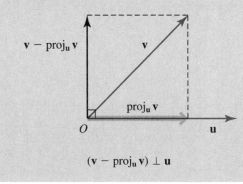

Figure 5.8 $(\mathbf{v} - \text{proj}_{\mathbf{u}}\, \mathbf{v}) \perp \mathbf{u}$

* Jorgen Pederson Gram (1850–1916) was a Danish actuary.

Erhard Schmidt (1876–1959) studied at Berlin and Göttingen. He taught at Bonn, Zurich, Erlangen, and Berlin. In 1946, he became the first director of the Research Institute for Mathematics for the German Academy of Sciences. He is considered a founder of the branch of modern mathematics called functional analysis.

Example 4   The set $\{(1, 2, 0, 3), (4, 0, 5, 8), (8, 1, 5, 6)\}$ is linearly independent in $\mathbf{R}^4$. The vectors form a basis for a three-dimensional subspace V of $\mathbf{R}^4$. Construct an orthonormal basis for V.

Solution   Let $\mathbf{v}_1 = (1, 2, 0, 3)$, $\mathbf{v}_2 = (4, 0, 5, 8)$, $\mathbf{v}_3 = (8, 1, 5, 6)$. We now use the Gram-Schmidt process to construct an orthogonal set $\{\mathbf{u}_1, \mathbf{u}_2, \mathbf{u}_3\}$ from these vectors.

Let $\mathbf{u}_1 = \mathbf{v}_1 = (1, 2, 0, 3)$

$$\text{Let } \mathbf{u}_2 = \mathbf{v}_2 - \text{proj}_{u_1}\, \mathbf{v}_2 = \mathbf{v}_2 - \frac{(\mathbf{v}_2 \cdot \mathbf{u}_1)}{(\mathbf{u}_1 \cdot \mathbf{u}_1)}\mathbf{u}_1$$

$$= (4, 0, 5, 8) - \frac{(4, 0, 5, 8) \cdot (1, 2, 0, 3)}{(1, 2, 0, 3) \cdot (1, 2, 0\ 3)}(1, 2, 0, 3)$$

$$= (4, 0, 5, 8) - 2(1, 2, 0, 3)$$

$$= (2, -4, 5, 2)$$

$$\text{Let } \mathbf{u}_3 = \mathbf{v}_3 - \text{proj}_{u_1}\, \mathbf{v}_3 - \text{proj}_{u_2}\, \mathbf{v}_3 = \mathbf{v}_3 - \frac{\mathbf{v}_3 \cdot \mathbf{u}_1}{\mathbf{u}_1 \cdot \mathbf{u}_1}\mathbf{u}_1 - \frac{\mathbf{v}_3 \cdot \mathbf{u}_2}{\mathbf{u}_2 \cdot \mathbf{u}_2}\mathbf{u}_2$$

$$= (8, 1, 5, 6) - \frac{(8, 1, 5, 6) \cdot (1, 2, 0, 3)}{(1, 2, 0, 3) \cdot (1, 2, 0, 3)}(1, 2, 0, 3)$$

$$- \frac{(8, 1, 5, 6) \cdot (2, -4, 5, 2)}{(2, -4, 5, 2) \cdot (2, -4, 5, 2)}(2, -4, 5, 2)$$

$$= (8, 1, 5, 6) - 2(1, 2, 0, 3) - 1(2, -4, 5, 2)$$

$$= (4, 1, 0, -2)$$

The set $\{(1, 2, 0, 3), (2, -4, 5, 2), (4, 1, 0, -2)\}$ is an orthogonal basis for V. (Check that the dot product of each pair of vectors is indeed zero.)

Let us now compute the norm of each vector and then normalize the vectors to get an orthonormal basis. We get

$$\|(1, 2, 0, 3)\| = \sqrt{1^2 + 2^2 + 0^2 + 3^2} = \sqrt{14}$$

$$\|(2, -4, 5, 2)\| = \sqrt{2^2 + (-4)^2 + 5^2 + 2^2} = 7$$

$$\|(4, 1, 0, -2)\| = \sqrt{4^2 + 1^2 + 0^2 + (-2)^2} = \sqrt{21}$$

Use these values to normalize the vectors, to arrive at the following orthonormal basis for V.

$$\left\{ \left(\frac{1}{\sqrt{14}}, \frac{2}{\sqrt{14}}, 0, \frac{3}{\sqrt{14}}\right), \quad \left(\frac{2}{7}, -\frac{4}{7}, \frac{5}{7}, \frac{2}{7}\right) \quad \left(\frac{4}{\sqrt{21}}, \frac{1}{\sqrt{21}}, 0, \frac{-2}{\sqrt{21}}\right) \right\}$$

Projection of a Vector onto a Subspace

We have defined the projection of a vector onto another vector. We now extend this concept to the projection of a vector onto a subspace. The projection of a vector onto a subspace tells us "how much" of the vector lies in the subspace.

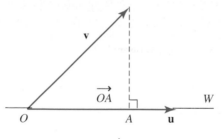

u unit vector

Figure 5.9 $\text{proj}_W \mathbf{v} = \overrightarrow{OA} = (\mathbf{v} \cdot \mathbf{u})\mathbf{u}$

Let $\overrightarrow{OA}$ be the projection of a vector $\mathbf{v}$ onto a vector $\mathbf{u}$ in $\mathbf{R}^n$. See Figure 5.9. Let W be the one-dimensional subspace of $\mathbf{R}^n$ consisting of all vectors that lie on the line defined by $\mathbf{u}$. Note that the projection of $\mathbf{v}$ onto any vector $\mathbf{u}$ that lies in this subspace is $\overrightarrow{OA}$. Thus we can define the vector projection of $\mathbf{v}$ onto the subspace W, written $\text{proj}_W \mathbf{v}$, to be $\overrightarrow{OA}$. The simplest expression for $\overrightarrow{OA}$ is obtained by taking $\mathbf{u}$ to be a unit vector. We then get

$$\text{proj}_W \mathbf{v} = \overrightarrow{OA} = (\mathbf{v} \cdot \mathbf{u})\mathbf{u} \qquad (\mathbf{u} \text{ a unit vector})$$

If W is a subspace of dimension m, then we extend this concept of projection by expressing the projection of $\mathbf{v}$ onto W as a linear combination of the vectors of an orthonormal basis of W as follows.

Definition   *Let W be a subspace of $\mathbf{R}^n$. Let $\{\mathbf{u}_1, \ldots, \mathbf{u}_m\}$ be an orthonormal basis for W. If $\mathbf{v}$ is a vector in $\mathbf{R}^n$, the **projection** of $\mathbf{v}$ onto W is denoted $\text{proj}_W \mathbf{v}$ and is defined by*

$$\text{proj}_W \mathbf{v} = (\mathbf{v} \cdot \mathbf{u}_1)\mathbf{u}_1 + (\mathbf{v} \cdot \mathbf{u}_2)\mathbf{u}_2 + \cdots + (\mathbf{v} \cdot \mathbf{u}_m)\mathbf{u}_m$$

See Figure 5.10.

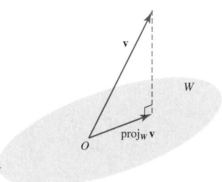

$\{\mathbf{u}_1, \ldots, \mathbf{u}_m\}$ orthonormal basis for W

Figure 5.10 $\text{proj}_W \mathbf{v} = (\mathbf{v} \cdot \mathbf{u}_1)\mathbf{u}_1 + \cdots + (\mathbf{v} \cdot \mathbf{u}_m)\mathbf{u}_m$

We say that a vector **v** is **orthogonal to a subspace** W if **v** is orthogonal to every vector in W. The following theorem tells us that if we have a vector **v** in $\mathbf{R}^n$ and if W is a subspace of $\mathbf{R}^n$, then we can **decompose** **v** into a vector that lies in the subspace W and a vector that is orthogonal to W.

Theorem 5.7   Let W be a subspace of $\mathbf{R}^n$. Every vector **v** in $\mathbf{R}^n$ can be written uniquely in the form

$$\mathbf{v} = \mathbf{w} + \mathbf{w}_\perp$$

where **w** is in W and $\mathbf{w}_\perp$ is orthogonal to W. The vectors **w** and $\mathbf{w}_\perp$ are

$$\mathbf{w} = \text{proj}_W \mathbf{v} \quad \text{and} \quad \mathbf{w}_\perp = \mathbf{v} - \text{proj}_W \mathbf{v}$$

See Figure 5.11.

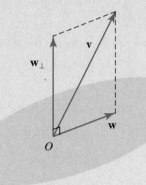

$\mathbf{v} = \mathbf{w} + \mathbf{w}_\perp$, where

Figure 5.11 $\mathbf{w} = \text{proj}_W \mathbf{v}$ and $\mathbf{w}_\perp = \mathbf{v} - \text{proj}_W \mathbf{v}$

Proof   Observe that

$$\mathbf{w} + \mathbf{w}_\perp = (\text{proj}_W \mathbf{v}) + (\mathbf{v} - \text{proj}_W \mathbf{v})$$
$$= \mathbf{v}$$

Further, **w** is in W. It remains to be shown that $(\mathbf{v} - \text{proj}_W \mathbf{v})$ is orthogonal to W. Let $\{\mathbf{u}_1, \ldots, \mathbf{u}_m\}$ be an orthonormal basis for W. We first show that $(\mathbf{v} - \text{proj}_W \mathbf{v})$ is orthogonal to each of these base vectors of W. We will then be able to show that $(\mathbf{v} - \text{proj}_W \mathbf{v})$ is orthogonal to an arbitrary vector in W. We get

$$\mathbf{u}_i \cdot (\mathbf{v} - \text{proj}_W \mathbf{v}) = \mathbf{u}_i \cdot (\mathbf{v} - (\mathbf{v} \cdot \mathbf{u}_1)\mathbf{u}_1 - \cdots - (\mathbf{v} \cdot \mathbf{u}_m)\mathbf{u}_m)$$
$$= \mathbf{u}_i \cdot \mathbf{v} - \mathbf{u}_i \cdot (\mathbf{v} \cdot \mathbf{u}_1)\mathbf{u}_1 - \cdots - \mathbf{u}_i \cdot (\mathbf{v} \cdot \mathbf{u}_m)\mathbf{u}_m$$
$$= \mathbf{u}_i \cdot \mathbf{v} - (\mathbf{v} \cdot \mathbf{u}_1)(\mathbf{u}_i \cdot \mathbf{u}_1) - \cdots - (\mathbf{v} \cdot \mathbf{u}_m)(\mathbf{u}_i \cdot \mathbf{u}_m)$$
$$= \mathbf{u}_i \cdot \mathbf{v} - (\mathbf{v} \cdot \mathbf{u}_i)(\mathbf{u}_i \cdot \mathbf{u}_i), \text{ since } \mathbf{u}_i \cdot \mathbf{u}_j = 0 \text{ unless } i = j$$
$$= \mathbf{u}_i \cdot \mathbf{v} - \mathbf{v} \cdot \mathbf{u}_i, \text{ since } \mathbf{u}_i \cdot \mathbf{u}_i = 1$$
$$= 0$$

(continues)

Thus $(\mathbf{v} - \text{proj}_W\,\mathbf{v})$ is orthogonal to each of the base vectors of W.

Let $\mathbf{w}' = c_1\mathbf{u}_1 + \cdots + c_m\mathbf{u}_m$ be an arbitrary vector in W. We get

$$\mathbf{w}' \cdot (\mathbf{v} - \text{proj}_W\,\mathbf{v}) = (c_1\mathbf{u}_1 + \cdots + c_m\mathbf{u}_m) \cdot (\mathbf{v} - \text{proj}_W\,\mathbf{v})$$

$$= c_1\mathbf{u}_1 \cdot (\mathbf{v} - \text{proj}_W\,\mathbf{v}) + \cdots + c_m\mathbf{u}_m \cdot (\mathbf{v} - \text{proj}_W\,\mathbf{v})$$

$$= 0$$

Thus $(\mathbf{v} - \text{proj}_W\,\mathbf{v})$ is orthogonal to W.

The proof of uniqueness is omitted.

Example 5   Consider the vector $\mathbf{v} = (3, 2, 6)$ in $\mathbf{R}^3$. Let W be the subspace of $\mathbf{R}^3$ consisting of all vectors of the form (a, b, b). Decompose $\mathbf{v}$ into the sum of a vector that lies in W and a vector orthogonal to W.

Solution   We need an orthonormal basis for W. We can write an arbitrary vector of W as follows:

$$(a, b, b) = a(1, 0, 0) + b(0, 1, 1)$$

The set $\{(1, 0, 0), (0, 1, 1)\}$ spans W and is linearly independent. It forms a basis for W. The vectors are orthogonal. Normalize each vector to get an orthonormal basis $\{\mathbf{u}_1, \mathbf{u}_2\}$ for W, where

$$\mathbf{u}_1 = (1, 0, 0), \qquad \mathbf{u}_2 = \left(0, \frac{1}{\sqrt{2}}, \frac{1}{\sqrt{2}}\right)$$

We get

$$\mathbf{w} = \text{proj}_W\,\mathbf{v} = (\mathbf{v} \cdot \mathbf{u}_1)\mathbf{u}_1 + (\mathbf{v} \cdot \mathbf{u}_2)\mathbf{u}_2$$

$$= ((3, 2, 6) \cdot (1, 0, 0))(1, 0, 0) + \left((3, 2, 6) \cdot \left(0, \frac{1}{\sqrt{2}}, \frac{1}{\sqrt{2}}\right)\right)\left(0, \frac{1}{\sqrt{2}}, \frac{1}{\sqrt{2}}\right)$$

$$= (3, 0, 0) + (0, 4, 4) = (3, 4, 4)$$

and

$$\mathbf{w}\perp = \mathbf{v} - \text{proj}_W\,\mathbf{v} = (3, 2, 6) - (3, 4, 4) = (0, -2, 2)$$

Thus the desired decomposition of $\mathbf{v}$ is

$$(3, 2, 6) = (3, 4, 4) + (0, -2, 2)$$

In this decomposition the vector $(3, 4, 4)$ lies in W and the vector $(0, -2, 2)$ is orthogonal to W.

Distance of a Point from a Subspace

We have discussed the distance of a point from another point in $\mathbf{R}^n$. We are now able to talk about the distance of a point from a subspace in $\mathbf{R}^n$. Let $\mathbf{x} = (x_1, \ldots, x_n)$ be a point in $\mathbf{R}^n$, W be a subspace of $\mathbf{R}^n$, and $\mathbf{y} = (y_1, \ldots, y_n)$ be a point in W. It is natural to define the distance of **x** from W, denoted $d(\mathbf{x}, W)$, to be the minimum of the distances from **x** to the points of W.

$$d(\mathbf{x}, W) = \min \{d(\mathbf{x}, \mathbf{y})\}, \text{ for all points } \mathbf{y} \text{ in } W$$

We now find an expression for $d(\mathbf{x}, W)$. We can write

$$\mathbf{x} - \mathbf{y} = (\mathbf{x} - \text{proj}_W \mathbf{x}) + (\text{proj}_W \mathbf{x} - \mathbf{y})$$

This is a decomposition of the vector $\mathbf{x} - \mathbf{y}$ into the sum of a vector $(\mathbf{x} - \text{proj}_W \mathbf{x})$ that is orthogonal to W and a vector $(\text{proj}_W \mathbf{x} - \mathbf{y})$ that lies in W. The Pythagorean Theorem thus gives

$$\|\mathbf{x} - \mathbf{y}\|^2 = \|\mathbf{x} - \text{proj}_W \mathbf{x}\|^2 + \|\text{proj}_W \mathbf{x} - \mathbf{y}\|^2$$

Therefore $\|\mathbf{x} - \mathbf{y}\|$ (which is equal to $d(\mathbf{x}, \mathbf{y})$) has a minimum value of $\|\mathbf{x} - \text{proj}_W \mathbf{x}\|$ when $\text{proj}_W \mathbf{x} - \mathbf{y} = 0$; that is, when $\mathbf{y} = \text{proj}_W \mathbf{x}$. Thus

$$d(\mathbf{x}, W) = \|\mathbf{x} - \text{proj}_W \mathbf{x}\|$$

The distance of a point from a subspace is the distance of the point from its projection in the subspace. See Figure 5.12.

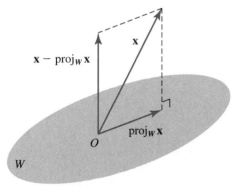

Figure 5.12

Distance of **x** from W
$d(\mathbf{x}, W) = \|\mathbf{x} - \text{proj}_W \mathbf{x}\|$

Example 6   Find the distance of the point $\mathbf{x} = (4, 1, -7)$ of $\mathbf{R}^3$ from the subspace W consisting of all vectors of the form (a, b, b).

Solution   The previous example tells us that the set $\{\mathbf{u}_1, \mathbf{u}_2\}$, where

$$\mathbf{u}_1 = (1, 0, 0), \qquad \mathbf{u}_2 = \left(0, \frac{1}{\sqrt{2}}, \frac{1}{\sqrt{2}}\right)$$

is an orthogonal basis for W. We can compute $\text{proj}_W \mathbf{x}$.

$$\text{proj}_W \mathbf{x} = (\mathbf{x} \cdot \mathbf{u}_1)\mathbf{u}_1 + (\mathbf{x} \cdot \mathbf{u}_2)\mathbf{u}_2$$

$$= ((4, 1, -7) \cdot (1, 0, 0))(1, 0, 0) + \left((4, 1, -7) \cdot \left(0, \frac{1}{\sqrt{2}}, \frac{1}{\sqrt{2}}\right)\right)\left(0, \frac{1}{\sqrt{2}}, \frac{1}{\sqrt{2}}\right)$$

$$= (4, 0, 0) + (0, -3, -3) = (4, -3, -3)$$

Thus

$$\|\mathbf{x} - \text{proj}_W \mathbf{x}\| = \|(4, 1, -7) - (4, -3, -3)\| = \|(0, 4, -4)\| = \sqrt{32}$$

The distance from $\mathbf{x}$ to W is $\sqrt{32}$.

We have seen the value of interpreting matrices as made up of columns and rows—this way of looking at matrices led to rank. We now introduce an important class of matrices whose columns and rows form orthonormal sets. We shall find that these matrices play an important computational role. In the next chapter we see that they also have special geometrical properties when used to define functions on a vector space.

Orthogonal Matrices

Definition   *A square matrix whose column vectors form an orthonormal set is called an **orthogonal matrix**.*

Example 7   Show that the following matrix A is an orthogonal matrix.

$$A = \begin{bmatrix} \dfrac{1}{\sqrt{2}} & \dfrac{1}{\sqrt{2}} \\ -\dfrac{1}{\sqrt{2}} & \dfrac{1}{\sqrt{2}} \end{bmatrix}$$

Solution   The column vectors of A are

$$\mathbf{a}_1 = \begin{bmatrix} \dfrac{1}{\sqrt{2}} \\ -\dfrac{1}{\sqrt{2}} \end{bmatrix} \quad \text{and} \quad \mathbf{a}_2 = \begin{bmatrix} \dfrac{1}{\sqrt{2}} \\ \dfrac{1}{\sqrt{2}} \end{bmatrix}$$

Observe that

$$\|\mathbf{a}_1\| = \sqrt{\left(\frac{1}{\sqrt{2}}\right)^2 + \left(-\frac{1}{\sqrt{2}}\right)^2} = 1, \quad \|\mathbf{a}_2\| = \sqrt{\left(\frac{1}{\sqrt{2}}\right)^2 + \left(\frac{1}{\sqrt{2}}\right)^2} = 1,$$

$$\mathbf{a}_1 \cdot \mathbf{a}_2 = \left(\frac{1}{\sqrt{2}}\right)\left(-\frac{1}{\sqrt{2}}\right) + \left(\frac{1}{\sqrt{2}}\right)\left(\frac{1}{\sqrt{2}}\right) = 0$$

The column vectors of A are therefore both unit vectors, and they are orthogonal. Thus A is an orthogonal matrix.

We now give four important properties of orthogonal matrices.

Theorem 5.8 Let A be an orthogonal matrix. Then

(a) the row vectors of A form an orthonormal set.

(b) A is invertible, with $A^{-1} = A^t$.

(c) A^{-1} is an orthogonal matrix.

(d) $|A| = 1$ or -1 (written $|A| = \pm 1$).

Proof We prove part (b) first; the other parts then follow.

(b) Consider the product $P = A^t A$. Computing this product involves taking the dot product of each row vector of A^t with each column vector of A. Since the rows of A^t are the columns of A, an orthonormal set, we get that $p_{ij} = 1$ if $i = j$ and $p_{ij} = 0$ if $i \neq j$. Thus $A^t A = I$, the identity matrix.

Let A be an $n \times n$ matrix. Since the column vectors of A form an orthonormal set, they are linearly independent and A is of rank n. Thus A^{-1} exists. Multiply both sides of $A^t A = I$ by A^{-1} to get $A^t = A^{-1}$.

(a) Since $A A^{-1} = I$ and $A^{-1} = A^t$, then $A A^t = I$. This implies that the row vectors of A form an orthonormal set.

(c) Since the rows of A form an orthonormal set, the columns of A^t form an orthonormal set. Thus A^t is an orthogonal matrix, implying that A^{-1} is orthogonal.

(d) Since $A A^t = I$, the properties of determinants give

$$|A A^t| = 1, \qquad |A||A^t| = 1, \qquad |A||A| = 1, \qquad |A|^2 = 1$$

Thus $|A| = \pm 1$.

Exercise Set 5.2

1. Which of the following are orthogonal sets of vectors?
   *(a) $\{(1, 2), (2, -1)\}$ *(b) $\{(3, -1), (0, 5)\}$
   (c) $\{(0, -2), (3, 0)\}$ *(d) $\{(4, 1), (2, -3)\}$
   (e) $\{(-3, 2), (2, 3)\}$

2. Which of the following are orthogonal sets of vectors?
   *(a) $\{(1, 2, 1), (4, -2, 0), (2, 4, -10)\}$
   *(b) $\{(3, -1, 1), (2, 0, 1), (1, 1, -2)\}$
   (c) $\{(1, 4, 2), (-2, -1, 3), (6, -1, -1)\}$
   (d) $\{(1, 2, -1, 1), (3, 1, 4, -1), (0, 1, -1, -3)\}$

3. Which of the following are orthonormal sets of vectors?
   *(a) $\left\{ \left(\frac{1}{3}, \frac{2}{3}, \frac{2}{3}\right), \left(\frac{2}{3}, -\frac{2}{3}, \frac{1}{3}\right), \left(\frac{2}{3}, \frac{1}{3}, -\frac{2}{3}\right) \right\}$

(b) $\left\{ \left(\frac{1}{\sqrt{10}}, \frac{3}{\sqrt{10}}\right), \left(\frac{-3}{\sqrt{10}}, \frac{1}{\sqrt{10}}\right) \right\}$

*(c) $\left\{ \left(\frac{1}{\sqrt{2}}, 0, \frac{1}{\sqrt{2}}\right), \left(\frac{1}{\sqrt{2}}, 0, \frac{-1}{\sqrt{2}}\right), (0, 1, 0) \right\}$

(d) $\left\{ \left(\frac{1}{\sqrt{6}}, \frac{-1}{\sqrt{6}}, \frac{2}{\sqrt{6}}\right), \left(0, \frac{2}{\sqrt{5}}, \frac{1}{\sqrt{5}}\right), \right.$
$\left. \left(\frac{5}{\sqrt{30}}, \frac{1}{\sqrt{30}}, \frac{-2}{\sqrt{30}}\right) \right\}$

*(e) $\left\{ \left(\frac{4}{\sqrt{20}}, \frac{2}{\sqrt{20}}, 0\right), \left(\frac{-1}{\sqrt{6}}, \frac{2}{\sqrt{6}}, \frac{1}{\sqrt{6}}\right), \right.$
$\left. \left(\frac{1}{\sqrt{32}}, \frac{-2}{\sqrt{32}}, \frac{5}{\sqrt{32}}\right) \right\}$

*4. The vectors $\mathbf{u}_1$, $\mathbf{u}_2$, and $\mathbf{u}_3$ that follow form an orthonormal basis for $\mathbf{R}^3$. Use the result of Theorem 5.5 to express the vector $\mathbf{v} = (2, -3, 1)$ as a linear combination of these basis vectors.

$$\mathbf{u}_1 = (0, -1, 0) \qquad \mathbf{u}_2 = \left(\tfrac{3}{5}, 0, -\tfrac{4}{5}\right), \qquad \mathbf{u}_3 = \left(\tfrac{4}{5}, 0, \tfrac{3}{5}\right)$$

5. The vectors $\mathbf{u}_1$, $\mathbf{u}_2$, and $\mathbf{u}_3$ that follow form an orthonormal basis for $\mathbf{R}^3$. Use the result of Theorem 5.5 to express the vector $\mathbf{v} = (7, 5, -1)$ as a linear combination of these basis vectors.

$$\mathbf{u}_1 = (1, 0, 0), \qquad \mathbf{u}_2 = \left(0, \frac{1}{\sqrt{2}}, \frac{1}{\sqrt{2}}\right),$$
$$\mathbf{u}_3 = \left(0, \frac{1}{\sqrt{2}}, \frac{-1}{\sqrt{2}}\right)$$

6. Determine the projection of the vector $\mathbf{v}$ onto the vector $\mathbf{u}$ for the following vectors.

*(a) $\mathbf{v} = (7, 4)$, $\mathbf{u} = (1, 2)$

(b) $\mathbf{v} = (-1, 5)$, $\mathbf{u} = (3, -2)$

*(c) $\mathbf{v} = (4, 6, 4)$, $\mathbf{u} = (1, 2, 3)$

(d) $\mathbf{v} = (6, -8, 7)$, $\mathbf{u} = (-1, 3, 0)$

*(e) $\mathbf{v} = (1, 2, 3, 0)$, $\mathbf{u} = (1, -1, 2, 3)$

7. Determine the projection of the vector $\mathbf{v}$ onto the vector $\mathbf{u}$ for the following vectors.

*(a) $\mathbf{v} = (1, 2)$, $\mathbf{u} = (2, 5)$

(b) $\mathbf{v} = (-1, 3)$, $\mathbf{u} = (2, 4)$

*(c) $\mathbf{v} = (1, 2, 3)$, $\mathbf{u} = (1, 2, 0)$

(d) $\mathbf{v} = (2, 1, 4)$, $\mathbf{u} = (-1, -3, 2)$

*(e) $\mathbf{v} = (2, -1, 3, 1)$, $\mathbf{u} = (-1, 2, 1, 3)$

8. The following vectors form a basis for $\mathbf{R}^2$. Use these vectors in the Gram-Schmidt process to construct an orthonormal basis for $\mathbf{R}^2$.

*(a) $(1, 2)$, $(-1, 3)$ (b) $(1, 1)$, $(6, 2)$

*(c) $(1, -1)$, $(4, -2)$

9. The following vectors form a basis for $\mathbf{R}^3$. Use these vectors in the Gram-Schmidt process to construct an orthonormal basis for $\mathbf{R}^3$.

*(a) $(1, 1, 1)$, $(2, 0, 1)$, $(2, 4, 5)$

(b) $(3, 2, 0)$, $(1, 5, -1)$, $(5, -1, 2)$

10. Construct an orthonormal basis for the subspace of $\mathbf{R}^3$ spanned by the following vectors.

*(a) $(1, 0, 2)$, $(-1, 0, 1)$ (b) $(1, -1, 1)$, $(1, 2, -1)$

11. Construct an othornormal basis for the subspace of $\mathbf{R}^4$ spanned by the following vectors.

(a) $(1, 2, 3, 4)$, $(-1, 1, 0, 1)$

*(b) $(3, 0, 0, 0)$, $(0, 1, 2, 1)$, $(0, -1, 3, 2)$

*12. Construct a vector in $\mathbf{R}^4$ that is orthogonal to the vector $(1, 2, -1, 1)$.

13. Construct a vector in $\mathbf{R}^4$ that is orthogonal to the vector $(2, 0, 1, 1)$.

14. Let W be the subspace of $\mathbf{R}^3$ having basis $\{(1, 1, 2), (0, -1, 3)\}$. Determine the projections of the following vectors onto W.

*(a) $(3, -1, 2)$ (b) $(1, 1, 1)$ *(c) $(4, 2, 1)$

15. Let V be the subspace of $\mathbf{R}^4$ having basis $\{(-1, 0, 2, 1), (1, -1, 0, 3)\}$. Determine the projections of the following vectors onto V.

*(a) $(1, -1, 1, -1)$ (b) $(2, 0, 1, -1)$

*(c) $(3, 2, 1, 0)$

*16. Consider the vector $\mathbf{v} = (1, 2, -1)$ in $\mathbf{R}^3$. Let V be the subspace of $\mathbf{R}^3$ consisting of all vectors of the form (a, a, b). Decompose $\mathbf{v}$ into the sum of a vector that lies in V and a vector orthogonal to V.

17. Consider a vector $\mathbf{v} = (4, 1, -2)$ in $\mathbf{R}^3$. Let V be the subspace of $\mathbf{R}^3$ consisting of all vectors of the form $(a, 2a, b)$. Decompose $\mathbf{v}$ into the sum of a vector that lies in V and a vector orthogonal to V.

*18. Consider the vector $\mathbf{v} = (3, 2, 1)$ in $\mathbf{R}^3$. Let W be the subspace of $\mathbf{R}^3$ consisting of all vectors of the form $(a, b, a + b)$. Decompose $\mathbf{v}$ into the sum of a vector that lies in W and a vector orthogonal to W.

*19. Find the distance of the point $\mathbf{x} = (1, 3, -2)$ of $\mathbf{R}^3$ from the subspace W consisting of all vectors of the form (a, a, b).

20. Find the distance of the point $\mathbf{x} = (2, 4, -1)$ of $\mathbf{R}^3$ from the subspace W consisting of all vectors of the form $(a, -2a, b)$.

*21. Find the distance of the point $\mathbf{x} = (1, 3, -2)$ of $\mathbf{R}^3$ from the subspace W consisting of all vectors of the form $(a, 2a, 3a)$. Note that this is finding the distance of a point from a line.

*22. Find an orthogonal basis for $\mathbf{R}^3$ that includes the vectors $(1, 2, -2)$ and $(6, 1, 4)$.

23. Let $\{\mathbf{u}_1, \ldots, \mathbf{u}_n\}$ be an orthonormal basis for a vector space V. Let $\mathbf{v}$ be a vector in V. Show that $\|\mathbf{v}\|$ can be written

$$\|\mathbf{v}\| = \sqrt{(\mathbf{v} \cdot \mathbf{u}_1)^2 + (\mathbf{v} \cdot \mathbf{u}_2)^2 + \cdots + (\mathbf{v} \cdot \mathbf{u}_n)^2}$$

*24. Let $\{\mathbf{u}_1, \ldots, \mathbf{u}_n\}$ be an orthonormal set in $\mathbf{R}^n$. Prove that this set is a basis for $\mathbf{R}^n$.

*25. Let $\mathbf{u}$ and $\mathbf{v}$ be vectors in $\mathbf{R}^n$. We discussed in the proof of the Gram-Schmidt theorem that $(\mathbf{v} - \text{proj}_\mathbf{u} \mathbf{v})$ is orthogonal to $\mathbf{u}$. Formally prove this result by showing that $(\mathbf{v} - \text{proj}_\mathbf{u} \mathbf{v}) \cdot \mathbf{u} = 0$, using the definition of $\text{proj}_\mathbf{u} \mathbf{v}$ and the properties of the dot product.

26. Prove that each of the following matrices is an orthogonal matrix.

*(a) $\begin{bmatrix} 1 & 0 \\ 0 & 1 \end{bmatrix}$ **(b)** $\begin{bmatrix} 0 & -1 \\ 1 & 0 \end{bmatrix}$

*(c) $\begin{bmatrix} \dfrac{\sqrt{3}}{2} & \dfrac{1}{2} \\ -\dfrac{1}{2} & \dfrac{\sqrt{3}}{2} \end{bmatrix}$ **(d)** $\begin{bmatrix} 1 & 0 & 0 \\ 0 & 0 & -1 \\ 0 & 1 & 0 \end{bmatrix}$

(e) $\begin{bmatrix} 0 & \dfrac{1}{\sqrt{2}} & -\dfrac{1}{\sqrt{2}} \\ -\dfrac{2}{\sqrt{6}} & \dfrac{1}{\sqrt{6}} & \dfrac{1}{\sqrt{6}} \\ \dfrac{1}{\sqrt{3}} & \dfrac{1}{\sqrt{3}} & \dfrac{1}{\sqrt{3}} \end{bmatrix}$

27. The following matrices are orthogonal matrices. Show that they have all the properties indicated in Theorem 5.8.

(a) $\begin{bmatrix} \dfrac{1}{\sqrt{2}} & \dfrac{1}{\sqrt{2}} \\ -\dfrac{1}{\sqrt{2}} & \dfrac{1}{\sqrt{2}} \end{bmatrix}$ **(b)** $\begin{bmatrix} \dfrac{\sqrt{3}}{2} & \dfrac{1}{2} \\ \dfrac{1}{2} & -\dfrac{\sqrt{3}}{2} \end{bmatrix}$

28. Let A and B be orthogonal matrices of the same size. Prove that the product matrix AB is also an orthogonal matrix.

29. Let $\mathbf{u}$ be a unit vector. Interpret $\mathbf{u}$ as a column matrix. Show that $A = I - 2\mathbf{u}\mathbf{u}^t$ is an orthogonal matrix.

*30. Prove that A is an orthogonal matrix if and only if $A^{-1} = A^t$. (This condition is sometimes taken as the definition of orthogonal matrix.)

*31. The complex analog of an orthogonal matrix is called a **unitary matrix**. A square matrix A is unitary if $A^{-1} = \overline{A}^t$. If A is unitary, prove that A^{-1} is also unitary.

32. Prove that a square matrix A is unitary if and only if its columns form a set of unit, mutually orthogonal vectors in $\mathbf{C}^n$.

*5.3 Cross Product†

In this section we introduce the cross product of two vectors in $\mathbf{R}^3$. The cross product is an important tool in many areas of science and engineering.

Definition   Let $\mathbf{u} = (u_1, u_2, u_3)$ and $\mathbf{v} = (v_1, v_2, v_3)$ be two vectors in $\mathbf{R}^3$. The **cross product** of $\mathbf{u}$ and $\mathbf{v}$ is denoted $\mathbf{u} \times \mathbf{v}$ and is the vector

$$\mathbf{u} \times \mathbf{v} = (u_2 v_3 - u_3 v_2, \, u_3 v_1 - u_1 v_3, \, u_1 v_2 - u_2 v_1)$$

Example 1   Determine the cross product $\mathbf{u} \times \mathbf{v}$ of the vectors

$$\mathbf{u} = (1, -2, 3) \quad \text{and} \quad \mathbf{v} = (0, 1, 4)$$

Solution   Using the above definition and letting $(u_1, u_2, u_3) = (1, -2, 3)$ and $(v_1, v_2, v_3) = (0, 1, 4)$, we get

$$\mathbf{u} \times \mathbf{v} = ((-2 \times 4) - (3 \times 1), \, (3 \times 0) - (1 \times 4), \, (1 \times 1) - (-2 \times 0))$$
$$= (-11, -4, 1)$$

---

† Some schools cover this material in the calculus sequence.

We shall use the notation $\mathbf{i} = (1, 0, 0)$, $\mathbf{j} = (0, 1, 0)$, and $\mathbf{k} = (0, 0, 1)$ for these **standard basis vectors** for $\mathbf{R}^3$. See Figure 5.13. This notation is commonly used in the sciences.

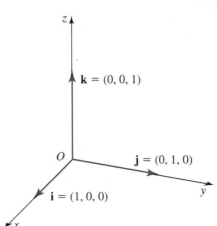

Figure 5.13

We can write $\mathbf{u} \times \mathbf{v}$ in the convenient form

$$\mathbf{u} \times \mathbf{v} = (u_2v_3 - u_3v_2)\mathbf{i} + (u_3v_1 - u_1v_3)\mathbf{j} + (u_1v_2 - u_2v_1)\mathbf{k}$$

It can be easily seen (by expanding the determinant) that

$$\mathbf{u} \times \mathbf{v} = \begin{vmatrix} \mathbf{i} & \mathbf{j} & \mathbf{k} \\ u_1 & u_2 & u_3 \\ v_1 & v_2 & v_3 \end{vmatrix}$$

Note that this is not an ordinary determinant since the elements of the first row are vectors, not scalars. The expansion of this determinant in terms of the first row gives a useful way of remembering the components of the vector $\mathbf{u} \times \mathbf{v}$. Furthermore, the algebraic properties of a determinant conveniently describe the algebraic properties of the cross product.

We shall find it convenient to use all three of the above forms of the cross product at various times.

Example 2   Use the determinant form of the cross product to compute $\mathbf{u} \times \mathbf{v}$ for the vectors

$$\mathbf{u} = (-2, 4, 1) \quad \text{and} \quad \mathbf{v} = (3, 1, 5)$$

Solution   Expanding the determinant in terms of the first row, we get

$$\mathbf{u} \times \mathbf{v} = \begin{vmatrix} \mathbf{i} & \mathbf{j} & \mathbf{k} \\ -2 & 4 & 1 \\ 3 & 1 & 5 \end{vmatrix}$$

$$= ((4 \times 5) - (1 \times 1))\mathbf{i} - ((-2 \times 5) - (1 \times 3))\mathbf{j} + ((-2 \times 1) - (4 \times 3))\mathbf{k}$$

$$= 19\mathbf{i} + 13\mathbf{j} - 14\mathbf{k}$$

$$= (19, 13, -14)$$

We now use this determinant form of the cross product to derive some of the properties of the cross product.

Theorem 5.9   Let $\mathbf{u}$ and $\mathbf{v}$ be vectors in $\mathbf{R}^3$. Then

$$\mathbf{u} \times \mathbf{u} = 0$$
$$\mathbf{u} \times \mathbf{v} = -(\mathbf{v} \times \mathbf{u})$$

Proof   Let $\mathbf{u} = (u_1, u_2, u_3)$ and $\mathbf{v} = (v_1, v_2, v_3)$. We get

$$\mathbf{u} \times \mathbf{u} = \begin{vmatrix} \mathbf{i} & \mathbf{j} & \mathbf{k} \\ u_1 & u_2 & u_3 \\ u_1 & u_2 & u_3 \end{vmatrix} = \mathbf{0}$$

since a determinant with two equal rows is zero. Note that since $\mathbf{u} \times \mathbf{u}$ is a vector, the zero is a vector, not a scalar.

Further, using the properties of determinants, we get

$$\mathbf{u} \times \mathbf{v} = \begin{vmatrix} \mathbf{i} & \mathbf{j} & \mathbf{k} \\ u_1 & u_2 & u_3 \\ v_1 & v_2 & v_3 \end{vmatrix} \underset{R1 \leftrightarrow R2}{=} -\begin{vmatrix} \mathbf{i} & \mathbf{j} & \mathbf{k} \\ v_1 & v_2 & v_3 \\ u_1 & u_2 & u_3 \end{vmatrix} = -(\mathbf{v} \times \mathbf{u})$$

Thus, in particular,

$$\mathbf{i} \times \mathbf{i} = 0, \qquad \mathbf{j} \times \mathbf{j} = 0, \qquad \mathbf{k} \times \mathbf{k} = 0$$
$$\mathbf{i} \times \mathbf{j} = -(\mathbf{j} \times \mathbf{i}), \qquad \mathbf{i} \times \mathbf{k} = -(\mathbf{k} \times \mathbf{i}), \qquad \mathbf{j} \times \mathbf{k} = -(\mathbf{k} \times \mathbf{j})$$

Theorem 5.10   If $\mathbf{i} = (1, 0, 0)$, $\mathbf{j} = (0, 1, 0)$, $\mathbf{k} = (0, 0, 1)$, then

$$\mathbf{i} \times \mathbf{j} = \mathbf{k}, \qquad \mathbf{j} \times \mathbf{k} = \mathbf{i}, \qquad \mathbf{k} \times \mathbf{i} = \mathbf{j}$$

See Figure 5.14.

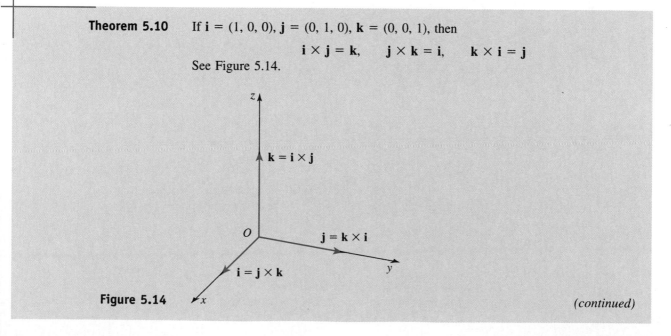

Figure 5.14

(continued)

Proof   To arrive at $\mathbf{i} \times \mathbf{j} = \mathbf{k}$, use $\mathbf{i} = (1, 0, 0)$ and $\mathbf{j} = (0, 1, 0)$ in the determinant form of the cross product as follows.

$$\mathbf{i} \times \mathbf{j} = \begin{vmatrix} \mathbf{i} & \mathbf{j} & \mathbf{k} \\ 1 & 0 & 0 \\ 0 & 1 & 0 \end{vmatrix}$$

$$= ((0 \times 0) - (0 \times 1))\mathbf{i} + ((1 \times 0) - (0 \times 0))\mathbf{j} + ((1 \times 1) - (0 \times 0))\mathbf{k}$$

$$= \mathbf{k}$$

$\mathbf{j} \times \mathbf{k} = \mathbf{i}$ and $\mathbf{k} \times \mathbf{i} = \mathbf{j}$ are proved similarly.

The Direction and Magnitude of the Vector u × v

Theorem 5.11   Let $\mathbf{u}$ and $\mathbf{v}$ be vectors in $\mathbf{R}^3$. Then

$$\mathbf{u} \cdot (\mathbf{u} \times \mathbf{v}) = 0$$
$$\mathbf{v} \cdot (\mathbf{u} \times \mathbf{v}) = 0$$

This result tells us that the vector $(\mathbf{u} \times \mathbf{v})$ is orthogonal to both $\mathbf{u}$ and $\mathbf{v}$. Therefore it is orthogonal to the space spanned by $\mathbf{u}$ and $\mathbf{v}$, namely the plane defined by $\mathbf{u}$ and $\mathbf{v}$. See Figure 5.15.

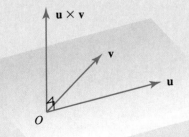

Figure 5.15 $\mathbf{u} \times \mathbf{v} \perp$ plane spanned by $\mathbf{u}$ and $\mathbf{v}$

Proof

$$\mathbf{u} \cdot (\mathbf{u} \times \mathbf{v}) = [u_1\mathbf{i} + u_2\mathbf{j} + u_3\mathbf{k}] \cdot \begin{vmatrix} \mathbf{i} & \mathbf{j} & \mathbf{k} \\ u_1 & u_2 & u_3 \\ v_1 & v_2 & v_3 \end{vmatrix}$$

$$= \begin{vmatrix} u_1 & u_2 & u_3 \\ u_1 & u_2 & u_3 \\ v_1 & v_2 & v_3 \end{vmatrix} = 0$$

It can be proved similarly that $\mathbf{v} \cdot (\mathbf{u} \times \mathbf{v}) = 0$.

Theorem 5.12   Let **u** and **v** be vectors in $\mathbf{R}^3$. Then $\|\mathbf{u} \times \mathbf{v}\| = \|\mathbf{u}\|\,\|\mathbf{v}\|\sin\theta$, where θ is the angle between **u** and **v**.

Proof   Using the definition of magnitude of a vector, we get

$$\|\mathbf{u} \times \mathbf{v}\|^2 = (u_2v_3 - u_3v_2)^2 + (u_3v_1 - u_1v_3)^2 + (u_1v_2 + u_2v_1)^2$$

On expanding the squares, this can be rewritten

$$= (u_1^2 + u_2^2 + u_3^2)(v_1^2 + v_2^2 + v_3^2) - (u_1v_1 + u_2v_2 - u_3v_3)^2$$
$$= \|\mathbf{u}\|^2\|\mathbf{v}\|^2 - (\mathbf{u} \cdot \mathbf{v})^2$$
$$= \|\mathbf{u}\|^2\|\mathbf{v}\|^2 - (\|\mathbf{u}\|\,\|\mathbf{v}\|\cos\theta)^2$$
$$= \|\mathbf{u}\|^2\|\mathbf{v}\|^2 - \|\mathbf{u}\|^2\|\mathbf{v}\|^2\cos^2\theta$$
$$= \|\mathbf{u}\|^2\|\mathbf{v}\|^2(1 - \cos^2\theta)$$
$$= \|\mathbf{u}\|^2\|\mathbf{v}\|^2\sin^2\theta$$

Thus

$$\|\mathbf{u} \times \mathbf{v}\| = \|\mathbf{u}\|\,\|\mathbf{v}\|\sin\theta$$

Area of a Triangle

The above result leads to the area of a triangle that is defined by two vectors. Consider the triangle whose edges are the vectors **u** and **v**. See Figure 5.16. We get

$$\text{Area of triangle} = \left(\tfrac{1}{2}\right)\text{base} \times \text{height}$$
$$= \left(\tfrac{1}{2}\right)\|\mathbf{u}\|\,\|\mathbf{v}\|\sin\theta$$
$$= \left(\tfrac{1}{2}\right)\|\mathbf{u} \times \mathbf{v}\|$$

$$\text{Area of a triangle with edges } \mathbf{u} \text{ and } \mathbf{v} = \left(\tfrac{1}{2}\right)\|\mathbf{u} \times \mathbf{v}\|$$

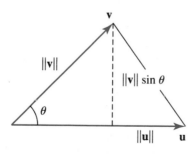

Length of base $= \|\mathbf{u}\|$
Height $= \|\mathbf{v}\|\sin\theta$

Figure 5.16

Example 3   Determine the area of the triangle having vertices $A(1, 1, 1)$, $B(-2, 3, 5)$, and $C(1, 7, 2)$.

Solution   We sketch this triangle in Figure 5.17. The points B and C define the following two edge vectors, starting from the point A.

$$\overrightarrow{AB} = (-2, 3, 5) - (1, 1, 1) = (-3, 2, 4)$$
$$\overrightarrow{AC} = (1, 7, 2) - (1, 1, 1) = (0, 6, 1)$$

Thus area of triangle $= \left(\tfrac{1}{2}\right)\|\overrightarrow{AB} \times \overrightarrow{AC}\|$

$$= \left(\tfrac{1}{2}\right)\|(-3, 2, 4) \times (0, 6, 1)\|$$
$$= \left(\tfrac{1}{2}\right)\|(-22, 3, -18)\|$$
$$= \left(\tfrac{1}{2}\right)\sqrt{22^2 + 3^2 + 18^2}$$
$$= \left(\tfrac{1}{2}\right)\sqrt{817}$$

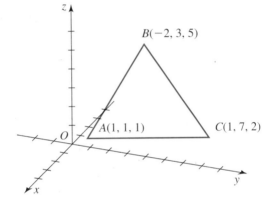

$$\overrightarrow{AB} = (-2, 3, 5) - (1, 1, 1) = (-3, 2, 4)$$
Figure 5.17 $\overrightarrow{AC} = (1, 7, 2) - (1, 1, 1) = (0, 6, 1)$

Volume of a Parallelepiped

Consider the parallelepiped whose adjacent edges are defined by the vectors **u**, **v**, and **w**. See Figure 5.18.

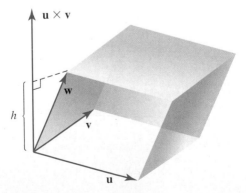

Figure 5.18

The area of the base is twice the area of the triangle defined by **u** and **v**. Thus area of base $= \|\mathbf{u} \times \mathbf{v}\|$. Further, volume $= \|\mathbf{u} \times \mathbf{v}\| \times h$, where h is the height.

Observe that

$$h = \text{magnitude of projection of } \mathbf{w} \text{ onto } \mathbf{u} \times \mathbf{v}$$

$$= \text{magnitude of } \left(\frac{\mathbf{w} \cdot (\mathbf{u} \times \mathbf{v})}{(\mathbf{u} \times \mathbf{v}) \cdot (\mathbf{u} \times \mathbf{v})} (\mathbf{u} \times \mathbf{v}) \right)$$

$$= \text{magnitude of } \left(\frac{\mathbf{w} \cdot (\mathbf{u} \times \mathbf{v})}{\|\mathbf{u} \times \mathbf{v}\|^2} (\mathbf{u} \times \mathbf{v}) \right)$$

$$= \frac{|\mathbf{w} \cdot (\mathbf{u} \times \mathbf{v})|}{\|\mathbf{u} \times \mathbf{v}\|}$$

Thus

Volume of a parallelepiped with adjacent edges **u**, **v**, and $\mathbf{w} = |\mathbf{w} \cdot (\mathbf{u} \times \mathbf{v})|$

The expression $\mathbf{w} \cdot (\mathbf{u} \times \mathbf{v})$ is called the **triple scalar product** of **u**, **v**, and **w**. It can be conveniently written as a determinant. Let

$$\mathbf{u} = (u_1, u_2, u_3), \quad \mathbf{v} = (v_1, v_2, v_3), \quad \text{and} \quad \mathbf{w} = (w_1, w_2, w_3)$$

Then

$$\mathbf{w} \cdot (\mathbf{u} \times \mathbf{v}) = [w_1\mathbf{i} + w_2\mathbf{j} + w_3\mathbf{k}] \cdot \begin{vmatrix} \mathbf{i} & \mathbf{j} & \mathbf{k} \\ u_1 & u_2 & u_3 \\ v_1 & v_2 & v_3 \end{vmatrix}$$

$$= \begin{vmatrix} w_1 & w_2 & w_3 \\ u_1 & u_2 & u_3 \\ v_1 & v_2 & v_3 \end{vmatrix} = \begin{vmatrix} u_1 & u_2 & u_3 \\ v_1 & v_2 & v_3 \\ w_1 & w_2 & w_3 \end{vmatrix}$$

Thus

Volume of a parallelepiped with edges

$$(u_1, u_2, u_3), (v_1, v_2, v_3), (w_1, w_2, w_3) = \text{absolute value of } \begin{vmatrix} u_1 & u_2 & u_3 \\ v_1 & v_2 & v_3 \\ w_1 & w_2 & w_3 \end{vmatrix}$$

Example 4   Find the volume of the parallelepiped having adjacent edges defined by the points $A(1, 1, 3)$, $B(3, 7, 1)$, $C(-2, 3, 3)$, $D(1, 2, 8)$.

Solution   We sketch the parallelepiped in Figure 5.19. This need be only a very rough sketch. The points A, B, C, and D define the following three adjacent edge vectors.

$$\overrightarrow{AB} = (3, 7, 1) - (1, 1, 3) = (2, 6, -2)$$
$$\overrightarrow{AC} = (-2, 3, 3) - (1, 1, 3) = (-3, 2, 0)$$
$$\overrightarrow{AD} = (1, 2, 8) - (1, 1, 3) = (0, 1, 5)$$

The volume of the parallelepiped is thus

$$= \text{absolute value of} \begin{vmatrix} 2 & 6 & -2 \\ -3 & 2 & 0 \\ 0 & 1 & 5 \end{vmatrix}$$

$$= \text{absolute value of } (116)$$

$$= 116$$

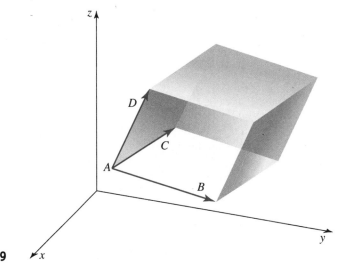

Figure 5.19

We have discussed numerous operations on vectors in this course. Addition, scalar multiplication, and dot product have been defined. In this section we have introduced yet another operation, namely the cross product. We now conveniently summarize the properties of the cross product. The summary includes properties already derived and further properties that tell us how the cross product interacts with addition, scalar multiplication, and dot product. You are asked to derive the new properties in the exercises that follow.

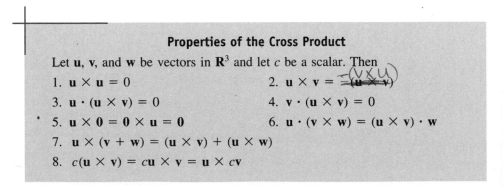

Properties of the Cross Product

Let **u**, **v**, and **w** be vectors in $\mathbf{R}^3$ and let c be a scalar. Then

1. $\mathbf{u} \times \mathbf{u} = 0$
2. $\mathbf{u} \times \mathbf{v} = -(\mathbf{u} \times \mathbf{v})$
3. $\mathbf{u} \cdot (\mathbf{u} \times \mathbf{v}) = 0$
4. $\mathbf{v} \cdot (\mathbf{u} \times \mathbf{v}) = 0$
5. $\mathbf{u} \times \mathbf{0} = \mathbf{0} \times \mathbf{u} = 0$
6. $\mathbf{u} \cdot (\mathbf{v} \times \mathbf{w}) = (\mathbf{u} \times \mathbf{v}) \cdot \mathbf{w}$
7. $\mathbf{u} \times (\mathbf{v} + \mathbf{w}) = (\mathbf{u} \times \mathbf{v}) + (\mathbf{u} \times \mathbf{w})$
8. $c(\mathbf{u} \times \mathbf{v}) = c\mathbf{u} \times \mathbf{v} = \mathbf{u} \times c\mathbf{v}$

If $\mathbf{i} = (1, 0, 0)$, $\mathbf{j} = (0, 1, 0)$, $\mathbf{k} = (0, 0, 1)$, then
9. $\mathbf{i} \times \mathbf{i} = \mathbf{0}$, $\mathbf{j} \times \mathbf{j} = \mathbf{0}$, $\mathbf{k} \times \mathbf{k} = \mathbf{0}$
10. $\mathbf{i} \times \mathbf{j} = \mathbf{k}$, $\mathbf{j} \times \mathbf{k} = \mathbf{i}$, $\mathbf{k} \times \mathbf{i} = \mathbf{j}$
11. $\mathbf{i} \times \mathbf{j} = -(\mathbf{j} \times \mathbf{i})$, $\mathbf{i} \times \mathbf{k} = -(\mathbf{k} \times \mathbf{i})$, $\mathbf{j} \times \mathbf{k} = -(\mathbf{k} \times \mathbf{j})$

Exercise Set 5.3

1. If $\mathbf{u} = (1, 2, 3)$, $\mathbf{v} = (-1, 0, 4)$, and $\mathbf{w} = (1, 2, -1)$, compute the following using the definition of the cross product.

    *(a) $\mathbf{u} \times \mathbf{v}$ (b) $\mathbf{v} \times \mathbf{u}$

    *(c) $\mathbf{u} \times \mathbf{w}$ (d) $\mathbf{v} \times \mathbf{w}$

    *(e) $(\mathbf{u} \times \mathbf{v}) \times \mathbf{w}$

2. If $\mathbf{u} = (-2, 2, 4)$, $\mathbf{v} = (3, 0, 5)$, and $\mathbf{w} = (4, -2, 1)$, compute the following using the determinant form of the cross product.

    *(a) $\mathbf{u} \times \mathbf{v}$ (b) $\mathbf{u} \times \mathbf{w}$

    *(c) $\mathbf{w} \times \mathbf{v}$ (d) $\mathbf{v} \times \mathbf{w}$

    *(e) $(\mathbf{w} \times \mathbf{v}) \times \mathbf{u}$

3. If $\mathbf{u} = 2\mathbf{i} + 3\mathbf{j} + \mathbf{k}$, $\mathbf{v} = -\mathbf{i} + 2\mathbf{j} + 4\mathbf{k}$, and $\mathbf{w} = 3\mathbf{i} - 7\mathbf{k}$, compute the following cross products.

    *(a) $\mathbf{u} \times \mathbf{v}$ (b) $\mathbf{u} \times \mathbf{w}$

    *(c) $\mathbf{w} \times \mathbf{u}$ (d) $\mathbf{v} \times \mathbf{w}$

    *(e) $(\mathbf{w} \times \mathbf{u}) \times \mathbf{v}$

4. If $\mathbf{u} = (3, 1, -2)$, $\mathbf{v} = (4, -1, 2)$, and $\mathbf{w} = (0, 3, -2)$, compute the following.

    *(a) $\mathbf{u} \times \mathbf{v}$ (b) $3\mathbf{v} \times 2\mathbf{w}$

    *(c) $(\mathbf{w} \times \mathbf{u}) \cdot \mathbf{v}$ *(d) $(\mathbf{w} + 2\mathbf{u}) \times \mathbf{v}$

    (e) $\mathbf{u} \cdot (\mathbf{v} \times \mathbf{w})$ *(f) $(\mathbf{v} \times \mathbf{u}) \cdot (\mathbf{w} \times \mathbf{v})$

5. Determine the areas of the triangles having the following vertices.

    *(a) $A(1, 2, 1)$, $B(-3, 4, 6)$, $C(1, 8, 3)$

    (b) $A(3, -1, 2)$, $B(0, 2, 6)$, $C(7, 1, 5)$

    *(c) $A(1, 0, 0)$, $B(0, 5, 2)$, $C(3, -4, 8)$

6. Find the volumes of the parallelepipeds having adjacent edges defined by the following points.

    *(a) $A(1, 2, 5)$, $B(4, 8, 1)$, $C(-3, 2, 3)$, $D(0, 3, 9)$

    (b) $A(3, 1, 6)$, $B(-2, 3, 4)$, $C(0, 2, -5)$, $D(3, -1, 4)$

    *(c) $A(0, 1, 2)$, $B(-3, 1, 4)$, $C(5, 2, 3)$, $D(-3, -2, 1)$

7. Prove the following.

    (a) $\mathbf{i} \cdot \mathbf{i} = 1$, $\mathbf{j} \cdot \mathbf{j} = 1$, $\mathbf{k} \cdot \mathbf{k} = 1$

    (b) $\mathbf{i} \cdot \mathbf{j} = 0$, $\mathbf{i} \cdot \mathbf{k} = 0$, $\mathbf{j} \cdot \mathbf{k} = 0$

*8. Prove that $(\mathbf{i} \times \mathbf{j}) \cdot \mathbf{k} = 1$.

9. Let $\mathbf{u}$ be a vector in $\mathbf{R}^3$. Prove that $\mathbf{u} \times \mathbf{0} = \mathbf{0} \times \mathbf{u} = \mathbf{0}$.

10. Let $\mathbf{u}$, $\mathbf{v}$, and $\mathbf{w}$ be vectors in $\mathbf{R}^3$. Prove that

$$\mathbf{u} \cdot (\mathbf{v} \times \mathbf{w}) = (\mathbf{u} \times \mathbf{v}) \cdot \mathbf{w}$$

11. Let $\mathbf{u}$, $\mathbf{v}$, and $\mathbf{w}$ be vectors in $\mathbf{R}^3$. Prove that

$$\mathbf{u} \times (\mathbf{v} \times \mathbf{w}) = (\mathbf{u} \cdot \mathbf{w})\mathbf{v} - (\mathbf{u} \cdot \mathbf{v})\mathbf{w}$$

*12. Let $\mathbf{u}$ and $\mathbf{v}$ be vectors in $\mathbf{R}^3$ and let c be a scalar. Prove that

$$c(\mathbf{u} \times \mathbf{v}) = c\mathbf{u} \times \mathbf{v} = \mathbf{u} \times c\mathbf{v}$$

13. Let $\mathbf{u}$ and $\mathbf{v}$ be nonzero vectors in $\mathbf{R}^3$. Prove that $\mathbf{u}$ and $\mathbf{v}$ are parallel if and only if $\mathbf{u} \times \mathbf{v} = \mathbf{0}$.

*14. Let $\mathbf{u}$, $\mathbf{v}$, and $\mathbf{w}$ be vectors in $\mathbf{R}^3$. Prove that

$$\mathbf{u} \times (\mathbf{v} \times \mathbf{w}) + \mathbf{v} \times (\mathbf{w} \times \mathbf{u}) + \mathbf{w} \times (\mathbf{u} \times \mathbf{v}) = \mathbf{0}$$

*15. Prove that the position vectors $\mathbf{u}$, $\mathbf{v}$, and $\mathbf{w}$ of $\mathbf{R}^3$ all lie in a plane if and only if $\mathbf{u} \cdot (\mathbf{v} \times \mathbf{w}) = 0$.

16. Let $\mathbf{t}$, $\mathbf{u}$, $\mathbf{v}$, and $\mathbf{w}$ be vectors in $\mathbf{R}^3$. Prove that

$$(\mathbf{t} \times \mathbf{u}) \cdot (\mathbf{v} \times \mathbf{w}) = \begin{vmatrix} \mathbf{t} \cdot \mathbf{v} & \mathbf{u} \cdot \mathbf{v} \\ \mathbf{t} \cdot \mathbf{w} & \mathbf{u} \cdot \mathbf{w} \end{vmatrix}$$

*5.4 Equations of Planes and Lines in Three-Space[†]

The tools that we develop in this section are useful for discussing planes and lines in three-space. We first look at planes.

Planes in R³

Let $P_0(x_0, y_0, z_0)$ be a point in a plane. Let (a, b, c) be a vector perpendicular to the plane, called a **normal** to the plane. See Figure 5.20. These two quantities, namely a point in a plane and a normal vector to the plane, characterize the plane. There is only one plane through a given point and having a given normal. We shall now derive the equation of a plane passing through the point $P_0(x_0, y_0, z_0)$ and having normal (a, b, c).

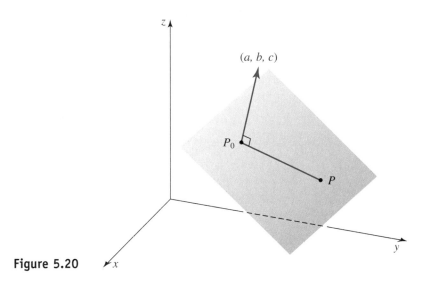

Figure 5.20

Let $P(x, y, z)$ be an arbitrary point in the plane. We get

$$\overrightarrow{P_0P} = (x, y, z) - (x_0, y_0, z_0)$$
$$= (x - x_0, y - y_0, z - z_0)$$

The vector $\overrightarrow{P_0P}$ lies in the plane. Thus the vectors (a, b, c) and $\overrightarrow{P_0P}$ are orthogonal. Their dot product is zero. This observation leads to an equation for the plane.

$$(a, b, c) \cdot \overrightarrow{P_0P} = 0,$$
$$(a, b, c) \cdot (x - x_0, y - y_0, z - z_0) = 0,$$
$$a(x - x_0) + b(y - y_0) + c(z - z_0) = 0$$

This is called the **point-normal form** of the equation of the plane.

[†] Some schools cover this material in the calculus sequence.

Let us now rewrite the equation. Multiplying we get

$$ax - ax_0 + by - by_0 + cz - cz_0 = 0,$$
$$ax + by + cz - ax_0 - by_0 - cz_0 = 0$$

The last three terms are constant. Combine them into a single constant d. We get

$$ax + by + cz + d = 0$$

This is called the **general form** of the equation of the plane.

Point-normal equation of a plane:
$$a(x - x_0) + b(y - y_0) + c(z - z_0) = 0$$
General form equation of a plane: $ax + by + cz + d = 0$

Example 1   Find the point-normal and general forms of the equation of the plane passing through the point $(1, 2, 3)$ and having normal $(-1, 4, 6)$.

Solution   Let $(x_0, y_0, z_0) = (1, 2, 3)$ and $(a, b, c) = (-1, 4, 6)$. The point-normal form is

$$-1(x - 1) + 4(y - 2) + 6(z - 3) = 0$$

Multiplying and simplifying,

$$-x + 1 + 4y - 8 + 6z - 18 = 0$$

The general form is

$$-x + 4y + 6z - 25 = 0$$

Example 2   Determine the equation of the plane through the three points $P_1(2, -1, 1)$, $P_2(-1, 1, 3)$, and $P_3(2, 0, -3)$.

Solution   The vectors $\overrightarrow{P_1P_2}$ and $\overrightarrow{P_1P_3}$ lie in the plane. Thus $\overrightarrow{P_1P_2} \times \overrightarrow{P_1P_3}$ will be normal to the plane. We get

$$\overrightarrow{P_1P_2} = (-1, 1, 3) - (2, -1, 1) = (-3, 2, 2)$$
$$\overrightarrow{P_1P_3} = (2, 0, -3) - (2, -1, 1) = (0, 1, -4)$$

Thus

$$\overrightarrow{P_1P_2} \times \overrightarrow{P_1P_3} = \begin{vmatrix} \mathbf{i} & \mathbf{j} & \mathbf{k} \\ -3 & 2 & 2 \\ 0 & 1 & -4 \end{vmatrix}$$
$$= (-8 - 2)\mathbf{i} - (12 - 0)\mathbf{j} + (-3 - 0)\mathbf{k}$$
$$= -10\mathbf{i} - 12\mathbf{j} - 3\mathbf{k}$$

Let $(x_0, y_0, z_0) = (2, -1, 1)$ and $(a, b, c) = (-10, -12, -3)$. The point-normal form gives

$$-10(x - 2) - 12(y + 1) - 3(z - 1) = 0,$$
$$-10x + 20 - 12y - 12 - 3z + 3 = 0$$

The equation of the plane is

$$-10x - 12y - 3z + 11 = 0$$

Observe that each of the given points $(2, -1, 1)$, $(-1, 1, 3)$, and $(2, 0, -3)$ satisfies this equation.

Example 3   Prove that the following system of equations has no solution.

$$2x - y + 3z = 6$$
$$-4x + 2y - 6z = 2$$
$$x + y + 4z = 5$$

Solution   Consider the general form of the equation of a plane.

$$ax + by + cz + d = 0$$

The vector (a, b, c) is normal to this plane. Interpret each of the given equations as defining a plane in $\mathbf{R}^3$. On comparison with the general form, it is seen that the following vectors are normals to these three planes.

$$(2, -1, 3), \quad (-4, 2, -6), \quad \text{and} \quad (1, 1, 4)$$

Observe that

$$(-4, 2, -6) = -2(2, -1, 3)$$

The normals to the first two planes are parallel. Thus these two planes are parallel, and they are distinct. The three planes can therefore have no points in common. The system of equations has no solution.

Lines in $\mathbf{R}^3$

Consider a line through the point $P_0(x_0, y_0, z_0)$ in the direction defined by the vector (a, b, c). See Figure 5.21. Let $P(x, y, z)$ be any other point on the line. We get

$$\overrightarrow{P_0P} = (x - x_0, y - y_0, z - z_0)$$

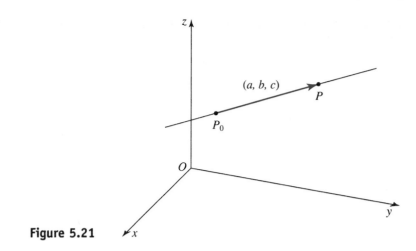

Figure 5.21

The vectors $\overrightarrow{P_0P}$ and (a, b, c) are parallel. Thus there exists a scalar t such that

$$\overrightarrow{P_0P} = t(a, b, c)$$

Equating the two expressions for $\overrightarrow{P_0P}$, we get

$$(x - x_0, y - y_0, z - z_0) = t(a, b, c)$$

This is called the **vector equation** of the line. Comparing the components of the vectors on the left and right of this equation gives

$$x - x_0 = ta, \qquad y - y_0 = tb, \qquad z - z_0 = tc$$

Rearranging these equations as follows gives the **parametric equations** of a line in $\mathbf{R}^3$.

Parametric Equations of a Line

$$x = x_0 + ta, \quad y = y_0 + tb, \quad z = z_0 + tc \qquad -\infty < t < \infty$$

As t varies, we get the points on the line.

Example 4 Find the parametric equations of the line through the point $(1, 2, 5)$ in the direction $(4, 3, 1)$. Determine any two points on the line.

Solution Let $(x_0, y_0, z_0) = (1, 2, 5)$ and $(a, b, c) = (4, 3, 1)$. The parametric equations of the line are

$$x = 1 + 4t, \quad y = 2 + 3t, \quad z = 5 + t \qquad -\infty < t < \infty$$

By letting t take on two convenient values, we can determine two other points on the line. For example, $t = 1$ leads to the point $(5, 5, 6)$, and $t = -2$ leads to the point $(-7, -4, 3)$.

Example 5   Find the parametric equations of the line through the points $(-1, 2, 6)$ and $(1, 5, 4)$.

Solution   Let (x_0, y_0, z_0) be the point $(-1, 2, 6)$. The direction of the line is given by the vector

$$(a, b, c) = (1, 5, 4) - (-1, 2, 6) = (2, 3, -2)$$

The parametric equations of the line are

$$x = -1 + 2t, \quad y = 2 + 3t, \quad z = 6 - 2t \qquad -\infty < t < \infty$$

The parametric equations of a line for which a, b, and c are all nonzero lead to another useful way of expressing the line, called the **symmetric equations** of the line. Isolating the variable t in each of the three parametric equations we get

$$t = \frac{x - x_0}{a}, \qquad t = \frac{y - y_0}{b}, \qquad t = \frac{z - z_0}{c}$$

Equating these expressions for t, we get the following symmetric equations.

Symmetric Equations of a Line

$$\frac{x - x_0}{a} = \frac{y - y_0}{b} = \frac{z - z_0}{c}$$

In this form, the line can be interpreted as the intersection of the planes

$$\frac{x - x_0}{a} = \frac{y - y_0}{b} \quad \text{and} \quad \frac{y - y_0}{b} = \frac{z - z_0}{c}$$

Example 6   Determine the symmetric equations of the line through the points $(-4, 2, -6)$ and $(1, 4, 3)$.

Solution   The direction of the line is given by the vector

$$(a, b, c) = (1, 4, 3) - (-4, 2, -6) = (5, 2, 9)$$

Let (x_0, y_0, z_0) be the point $(-4, 2, -6)$. (Either point can be used.) The symmetric equations of the line are

$$\frac{x - (-4)}{5} = \frac{y - 2}{2} = \frac{z - (-6)}{9}$$

Write these equations in the form

$$\frac{x + 4}{5} = \frac{y - 2}{2} \quad \text{and} \quad \frac{y - 2}{2} = \frac{z + 6}{9}$$

Cross multiply these equations and simplify.

$$2(x + 4) = 5(y - 2) \quad \text{and} \quad 9(y - 2) = 2(z + 6),$$
$$2x - 5y + 18 = 0 \quad \text{and} \quad 9y - 2z - 30 = 0$$

The line through the points $(-4, 2, -6)$ and $(1, 4, 3)$ is thus the intersection of the planes

$$2x - 5y + 18 = 0 \quad \text{and} \quad 9y - 2z - 30 = 0$$

Exercise Set 5.4

1. Determine the point-normal form and the general form of the equations of the plane through each of the following points and having the given normals.

   *(a) point $(1, -2, 4)$; normal $(1, 1, 1)$

   (b) point $(-3, 5, 6)$; normal $(-2, 4, 5)$

   *(c) point $(0, 0, 0)$; normal $(1, 2, 3)$

   (d) point $(4, 5, -2)$; normal $(-1, 4, 3)$ $\longrightarrow$ SCALAR

2. Determine a general form of the equation of the plane through each of the following sets of three points.

   *(a) $P_1(1, -2, 3)$, $P_2(1, 0, 2)$, and $P_3(-1, 4, 6)$

   (b) $P_1(0, 0, 0)$, $P_2(1, 2, 4)$, and $P_3(-3, 5, 1)$

   *(c) $P_1(-1, -1, 2)$, $P_2(3, 5, 4)$, and $P_3(1, 2, 5)$

   (d) $P_1(7, 1, 3)$, $P_2(-2, 4, -3)$, and $P_3(5, 4, 1)$

*3. Prove that the planes $3x - 2y + 4z - 3 = 0$ and $-6x + 4y - 8z + 7 = 0$ are parallel.

4. Use the general form of the equation of a plane to prove that the following system of equations has no solutions.

   $$x + y - 3z = 7$$
   $$3x - 6y + 9z = 6$$
   $$-x + 2y - 3z = 2$$

*5. Find the equation of the plane parallel to the plane $2x - 3y + z + 4 = 0$, passing through the point $(1, 2, -3)$.   POINT NORMAL EQN.

6. Determine parametric equations and symmetric equations of the lines through the following points, in the given directions.

   *(a) point $(1, 2, 3)$; direction $(-1, 2, 4)$

   (b) point $(-3, 1, 2)$; direction $(1, 1, 1)$

   *(c) point $(0, 0, 0)$; direction $(-2, -3, 5)$

   (d) point $(-2, -4, 1)$; direction $(2, -2, 4)$

*7. Find the equation of the line through the point $(1, 2, -4)$, parallel to the line $x = 4 + 2t$, $y = -1 + 3t$, $z = 2 + t$, where $-\infty < t < \infty$.

8. Find the equation of the line through the point $(2, -3, 1)$ in a direction orthogonal to the line

   $$\frac{x + 1}{3} = \frac{y - 1}{2} = \frac{z + 2}{5}$$

*9. Determine the equation of the line through the point $(4, -1, 3)$, perpendicular to the plane.

   $$2x - y + 4z + 7 = 0$$

10. Give general forms for the equations of the xy plane, xz plane, and yz plane.

*11. Show that there are many planes that contain the three points $P_1(3, -5, 5)$, $P_2(-1, 1, 3)$, and $P_3(5, -8, 6)$. Interpret your conclusion geometrically.

*12. Find an equation for the plane through the point $(4, -1, 5)$, in a direction perpendicular to the line $x = 1 - t$, $y = 3 + 2t$, $z = 5 - 4t$, where $-\infty < t < \infty$.

*13. Show that the line $x = 1 + t$, $y = 14 - t$, $z = 2 - t$, where $-\infty < t < \infty$, lies in the plane

   $$2x - y + 3z + 6 = 0$$

$$2(1+t) - (14-t) + 3(2-t) + 6 = 0$$
$$2 + 2t - 14 + t + 6 - 3t + 6 = 0$$
$$0 = 0 \quad \therefore \text{yes}$$

14. Prove that the line $x = 4 + 2t$, $y = 5 + t$, $z = 7 + 2t$, where $-\infty < t < \infty$, never intersects the plane

$$3x + 2y - 4z + 7 = 0$$

15. Find an equation of the line through the point $(5, -1, 2)$ in a direction perpendicular to the line $x = 5 - 2t$, $y = 2 + 3t$, $z = 2t$, where $-\infty < t < \infty$.

*16. Prove that the lines $x = -1 - 4t$, $y = 4 + 4t$, $z = 7 + 8t$, where $-\infty < t < \infty$, and $x = 4 + 3h$, $y = 1 - h$, $z = 5 + 2h$, where $-\infty < h < \infty$, intersect at right angles. Find the point of intersection.

*17. Two lines are said to be **skew** if they are not parallel and do not intersect. Show that the following two lines are skew.

$$x = 1 + 4t, \quad y = 2 + 5t, \quad z = 3 + 3t,$$
$$\text{where } -\infty < t < \infty$$
$$x = 2 + h, \quad y = 1 - 3h, \quad z = -1 - 2h,$$
$$\text{where } -\infty < h < \infty$$

18. ***(a)** Show that the set of all points on the plane $2x + 3y - 4z = 0$ forms a subspace of $\mathbf{R}^3$. Find a basis for this subspace.

    *(b) Show that the set of all points on the plane $3x - 4y + 2z - 6 = 0$ does not form a subspace of $\mathbf{R}^3$.

    (c) Prove that the set of all points on the plane $ax + by + cz + d = 0$ forms a subspace of $\mathbf{R}^3$ if and only if $d = 0$.

19. Prove that a line can be written in symmetric form if and only if it is not orthogonal to the x axis, y axis, or z axis.

Review Exercises Chapter 5

1. Determine the dot products of the following pairs of vectors.

    (a) $(1, 2)$, $(3, -4)$ **(b)** $(1, -2, 3)$, $(4, 2, -7)$

    (c) $(2, 2, -5)$, $(3, 2, -1)$

2. Find the norms of the following vectors.

    (a) $(1, -4)$ **(b)** $(-2, 1, 3)$

    (c) $(1, -2, 3, 4)$

3. Determine the cosines of the angles between the following vectors.

    (a) $(-1, 1)$, $(2, 3)$ **(b)** $(1, 2, -3)$, $(4, 1, 2)$

4. Find a vector orthogonal to $(-2, 1, 5)$.

5. Determine the distances between the following pairs of points.

    (a) $(1, -2)$, $(5, 3)$ **(b)** $(3, 2, 1)$, $(7, 1, 2)$

    (c) $(3, 1, -1, 2)$, $(4, 1, 6, 2)$

6. Find all the values of c such that $\|c(1, 2, 3)\| = 196$.

7. Determine the projection of the vector $\mathbf{v}$ onto the vector $\mathbf{u}$ for the following vectors.

    (a) $\mathbf{v} = (1, 3)$, $\mathbf{u} = (2, 4)$

    (b) $\mathbf{v} = (-1, 3, 4)$, $\mathbf{u} = (-1, 2, 4)$

8. Construct an orthonormal basis for the subspace of $\mathbf{R}^4$ spanned by the vectors $(1, 2, 3, -1)$, $(2, 0, -1, 1)$, $(3, 2, 0, 1)$.

9. Determine an orthonormal basis for the subspace of $\mathbf{R}^3$ consisting of vectors of the form $(x, y, x + 2y)$.

10. Let W be the subspace of $\mathbf{R}^3$ having basis $\{(1, -1, 3), (2, 1, 1)\}$. Find the projection of $(3, 1, -2)$ onto W.

11. Find the distance of the point $(1, 2, -4)$ in $\mathbf{R}^3$ from the subspace of vectors of the form $(a, 3a, b)$.

12. Prove that a square matrix A is orthogonal if and only if A^t is orthogonal.

13. Consider the vector $\mathbf{v} = (1, 3, -1)$ in $\mathbf{R}^3$. Let W be the subspace of $\mathbf{R}^3$ consisting of all vectors of the form $(a, b, a - 2b)$. Decompose $\mathbf{v}$ into the sum of a vector that lies in W and a vector orthogonal to W.

14. Prove that if two vectors $\mathbf{u}$ and $\mathbf{v}$ are orthogonal, then they are linearly independent.

15. Prove that if $\mathbf{u}$ is orthogonal to both the vectors $\mathbf{v}$ and $\mathbf{w}$, then $\mathbf{u}$ is orthogonal to every vector in the subspace generated by $\mathbf{v}$ and $\mathbf{w}$.

16. If $\mathbf{u} = (1, 1, -3)$, $\mathbf{v} = (2, -1, 4)$ and $\mathbf{w} = (2, 0, -2)$, compute the following.

    (a) $\mathbf{u} \times \mathbf{v}$ **(b)** $2\mathbf{v} \times 3\mathbf{w}$

    (c) $(\mathbf{w} \times \mathbf{u}) \cdot \mathbf{v}$ **(d)** $(\mathbf{w} - 3\mathbf{u}) \times \mathbf{v}$

    (e) $\mathbf{u} \cdot (\mathbf{v} \times \mathbf{w})$ **(f)** $(\mathbf{v} \times \mathbf{u}) \cdot (\mathbf{w} \times \mathbf{v})$

17. Determine the area of the triangle having vertices $(1, -2, 3)$, $(4, 1, 0)$, $(5, -2, 1)$.

18. Let $\mathbf{u}$, $\mathbf{v}$, and $\mathbf{w}$ be vectors in $\mathbf{R}^3$. Prove that

$$\mathbf{u} \times (\mathbf{v} + \mathbf{w}) = (\mathbf{u} \times \mathbf{v}) + (\mathbf{u} \times \mathbf{w})$$

19. Let $\mathbf{u}$ and $\mathbf{v}$ be vectors in $\mathbf{R}^3$. Prove that $\mathbf{u} \times \mathbf{v} = \mathbf{v} \times \mathbf{u}$ if and only if $\mathbf{u} \times \mathbf{v} = \mathbf{0}$.

20. Let $\mathbf{u}$ and $\mathbf{v}$ be vectors in $\mathbf{R}^3$. Prove that

$$(\mathbf{u} + \mathbf{v}) \times (\mathbf{u} - \mathbf{v}) = 2\mathbf{v} \times \mathbf{u}$$

21. Determine the point-normal form and the general form of the equation of the plane through the point $(1, -3, 2)$ and having normal $(1, -1, 3)$.

22. Determine a general form of the equation of the plane through the three points $(1, -3, 1)$, $(2, 1, 4)$, and $(-1, 4, 2)$.

23. Find the parametric equations and symmetric equations of the line through the point $(1, -2, 3)$ in the direction $(2, -1, 4)$.

24. Find the equation of the line through the point $(-2, 1, 3)$, perpendicular to the plane $x - 3y + 2z = 4$.

25. Show that the line $x = 2 + t$, $y = 4 - 3t$, $z = 2 - 5t$, where t is a real number, lies in the plane $2x - y + z = 2$.

MATLAB Discussion

M.15   Dot Product, Norm, Angle, Distance, Projection (Section 5.1)

The dot product on n-dimensional Euclidean space is

$$\mathbf{u} \cdot \mathbf{v} = u_1 v_1 + \cdots + u_n v_n$$

The geometry is defined in terms of the dot product.

norm	*angle*	*proj* $\mathbf{u}$ *onto* $\mathbf{v}$	*distance X to Y*
$\|\mathbf{u}\| = \sqrt{\mathbf{u} \cdot \mathbf{u}}$	$\cos \theta = \dfrac{\mathbf{u} \cdot \mathbf{v}}{\|\mathbf{u}\|\|\mathbf{v}\|}$	$\text{proj}_{\mathbf{v}}\, \mathbf{u} = \dfrac{\mathbf{u} \cdot \mathbf{v}}{\mathbf{v} \cdot \mathbf{v}} \mathbf{v}$	$d(X, Y) = \|X - Y\|$

The following functions have been written to compute these quantities.

dot(u, v),   **mag**(u),   **angle**(u, v),   **proj**(u, v),   **dist**(X, Y)

Example 1   Let $\mathbf{u} = (1, -2, 4)$, $\mathbf{v} = (3, 1, 2)$, $X = (1, 2, 3)$, and $Y = (-4, 2, 3)$. Find $\mathbf{u} \cdot \mathbf{v}$, $\|\mathbf{u}\|$, the angle between $\mathbf{u}$ and $\mathbf{v}$, $\text{proj}_{\mathbf{v}}\, \mathbf{u}$, and $d(X, Y)$.

We enter the data and use the functions, one at a time.

»u = [1   2   4];   v = [3   1   2];   X = [1   2   3];   Y = [−4   2   3];

dot prod	*magnitude*	*angle*	*projection*	*distance*
»**dot(u,v)**	»**mag(u)**	»**angle(u,v)**	»**proj(u,v)**	»**dist(X,Y)**
ans =	ans =	ans =	ans =	ans =
13	4.5826	40.6964 {degrees}	2.7857 0.9286 1.8571	5

*Exercise 1   Let $\mathbf{u} = (2, 0, 6)$, $\mathbf{v} = (4, -5, -1)$, $X = (3, 3, 3)$, $Y = (-2, 5, 1)$. Find $\mathbf{u} \cdot \mathbf{v}$, $\|\mathbf{u}\|$, the angle between $\mathbf{u}$ and $\mathbf{v}$, $\text{proj}_{\mathbf{v}}\, \mathbf{u}$, and $d(X, Y)$.

Exercise 2   Let $\mathbf{u} = (1, 2, -5, 3)$, $\mathbf{v} = (0, 4, 2, -8)$, $X = (6, 3, 2, 7)$, $Y = (9, -4, -3, 2)$. Find $\mathbf{u} \cdot \mathbf{v}$, $\|\mathbf{u}\|$, the angle between $\mathbf{u}$ and $\mathbf{v}$, $\text{proj}_{\mathbf{v}}\, \mathbf{u}$, and $d(X, Y)$.

Exercise 3   Let $\mathbf{u} = (3, 2, -1)$ and $\mathbf{v} = (-5, 9, 4)$. Compute $\|\mathbf{u}\|$, $\|\mathbf{v}\|$, and $\|\mathbf{u} + \mathbf{v}\|$, using MATLAB, and verify that the triangle inequality holds.

Exercise 4   Use MATLAB to investigate the following matrix A. Determine as many properties as you can. (*Hint:* Look at the vectors that make up the rows of A, det(A), etc.) Construct two other matrices, one 2×2, the other 3×3, that have the same properties as A.

$$A = \begin{bmatrix} \dfrac{1}{\sqrt{2}} & \dfrac{1}{\sqrt{2}} \\ -\dfrac{1}{\sqrt{2}} & \dfrac{1}{\sqrt{2}} \end{bmatrix}$$

M.16   Gram-Schmidt Orthogonalization (Section 5.2)

The Gram-Schmidt orthogonalization process provides a way in which an orthonormal basis for a vector space or subspace can be constructed from a given set of vectors.

Example 1   Find an orthonormal basis for the vector space spanned by the set of vectors $\{(1, 2, 2), (3, 2, 0),$ and $(0, 4, 2)\}$.

»Gram   {author-defined function for
Gram-Schmidt Orthogonalization   G/S orthogonalization}

Number of vectors to be orthonormalized: 3
Number of elements in each vector: 3

Vector 1:   {enter the vectors}
[1   2   2]
          :

All steps? y/n: y   {select "All steps" option}
[press return at each step to continue]
Initial vectors:
Vector 1   {displays original vectors}
  1   2   2
          :

Perform orthogonalization

Orthogonalizing vector 1
Orthogonal basis vector   {gives orthogonal vectors}
  1   2   2
          :

Normalizing vectors

Normalize vector 1   {gives orthonormal set}
  0.3333   0.6667   0.6667
          :

Orthonormal basis
  0.3333   0.6667   0.6667
  0.8085   0.1617   −0.5659
  −0.4851   0.7276   −0.4851

When the given vectors are not linearly independent, the routine **Gram** will warn you of this fact and find an orthonormal basis for the subspace that is spanned by the vectors.

Exercise 1   Find orthonormal bases for the spaces spanned by the following sets.

*(a) $\{(1, 2, 0), (1, 0, 1), (0, 2, -1)\}$ (b) $\{(4, 2, 1), (1, 5, 2), (3, 10, 6)\}$

M.17   Cross Product (Section 5.3)

Let $\mathbf{A} = (a_1, a_2, a_3)$ and $\mathbf{B} = (b_1, b_2, b_3)$. The cross product of $\mathbf{A}$ and $\mathbf{B}$ is

$$\mathbf{A} \times \mathbf{B} = (a_2 b_3 - b_2 a_3, b_1 a_3 - a_1 b_3, a_1 b_2 - b_1 a_2)$$

The cross product of $\mathbf{A}$ and $\mathbf{B}$ is also given by the following determinant.

$$\mathbf{A} \times \mathbf{B} = \begin{vmatrix} \mathbf{i} & \mathbf{j} & \mathbf{k} \\ a_1 & a_2 & a_3 \\ b_1 & b_2 & b_3 \end{vmatrix}$$

Here $\mathbf{i}$, $\mathbf{j}$, and $\mathbf{k}$ are the unit vectors in the x, y, and z directions.

Example 1   Find the cross product of $(1, 2, 1)$ and $(1, -2, 1)$.

```
»a = [1   2   1];   b = [1   −2   1];        {enter the vectors}
»c = cross(a, b)                             {author-defined function}
c =
    4    0   −4
```
The cross product of the vectors is $(4, 0, -4)$.

Exercise 1   Find the cross product of

*(a) $(3, 4, 0), (5, -2, 9)$ (b) $(3, -1, 2), (6, 8, -3)$.

Triple Scalar Product

Let $\mathbf{A} = (a_1, a_2, a_3)$, $\mathbf{B} = (b_1, b_2, b_3)$, and $\mathbf{C} = (c_1, c_2, c_3)$. The triple scalar product of $\mathbf{A}$, $\mathbf{B}$, and $\mathbf{C}$ is

$$\mathbf{A} \cdot (\mathbf{B} \times \mathbf{C}) = \begin{vmatrix} a_1 & a_2 & a_3 \\ b_1 & b_2 & b_3 \\ c_1 & c_2 & c_3 \end{vmatrix}$$

Example 2   Find the triple scalar product of $(1, 2, 1)$, $(1, -2, 1)$, and $(3, 0, 6)$.

```
»a = [1   2   1];   b = [1   −2   1];   c = [3   0   6];    {enter the vectors}
»d = tdot(a, b, c)                                          {author-defined function}
d =
   −12
```
The triple scalar product is -12.

Exercise 2   Let $\mathbf{A} = (3, 6, 0)$, $\mathbf{B} = (0, 1, 5)$, and $\mathbf{C} = (2, -2, 6)$. Find

*(a) $\mathbf{A} \cdot (\mathbf{B} \times \mathbf{C})$ (b) $\mathbf{A} \cdot (\mathbf{A} \times \mathbf{B})$ (c) $\mathbf{A} \cdot (\mathbf{B} \times \mathbf{B})$

6

c h a p t e r

General Vector Spaces

In this chapter we generalize the concept of a vector space. We discuss the underlying algebraic structure of the vector space $\mathbf{R}^n$. Any set with this structure has the same mathematical properties as $\mathbf{R}^n$ and will be called a vector space. The results that were developed for the vector space $\mathbf{R}^n$ will also apply to such vector spaces. We shall, for example, find that certain spaces of functions have the same mathematical properties as the vector space $\mathbf{R}^n$.

6.1 General Vector Spaces

The vector space $\mathbf{R}^n$ is a set of elements called vectors, on which two operations, namely addition and scalar multiplication, have been defined. We have seen that the vector space $\mathbf{R}^n$ is closed under these operations; the sum of two vectors in $\mathbf{R}^n$ lies in $\mathbf{R}^n$, and the scalar multiple of a vector in $\mathbf{R}^n$ also lies in $\mathbf{R}^n$. The vector space $\mathbf{R}^n$ also has other algebraic properties. For example, we have seen that vectors in $\mathbf{R}^n$ are commutative and associative under addition:

$$\mathbf{u} + \mathbf{v} = \mathbf{v} + \mathbf{u}$$
$$\mathbf{u} + (\mathbf{v} + \mathbf{w}) = (\mathbf{u} + \mathbf{v}) + \mathbf{w}$$

Our aim in this section will be to focus on these and other algebraic properties of $\mathbf{R}^n$. We shall draw up a set of **axioms** based on the properties of $\mathbf{R}^n$. Any set that satisfies these axioms will have similar algebraic properties to the vector space $\mathbf{R}^n$. Such a set will be called a **vector space** and its elements will be called **vectors**. The strength of this approach lies in the fact that the concepts and results developed for the vector space $\mathbf{R}^n$ will apply to other vector spaces.

We now give the formal definition of a vector space, based on the algebraic properties of $\mathbf{R}^n$ that were discussed in Section 5.1. Observe that the axioms can be separated into three convenient groups.

Definition   *A **vector space** is a set V of elements called **vectors**, having operations of addition and scalar multiplication defined on it that satisfy the following conditions. (**u**, **v**, and **w** are arbitrary elements of V, and c and d are scalars.)*

Closure Axioms

    1. The sum $\mathbf{u} + \mathbf{v}$ exists and is an element of V. (closed under addition)

    *2. c**u** is an element of V. (closed under scalar multiplication)*

Addition Axioms

    3. $\mathbf{u} + \mathbf{v} = \mathbf{v} + \mathbf{u}$ (commutative property)

    4. $\mathbf{u} + (\mathbf{v} + \mathbf{w}) = (\mathbf{u} + \mathbf{v}) + \mathbf{w}$ (associative property)

    *5. There exists an element of V, called the **zero vector**, denoted $\mathbf{0}$, such that $\mathbf{u} + \mathbf{0} = \mathbf{u}$.*

    *6. For every element **u** of V, there exists an element called the **negative** of **u**, denoted $-\mathbf{u}$, such that $\mathbf{u} + (-\mathbf{u}) = \mathbf{0}$.*

Scalar Multiplication Axioms

    7. $c(\mathbf{u} + \mathbf{v}) = c\mathbf{u} + c\mathbf{v}$

    8. $(c + d)\mathbf{u} = c\mathbf{u} + d\mathbf{u}$

    9. $(cd)\mathbf{u} = c(d\mathbf{u})$

    10. $1\mathbf{u} = \mathbf{u}$

The two most common sets of scalars used in vector spaces are the **set of real numbers** and the **set of complex numbers**. The vector spaces are then called *real vector spaces* and *complex vector spaces*. We shall focus primarily on real vector spaces, mentioning briefly, when appropriate, results for complex vector spaces. We now give examples of vector spaces.

Vector Spaces of Matrices

Consider the set of real 2×2 matrices. Denote this set M_{22}. We have defined operations of addition and scalar multiplication on this set. The set does in fact form a vector space. We shall discuss axioms 1, 3, 4, 5, and 6, leaving the remaining axioms for you to check.

Use vector notation for the elements of M_{22}. Let

$$\mathbf{u} = \begin{bmatrix} a & b \\ c & d \end{bmatrix} \quad and \quad \mathbf{v} = \begin{bmatrix} e & f \\ g & h \end{bmatrix}$$

be two arbitrary 2×2 matrices. We get

Axiom 1:

$$\mathbf{u} + \mathbf{v} = \begin{bmatrix} a & b \\ c & d \end{bmatrix} + \begin{bmatrix} e & f \\ g & h \end{bmatrix} = \begin{bmatrix} a + e & b + f \\ c + g & d + h \end{bmatrix}$$

$\mathbf{u} + \mathbf{v}$ is a 2×2 matrix. Thus M_{22} is closed under addition.

Axioms 3 and 4:

From our previous discussions we know that 2×2 matrices are commutative and associative under addition (Theorem 2.2).

Axiom 5:

The 2×2 zero matrix is $\mathbf{0} = \begin{bmatrix} 0 & 0 \\ 0 & 0 \end{bmatrix}$, since

$$\mathbf{u} + \mathbf{0} = \begin{bmatrix} a & b \\ c & d \end{bmatrix} + \begin{bmatrix} 0 & 0 \\ 0 & 0 \end{bmatrix} = \begin{bmatrix} a & b \\ c & d \end{bmatrix} = \mathbf{u}$$

Axiom 6:

If $\mathbf{u} = \begin{bmatrix} a & b \\ c & d \end{bmatrix}$, then $-\mathbf{u} = \begin{bmatrix} -a & -b \\ -c & -d \end{bmatrix}$, since

$$\mathbf{u} + (-\mathbf{u}) = \begin{bmatrix} a & b \\ c & d \end{bmatrix} + \begin{bmatrix} -a & -b \\ -c & -d \end{bmatrix} = \begin{bmatrix} a - a & b - b \\ c - c & d - d \end{bmatrix} = \begin{bmatrix} 0 & 0 \\ 0 & 0 \end{bmatrix} = \mathbf{0}$$

The set M_{22} of 2×2 matrices is a vector space. The algebraic properties of M_{22} are similiar to those of $\mathbf{R}^n$. Similarly, we get

M_{mn}, the set of $\mathbf{m} \times \mathbf{n}$ matrices, is a vector space.

Vector Spaces of Functions

Let V be the set of all functions having the real numbers as their domain. Each element of V, such as f, will map the real line into the real line. We shall now introduce operations of addition and scalar multiplication on V that will make it into a vector space.

Let f and g be arbitrary elements of V. Define their sum, $f + g$, to be the function such that

$$(f + g)(x) = f(x) + g(x)$$

This defines $f + g$ to be a function with domain the set of real numbers. To find the value of $f + g$ for any real number x, we add the value of f at x and the value of g at x. This operation is called **pointwise addition**.

We next define scalar multiplication on the elements of V. Let c be an arbitrary scalar. The scalar multiple of f, cf is the function such that

$$(cf)(x) = c[f(x)]$$

This defines c to be a function with domain the set of real numbers. To find the value of cf for any real number x, multiply the value of f at x by c. This operation is called **pointwise scalar multiplication**.

To get a geometrical feel for these two operations on functions, let us consider two specific functions:

$$f(x) = x \quad \text{and} \quad g(x) = x^2$$

Then $f + g$ is the function defined by

$$(f + g)(x) = x + x^2$$

The scalar multiple of f by a scalar such as 3 is the function $3f$ defined by

$$(3f)(x) = 3x$$

Having defined operations of addition and scalar multiplication on this function space V, let us now check that V is a vector space. We shall look at axioms 1, 2, 5, and 6, leaving the remaining axioms for you to check.

Axiom 1:

$f + g$ is defined by $(f + g)(x) = f(x) + g(x)$. $f + g$ is thus a function with domain the set of real numbers. $f + g$ is an element of V. Thus V is closed under addition.

Axiom 2:

cf is defined by $(cf) = c[f(x)]$. cf is thus a function with domain the set of real numbers. cf is an element of V. Thus V is closed under scalar multiplication.

Axiom 5:

Let $\mathbf{0}$ be the function such that $\mathbf{0}(x) = 0$ for every real number x. $\mathbf{0}$ is called the **zero function**. We get

$$(f + \mathbf{0})(x) = f(x) + \mathbf{0}(x) = f(x) + 0 = f(x) \qquad \text{for every real number } x$$

The value of the function $f + \mathbf{0}$ is the same as the value of f at every x. Thus $f + \mathbf{0} = f$. $\mathbf{0}$ is the **zero vector**.

Axiom 6:

Consider the function $-f$ defined by $(-f)(x) = -[f(x)]$. We shall show that $-f$ is the negative of f.

$$[f + (-f)](x) = f(x) + (-f)(x)$$
$$= f(x) - [f(x)]$$
$$= 0$$
$$= \mathbf{0}(x)$$

The value of the function $[f + (-f)]$ is the same as the value of $\mathbf{0}$ at every x. Thus $[f + (-f)] = \mathbf{0}$. Therefore $-f$ is the negative of f.

There are many vector spaces of functions. For example, the set of all functions having domain $[-\pi, \pi]$ with operations of pointwise addition and pointwise scalar multiplication is a vector space. You will investigate other function vector spaces in the exercises that follow.

*The Complex Vector Space $\mathbf{C}^n$

We now extend the concept of the real vector space $\mathbf{R}^n$ to a complex vector space $\mathbf{C}^n$. Let $(u_1, \ldots, u_n)$ be a sequence of n complex numbers. The set of all such sequences is denoted $\mathbf{C}^n$. For example, $(2 + 3i, 4 - 6i)$ is an element of $\mathbf{C}^2$, while $(3i, 1 - 5i, 2)$ is an element of $\mathbf{C}^3$.

Let operations of addition and scalar multiplication (by a complex scalar c) be defined on $\mathbf{C}^n$ as follows:

$$(u_1, \ldots, u_n) + (v_1, \ldots, v_n) = (u_1 + v_1, \ldots, u_n + v_n)$$
$$c(u_1, \ldots, u_n) = (cu_1, \ldots, cu_n)$$

For example, consider the two vectors $\mathbf{u} = (2 + i, 3 - 4i)$ and $\mathbf{v} = (1 - 3i, 5 + 3i)$ of $\mathbf{C}^2$ and the scalar $c = 4 + 3i$. Then

$$\mathbf{u} + \mathbf{v} = (2 + i, 3 - 4i) + (1 - 3i, 5 + 3i) = (3 - 2i, 8 - i)$$
$$c\mathbf{u} = (4 + 3i)(2 + i, 3 - 4i) = (5 + 10i, 24 - 7i)$$

It can be shown that $\mathbf{C}^n$ with these two operations is a complex vector space (see the following exercises).

We now give a theorem that contains useful properties of vectors. These are properties that were immediately apparent for $\mathbf{R}^n$ and were taken almost for granted. They are not, however, so apparent for all vector spaces.

Theorem 6.1   Let V be a vector space, $\mathbf{v}$ a vector in V, $\mathbf{0}$ the zero vector of V, c a scalar, and 0 the zero scalar. Then

(a) $0\mathbf{v} = \mathbf{0}$

(b) $c\mathbf{0} = \mathbf{0}$

(c) $(-1)\mathbf{v} = -\mathbf{v}$

(d) If $c\mathbf{0} = \mathbf{0}$, then either $c = 0$ or $\mathbf{v} = \mathbf{0}$.

Proof   We shall prove (a) and (c), leaving the proofs of (b) and (d) for you to complete in the exercises that follow.

(a) $0\mathbf{v} + 0\mathbf{v} = (0 + 0)\mathbf{v}$ (axiom 8)

$\qquad\qquad = 0\mathbf{v}$ (property of the scalar 0)

Add the negative of $0\mathbf{v}$, namely $-0\mathbf{v}$, to both sides of this equation.

$$(0\mathbf{v} + 0\mathbf{v}) + (-0\mathbf{v}) = 0\mathbf{v} + (-0\mathbf{v})$$

$$0\mathbf{v} + [(0\mathbf{v} + (-0\mathbf{v})] = \mathbf{0} \quad \text{(axioms 4 and 6)}$$

$$0\mathbf{v} + \mathbf{0} = \mathbf{0} \quad \text{(axiom 6)}$$

$$0\mathbf{v} = \mathbf{0} \quad \text{(axiom 5)}$$

(c) $(-1)\mathbf{v} + \mathbf{v} = (-1)\mathbf{v} + 1\mathbf{v}$ (axiom 10)

$\qquad\qquad = [(-1) + 1]\mathbf{v}$ (axiom 8)

$\qquad\qquad = 0\mathbf{v}$ (property of scalar 0)

$\qquad\qquad = \mathbf{0}$ (part (a) above)

Thus $(-1)\mathbf{v}$ is the negative of $\mathbf{v}$ (axiom 6).

Subspaces

We have seen how certain subsets of the vector space $\mathbf{R}^n$ that were closed under addition and under scalar multiplication form vector spaces in their own right. Such subsets were called subspaces. We now extend this concept to general vector spaces.

Definition   *Let V be a vector space and U be a nonempty subset of V. U is said to be a **subspace** of V if it is closed under addition and under scalar multiplication.*

To see that such a subset U is a vector space, observe that U will satisfy the closure axioms 1 and 2 from the above definition and that it will "inherit" the requirements of axioms 3 through 10 from the larger space V.

A subspace is a vector space embedded in a larger vector space.

Example 1   Prove that the set U of 2×2 diagonal matrices is a subspace of the vector space M_{22} of 2×2 matrices.

Solution   We have to show that U is closed under addition and under scalar multiplication. Consider the following two elements of U.

$$\mathbf{u} = \begin{bmatrix} a & 0 \\ 0 & b \end{bmatrix} \quad \text{and} \quad \mathbf{v} = \begin{bmatrix} p & 0 \\ 0 & q \end{bmatrix}$$

We get

$$\mathbf{u} + \mathbf{v} = \begin{bmatrix} a & 0 \\ 0 & b \end{bmatrix} + \begin{bmatrix} p & 0 \\ 0 & q \end{bmatrix} = \begin{bmatrix} a + p & 0 \\ 0 & b + q \end{bmatrix}$$

Observe that $\mathbf{u} + \mathbf{v}$ is a 2×2 diagonal matrix and is thus an element of U. U is closed under addition.

Let c be a scalar. We get

$$c\mathbf{u} = c\begin{bmatrix} a & 0 \\ 0 & b \end{bmatrix} = \begin{bmatrix} ca & 0 \\ 0 & cb \end{bmatrix}$$

$c\mathbf{u}$ is a 2×2 diagonal matrix. Thus U is closed under scalar multiplication.

U is a subspace of M_{22}. It is a vector space of matrices, embedded in M_{22}.

Example 2   Let P_n denote the set of real polynomial functions of degree $\leq n$. Prove that P_n is a vector space if addition and scalar multiplication are defined on polynomials in a pointwise manner.

Solution   We could prove that P_n, with these two operations, satisfies all the axioms of a vector space. There is, however, an easier way. P_n is a subset of the vector space V of functions having domain the set of real numbers. Let us show that this subset is a subspace of V; it will then be a vector space in its own right. We show that P_n is closed under addition and scalar multiplication.

Let f and g be two elements of P_n defined by

$$f(x) = a_n x^n + a_{n-1} x^{n-1} + \cdots + a_1 x + a_0$$

$$\text{and}$$

$$g(x) = b_n x^n + b_{n-1} x^{n-1} + \cdots + b_1 x + b_0$$

We first consider addition. $f + g$ is defined by

$$\begin{aligned} (f + g)(x) &= f(x) + g(x) \\ &= [a_n x^n + a_{n-1} x^{n-1} + \cdots + a_1 x + a_0] \\ &\quad + [b_n x^n + b_{n-1} x^{n-1} + \cdots + b_1 x + b_0] \\ &= (a_n + b_n)x^n + (a_{n-1} + b_{n-1})x^{n-1} \\ &\quad + \cdots + (a_1 + b_1)x + (a_0 + b_0) \end{aligned}$$

$(f + g)(x)$ is a polynomial of degree $\leq n$. Thus $f + g$ is an element of P_n. P_n is closed under addition.

We now look at scalar multiplication. cf is defined by

$$(cf)(x) = c[f(x)]$$
$$= c[a_n x^n + a_{n-1} x^{n-1} + \cdots + a_1 x + a_0]$$
$$= ca_n x^n + ca_{n-1} x^{n-1} + \cdots + ca_1 x + ca_0$$

$(cf)(x)$ is a polynomial of degree $\leq n$. Thus cf is an element of P_n. P_n is closed under scalar multiplication.

We have proved that P_n, a subset of the vector space V of functions, is closed under addition and scalar multiplication. It is thus a subspace of V and therefore a vector space.

We now extend the concepts of linear combinations, spanning sets, linear dependence/independence, basis, and dimension to general vector spaces.

A vector $\mathbf{v}$ in a vector space V is said to be a **linear combination** of the vectors $\mathbf{v}_1, \mathbf{v}_2, \ldots, \mathbf{v}_m$ if there exist scalars $a_1, a_2, \ldots, a_m$ such that

$$\mathbf{v} = a_1 \mathbf{v}_1 + a_2 \mathbf{v}_2 + \cdots + a_m \mathbf{v}_m$$

Example 3   Determine whether the matrix $\begin{bmatrix} -1 & 7 \\ 8 & -1 \end{bmatrix}$ is a linear combination of the matrices $\begin{bmatrix} 1 & 0 \\ 2 & 1 \end{bmatrix}, \begin{bmatrix} 2 & -3 \\ 0 & 2 \end{bmatrix}$, and $\begin{bmatrix} 0 & 1 \\ 2 & 0 \end{bmatrix}$.

Solution   In this example we are looking at the vector space M_{22} of 2×2 matrices. The given matrix will be a linear combination of the other three matrices if there exist scalars a_1, a_2, a_3 such that we can write

$$\begin{bmatrix} -1 & 7 \\ 8 & -1 \end{bmatrix} = a_1 \begin{bmatrix} 1 & 0 \\ 2 & 1 \end{bmatrix} + a_2 \begin{bmatrix} 2 & -3 \\ 0 & 2 \end{bmatrix} + a_3 \begin{bmatrix} 0 & 1 \\ 2 & 0 \end{bmatrix}$$

This gives

$$\begin{bmatrix} -1 & 7 \\ 8 & -1 \end{bmatrix} = \begin{bmatrix} a_1 + 2a_2 & -3a_2 + a_3 \\ 2a_1 + 2a_3 & a_1 + 2a_2 \end{bmatrix}$$

On equating corresponding elements, we get the following system of linear equations.

$$\begin{aligned} a_1 + 2a_2 \quad\quad\quad &= -1 \\ -3a_2 + a_3 &= 7 \\ 2a_1 \quad\quad + 2a_3 &= 8 \\ a_1 + 2a_2 \quad\quad\quad &= -1 \end{aligned}$$

This system can be shown to have a unique solution $a_1 = 3$, $a_2 = -2$, $a_3 = 1$. The given matrix is thus the following linear combination of the other three matrices.

$$\begin{bmatrix} -1 & 7 \\ 8 & -1 \end{bmatrix} = 3\begin{bmatrix} 1 & 0 \\ 2 & 1 \end{bmatrix} - 2\begin{bmatrix} 2 & -3 \\ 0 & 2 \end{bmatrix} + \begin{bmatrix} 0 & 1 \\ 2 & 0 \end{bmatrix}$$

If it had turned out that the preceding system of equations had no solution, then the given matrix would not have been a linear combination of the other matrices.

≡≡≡

The vectors $\mathbf{v}_1$, $\mathbf{v}_2$, . . . , $\mathbf{v}_m$ **span** a vector space V if every vector in V can be expressed as a linear combination of these vectors.

Example 4   Show that the following matrices span the vector space M_{22} of 2×2 matrices.

$$\begin{bmatrix} 1 & 0 \\ 0 & 0 \end{bmatrix}, \quad \begin{bmatrix} 0 & 1 \\ 0 & 0 \end{bmatrix}, \quad \begin{bmatrix} 0 & 0 \\ 1 & 0 \end{bmatrix}, \quad \begin{bmatrix} 0 & 0 \\ 0 & 1 \end{bmatrix}$$

Solution   Let $\begin{bmatrix} a & b \\ c & d \end{bmatrix}$ be an arbitrary element of M_{22}. We can express this matrix as follows:

$$\begin{bmatrix} a & b \\ c & d \end{bmatrix} = a\begin{bmatrix} 1 & 0 \\ 0 & 0 \end{bmatrix} + b\begin{bmatrix} 0 & 1 \\ 0 & 0 \end{bmatrix} + c\begin{bmatrix} 0 & 0 \\ 1 & 0 \end{bmatrix} + d\begin{bmatrix} 0 & 0 \\ 0 & 1 \end{bmatrix}$$

proving the result.

≡≡≡

The set of vectors $\{\mathbf{v}_1, \mathbf{v}_2, \ldots, \mathbf{v}_m\}$ is **linearly dependent** if there exist scalars $a_1, a_2, \ldots, a_m$, not all zero, such that

$$a_1\mathbf{v}_1 + a_2\mathbf{v}_2 + \cdots + a_m\mathbf{v}_m = 0$$

If the set is not linearly dependent, it is **linearly independent**.

Example 5   Consider the functions $f(x) = x^2 + 1$, $g(x) = 3x - 1$, $h(x) = -4x + 1$ of the vector space P_2 of polynomials of degree ≤ 2. Show that $\{f, g, h\}$ is linearly independent.

Solution   Let us examine the identity

$$a_1 f + a_2 g + a_3 h = \mathbf{0}$$

$\{f, g, h\}$ will be linearly independent if we can show that this identity implies that $a_1 = 0$, $a_2 = 0$, $a_3 = 0$. We can write the identity as

$$a_1(x^2 + 1) + a_2(3x - 1) + a_3(-4x + 1) = 0$$

where x is any real number. Consider three convenient values of x. We get

$$\begin{aligned} x = \quad 0: \quad & a_1 - a_2 + a_3 = 0 \\ x = \quad 1: \quad & 2a_1 + 2a_2 + 3a_3 = 0 \\ x = -1: \quad & 2a_1 - 4a_2 + 5a_3 = 0 \end{aligned}$$

It can be shown that this system of three equations has the unique solution

$$a_1 = 0, \qquad a_2 = 0, \qquad a_3 = 0$$

Thus $a_1 f + a_2 g + a_3 h = \mathbf{0}$ implies that $a_1 = 0$, $a_2 = 0$, $a_3 = 0$. $\{f, g, h\}$ is linearly independent.

The set $\{\mathbf{v}_1, \mathbf{v}_2, \ldots, \mathbf{v}_m\}$ is a **basis** for a vector space V if the vectors span V and are linearly independent.

Example 6   Show that $\{f, g, h\}$, where $f(x) = x^2 + 1$, $g(x) = 3x - 1$, and $h(x) = -4x + 1$ is a basis for P_2.

Solution $\{f, g, h\}$ will form a basis for P_2 if these functions span P_2 and are linearly independent. We have seen in the previous example that the functions are linearly independent. It thus remains to show that they span P_2. Let p be an arbitrary function in P_2. p is thus a polynomial of the form

$$p(x) = bx^2 + cx + d$$

The functions f, g, and h will span P_2 if there exist scalars a_1, a_2, a_3 such that we can write

$$p(x) = a_1 f(x) + a_2 g(x) + a_3 h(x)$$

This gives

$$bx^2 + cx + d = a_1(x^2 + 1) + a_2(3x - 1) + a_3(-4x + 1)$$
$$= a_1 x^2 + (3a_2 - 4a_3)x + (a_1 - a_2 + a_3)$$

Comparing the coefficients leads to the following system of equations.

$$
\begin{aligned}
a_1 & & & = b \\
& 3a_2 & - 4a_3 & = c \\
a_1 & - a_2 & + a_3 & = d
\end{aligned}
$$

It can be shown that this system of equations has solution

$$a_1 = b, \qquad a_2 = 4b - 4d - c, \qquad a_3 = 3b - 3d - c$$

Thus the polynomial p can be expressed

$$p(x) = a_1 f(x) + a_2 g(x) + a_3 h(x)$$

The functions f, g, and h span P_2.

These functions span P_2 and are linearly independent. They form a basis for P_2.

We are now in a position to appreciate the full power of the approach we have taken, namely that of gradually generalizing the algebraic concepts of the vector space $\mathbf{R}^n$. Theorems that were developed for $\mathbf{R}^n$ hold also for more general vector spaces since the algebraic concepts are identical. To illustrate the point, we now focus on one important theorem and the definition that arises from that theorem.

Any two bases for a given vector space have the same number of vectors. *The **dimension** of a vector space is the number of vectors in a basis.*

We are now able to talk about dimensions of spaces of matrices and dimensions of certain spaces of functions in the same way as we can talk about the dimension of $\mathbf{R}^n$.

For example, consider the vector space M_{22} of 2×2 matrices. The matrices

$$\begin{bmatrix} 1 & 0 \\ 0 & 0 \end{bmatrix}, \quad \begin{bmatrix} 0 & 1 \\ 0 & 0 \end{bmatrix}, \quad \begin{bmatrix} 0 & 0 \\ 1 & 0 \end{bmatrix}, \quad \begin{bmatrix} 0 & 0 \\ 0 & 1 \end{bmatrix}$$

span M_{22} and are linearly independent. They thus form a basis for M_{22}. The dimension of the vector space M_{22} is 4. Similarly, the dimension of the vector space M_{mn} is mn.

Consider the vector space of polynomials of degree ≤ 2, P_2. The functions $x^2, x, 1$ span P_2 since any polynomial $ax^2 + bx + c$ can be written

$$ax^2 + bx + c = a(x^2) + b(x) + c(1)$$

$x^2, x, 1$ are also linearly independent since

$$ax^2 + bx + c = 0 \qquad \text{for all values of } x$$

implies that $a = 0$, $b = 0$, $c = 0$. $\{x^2, x, 1\}$ is thus a basis for P_2. The dimension of the vector space P_2 is 3. Similarly, the set $\{x^n, x^{n-1}, \ldots, x, 1\}$ is a basis for P_n and the dimension of P_n is $n + 1$. This basis is called the **standard basis** for P_n.

Consider the vector space $\mathbf{C}^2$. The vectors $(1, 0)$ and $(0, 1)$ span $\mathbf{C}^2$ since an arbitrary vector $(a + bi, c + d)$ can be written

$$(a + bi, c + di) = (a + bi)(1, 0) + (c + di)(0, 1)$$

$(1, 0)$ and $(0, 1)$ are also linearly independent in $\mathbf{C}^2$. Thus $\{(1, 0), (0, 1)\}$ is a basis for $\mathbf{C}^2$, and the dimension of $\mathbf{C}^2$ is 2. Similarly, $\{(1, \ldots, 0), \ldots, (0, \ldots, 1)\}$ is a basis for $\mathbf{C}^n$, and its dimension is n. This basis is the **standard basis** for $\mathbf{C}^n$.

We mention at this time that vector spaces can be **infinite-dimensional**; there may not be a finite set of vectors that forms a basis. The set of functions having as their domain the real numbers is such a vector space. We shall be primarily interested in finite dimensional vector spaces in this course.

Exercise Set 6.1

*1. We have discussed the fact that M_{22}, the set of 2×2 matrices with the usual operations of matrix addition and scalar multiplication, satisfies axioms 1, 3, 4, 5, and 6 of a vector space. Prove that M_{22} also satisfies the remaining vector space axioms, completing the proof that M_{22} is a vector space.

2. *(a) Let $f(x) = x + 2$ and $g(x) = x^2 - 1$. Compute the functions $f + g$, $2f$, and $3g$.

  (b) Let $f(x) = 2x$ and $g(x) = 4 - 2x$. Compute $f + g$, $3f$, and $-g$.

*3. We have shown that the set of functions with operations of pointwise addition and scalar multiplication, with domain the set of real numbers, satisfies axioms 1, 2, 5, and 6 of a vector space. Prove that this set also satisfies the remaining vector space axioms, completing the proof that this set is a vector space.

4. Prove that the set C^n with the operations of addition and scalar multiplication defined as follows is a vector space.

$$(u_1, \dots, u_n) + (v_1, \dots, v_n) = (u_1 + v_1, \dots, u_n + v_n)$$

$$c(u_1, \dots, u_n) = (cu_1, \dots, cu_n)$$

Determine $\mathbf{u} + \mathbf{v}$ and $c\mathbf{u}$ for the following vectors and scalars in C^2.

*(a) $\mathbf{u} = (2 - i, 3 + 4i)$, $\mathbf{v} = (5, 1 + 3i)$, $c = 3 - 2i$

(b) $\mathbf{u} = (1 + 5i, -2 - 3i)$, $\mathbf{v} = (2i, 3 - 2i)$,
$c = 4 + i$

5. (a) Consider the set of all continuous functions with operations of pointwise addition and scalar multiplication, having domain [0, 1]. Is this set a vector space?

*(b) Consider the set of all discontinuous functions with operations of pointwise addition and scalar multiplication, having domain [0, 1]. Is this set a vector space?

6. Determine which of the following subsets of M_{22} form subspaces.

*(a) The subset having diagonal elements zero

(b) The subset consisting of matrices the sum of whose elements is 6 $\left(\text{For example,} \begin{bmatrix} 2 & -1 \\ 0 & 5 \end{bmatrix} \text{would be such a matrix.} \right)$

*(c) The subset of matrices of the form $\begin{bmatrix} a & a^2 \\ b & b^2 \end{bmatrix}$

(d) The subset of matrices of the form $\begin{bmatrix} a & a + 2 \\ b & c \end{bmatrix}$

(e) The subset of 2×2 symmetric matrices

*(f) The subset of 2×2 invertible matrices

7. Which of the following subsets of M_{23} form subspaces?

*(a) The subset of matrices of the form $\begin{bmatrix} a & b & 0 \\ c & d & 0 \end{bmatrix}$

(b) The subset of matrices of the form $\begin{bmatrix} a & 2a & 3a \\ b & 2b & 3b \end{bmatrix}$

*8. P_3 is the vector space of polynomials of degree ≤ 3, and P_2 is the vector space of polynomials of degree ≤ 2. Prove that P_2 is a subspace of P_3.

9. Consider the set of polynomials of degree 2. Prove that this set is not closed under addition or under scalar multiplication. It is thus not a vector space.

*10. Let S be the set of all functions of the form $f(x) = ax^2 + bx + 3$, where a and b are real numbers. Is S a subspace of P_2?

*11. Is R^n a subspace of C^n?

12. Prove that every subspace of a vector space V has to contain the zero vector of V. This is often a quick way of deciding that some subsets cannot be subspaces. Use this criterion to prove the following:

*(a) The set of matrices of the form $\begin{bmatrix} a & 1 \\ b & c \end{bmatrix}$ is not a subspace of M_{22}.

(b) The set of all functions of the form $f(x) = ax + 2$ is not a subspace of P_2.

13. Consider the vector space of functions defined on the set of real numbers. Which of the following are subspaces of this vector space?

*(a) The subset consisting of all functions f such that $f(0) = 0$

*(b) The subset consisting of all functions f such that $f(0) = 3$

(c) The subset of all constant functions

14. Let V be a vector space $\mathbf{u}$, $\mathbf{v}$ and $\mathbf{w}$ be vectors in V, and a, b, and c be scalars. Use the axioms of a vector space to prove the following.

*(a) $c\mathbf{0} = \mathbf{0}$

(b) If $c\mathbf{v} = \mathbf{0}$, then either $c = 0$ or $\mathbf{v} = \mathbf{0}$.

(c) $-(-\mathbf{v}) = \mathbf{v}$

(d) If $\mathbf{u} + \mathbf{w} = \mathbf{v} + \mathbf{w}$, then $\mathbf{u} = \mathbf{v}$.

(e) If $\mathbf{u} \ne \mathbf{0}$ and $a\mathbf{u} = b\mathbf{u}$, then $a = b$.

15. In each of the following, determine whether the first matrix is a linear combination of the matrices that follow.

*(a) $\begin{bmatrix} 5 & 7 \\ 5 & -10 \end{bmatrix}$; $\begin{bmatrix} 1 & 2 \\ 3 & -4 \end{bmatrix}$, $\begin{bmatrix} 0 & 3 \\ 1 & 2 \end{bmatrix}$, $\begin{bmatrix} 1 & 2 \\ 0 & 0 \end{bmatrix}$

(b) $\begin{bmatrix} 7 & 6 \\ -5 & -3 \end{bmatrix}$; $\begin{bmatrix} 3 & 0 \\ 1 & 1 \end{bmatrix}$, $\begin{bmatrix} 0 & 1 \\ 3 & 4 \end{bmatrix}$, $\begin{bmatrix} 1 & 2 \\ 0 & 1 \end{bmatrix}$

*(c) $\begin{bmatrix} 4 & 1 \\ 7 & 10 \end{bmatrix}$; $\begin{bmatrix} 1 & 1 \\ 1 & 1 \end{bmatrix}$, $\begin{bmatrix} 3 & 1 \\ 0 & 0 \end{bmatrix}$, $\begin{bmatrix} -1 & -1 \\ 2 & 3 \end{bmatrix}$

16. In each of the following, determine whether the first function is a linear combination of the functions that follow.

*(a) $f(x) = 3x^2 + 2x + 9$; $g(x) = x^2 + 1$, $h(x) = x + 3$

*(b) $f(x) = 2x^2 + x - 3$; $g(x) = x^2 - x + 1$,
$h(x) = x^2 + 2x - 2$

(c) $f(x) = x^2 + 4x + 5$; $g(x) = x^2 + x - 1$,
$h(x) = x^2 + 2x + 1$

17. Determine whether or not the following sets are linearly dependent.

*(a) $\{f, g, h\}$ where $f(x) = 2x^2 + 1$, $g(x) = x^2 + 4x$,
$h(x) = x^2 - 4x + 1$

(b) $\{f, g, h\}$ where $f(x) = x^2 + 3$, $g(x) = x + 1$,
$h(x) = 2x^2 - 3x + 3$

*(c) $\{f, g, h\}$ where $f(x) = x^2 + 3x - 1$, $g(x) = x + 3$,
$h(x) = 2x^2 - x + 1$

(d) $\left\{ \begin{bmatrix} 1 & 0 \\ 0 & 0 \end{bmatrix}, \begin{bmatrix} 0 & 2 \\ 0 & 0 \end{bmatrix}, \begin{bmatrix} 0 & 0 \\ 3 & 0 \end{bmatrix}, \begin{bmatrix} 0 & 0 \\ 0 & 4 \end{bmatrix} \right\}$

*(e) $\left\{ \begin{bmatrix} 1 & 2 \\ 3 & 1 \end{bmatrix}, \begin{bmatrix} 1 & 1 \\ 1 & 1 \end{bmatrix}, \begin{bmatrix} 2 & 1 \\ 4 & 2 \end{bmatrix} \right\}$

(f) $\left\{ \begin{bmatrix} 1 & 2 \\ -1 & 0 \end{bmatrix}, \begin{bmatrix} 1 & 2 \\ 1 & 1 \end{bmatrix}, \begin{bmatrix} 1 & 2 \\ 5 & 3 \end{bmatrix} \right\}$

18. Determine a basis for each of the following vector spaces and give the dimension of the space.

*(a) P_3 (b) M_{33} *(c) M_{23}

(d) The subspace of M_{22} consisting of all diagonal matrices

*(e) The subspace of M_{22} consisting of all symmetric matrices

19. (a) Is the function $f(x) = x + 5$ in the subspace spanned by $g(x) = x + 1$ and $h(x) = x + 3$?

*(b) Is the function $f(x) = 3x^2 + 5x + 1$ in the subspace spanned by $g(x) = 2x^2 + 3$ and $h(x) = x^2 + 3x - 1$?

(c) Give three other functions in the space spanned by $g(x) = 2x^2 + 3$ and $h(x) = x^2 + 3x - 1$.

*(d) Give a basis for the space spanned by $f(x) = 2x + 3$, $g(x) = x - 1$, and $h(x) = -x - 4$.

20. Are the following sets bases for the given vector spaces?

(a) $\{f, g, h\}$ where $f(x) = x^2 + 2x - 1$, $g(x) = x + 3$,
$h(x) = x^2 + 3x + 2$, for P_2

(b) $\{f, g, h\}$ where $f(x) = x^2 + x - 3$,
$g(x) = x^2 - x + 1$, $h(x) = x^2 + x + 1$, for P_2

*(c) $\left\{ \begin{bmatrix} 1 & 2 \\ 0 & 1 \end{bmatrix}, \begin{bmatrix} 3 & 4 \\ 1 & 1 \end{bmatrix}, \begin{bmatrix} 1 & 2 \\ 1 & 1 \end{bmatrix}, \begin{bmatrix} 0 & 2 \\ 1 & 2 \end{bmatrix} \right\}$, for M_{22}

(d) $\left\{ \begin{bmatrix} 1 & 0 \\ 0 & 0 \end{bmatrix}, \begin{bmatrix} 1 & 2 \\ 0 & 0 \end{bmatrix}, \begin{bmatrix} 1 & 2 \\ 3 & 0 \end{bmatrix}, \begin{bmatrix} 1 & 2 \\ 3 & 4 \end{bmatrix} \right\}$, for M_{22}

*(e) $\{(2 - 3i, 1 + 4i), (1 + i, 2)\}$, for $\mathbf{C}^2$

(f) $\{(1 + 2i, 3 - i, 1), (4 + i, 3i, 1 + i),$
$(-2 + 3i, 6 - 5i, 1 - i)\}$, for $\mathbf{C}^3$

*21. Let V be the vector space of functions defined on the set of real numbers. Let f, g, and h be elements of V that are twice differentiable. The function $W(x)$ defined below is called the **Wronskian** of f, g, and h. $f'(x)$ is the derivative of f, and $f''(x)$ is the second derivative, and so on. Prove that f, g, and h are linearly independent if the Wronskian is not the zero function. [If $W(x) = 0$, we cannot say anything about the linear dependence or independence of f, g, and h.] Readers who go on to study differential equations will make use of the Wronskian in that field.

$$W(x) = \begin{vmatrix} f(x) & g(x) & h(x) \\ f'(x) & g'(x) & h'(x) \\ f''(x) & g''(x) & h''(x) \end{vmatrix}$$

22. Use the Wronskian, introduced in the previous exercise, to show that the following functions are linearly independent.

(a) $1, x, x^2$ (b) $1, x, e^x$

*(c) $\sin x, \cos x, x \sin x$ (d) e^x, xe^x, x^2e^x

*23. If U is a subspace of a vector space V, and if $\mathbf{u}$ and $\mathbf{v}$ are elements of V, but one or both not in U, can $\mathbf{u} + \mathbf{v}$ be in U? Can $c\mathbf{u}$ be in U for some nonzero scalar c if $\mathbf{u}$ is not in U?

24. Prove that a necessary and sufficient condition for a subset U of a vector space V to be a subspace is that $a\mathbf{u} + b\mathbf{v}$ be in U for all scalars a and b and all vectors U.

25. Prove the following:

*(a) The union of two subspaces need not be a subspace.

(b) The intersection of two subspaces is a subspace.

6.2   Inner Product Spaces

In the previous section we generalized the vector space $\mathbf{R}^n$. We drew up a set of axioms based on the properties of $\mathbf{R}^n$, and any set that satisfied those axioms was called a vector space. Such a space had similar algebraic properties to $\mathbf{R}^n$. We now proceed one stage further in this process of generalization; we extend the concepts of dot product of two vectors, norm of a vector, angle between vectors, and distance between points to general vector spaces.

The dot product was a key concept on $\mathbf{R}^n$ that led to definitions of norm, angle, and distance. Our approach will be to generalize the dot product of $\mathbf{R}^n$ to a general vector space with a mathematical structure called an **inner product**. This will be used to define norm, angle, and distance for a general vector space. The following definition of inner product is based on the properties of the dot product given in Section 5.1.

Definition   *An inner product on a real vector space V is a function that associates a number, denoted $\langle \mathbf{u}, \mathbf{v} \rangle$, with each pair of vectors $\mathbf{u}$ and $\mathbf{v}$ of V. This function has to satisfy the following conditions for vectors $\mathbf{u}$, $\mathbf{v}$, and $\mathbf{w}$ and scalar c.*

1. $\langle \mathbf{u}, \mathbf{v} \rangle = \langle \mathbf{v}, \mathbf{u} \rangle$ *(symmetry axiom)*
2. $\langle \mathbf{u} + \mathbf{v}, \mathbf{w} \rangle = \langle \mathbf{u}, \mathbf{w} \rangle + \langle \mathbf{v}, \mathbf{w} \rangle$ *(additive axiom)*
3. $\langle c\mathbf{u}, \mathbf{v} \rangle = c\langle \mathbf{u}, \mathbf{v} \rangle$ *(homogeneity axiom)*
4. $\langle \mathbf{u}, \mathbf{u} \rangle \geq 0$, *and* $\langle \mathbf{u}, \mathbf{u} \rangle = 0$ *if and only if* $\mathbf{u} = \mathbf{0}$ *(positive definite axiom)*

A vector space V on which an inner product is defined is called an **inner product space**.

Any function on a vector space that satisfies the axioms of an inner product defines an inner product on that space. There can be many inner products on a given vector space. The dot product on $\mathbf{R}^2$ is an inner product on that space. The following example illustrates another inner product on $\mathbf{R}^2$.

Example 1   Let $\mathbf{u} = (x_1, x_2)$, $\mathbf{v} = (y_1, y_2)$, and $\mathbf{w} = (z_1, z_2)$ be arbitrary vectors in $\mathbf{R}^2$. Prove that $\langle \mathbf{u}, \mathbf{v} \rangle$ defined as follows is an inner product on $\mathbf{R}^2$.

$$\langle \mathbf{u}, \mathbf{v} \rangle = x_1 y_1 + 4x_2 y_2$$

Determine the inner product of the vectors $(-2, 5)$, $(3, 1)$ under this inner product.

Solution   We check each of the four inner product axioms. Use the algebraic properties of real numbers and vectors and apply the above inner product at appropriate times.

Axiom 1:
$$\begin{aligned}\langle \mathbf{u}, \mathbf{v} \rangle &= x_1 y_1 + 4x_2 y_2 \\ &= y_1 x_1 + 4y_2 x_2 \\ &= \langle \mathbf{v}, \mathbf{u} \rangle\end{aligned}$$

Axiom 2:
$$\langle \mathbf{u} + \mathbf{v}, \mathbf{w} \rangle = \langle (x_1, x_2) + (y_1, y_2), (z_1, z_2) \rangle$$
$$= \langle (x_1 + y_1, x_2 + y_2), (z_1, z_2) \rangle$$
$$= (x_1 + y_1)z_1 + 4(x_2 + y_2)z_2$$
$$= x_1 z_1 + y_1 z_1 + 4x_2 z_2 + 4y_2 z_2$$
$$= x_1 z_1 + 4x_2 z_2 + y_1 z_1 + 4y_2 z_2$$
$$= \langle (x_1, x_2), (z_1, z_2) \rangle + \langle (y_1, y_2), (z_1, z_2) \rangle$$
$$= \langle \mathbf{u}, \mathbf{w} \rangle + \langle \mathbf{v}, \mathbf{w} \rangle$$

Axiom 3:
$$\langle c\mathbf{u}, \mathbf{v} \rangle = \langle c(x_1, x_2), (y_1, y_2) \rangle = \langle (cx_1, cx_2), (y_1, y_2) \rangle$$
$$= cx_1 y_1 + 4cx_2 y_2 = c(x_1 y_1 + 4x_2 y_2)$$
$$= c\langle \mathbf{u}, \mathbf{v} \rangle$$

Axiom 4:
$$\langle \mathbf{u}, \mathbf{u} \rangle = \langle (x_1, x_2), (x_1, x_2) \rangle$$
$$= x_1^2 + 4x_2^2 \geq 0$$

Further, $x_1^2 + 4x_2^2 = 0$ if and only if $x_1 = 0$ and $x_2 = 0$. That is, $\mathbf{u} = \mathbf{0}$. Thus $\langle \mathbf{u}, \mathbf{u} \rangle \geq 0$, and $\langle \mathbf{u}, \mathbf{u} \rangle = 0$ if and only if $\mathbf{u} = 0$.

The four inner product axioms are satisfied. $\langle \mathbf{u}, \mathbf{v} \rangle = x_1 y_1 + 4x_2 y_2$ is an inner product on $\mathbf{R}^2$.

The inner product of the vectors $(-2, 5)$, $(3, 1)$ is

$$\langle (-2, 5), (3, 1) \rangle = (-2 \times 3) + 4(5 \times 1) = 14$$

There are many other inner products on $\mathbf{R}^2$. The dot product is, however, the most important inner product on $\mathbf{R}^2$ because it leads to angles and distances of Euclidean geometry. We shall discuss a non-Euclidean geometry based on the inner product of this example later in this section.

We now illustrate inner products on the real vector spaces of matrices and functions.

Example 2 Consider the vector space M_{22} of 2×2 matrices. Let $\mathbf{u}$ and $\mathbf{v}$ defined as follows be arbitrary 2×2 matrices.

$$\mathbf{u} = \begin{bmatrix} a & b \\ c & d \end{bmatrix}, \qquad \mathbf{v} = \begin{bmatrix} e & f \\ g & h \end{bmatrix}$$

Prove that the following function is an inner product on M_{22}.

$$\langle \mathbf{u}, \mathbf{v} \rangle = ae + bf + cg + dh$$

(One multiplies corresponding elements of the matrices and adds.)

Determine the inner product of the matrices

$$\begin{bmatrix} 2 & -3 \\ 0 & 1 \end{bmatrix} \quad \text{and} \quad \begin{bmatrix} 5 & 2 \\ 9 & 0 \end{bmatrix}$$

Solution   We shall verify axioms 1 and 3 of an inner product, leaving axioms 2 and 4 for you to check in the exercises that follow.

Axiom 1: $\langle \mathbf{u}, \mathbf{v} \rangle = ae + bf + cg + dh = ea + fb + gc + hd$

$$= \langle \mathbf{v}, \mathbf{u} \rangle$$

Axiom 3: Let k be a scalar. Then

$$\langle k\mathbf{u}, \mathbf{v} \rangle = kae + kbf + kcg + kdh = k(ae + bf + cg + dh)$$
$$= k \langle \mathbf{u}, \mathbf{v} \rangle$$

We now compute the inner product of the given matrices. We get

$$\left\langle \begin{bmatrix} 2 & -3 \\ 0 & 1 \end{bmatrix}, \begin{bmatrix} 5 & 2 \\ 9 & 0 \end{bmatrix} \right\rangle = (2 \times 5) + (-3 \times 2) + (0 \times 9) + (1 \times 0) = 4$$

Example 3   Consider the vector space P_n, of polynomials of degree $\leq n$. Let f and g be elements of P_n. Prove that the following function defines an inner product on P_n.

$$\langle f, g \rangle = \int_0^1 f(x)g(x) \, dx$$

Determine the inner product of the polynomials

$$f(x) = x^2 + 2x - 1 \quad \text{and} \quad g(x) = 4x + 1$$

Solution   We shall verify axioms 1 and 2 of an inner product, leaving axioms 3 and 4 for you to check in the exercises that follow. We use the properties of integrals.

Axiom 1: $\langle f, g \rangle = \int_0^1 f(x)g(x) \, dx = \int_0^1 g(x)f(x) \, dx$

$$= \langle g, f \rangle$$

Axiom 2: $\langle f + g, h \rangle = \int_0^1 [f(x) + g(x)]h(x) \, dx$

$$= \int_0^1 [f(x)h(x) + g(x)h(x)] \, dx$$
$$= \int_0^1 f(x)h(x) \, dx + \int_0^1 g(x)h(x) \, dx$$
$$= \langle f, h \rangle + \langle g, h \rangle$$

We now find the inner product of the functions $f(x) = x^2 + 2x - 1$ and $g(x) = 4x + 1$.

$$\langle x^2 + 2x - 1, 4x + 1 \rangle = \int_0^1 (x^2 + 2x - 1)(4x + 1)\, dx$$
$$= \int_0^1 (4x^3 + 9x^2 - 2x - 1)\, dx = 2$$

(We shall not include steps of integration in this section. We assume that you have the necessary expertise to integrate the functions involved, and to arrive at the given answers.)

Norm of a Vector

The norm of a vector in $\mathbf{R}^n$ can be expressed in terms of the dot product as follows:

$$\|(x_1, \ldots, x_n)\| = \sqrt{(x_1^2 + \cdots + x_n^2)}$$
$$= \sqrt{(x_1, \ldots, x_n) \cdot (x_1, \ldots, x_n)}$$

This definition of norm gave the norms we expected for position vectors in $\mathbf{R}^2$ and $\mathbf{R}^3$ in Euclidean geometry. To get the norm of a vector in a general vector space, we generalize this definition, using the inner product in place of the dot product. The norms that we thus obtain do not necessarily have geometric interpretations, but are often important in numerical work.

Definition　*Let V be an inner product space. The norm of a vector* **v** *is denoted* $\|\mathbf{v}\|$ *and is defined by*

$$\|\mathbf{v}\| = \sqrt{\langle \mathbf{v}, \mathbf{v} \rangle}$$

Example 4　Consider the vector space P_n of polynomials with inner product

$$\langle f, g \rangle = \int_0^1 f(x)g(x)\, dx$$

The norm of the function f **generated** by this inner product is

$$\|f\| = \sqrt{\langle f, f \rangle} = \sqrt{\int_0^1 [f(x)]^2\, dx}$$

Determine the norm of the function $f(x) = 5x^2 + 1$.

Solution　Using the above definition of norm, we get

$$\|5x^2 + 1\| = \sqrt{\int_0^1 [5x^2 + 1]^2\, dx}$$
$$= \sqrt{\int_0^1 [25x^4 + 10x^2 + 1]\, dx}$$
$$= \sqrt{\tfrac{28}{3}}$$

The norm of the function $f(x) = 5x^2 + 1$ is $\sqrt{\frac{28}{3}}$.

Angle

The dot product in $\mathbf{R}^n$ was used to define angles between vectors. The angle θ between vectors $\mathbf{u}$ and $\mathbf{v}$ in $\mathbf{R}^n$ is defined by

$$\cos \theta = \frac{\mathbf{u} \cdot \mathbf{v}}{\|\mathbf{u}\| \, \|\mathbf{v}\|}$$

We adopt this as our definition of angle for a real vector space, using the inner product generalization of the dot product.

Definition   *Let V be a real inner product space. The angle θ between two nonzero vectors $\mathbf{u}$ and $\mathbf{v}$ is given by*

$$\cos \theta = \frac{\langle \mathbf{u}, \mathbf{v} \rangle}{\|\mathbf{u}\| \, \|\mathbf{v}\|}$$

Example 5   Consider the product space P_n of polynomials with inner product

$$\langle f, g \rangle = \int_0^1 f(x)g(x) \, dx$$

The angle between two nonzero functions f and g is given by

$$\cos \theta = \frac{\langle f, g \rangle}{\|f\| \, \|g\|} = \frac{\int_0^1 f(x)g(x) \, dx}{\|f\| \, \|g\|}$$

Determine the cosine of the angle between the functions

$$f(x) = 5x^2 \quad \text{and} \quad g(x) = 3x$$

Solution   We first compute $\|f\|$ and $\|g\|$.

$$\|5x^2\| = \sqrt{\int_0^1 [5x^2]^2 \, dx} = \sqrt{5} \quad \text{and} \quad \|3x\| = \sqrt{\int_0^1 [3x]^2 \, dx} = \sqrt{3}$$

Thus

$$\cos \theta = \frac{\int_0^1 f(x)g(x) \, dx}{\|f\| \, \|g\|} = \frac{\int_0^1 (5x^2)(3x) \, dx}{\sqrt{5}\sqrt{3}} = \frac{\sqrt{15}}{4}$$

Orthogonal Vectors

Let V be an inner product space. Two vectors $\mathbf{u}$ and $\mathbf{v}$ in V are said to be orthogonal if

$$\langle \mathbf{u}, \mathbf{v} \rangle = 0$$

Example 6   Show that the functions $f(x) = 3x - 2$ and $g(x) = x$ are orthogonal in P_n with inner product $\langle f, g \rangle = \int_0^1 f(x)g(x) \, dx$.

Solution   We get

$$\langle 3x - 2, x \rangle = \int_0^1 (3x - 2)(x) \, dx = [x^3 - x^2]_0^1 = 0$$

Thus the functions f and g are orthogonal in this inner product space.

Distance

Our final task is to extend the Euclidean concept of distance to general vector spaces. As for norm, the concept of distance that we arrive at will not in general have direct geometrical interpretation. It is, however, useful in numerical mathematics to be able to discuss how far apart various functions are.

Definition   *Let V be an inner product space with vector norm defined by*

$$\|\mathbf{v}\| = \sqrt{\langle \mathbf{v}, \mathbf{v} \rangle}$$

The distance between two vectors (points) $\mathbf{u}$ and $\mathbf{v}$ is denoted $d(\mathbf{u}, \mathbf{v})$ and is defined by

$$d(\mathbf{u}, \mathbf{v}) = \|\mathbf{u} - \mathbf{v}\| \quad (= \sqrt{\langle \mathbf{u} - \mathbf{v}, \mathbf{u} - \mathbf{v} \rangle})$$

Example 7   Consider the inner product space P_n of polynomials discussed earlier. Determine which of the functions $g(x) = x^2 - 3x + 5$ or $h(x) = x^2 + 4$ is closest to $f(x) = x^2$.

Solution   We compute the distances between f and g and between f and h. It is easiest to work initially in terms of the squares of the distances.

$$[d(f, g)]^2 = \langle f - g, f - g \rangle = \langle 3x - 5, 3x - 5 \rangle = \int_0^1 (3x - 5)^2 \, dx = 13$$

$$[d(f, h)]^2 = \langle f - h, f - h \rangle = \langle -4, -4 \rangle = \int_0^1 (-4)^2 \, dx = 16$$

Thus $d(f, g) = \sqrt{13}$ and $d(f, h) = 4$. The distance between f and h is 4, as we might suspect. g is closer than h to f.

In practice, relative distances between functions are often more significant than absolute distances.

The complex vector space $\mathbf{C}^n$ that was introduced in the previous section is important in many branches of mathematics, physics, and engineering. We now discuss the inner product that is most often used on $\mathbf{C}^n$.

*Inner Product on C^n

For a complex vector space, the first axiom of inner product is modified to read $\langle \mathbf{u}, \mathbf{v} \rangle = \overline{\langle \mathbf{v}, \mathbf{u} \rangle}$. An inner product can then be used to define norm, orthogonality, and distance, as for a real vector space.

Let $\mathbf{u} = (x_1, \ldots, x_n)$ and $\mathbf{v} = (y_1, \ldots, y_n)$ be elements of $\mathbf{C}^n$. The most useful inner product for $\mathbf{C}^n$ is

$$\langle \mathbf{u}, \mathbf{v} \rangle = x_1 \bar{y}_1 + \cdots + x_n \bar{y}_n$$

It can be shown that this definition satisfies the inner product axioms for a complex vector space (see following exercises). This inner product leads to the following definitions of orthogonality, norm, and distance for $\mathbf{C}^n$.

$$\mathbf{u} \perp \mathbf{v} \text{ if } \langle \mathbf{u}, \mathbf{v} \rangle = 0 \qquad \|\mathbf{u}\| = \sqrt{x_1 \bar{x}_1 + \cdots + x_n \bar{x}_n} \qquad d(\mathbf{u}, \mathbf{v}) = \|\mathbf{u} - \mathbf{v}\|$$

Example 8   Consider the vectors $\mathbf{u} = (2 + 3i, -1 + 5i)$, $\mathbf{v} = (1 + i, -i)$ in $\mathbf{C}^2$. Compute

(a) $\langle \mathbf{u}, \mathbf{v} \rangle$, and show that $\mathbf{u}$ and $\mathbf{v}$ are orthogonal.

(b) $\|\mathbf{u}\|$ and $\|\mathbf{v}\|$

(c) $d(\mathbf{u}, \mathbf{v})$

Solution   (a) $\langle \mathbf{u}, \mathbf{v} \rangle = (2 + 3i)(1 - i) + (-1 + 5i)(i) = 5 + i - i - 5 = 0$. Thus $\mathbf{u}$ and $\mathbf{v}$ are orthogonal.

(b) $\|\mathbf{u}\| = \sqrt{(2 + 3i)(2 - 3i) + (-1 + 5i)(-1 - 5i)} = \sqrt{13 + 26} = \sqrt{39}$
$\|\mathbf{v}\| = \sqrt{(1 + i)(1 + i) + (-i)(i)} = \sqrt{3}$

(c) $d(\mathbf{u}, \mathbf{v}) = \|\mathbf{u} - \mathbf{v}\| = \|(2 + 3i, -1 + 5i) - (1 + i, -i)\|$
$= \|(1 + 2i, -1 + 6i\|$
$= \sqrt{(1 + 2i)(1 - 2i) + (-1 + 6i)(-1 - 6i)} = \sqrt{5 + 37} = \sqrt{42}$

Exercise Set 6.2

1. Let $\mathbf{u} = (x_1, x_2)$ and $\mathbf{v} = (y_1, y_2)$ be elements of $\mathbf{R}^2$. Prove that the following function defines an inner product on $\mathbf{R}^2$.

$$\langle \mathbf{u}, \mathbf{v} \rangle = 4x_1 y_1 + 9x_2 y_2$$

2. Let $\mathbf{u} = (x_1, x_2, x_3)$ and $\mathbf{v} = (y_1, y_2, y_3)$ be elements of $\mathbf{R}^3$. Prove that the following function defines an inner product on $\mathbf{R}^3$.

$$\langle \mathbf{u}, \mathbf{v} \rangle = x_1 y_1 + 2x_2 y_2 + 4x_3 y_3$$

*3. Let $\mathbf{u} = (x_1, x_2)$ and $\mathbf{v} = (y_1, y_2)$ be elements of $\mathbf{R}^2$. Prove that the following function does not define an inner product on $\mathbf{R}^2$.

$$\langle \mathbf{u}, \mathbf{v} \rangle = 2x_1 y_1 - x_2 y_2$$

*4. Consider the vector space M_{22} of 2×2 matrices. Let $\mathbf{u}$ and $\mathbf{v}$ defined as follows be arbitrary 2×2 matrices.

$$\mathbf{u} = \begin{bmatrix} a & b \\ c & d \end{bmatrix}, \qquad \mathbf{v} = \begin{bmatrix} e & f \\ g & h \end{bmatrix}$$

Prove that the following function satisfies axioms 2 and 4 of the inner product.

$$\langle \mathbf{u}, \mathbf{v} \rangle = ae + bf + cg + dh$$

(This exercise completes Example 2 of this section, showing that this function is an inner product.)

5. Let **u** and **v** defined as follows be arbitrary elements of M_{22}.

$$\mathbf{u} = \begin{bmatrix} a & b \\ c & d \end{bmatrix}, \quad \mathbf{v} = \begin{bmatrix} e & f \\ g & h \end{bmatrix}$$

Prove that the following function is an inner product.

$$\langle \mathbf{u}, \mathbf{v} \rangle = ae + 2bf + 3cg + 4dh$$

Determine the inner products of the following pairs of matrices.

*(a) $\mathbf{u} = \begin{bmatrix} 1 & 2 \\ 0 & -3 \end{bmatrix}$, $\mathbf{v} = \begin{bmatrix} 4 & 1 \\ -3 & 2 \end{bmatrix}$

(b) $\mathbf{u} = \begin{bmatrix} -2 & 4 \\ 1 & 0 \end{bmatrix}$, $\mathbf{v} = \begin{bmatrix} 5 & -2 \\ 0 & -3 \end{bmatrix}$

6. Consider the vector space P_n, of polynomials of degree $\leq n$. Let f and g be elements of this space. Prove that the following function satisfies axioms 3 and 4 of the inner product.

$$\langle f, g \rangle = \int_0^1 f(x)g(x)\, dx$$

(This exercise completes Example 3 of this section, showing that this function is an inner product.)

7. Let f and g be arbitrary elements of the vector space P_n. Prove that the following function defines an inner product on P_n if a and b are real numbers with $a < b$. Show that it does not define an inner product if $a \leq b$.

$$\langle f, g \rangle = \int_a^b f(x)g(x)\, dx$$

In Exercises 8–15, all functions are in the inner product space P_n, with inner product defined by

$$\langle f, g \rangle = \int_0^1 f(x)g(x)\, dx$$

8. Determine the inner products of the following functions.
   *(a) $f(x) = 2x + 1$, $g(x) = 3x - 2$
   (b) $f(x) = x^2 + 2$, $g(x) = 3$
   *(c) $f(x) = x^2 + 3x - 2$, $g(x) = x + 1$
   (d) $f(x) = x^3 + 2x - 1$, $g(x) = 3x^2 - 4x + 2$

9. Determine the norms of the following functions.
   *(a) $f(x) = 4x - 2$ (b) $f(x) = 7x^3$
   *(c) $f(x) = 3x^2 + 2$ (d) $g(x) = x^2 + x + 1$

*10. Prove that the functions $f(x) = x^2$ and $g(x) = 4x - 3$ are orthogonal.

11. Prove that the functions $f(x) = 1$ and $g(x) = \frac{1}{2} - x$ are orthogonal.

*12. Compute the cosine of the angle between the functions

$$f(x) = 5x^2 \quad \text{and} \quad g(x) = 9x$$

*13. Determine a linear function that is orthogonal to

$$f(x) = 6x + 12$$

*14. Compute the distance between the functions

$$f(x) = x^2 + 3x + 1 \quad \text{and} \quad g(x) = x^2 + x - 3$$

*15. Determine which of the functions $g(x) = x^2 + 2x - 3$ or $h(x) = x^2 - 3x + 4$ is closest to $f(x) = x^2$.

In Exercises 16–20, all the matrices are elements of the inner product space M_{22} with inner product defined by

$$\left\langle \begin{bmatrix} a & b \\ c & d \end{bmatrix}, \begin{bmatrix} e & f \\ g & h \end{bmatrix} \right\rangle = ae + bf + cg + dh$$

16. Determine the inner products of the following matrices.

*(a) $\begin{bmatrix} 1 & 2 \\ 3 & 4 \end{bmatrix}$, $\begin{bmatrix} -2 & 0 \\ -3 & 5 \end{bmatrix}$

(b) $\begin{bmatrix} 0 & -3 \\ 2 & 5 \end{bmatrix}$, $\begin{bmatrix} 3 & 6 \\ -2 & -7 \end{bmatrix}$

17. Compute the norms of the following matrices.

*(a) $\begin{bmatrix} 1 & 2 \\ 3 & 4 \end{bmatrix}$ (b) $\begin{bmatrix} 0 & 1 \\ -1 & 3 \end{bmatrix}$

*(c) $\begin{bmatrix} 5 & -2 \\ -1 & 6 \end{bmatrix}$ (d) $\begin{bmatrix} 4 & -2 \\ -1 & -3 \end{bmatrix}$

18. Prove that the following pairs of matrices are orthogonal.

*(a) $\begin{bmatrix} 1 & 2 \\ -1 & 1 \end{bmatrix}$, $\begin{bmatrix} 2 & 4 \\ 3 & -7 \end{bmatrix}$

(b) $\begin{bmatrix} 5 & 2 \\ -3 & 2 \end{bmatrix}$, $\begin{bmatrix} -1 & 6 \\ 1 & -2 \end{bmatrix}$

*19. Determine a matrix orthogonal to the matrix

$$\begin{bmatrix} 1 & 2 \\ 3 & 4 \end{bmatrix}$$

20. Compute the distances between the following pairs of matrices.

*(a) $\begin{bmatrix} 4 & 0 \\ -1 & 3 \end{bmatrix}$, $\begin{bmatrix} 1 & 1 \\ 1 & 1 \end{bmatrix}$

(b) $\begin{bmatrix} 2 & -3 \\ -1 & 4 \end{bmatrix}$, $\begin{bmatrix} -3 & 2 \\ 1 & 0 \end{bmatrix}$

21. Consider the following vectors **u** and **v** in the complex inner product space $\mathbf{C}^2$. Compute $\langle \mathbf{u}, \mathbf{v} \rangle$, $\|\mathbf{u}\|$ and $\|\mathbf{v}\|$, $d(\mathbf{u}, \mathbf{v})$. Determine whether **u** and **v** are orthogonal.

*(a) $\mathbf{u} = (2 - i, 3 + 2i)$, $\mathbf{v} = (3 - 2i, 2 + i)$

(b) $\mathbf{u} = (4 + 3i, 1 - i)$, $\mathbf{v} = (2 + i, 4 - 5i)$

(c) $\mathbf{u} = (2 + 3i, -1)$, $\mathbf{v} = (-i, 3 + 2i)$

*(d) $\mathbf{u} = (2 - 3i, -2 + 3i)$, $\mathbf{v} = (1, 1)$

22. Consider the following vectors **u** and **v** in the complex inner product space $\mathbf{C}^2$. Compute $\langle \mathbf{u}, \mathbf{v} \rangle$, $\|\mathbf{u}\|$ and $\|\mathbf{v}\|$, $d(\mathbf{u}, \mathbf{v})$. Determine whether **u** and **v** are orthogonal.

*(a) $\mathbf{u} = (1 + 4i, 1 + i)$, $\mathbf{v} = (1 + i, -4 + i)$

(b) $\mathbf{u} = (2 + 7i, 1 + i)$, $\mathbf{v} = (3 - 4i, 2 + 5i)$

*(c) $\mathbf{u} = (1 - 3i, 1 + i)$, $\mathbf{v} = (2 - i, 5i)$

(d) $\mathbf{u} = (3 + i, 2 + 2i)$, $\mathbf{v} = \left(2 + i, -2 + \frac{2}{3}i\right)$

*23. Let $\mathbf{u} = (x_1, \ldots, x_n)$ and $\mathbf{v} = (y_1, \ldots, y_n)$ be elements of $\mathbf{C}^n$. Prove that

$$\langle \mathbf{u}, \mathbf{v} \rangle = x_1 \bar{y}_1 + \cdots + x_n \bar{y}_n$$

satisfies the inner product axioms for a complex vector space.

*24. Prove that if **u** and **v** are vectors in the complex inner product space $\mathbf{C}^n$ and k is a scalar, then

$$\langle \mathbf{u}, k\mathbf{0} \rangle = \bar{k}\langle \mathbf{u}, \mathbf{v} \rangle$$

25. Let **u** and **v** be vectors in a real inner product space and let c be a scalar. Prove that (a) $\langle \mathbf{0}, \mathbf{v} \rangle = \langle \mathbf{v}, \mathbf{0} \rangle = 0$, (b) $\langle \mathbf{u}, \mathbf{v} + \mathbf{w} \rangle = \langle \mathbf{u}, \mathbf{v} \rangle + \langle \mathbf{u}, \mathbf{w} \rangle$, and (c) $\langle \mathbf{u}, c\mathbf{0} \rangle = c\langle \mathbf{u}, \mathbf{v} \rangle$.

26. Let **u** and **v** be vectors in $\mathbf{R}^n$. Let A be an $n \times n$ symmetric matrix such that $\mathbf{w}A\mathbf{w}^t > 0$ for every nonzero vector **w** in $\mathbf{R}^n$. Such a matrix is called **positive definite**.

*(a) Show that $\langle \mathbf{u}, \mathbf{v} \rangle = \mathbf{u}A\mathbf{v}^t$ satisfies all the axioms of an inner product and thus defines an inner product on $\mathbf{R}^n$. Thus any positive definite matrix can be used to define an inner product on $\mathbf{R}^n$.

(b) Show that the identity matrix I_n defines the dot product on $\mathbf{R}^n$.

*(c) Let $\mathbf{u} = (1, 0)$, $\mathbf{v} = (0, 1)$. Find $\langle \mathbf{u}, \mathbf{v} \rangle$, $\|\mathbf{u}\|$, $\|\mathbf{v}\|$, and $d(\mathbf{u}, \mathbf{v})$ for each of the inner products defined by the following positive definite matrices.

$$A = \begin{bmatrix} 2 & 0 \\ 0 & 3 \end{bmatrix}, \quad B = \begin{bmatrix} 1 & 2 \\ 2 & 3 \end{bmatrix}, \quad C = \begin{bmatrix} 2 & 3 \\ 3 & 5 \end{bmatrix}$$

*(d) The inner product of Example 1 of this section can be defined by a positive definite matrix A. Determine A.

*6.3 Non-Euclidean Geometry

In this section we explore the significance of the mathematical freedom we have in the many inner products that can be placed on $\mathbf{R}^n$. The inner product is a very powerful tool that enables us to control the geometry of $\mathbf{R}^n$. The inner product controls the norms of vectors, angles between vectors, and distances between points, since each of these is defined in terms of the inner product.

$$\|\mathbf{v}\| = \sqrt{\langle \mathbf{v}, \mathbf{v} \rangle}, \quad \cos \theta = \frac{\langle \mathbf{u}, \mathbf{v} \rangle}{\|\mathbf{u}\| \|\mathbf{v}\|}, \quad d(\mathbf{u}, \mathbf{v}) = \|\mathbf{u} - \mathbf{v}\| = \sqrt{\langle \mathbf{u} - \mathbf{v}, \mathbf{u} - \mathbf{v} \rangle}$$

Different inner products on $\mathbf{R}^n$ lead to different measures of vector norm, angle, and distance—that is, to different geometries. The dot product is a particular inner product that leads to Euclidean geometry. Let us now look at some of the other geometries that we can get by putting other inner products on $\mathbf{R}^n$. These **non-Euclidean geometries** have been studied by mathematicians and are used by scientists. We shall see how one of these geometries is used to describe space-time in the theory of special relativity.

We first look at a non-Euclidean geometry on $\mathbf{R}^2$. Consider the inner product that was discussed in Example 1 of the previous section. Let (x_1, x_2) and (y_1, y_2) be arbitrary vectors in $\mathbf{R}^2$, and let the inner product be defined by

$$\langle (x_1, x_2), (y_1, y_2) \rangle = x_1 y_1 + 4 x_2 y_2$$

This inner product differs from the dot product in the appearance of a 4. Consider the vector $(0, 1)$ in this space. The norm of this vector is

$$\|(0, 1)\| = \sqrt{\langle (0, 1), (0, 1) \rangle}$$
$$= \sqrt{(0 \times 0) + 4(1 \times 1)}$$
$$= 2$$

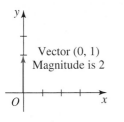

Vector (0, 1)
Magnitude is 2

Figure 6.1

We illustrate this vector in Figure 6.1. The norm of this vector in Euclidean geometry is, of course, 1; in our new geometry, however, the norm is 2.

Consider the vectors $(1, 1)$ and $(-4, 1)$. The inner product of these vectors is

$$\langle (1, 1), (-4, 1) \rangle = (1 \times -4) + 4(1 \times 1)$$
$$= 0$$

Thus the vectors $(1, 1)$ and $(-4, 1)$ are orthogonal. We illustrate these vectors in Figure 6.2. The vectors do not appear to be at right angles; they are not orthogonal in the Euclidean geometry of our everyday experience. However, the angle between these vectors is a right angle in our new geometry.

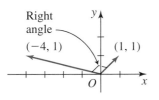

Right angle
(−4, 1)   (1, 1)

Figure 6.2

Finally, let us use the definition of distance based on this inner product to compute the distance between the points $(1, 0)$ and $(0, 1)$. We have that

$$d((1, 0), (0, 1)) = \|(1, 0) - (0, 1)\|$$
$$= \|(1, -1)\|$$
$$= \sqrt{\langle (1, -1), (1, -1) \rangle}$$
$$= \sqrt{(1 \times 1) + 4(-1 \times -1)}$$
$$= \sqrt{5}$$

This distance is illustrated in Figure 6.3. The distance between these same two points in Euclidean space, on applying the Pythagorean Theorem, is $\sqrt{2}$.

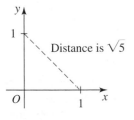

Distance is $\sqrt{5}$

Figure 6.3

You should appreciate that the geometry that we have developed in this example is correct mathematics, even though it does not describe the everyday world in which we live! Mathematics exists as an elegant field in itself, without having to be useful or related to real-world concepts. Even though an important aim of this course is to show how the mathematics that we develop is applied, the reader should be aware that much fine mathematics is not oriented toward applications. Of course, one never knows when such pure mathematics will turn out to be useful. This, for example, happened to be the case for non-Euclidean geometry. Much of non-Euclidean geometry was developed by a German mathematician named Bernhard Riemann in the mid-nineteenth century. In the early part of this century, Albert Einstein found that this was the type of geometry he needed for describing space-time. In space-time one uses $\mathbf{R}^4$ to represent three spatial dimensions and one time dimension. Euclidean geometry is not the appropriate geometry when space and time are "mixed."

We also mention that matrix inner product spaces and function inner product spaces form the mathematical framework of the theory of quantum mechanics. Quantum mechanics is the mathematical model that was developed in the early part of this century by such physicists as Neils Bohr, Max Planck, and Werner Heisenberg to describe the behavior of atoms and electrons. Thus the theory of inner product spaces forms the cornerstone of the two foremost physical theories of the twentieth century. These theories have had a great impact on all of our lives; they have, for example, led to the development of both atomic energy and atomic weapons.

Special Relativity

Special relativity was developed by Albert Einstein* to describe the physical world in which we live. At the time, Newtonian mechanics was the theory used to describe the motions of bodies. However, experiments had led scientists to believe that large-scale motions of bodies, such as planetary motions, were not accurately described by Newtonian mechanics. One of Einstein's contributions to science was the development of more accurate mathematical models. He first developed special relativity, which did not incorporate gravity; later he included gravity in his general theory of relativity. Here we introduce the mathematical model of special relativity by describing the predictions of the theory and then showing how they arise out of the mathematics.

The nearest star to Earth, other than the Sun, is Alpha Centauri; it is about 4 light-years away. (A light-year is the distance light travels in 1 year, 5.88×10^{12} miles.) Consider a pair of twins, who are separated immediately after birth. Twin 1 remains on Earth, and Twin 2 is flown off to Alpha Centauri in a spaceship at 0.8 the speed of light. On arriving at Alpha Centauri, he immediately flies back to Earth at the same high speed. Special relativity predicts that when the twins get together, the twin who remained on Earth will be 10 years old, while the one who went off to Alpha Centauri will be 6 years old. We present this scenario in terms of twins, since it clearly displays the phenomenon involved, namely the difference in the aging process. The one twin will be a 10-year-old child, the other twin a 6-year-old child. The same type of phenomenon is predicted for any two people, one of whom stays on Earth while the other travels to a distant planet and back. The numbers will vary according to how far and how fast the traveler goes. For example, this phenomenon was experienced to a very small degree by the astronauts who went to the Moon. On their arrival back on Earth, they were fractionally younger than they would have been had they remained on Earth.

Numerous experiments have been constructed to test this hypothesis that time on Earth varies from time recorded on an object moving relative to Earth. We mention one experiment conducted by Professor J. C. Hafele, Department of Physics, Washington

---

* Albert Einstein (1879–1955) was one of the greatest scientists of all time. He studied in Zurich, and taught in Zurich, Prague, Berlin, and Princeton. He is best known for his theories of relativity, which revolutionized scientific thought with new conceptions of time, space, mass, motion, and gravitation, but he received the Nobel Prize for his work on the photoelectric effect. Einstein moved to the United States when Hitler came to power. He renounced his pacifist stand, being convinced that Hitler's regime could be put down only by force. He wrote a letter with the physicists Teller and Wigner to Roosevelt that helped initiate the efforts that produced the atomic bomb (although Einstein knew nothing of these efforts). Einstein was fond of classical music and played the violin. Although not associated with any orthodox religion, he was by nature deeply religious. He never believed the universe was one of chance or chaos.

University, St. Louis, and Dr. Richard E. Keating, U.S. Naval Observatory, Washington, D.C., and reported in the journal **Science**, Volume 177, 1972. During October 1971, four atomic clocks were flown on regularly scheduled jet flights around the world twice. There were slight differences in the times recorded by the clocks and clocks on Earth. These differences were in accord with the predictions of relativity theory. This aging variation will not, however, be realized by humans to the extent illustrated by the trip to Alpha Centauri in the near future as the energies required to produce such high speeds in a macroscopic body are prohibitive. This is probably fortunate from a sociological point of view. Theoretically, a man who went off on such a trip could be of an age to marry his great-granddaughter when he returned!

We now introduce the mathematical model of special relativity and see how the predictions arise out of this model. Special relativity is a model of space-time and hence involves four coordinates: three space coordinates—x_1, x_2, x_3—and a time coordinate, x_4. We use the vector space $\mathbf{R}^4$ to represent space-time. There are many inner products that we can place on $\mathbf{R}^4$; each would lead to a geometry for $\mathbf{R}^4$. None of the inner products, however, leads to a geometry that conforms to experimental results. If one drops the fourth inner product axiom, one comes up with a pseudo inner product that leads to a geometry that fits experimental results. This geometry of special relativity is called **Minkowski geometry** after Hermann Minkowski,* who gave this geometrical interpretation of special relativity.

Let $X = (x_1, x_2, x_3, x_4)$ and $Y = (y_1, y_2, y_3, y_4)$ be arbitrary elements of $\mathbf{R}^4$. The following function $\langle X, Y \rangle$ plays the role of an inner product in this geometry.

$$\langle X, Y \rangle = -x_1 y_1 - x_2 y_2 - x_3 y_3 + x_4 y_4$$

Note that $\langle X, Y \rangle$ is not a true inner product; the negative signs cause it to violate axiom 4 of an inner product. See the following exercises.

This pseudo inner product is used as follows to define norms of vectors and distances between points. Observe the use of absolute value signs in these definitions. The justification for these definitions is simple—they are the ones that work for the geometry to describe space-time!

$$\|X\| = \sqrt{|\langle X, X \rangle|}$$

and

$$d(X, Y) = \|X - Y\| = \|(x_1 - y_1, x_2 - y_2, x_3 - y_3, x_4 - y_4)\|$$
$$= \sqrt{\left| -(x_1 - y_1)^2 - (x_2 - y_2)^2 - (x_3 - y_3)^2 + (x_4 - y_4)^2 \right|}$$

---

* Hermann Minkowski (1864–1909) was educated in Königsberg and taught at Bonn, Königsberg, Zurich, and Göttingen. His interests were in number theory, geometry, and algebra. He became interested in relativity in his later years. He was the first to conceive that the principle of relativity formulated by Einstein led to the abandonment of the concept of space and time as separate entities and to their replacement by a four-dimensional space-time. This became the framework for the later developments of the theory and led Einstein to his bolder conception of general relativity.

Let us now see how this geometry is used to describe the behavior of time for the voyage to Alpha Centauri. We draw a **space-time diagram**. For convenience assume that Alpha Centauri lies in the direction of the x_1 axis from Earth. The twin on Earth advances in time, x_4, while the twin on board the spaceship advances in time and also moves in the direction of increasing x_1 on the outward voyage and then in a direction of decreasing x_1 on the homeward voyage. The space-time diagram is shown in Figure 6.4. The path of Twin 1 is PQ. The path of Twin 2 is PR to Alpha Centauri and then RQ back to Earth. There is no motion in either the x_2 or x_3 directions; hence we suppress these axes in the diagram.

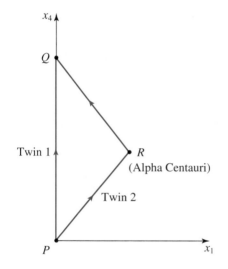

Figure 6.4

We now look at the situation from the point of view of each twin.

Twin 1   Distance to Alpha Centauri and back = 8 light-years; speed of spaceship = 0.8 speed of light. Thus

$$\text{Duration of voyage relative to Earth} = \frac{\text{distance}}{\text{speed}} = \frac{8}{0.8} = 10 \text{ years}$$

Geometrically, this means that the length of PQ is 10.

Twin 2   Let us now arrive at coordinates for the various points in the diagram. See Figure 6.5. Let P be the origin, $(0, 0, 0, 0)$, in $\mathbf{R}^4$. Since $PQ = 10$, Q is the point $(0, 0, 0, 10)$ and S is the point $(0, 0, 0, 5)$.

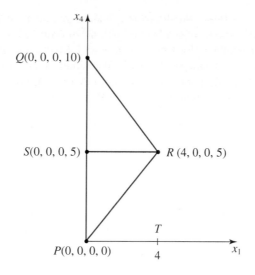

Figure 6.5

PT is the spatial distance of Alpha Centauri from Earth, namely 4 light-years. Thus *R* is the point (4, 0, 0, 5). We now use Minkowski geometry to compute the distance between *R* and *P*.

$$d(R, P) = \|(4, 0, 0, 5) - (0, 0, 0, 0)\|$$
$$= \sqrt{\left| -(4 - 0)^2 - (0 - 0)^2 - (0 - 0)^2 + (5 - 0)^2 \right|}$$
$$= \sqrt{|-16 + 25|} = \sqrt{9} = 3$$

Similarly, in this geometry,

$$d(R, Q) = 3$$

We have completed the mathematics. Now comes the very interesting part of relating the geometrical results to the physical situation. The following is a postulate of special relativity that gives the geometry physical meaning.

The distance between two points on the path of an observer, such as Twin 2, in Minkowski geometry, corresponds to the time recorded by that observer in traveling between the two points.

Thus *d(R, P)* implies that Twin 2 ages 3 years in traveling from *P* to *R*. Similarly, *d(R, Q)* = 3 implies that Twin 2 ages 3 years in traveling from *R* to *Q*. The total duration of the voyage for Twin 2 is thus 6 years.

Therefore, when the twins meet at *Q*, Twin 1 will have aged 10 years while Twin 2 will have aged 6 years.

Note that this model introduces a new kind of geometry in which the straight-line distance is not the shortest distance between two points. In Figure 6.5 the straight-line distance *PQ* from *P* to *Q* is 10, whereas the distance *PR* + *RQ* from *P* to *Q* is 6, smaller! Thus, not only is the physical interpretation of the model fascinating, it opens up a new trend in geometrical thinking.

This example illustrates the flexibility that one has in applying mathematics. If the standard body of mathematics (inner product axioms in this case) does not fit the situation, maybe a slight modification will. Mathematicians have molded and developed mathematics to suit their needs. Mathematics is not as rigid and absolute as it is sometimes made out to be; it is a field that is continually being developed and applied in the spirit presented here.

Exercise Set 6.3

*1. Consider $\mathbf{R}^2$ with the inner product of Example 1 of the previous section.

$$\langle(x_1, x_2), (y_1, y_2)\rangle = x_1 y_1 + 4x_2 y_2$$

Determine the equation of the circle with center at the origin and radius 1 in this space. Sketch the circle.

2. Consider $\mathbf{R}^2$ with the inner product

$$\langle(x_1, x_2), (y_1, y_2)\rangle = 4x_1 y_1 + 9x_2 y_2$$

*(a) Determine the norms of the following vectors.

$$(1, 0), \quad (0,1), \quad (1,1), \quad (2,3)$$

(b) Prove that the vectors $(2, 1)$ and $(-9, 8)$ are orthogonal in this space. Sketch these vectors. Observe that they are not orthogonal in Euclidean space.

*(c) Determine the distance between the points $(1, 0)$ and $(0, 1)$ in this space.

(d) Determine the equation of the circle with center the origin and radius 1 in this space. Sketch the circle.

3. Consider $\mathbf{R}^2$ with the inner product

$$\langle(x_1, x_2), (y_1, y_2)\rangle = x_1 y_1 + 16x_2 y_2$$

(a) Determine the norms of the vectors $(1, 0)$, $(0, 1)$ and $(1, 1)$ in this space.

(b) Prove that the vectors $(1, 1)$ and $(-16, 1)$ are orthogonal.

*(c) Determine the distances between the points $(5, 0)$ and $(0, 4)$, both in this space and in Euclidean space.

*(d) Determine the equation of the circle with center the origin and radius 1 in this space. Sketch the circle.

*4. Determine the inner product that must be placed on $\mathbf{R}^2$ for the equation of the circle with center the origin and radius 1 to be $x^2/4 + y^2/25 = 1$.

5. Prove that the pseudo inner product of Minkowski geometry violates axiom 4 of an inner product. (*Hint:* Find a nonzero element X of $\mathbf{R}^4$ such that $\langle X, X\rangle = 0$.)

*6. Use the definition of distance between two points in Minkowski space to show that the distance between the points R and Q in Figure 6.5 is 3.

7. Determine the distance between points $P(0, 0, 0, 0)$ and $M(1, 0, 0, 1)$ in Minkowski geometry. This reveals another interesting aspect of this geometry, namely that noncoincident points can be zero distance apart. The special theory of relativity gives a physical interpretation to such a line PM, all of whose points are zero distance apart; it is the path of a **photon**, a light particle.

*8. Prove that the vectors $(2, 0, 0, 1)$ and $(1, 0, 0, 2)$ are orthogonal in Minkowski geometry. Sketch these vectors, and observe that they are symmetrical about the "45°" vector $(1, 0, 0, 1)$. Any pair of vectors that are thus symmetrical about the vector $(1, 0, 0, 1)$ will be orthogonal in Minkowski geometry. Prove this result by demonstrating that the vectors $(a, 0, 0, b)$ and $(b, 0, 0, a)$ are orthogonal.

*9. Determine the equations of the circles with center the origin and radii 1, 2, and 3 in the $x_1 x_4$ plane of Minkowski space. Sketch these circles.

*10. The star Sirius is 8 light-years from Earth. Sirius is the nearest star, other than the Sun, that is visible with the naked eye from most parts of North America. It is the brightest appearing of all stars. Light reaches us from the Sun in 8 minutes, while it takes 8 years to reach us from Sirius. Suppose a spaceship that leaves for Sirius returns to Earth 20 years later. What is the duration of the voyage for a person on the spaceship? What is the speed of the spaceship?

11. A spaceship makes a round-trip to the bright star Capella, which is 45 light-years from Earth. If the duration of the voyage relative to Earth is 120 years, what is the duration of the voyage from the traveler's viewpoint?

*12. The star cluster Pleiades in the constellation Taurus is 410 light-years from Earth. A traveler to the cluster ages 40 years on a round-trip. By the time he returns to Earth, how many centuries will have passed on Earth?

13. The star cluster Praesepe in the constellation Cancer is 515 light-years from Earth. A traveler flies to Praesepe and back at 0.99999 the speed of light. Determine the durations of the voyage from both Earth and traveler viewpoints.

*6.4 Approximation of Functions

Many problems in the physical sciences and engineering involve approximating given functions by polynomials or by trigonometric functions. For example, it may be necessary to approximate $f(x) = e^x$ by a linear function of the form $g(x) = a + bx$ over the interval $[0, 1]$, or by a trigonometric function of the form $h(x) = a + b \sin(x) + c \cos(x)$ over the interval $[-\pi, \pi]$. Furthermore, the problem of approximating functions by polynomials is central to software development for computers since computers are essentially only arithmetic devices that can compute polynomials and not much more. We now introduce the techniques for approximating functions.

Let $C[a, b]$ be the inner product space of continuous functions over the interval $[a, b]$ with inner product $\langle f, g \rangle = \int_a^b f(x)g(x)\,dx$, and geometry defined by this inner product. Let W be a subspace of $C[a, b]$. Suppose f is in $C[a, b]$, but outside W, and we want to find the "best" approximation to f that lies in W. We define the "best" approximation to be the function g in W such that the distance $\|f - g\|$ between f and g is a minimum.

Definition   *Let $C[a, b]$ be the vector space of continuous functions over the interval $[a, b]$, f be an element of $C[a, b]$, and W be a subspace of $C[a, b]$. The function g in W such that $\int_a^b [f(x) - g(x)]^2\,dx$ is a minimum is called the **least squares approximation** to f.*

This approximation is called "the least squares approximation" since the distance formula is based on squaring.

We now give a method for finding the least squares approximation $g(x)$. We can sit back and really appreciate the approach of extending geometrical structures and results to more abstract surroundings! We have already looked at a situation like this in $\mathbf{R}^n$. We know from Section 5.2 that if W is a subspace of $\mathbf{R}^n$ and if $\mathbf{x}$ defines a point in $\mathbf{R}^n$, then the element of W that is closest to $\mathbf{x}$ is $\text{proj}_W \mathbf{x}$. This result, although derived in a geometrical environment, was developed in terms of inner product space concepts that apply in this function space environment. It is very helpful to picture functions as being geometrical vectors. See Figure 6.6. The least squares approximation to f in the subspace W is $g = \text{proj}_W f$.

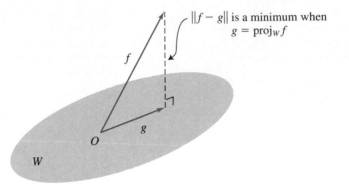

$\|f - g\|$ is a minimum when
$g = \text{proj}_W f$

Figure 6.6

We can again use our geometrical results to compute $\text{proj}_W f$. Let $\{g_1, \ldots, g_n\}$ be an orthonormal basis for W. Then we know from Section 5.2 that

$$\text{proj}_W f = \langle f, g_1 \rangle g_1 + \cdots + \langle f, g_n \rangle g_n$$

Example 1   Find the least squares linear approximation to $f(x) = e^x$ over the interval $[-1, 1]$.

Solution   Let the linear approximation be $g(x) = a + bx$. f is an element of $C[-1, 1]$, and g is an element of the subspace $P_1[-1, 1]$ of polynomials of degree less than or equal to 1 over $[-1, 1]$. The set $\{1, x\}$ is a basis for $P_1[-1, 1]$. We get

$$\langle 1, x \rangle = \int_{-1}^{1} (1 \cdot x)\, dx = 0$$

Thus the functions are orthogonal. The magnitudes of these vectors are given by

$$\|1\|^2 = \int_{-1}^{1} (1 \cdot 1)\, dx = 2 \quad \text{and} \quad \|x\|^2 = \int_{-1}^{1} (x \cdot x)\, dx = \frac{2}{3}$$

Thus the set $\left\{ \dfrac{1}{\sqrt{2}}, \sqrt{\dfrac{3}{2}} x \right\}$ is an orthonormal basis for $P_1[-1, 1]$. We now get

$$
\begin{aligned}
\text{proj}_W f &= \langle f, g_1 \rangle g_1 + \langle f, g_2 \rangle g_2 \\
&= \int_{-1}^{1} \left(e^x \sqrt{\frac{1}{2}} \right) dx\, \frac{1}{\sqrt{2}} + \int_{-1}^{1} \left(e^x \sqrt{\frac{3}{2}} x \right) dx\, \sqrt{\frac{3}{2}} x \\
&= \frac{1}{2}(e - e^{-1}) + 3e^{-1}x
\end{aligned}
$$

The least squares linear approximation to $f(x) = e^x$ over the interval $[-1, 1]$ is $g(x) = \frac{1}{2}(e - e^{-1}) + 3e^{-1}x$. This gives $g(x) = 1.18 + 1.1x$, to two decimal places. See Figure 6.7.

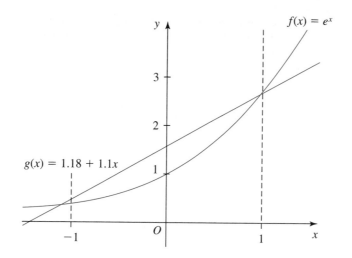

Figure 6.7

In the preceding example we found the linear approximation to f in $P_1[-1, 1]$. Higher-degree polynomial approximations can be found in the space $P_n[-1, 1]$ of polynomials of degree less than or equal to n. To arrive at the approximations, an orthogonal basis has to be constructed in $P_n[-1, 1]$ using the Gram-Schmidt orthogonalization process. The orthogonal functions found in this manner are called **Legendre polynomials**. The first six Legendre polynomials are

$$1, \qquad x, \qquad x^2 - \tfrac{1}{3}, \qquad x^3 - \tfrac{3}{5}x, \qquad x^4 + \tfrac{6}{7}x^2 + \tfrac{3}{35}, \qquad x^5 - \tfrac{10}{9}x^3 + \tfrac{5}{21}x$$

We next look at approximations of functions in terms of trigonometric functions. Such approximations are widely used in fields such as heat conduction, electromagnetism, electric circuits, and mechanical vibrations. The initial work in this area was undertaken by Jean Baptiste Fourier,* a French mathematician who developed the methods to analyze conduction in an insulated bar. The approximations are often used in discussing solutions to partial differential equations that describe physical situations.

Fourier Approximations

Let f be a function in $C[-\pi, \pi]$ with inner product $\langle f, g \rangle = \int_{-\pi}^{\pi} f(x)g(x)\, dx$, and geometry defined by this inner product. Let us find the least squares approximation of f in the space $T[-\pi, \pi]$ of trigonometric polynomials spanned by the set $\{1, \cos x, \sin x, \ldots, \cos nx, \sin nx\}$, where n is a positive integer.

---

* Jean Baptiste Fourier (1768–1830) lost his parents at an early age and was placed in a military school, where he developed his passion for mathematics. He was jailed during the French Revolution for his defense of the victims of the Terror. He later held diplomatic positions under Napoleon in Egypt and France, who made him both a baron and a count. Fourier is best known for the important mathematical techniques that he developed in his work on diffusion of heat.

It can be shown that the vectors 1, cos x, sin x, . . . , cos nx, sin nx are mutually orthogonal in this space (see the following exercises). The magnitudes of these vectors are given by

$$\|1\|^2 = \int_{-\pi}^{\pi} (1 \cdot 1) \, dx = 2\pi, \qquad \|\cos nx\|^2 = \int_{-\pi}^{\pi} (\cos nx \cdot \cos nx) \, dx = \pi,$$

$$\|\sin nx\|^2 = \int_{-\pi}^{\pi} (\sin nx \cdot \sin nx) \, dx = \pi$$

Thus the following set is an orthonormal basis for $T[-\pi, \pi]$.

$$\{g_0, \ldots, g_{2n}\} = \left\{ \frac{1}{\sqrt{2\pi}}, \frac{1}{\sqrt{\pi}} \cos x, \frac{1}{\sqrt{\pi}} \sin x, \ldots, \frac{1}{\sqrt{\pi}} \cos nx, \frac{1}{\sqrt{\pi}} \sin nx \right\}$$

Let us use this orthonormal basis in the following formula to find the least squares approximation g of f.

$$g(x) = \text{proj}_T f = \langle f, g_0 \rangle g_0 + \cdots + \langle f, g_{2n} \rangle g_{2n}$$

We get

$$g(x) = \left\langle f, \frac{1}{\sqrt{2\pi}} \right\rangle \frac{1}{\sqrt{2\pi}} + \left\langle f, \frac{1}{\sqrt{\pi}} \cos x \right\rangle \frac{1}{\sqrt{\pi}} \cos x + \left\langle f, \frac{1}{\sqrt{\pi}} \sin x \right\rangle \frac{1}{\sqrt{\pi}} \sin x + \cdots +$$

$$\left\langle f, \frac{1}{\sqrt{\pi}} \cos nx \right\rangle \frac{1}{\sqrt{\pi}} \cos nx + \left\langle f, \frac{1}{\sqrt{\pi}} \sin nx \right\rangle \frac{1}{\sqrt{\pi}} \sin nx$$

Let us introduce the following convenient notation.

$$a_0 = \left\langle f, \frac{1}{\sqrt{2\pi}} \right\rangle \frac{1}{\sqrt{2\pi}} = \int_{-\pi}^{\pi} \left(f(x) \cdot \frac{1}{\sqrt{2\pi}} \right) dx \frac{1}{\sqrt{2\pi}} = \frac{1}{2\pi} \int_{-\pi}^{\pi} f(x) \, dx$$

$$a_k = \left\langle f, \frac{1}{\sqrt{\pi}} \cos kx \right\rangle \frac{1}{\sqrt{\pi}} = \int_{-\pi}^{\pi} \left(f(x) \cdot \frac{1}{\sqrt{\pi}} \cos kx \right) dx \frac{1}{\sqrt{\pi}}$$

$$= \frac{1}{\pi} \int_{-\pi}^{\pi} f(x) \cos kx \, dx$$

$$b_k = \left\langle f, \frac{1}{\sqrt{\pi}} \sin kx \right\rangle \frac{1}{\sqrt{\pi}} = \int_{-\pi}^{\pi} \left(f(x) \cdot \frac{1}{\sqrt{\pi}} \sin kx \right) dx \frac{1}{\sqrt{\pi}}$$

$$= \frac{1}{\pi} \int_{-\pi}^{\pi} f(x) \sin kx \, dx$$

The trigonometric approximation of $f(x)$ can now be written:

$$g(x) = a_0 + \sum_{k=1}^{n} (a_k \cos kx + b_k \sin kx)$$

$g(x)$ is called the **nth-order Fourier approximation** of $f(x)$. The coefficients $a_0, a_1, b_1, \ldots, a_n, b_n$ are called **Fourier coefficients**.

As n increases, this approximation naturally becomes an increasingly better approximation in the sense that $\|f - g\|$ gets smaller. The infinite sum

$$g(x) = a_0 + \sum_{k=1}^{\infty} (a_k \cos kx + b_k \sin kx)$$

is known as the **Fourier series** of f on the interval $[-\pi, \pi]$.

Example 2   Find the fourth-order Fourier approximation to $f(x) = x$ over the interval $[-\pi, \pi]$.

Solution   Using the above Fourier coefficients with $f(x) = x$, and using integration by parts, we get

$$a_0 = \frac{1}{2\pi} \int_{-\pi}^{\pi} f(x)\, dx = \frac{1}{2\pi} \int_{-\pi}^{\pi} x\, dx = \frac{1}{2\pi}\left[\frac{x^2}{2}\right]_{-\pi}^{\pi} = 0$$

$$a_k = \frac{1}{\pi} \int_{-\pi}^{\pi} (f(x) \cdot \cos kx)\, dx = \frac{1}{\pi} \int_{-\pi}^{\pi} (x \cdot \cos kx)\, dx$$

$$= \frac{1}{\pi}\left[\frac{x}{k} \sin kx + \frac{1}{k^2} \cos kx\right]_{-\pi}^{\pi} = 0$$

$$b_k = \frac{1}{\pi} \int_{-\pi}^{\pi} (f(x)\sin kx)\, dx = \frac{1}{\pi} \int_{-\pi}^{\pi} (x \cdot \sin kx)\, dx$$

$$= \frac{1}{\pi}\left[-\frac{x}{k} \cos kx + \frac{1}{k^2} \sin kx\right]_{-\pi}^{\pi} = \frac{2(-1)^{k+1}}{k}$$

The Fourier approximation of f is

$$g(x) = \sum_{k=1}^{n} \frac{2(-1)^{k+1}}{k} \sin kx$$

Taking $k = 1, \ldots, 4$ we get fourth-order approximation

$$g(x) = 2\left(\sin x - \tfrac{1}{2} \sin 2x + \tfrac{1}{3} \sin 3x - \tfrac{1}{4} \sin 4x\right)$$

We have been primarily interested in real vector spaces in this course—that is, vector spaces where the scalars are real numbers. We have occasionally mentioned complex vector spaces, where the scalars are complex numbers. Our introduction to vector spaces would not be complete without illustrating an application of vector spaces where the scalars are not real or complex numbers. Furthermore, this example is interesting in that there are only a finite number of vectors in the spaces.

Coding Theory

Messages are sent electronically as sequences of 0s and 1s. Errors can occur in messages due to noise or interference. For example, the message (1, 1, 0, 1, 1, 0, 1) could be sent, and the message (1, 1, 0, 1, 0, 0, 1) might be received. An error has occurred in the fifth entry. We shall now look at methods that are used to detect and correct such errors.

The scalars in a vector field can be sets other than real or complex numbers. The requirement is that the scalars form an algebraic **field**. A field is a set of elements with two operations that satisfy certain axioms. Readers who go on to take a course in modern algebra will study fields. In this example we shall use the field $\{0, 1\}$ of scalars—having only these two elements—with operations of addition and multiplication defined as follows:

$$0 + 0 = 0, \qquad 0 + 1 = 1, \qquad 1 + 0 = 1, \qquad 1 + 1 = 0$$
$$0 \cdot 0 = 0, \qquad 0 \cdot 1 = 0, \qquad 1 \cdot 0 = 0, \qquad 1 \cdot 1 = 1$$

Let V_7 be the vector space of 7-tuples of 0s and 1s over this field of scalars, where addition and scalar multiplication are defined in the usual componentwise manner. For example,

$$(1, 0, 0, 1, 1, 0, 1) + (0, 1, 1, 1, 0, 0, 1)$$
$$= (1 + 0, 0 + 1, 0 + 1, 1 + 1, 1 + 0, 0 + 0, 1 + 1)$$
$$= (1, 1, 1, 0, 1, 0, 0)$$
$$0(1, 0, 0, 1, 1, 0, 1) = (0 \cdot 1, 0 \cdot 0, 0 \cdot 0, 0 \cdot 1, 0 \cdot 1, 0 \cdot 0, 0 \cdot 1) = (0, 0, 0, 0, 0, 0, 0)$$
$$1(1, 0, 0, 1, 1, 0, 1) = (1 \cdot 1, 1 \cdot 0, 1 \cdot 0, 1 \cdot 1, 1 \cdot 1, 1 \cdot 0, 1 \cdot 1) = (1, 0, 0, 1, 1, 0, 1)$$

Since each vector in V_7 has seven components, and each of these components can be either 0 or 1, there are 2^7 vectors in this space. The four-dimensional subspace of V_7 having basis

$$\{(1, 0, 0, 0, 0, 1, 1), (0, 1, 0, 0, 1, 0, 1), (0, 0, 1, 0, 1, 1, 0), (0, 0, 0, 1, 1, 1, 1)\}$$

is called a **Hamming code** and is denoted $C_{7,4}$. The vectors in $C_{7,4}$ can be used as messages. Each vector in $C_{7,4}$ can be written

$$\mathbf{v}_i = a_1(1, 0, 0, 0, 0, 1, 1) + a_2(0, 1, 0, 0, 1, 0, 1)$$
$$+ a_3(0, 0, 1, 0, 1, 1, 0) + a_4(0, 0, 0, 1, 1, 1, 1)$$

Since each of the four scalars a_1, a_2, a_3, and a_4 can take one of the values 1 or 0, we have 2^4—that is, 16—vectors in $C_{7,4}$. The Hamming code can thus be used to send 16 different messages, $\mathbf{v}_1, \ldots, \mathbf{v}_{16}$. You are asked to list these vectors in the following exercises.

When an error occurs in one location of a transmitted message, the resulting incorrect vector lies in V_7, outside the subspace $C_{7,4}$. It can be proved that there is exactly one vector in $C_{7,4}$ that differs from this incorrect vector in one location. Thus the error can be detected and corrected. For example, suppose the received vector is $(1, 0, 1, 1, 1, 0, 1)$. This vector cannot be expressed as a linear combination of the above base vectors; it is not in $C_{7,4}$. There is a single vector in $C_{7,4}$ that differs from this vector in one entry, namely $(1, 0, 1, 0, 1, 0, 1)$. The correct message is therefore $(1, 0, 1, 0, 1, 0, 1)$. The Hamming code is called an **error-correcting code** because of this facility to detect and correct errors.

Let us look at the geometry underlying this code. The distance between two vectors **u** and **w** in V_7, denoted $d(\mathbf{u}, \mathbf{w})$, is defined to be the number of components that differ in **u** and **w**. Thus, for example,

$$d((1, 0, 0, 1, 0, 1, 1), (1, 1, 0, 0, 1, 1, 1)) = 3$$

since the second, fourth, and fifth components of these vectors differ. A sphere of radius 1 about a vector in V_7 contains all those vectors that differ in one component from the vector. It can be shown that the spheres of radii 1, centered at the vectors of $C_{7,4}$, will be disjoint and that every element of V_7 lies in one such sphere. See Figure 6.8. Thus, if a vector **u** is received that is not in $C_{7,4}$, then it will be in a single sphere having center $\mathbf{v}_i$, the correct message. In practice, electrical circuits called gates are used to test whether the received message is in $C_{7,4}$, and if it is not, to determine the center of the sphere in which it lies, giving the correct message.

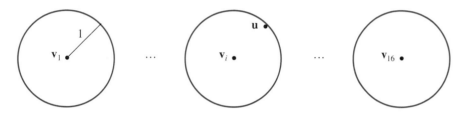

Figure 6.8

Hamming code $\{\mathbf{v}_1, \ldots, \mathbf{v}_{16}\}$
Incorrect message **u**; correct message $\mathbf{v}_i$

Other error-detecting codes exist. The **Golay code**, for example, is a twelve-dimensional subspace of V_{23} that is denoted $C_{23,12}$. The code space has 2^{12} elements that can be used to represent 4096 messages. This code can detect and correct errors in one, two, or three locations. It can be shown that the spheres of radii 3, centered at the vectors of $C_{23,12}$, are disjoint and that every element of V_{23} lies in one such sphere. If a vector **u** that is received has an error in one, two, or three locations, it will lie in a sphere centered at the correct message $\mathbf{v}_i$, and $\mathbf{v}_i$ can be retrieved.

A good introduction to coding theory can be found in *A Common Sense Approach to the Theory of Error Correcting Codes* by Benjamin Arazi, The MIT Press, 1988.

Exercise Set 6.4

Least Squares Approximations

In Exercises 1–4, find the least squares linear approximations, $g(x) = a + bx$, to the given functions over the stated intervals.

*1. $f(x) = x^2$ over the interval $[-1, 1]$

*2. $f(x) = e^x$ over $[0, 1]$

3. $f(x) = \sqrt{x}$ over $[0, 1]$

*4. $f(x) = \cos x$ over $[0, \pi]$

In Exercises 5–8, find the least squares quadratic approximations, $g(x) = a + bx + cx^2$, to the given functions over the stated intervals.

5. $f(x) = e^x$ over *(**a**) $[-1, 1]$ and (**b**) $[0, 1]$

*6. $f(x) = \sqrt{x}$ over $[0, 1]$

7. $f(x) = x^2$ over $[0, 1]$

*8. $f(x) = \sin x$ over $[0, \pi]$

In Exercises 9–11, find the fourth-order Fourier approximations to the given functions over the stated intervals.

*9. $f(x) = x$ over $[0, 2\pi]$

10. $f(x) = 1 + x$ over $[-\pi, \pi]$

*11. $f(x) = x^2$ over $[0, 2\pi]$

12. Show that the vectors $1, \cos x, \sin x, \dots, \cos nx, \sin nx$ are mutually orthogonal in the space $C[-\pi, \pi]$.

Coding Theory

*13. List all the vectors in the Hamming code $C_{7,4}$.

14. List all the vectors of V_7 that lie in the sphere having radius 1 and the following centers.

*(a) $(0, 0, 1, 0, 1, 1, 0)$ ***(b)** $(1, 1, 0, 1, 0, 0, 1)$

(c) $(1, 1, 1, 1, 1, 1, 1)$ **(d)** $(0, 0, 0, 0, 0, 0, 0)$

15. The following messages are received using the Hamming code. Find the correct message in each case.

*(a) $(0, 0, 0, 0, 1, 0, 0)$ **(b)** $(0, 1, 1, 1, 0, 0, 1)$

*(c) $(1, 0, 1, 0, 1, 1, 0)$ **(d)** $(1, 0, 1, 1, 0, 1, 0)$

16. Consider the vector space V_{23} and the Golay code subspace $C_{23,12}$. Prove that there are ***(a)** 2^{23} vectors in V_{23}, **(b)** 4096 vectors in $C_{23,12}$, ***(c)** 2048 vectors in each sphere of radius 3 about a vector in $C_{23,12}$ (given that each element of V_{23} is in one sphere). How many vectors of distance 1, 2, and 3 are in each sphere of radius 3?

Review Exercises Chapter 6

1. Let $f(x) = 3x - 1$ and $g(x) = 2x^2 + 3$. Compute the functions $f + g$, $3f$, and $2f - 3g$.

2. Let S be the set of all functions of the form

$$f(x) = 3x^2 + ax - b$$

where a and b are real numbers. Is S a subspace of P_2?

3. Prove that $f(x) = 13x^2 + 8x - 21$ is a linear combination of $g(x) = 2x^2 + x - 3$ and $h(x) = -3x^2 - 2x + 5$.

4. Find a basis for the vector space of upper triangular 3×3 matrices.

5. Show that $\{x^2 + 2x - 3, 3x^2 + x - 1, 4x^2 + 3x - 3\}$ is a basis for P_2.

6. Let $\mathbf{u} = (x_1, x_2)$ and $\mathbf{v} = (y_1, y_2)$ be elements of $\mathbf{R}^2$. Prove that the following function defines an inner product on $\mathbf{R}^2$.

$$\langle \mathbf{u}, \mathbf{v} \rangle = 2x_1 y_1 + 3x_2 y_2$$

7. Determine the inner product, the norms, and the distance between the functions $f(x) = 3x - 1$, $g(x) = 5x + 3$ in the inner product space P_2, with inner product defined by $\langle f, g \rangle = \int_0^1 f(x)g(x)\, dx$.

8. Let $\mathbf{R}^2$ have an inner product defined by $\langle (x_1, x_2), (y_1, y_2) \rangle = 2x_1 y_1 + 3x_2 y_2$. Determine the norms and the distance between $(1, -2)$, $(3, 2)$ in this space.

9. Is the condition that a subset of a vector space contain the zero vector a necessary and sufficient condition for the subset to be a subspace?

10. Consider $\mathbf{R}^2$ with the inner product

$$\langle (x_1, x_2), (y_1, y_2) \rangle = 2x_1 y_1 + 3x_2 y_2$$

Determine the equation of the circle with center the origin and radius 1 in this space. Sketch the circle.

11. Find the least squares linear approximation to

$$f(x) = x^2 + 2x - 1$$

over the interval $[-1, 1]$.

12. Find the fourth-order Fourier approximation to $f(x) = 2x - 1$ over $[0, 2\pi]$.

13. List all the vectors of V_7 that lie in the sphere having radius 1 and center $(0, 1, 1, 0, 0, 1, 1)$.

MATLAB Discussion

M.18   Inner Product, Non-Euclidean Geometry
(Sections 6.2, 6.3)

Let **u** and **v** be row vectors in $\mathbf{R}^n$. Let A be an $n \times n$ real symmetric matrix. Suppose A is such that $\mathbf{u}A\mathbf{u}^t > 0$ for all nonzero vectors **u**. Such a matrix is said to be **positive definite**. An inner product on $\mathbf{R}^n$ is defined by $\langle \mathbf{u}, \mathbf{v} \rangle = \mathbf{u}A\mathbf{v}^t$. This inner product leads to non-Euclidean geometries, where

$$\|\mathbf{u}\| = \sqrt{\langle \mathbf{u}, \mathbf{u} \rangle}, \quad \cos \theta = \frac{\langle \mathbf{u}, \mathbf{v} \rangle}{\|\mathbf{u}\| \, \|\mathbf{v}\|}, \quad d(X, Y) = \|X - Y\|$$

MATLAB functions have been written to explore these non-Euclidean spaces where

$$\langle \mathbf{u}, \mathbf{v} \rangle = \mathbf{u}A\mathbf{v}^t$$

The "g" in each of these functions stands for "generalized."

> **ginner**(u,v,A): inner product of **u** and **v**
>
> **gnorm**(u,A): norm of **u**
>
> **gangle**(u,v,A): angle between **u** and **v** in degrees
>
> **gdist**(X,Y,A): distance between the points X and Y
>
> **circles**(A): sketches the circles of radii 1, 2, and 3 in $\mathbf{R}^2$

Example 1   Consider $\mathbf{R}^2$ with an inner product defined by the matrix $\mathbf{A} = \begin{bmatrix} 1 & 0 \\ 0 & 4 \end{bmatrix}$. This inner product is positive definite. Let us use MATLAB to determine

(a) the norm of the vector (0, 1).

(b) the angle between the vectors (1, 1) and (−4, 1).

(c) the distance between the points (1, 0) and (0, 1).

(d) Graph the circles having centers at the origin and radii 1, 2, and 3 in this space.

Using the above functions, we get

```
»A = [1  0;0  4];   u = [0  1];
»gnorm(u,A)
ans =
        2            {the magnitude of (0, 1) is 2 in this space; it is 1 in Euclidean space}
»u = [1  1];   v = [−4  1];
»gangle(u,v,A)
ans =
       90            {units are degrees—the vectors are at right angles in this space}
```

»X = [1 0];   Y = [0 1];
»gdist(X,Y,A)
ans =
    2.2361   {the distance between the points (1, 0) and (0, 1) is $\sqrt{5}$ in this space}
                  {it is $\sqrt{2}$ in Euclidean space}
»**circles(A)**
★★**computing pts for circle radius 1**★★
★★**computing pts for circle radius 2**★★
★★**computing pts for circle radius 3**★★

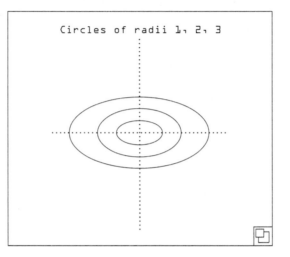

Circles of radii 1, 2, 3

Note: Use "Enter" and "shg" (or "figure(gcf)") to alternate between command and graph windows on IBMs.

Observe that the circles appear as ellipses in this geometry. The reader is asked to determine the equations of these circles in the exercises that follow.

*Exercise 1   Consider the above non-Euclidean geometry on $\mathbf{R}^2$, having an inner product defined by the matrix $A = \begin{bmatrix} 1 & 0 \\ 0 & 4 \end{bmatrix}$. Determine the following using mathematics, and compare your results with the above computer output.

(a)   The norm of the vector (0, 1)

(b)   The angle between the vectors (1, 1) and (−4, 1)

(c)   The distance between the points (1, 0) and (0, 1)

(d)   The equations of the circles having centers at the origin and radii 1, 2, and 3

(e)   Prove that the matrix A is positive definite

Exercise 2   Find the matrix that defines the dot product, and hence Euclidean geometry on $\mathbf{R}^2$. Use this matrix to sketch the circles of radii 1, 2, and 3, having centers at the origin, in Euclidean space using MATLAB.

Exercise 3   Show that the following matrix A is positive definite and hence can be used to define a non-Euclidean geometry on $\mathbf{R}^2$.

$$A = \begin{bmatrix} 6 & 2 \\ 2 & 9 \end{bmatrix}$$

Use the function **circles**(A) to sketch the circles having centers at the origin and radii 1, 2, and 3 in this space. Find the equations of these circles.

Exercise 4   The following matrix A is not positive definite. It can, however, be used to define a pseudo inner product $\langle \mathbf{u}, \mathbf{v} \rangle = \mathbf{u}A\mathbf{v}^t$ on $\mathbf{R}^2$ (the positive definite axiom is violated), with $\|\mathbf{u}\| = \sqrt{|\langle \mathbf{u}, \mathbf{u} \rangle|}$.

$$A = \begin{bmatrix} -1 & 0 \\ 0 & 1 \end{bmatrix}$$

(a)   Find a vector $\mathbf{u}$ such that $\|\mathbf{u}\| = 0$, but $\mathbf{u} \neq \mathbf{0}$. (This proves that A is not positive definite.)

(b)   Find all vectors $\mathbf{u}$ in $\mathbf{R}^2$ such that $\|\mathbf{u}\| = 0$, $\mathbf{u} \neq \mathbf{0}$. Sketch these vectors.

(c)   Find two distinct points in this space that are zero distance apart. Can you generalize—that is, find the relationship between points that are zero distance apart?

(d)   Use **circles**(A) to sketch the circles center the origin and radii 1, 2, and 3 in this space. Find the equations of these circles.

This space is two-dimensional **Minkowski space**. Four-dimensional Minkowski space is the space of special relativity. It is discussed in the following section.

Exercise 5   Use **circles**(A) to sketch circles having centers at the origin and radii 1, 2, and 3 in the spaces with inner products defined by the following matrices

(a) $A = \begin{bmatrix} 1 & 0 \\ 0 & 0 \end{bmatrix}$,   **(b)** $B = \begin{bmatrix} 0 & 0 \\ 0 & -1 \end{bmatrix}$,   **(c)** $C = \begin{bmatrix} 1 & 1 \\ 1 & 1 \end{bmatrix}$

Determine the equations of these circles.

Exercise 6   Minkowski geometry on $\mathbf{R}^3$ is defined using the pseudo inner product $\langle \mathbf{u}, \mathbf{v} \rangle = \mathbf{u}A\mathbf{v}^t$, where

$$A = \begin{bmatrix} -1 & 0 & 0 \\ 0 & -1 & 0 \\ 0 & 0 & 1 \end{bmatrix}$$

(a)   Find all vectors $\mathbf{u}$ in $\mathbf{R}^3$ such that $\langle \mathbf{u}, \mathbf{u} \rangle = 0$, $\mathbf{u} \neq \mathbf{0}$.

(b)   Determine the spheres with their centers at the origin and radii 1, 2, and 4.

M.19   Space-Time Travel (Section 6.3)

The geometry of Minkowski space can be used to describe travel through space-time. An astronaut goes to Alpha Centauri, which is 4 light-years from Earth, with 0.8 the speed of light and returns to Earth. We can use the function **space**, based on the pseudo inner product of Minkowski space, to simulate a space voyage and predict the duration of the voyage, both from the point of view of Earth and the astronaut.

»**space**
Give the speed: .8
  the distance: 4
  name of astronaut: Geraint
Nationality
English(E), French(F), German(G), Spanish(S), Turkish(T), Welsh(W): W

	Times
Earth	**Spaceship**
E-time	**S-time**
1.0000	**0.6000**
2.0000	**1.2000**
3.0000	**1.8000**
4.0000	**2.4000**
5.0000	**3.0000**
6.0000	**3.6000**
7.0000	**4.2000**
8.0000	**4.8000**
9.0000	**5.4000**
10.0000	**6.0000**

times=
10.0000   **6.0000**

Graphics window gives picture of voyage

```
                Space-Time Voyage
Croeso adref Geraint!
        o <                                      *
        o        <                               *
        o              <                          *
        o                   <                     *
        o                        <              * 
Earth   o                             < >* star
path    o                               > * path
        o                             >    *
        o                        >         *
        o              >         Spaceship *
        o >        >              path     *
```

The time lapse on Earth is 10 years; on the spaceship it is 6 years.

Note: Use "Enter" and "shg" (or "figure(gcf)") to alternate between command and graph windows on IBMs.

*Exercise 1   A spaceship goes off to Capella, 45 light-years from Earth, at 0.75 the speed of light and returns to Earth. How long will the trip take both from Earth's point of view and from the point of view of a person on the spaceship?

Exercise 2   A spaceship goes on a round-trip to Pleiades, 410 light-years from Earth at 0.9999 the speed of light. How many years will have passed on Earth, and how many on the spaceship when the ship arrives back?

chapter

7

Linear Transformations

A function is a rule that

assigns to each element of one set exactly
one element of another set. Functions are
used in most areas of mathematics, and are
important in applications for describing the
dependency of one variable upon another.
In this chapter we introduce an important
class of functions between vector spaces,
called linear transformations. Linear transformations will give us further insight into the
behavior of systems of linear equations. We
shall see how linear transformations are
used in computer graphics to manipulate
pictures.

7.1 Introduction to Linear Transformations

Definition   *Let U and V be vector spaces. A **transformation** T of U into V, written T: U → V, is a rule that assigns to each vector **u** in U, a unique vector **v** in V. U is called the **domain** of T.*

*We write T(**u**) = **v**; **v** is called the **image** of **u** under T.*

*The terms **mapping** and **function** are also used for a transformation.*

For example, consider the transformation $T: \mathbf{R}^3 \to \mathbf{R}^2$, defined by

$$T(x, y, z) = (2x, y - z)$$

The domain of T is $\mathbf{R}^3$. The image of the vector $(1, 4, -2)$ under T is $(2, 6)$.

A vector space has two operations defined on it, namely addition and scalar multiplication. The most important transformations between vector spaces are those that **preserve** these linear structures in the following sense.

Definition   *Let U and V be vector spaces. Let **u**₁ and **u**₂ be vectors in U, and let c be a scalar. A transformation T: U → V is said to be **linear** if*

$$T(\mathbf{u}_1 + \mathbf{u}_2) = T(\mathbf{u}_1) + T(\mathbf{u}_2)$$
$$T(c\mathbf{u}_1) = cT(\mathbf{u}_1)$$

The first condition implies that T maps the sum of two vectors into the sum of the images of those vectors. The second condition implies that T maps the scalar multiple of a vector into the same scalar multiple of the image. Thus the operations of addition and scalar multiplication are preserved under the mapping. These structure-preserving ideas of a linear transformation are illustrated in Figure 7.1.

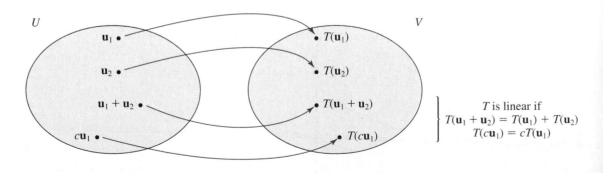

Figure 7.1

Example 1   Prove that the following transformation $T: \mathbf{R}^2 \to \mathbf{R}^2$ is linear.

$$T(x, y) = (x - y, 3x)$$

Solution   We first show that T preserves addition. Let (x_1, y_1) and (x_2, y_2) be elements of $\mathbf{R}^2$. Then

$$
\begin{aligned}
T((x_1, y_1) + (x_2, y_2)) &= T(x_1 + x_2, y_1 + y_2) && \text{by vector addition} \\
&= (x_1 + x_2 - y_1 - y_2, 3x_1 + 3x_2) && \text{by definition of } T \\
&= (x_1 - y_1, 3x_1) + (x_2 - y_2, 3x_2) && \text{by vector addition} \\
&= T(x_1, y_1) + T(x_2, y_2) && \text{by definition of } T
\end{aligned}
$$

Thus T preserves vector addition.

We now show that T preserves scalar multiplication. Let c be a scalar. Then

$$
\begin{aligned}
T(c(x_1, y_1)) &= T(cx_1, cy_1) && \text{by scalar multiplication of a vector} \\
&= (cx_1 - cy_1, 3cx_1) && \text{by definition of } T \\
&= c(x_1 - y_1, 3x_1) && \text{by scalar multiplication of a vector} \\
&= cT(x_1, y_1) && \text{by definition of } T
\end{aligned}
$$

Thus T preserves scalar multiplication. T is linear.

*A linear transformation of a vector space into itself is called a **linear operator**.* $T: \mathbf{R}^2 \to \mathbf{R}^2$ defined by $T(x, y) = (x - y, 3x)$ is an example of a linear operator.

The following example illustrates a transformation that is not linear.

Example 2   Show that the transformation $T: \mathbf{R}^3 \to \mathbf{R}^2$ is not linear.

$$T(x, y, z) = (xy, z)$$

Solution   Let us first test addition. Let (x_1, y_1, z_1) and (x_2, y_2, z_2) be elements of $\mathbf{R}^3$. Then

$$
\begin{aligned}
T((x_1, y_1, z_1) + (x_2, y_2, z_2)) &= T(x_1 + x_2, y_1 + y_2, z_1 + z_2) \\
&= ((x_1 + x_2)(y_1 + y_2), z_1 + z_2) \\
&= (x_1 y_1 + x_2 y_2 + x_1 y_2 + x_2 y_1, z_1 + z_2)
\end{aligned}
$$

and

$$
\begin{aligned}
T(x_1, y_1, z_1) + T(x_2, y_2, z_2) &= (x_1 y_1, z_1) + (x_2 y_2, z_2) \\
&= (x_1 y_1 + x_2 y_2, z_1 + z_2)
\end{aligned}
$$

Thus, in general,

$$T((x_1, y_1, z_1) + (x_2, y_2, z_2)) \neq T(x_1, y_1, z_1) + T(x_2, y_2, z_2)$$

Since vector addition is not preserved, T is not linear.

Example 3   Let P_n be the vector space of real polynomial functions of degree $\leq n$. Show that the following transformation $T: P_2 \rightarrow P_1$ is linear.

$$T(a_2 x^2 + a_1 x + a_0) = (a_2 + a_1)x + a_0$$

Solution   Let $a_2 x^2 + a_1 x + a_0$ and $b_2 x^2 + b_1 x + b_0$ be elements of P_2, and let c be a scalar. Then

$$
\begin{aligned}
T((a_2 x^2 &+ a_1 x + a_0) + (b_2 x^2 + b_1 x + b_0)) \\
&= T((a_2 + b_2)x^2 + (a_1 + b_1)x + (a_0 + b_0)) \\
&= (a_2 + b_2 + a_1 + b_1)x + (a_0 + b_0) \\
&= (a_2 + a_1)x + a_0 + (b_2 + b_1)x + b_0 \\
&= T(a_2 x^2 + a_1 x + a_0) + T(b_2 x^2 + b_1 x + b_0)
\end{aligned}
$$

Thus T preserves addition.

$$
\begin{aligned}
T(c(a_2 x^2 + a_1 x + a_0)) &= T(ca_2 x^2 + ca_1 x + ca_0) \\
&= (ca_2 + ca_1)x + ca_0 \\
&= c((a_2 + a_1)x + a_0) \\
&= cT(a_2 x^2 + a_1 x + a_0)
\end{aligned}
$$

T preserves scalar multiplication. Therefore T is linear.

Example 4   Let D be the operation of taking the derivative. D can be interpreted as a mapping of P_n into itself. For example,

$$D(4x^3 - 3x^2 + 2x + 1) = 12x^2 - 6x + 2$$

D maps the element $4x^3 - 3x^2 + 2x + 1$ of P_3 into the element $12x^2 - 6x + 2$ of P_3. Let us show that D is a linear operator on P_n.

Let f and g be elements of P_n and c be a scalar. The following properties of the derivative imply that D is a linear operator.

$$D(f + g) = Df + Dg$$
$$D(cf) = cD(f)$$

We now find that *every matrix defines a linear transformation*. We shall use this result a great deal.

Theorem 7.1 Let A be an $m \times n$ matrix. Let $\mathbf{x}$ be an element of $\mathbf{R}^n$, written as a column matrix. The transformation $T: \mathbf{R}^n \to \mathbf{R}^m$, defined by $T(\mathbf{x}) = A\mathbf{x}$, is linear. Such a linear transformation is called a **matrix transformation**.

Proof Observe that since A is an $m \times n$ matrix and $\mathbf{x}$ is an $n \times 1$ matrix, then $A\mathbf{x}$ is an $m \times 1$ matrix. Thus $T(\mathbf{x}) = A\mathbf{x}$ defines a transformation of $\mathbf{R}^n$ into $\mathbf{R}^m$. It remains to prove that T is linear. Let $\mathbf{x}$ and $\mathbf{y}$ be elements of $\mathbf{R}^n$, written as column matrices, and c be a scalar. We know from the definition of T and the properties of matrices that

$$T(\mathbf{x} + \mathbf{y}) = A(\mathbf{x} + \mathbf{y})$$
$$= A\mathbf{x} + A\mathbf{y}$$
$$= T(\mathbf{x}) + T(\mathbf{y}) \quad \text{addition is preserved}$$

$$T(c\mathbf{x}) = A(c\mathbf{x})$$
$$= cA\mathbf{x}$$
$$= cT(\mathbf{x}) \quad \text{scalar multiplication is preserved}$$

Thus the transformation is linear.

Example 5 Consider the linear transformation T, defined by the following 3×2 matrix A. Find the image of an arbitrary vector under T, and use this result to determine the image of the given vector $\mathbf{x}$.

$$A = \begin{bmatrix} 1 & -1 \\ 0 & 2 \\ 1 & 3 \end{bmatrix}, \quad \mathbf{x} = \begin{bmatrix} 5 \\ -1 \end{bmatrix}$$

Solution Since A is a 3×2 matrix, it defines a linear transformation $T: \mathbf{R}^2 \to \mathbf{R}^3$. We get

$$T\left(\begin{bmatrix} x \\ y \end{bmatrix}\right) = \begin{bmatrix} 1 & -1 \\ 0 & 2 \\ 1 & 3 \end{bmatrix} \begin{bmatrix} x \\ y \end{bmatrix} = \begin{bmatrix} x - y \\ 2y \\ x + 3y \end{bmatrix}$$

Letting $x = 5$, $y = -1$ we get

$$T\left(\begin{bmatrix} 5 \\ -1 \end{bmatrix}\right) = \begin{bmatrix} 6 \\ -2 \\ 2 \end{bmatrix}$$

See Figure 7.2.

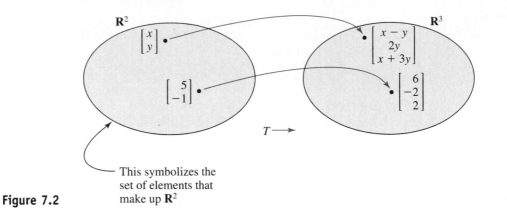

Figure 7.2

This symbolizes the
set of elements that
make up $\mathbf{R}^2$

We are familiar with the concept of combining functions into composite functions. Linear transformations can be similarly combined.

Composition of Linear Transformations

Let U, V, and W be vector spaces and $T_1: U \rightarrow V$, $T_2: V \rightarrow W$ be linear transformations between these spaces. These two transformations can be combined into a single **composite transformation** $T: U \rightarrow W$ as follows. Let $\mathbf{u}$ be a vector in U. Then

$$T(\mathbf{u}) = T_2[T_1(\mathbf{u})]$$

This composite transformation is denoted by $T = T_2 \circ T_1$. See Figure 7.3.

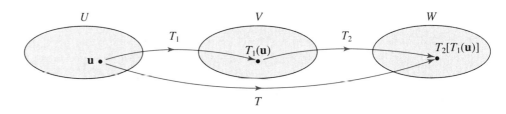

Figure 7.3   The composite transformation $T = T_2 \circ T_1$

Theorem 7.2   The composite of two linear transformations is itself a linear transformation.

Proof   Let U, V, and W be vector spaces, and let $T_1: U \rightarrow V$, $T_2: V \rightarrow W$ be linear transformations. Let T be the composite transformation $T = T_2 \circ T_1$. Let $\mathbf{u}$ and $\mathbf{v}$ be vectors in U, and c be a scalar. We use the linearity of T_1 and T_2 to get

$$T(\mathbf{u}_1 + \mathbf{u}_2) = T_2[T_1(\mathbf{u}_1 + \mathbf{u}_2)] = T_2[T_1(\mathbf{u}_1) + T_1(\mathbf{u}_2)]$$
$$= T_2[T_1(\mathbf{u}_1)] + T_2[T_1(\mathbf{u}_2)] = T(\mathbf{u}_1) + T(\mathbf{u}_2)$$

$$T(c\mathbf{u}_1) = T_2[T_1(c\mathbf{u}_1)] = T_2[cT_1(\mathbf{u}_1)] = cT_2[T_1(\mathbf{u}_1)] = cT(\mathbf{u}_1)$$

Thus T is linear.

Linear transformations defined by matrices are very important. We now discuss the composition of two matrix transformations.

Let $T_1: \mathbf{R}^n \to \mathbf{R}^m$ and $T_2: \mathbf{R}^m \to \mathbf{R}^s$ be matrix transformations defined by $T_1(\mathbf{x}) = A_1\mathbf{x}$ and $T_2(\mathbf{x}) = A_2\mathbf{x}$. Let T be the composite transformation $T = T_2 \circ T_1$. Then

$$T(\mathbf{x}) = T_2[T_1(\mathbf{x})] = T_2[A_1\mathbf{x}] = A_2A_1\mathbf{x}$$

Thus T is defined by the product matrix A_2A_1.

$$T(\mathbf{x}) = A_2A_1\mathbf{x}$$

Example 6 Let $T_1(\mathbf{x}) = A_1\mathbf{x}$ and $T_2(\mathbf{x}) = A_2\mathbf{x}$ be defined by the following matrices A_1 and A_2. Let $T = T_2 \circ T_1$. Find the image of the vector $\mathbf{x}$ under T.

$$A_1 = \begin{bmatrix} 3 & 0 & -1 \\ 4 & 2 & 0 \end{bmatrix}, \qquad A_2 = \begin{bmatrix} 1 & -2 \\ 4 & 0 \end{bmatrix}, \qquad \mathbf{x} = \begin{bmatrix} 1 \\ 4 \\ 2 \end{bmatrix}$$

Solution T is defined by the product matrix A_2A_1. We get

$$A_2A_1 = \begin{bmatrix} 1 & -2 \\ 4 & 0 \end{bmatrix} \begin{bmatrix} 3 & 0 & -1 \\ 4 & 2 & 0 \end{bmatrix} = \begin{bmatrix} -5 & -4 & -1 \\ 12 & 0 & -4 \end{bmatrix}$$

Thus

$$T(\mathbf{x}) = \begin{bmatrix} -5 & -4 & -1 \\ 12 & 0 & -4 \end{bmatrix} \begin{bmatrix} 1 \\ 4 \\ 2 \end{bmatrix} = \begin{bmatrix} -23 \\ 4 \end{bmatrix}$$

We can extend the results of this discussion in a natural way. Let $T_1, \ldots, T_n$, be a sequence of linear mappings on $\mathbf{R}^p, \ldots, \mathbf{R}^s$ defined by matrices $A_1, \ldots, A_n$. The composite $T = T_n \circ \cdots \circ T_1$ is given by the matrix product $A_n \ldots A_1$ (if this product exists).

Exercise Set 7.1

*1. Prove that the transformation $T: \mathbf{R}^2 \to \mathbf{R}^2$, defined by $T(x, y) = (2x, x - y)$, is linear. Find the images of the elements $(1, 2)$ and $(-1, 4)$ under this transformation.

2. Prove that $T(x, y) = (3x + y, 2y, x - y)$ defines a linear transformation $T: \mathbf{R}^2 \to \mathbf{R}^3$. Find the images of $(1, 2)$ and $(2, -5)$.

*3. Prove that $T: \mathbf{R}^3 \to \mathbf{R}^3$, defined by $T(x, y, z) = (0, y, 0)$, is linear. This transformation is also called a projection. Why is this term appropriate?

4. Prove that the following transformations $T: \mathbf{R}^3 \to \mathbf{R}^2$ are not linear.

   *(a) $T(x, y, z) = (3x, y^2)$

   (b) $T(x, y, z) = (x + 2, 4y)$

5. Prove that $T: \mathbf{R}^2 \to \mathbf{R}$, defined by $T(x, y) = x + a$, where a is a nonzero scalar, is not linear.

In Exercises 6–11, determine whether the given transformations are linear.

*6. $T(x, y, z) = (2x, y)$ of $\mathbf{R}^3 \to \mathbf{R}^2$

7. $T(x, y) = x - y$ of $\mathbf{R}^2 \to \mathbf{R}$

8. $T(x, y) = (x, y, z)$ of $\mathbf{R}^2 \to \mathbf{R}^3$ when ***(a)** $z = 0$, **(b)** $z = 1$

9. $T(x) = (x, 2x, 3x)$ of $\mathbf{R} \to \mathbf{R}^3$

*10. $T(x, y) = (x^2, y)$ of $\mathbf{R}^2 \to \mathbf{R}^2$

*11. $T(x, y, z) = (x + 2y, x + y + z, 3z)$ of $\mathbf{R}^3 \to \mathbf{R}^3$

*12. Consider the linear transformation $T: \mathbf{R}^2 \to \mathbf{R}^3$, defined by the following matrix A. Determine the images of the given vectors $\mathbf{x}$, $\mathbf{y}$, and $\mathbf{z}$.

$$A = \begin{bmatrix} 1 & 2 \\ -1 & 3 \\ 1 & 2 \end{bmatrix};$$

$$\mathbf{x} = \begin{bmatrix} -1 \\ 1 \end{bmatrix}, \quad \mathbf{y} = \begin{bmatrix} 2 \\ 3 \end{bmatrix}, \quad \mathbf{z} = \begin{bmatrix} 5 \\ 2 \end{bmatrix}$$

13. Consider the linear operator T on $\mathbf{R}^2$, defined by the following matrix A. Determine the images of the given vectors $\mathbf{x}$, $\mathbf{y}$, and $\mathbf{z}$.

$$A = \begin{bmatrix} 0 & -4 \\ 1 & 2 \end{bmatrix};$$

$$\mathbf{x} = \begin{bmatrix} 5 \\ -2 \end{bmatrix}, \quad \mathbf{y} = \begin{bmatrix} 4 \\ 0 \end{bmatrix}, \quad \mathbf{z} = \begin{bmatrix} 3 \\ 2 \end{bmatrix}$$

*14. Let A be an $m \times n$ matrix and $\mathbf{c}$ be a column vector in $\mathbf{R}^n$. Prove that the following transformation $T: \mathbf{R}^n \to \mathbf{R}^m$ is not linear if $\mathbf{c} \ne \mathbf{0}$: $T(\mathbf{x}) = A\mathbf{x} + \mathbf{c}$.

15. Let T be a linear transformation with domain U. If $\mathbf{v}$ and $\mathbf{w}$ are vectors in U, prove that **(a)** $T(-\mathbf{v}) = -T(\mathbf{v})$, **(b)** $T(\mathbf{v} - \mathbf{w}) = T(\mathbf{v}) - T(\mathbf{w})$.

16. Let T be a linear transformation with domain U. Let $\mathbf{v}_1, \ldots, \mathbf{v}_m$ be vectors in U, and $a_1, \ldots, a_m$ be scalars. Prove that

$$T(a_1\mathbf{v}_1 + \cdots + a_m\mathbf{v}_m) = a_1T(\mathbf{v}_1) + \cdots + a_mT(\mathbf{v}_m)$$

(Can you prove this result by induction?)

*17. Let $\mathbf{x}$ be an arbitrary vector in $\mathbf{R}^n$ and $\mathbf{y}$ a fixed vector. Prove that the transformation $T(\mathbf{x}) = \mathbf{x} \cdot \mathbf{y}$ is a linear transformation of $\mathbf{R}^n \to \mathbf{R}$.

18. Let J be the operation of taking a definite integral from 0 to 1. J can be used to define a mapping of P_n into $\mathbf{R}$. For example,

$$J(4x^3 - 6x^2 + 2x + 3) = \int_0^1 (4x^3 - 6x^2 + 2x + 3) \, dx$$
$$= [x^4 - 2x^3 + x^2 + 3x]_0^1 = 3$$

J maps the element $4x^3 - 6x^2 + 2x + 3$ of P_n into the element 3 of $\mathbf{R}$. Show that J is a linear mapping of $P_n \to \mathbf{R}$.

19. Let $T_1(\mathbf{x}) = A_1\mathbf{x}$ and $T_2(\mathbf{x}) = A_2\mathbf{x}$ be defined by the following matrices A_1 and A_2. Let $T = T_2 \circ T_1$. Find the matrix that defines T and use it to determine the image of the vector $\mathbf{x}$ under T.

*(a) $A_1 = \begin{bmatrix} 1 & 2 \\ 3 & 0 \end{bmatrix}$, $\quad A_2 = \begin{bmatrix} -1 & 0 \\ 1 & 5 \end{bmatrix}$, $\quad \mathbf{x} = \begin{bmatrix} 5 \\ 2 \end{bmatrix}$

(b) $A_1 = \begin{bmatrix} 0 & 1 & 2 \\ 3 & 4 & -1 \end{bmatrix}$,

$A_2 = \begin{bmatrix} 2 & 2 \\ 1 & -1 \end{bmatrix}$, $\quad \mathbf{x} = \begin{bmatrix} 0 \\ 1 \\ 3 \end{bmatrix}$

*(c) $A_1 = \begin{bmatrix} 3 & -2 \\ 0 & 1 \end{bmatrix}$,

$A_2 = \begin{bmatrix} 2 & 2 \\ 1 & -1 \\ 0 & 4 \end{bmatrix}$, $\quad \mathbf{x} = \begin{bmatrix} -3 \\ 2 \end{bmatrix}$

*20. Let $T_1(x, y) = (2x, -y)$ and $T_2(x, y) = (0, x + y)$. Let $T = T_2 \circ T_1$. Find an equation for T and use it to determine the image of $(2, -3)$ under T.

21. Let $T_1(x, y) = (3x + y, 4y)$ and $T_2(x, y) = (2y, x - y)$. Let $T = T_2 \circ T_1$. Find an equation for T and use it to determine the image of $(1, 2)$ under T.

*22. Let $T_1(x, y) = (x + y, 2x, 3y)$ and $T_2(x, y, z) = (x, x + y, 2x - z)$. Let $T = T_2 \circ T_1$. Find an equation for T and use it to determine the image of $(-2, 5)$ under T.

*23. Let U, V, and W be vector spaces. Let T_1 be a linear transformation of $U \to V$, and let T_2 be a linear transformation of $V \to W$. Is $T_2 \circ T_1 = T_1 \circ T_2$?

24. In this section we have defined a way of combining certain transformations T_1 and T_2 to get a composite transformation $T_2 \circ T_1$. We can also add certain transformations and multiply a transformation by a scalar. Let T_1 and T_2 both be linear transformations of $U \to V$, and c be a scalar. Let transformations $T_1 + T_2$ and cT_1 be defined by

$$(T_1 + T_2)(\mathbf{u}) = T_1(\mathbf{u}) + T_2(\mathbf{u}) \quad \text{and} \quad (cT_1)(\mathbf{u}) = cT_1(\mathbf{u})$$

(a) Prove that $T_1 + T_2$ and cT_1 are both linear transformations.

(b) Prove that if T_1 and T_2 are matrix transformations defined by the matrices A_1 and A_2, then $(T_1 + T_2)$ is defined by the matrix $(A_1 + A_2)$, and cT_1 is defined by the matrix cA_1.

(c) Prove that the set of transformations under these operations of addition and scalar multiplication is a vector space.

*25. Let D be a mapping of $M_{nn} \to \mathbf{R}$, defined by $D(A) = |A|$. (D maps the square matrix into its determinant.) Is D a linear mapping?

7.2   Matrix Transformations, Computer Graphics, and Fractals

In the previous section we saw that a matrix defines a linear transformation. In this section we find that any linear transformation $T: \mathbf{R}^n \to \mathbf{R}^m$ can be defined by a matrix A. Such a matrix representation of a linear mapping is an important tool for discussing linear mappings. We shall determine the matrices that describe a number of important linear transformations such as rotations and expansions. Applications will be seen in fractal geometry.

Rotation about the Origin

Consider a rotation $T: \mathbf{R}^2 \to \mathbf{R}^2$ about the origin. See Figure 7.4(a). Let

$$T(\mathbf{u}) = \mathbf{u}' \quad \text{and} \quad T(\mathbf{v}) = \mathbf{v}'$$

Parallelogram ① is transformed into parallelogram ②. Since the diagonal of parallelogram ① is mapped into the diagonal of parallelogram ②,

$$T(\mathbf{u} + \mathbf{v}) = \mathbf{u}' + \mathbf{v}'$$
$$= T(\mathbf{u}) + T(\mathbf{v}) \quad \text{addition is preserved}$$

Further, in Figure 7.4(b), $T(\mathbf{w}) = \mathbf{w}'$. Since the image of the vector $c\mathbf{w}$ is $c\mathbf{w}'$,

$$T(c\mathbf{w}) = c\mathbf{w}'$$
$$= cT(\mathbf{w}) \quad \text{scalar multiplication is preserved}$$

Thus a rotation about the origin is a linear operator.

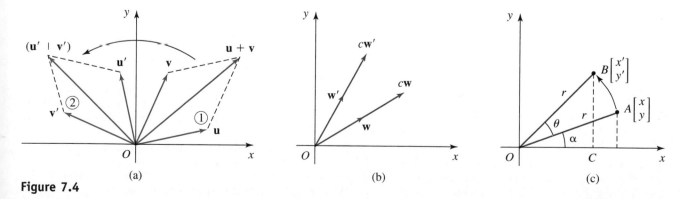

Figure 7.4

Having seen that a rotation is a linear operator, let us now develop a working expression for such a rotation—an expression that can be used to give the coordinates of the image of any given point under a rotation. Consider a counterclockwise rotation through an angle θ about the origin. See Figure 7.4(c). Such a rotation would map the point A into the point B. The distance OA is equal to OB; let it be r. Let the angle AOC be α. We get

$$x' = OC = r\cos(\alpha + \theta) = r\cos\alpha\cos\theta - r\sin\alpha\sin\theta$$
$$= x\cos\theta - y\sin\theta$$

$$y' = BC = r\sin(\alpha + \theta) = r\sin\alpha\cos\theta + r\cos\alpha\sin\theta$$
$$= y\cos\theta + x\sin\theta$$
$$= x\sin\theta + y\cos\theta$$

These expressions for x' and y' can be combined into a single matrix equation.

$$\begin{bmatrix} x' \\ y' \end{bmatrix} = \begin{bmatrix} \cos\theta & -\sin\theta \\ \sin\theta & \cos\theta \end{bmatrix} \begin{bmatrix} x \\ y \end{bmatrix}$$

Thus

$$T\left(\begin{bmatrix} x' \\ y' \end{bmatrix}\right) = \begin{bmatrix} \cos\theta & -\sin\theta \\ \sin\theta & \cos\theta \end{bmatrix} \begin{bmatrix} x \\ y \end{bmatrix}$$

A rotation about the origin is defined by matrix multiplication, confirming that a rotation is indeed a linear operator. Note that θ is positive for a counterclockwise rotation, negative for a clockwise rotation.

For example, let us find the image of the point $\begin{bmatrix} 3 \\ 2 \end{bmatrix}$ under a rotation of $\pi/2$ about the origin. Using $\cos(\pi/2) = 0$ and $\sin(\pi/2) = 1$ in the rotation matrix, we get

$$T\left(\begin{bmatrix} 3 \\ 2 \end{bmatrix}\right) = \begin{bmatrix} 0 & -1 \\ 1 & 0 \end{bmatrix} \begin{bmatrix} 3 \\ 2 \end{bmatrix} = \begin{bmatrix} -2 \\ 3 \end{bmatrix}$$

The image of the point $\begin{bmatrix} 3 \\ 2 \end{bmatrix}$ under a rotation of $\pi/2$ about the origin is $\begin{bmatrix} -2 \\ 3 \end{bmatrix}$.

As in the case of a rotation, each linear transformation can be defined by a matrix. While it is possible to use ad hoc ways of arriving at matrix representations of linear transformations, as for the rotation matrix, it is desirable to have a standard method that can be applied in all cases. We now develop such a method for linear mappings on $\mathbf{R}^n$.

Matrix Representation

Let $T: \mathbf{R}^n \to \mathbf{R}^m$ be a linear transformation, $\{\mathbf{e}_1, \ldots, \mathbf{e}_n\}$ be the standard basis for $\mathbf{R}^n$, and $\mathbf{u}$ be a vector in $\mathbf{R}^n$. Write $\mathbf{e}_1, \ldots, \mathbf{e}_n$ and $\mathbf{u}$ as column matrices.

$$\mathbf{e}_1 = \begin{bmatrix} 1 \\ \vdots \\ 0 \end{bmatrix}, \quad \mathbf{e}_n = \begin{bmatrix} 0 \\ \vdots \\ 1 \end{bmatrix}, \quad \mathbf{u} = \begin{bmatrix} a_1 \\ \vdots \\ a_n \end{bmatrix}$$

We can express **u** in terms of the basis:

$$\mathbf{u} = a_1\mathbf{e}_1 + \cdots + a_n\mathbf{e}_n$$

Since T is a linear transformation,

$$T(\mathbf{u}) = T(a_1\mathbf{e}_1 + \cdots + a_n\mathbf{e}_n)$$
$$= a_1T(\mathbf{e}_1) + \cdots + a_nT(\mathbf{e}_n)$$
$$= [T(\mathbf{e}_1) \quad \cdots \quad T(\mathbf{e}_n)] \begin{bmatrix} a_1 \\ \vdots \\ a_n \end{bmatrix}$$

Thus the linear transformation T is defined by the matrix

$$A = [T(\mathbf{e}_1) \quad \cdots \quad T(\mathbf{e}_n)]$$

A is called the **standard matrix** of T.

Note that the vectors of $\mathbf{R}^n$ must be written in column form to make use of this matrix way of expressing a linear transformation. You are asked to derive the rotation matrix in this manner in the following exercises.

Example 1   Determine the standard matrix of $T\left(\begin{bmatrix} x \\ y \end{bmatrix}\right) = \begin{bmatrix} 2x + y \\ 3y \end{bmatrix}$.

Solution   We find the effect of T on the standard basis.

$$T\left(\begin{bmatrix} 1 \\ 0 \end{bmatrix}\right) = \begin{bmatrix} 2 \\ 0 \end{bmatrix}, \quad T\left(\begin{bmatrix} 0 \\ 1 \end{bmatrix}\right) = \begin{bmatrix} 1 \\ 3 \end{bmatrix}$$

These matrices are the columns of the standard matrix A.

$$A = \begin{bmatrix} 2 & 1 \\ 0 & 3 \end{bmatrix}$$

T can be written as a matrix transformation:

$$T\left(\begin{bmatrix} x \\ y \end{bmatrix}\right) = \begin{bmatrix} 2 & 1 \\ 0 & 3 \end{bmatrix} \begin{bmatrix} x \\ y \end{bmatrix}$$

Let us now find the matrices that describe a number of important transformations.

Dilation and Contraction   *Scalar mult.*

Consider the operator $T\colon \mathbf{R}^2 \to \mathbf{R}^2$, defined by $T\left(\begin{bmatrix} x \\ y \end{bmatrix}\right) = r\begin{bmatrix} x \\ y \end{bmatrix}$, where r is a scalar. If $r > 1$, then T moves the points *outward* from the origin; it is called a **dilation of** factor r. If $0 < r < 1$, then T moves points *inward* toward the origin; it is called a **contraction of factor** r. See Figure 7.5. It can be easily shown that T is linear. Let us find the standard matrix of T.

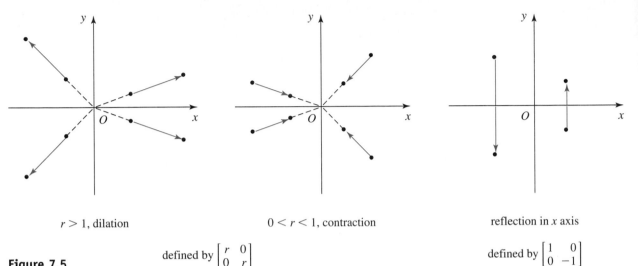

$r > 1$, dilation $\qquad\qquad$ $0 < r < 1$, contraction $\qquad\qquad$ reflection in x axis

Figure 7.5 $\qquad$ defined by $\begin{bmatrix} r & 0 \\ 0 & r \end{bmatrix}$ $\qquad\qquad\qquad\qquad\qquad$ defined by $\begin{bmatrix} 1 & 0 \\ 0 & -1 \end{bmatrix}$

We find the effect of T on the standard basis.

$$T\left(\begin{bmatrix} 1 \\ 0 \end{bmatrix}\right) = \begin{bmatrix} r \\ 0 \end{bmatrix}, \qquad T\left(\begin{bmatrix} 0 \\ 1 \end{bmatrix}\right) = \begin{bmatrix} 0 \\ r \end{bmatrix}$$

The standard matrix is

$$A = \begin{bmatrix} r & 0 \\ 0 & r \end{bmatrix}$$

T can be written as a matrix transformation:

$$T\left(\begin{bmatrix} x \\ y \end{bmatrix}\right) = \begin{bmatrix} r & 0 \\ 0 & r \end{bmatrix} \begin{bmatrix} x \\ y \end{bmatrix}$$

For example, consider the dilation T defined by the matrix $\begin{bmatrix} 3 & 0 \\ 0 & 3 \end{bmatrix}$. We get

$$T\left(\begin{bmatrix} 2 \\ 1 \end{bmatrix}\right) = \begin{bmatrix} 3 & 0 \\ 0 & 3 \end{bmatrix} \begin{bmatrix} 2 \\ 1 \end{bmatrix} = \begin{bmatrix} 6 \\ 3 \end{bmatrix}$$

Thus T maps the point $\begin{bmatrix} 2 \\ 1 \end{bmatrix}$ into the point $\begin{bmatrix} 6 \\ 3 \end{bmatrix}$. The image is in the same direction as the original point from the origin, but is three times its distance from the origin.

Reflection

Consider the operator $T: \mathbf{R}^2 \to \mathbf{R}^2$, defined by $T\left(\begin{bmatrix} x \\ y \end{bmatrix}\right) = \begin{bmatrix} x \\ -y \end{bmatrix}$, that maps a point into its mirror image in the x axis. See Figure 7.5. T is called a **reflection**.

It can be shown that T is linear. Let us find its standard matrix. The effect of T on the standard basis is

$$T\left(\begin{bmatrix} 1 \\ 0 \end{bmatrix}\right) = \begin{bmatrix} 1 \\ 0 \end{bmatrix}, \qquad T\left(\begin{bmatrix} 0 \\ 1 \end{bmatrix}\right) = \begin{bmatrix} 0 \\ -1 \end{bmatrix}$$

The standard matrix is thus

$$A = \begin{bmatrix} 1 & 0 \\ 0 & -1 \end{bmatrix}$$

A reflection about the x axis can therefore be written:

$$T\left(\begin{bmatrix} x \\ y \end{bmatrix}\right) = \begin{bmatrix} 1 & 0 \\ 0 & -1 \end{bmatrix} \begin{bmatrix} x \\ y \end{bmatrix}$$

Applications often involve a sequence of linear transformations. A single composite transformation $T = T_n \circ \cdots \circ T_1$ is described by the product of the matrices of the component transformations $A_n \cdots A_1$.

Example 2   Determine the matrix that describes a reflection in the x axis, followed by a rotation through $\pi/2$, followed by a dilation of factor 3. Find the image of the point $\begin{bmatrix} 4 \\ 1 \end{bmatrix}$ under this sequence of mappings.

Solution   The matrices that define the reflection, rotation, and dilation are

$$\begin{bmatrix} 1 & 0 \\ 0 & -1 \end{bmatrix}, \qquad \begin{bmatrix} \cos(\pi/2) & -\sin(\pi/2) \\ \sin(\pi/2) & \cos(\pi/2) \end{bmatrix}, \qquad \begin{bmatrix} 3 & 0 \\ 0 & 3 \end{bmatrix}$$

The composite mapping is thus defined by the single product matrix

$$\begin{bmatrix} 3 & 0 \\ 0 & 3 \end{bmatrix} \begin{bmatrix} \cos(\pi/2) & -\sin(\pi/2) \\ \sin(\pi/2) & \cos(\pi/2) \end{bmatrix} \begin{bmatrix} 1 & 0 \\ 0 & -1 \end{bmatrix} = \begin{bmatrix} 3 & 0 \\ 0 & 3 \end{bmatrix} \begin{bmatrix} 0 & -1 \\ 1 & 0 \end{bmatrix} \begin{bmatrix} 1 & 0 \\ 0 & -1 \end{bmatrix}$$

$$= \begin{bmatrix} 0 & 3 \\ 3 & 0 \end{bmatrix}$$

The image of the point $\begin{bmatrix} 4 \\ 1 \end{bmatrix}$ is

$$\begin{bmatrix} 0 & 3 \\ 3 & 0 \end{bmatrix} \begin{bmatrix} 4 \\ 1 \end{bmatrix} = \begin{bmatrix} 3 \\ 12 \end{bmatrix}$$

Transformations Defined by Nonsingular Matrices

A **nonsingular transformation** is a mapping $T: \mathbf{R}^n \to \mathbf{R}^n$ defined by

$$T(\mathbf{u}) = A\mathbf{u}$$

where A is a nonsingular matrix.

Nonsingular transformations are important in that they **preserve the linearity** of a vector space in the following sense.

Theorem 7.3   A nonsingular linear transformation T maps

(a)   lines into lines.

(b)   segments of lines into segments of lines.

(c)   parallel lines into parallel lines.

(d)   lines through the origin into lines through the origin.

We ask you to discuss these results in the exercises that follow.

Example 3   Consider the operator $T\colon \mathbf{R}^2 \to \mathbf{R}^2$ defined by the matrix $A = \begin{bmatrix} 4 & 2 \\ 2 & 3 \end{bmatrix}$. Determine the image of the unit square under this transformation.

Solution   The unit square in $\mathbf{R}^2$ is the square whose vertices are the points

$$P\begin{bmatrix} 1 \\ 0 \end{bmatrix}, \qquad Q\begin{bmatrix} 1 \\ 1 \end{bmatrix}, \qquad R\begin{bmatrix} 0 \\ 1 \end{bmatrix}, \qquad O\begin{bmatrix} 0 \\ 0 \end{bmatrix}$$

See Figure 7.6a. Let us compute the images of these points under the matrix transformation. It is convenient to use the notation $\mathbf{u} \mapsto A\mathbf{u}$ to discuss the images of specific points. Multiplying each point by the matrix, we get

$$\begin{array}{cccccccc} P & P' & Q & Q' & R & R' & O & O \end{array}$$

$$\begin{bmatrix} 1 \\ 0 \end{bmatrix} \mapsto \begin{bmatrix} 4 \\ 2 \end{bmatrix}, \qquad \begin{bmatrix} 1 \\ 1 \end{bmatrix} \mapsto \begin{bmatrix} 6 \\ 5 \end{bmatrix}, \qquad \begin{bmatrix} 0 \\ 1 \end{bmatrix} \mapsto \begin{bmatrix} 2 \\ 3 \end{bmatrix}, \qquad \begin{bmatrix} 0 \\ 0 \end{bmatrix} \mapsto \begin{bmatrix} 0 \\ 0 \end{bmatrix}$$

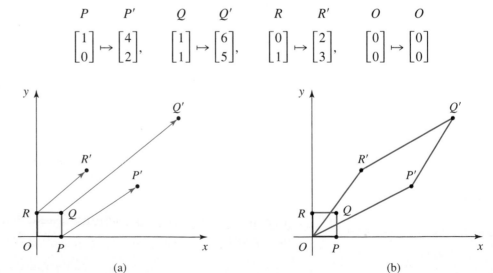

Figure 7.6   (a)   (b)

Further, $|A| = (4 \times 3) - (2 \times 2) = 8 \neq 0$. The matrix A is nonsingular; thus line segments are mapped into line segments. We get

$$OP \to OP', \qquad PQ \to P'Q', \qquad QR \to Q'R', \qquad OR \to OR'$$

The square $PQRO$ is transformed into the parallelogram $P'Q'R'O$; see Figure 7.6b.

These operators on $\mathbf{R}^2$ are used in computer software to rotate, stretch, or contract figures. We discuss this application later in this section.

Solid bodies can be described geometrically. When loads are applied to bodies, changes in shape called **deformations** occur. For example, the square $PQRO$ in Figure 7.6 could represent a physical body that is deformed into the shape $P'Q'R'O$. Such deformations can be modeled and analyzed on computers using these mathematical techniques. The fields of science that investigate such problems are called **elasticity** and **plasticity**.

Nonsingular transformations are also important because they have an inverse transformation. Let T be a nonsingular transformation defined by a matrix A, and let $A\mathbf{u} = \mathbf{v}$. Since A is nonsingular, A^{-1} exists, and $A^{-1}\mathbf{v} = \mathbf{u}$. A^{-1} thus defines a transformation, called the **inverse** of T, denoted T^{-1}.

Rotations, dilations, contractions, and reflections are nonsingular transformations. They thus have inverses. Let us show that the inverse of a dilation of factor r is a contraction of factor $1/r$. The dilation matrix A and its inverse are

$$A = \begin{bmatrix} r & 0 \\ 0 & r \end{bmatrix}, \qquad A^{-1} = \begin{bmatrix} \frac{1}{r} & 0 \\ 0 & \frac{1}{r} \end{bmatrix}$$

Thus A^{-1} defines a contraction of factor $1/r$. We ask you to discuss the inverses of rotations and reflections in the following exercises.

The next class of transformations are important in that they **preserve the geometry** of Euclidean space.

Orthogonal Transformations

vecto · orthog. matri

An **orthogonal transformation** is a mapping $T: \mathbf{R}^n \to \mathbf{R}^n$, defined by

$$T(\mathbf{u}) = A\mathbf{u}$$

where A is an orthogonal matrix.

Since an orthogonal matrix is nonsingular, then orthogonal transformations preserve linearity. They have the following additional properties.

Theorem 7.4   Let T be an orthogonal transformation on $\mathbf{R}^n$. Let $\mathbf{u}$ and $\mathbf{v}$ be elements of $\mathbf{R}^n$. Let P and Q be the points in $\mathbf{R}^n$ defined by $\mathbf{u}$ and $\mathbf{v}$, and let R and S be their images.

(continues)

Then
$$\|\mathbf{u}\| = \|T(\mathbf{u})\|$$
$$\text{angle between } \mathbf{u} \text{ and } \mathbf{v} = \text{angle between } T(\mathbf{u}) \text{ and } T(\mathbf{v})$$
$$d(P, Q) = d(R, S)$$

Orthogonal transformations preserve norms, angles, and distances. See Figure 7.7. Orthogonal transformations preserve the shapes of rigid bodies—they are often referred to as **rigid motions**.

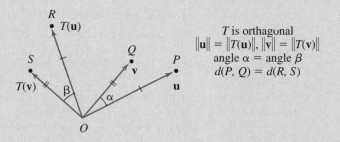

T is orthogonal
$\|\mathbf{u}\| = \|T(\mathbf{u})\|, \|\mathbf{v}\| = \|T(\mathbf{v})\|$
angle α = angle β
$d(P, Q) = d(R, S)$

Figure 7.7

Proof   We show that orthogonal transformations preserve dot products. Since norms, angles, and distances are defined in terms of dot products, this leads to their preservation. Let the orthogonal transformation T be defined by an orthogonal matrix A. We get

$$T(\mathbf{u}) \cdot T(\mathbf{v}) = (A\mathbf{u}) \cdot (A\mathbf{v}) = (A\mathbf{u})^t(A\mathbf{v}) = (\mathbf{u}^t A^t)(A\mathbf{v})$$
$$= \mathbf{u}^t A^t A \mathbf{v} = \mathbf{u}^t I \mathbf{v} = \mathbf{u}^t \mathbf{v} = \mathbf{u} \cdot \mathbf{v}$$

Thus orthogonal transformations preserve dot products.

We now look at norms.

$$\|T(\mathbf{u})\| = \|A\mathbf{u}\| = \sqrt{(A\mathbf{u}) \cdot (A\mathbf{u})}$$
$$= \sqrt{\mathbf{u} \cdot \mathbf{u}} \quad \text{since dot product is preserved}$$
$$= \|\mathbf{u}\|$$

Thus norm is preserved.

We leave it for you to show that orthogonal transformations also preserve angles and distances in the exercises that follow.

Example 4   Let T be the orthogonal transformation defined by the following orthogonal matrix A. Show that T preserves the norms of the given vectors $\mathbf{u}$ and $\mathbf{v}$, preserves the angle between these vectors, and also preserves the distance between the points defined by the vectors.

$$A = \begin{bmatrix} \dfrac{1}{\sqrt{2}} & \dfrac{1}{\sqrt{2}} \\ -\dfrac{1}{\sqrt{2}} & \dfrac{1}{\sqrt{2}} \end{bmatrix}, \quad \mathbf{u} = \begin{bmatrix} 2 \\ 0 \end{bmatrix}, \quad \mathbf{v} = \begin{bmatrix} 3 \\ 4 \end{bmatrix}$$

Solution Let us first show that T preserves the norms of $\mathbf{u}$ and $\mathbf{v}$. We get

$$T(\mathbf{u}) = A\mathbf{u} = \begin{bmatrix} \sqrt{2} \\ -\sqrt{2} \end{bmatrix} \quad \text{thus} \quad \|T(\mathbf{u})\| = 2 = \|\mathbf{u}\|$$

$$T(\mathbf{v}) = A\mathbf{v} = \begin{bmatrix} \dfrac{7}{\sqrt{2}} \\ \dfrac{1}{\sqrt{2}} \end{bmatrix} \quad \text{and} \quad \|T(\mathbf{v})\| = 5 = \|\mathbf{v}\|$$

Thus T preserves the norms of $\mathbf{u}$ and $\mathbf{v}$.

Next we show that T preserves the angle between $\mathbf{u}$ and $\mathbf{v}$. We get

$$\frac{\mathbf{u} \cdot \mathbf{v}}{\|\mathbf{u}\| \, \|\mathbf{v}\|} = \frac{6}{2 \times 5} = 0.6 \quad \text{and} \quad \frac{T(\mathbf{u}) \cdot T(\mathbf{v})}{\|T(\mathbf{u})\| \, \|T(\mathbf{u})\|} = \frac{6}{2 \times 5} = 0.6$$

Thus the angle between $\mathbf{u}$ and $\mathbf{v}$ is equal to the angle between $T(\mathbf{u})$ and $T(\mathbf{v})$.

Finally, we show that T preserves the distances between the points defined by the vectors. The distance between the original points is

$$d\left(\begin{bmatrix} 2 \\ 0 \end{bmatrix}, \begin{bmatrix} 3 \\ 4 \end{bmatrix} \right) = \left\| \begin{bmatrix} 2 \\ 0 \end{bmatrix} - \begin{bmatrix} 3 \\ 4 \end{bmatrix} \right\| = \left\| \begin{bmatrix} -1 \\ -4 \end{bmatrix} \right\| = \sqrt{17}$$

The distance between the images is

$$d\left(\begin{bmatrix} \sqrt{2} \\ -\sqrt{2} \end{bmatrix}, \begin{bmatrix} \dfrac{7}{\sqrt{2}} \\ \dfrac{1}{\sqrt{2}} \end{bmatrix} \right) = \left\| \begin{bmatrix} \sqrt{2} \\ -\sqrt{2} \end{bmatrix} - \begin{bmatrix} \dfrac{7}{\sqrt{2}} \\ \dfrac{1}{\sqrt{2}} \end{bmatrix} \right\| = \left\| \begin{bmatrix} -\dfrac{5}{\sqrt{2}} \\ -\dfrac{3}{\sqrt{2}} \end{bmatrix} \right\| = \sqrt{17}$$

Thus the distance is preserved.

Observe that T defines a rotation of points in a plane through an angle of $\pi/4$ in a clockwise direction about the origin. Intuitively we expect all rotation matrices to preserve norms, angles, and distances.

Translations and Affine Transformations

We complete this section with a discussion of transformations that, even though they are not linear, are important in mathematics and in applications.

A **translation** is a transformation $T: \mathbf{R}^n \to \mathbf{R}^n$, defined by

$$T(\mathbf{u}) = \mathbf{u} + \mathbf{v}$$

where $\mathbf{v}$ is a fixed vector.

A translation slides points in a direction and distance defined by the vector $\mathbf{v}$. Thus, quite naturally, *translations preserve lines, angles, and distances*. For example, consider the following translation on $\mathbf{R}^2$.

$$T\left(\begin{bmatrix} x \\ y \end{bmatrix}\right) = \begin{bmatrix} x \\ y \end{bmatrix} + \begin{bmatrix} 4 \\ 2 \end{bmatrix}$$

Let us determine the effect of T on the triangle PQR having vertices $\begin{bmatrix} 1 \\ 2 \end{bmatrix}$, $\begin{bmatrix} 2 \\ 8 \end{bmatrix}$, $\begin{bmatrix} 3 \\ 2 \end{bmatrix}$. We see that

$$\begin{array}{cccccc} P & P' & Q & Q' & R & R' \end{array}$$

$$\begin{bmatrix} 1 \\ 2 \end{bmatrix} \mapsto \begin{bmatrix} 5 \\ 4 \end{bmatrix}, \quad \begin{bmatrix} 2 \\ 8 \end{bmatrix} \mapsto \begin{bmatrix} 6 \\ 10 \end{bmatrix}, \quad \begin{bmatrix} 3 \\ 2 \end{bmatrix} \mapsto \begin{bmatrix} 7 \\ 4 \end{bmatrix}$$

The triangle PQR is transformed into the triangle $P'Q'R'$ in Figure 7.8.

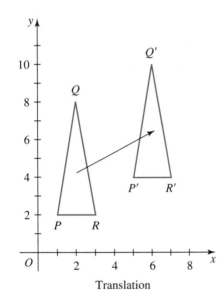

Figure 7.8   Translation

Example 5   Find the equation of the image of the line $y = 2x + 3$ under the translation $T\left(\begin{bmatrix} x \\ y \end{bmatrix}\right) = \begin{bmatrix} x \\ y \end{bmatrix} + \begin{bmatrix} 2 \\ 1 \end{bmatrix}$.

Solution   The equation $y = 2x + 3$ describes points on the line of slope 2 and y intercept 3. T will slide this line into another line. We want to find the equation of this image line. We get

$$T\left(\begin{bmatrix} x \\ y \end{bmatrix}\right) = \begin{bmatrix} x \\ y \end{bmatrix} + \begin{bmatrix} 2 \\ 1 \end{bmatrix} = \begin{bmatrix} x \\ 2x + 3 \end{bmatrix} + \begin{bmatrix} 2 \\ 1 \end{bmatrix} = \begin{bmatrix} x + 2 \\ 2x + 4 \end{bmatrix} = \begin{bmatrix} x' \\ y' \end{bmatrix}$$

We see that $y' = 2x'$ for the image point. Thus the equation of the image line is $y = 2x$.

An **affine transformation** is a transformation $T: \mathbf{R}^n \to \mathbf{R}^n$, defined by

$$T(\mathbf{u}) = A\mathbf{u} + \mathbf{v}$$

where A is a matrix and $\mathbf{v}$ is a fixed vector.

An affine transformation can be interpreted as a matrix transformation followed by a translation.

For example, consider the following affine transformation on $\mathbf{R}^2$.

$$T\left(\begin{bmatrix} x \\ y \end{bmatrix}\right) = \begin{bmatrix} 2 & 1 \\ 1 & 1 \end{bmatrix} \begin{bmatrix} x \\ y \end{bmatrix} + \begin{bmatrix} 1 \\ 2 \end{bmatrix}$$

Let us find the image of the unit square in Figure 7.9. We get

$$
\begin{array}{cccccccc}
P & P' & Q & Q' & R & R' & O & O' \\
\begin{bmatrix} 1 \\ 0 \end{bmatrix} \mapsto \begin{bmatrix} 3 \\ 3 \end{bmatrix}, & & \begin{bmatrix} 1 \\ 1 \end{bmatrix} \mapsto \begin{bmatrix} 4 \\ 4 \end{bmatrix}, & & \begin{bmatrix} 0 \\ 1 \end{bmatrix} \mapsto \begin{bmatrix} 2 \\ 3 \end{bmatrix}, & & \begin{bmatrix} 0 \\ 0 \end{bmatrix} \mapsto \begin{bmatrix} 1 \\ 2 \end{bmatrix}
\end{array}
$$

Since the matrix is nonsingular, line segments are mapped into line segments. We get

$$OP \to O'P', \qquad PQ \to P'Q', \qquad QR \to Q'R', \qquad OR \to O'R'$$

The square $PQRO$ is transformed into the parallelogram $P'Q'R'O'$.

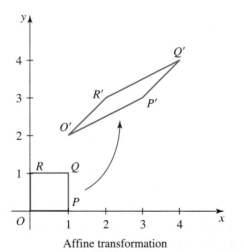

Figure 7.9   Affine transformation

You are asked to show in the following exercises that translations and affine transformations are not linear transformations.

We have introduced a number of fundamental transformations in this section. You will meet other transformations—namely **projection**, **scaling**, and **shear**—in the exercises that follow. These basic transformations are the building blocks for creating other transformations, using composition. We now discuss the use of these transformations in computer graphics.

*Transformations in Computer Graphics

Computer graphics is the field that studies the creation and manipulation of pictures with the aid of computers. The impact of computer graphics is felt in many homes through video games; its uses in research, industry, and business are vast and ever expanding. Architects use computer graphics to explore designs, molecular biologists display and manipulate pictures of molecules to gain insight into their structure, pilots are trained using graphics flight simulators, and transportation engineers use computer-generated transforms in their planning work—to mention only a few applications.

The manipulation of pictures in computer graphics is carried out using sequences of transformations. Rotations, reflections, dilations, and contractions are defined by matrices, using matrix multiplication. A sequence of such transformations can be performed by a single linear transformation defined by the product of the matrices. Unfortunately, translation, as it now stands, uses matrix addition, and any sequence of transformations involving translations cannot be combined in this manner into a single matrix. However, if coordinates called **homogeneous coordinates** are used to describe points in a plane, then translations can also be accomplished through matrix multiplication, and any sequence of these transformations can be defined in terms of a single matrix. In homogeneous coordinates, a third component of 1 is added to each coordinate, and rotation, reflection, dilation/contraction, and translation—R, R_e, D, and T, respectively—are defined by the following matrices.

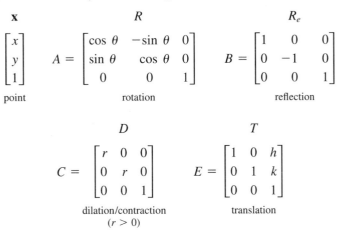

$$\mathbf{x} \qquad\qquad\qquad R \qquad\qquad\qquad R_e$$

$$\begin{bmatrix} x \\ y \\ 1 \end{bmatrix} \qquad A = \begin{bmatrix} \cos\theta & -\sin\theta & 0 \\ \sin\theta & \cos\theta & 0 \\ 0 & 0 & 1 \end{bmatrix} \qquad B = \begin{bmatrix} 1 & 0 & 0 \\ 0 & -1 & 0 \\ 0 & 0 & 1 \end{bmatrix}$$

point   rotation   reflection

$$\qquad\qquad D \qquad\qquad\qquad\qquad T$$

$$C = \begin{bmatrix} r & 0 & 0 \\ 0 & r & 0 \\ 0 & 0 & 1 \end{bmatrix} \qquad E = \begin{bmatrix} 1 & 0 & h \\ 0 & 1 & k \\ 0 & 0 & 1 \end{bmatrix}$$

dilation/contraction   translation
($r > 0$)

Thus, for example, a dilation D followed by a translation T and then a rotation R would be defined by $RTD(\mathbf{x}) = AEC(\mathbf{x})$. The composite transformation RTD would be described by the single matrix AEC.

The programming language True BASIC has built-in subroutines for rotation, translation, and dilation/contraction (and also scale and shear, see Exercises 15 and 18 following) that can be used to move pictures on the screen. To accomplish this movement, the subroutines convert screen coordinates into homogeneous coordinates and use the matrices that define these transformations in homogeneous coordinates. For further information on the way these matrices are used in the subroutines, see *True BASIC Reference Manual*, by John G. Kemeney and Thomas E. Kurtz, True BASIC, Inc., pages 291–294, 1988.

We now illustrate how the transformations are used to rotate a geometrical figure about a point other than the origin.

Example 6   Determine the matrix that defines a rotation of a plane through an angle θ about a point $P(h, k)$. Use this general result to find the matrix that defines a rotation of the plane through an angle of $\pi/2$ about the point $(5, 4)$. Find the image of the triangle having the vertices $A(1, 2)$, $B(2, 8)$, and $C(3, 2)$ under this rotation. See Figure 7.10.

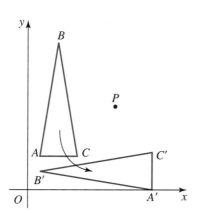

Figure 7.10   Rotation about P

Solution   The rotation about P can be accomplished by a sequence of three of the above transformations:

(a)   A translation T_1 of the plane that takes P to the origin O.

(b)   A rotation R of the plane about the origin through an angle θ.

(c)   A translation T_2 of the plane that takes O back to P.

The matrices that describe these transformations are as follows.

$$
\begin{array}{ccc}
T_1 & R & T_2
\end{array}
$$

$$
\begin{bmatrix} 1 & 0 & -h \\ 0 & 1 & -k \\ 0 & 0 & 1 \end{bmatrix}
\quad
\begin{bmatrix} \cos\theta & -\sin\theta & 0 \\ \sin\theta & \cos\theta & 0 \\ 0 & 0 & 1 \end{bmatrix}
\quad
\begin{bmatrix} 1 & 0 & h \\ 0 & 1 & k \\ 0 & 0 & 1 \end{bmatrix}
$$

The rotation R_p about P can be accomplished as follows.

$$
R_p\left(\begin{bmatrix} x \\ y \\ 1 \end{bmatrix}\right) = T_2 R T_1 \left(\begin{bmatrix} x \\ y \\ 1 \end{bmatrix}\right) = \begin{bmatrix} 1 & 0 & h \\ 0 & 1 & k \\ 0 & 0 & 1 \end{bmatrix} \begin{bmatrix} \cos\theta & -\sin\theta & 0 \\ \sin\theta & \cos\theta & 0 \\ 0 & 0 & 1 \end{bmatrix} \begin{bmatrix} 1 & 0 & -h \\ 0 & 1 & -k \\ 0 & 0 & 1 \end{bmatrix} \begin{bmatrix} x \\ y \\ 1 \end{bmatrix}
$$

$$
= \begin{bmatrix} \cos\theta & -\sin\theta & -h\cos\theta + k\sin\theta + h \\ \sin\theta & \cos\theta & -h\sin\theta - k\cos\theta + k \\ 0 & 0 & 1 \end{bmatrix} \begin{bmatrix} x \\ y \\ 1 \end{bmatrix}
$$

To get the specific matrix that defines the rotation of the plane through an angle $\pi/2$ about the point $P(5, 4)$, for example, let $h = 5$, $k = 4$, and $\theta = \pi/2$. The rotation matrix is

$$M = \begin{bmatrix} 0 & -1 & 9 \\ 1 & 0 & -1 \\ 0 & 0 & 1 \end{bmatrix}$$

To find the images of the vertices of the triangle ABC, write these vertices in column form as homogeneous coordinates and multiply by M. On performing the matrix multiplication we get

$$
\begin{matrix}
A & & A' & & B & & B' & & C & & C' \\
\begin{bmatrix} 1 \\ 2 \\ 1 \end{bmatrix} & \mapsto & \begin{bmatrix} 7 \\ 0 \\ 1 \end{bmatrix}, & & \begin{bmatrix} 2 \\ 8 \\ 1 \end{bmatrix} & \mapsto & \begin{bmatrix} 1 \\ 1 \\ 1 \end{bmatrix}, & & \begin{bmatrix} 3 \\ 2 \\ 1 \end{bmatrix} & \mapsto & \begin{bmatrix} 7 \\ 2 \\ 1 \end{bmatrix}
\end{matrix}
$$

The triangle with vertices $A(1, 2)$, $B(2, 8)$, and $C(3, 2)$ is transformed into the triangle with vertices $A'(7, 0)$, $B'(1, 1)$, and $C'(7, 2)$. See Figure 7.10.

*Fractal Pictures of Nature

Computer graphics systems based on traditional Euclidean geometry are suitable for creating pictures of man-made objects such as machinery, buildings, and airplanes. Images of such objects can be created using lines, circles, and so on. However, these techniques are not appropriate when it comes to constructing images of natural objects such as animals, trees, and landscapes. In the words of mathematician Benoit B. Mandelbrot, "Clouds are not spheres, mountains are not cones, coastlines are not circles, and bark is not smooth, nor does lightning travel in straight lines." However, nature does wear its irregularities in an unexpectedly orderly fashion; it is full of shapes that repeat themselves on different scales within the same object. In 1975 Mandelbrot introduced a new geometry, which he called **fractal geometry**, that can be used to describe natural phenomena. A fractal is a convenient label for irregular and fragmented self-similar shapes. Fractal objects contain structures nested within one another, each smaller structure being a miniature, though not necessarily identical, version of the larger form. The story behind the word *fractal* is interesting. Mandelbrot came across the Latin adjective *fractus*, from the verb *frangere*, to break, in his son's Latin book. The resonance of the main English cognates *fracture* and *fraction* seemed appropriate, and he coined the word *fractal*!

We now discuss methods that are being developed by a research team at the Georgia Institute of Technology for forming images of natural objects using fractals. These fractal images of nature are generated using affine transformations. Figure 7.11 shows a fractal image of a fern being gradually generated. Let us see how this is done.

Figure 7.11

Consider the following four affine transformations $T_1, \ldots, T_4$. Associate probabilities $p_1, \ldots, p_4$ with these transformations.

$$T_1\left(\begin{bmatrix} x \\ y \end{bmatrix}\right) = \begin{bmatrix} 0.86 & 0.03 \\ -0.03 & 0.86 \end{bmatrix} \begin{bmatrix} x \\ y \end{bmatrix} + \begin{bmatrix} 0 \\ 1.5 \end{bmatrix}, p_1 = 0.83$$

$$T_2\left(\begin{bmatrix} x \\ y \end{bmatrix}\right) = \begin{bmatrix} 0.2 & -0.25 \\ 0.21 & 0.23 \end{bmatrix} \begin{bmatrix} x \\ y \end{bmatrix} + \begin{bmatrix} 0 \\ 1.5 \end{bmatrix}, p_2 = 0.08$$

$$T_3\left(\begin{bmatrix} x \\ y \end{bmatrix}\right) = \begin{bmatrix} -0.15 & 0.27 \\ 0.25 & 0.26 \end{bmatrix} \begin{bmatrix} x \\ y \end{bmatrix} + \begin{bmatrix} 0 \\ 0.45 \end{bmatrix}, p_3 = 0.08$$

$$T_4\left(\begin{bmatrix} x \\ y \end{bmatrix}\right) = \begin{bmatrix} 0 & 0 \\ 0 & 0.17 \end{bmatrix} \begin{bmatrix} x \\ y \end{bmatrix} + \begin{bmatrix} 0 \\ 0 \end{bmatrix}, p_4 = 0.01$$

The following algorithm is used on a computer to produce the image of the fern.

1. Let $x = 0$, $y = 0$.

2. Use a random generator to select an affine transformation T_i according to the given probabilities.

3. Let $(x', y') = T_i(x, y)$.

4. Plot (x', y').

5. Let $(x, y) = (x', y')$.

6. Repeat steps 2, 3, 4, and 5 five thousand times.

As step 4 is executed, each of five thousand times, the image of the fern gradually appears. A computer program for producing such fractals is given at the end of this discussion.

Each affine transformation T_i involves six parameters a, b, c, d, e, f, and a probability p_i, as follows:

$$T_i\left(\begin{bmatrix} x \\ y \end{bmatrix}\right) = \begin{bmatrix} a & b \\ c & d \end{bmatrix} \begin{bmatrix} x \\ y \end{bmatrix} + \begin{bmatrix} e \\ f \end{bmatrix}, p_i$$

The affine transformations and corresponding probabilities that generate a fractal are written as rows of matrix, called an *iterated function system* (IFS). The IFS for the fern is as follows.

IFS for a Fern

T	a	b	c	d	e	f	p
1	0.86	0.03	−0.03	0.86	0	1.5	0.83
2	0.2	−0.25	0.21	0.23	0	1.5	0.08
3	−0.15	0.27	0.25	0.26	0	0.45	0.08
4	0	0	0	0.17	0	0	0.01

The appropriate affine transformations that produce a given fractal object are found by determining transformations that map the object (called the **attractor**) into various disjoint images, the union of which is the whole fractal. A theorem called the **collage theorem** then guarantees that the transformations can be grouped into an IFS that produces the fractal.

Different probabilities do not, in general, lead to different images, but they do affect the rate at which the image is produced. Appropriate probabilities are

$$p_i = \frac{\text{area of the image under transformation } T_i}{\text{area of image of object}}$$

Readers who wish to know more about this fascinating topic are referred to an excellent survey article: "A Better Way to Compress Images," by Michael F. Barnsley and Alan D. Sloan, *BYTE*, vol. 13, no. 1, 1988, pages 215–223. The method developed at the Georgia Institute of Technology is extremely important because it can be used to produce an image to any desired degree of accuracy using a highly compressed data set. A fractal image containing infinitely many points, whose organization is too complicated to describe directly, can be reproduced using mathematical formulas. For a more in-depth mathematical introduction to fractal geometry, containing details of how the affine transformations are constructed, we suggest *Fractals Everywhere*, by Michael F. Barnsley, Academic Press Inc., 1988. The software package *The Desktop Fractal Design Handbook and System*, by Michael F. Barnsley, Academic Press Inc., 1989, can be used to produce fractals.

```
!True BASIC program for a fractal fern
RANDOMIZE
!the IFS description of the fern.
!Change these statements to get other fractals.
DIM a(4), b(4), c(4), d(4), e(4), f(4), p(4)
DATA 4
DATA .86, .03, −.03, .86, 0, 1.5, .83
DATA .2, −.25, .21, .23, 0, 1.5, .08
DATA −.15, .27, .25, .26, 0, .45, .08
DATA 0, 0, 0, .17, 0, 0, .04
!read in the IFS description
READ m    !the number of affine transformations
LET pt=0    !pt will be accumulated probabilities
```

```
FOR j=1 to m
    READ a(j), b(j), c(j), d(j), e(j), f(j), pk
    LET pt=pt+pk
    LET p(j)=pt
NEXT j
SET WINDOW −10, 12, −1, 12   !set window size
LET x=0   !initialize x & y
LET y=0
FOR n=1 to 5000   !5000 iterations
    LET pk-rnd   !select a random number
    IF pk<=p(1) then
        LET k=1
    ELSEIF pk<=p(2) then
        LET k=2
    ELSEIF pk<=p(3) then
        LET k=3
    ELSE
        LET k=4
    END IF
    LET newx= a(k)*x+b(k)*y+e(k)
    LET newy= c(k)*x+d(k)*y+f(k)
    PLOT newx,newy   !plot the point
    LET x=newx
    LET y=newy
NEXT n
END
```

Exercise Set 7.2

1. Find the matrix that defines a rotation of a plane about the origin through each of the following angles. Determine the image of the point $\begin{bmatrix} 2 \\ 1 \end{bmatrix}$ under each transformation.

   *(a) $\pi/2$ (b) $-\pi/2$ *(c) $\pi/4$

   *(d) π (e) $-3\pi/2$ *(f) $\pi/6$

   (g) $-\pi/3$

2. Derive the rotation matrix by finding the effect of a rotation on the standard basis of $\mathbf{R}^2$.

*3. Find the equation of the image of the ellipse

$$\frac{x^2}{4} + \frac{y^2}{9} = 1$$

under a rotation through an angle of $\pi/2$.

*4. Find a single matrix that defines a rotation of the plane through an angle of $\pi/2$ about the origin, while at the same time moves points to twice their original distance from the origin. (*Hint:* Construct a composite transformation.)

5. Determine a single matrix that defines both a rotation about the origin through an angle θ and a dilation of factor r.

*6. Find the equation of the image of the unit circle $x^2 + y^2 = 1$ under a dilation of factor 3.

*7. Consider the sequence of transformations

$$T_n(\mathbf{u}) = (A)^n\mathbf{u}, \; n = 1, \ldots, 4$$

where A is a dilation matrix of factor 1.5. Sketch the images of the unit circle, in one coordinate system, under this sequence of transformations.

*8. Determine the matrix that defines a reflection in the y axis. Find the image of the point $\begin{bmatrix} 2 \\ 1 \end{bmatrix}$ under this transformation.

9. Determine the matrix that defines a reflection in the line $y = x$.

10. Consider the transformations on $\mathbf{R}^2$ defined by each of the following matrices. Find the image of the unit square under each transformation.

*(a) $\begin{bmatrix} 0 & -1 \\ 1 & 0 \end{bmatrix}$ **(b)** $\begin{bmatrix} 2 & 0 \\ 0 & 2 \end{bmatrix}$ ***(c)** $\begin{bmatrix} 3 & 0 \\ 1 & 4 \end{bmatrix}$

(d) $\begin{bmatrix} 4 & -1 \\ 1 & 5 \end{bmatrix}$ ***(e)** $\begin{bmatrix} -2 & -3 \\ 0 & 4 \end{bmatrix}$ **(f)** $\begin{bmatrix} -2 & -4 \\ -4 & -1 \end{bmatrix}$

*(g) $\begin{bmatrix} 0 & -2 \\ 2 & 0 \end{bmatrix}$ **(h)** $\begin{bmatrix} 0 & 3 \\ -3 & 0 \end{bmatrix}$

11. Find the standard matrix of each of the following linear operations on $\mathbf{R}^2$.

*(a) $T\left(\begin{bmatrix} x \\ y \end{bmatrix} \right) = \begin{bmatrix} 2x \\ x - y \end{bmatrix}$ **(b)** $T\left(\begin{bmatrix} x \\ y \end{bmatrix} \right) = \begin{bmatrix} x - y \\ x + y \end{bmatrix}$

*(c) $T\left(\begin{bmatrix} x \\ y \end{bmatrix} \right) = \begin{bmatrix} 2x - 5y \\ 3y \end{bmatrix}$

(d) $T\left(\begin{bmatrix} x \\ y \end{bmatrix} \right) = \begin{bmatrix} 2y \\ -3x \end{bmatrix}$

Projection

12. Find the standard matrix of the operator

$$T\left(\begin{bmatrix} x \\ y \end{bmatrix} \right) = \begin{bmatrix} x \\ 0 \end{bmatrix}$$

Observe that this transformation projects all points onto the x axis. It is called a **projection operator**.

*13. Determine the matrix that defines projection onto the y axis.

*14. Determine the matrix that defines projection onto the line $y = x$. Find the image of $\begin{bmatrix} 4 \\ 2 \end{bmatrix}$ under this projection.

Scaling

*15. Find the standard matrix of the operator

$$T\left(\begin{bmatrix} x \\ y \end{bmatrix} \right) = \begin{bmatrix} ax \\ by \end{bmatrix}$$

where a and b are positive scalars. T is called a **scaling of factor a in the x direction and factor b in the y direction**. A scaling distorts a figure since x and y do not change in the same manner. Sketch the image of the unit square under this transformation when $a = 3$ and $b = 2$.

*16. Find the equation of the image of the line $y = 2x$ under a scaling of factor 2 in the x direction and factor 3 in the y direction.

17. Find the equation of the image of the unit circle, $x^2 + y^2 = 1$, under a scaling of factor 4 in the x direction and factor 3 in the y direction.

Shear

*18. Find the standard matrix of the operator

$$T\left(\begin{bmatrix} x \\ y \end{bmatrix} \right) = \begin{bmatrix} x + cy \\ y \end{bmatrix}$$

where c is a scalar. T is called a **shear of factor c in the x direction**. Sketch the image of the unit square under a shear of factor 2 in the x direction. Observe how the x value of each point is increased by a factor of $2y$, causing a shearing of the figure.

19. Sketch the image of the unit square under a shear of factor 0.5 in the y direction.

*20. Find the equation of the image of the line $y = 3x$ under a shear of factor 5 in the x direction.

21. Prove that a transformation defined by a 2×2 nonsingular matrix always maps straight lines into straight lines.

22. Show that the following matrix A is orthogonal. Show that the transformation defined by A preserves the norms of the vectors $\mathbf{u}$ and $\mathbf{v}$, preserves the angle between the vectors, and also preserves the distance between the points defined by the vectors.

$$A = \begin{bmatrix} 0 & -1 \\ 1 & 0 \end{bmatrix}, \quad \mathbf{u} = \begin{bmatrix} 3 \\ 3 \end{bmatrix}, \quad \mathbf{v} = \begin{bmatrix} 1 \\ 5 \end{bmatrix}$$

*23. Prove that a rotation matrix is an orthogonal matrix.

24. Let A be an $n \times n$ orthogonal matrix. A defines a transformation of $\mathbf{R}^n$ into itself. Let $\mathbf{u}$ and $\mathbf{v}$ be elements of $\mathbf{R}^n$. Let P and Q be the points defined by $\mathbf{u}$ and $\mathbf{v}$, and let R and S be the points defined by $A\mathbf{u}$ and $A\mathbf{v}$. See Figure 7.7. Prove that

(a) the angle between $\mathbf{u}$ and $\mathbf{v}$ is equal to the angle between $A\mathbf{u}$ and $A\mathbf{v}$.

(b) the distance between P and Q is equal to the distance between S and T.

Thus the transformation defined by A preserves angles and distances.

25. Show that translations and affine transformations are not linear.

26. Find the image of the line $y = 3x + 1$ under the translation

$$T\begin{bmatrix} x \\ y \end{bmatrix} = \begin{bmatrix} x \\ y \end{bmatrix} + \begin{bmatrix} p \\ q \end{bmatrix}$$

where *(a) $p = 2$, $q = 5$ and (b) $p = -1$, $q = 1$.

27. Find and sketch the image of the unit square and the unit circle under the affine transformations $T(\mathbf{u}) = A\mathbf{u} + \mathbf{v}$ defined by the following matrices and vectors.

*(a) $A = \begin{bmatrix} 2 & 0 \\ 0 & 2 \end{bmatrix}$, $\mathbf{v} = \begin{bmatrix} 4 \\ 4 \end{bmatrix}$

(b) $A = \begin{bmatrix} 3 & 0 \\ 0 & 2 \end{bmatrix}$, $\mathbf{v} = \begin{bmatrix} 4 \\ 1 \end{bmatrix}$

*(c) $A = \begin{bmatrix} \frac{1}{\sqrt{2}} & -\frac{1}{\sqrt{2}} \\ \frac{1}{\sqrt{2}} & \frac{1}{\sqrt{2}} \end{bmatrix}$, $\mathbf{v} = \begin{bmatrix} 3 \\ 1 \end{bmatrix}$

(d) $A = \begin{bmatrix} \frac{1}{2} & 0 \\ 0 & \frac{1}{4} \end{bmatrix}$, $\mathbf{v} = \begin{bmatrix} 3 \\ 3 \end{bmatrix}$

28. (a) Show that the inverse of a rotation through an angle θ is a rotation through an angle $-\theta$.

(b) Show that the inverse of a reflection in the x axis is the reflection in the x axis.

***29.** Determine the inverse of the affine transformation $T(\mathbf{u}) = A\mathbf{u} + \mathbf{v}$, where A is nonsingular.

30. Construct single 2×2 matrices that define the following transformations on $\mathbf{R}^2$. Find the image of the point $\begin{bmatrix} 2 \\ 1 \end{bmatrix}$ under each transformation.

*(a) A rotation through $\pi/2$ counterclockwise, then a dilation of factor 2.

(b) A dilation of factor 4, then a reflection in the x axis.

*(c) A reflection about the line $y = x$, then a rotation through π.

31. Construct single 2×2 matrices that define the following transformations on $\mathbf{R}^2$. Find the image of the point $\begin{bmatrix} 3 \\ 2 \end{bmatrix}$ under each transformation.

*(a) A dilation of factor 3, then a shear of factor 2 in the x direction.

*(b) A scaling of factor 3 in the x direction, of factor 2 in the y direction, then a reflection in the line $y = x$.

(c) A dilation of factor 2, then a shear of factor 3 in the x direction, then a rotation through $\pi/2$ counterclockwise.

32. Consider the mapping $T: \mathbf{R}^n \rightarrow \mathbf{R}^n$, defined by

$$T(\mathbf{u}) = A\mathbf{u}$$

T will also have a standard matrix A'. Show that A and A' are identical.

33. Let A be the rotation matrix for $\pi/4$. Show that $A^8 = I$, the identity 2×2 matrix. Give a geometrical reason for expecting this result.

34. Show, for the following matrices A and B, that $A^2BA^2 = I$. Give a geometrical reason for expecting this result.

$$A = \begin{bmatrix} \frac{1}{\sqrt{2}} & -\frac{1}{\sqrt{2}} \\ \frac{1}{\sqrt{2}} & \frac{1}{\sqrt{2}} \end{bmatrix}, \quad B = \begin{bmatrix} -1 & 0 \\ 0 & -1 \end{bmatrix}$$

35. Consider the sequence of transformations $T_n(\mathbf{u}) = (BA)^n\mathbf{u}$, where A is a rotation matrix through $1°$, B is a dilation of factor 1.005, and $\mathbf{u}$ is the point $x = 1$, $y = 0$. What would the set of image points look like as n goes from 1 to 720? (If you have a computer, try it!)

36. Find and sketch (in one coordinate system) the image of the unit square under the affine transformations $T_n(\mathbf{u}) = A^n\mathbf{u} + \mathbf{v}$, for $n = 1, 2, 3$, defined by the following matrix A and vector $\mathbf{v}$.

*(a) $A = \begin{bmatrix} \frac{1}{2} & 0 \\ 0 & \frac{1}{2} \end{bmatrix}$, $\mathbf{v} = \begin{bmatrix} 2 \\ 2 \end{bmatrix}$

(b) $A = \begin{bmatrix} \frac{1}{\sqrt{2}} & -\frac{1}{\sqrt{2}} \\ \frac{1}{\sqrt{2}} & \frac{1}{\sqrt{2}} \end{bmatrix}$, $\mathbf{v} = \begin{bmatrix} 3 \\ 2 \end{bmatrix}$

37. Find the image of the triangle having vertices (1, 2), (3, 4), and (4, 6) under the translation that takes the point (1, 2) to (2, −3).

***38.** Transformations T_1 and T_2 are commutative if

$$T_2 \circ T_1(\mathbf{u}) = T_1 \circ T_2(\mathbf{u})$$

for all vectors $\mathbf{u}$. Let R be a rotation, D a dilation, F a reflection, S a scaling, H a shear, and A an affine transformation. Which pairs of transformations are commutative?

***39.** Find the matrix that defines a rotation of 3-space through an angle of $\pi/2$ about the z axis. (You may consider either direction.)

40. Find the matrix that defines an expansion of three-dimensional space outward from the origin, so that each point moves to three times as far away.

*41. Determine the matrix that can be used to define a rotation through $\pi/2$ about the point $(5, 1)$. Find the image of the unit square under this rotation.

*42. Consider the general translation defined by the following matrix T. Does this transformation have an inverse? If so, find it.

$$T = \begin{bmatrix} 1 & 0 & h \\ 0 & 1 & k \\ 0 & 0 & 1 \end{bmatrix}$$

43. Consider the general scaling defined by the following matrix S. Does this transformation have an inverse? If so, find it.

$$S = \begin{bmatrix} c & 0 & 0 \\ 0 & d & 0 \\ 0 & 0 & 1 \end{bmatrix}$$

*44. Find the image of the triangle having the following vertices A, B, and C (in homogeneous coordinates), under the sequence of transformation T followed by R, followed by S. Sketch the original and final triangle.

$$A\begin{bmatrix} 1 \\ 6 \\ 1 \end{bmatrix}, \quad B\begin{bmatrix} 3 \\ 0 \\ 1 \end{bmatrix}, \quad C\begin{bmatrix} 4 \\ 16 \\ 1 \end{bmatrix};$$

$$T = \begin{bmatrix} 1 & 0 & 4 \\ 0 & 1 & -3 \\ 0 & 0 & 1 \end{bmatrix}, \quad R = \begin{bmatrix} 0 & 1 & 0 \\ -1 & 0 & 0 \\ 0 & 0 & 1 \end{bmatrix}, \quad S = \begin{bmatrix} 3 & 0 & 0 \\ 0 & 5 & 0 \\ 0 & 0 & 1 \end{bmatrix}$$

Computer Exercises

45. The IFS code for a fractal called the **Sierpinski triangle** is

T	a	b	c	d	e	f	p
1	0.5	0	0	0.5	0.5	0.5	0.34
2	0.5	0	0	0.5	1	0	0.33
3	0.5	0	0	0.5	0	0	0.33

Use the computer program with window 0, 2, 0, 1 to find out what a Sierpinski triangle looks like. How many different-sized triangles have you been able to produce in your image?

46. The IFS codes for a fractal tree and "dragon" are as follows. Plot these images.

Tree

T	a	b	c	d	e	f	p
1	0.42	0.42	-0.42	0.42	0	0.2	0.4
2	0.42	-0.42	0.42	0.42	0	0.2	0.4
3	0.1	0	0	0.1	0	0.2	0.15
4	0	0	0	0.5	0	0	0.05

Window $-0.5, 0.5, 0, 0.5$

Dragon

T	a	b	c	d	e	f	p
1	0.5	0.5	-0.5	0.5	0.125	0.625	0.5
2	0.5	0.5	-0.5	0.5	-0.125	0.375	0.5

Window 0, 1, 0, 1

7.3   Kernel and Range

A linear transformation is a function from one vector space, the **domain**, into another vector space. Two further vector spaces, called **kernel** and **range**, are associated with every linear transformation. In this section we introduce and discuss the properties of these spaces.

The following theorem gives an important property of all linear transformations. It paves the way for the introduction of kernel and range.

Theorem 7.5   Let $T: U \to V$ be a linear transformation. Let $\mathbf{0}_V$ and $\mathbf{0}_U$ be the zero vectors of U and V. Then

$$T(\mathbf{0}_U) = \mathbf{0}_V$$

That is, a linear transformation maps a zero vector into a zero vector.

Proof   Let $\mathbf{u}$ be a vector in U and let $T(\mathbf{u}) = \mathbf{v}$. Let 0 be the zero scalar. Since $0\mathbf{u} = \mathbf{0}_U$ and $0\mathbf{v} = \mathbf{0}_V$ and T is linear, we get

$$T(\mathbf{0}_U) = T(0\mathbf{u}) = 0T(\mathbf{u}) = 0\mathbf{v} = \mathbf{0}_V$$

Definition   *Let $T: U \to V$ be a linear transformation.*
The set of vectors in U that are mapped into the zero vector of V is called the **kernel** *of T. The kernel is denoted* ker(T).
The set of vectors in V that are the images of vectors in U is called the **range** *of T. The range is denoted* range(T).

We illustrate these sets in Figure 7.12. Whenever we introduce sets in linear algebra, we are interested in knowing whether they are vector spaces or not! We now find that the kernel and range are indeed vector spaces.

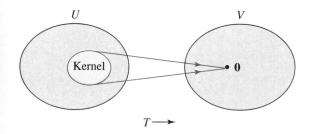

Kernel
All vectors in U that are mapped into $\mathbf{0}$

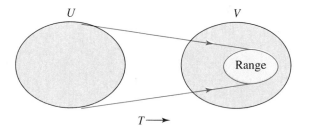

Range
All vectors in V that are images of vectors in U

Figure 7.12

Theorem 7.6   Let $T: U \to V$ be a linear transformation.
(a)   The kernel of T is a subspace of U.
(b)   The range of T is a subspace of V.

(continues)

Proof

(a) From the previous theorem we know that the kernel is nonempty since it contains the zero vector of U. To prove that the kernel is a subspace of U, it remains to show that it is closed under addition and scalar multiplication.

First we prove closure under addition. Let $\mathbf{u}_1$ and $\mathbf{u}_2$ be elements of $\ker(T)$. Thus $T(\mathbf{u}_1) = \mathbf{0}$ and $T(\mathbf{u}_2) = \mathbf{0}$. Using the linearity of T, we get

$$T(\mathbf{u}_1 + \mathbf{u}_2) = T(\mathbf{u}_1) + T(\mathbf{u}_2) = \mathbf{0} + \mathbf{0} = \mathbf{0}$$

The vector $\mathbf{u}_1 + \mathbf{u}_2$ is mapped into $\mathbf{0}$. Thus $\mathbf{u}_1 + \mathbf{u}_2$ is in $\ker(T)$.

Let us now show that $\ker(T)$ is closed under scalar multiplication. Let c be a scalar. Again using the linearity of T, we get

$$T(c\mathbf{u}_1) = cT(\mathbf{u}_1) = c\mathbf{0} = \mathbf{0}$$

Thus $c\mathbf{u}_1$ is in $\ker(T)$.

The kernel is closed under addition and under scalar multiplication. It is a subspace of U.

(b) The previous theorem tells us that the range is nonempty since it contains the zero vector of V. To prove that the range is a subspace of V, it remains to show that it is closed under addition and scalar multiplication. Let $\mathbf{v}_1$ and $\mathbf{v}_2$ be elements of $\text{range}(T)$. Thus there exist vectors $\mathbf{w}_1$ and $\mathbf{w}_2$ in the domain U such that

$$T(\mathbf{w}_1) = \mathbf{v}_1 \quad \text{and} \quad T(\mathbf{w}_2) = \mathbf{v}_2$$

Using the linearity of T,

$$T(\mathbf{w}_1 + \mathbf{w}_2) = T(\mathbf{w}_1) + T(\mathbf{w}_2) = \mathbf{v}_1 + \mathbf{v}_2$$

The vector $\mathbf{v}_1 + \mathbf{v}_2$ is the image of $\mathbf{w}_1 + \mathbf{w}_2$. Thus $\mathbf{v}_1 + \mathbf{v}_2$ is in the range.

Let c be a scalar. By the linearity of T,

$$T(c\mathbf{w}_1) = cT(\mathbf{w}_1) = c\mathbf{v}_1$$

The vector $c\mathbf{v}_1$ is the image of $c\mathbf{w}_1$. Thus $c\mathbf{v}_1$ is in the range.

The range is closed under addition and under scalar multiplication. It is a subspace of V.

Example 1   Find the kernel and range of the linear operator

$$T(x, y, z) = (x, y, 0)$$

Solution   Since the linear operator T maps $\mathbf{R}^3$ into $\mathbf{R}^3$, the kernel and range will both be subspaces of $\mathbf{R}^3$.

Kernel: ker(T) is the subset that is mapped into (0, 0, 0). We see that

$$T(x, y, z) = (x, y, 0)$$
$$= (0, 0, 0), \quad \text{if } x = 0, y = 0$$

Thus ker(T) is the set of all vectors of the form (0, 0, z). We express this as

$$\text{ker}(T) = \{(0, 0, z)\}$$

Geometrically, ker(T) is the set of all vectors that lie on the z axis.

Range: The range of T is the set of all vectors of the form (x, y, 0). Thus

$$\text{range}(T) = \{(x, y, 0)\}$$

Range(T) is the set of all vectors that lie in the xy plane.

  We illustrate this transformation in Figure 7.13. Observe that T projects the vector (x, y, z) into the vector (x, y, 0) in the xy plane. T projects all vectors onto the xy plane. T is an example of **projection operator**.

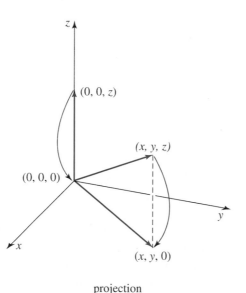

Figure 7.13

projection
$f(x, y, z) = (x, y, 0)$

  Projections are important in applications. The world in which we live has three spatial dimensions. When we observe an object, however, we get a two-dimensional impression of that object, the view changing from location to location. Projections can be used to illustrate what three-dimensional objects look like from various locations. Such transformations are used in architecture, the auto industry, and the aerospace industry. The outline of the object of interest, relative to a suitable coordinate system, is fed into a computer. The computer program contains an appropriate projection transformation that maps the object onto a plane. The output gives a two-dimensional view of the object, the outline being graphed by the computer. In this manner, various transformations can be used to lead to various perspectives of an object. The General Electric Plant at Daytona Beach, Florida, uses such a computer graphics system for simulating aircraft. The Grumman Aerospace Corporation in Bethpage, New York, uses a graphics system in designing aircraft. We illustrate these concepts in Figure 7.14 for an aircraft. The projection used is onto the *xz* plane. The image represents the view an observer at *A* has of the aircraft; it would be graphed out by a computer.

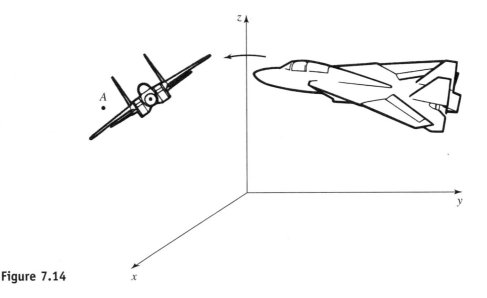

Figure 7.14

We now discuss kernels and ranges of matrix transformations. The following theorem gives us information about the range of a matrix transformation.

Theorem 7.7   The range of $T: \mathbf{R}^n \rightarrow \mathbf{R}^m$, defined by $T(\mathbf{u}) = A\mathbf{u}$, is spanned by the column vectors of A.

Proof   Let $\mathbf{v}$ be a vector in the range. There exists a vector $\mathbf{u}$ such that $T(\mathbf{u}) = \mathbf{v}$. Express $\mathbf{u}$ in terms of the standard basis of $\mathbf{R}^n$.

$$\mathbf{u} = a_1\mathbf{e}_1 + \cdots + a_n\mathbf{e}_n$$

Thus

$$\mathbf{v} = T(a_1\mathbf{e}_1 + \cdots + a_n\mathbf{e}_n)$$
$$= a_1 T(\mathbf{e}_1) + \cdots + a_n T(\mathbf{e}_n)$$

Therefore the column vectors of A, namely, $T(\mathbf{e}_1), \ldots, T(\mathbf{e}_n)$, span the range of A.

Example 2   Determine the kernel and the range of the transformation defined by the following matrix.

$$A = \begin{bmatrix} 1 & 2 & 3 \\ 0 & -1 & 1 \\ 1 & 1 & 4 \end{bmatrix}$$

Solution A is a 3×3 matrix. Thus A defines a linear operator $T: \mathbf{R}^3 \to \mathbf{R}^3$,

$$T(\mathbf{x}) = A(\mathbf{x})$$

The elements of $\mathbf{R}^3$ are written in column matrix form for the purpose of matrix multiplication. For convenience, we express the elements of $\mathbf{R}^3$ in row form at all other times.

Kernel: The kernel will consist of all vectors $\mathbf{x} = (x_1, x_2, x_3)$ in $\mathbf{R}^3$ such that

$$T(\mathbf{x}) = \mathbf{0}$$

Thus

$$\begin{bmatrix} 1 & 2 & 3 \\ 0 & -1 & 1 \\ 1 & 1 & 4 \end{bmatrix} \begin{bmatrix} x_1 \\ x_2 \\ x_3 \end{bmatrix} = \begin{bmatrix} 0 \\ 0 \\ 0 \end{bmatrix}$$

This matrix equation corresponds to the following system of linear equations.

$$x_1 + 2x_2 + 3x_3 = 0$$
$$- x_2 + x_3 = 0$$
$$x_1 + x_2 + 4x_3 = 0$$

On solving this system, we get many solutions, $x_1 = -5r$, $x_2 = r$, $x_3 = r$. The kernel is thus the set of vectors of the form $(-5r, r, r)$.

$$\ker(T) = \{(-5r, r, r)\}$$

Ker(T) is a one-dimensional subspace of $\mathbf{R}^3$ with basis $(-5, 1, 1)$.

Range: The range is spanned by the column vectors of A. Write these column vectors as rows of a matrix and compute the reduced echelon form of the matrix. The nonzero row vectors will give a basis for the range. We get

$$\begin{bmatrix} 1 & 0 & 1 \\ 2 & -1 & 1 \\ 3 & 1 & 4 \end{bmatrix} \approx \begin{bmatrix} 1 & 0 & 1 \\ 0 & -1 & -1 \\ 0 & 1 & 1 \end{bmatrix} \approx \begin{bmatrix} 1 & 0 & 1 \\ 0 & 1 & 1 \\ 0 & 1 & 1 \end{bmatrix} \approx \begin{bmatrix} 1 & 0 & 1 \\ 0 & 1 & 1 \\ 0 & 0 & 0 \end{bmatrix}$$

The vectors $(1, 0, 1)$ and $(0, 1, 1)$ span the range of T. An arbitrary vector in the range will be a linear combination of these vectors.

$$s(1, 0, 1) + t(0, 1, 1)$$

Thus the range of T is

$$\text{range}(T) = \{(s, t, s + t)\}$$

Range (T) is a two-dimensional subspace of $\mathbf{R}^3$ with basis $\{(1, 0, 1), (0, 1, 1)\}$.

The following theorem gives an important relationship between the "sizes" of the kernel and the range of a linear transformation.

Theorem 7.8   Let $T: U \to V$ be a linear transformation. Then

$$\dim \ker(T) + \dim \text{range}(T) = \dim \text{domain}(T)$$

[Observe that this result holds for the linear transformation T of the above example: $\dim \ker(T) = 1$, $\dim \text{range}(T) = 2$, $\dim \text{domain}(T) = 3$.]

Proof   Let us assume that the kernel consists of more than the zero vector, and that it is not the whole of U. (You are asked to prove the result for these two special cases in the exercises that follow.)

Let $\mathbf{u}_1, \ldots, \mathbf{u}_m$ be a basis for $\ker(T)$. Add vectors $\mathbf{u}_{m+1}, \ldots, \mathbf{u}_n$ to this set to get a basis $\mathbf{u}_1, \ldots, \mathbf{u}_n$ for U. We shall show that $T(\mathbf{u}_{m+1}), \ldots, T(\mathbf{u}_n)$ form a basis for the range, thus proving the theorem.

Let $\mathbf{u}$ be a vector in U. $\mathbf{u}$ can be expressed as a linear combination of the basis vectors as follows.

$$\mathbf{u} = a_1\mathbf{u}_1 + \cdots + a_m\mathbf{u}_m + a_{m+1}\mathbf{u}_{m+1} + \cdots a_n\mathbf{u}_n$$

Thus

$$T(\mathbf{u}) = T(a_1\mathbf{u}_1 + \cdots + a_m\mathbf{u}_m + a_{m+1}\mathbf{u}_{m+1} + \cdots + a_n\mathbf{u}_n)$$

The linearity of T gives

$$T(\mathbf{u}) = a_1T(u_1) + \cdots + a_mT(u_m) + a_{m+1}T(u_{m+1}) + a_nT(u_n)$$

Since $\mathbf{u}_1, \ldots, \mathbf{u}_m$ are in this kernel, this reduces to

$$T(\mathbf{u}) = a_{m+1}T(\mathbf{u}_{m+1}) + \cdots + a_nT(\mathbf{u}_n)$$

$T(\mathbf{u})$ represents an arbitrary vector in the range of T. Thus the vectors $T(\mathbf{u}_{m+1}), \ldots, T(\mathbf{u}_n)$ span the range.

It remains to prove that these vectors are also linearly independent. Consider the identity

$$b_{m+1}T(\mathbf{u}_{m+1}) + \cdots + b_nT(\mathbf{u}_n) = \mathbf{0} \qquad (1)$$

where the scalars have been labeled thus for convenience. The linearity of T implies that

$$T(b_{m+1}\mathbf{u}_{m+1} + \cdots + b_n\mathbf{u}_n) = \mathbf{0}$$

This means that the vector $b_{m+1}\mathbf{u}_{m+1} + \cdots + b_n\mathbf{u}_n$ is in the kernel. Thus it can be expressed as a linear combination of the basis of the kernel. Let

$$b_{m+1}\mathbf{u}_{m+1} + \cdots + b_n\mathbf{u}_n = c_1\mathbf{u}_1 + \cdots + c_m\mathbf{u}_m$$

Thus

$$c_1\mathbf{u}_1 + \cdots + c_m\mathbf{u}_m - b_{m+1}\mathbf{u}_{m+1} - \cdots - b_n\mathbf{u}_n = \mathbf{0}$$

Since the vectors $\mathbf{u}_1, \ldots, \mathbf{u}_m, \mathbf{u}_{m+1}, \ldots, \mathbf{u}_n$ are a basis, they are linearly independent. Therefore the coefficients are all zero.

$$c_1 = 0, \ldots, c_m = 0, \qquad b_{m+1} = 0, \ldots, b_n = 0$$

Returning to identity (1), this implies that $T(\mathbf{u}_{m+1}), \ldots, T(\mathbf{u}_n)$ are linearly independent. The set of vectors $T(\mathbf{u}_{m+1}), \ldots, T(\mathbf{u}_n)$ is a basis for the range. The theorem is proven.

Note that the "bigger" the kernel, the "smaller" the range, and vice versa.

We remind the reader that the subspace spanned by the column vectors of a matrix is called the **column space** of the matrix, and that the dimension of the column space is called the **rank** of the matrix. We now have the following result.

The dimension of the range of a matrix transformation is the rank of the matrix.

Example 3 Find the dimensions of the kernel and range of the linear transformation T defined by the matrix

$$A = \begin{bmatrix} 1 & 0 & 3 \\ 0 & 1 & 5 \\ 0 & 0 & 0 \end{bmatrix}$$

Solution Observe that the matrix A is in reduced echelon form. The row vectors are linearly independent. Thus $\operatorname{rank}(A) = 2$, implying that dim range$(T) = 2$.

The domain of T is $\mathbf{R}^3$; dim domain$(T) = 3$. Therefore dim ker$(T) = 1$.

Terminology

The kernel of a linear mapping T is often called the **null space**. Dim ker(T) is called the **nullity**, and dim range(T) is called the **rank** of the transformation. This latter term comes from our observation that the dimension of the range of a matrix mapping is the rank of the matrix. The previous theorem is often referred to as the **rank/nullity theorem** and is written in the following form:

$$\operatorname{rank}(T) + \operatorname{nullity}(T) = \operatorname{dim\ domain}(T)$$

We close this section by developing a relationship between the kernel and the one-to-one property of a linear transformation, and showing a theoretical application of this relationship.

Definition *A transformation T is said to be **one-to-one** if each element in the range of T corresponds to just one element in the domain of T. See Figure 7.15. This means that T is one-to-one if $T(\mathbf{u}) = T(\mathbf{v})$ implies that $\mathbf{u} = \mathbf{v}$.*

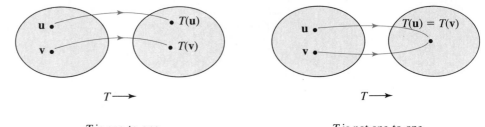

Figure 7.15 T is one-to-one T is not one-to-one

Theorem 7.9   A linear transformation T is one-to-one if and only if the kernel is the zero vector.

Proof   Let us first assume that T is one-to-one. The kernel consists of all the vectors that are mapped into zero. Since T is one-to-one, the kernel must consist of a single vector. However, we know that the zero vector must be in the kernel. Thus the kernel is the zero vector.

Conversely, let us assume that the kernel is the zero vector. Let $\mathbf{u}$ and $\mathbf{v}$ be vectors such that $T(\mathbf{u}) = T(\mathbf{v})$. Using the linear properties of T, we get

$$T(\mathbf{u}) - T(\mathbf{v}) = \mathbf{0}$$
$$T(\mathbf{u} - \mathbf{v}) = \mathbf{0}$$

Thus $\mathbf{u} - \mathbf{v}$ is in the kernel. But the kernel is the zero vector. Therefore

$$\mathbf{u} - \mathbf{v} = \mathbf{0}$$
$$\mathbf{u} = \mathbf{v}$$

The transformation is thus one-to-one.

Theorem 7.10   The transformation $T: \mathbf{R}^n \to \mathbf{R}^n$, defined by $T(\mathbf{x}) = A\mathbf{x}$, is one-to-one if and only if A is nonsingular.

Proof   Let T be one-to-one. Thus $\ker(T) = \mathbf{0}$. The rank/nullity theorem implies that dim range$(T) = n$. Thus the n columns of A are linearly independent, implying that the rank is n and $|A| \neq 0$.

Conversely, assume that A is nonsingular. This implies that the rank of A is n and that its n column vectors are linearly independent. This means that dim range$(T) = n$. The rank nullity theorem now implies that $\ker(T) = \mathbf{0}$. Thus T is one-to-one.

Example 4   Determine whether or not the linear transformations T_A and T_B defined by the following matrices are one-to-one.

$$\text{(a)} \quad A = \begin{bmatrix} 1 & -2 & 5 & 7 \\ 0 & 1 & 9 & 8 \\ 0 & 0 & 1 & 3 \end{bmatrix} \qquad \text{(b)} \quad B = \begin{bmatrix} 2 & 0 & -1 \\ 3 & 4 & 2 \\ 0 & 7 & 5 \end{bmatrix}$$

Solution

(a) Since the rows of A are linearly independent, rank$(A) = 3$, implying that dim range$(T_A) = 3$. The domain of T is $\mathbf{R}^4$, thus dim domain$(T_A) = 4$. By the rank/nullity theorem dim ker$(T_A) = 1$. Since dim ker$(T_A) \ne 0$, the kernel is not the zero vector. Therefore the mapping is not one-to-one.

(b) It can be shown that $|B| = -9 \ne 0$. The matrix B is nonsingular. Therefore the mapping T_B is one-to-one.

Theorem 7.11

Let $T: U \to V$ be a one-to-one linear transformation. If the set $\{\mathbf{u}_1, \dots, \mathbf{u}_n\}$ is linearly independent in U, then $\{T(\mathbf{u}_1), \dots, T(\mathbf{u}_n)\}$ is linearly independent in V. That is, one-to-one linear transformations preserve linear independence.

Proof Consider the identity

$$a_1 T(\mathbf{u}_1) + \cdots + a_n T(\mathbf{u}_n) = \mathbf{0} \tag{1}$$

for scalars $a_1, \dots, a_n$. Since T is linear, this can be written

$$T(a_1 \mathbf{u}_1 + \cdots + a_n \mathbf{u}_n) = \mathbf{0}$$

T is one-to-one, thus the kernel is the zero vector. Therefore

$$a_1 \mathbf{u}_1 + \cdots + a_n \mathbf{u}_n = \mathbf{0}$$

But the set $\{\mathbf{u}_1, \dots, \mathbf{u}_n\}$ is linearly independent. Thus

$$a_1 = 0, \dots, a_n = 0$$

Returning to identity (1), this means that $\{T(\mathbf{u}_1), \dots, T(\mathbf{u}_n)\}$ is linearly independent.

Exercise Set 7.3

1. Consider the linear transformation T defined by each of the following matrices. Determine the kernel and range of each transformation. Show that dim ker(T) + dim range(T) = dim domain(T) for each transformation.

*(a) $\begin{bmatrix} 1 & 2 \\ 3 & 0 \end{bmatrix}$

(b) $\begin{bmatrix} 2 & 0 \\ 3 & 0 \end{bmatrix}$

*(c) $\begin{bmatrix} 2 & 4 \\ 4 & 8 \end{bmatrix}$

(d) $\begin{bmatrix} 1 & 2 \\ -1 & 3 \end{bmatrix}$

*(e) $\begin{bmatrix} 1 & 2 & 3 \\ 0 & 1 & 2 \end{bmatrix}$

(f) $\begin{bmatrix} 1 & 0 & 0 \\ 0 & 2 & 0 \\ 0 & 0 & 3 \end{bmatrix}$

*(g) $\begin{bmatrix} 0 & 1 & 0 \\ 0 & 2 & 0 \\ 0 & 0 & 4 \end{bmatrix}$

*(h) $\begin{bmatrix} 1 & 2 & 1 \\ -1 & -2 & 0 \\ 2 & 4 & 1 \end{bmatrix}$

(i) $\begin{bmatrix} 1 & 1 & 5 \\ 0 & 1 & 3 \\ 2 & 1 & 7 \end{bmatrix}$

2. Determine the kernel and range of each of the following transformations. Show that dim ker(T) + dim range(T) = dim domain(T) for each transformation.

*(a) $T(x, y, z) = (x, 0, 0)$ of $\mathbf{R}^3 \to \mathbf{R}^3$

*(b) $T(x, y, z) = (x + y, z)$ for $\mathbf{R}^3 \to \mathbf{R}^2$

(c) $T(x, y, z) = x + y + z$ of $\mathbf{R}^3 \to \mathbf{R}$

(d) $T(x, y) = (x, 2x, 3x)$ of $\mathbf{R}^2 \to \mathbf{R}^3$

*(e) $T(x, y) = (3x, x - y, y)$ of $\mathbf{R}^2 \to \mathbf{R}^3$

3. Let $T: U \to U$. Let $\mathbf{u}$ be an arbitrary vector in U, $\mathbf{v}$ a fixed vector in U, and $\mathbf{0}$ the zero vector of U. Determine which of the following transformations are linear. Find the kernel and range of each linear transformation.

   *(a) $T(\mathbf{u}) = 5\mathbf{u}$ *(b) $T(\mathbf{u}) = 2\mathbf{u} + 3\mathbf{v}$

   (c) $T(\mathbf{u}) = \mathbf{u}$ (This is called the **identity transformation** on U.)

   (d) $T(\mathbf{0}) = \mathbf{0}$ (This is called the **zero transformation** on U.)

4. Use the rank/nullity theorem to find the dimensions of the kernels and ranges of the linear transformations defined by the following matrices. State whether or not the transformations are one-to-one.

   *(a) $A = \begin{bmatrix} 1 & 8 & 2 \\ 0 & 1 & -4 \\ 0 & 0 & 0 \end{bmatrix}$ (b) $B = \begin{bmatrix} 1 & 4 & 2 \\ 0 & 1 & 9 \\ 0 & 0 & 1 \end{bmatrix}$

   *(c) $C = \begin{bmatrix} 1 & 7 & 3 & 2 \\ 0 & 1 & 5 & 4 \\ 0 & 0 & 1 & 7 \end{bmatrix}$ (d) $D = \begin{bmatrix} 1 & 7 & 3 \\ 0 & 1 & 5 \\ 0 & 0 & 0 \end{bmatrix}$

   (e) $E = \begin{bmatrix} 1 & -2 & 3 & 5 \\ 1 & -1 & 8 & 7 \\ 2 & -4 & 6 & 10 \end{bmatrix}$

   *(f) $F = \begin{bmatrix} 1 & 2 & -3 \\ -2 & 4 & 6 \\ 0 & 2 & -1 \\ 0 & -4 & 2 \end{bmatrix}$

5. Use determinants to decide whether or not the linear transformations defined by the following matrices are one-to-one.

   *(a) $A = \begin{bmatrix} 1 & 2 & 5 \\ 0 & 3 & 6 \\ 0 & 0 & 4 \end{bmatrix}$ (b) $B = \begin{bmatrix} 3 & 2 & 4 \\ -2 & 0 & 0 \\ 5 & 1 & 2 \end{bmatrix}$

   *(c) $C = \begin{bmatrix} 1 & 2 & 3 \\ 2 & 4 & 6 \\ 3 & 6 & 9 \end{bmatrix}$ (d) $D = \begin{bmatrix} -1 & 2 & 4 \\ 3 & 1 & 2 \\ 1 & 5 & 10 \end{bmatrix}$

   (e) $E = \begin{bmatrix} 2 & 0 & 1 & 3 \\ 0 & 1 & 4 & 7 \\ 0 & 0 & 3 & 8 \\ 0 & 0 & 0 & -4 \end{bmatrix}$

*(f) $F = \begin{bmatrix} 2 & 1 & 3 & 0 \\ 4 & 3 & 0 & 2 \\ -1 & 7 & 8 & 1 \\ 0 & -2 & 4 & 0 \end{bmatrix}$

6. Let $T: U \to U$. A vector $\mathbf{u}$ is said to be a fixed point of T if $T(\mathbf{u}) = \mathbf{u}$. Determine the fixed points (if any) of the following transformations.

   *(a) $T(x, y) = (x, 3y)$ (b) $T(x, y) = (x, 2)$

   *(c) $T(x, y) = (x, y + 1)$ (d) $T(x, y) = (x, y)$

   (e) $T(x, y) = (y, x)$

   *(f) $T(x, y) = (x + y, x - y)$

   (g) Prove that if T is linear, the set of fixed points is a subspace.

7. Let $T: U \to V$ be linear. Prove that

   $$\dim \ker(T) + \dim \text{range}(T) = \dim \text{domain}(T)$$

   (a) when $\ker(T) = \mathbf{0}$, (b) when $\ker(T) = U$. These were the special cases mentioned in the proof of Theorem 7.8.

8. Let $T: U \to V$ be a linear transformation.

   (a) Prove that when $\dim(U) = \dim(V)$, then $\text{range}(T) = V$ if and only if $\ker(T) = \mathbf{0}$.

   (b) Prove that $\ker(T) = U$ if and only if $\text{range}(T) = \mathbf{0}$.

*9. Let $T: U \to V$ be a linear transformation. Prove that the dimension of the range of T can never be greater than the dimension of the domain.

10. Let an $m \times n$ matrix A define a mapping $T: \mathbf{R}^n \to \mathbf{R}^m$. A^t, being an $n \times m$ matrix, will define a mapping $T^t: \mathbf{R}^m \to \mathbf{R}^n$. Show that

    $$\dim \text{range}(T) = \dim \text{range}(T^t)$$

*11. Consider the projection $T(x, y, z) = (x, y, 0)$ of $\mathbf{R}^3 \to \mathbf{R}^3$. Find the set of vectors in $\mathbf{R}^3$ that are mapped by T into the vector $(1, 2, 0)$. Sketch this set.

12. Consider the linear operator $T(x, y) = (x - y, 2y - 2x)$ of $\mathbf{R}^2 \to \mathbf{R}^2$. Find and sketch the set of vectors in $\mathbf{R}^2$ that are mapped by T into the vector $(2, -4)$.

*13. Consider the linear operator $T: \mathbf{R}^2 \to \mathbf{R}^2$, defined by $T(x, y) = (2x, 3x)$. Find and sketch the set of vectors that are mapped by T into the vector $(4, 6)$. Is this set a subspace of $\mathbf{R}^2$?

14. Consider the linear transformation $T: \mathbf{R}^3 \to \mathbf{R}^2$, defined by $T(x, y, z) = (x - y, x + z)$. Find and sketch the set of vectors that is mapped by T into the vector $(1, 4)$. Is this set a subspace?

*15. Let $T: U \to V$ be a linear mapping. Let $\mathbf{v}$ be a nonzero vector in V. Let W be the set of vectors in U such that $T(\mathbf{w}) = \mathbf{v}$. Is W a subspace of U?

*16. Prove that $T: P_2 \to P_2$, defined as follows, is linear. Find the kernel and range of T. Give bases for these subspaces.

$$T(a_2 x^2 + a_1 x + a_0) = (a_2 + a_1)x^2 + a_1 x + 2a_0$$

17. Prove that $T: P_3 \to P_2$, defined as follows, is linear. Find the kernel and range of T. Give bases for these subspaces.

$$T(a_3 x^3 + a_2 x^2 + a_1 x + a_0) = a_3 x^2 - a_0$$

*18. Prove that $g: P_2 \to P_3$, defined as follows, is linear. Find the kernel and range of g. Give bases for these subspaces.

$$g(a_2 x^2 + a_1 x + a_0) = 2a_2 x^3 + a_1 x + 3a_0$$

*19. Prove that $T: P_1 \to P_1$, defined as follows, is not linear.

$$T(a_1 x + a_0) = a_1 x + 4$$

20. Determine whether $g: P_2 \to P_2$, defined as follows, is linear.

$$g(a_2 x^2 + a_1 x + a_0) = a_2 x^2 + (a_1)^2 x + 2a_0$$

*21. Let D be the operation of taking the derivative. Interpret D as a linear operator on P_n. Find the image of $x^3 - 3x^2 + 2x + 1$ under D. Determine the kernel and range of D.

22. Let D^2 be the operation of taking the second derivative.

*(a) Find the image of $2x^3 + 3x^2 - 5x + 4$ under D^2. Prove that D^2 is a linear operator on P_n. Determine the kernel and range of D^2.

(b) Consider the transformation $D^2 + D + 3$ of P_n into itself. Find the image of $x^3 - 2x^2 + 6x + 1$ under this transformation. Prove that the transformation is linear. Determine the kernel and range of the transformation.

*23. Let D be the operation of taking the derivative. Interpret D as an operator on P_n. Find the set of polynomials that are mapped by D into $3x^2 - 4x + 7$.

24. Let D^2 be the operation of taking the second derivative. Interpret D^2 as an operator on P_n. Find the set of polynomials that are mapped by D^2 into $4x^3 + 6x^2 - 2x + 3$.

*25. Let p be a polynomial in P_n and let $T(p) = \int_0^1 p(x)\, dx$ be a linear transformation of P_n into $\mathbf{R}$. Find

$$T(8x^3 + 6x^2 + 4x + 1)$$

Determine the kernel and range of T. Find the set of elements of P_n that are mapped into 2.

26. Let U be the vector space of 2×2 matrices. Let A and B be 2×2 matrices. Determine whether the following transformations are linear. Find the kernel and range of each linear transformation.

(a) $T(A) = A^t$ ***(b)** $T(A) = |A|$

(c) $T(A) = \text{trace}(A)$ ***(d)** $T(A) = A^2$

(e) $T(A) = A + B$ ***(f)** $T(A) = a_{11}$

(g) $T(A) = 0$ **(h)** $T(A) = I_2$

*(i) $T(A) = A + A^t$

*(j) $T(A) = \text{reduced echelon form}(A)$

27. Let U be a vector space with an inner product, and let $\mathbf{u}$ be an arbitrary vector in the space and $\mathbf{v}$ be a fixed vector. Determine which of the following transformations are linear. Find the kernel and range of each linear transformation.

*(a) $T(\mathbf{u}) = \|\mathbf{u}\|$ ***(b)** $T(\mathbf{u}) = \langle \mathbf{u}, \mathbf{v} \rangle$

(c) $T(\mathbf{u}) = d(\mathbf{u} - \mathbf{v})$

*28. Let $T: U \to V$ be a linear transformation. Prove that

$$\dim \text{range}(T) = \dim \text{domain}(T)$$

if and only if T is one-to-one.

29. Show that no one-to-one linear transformation can exist from $\mathbf{R}^3 \to \mathbf{R}^2$. Generalize this result by proving that no one-to-one linear transformation can exist from $U \to V$ if the dimension of U is greater than the dimension of V.

30. Let $T: U \to V$ be a linear transformation. Prove that T is one-to-one if and only if it preserves linear independence.

*7.4 Transformations and Systems of Linear Equations

Linear transformations and the concepts of kernel and range play an important role in the analyses of systems of linear equations. We shall find that they enable us to "visualize" the sets of solutions.

We have seen that a system of m linear equations in n variables can be written in the matrix form

$$A\mathbf{x} = \mathbf{y}$$

A is an $m \times n$ matrix; it is the matrix of coefficients of the system. The set of solutions is the set of all $\mathbf{x}$'s that satisfy this equation. We now have a very elegant way of looking at this solution set. Let $T: \mathbf{R}^n \to \mathbf{R}^m$ be the linear transformation defined by A. The system of equations can now be written

$$T(\mathbf{x}) = y$$

The set of solutions is thus the set of vectors in $\mathbf{R}^n$ that are mapped by T into the vector $\mathbf{y}$. If $\mathbf{y}$ is not in the range of T, then the system has no solution. See Figure 7.16.

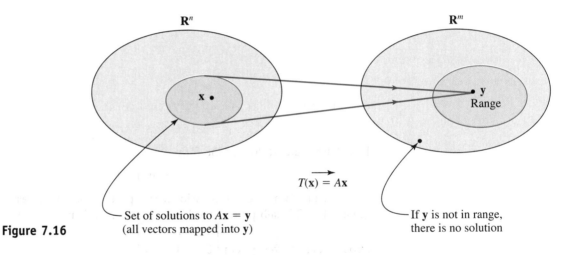

Figure 7.16   Set of solutions to $A\mathbf{x} = \mathbf{y}$ (all vectors mapped into $\mathbf{y}$)   If $\mathbf{y}$ is not in range, there is no solution

Homogeneous Equations

The above way of looking at systems of linear equations leads directly to the following result.

Theorem 7.12   The set of solutions to a homogeneous system of m linear equations in n variables, $A\mathbf{x} = \mathbf{0}$, is a subspace of $\mathbf{R}^n$.

Proof   Let T be the linear transformation of $\mathbf{R}^n$ into $\mathbf{R}^m$, defined by A. The set of solutions is the set of vectors in $\mathbf{R}^n$ that are mapped by T into the zero vector. The set of solutions is the kernel of the transformation and is thus a subspace.

Example 1   Solve the following homogeneous system of linear equations. Interpret the set of solutions as a subspace. Sketch the subspace of solutions.

$$x_1 + 2x_2 + 3x_3 = 0$$
$$- x_2 + x_3 = 0$$
$$x_1 + x_2 + 4x_3 = 0$$

Solution Using Gauss-Jordan elimination, we get

$$\begin{bmatrix} 1 & 2 & 3 & 0 \\ 0 & -1 & 1 & 0 \\ 1 & 1 & 4 & 0 \end{bmatrix} \approx \begin{bmatrix} 1 & 2 & 3 & 0 \\ 0 & -1 & 1 & 0 \\ 0 & -1 & 1 & 0 \end{bmatrix} \approx \begin{bmatrix} 1 & 2 & 3 & 0 \\ 0 & 1 & -1 & 0 \\ 0 & -1 & 1 & 0 \end{bmatrix}$$

$$\approx \begin{bmatrix} 1 & 0 & 5 & 0 \\ 0 & 1 & -1 & 0 \\ 0 & 0 & 0 & 0 \end{bmatrix}$$

Thus

$$x_1 \qquad + 5x_3 = 0$$
$$x_2 - x_3 = 0$$

giving

$$x_1 = 5x_3 \qquad x_2 = x_3$$

Assign the value r to x_3. An arbitrary solution is thus

$$x_1 = -5r, \qquad x_2 = r, \qquad x_3 = r$$

The solutions are vectors of the form

$$(-5r, r, r)$$

These vectors form a one-dimensional subspace of $\mathbf{R}^3$, with basis $(-5, 1, 1)$. See Figure 7.17. This subspace is the kernel of the transformation defined by the matrix

of coefficients of the system $\begin{bmatrix} 1 & 2 & 3 \\ 0 & -1 & 1 \\ 1 & 4 & 4 \end{bmatrix}$.

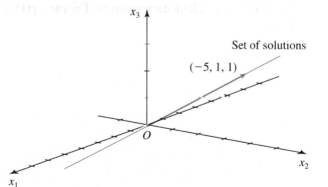

Figure 7.17

Nonhomogeneous Equations

We now find that the set of solutions to be a nonhomogeneous system of linear equations does not form a subspace.

Let $A\mathbf{x} = \mathbf{y}(\mathbf{y} \neq 0)$ be a nonhomogeneous system of linear equations. Let $\mathbf{x}_1$ and $\mathbf{x}_2$ be solutions. Thus

$$A\mathbf{x}_1 = \mathbf{y} \quad \text{and} \quad A\mathbf{x}_2 = \mathbf{y}$$

Adding these equations gives

$$A\mathbf{x}_1 + A\mathbf{x}_2 = 2\mathbf{y}$$
$$A(\mathbf{x}_1 + \mathbf{x}_2) = 2\mathbf{y}$$

Therefore $\mathbf{x}_1 + \mathbf{x}_2$ does not satisfy $A\mathbf{x} = \mathbf{y}$. It is not a solution. The set of solutions is not closed under addition; it is not a subspace.

All is not lost, however! We shall now find that though the set of solutions to a nonhomogeneous system is not itself a subspace, it can be obtained by sliding a certain subspace. This result will enable us to picture the set of solutions.

Theorem 7.13

Let $A\mathbf{x} = \mathbf{y}$ be a nonhomogeneous system of m linear equations in n variables. Let $\mathbf{x}_1$ be a particular solution. Every other solution can be written in the form $\mathbf{x} = \mathbf{z} + \mathbf{x}_1$, where $\mathbf{z}$ is an element of the kernel of the transformation T defined by A. The solution is unique if the kernel consists of the zero vector only.

Proof $\mathbf{x}_1$ is a solution. Thus $A\mathbf{x}_1 = \mathbf{y}$. Let $\mathbf{x}$ be an arbitrary solution. Thus $A\mathbf{x} = \mathbf{y}$. Equating $A\mathbf{x}_1$ and $A\mathbf{x}$, we get

$$A\mathbf{x}_1 = A\mathbf{x}$$
$$A\mathbf{x} - A\mathbf{x}_1 = \mathbf{0}$$
$$A(\mathbf{x} - \mathbf{x}_1) = \mathbf{0}$$
$$T(\mathbf{x} - \mathbf{x}_1) = \mathbf{0}$$

Thus $\mathbf{x} - \mathbf{x}_1$ is an element of the kernal of T; call it $\mathbf{z}$.

$$\mathbf{x} - \mathbf{x}_1 = \mathbf{z}$$

We can write

$$\mathbf{x} = \mathbf{z} + \mathbf{x}_1$$

Note that the solution is unique if the only value of $\mathbf{z}$ is $\mathbf{0}$—that is, if the kernel is the zero vector.

This result implies that the set of solutions to a nonhomogeneous system of linear equations $A\mathbf{x} = \mathbf{y}$ can be generated from the kernel of the transformation defined by the matrix of coefficients and a particular solution $\mathbf{x}_1$. If we take any vector $\mathbf{z}$ in the kernel and add $\mathbf{x}_1$ to it, we get a solution. Geometrically, this means that the set of solutions is obtained by sliding the kernel in the direction and distance defined by the vector $\mathbf{x}_1$. See Figure 7.18.

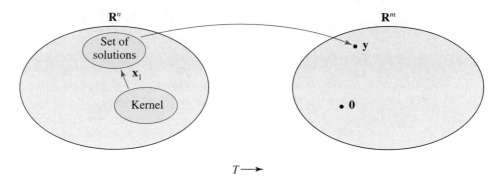

Figure 7.18   Set of solutions to $A\mathbf{x} = \mathbf{y}$

Example 2 Solve the following system of linear equations. Sketch the set of solutions.

$$x_1 + 2x_2 + 3x_3 = 11$$
$$- x_2 + x_3 = -2$$
$$x_1 + x_2 + 4x_3 = 9$$

Solution Using Gauss-Jordan elimination, we get

$$\begin{bmatrix} 1 & 2 & 3 & 11 \\ 0 & -1 & 1 & -2 \\ 1 & 1 & 4 & 9 \end{bmatrix} \approx \begin{bmatrix} 1 & 2 & 3 & 11 \\ 0 & -1 & 1 & -2 \\ 0 & -1 & 1 & -2 \end{bmatrix} \approx \begin{bmatrix} 1 & 2 & 3 & 11 \\ 0 & 1 & -1 & 2 \\ 0 & -1 & 1 & -2 \end{bmatrix}$$

$$\approx \begin{bmatrix} 1 & 0 & 5 & 7 \\ 0 & 1 & -1 & 2 \\ 0 & 0 & 0 & 0 \end{bmatrix}$$

Thus

$$x_1 + 5x_3 = 7$$
$$x_2 - x_3 = 2$$

giving

$$x_1 = -5x_3 + 7, \qquad x_2 = x_3 + 2$$

Assign the value r to x_3. An arbitrary solution is thus

$$x_1 = -5r + 7, \qquad x_2 = r + 2, \qquad x_3 = r$$

The solutions are vectors of the form

$$(-5r + 7, r + 2, r)$$

Let us "pull this solution apart." We separate the part involving the parameter r from the constant part. The part involving r will be in the kernel of the transformation defined by A, while the constant part will be a particular solution $\mathbf{x}_1$ to $A\mathbf{x} = \mathbf{y}$.

$$(-5r + 7, r + 2, r) = \quad r(-5, 1, 1) \quad + \quad (7, 2, 0)$$

arbitrary solution   element of kernel   a particular solution
to $A\mathbf{x} = \mathbf{y}$ to $A\mathbf{x} = \mathbf{y}$

We found in Example 1 that the kernel of the mapping defined by this matrix of coefficients is indeed the set of vectors of the form $r(-5, 1, 1)$. The reader can verify by substitution that $(7, 2, 0)$ is indeed a particular solution to the given system.

The set of solutions can be represented geometrically by sliding the kernel, namely the line defined by the vector $(-5, 1, 1)$, in the direction and distance defined by the vector $(7, 2, 0)$. See Figure 7.19.

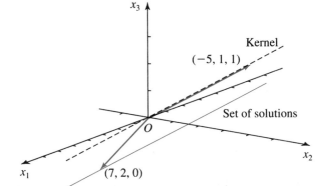

Figure 7.19

Many Systems

Consider a number of linear systems, $A\mathbf{x} = \mathbf{y}_1$, $A\mathbf{x} = \mathbf{y}_2$, $A\mathbf{x} = \mathbf{y}_3$, . . . , all having the same matrix of coefficients A. Let T be the linear transformation defined by A. Let $\mathbf{x}_1, \mathbf{x}_2, \mathbf{x}_3, \ldots$ be particular solutions to these systems. Then the sets of solutions to these systems are

$$\ker(T) + \mathbf{x}_1, \qquad \ker(T) + \mathbf{x}_2, \qquad \ker(T) + \mathbf{x}_3, \ldots$$

These sets are "parallel" sets, each being the $\ker(T)$ translated by the amounts $\mathbf{x}_1, \mathbf{x}_2, \mathbf{x}_3, \ldots$. Thus, for example, the solutions to the systems

$$x_1 + 2x_2 + 3x_3 = a_1$$
$$- x_2 + x_3 = a_2$$
$$x_1 + x_2 + 4x_3 = a_3$$

will all be straight lines parallel to the line defined by the vector $(-5, 1, 1)$.

Example 3   Analyze the solutions to the following system of equations.

$$x_1 - 2x_2 + 3x_3 + x_4 = 1$$
$$2x_1 - 3x_2 + 2x_3 - x_4 = 4$$
$$3x_1 - 5x_2 + 5x_3 \quad\quad = 5$$
$$x_1 - x_2 - x_3 - 2x_4 = 3$$

Solution   Solve using Gauss-Jordan elimination.

$$\begin{bmatrix} 1 & -2 & 3 & 1 & 1 \\ 2 & -3 & 2 & -1 & 4 \\ 3 & -5 & 5 & 0 & 5 \\ 1 & -1 & -1 & -2 & 3 \end{bmatrix} \approx \begin{bmatrix} 1 & -2 & 3 & 1 & 1 \\ 0 & 1 & -4 & -3 & 2 \\ 0 & 1 & -4 & -3 & 2 \\ 0 & 1 & -4 & -3 & 2 \end{bmatrix} \approx \begin{bmatrix} 1 & 0 & -5 & -5 & 5 \\ 0 & 1 & -4 & -3 & 2 \\ 0 & 0 & 0 & 0 & 0 \\ 0 & 0 & 0 & 0 & 0 \end{bmatrix}$$

We get

$$x_1 \qquad - 5x_3 - 5x_4 = 5$$
$$x_2 - 4x_3 - 3x_4 = 2$$

Express the leading variables in terms of the remaining variables.

$$x_1 = 5x_3 + 5x_4 + 5, \qquad x_2 = 4x_3 + 3x_4 + 2$$

The arbitrary solution is

$$(5r + 5s + 5, \, 4r + 3s + 2, \, r, \, s)$$

Separate the parts of this vector as follows.

$$(5r + 5s + 5, \, 4r + 3s + 2, \, r, \, s) = r(5, 4, 1, 0) + s(5, 3, 0, 1) \quad + \quad (5, 2, 0, 0)$$

arbitrary solution
to $A\mathbf{x} = \mathbf{y}$ kernel of mapping defined by A a particular solution
to $A\mathbf{x} = \mathbf{y}$

Observe that the kernel is a two-dimensional subspace of $\mathbf{R}^4$ with basis $(5, 4, 1, 0)$, $(5, 3, 0, 1)$. It is a plane through the origin. The set of solutions to the given system is this plane translated in a manner described by the vector $(5, 2, 0, 0)$.

It will be of interest to students who have studied differential equations to now see that these concepts relate to methods used in solving differential equations. The following example illustrates the ideas involved.

Example 4   Solve the differential equation $d^2y/dx^2 - 9y = 2x$.

Solution   Let D be the operation of taking the derivative. We can write the differential equation in the form

$$D^2y - 9y = 2x$$
$$(D^2 - 9)y = 2x$$

Let V be the vector space of functions that have second derivatives. $D^2 - 9$ is a linear transformation of V into itself. The solution y to the differential equation is the function that is mapped by $D^2 - 9$ into $2x$. The situation is analogous to solving a system of nonhomogeneous linear equations. We get

$$\text{arbitrary solution} = \frac{\text{arbitrary element of kernel}}{\text{of } D^2 - 9} + \frac{\text{particular solution to}}{D^2y - 9y = 2x}$$

Let us first find the kernel of $D^2 - 9$. The general theory tells us that since the operator is of order two, the kernel will be of dimension two, with basis vectors of the form e^{mx}. Since these vectors are in the kernel,

$$(D^2 - 9)e^{mx} = 0$$
$$(m^2 - 9)e^{mx} = 0$$
$$(m - 3)(m + 3)e^{mx} = 0$$
$$m = 3 \text{ or } -3$$

Basis vectors of the kernel are thus e^{3x} and e^{-3x}. An arbitrary vector in the kernel can be written $re^{3x} + se^{-3x}$.

It remains to find a particular solution to $D^2y - 9y = 2x$. The simplest solution to find is the one whose second derivative is zero. We get $-9y = 2x$; $y = -2x/9$. Thus an arbitrary solution to the differential equation can be written

$$y = re^{3x} + se^{-3x} - 2x/9$$

Exercise Set 7.4

Consider the following systems of linear equations. For convenience the solutions are given. Analyze the solutions by representing each set of solutions as the sum of the kernel of the transformation defined by the matrix of coefficients and a particular solution.

*1.
$$x_1 - x_2 + x_3 = 1$$
$$2x_1 - 2x_2 + 3x_3 = 3$$
$$x_1 - x_2 - x_3 = -1$$
Solutions: $x_1 = r, x_2 = r, x_3 = 1$

2.
$$x_1 + x_2 + x_3 = 3$$
$$2x_1 + 3x_2 + x_3 = 5$$
$$x_1 - x_2 - 2x_3 = -5$$
Solutions: $x_1 = 0, x_2 = 1, x_3 = 2$

*3.
$$x_1 - 2x_2 + 3x_3 = 1$$
$$3x_1 - 4x_2 + 5x_3 = 3$$
$$2x_1 - 3x_2 + 4x_3 = 2$$
Solutions: $x_1 = r + 1, x_2 = 2r, x_3 = r$

4.
$$x_1 - x_2 + x_3 = 3$$
$$-2x_1 + 2x_2 - 2x_3 = -6$$
Solutions: $x_1 = r - s + 3, x_2 = r, x_3 = s$

*5.
$$x_1 - x_2 - x_3 + 2x_4 = 4$$
$$2x_1 - x_2 + 3x_3 - x_4 = 2$$
Solutions: $x_1 = -4r + 3s - 2, x_2 = -5r + 5s - 6$
$$x_3 = r, x_4 = s$$

6.
$$x_1 + 2x_2 - x_3 + x_4 + 2x_5 = 0$$
$$x_2 + 2x_3 - 3x_4 - 3x_5 = 2$$
$$x_3 + 5x_4 + 3x_5 = -3$$

Solutions: $x_1 = -32r - 23s - 19$
$$x_2 = 13r + 9s + 8$$
$$x_3 = -5r - 3s - 3$$
$$x_4 = r, x_5 = s$$

7. Consider the transformation $T: \mathbf{R}^3 \to \mathbf{R}^3$, defined by the following matrix A. The range of this transformation is the set of vectors of the form $(x, y, x + y)$—that is, the plane $z = x + y$. Use this information to determine whether solutions exist to the following system of equations $A\mathbf{x} = \mathbf{y}$, $\mathbf{y}$ taking on the various values. (You need not determine the solutions, if they exist.)

*(a) $\mathbf{y} = (1, 1, 2)$ (b) $\mathbf{y} = (-1, 2, 3)$

(c) $\mathbf{y} = (3, 2, 5)$ *(d) $\mathbf{y} = (2, 4, 5)$

$$A = \begin{bmatrix} 1 & 2 & 3 \\ 0 & -1 & 1 \\ 1 & 1 & 4 \end{bmatrix}$$

*8. In this section we solved nonhomogeneous systems of equations, arriving at an arbitrary solution. The solution was decomposed into the kernel of the transformation defined by the matrix of coefficients and a particular solution to the system. Is the particular solution $\mathbf{x}_1$ unique? If $\mathbf{x}_1$ is not unique, determine a vector other than $(7, 2, 0)$ that can be used in Example 2 of this section.

9. Determine a vector other than $(5, 2, 0, 0)$ that can be used as a particular solution in the analysis of the set of solutions for Example 3 of this section.

*10. Construct a nonhomogeneous system of linear equations $A\mathbf{x} = \mathbf{y}$ that has the following matrix of coefficients A and particular solution $\mathbf{x}_1$.

$$A = \begin{bmatrix} 1 & 2 & 1 \\ 0 & 1 & 2 \\ 1 & 1 & -1 \end{bmatrix}, \qquad \mathbf{x}_1 = (1, -1, 4)$$

11. Construct a nonhomogeneous system of linear equations $A\mathbf{x} = \mathbf{y}$ that has the following matrix of coefficients A and particular solution $\mathbf{x}_1$.

$$A = \begin{bmatrix} 2 & 1 & 0 \\ 3 & 3 & 1 \\ 0 & 1 & 1 \end{bmatrix}, \qquad \mathbf{x}_1 = (2, 0, 3)$$

12. Prove that the solution to a nonhomogeneous system of linear equations is unique if and only if the kernel of the mapping defined by the matrix of coefficients is the zero vector.

Exercises 13 and 14 are intended for students who have a knowledge of differential equations.

13. Find a basis for the kernel of each of the following operators. Give an arbitrary function in the kernel.

   *(a) $D^2 + D - 2$ **(b)** $D^2 + 4D + 3$

   *(c) $D^2 + 2D - 8$

14. Solve the following differential equations.

   (a) $d^2y/dx^2 + 5dy/dx + 6y = 8$

   *(b) $d^2y/dx^2 - 3dy/dx = 8$

   (c) $d^2y/dx^2 - 7dy/dx + 12y = 24$

   *(d) $d^2y/dx^2 + 8dy/dx = 3e^{2x}$

*7.5 Least Squares Curves

This section is possibly the most attractive in the book! It illustrates mathematics at its best. Theoretical results developed earlier come together in a very elegant, powerful way to arrive at an important computational tool. We derive the method of finding a polynomial that best fits given data points—a method that is extremely important to the natural sciences, social sciences, and engineering.

We have seen that a system $A\mathbf{x} = \mathbf{y}$ of n equations in n variables, where A is invertible, has the unique solution $\mathbf{x} = A^{-1}\mathbf{y}$. However, if $A\mathbf{x} = \mathbf{y}$ is a system of n equations in m variables, where $n > m$, the system does not, in general, have a solution and it is then said to be **overdetermined**. A is not a square matrix for such a system, and A^{-1} does not exist. We shall introduce a matrix called the **pseudoinverse** of A, denoted pinv(A), that leads to a **least squares solutions**, $\mathbf{x} = \text{pinv}(A)\mathbf{y}$, for an overdetermined system. This is not a true solution, but is in some sense the closest we get to a true solution for the system. We shall see an application of overdetermined systems in finding curves that "best" fit data.

Definition   *Let A be a matrix. The matrix $(A^tA)^{-1}A^t$ is called the **pseudoinverse** of A and is denoted* pinv(A).

We saw that not every matrix has an inverse. Similarly, not every matrix has a pseudoinverse. The matrix A has a pseudoinverse if $(A^tA)^{-1}$ exists.

Example 1   Find the pseudoinverse of $A = \begin{bmatrix} 1 & 2 \\ -1 & 3 \\ 2 & 4 \end{bmatrix}$.

Solution   We compute the pseudoinverse of A in stages.

$$A'A = \begin{bmatrix} 1 & -1 & 2 \\ 2 & 3 & 4 \end{bmatrix} \begin{bmatrix} 1 & 2 \\ -1 & 3 \\ 2 & 4 \end{bmatrix} = \begin{bmatrix} 6 & 7 \\ 7 & 29 \end{bmatrix}$$

$$(A'A)^{-1} = \frac{1}{|A'A|} \text{adj}(A'A) = \frac{1}{125} \begin{bmatrix} 29 & -7 \\ -7 & 6 \end{bmatrix}$$

$$\text{pinv}(A) = (A'A)^{-1}A' = \frac{1}{125} \begin{bmatrix} 29 & -7 \\ -7 & 6 \end{bmatrix} \begin{bmatrix} 1 & -1 & 2 \\ 2 & 3 & 4 \end{bmatrix} = \frac{1}{25} \begin{bmatrix} 3 & -10 & 6 \\ 1 & 5 & 2 \end{bmatrix}$$

We now use the concept of pseudoinverse to further our understanding of systems of linear equations. Let $A\mathbf{x} = \mathbf{y}$ be a system of n linear equations in m variables with $n > m$, where A is of rank m. Multiply each side of this matrix equation by A' to get

$$A'A\mathbf{x} = A'\mathbf{y}$$

The matrix $A'A$ can be shown to be invertible for such a system. Multiple each side of this equation by $(A'A)^{-1}$ to get

$$\mathbf{x} = [(A'A)^{-1}A']\mathbf{y}$$
$$= \text{pinv}(A)\mathbf{y}$$

This value of $\mathbf{x}$ is called the least squares solution to the system of equations.

System	Least Squares Solution
$A\mathbf{x} = \mathbf{y}$	$\mathbf{x} = \text{pinv}(A)\mathbf{y}$

Let $A\mathbf{x} = \mathbf{y}$ be a system of n linear equations in m variables with $n > m$, where A is of rank m. This system has a least squares solution. If the system has a unique solution, the least squares solution is that unique solution. The system cannot have many solutions.

Example 2   Find the least squares solution of the following overdetermined system of equations. Sketch the solution.

$$\begin{aligned} x + y &= 6 \\ -x + y &= 3 \\ 2x + 3y &= 9 \end{aligned}$$

Solution   The matrix of coefficients is

$$A = \begin{bmatrix} 1 & 1 \\ -1 & 1 \\ 2 & 3 \end{bmatrix} \quad \text{and} \quad y = \begin{bmatrix} 6 \\ 3 \\ 9 \end{bmatrix}$$

The column vectors of A are linearly independent. Thus the rank of A is 2. This system has a least squares solution. We compute pinv(A).

$$A'A = \begin{bmatrix} 1 & -1 & 2 \\ 1 & 1 & 3 \end{bmatrix} \begin{bmatrix} 1 & 1 \\ -1 & 1 \\ 2 & 3 \end{bmatrix} = \begin{bmatrix} 6 & 6 \\ 6 & 11 \end{bmatrix}$$

$$(A'A)^{-1} = \frac{1}{|A'A|} \text{adj}(A'A) = \frac{1}{30} \begin{bmatrix} 11 & -6 \\ -6 & 6 \end{bmatrix}$$

$$\text{pinv}(A) = (A'A)^{-1}A' = \frac{1}{30} \begin{bmatrix} 11 & -6 \\ -6 & 6 \end{bmatrix} \begin{bmatrix} 1 & -1 & 2 \\ 1 & 1 & 3 \end{bmatrix} = \frac{1}{30} \begin{bmatrix} 5 & -17 & 4 \\ 0 & 12 & 6 \end{bmatrix}$$

The least squares solution is

$$\text{pinv}(A)y = \frac{1}{30} \begin{bmatrix} 5 & -17 & 4 \\ 0 & 12 & 6 \end{bmatrix} \begin{bmatrix} 6 \\ 3 \\ 9 \end{bmatrix} = \begin{bmatrix} \frac{1}{2} \\ 3 \end{bmatrix}$$

The least squares solution is the point $P\left(\frac{1}{2}, 3\right)$ in Figure 7.20.

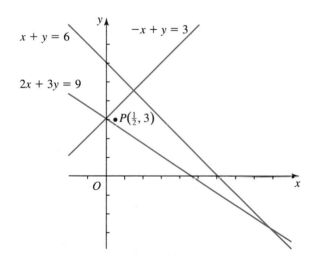

Figure 7.20

Let us now see how least squares solutions can be used to find curves that best fit given data.

Least Squares Curves

Many branches of science and business use equations based on data that has been determined from experimental results. In Section 1.4 we discussed the problem of finding a unique polynomial of degree two that passes through three data points. In many applications, however, there is too much data to lead to an equation that can exactly fit all the data. One then uses the equation of a line or curve that in some sense "best" fits the data. For example, suppose the data consists of the points $(x_1, y_1), \ldots, (x_n, y_n)$, shown in Figure 7.21(a). These points lie approximately on a line. We would want the equation of the line that best fits these points. On the other hand, the points might closely fit a parabola, as shown in Figure 7.21(b). We would then want to find the parabola that most closely fits these points.

Many criteria can be used for the best fit in such cases. The one that has generally been found to be most satisfactory is called the **least squares** line or curve—found by solving an overdetermined system of equations. The least squares line or curve is such that $d_1^2 + \cdots + d_n^2$ in Figure 7.21 is a minimum. In Section 6.4 we discussed least squares approximations to functions. This discussion is the discrete analogue of that problem. In the function situation we had to find a curve of a certain type that best fitted a continuous set of data points over a given interval (the graph of the function); here we want the best fit to a discrete set of data points over a given interval. Although the techniques that we develop to arrive at results are different, look for certain similarities in concepts in the two situations. In both situations the "best" result is one obtained by minimizing certain squares—hence the term *least squares*.

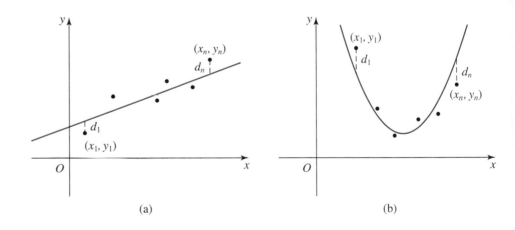

(a)   (b)

Least squares line or curve minimizes
$$d_1^2 + \cdots + d_n^2$$

Figure 7.21

We now illustrate how to fit a least squares polynomial to given data. The method involves constructing a system of linear equations. The least squares solution to this system of equations gives the coefficients of the polynomial. We shall justify the method after seeing how it works.

Example 3   Find the least squares line for the following data points.

$$(1, 1), (2, 4), (3, 2), (4, 4)$$

Solution   Let the equation of the line be $y = a + bx$. Substituting for these points into the equation of the line, we get the overdetermined system

$$a + b = 1$$
$$a + 2b = 4$$
$$a + 3b = 2$$
$$a + 4b = 4$$

We find the least squares solution. The matrix of coefficients A and column vector $\mathbf{y}$ are as follows.

$$A = \begin{bmatrix} 1 & 1 \\ 1 & 2 \\ 1 & 3 \\ 1 & 4 \end{bmatrix} \quad \text{and} \quad \mathbf{y} = \begin{bmatrix} 1 \\ 4 \\ 2 \\ 4 \end{bmatrix}$$

It can be shown that

$$\text{pinv}(A) = (A^t A)^{-1} A^t = \frac{1}{20} \begin{bmatrix} 20 & 10 & 0 & -10 \\ -6 & -2 & 2 & 6 \end{bmatrix}$$

The least squares solution is

$$[(A^t A)^{-1} A^t]\mathbf{y} = \frac{1}{20} \begin{bmatrix} 20 & 10 & 0 & -10 \\ -6 & -2 & 2 & 6 \end{bmatrix} \begin{bmatrix} 1 \\ 4 \\ 2 \\ 4 \end{bmatrix} = \begin{bmatrix} 1 \\ 0.7 \end{bmatrix}$$

Thus $a = 1$, $b = 0.7$.

The equation of the least squares line for this data is

$$y = 1 + 0.7x$$

This is the line that is generally thought to be the line of best fit for these points. See Figure 7.22.

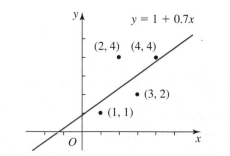

Figure 7.22

Example 4   Find the least squares parabola for the following data points.

$$(1, 7), (2, 2), (3, 1), (4, 3)$$

Solution   Let the equation of the parabola be $y = a + bx + cx^2$. Substituting for these points into the equation of the parabola, we get the system

$$a + b + c = 7$$
$$a + 2b + 4c = 2$$
$$a + 3b + 9c = 1$$
$$a + 4b + 16c = 3$$

We find the least squares solution. The matrix of coefficients A and column vector $\mathbf{y}$ are as follows.

$$A = \begin{bmatrix} 1 & 1 & 1 \\ 1 & 2 & 4 \\ 1 & 3 & 9 \\ 1 & 4 & 16 \end{bmatrix} \quad \text{and} \quad \mathbf{y} = \begin{bmatrix} 7 \\ 2 \\ 1 \\ 3 \end{bmatrix}$$

It can be shown that

$$\text{pinv}(A) = (A'A)^{-1}A' = \frac{1}{20}\begin{bmatrix} 45 & -15 & -25 & 15 \\ -31 & 23 & 27 & -19 \\ 5 & -5 & -5 & 5 \end{bmatrix}$$

The least squares solution is

$$[(A'A)^{-1}A']\mathbf{y} = \frac{1}{20}\begin{bmatrix} 45 & -15 & -25 & 15 \\ -31 & 23 & 27 & -19 \\ 5 & -5 & -5 & 5 \end{bmatrix}\begin{bmatrix} 7 \\ 2 \\ 1 \\ 3 \end{bmatrix} = \begin{bmatrix} 15.25 \\ -10.05 \\ 1.75 \end{bmatrix}$$

Thus $a = 15.25$, $b = -10.05$, $c = 1.75$.

The equation of the least squares parabola for these data points is

$$y = 15.25 - 10.05x + 1.75x^2$$

We illustrate this parabola in Figure 7.23.

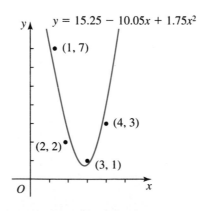

Figure 7.23

We now give the proof of the method used in the above examples to arrive at the least squares polynomials. The proof uses the concept of vector projection onto the range of a matrix. We see geometry at its best in deriving this important result. The proof is challenging and very rewarding!

Theorem 7.14   Let $(x_1, y_1), \ldots, (x_n, y_n)$ be a set of n data points. Let $y = a_0 + \cdots + a_m x^m$ be a polynomial of degree m $(n > m)$ that is to be fitted to these points. Substituting these points into the polynomial leads to a system $A\mathbf{x} = \mathbf{y}$ of n linear equations in the m variables $a_0, \ldots, a_m$, where

$$A = \begin{bmatrix} 1 & x_1 & \cdots & x_1^m \\ \vdots & \vdots & \vdots & \vdots \\ 1 & x_n & \cdots & x_n^m \end{bmatrix} \quad \text{and} \quad \mathbf{y} = \begin{bmatrix} y_1 \\ \vdots \\ y_n \end{bmatrix}$$

The least squares solution to this system gives the coefficients of the least squares polynomial for these data points.

Proof   The system $A\mathbf{x} = \mathbf{y}$ of n linear equations in m variables $(n > m)$ is, in general, overdetermined. It has no solution. (This is the case, for example, when all the points do not lie on a vertical line.) Thus $\mathbf{y}$ is not in the range of A. Let $(x_1, y'_1), \ldots, (x_n, y'_n)$ be the "data points" that lie on the least squares curve. Let the corresponding system of equations be $A\mathbf{x} = \mathbf{y}'$, where $\mathbf{y}' = \begin{bmatrix} y'_1 \\ \vdots \\ y'_n \end{bmatrix}$. Since the curve passes through these points, this system of n equations in m variables does have a unique solution $\mathbf{x}'$, the components of $\mathbf{x}'$ being the coefficients of the least squares polynomial. Thus $\mathbf{y}'$ is in the range of A. The least squares condition that $d_1^2 + \cdots + d_n^2$ is a minimum is equivalent to $\|\mathbf{y}' - \mathbf{y}\|$ being a minimum. Thus $\mathbf{y}'$ is the projection of $\mathbf{y}$ onto the range of A. See Figure 7.24. This implies that $\mathbf{y} - \mathbf{y}'$ is orthogonal to the range of A; $\mathbf{y} - \mathbf{y}'$ is orthogonal to the column vectors of A. The dot product of $\mathbf{y} - \mathbf{y}'$ with each column vector of A is zero. Interpreting the column vectors of A as the row vectors of A^t, this orthogonality condition can be expressed as the matrix product $A^t(\mathbf{y} - \mathbf{y}') = \mathbf{0}$. Since $A\mathbf{x}' = \mathbf{y}'$, we get

$$A^t(\mathbf{y} - A\mathbf{x}') = \mathbf{0}$$
$$-A^t A\mathbf{x}' = -A^t\mathbf{y}$$
$$\mathbf{x}' = (A^t A)^{-1} A^t\mathbf{y}$$
$$\mathbf{x}' = \text{pinv}(A)\mathbf{y}$$

The theorem is proved.

(continues)

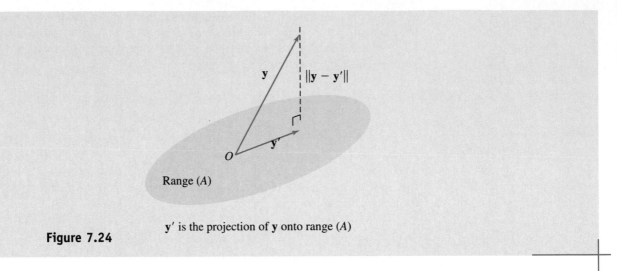

Figure 7.24

y′ is the projection of **y** onto range (A)

Example 5   **Hooke's law** states that when a force is applied to a spring, the length of the spring will be a linear function of the force. If L is the length of the spring when the force is F, this means that there exist (spring) constants a and b such that

$$L = a + bF$$

We shall now see that the spring constants, and hence the relationship between length and force for a spring, can be determined from experimental results using the method of least squares. Let various weights be suspended from the spring, and the length of the spring be measured in each case. Let the results be as follows.

Force, F (in ounces)	2	4	6	8
Length, L (in inches)	8.2	11.6	14.3	17.5

Write these statistics as points, where the first component is F and the second component is L. We get

$$(2, 8.2), \quad (4, 11.6), \quad (6, 14.3), \quad (8, 17.5)$$

In theory these points should all lie on a straight line. In practice, as in most experiments where measurements are taken, the readings are not completely accurate. The least squares line through these points will give the most satisfactory equation for the line. We get the system

$$a + 2b = 8.2$$
$$a + 4b = 11.6$$
$$a + 6b = 14.3$$
$$a + 8b = 17.5$$

The matrix of coefficients A and constant column matrix $\mathbf{y}$ are as follows.

$$A = \begin{bmatrix} 1 & 2 \\ 1 & 4 \\ 1 & 6 \\ 1 & 8 \end{bmatrix} \quad \text{and} \quad \mathbf{y} = \begin{bmatrix} 8.2 \\ 11.6 \\ 14.3 \\ 17.5 \end{bmatrix}$$

We get

$$\text{pinv}(A) = (A^tA)^{-1}A^t = \begin{bmatrix} 1 & 0.5 & 0 & -0.05 \\ -0.15 & -0.05 & 0.05 & 0.15 \end{bmatrix}$$

The least squares solution is

$$[(A^tA)^{-1}A^t]\mathbf{y} = \begin{bmatrix} 1 & 0.5 & 0 & -0.5 \\ -0.15 & -0.05 & 0.05 & 0.15 \end{bmatrix} \begin{bmatrix} 8.2 \\ 11.6 \\ 14.3 \\ 17.5 \end{bmatrix} = \begin{bmatrix} 5.25 \\ 1.53 \end{bmatrix}$$

Thus the spring constants are $a = 5.25$ and $b = 1.53$.
   The equation for the spring is

$$L = 5.25 + 1.53F$$

Thus, for example, when a weight of 20 ounces is attached to the spring, we can expect the length of the spring to be approximately $5.25 + (1.53 \times 20)$; that is, 35.85 inches.

   The following application illustrates a situation that frequently arises in analyses involving a least squares fit. One wants to compare the results from two sets of data points that have the same matrix of coefficients. The techniques of handling many systems having a common matrix of coefficients are useful.

Example 6   The following statistics give the estimated public expenditures on education, in U.S. dollars per inhabitant, for North America and Europe (*UNESCO Statistical Yearbook,* 1987). Let us construct least squares lines for these statistics and arrive at some conclusions.

	1970	1975	1980	1985
Europe	90	197	335	394
North America	317	474	816	1101

Solution   For convenience let us measure years from 1970 so that 1970 is 0, 1975 is 5, and so on. The above statistics give the following data points.

| Europe | (0, 90), (5, 197), (10, 335), (15, 394) |
| North America | (0, 317), (5, 474), (10, 816), (15, 1101) |

These two sets of data points lead to two systems of equations having the following common matrix of coefficients A and constant vectors $\mathbf{y}_1$ and $\mathbf{y}_2$.

$$A = \begin{bmatrix} 1 & 0 \\ 1 & 5 \\ 1 & 10 \\ 1 & 15 \end{bmatrix}, \quad \mathbf{y}_1 = \begin{bmatrix} 90 \\ 197 \\ 335 \\ 394 \end{bmatrix}, \quad \text{and} \quad \mathbf{y}_2 = \begin{bmatrix} 317 \\ 474 \\ 816 \\ 1101 \end{bmatrix}$$

It can be shown that

$$\text{pinv}(A) = (A^tA)^{-1}A^t = \frac{1}{50}\begin{bmatrix} 35 & 20 & 5 & -10 \\ -3 & -1 & 1 & 3 \end{bmatrix}$$

The least squares solutions are

$$[(A^tA)^{-1}A^t][\mathbf{y}_1 \quad \mathbf{y}_2] = \frac{1}{50}\begin{bmatrix} 35 & 20 & 5 & -10 \\ -3 & -1 & 1 & 3 \end{bmatrix}\begin{bmatrix} 90 & 317 \\ 197 & 474 \\ 335 & 816 \\ 394 & 1101 \end{bmatrix} = \begin{bmatrix} 96.50 & 272.90 \\ 21.00 & 53.88 \end{bmatrix}$$

The least squares lines are thus

$$\begin{aligned} \text{Europe:} \quad & y = 96.50 + 21.00x \\ \text{North America:} \quad & y = 272.90 + 53.88x \end{aligned}$$

One can use these lines to predict future expenditures. For example, predictions for 1995 ($x = 25$) are 621.50 (Europe) and 1619.90 (North America). However, the most interesting observation is that the slope of the North American line, namely 53.88, is more than twice the slope of the European line, 21. This indicates that the current gap between expenditures is likely to increase. It seems that total expenditures of money is not the problem with North American education. Part of the problem may be in the allocation or implementation of the funds.

We complete this section with a geometrical spinoff from the proof of the above theorem. We find that a pseudoinverse can be used to give the projection of a vector onto a subspace.

Theorem 7.15

Let W be the subspace of $\mathbf{R}^n$ generated by linearly independent vectors $\mathbf{u}_1, \ldots, \mathbf{u}_m$. Let $A = [\mathbf{u}_1 \quad \cdots \quad \mathbf{u}_m]$. The projection of a vector $\mathbf{y}$ onto W is given by

$$\text{proj}_W \mathbf{y} = A \,\text{pinv}(A)\mathbf{y}$$

$A \,\text{pinv}(A)$ is called a **projection matrix**.

Proof Since the vectors are linearly independent, pinv(A) exists. The result now follows directly from the proof of the previous theorem. Let $\mathbf{x}'$ be the least squares solution to $A\mathbf{x} = \mathbf{y}$. Thus

$$\mathbf{x}' = \text{pinv}(A)\mathbf{y}$$

Multiply both sides by A.

$$A\mathbf{x}' = A\,\text{pinv}(A)\mathbf{y}$$

But $A\mathbf{x}'$ is the projection of $\mathbf{y}$ onto range(A). Therefore

$$\text{proj}_W\,\mathbf{y} = A\,\text{pinv}(A)\mathbf{y}$$

Example 7 Find the projection matrix for the plane $x - 2y - z = 0$ in $\mathbf{R}^3$. Use this matrix to find the projection of the vector $(1, 2, 3)$ onto this plane.

Solution Let W be the subspace of vectors that lie in this plane. W consists of vectors of the form (x, y, z), where $x = 2y + z$. Thus $W = \{(2y + z, y, z)\}$. We can write $W = \{y(2, 1, 0) + z(1, 0, 1)\}$. Therefore W is the space generated by the vectors $(2, 1, 0)$ and $(1, 0, 1)$. Let A be the matrix having these vectors as columns.

$$A = \begin{bmatrix} 2 & 1 \\ 1 & 0 \\ 0 & 1 \end{bmatrix}$$

It can be shown that

$$\text{pinv}(A) = \frac{1}{6}\begin{bmatrix} 2 & 2 & -2 \\ 1 & -2 & 5 \end{bmatrix}$$

The projection matrix is

$$A\,\text{pinv}(A) = \frac{1}{6}\begin{bmatrix} 5 & 2 & 1 \\ 2 & 2 & -2 \\ 1 & -2 & 5 \end{bmatrix}$$

The projection of $(1, 2, 3)$ onto W is computed by multiplying this vector, in column form, by $A\,\text{pinv}(A)$. We get

$$\frac{1}{6}\begin{bmatrix} 5 & 2 & 1 \\ 2 & 2 & -2 \\ 1 & -2 & 5 \end{bmatrix}\begin{bmatrix} 1 \\ 2 \\ 3 \end{bmatrix} = \begin{bmatrix} 2 \\ 0 \\ 2 \end{bmatrix}$$

Thus the projection of $(1, 2, 3)$ onto the plane $x - 2y - z = 0$ is $(2, 0, 2)$.

Exercise Set 7.5

Determine the pseudoinverses of the matrices in Exercises 1–4, if they exist.

*1. $\begin{bmatrix} 1 & 3 \\ 0 & 1 \\ -1 & 2 \end{bmatrix}$
2. $\begin{bmatrix} 1 & 1 \\ 2 & 3 \\ 0 & 1 \end{bmatrix}$

*3. $\begin{bmatrix} 1 & 2 \\ 2 & 4 \\ 3 & 6 \end{bmatrix}$
4. $\begin{bmatrix} 1 & 2 \\ 3 & 4 \\ 5 & 6 \end{bmatrix}$

Determine the pseudoinverses of the matrices in Exercises 5–8, if they exist.

5. $\begin{bmatrix} 1 & 3 \\ 2 & 0 \\ 1 & -1 \end{bmatrix}$
*6. $\begin{bmatrix} 4 & 1 & 0 \\ 2 & 3 & 1 \end{bmatrix}$

7. $\begin{bmatrix} 1 & 2 & 3 \\ -2 & 4 & 6 \end{bmatrix}$
*8. $\begin{bmatrix} 1 & 2 \\ -3 & 5 \end{bmatrix}$

Determine the least squares lines for the data points in Exercises 9–14. (Hint: Note that the same pseudoinverse is involved for each set of data.)

*9. (1, 0), (2, 4), (3, 7)
10. (1, 1), (2, 3), (3, 7)
*11. (1, 1), (2, 5), (3, 9)
12. (1, 1), (2, 6), (3, 9)
*13. (1, 7), (2, 2), (3, 0)
14. (1, 12), (2, 5), (3, 1)

Determine the least squares lines for the data points in Exercises 15–18. (Hint: Note that the same pseudoinverse is involved for each set of data.)

*15. (1, 0), (2, 3), (3, 5), (4, 6)
16. (1, 1), (2, 3), (3, 5), (4, 9)
*17. (1, 9), (2, 7), (3, 3), (4, 2)
18. (1, 21), (2, 17), (3, 12), (4, 7)

Determine the least squares lines for the data points in Exercises 19–20.

*19. (1, 0), (2, 1), (3, 4), (4, 7), (5, 9)
20. (1, 1), (2, 3), (3, 4), (4, 8), (5, 11)

Determine the least squares parabolas for the data points in Exercises 21–24.

*21. (1, 5), (2, 2), (3, 3), (4, 8)
22. (1, 4), (2, 0), (3, 3), (4, 5)

*23. (1, 2), (2, 5), (3, 7), (4, 1)
24. (1, 23), (2, 4), (3, 7), (4, 17)

25. The following loads (in ounces) are suspended from a spring and the stated lengths (in inches) obtained. Determine the spring constants and the equation that describes each spring. Predict the length of each spring when a weight of 15 ounces is suspended.

*(a)

Force F	2	4	6	8
Length L	5.9	8	10.3	12.2

(b)

Force F	2	4	6	8
Length L	7.4	10.3	13.7	16.7

*26. The rate of gasoline consumption of a car depends upon its speed. To determine the effect of speed on fuel consumption, the Federal Highway Administration tested cars of various weights at various speeds. One car weighing 3980 pounds had the following results. Determine the least squares line corresponding to these data. Use this equation to predict the mileage per gallon of this car at 55 mph.

Speed (miles per hour)	30	40	50	60	70
Gas used (miles per gallon)		18.25	20.00	16.32	15.77	13.61

27. A state highway patrol has the following statistics on age of driver and traffic fatalities. Determine the least squares line and use it to predict the number of driver fatalities at age 22.

Age	20	25	30	35	40	45	50
Number of fatalities	101	115	92	64	60	50	49

*28. An advertising company has arrived at the following statistics relating the amount of money spent on advertising a certain product to the realized sale of that product. Find the least squares line and use it to predict the sales when $5000 is spent on advertising.

Dollars spent	1000	1500	2000	2500	3000	3500
Units sold	520	540	582	600	610	615

29. A College of Arts and Sciences has compiled the following data on the verbal SAT scores of ten entering freshmen and their graduating GPA scores four years later. Determine the least squares line and use it to predict the graduating GPA of a student entering with an SAT of 670.

SAT 490 450 640 510 680 610 480 450 600 650
GPA 3.2 2.8 3.9 2.4 4.0 3.5 2.7 3.6 3.5 2.4

*30. An insurance company has computed the following percentages of people dying at various ages from 20 to 50. Predict the percentage of deaths at 23 years of age.

Age in years	20	25	30	35	40	45	50
% deaths	18	19	21	25	35	54	83

31. The production numbers of a company for the last six-month period were as follows. Determine a least squares line (called the **cost-volume formula**) and use it to predict the total costs for the next month if production was planned to rise sharply to 165 units.

	July	Aug	Sept	Oct	Nov	Dec
Units of production	120	140	155	135	110	105
Total cost (in dollars)	6300	6500	6670	6450	6100	5950

*32. The following table gives the numbers of students, per 100,000 inhabitants, attending college in various countries. (*Statistical Yearbook of the United Nations 1983/84*, p. 296.) Find the least squares lines and use them to predict the numbers for the year 2000.

	1975	1980	1983	1984
U.K.	1308	1478	1784	1795
U.S.A.	5179	5313	5322	5185
Japan	2017	2065	2026	2006

33. The numbers of tourist arrivals in units of 1000 in various countries are given in the following table. (*Statistical Yearbook of the United Nations 1983/84*, p. 985.) Predict the number of tourists for these countries in 1990.

	1980	1981	1982	1983
Canada	12876	12811	12183	12854
China	5703	7767	7924	9477
U.K.	12420	11453	11636	12499
U.S.A.	22500	23050	21916	20441

34. The following table gives the number of passenger cars in units of 1000 in various countries. (*Statistical Yearbook of the United Nations 1983/84*, p. 1011.) Predict the number of passenger cars in these countries in 1990.

	1975	1978	1979	1980
Canada	8692	9745	9985	10255
Japan	7236	21280	22667	23659
U.K.	13949	14663	15206	15055
U.S.A.	100076	116575	116573	118459

	1981	1982	1983
Canada	10199	10530	10731
Japan	24613	25539	26386
U.K.	15227	15591	15885
U.S.A.	123461	123698	126728

*35. Population growth is described in units of one million by the following table. (*Statistical Yearbook of the United Nations 1983/84*, p. 6.) Use a least squares polynomial of degree two to predict populations in the year 2000.

	1950	1960	1965	1970
China	554	801	873	984
Japan	84	94	99	104
Western Europe	122	135	143	148
North America	166	199	214	227
World	2504	3014	3324	3683

	1975	1980	1983	1984
China	1102	1183	1225	1239
Japan	112	117	119	119
Western Europe	152	154	154	154
North America	239	252	259	261
World	4076	4453	4685	4763

36. Find the projection matrices for the following planes in $\mathbf{R}^3$. Use the matrices to find the projections of the given vectors onto those planes.

  *(a) plane $x - y - z = 0$; vector $(1, 2, 0)$

  (b) plane $x - y + z = 0$; vector $(1, 1, 1)$

  *(c) plane $x - 2y + z = 0$; vector $(0, 3, 0)$

37. Consider the subspaces of $\mathbf{R}^3$ generated by the following vectors. Find the projection matrix for each subspace and use it to find the projection of the given vector $\mathbf{y}$ onto W.

  *(a) W generated by $(1, -1, 1)$ and $(1, 1, -1)$; $\mathbf{y} = (-2, 1, 3)$

  (b) W generated by $(1, 0, 1)$ and $(2, 1, 0)$; $\mathbf{y} = (6, 0, 12)$

38. Consider the subspace W generated by the vectors $(1, 1, 1)$ and $(2, 1, 0)$. Find the projection matrix for W. Use this matrix to find the projection of the vector $(3, 0, 1)$ onto W. Show that the projection of $(6, 3, 0)$ onto W is, in fact, $(6, 3, 0)$. What does this mean?

*39. Trace the steps used in this section to derive the least squares solution to a system of linear equations. These steps follow logically, one after the other—why is the least squares solution not an actual solution?

40. Let A be an invertible matrix. Prove that $\text{pinv}(A) = A^{-1}$.

*41. Let A and B be two matrices for which the product AB exists.

  (a) Is $\text{pinv}(AB) = \text{pinv}(A)\text{pinv}(B)$?

  (b) Is $\text{pinv}(AB) = \text{pinv}(B)\text{pinv}(A)$, as for a matrix inverse?

42. Prove the following identities.

  (a) $\text{pinv}(cA) = \dfrac{1}{c}\text{pinv}(A)$, $c \neq 0$

  *(b) $\text{pinv}(\text{pinv}(A)) = A$, if A is a square matrix

  *(c) $\text{pinv}(A^t) = (\text{pinv}(A))^t$, if A is a square matrix

43. Let A be an $m \times n$ matrix. Prove that $\text{pinv}(A)$ exists if and only if A is of rank n. Show that one consequence of this result is that $\text{pinv}(A)$ does not exist if $m < n$. (*Hint:* Make use of the rank/nullity theorem.)

44. Let P be a projection matrix. Prove that *(a) P is symmetric and (b) $P^2 = P$. (P is idempotent.)

Review Exercises Chapter 7

1. Determine whether the following transformations are linear.

  (a) $T(x, y) = (2x, y, y - x)$ of $\mathbf{R}^2 \to \mathbf{R}^3$

  (b) $T(x, y) = (x + y, 2y + 3)$ of $\mathbf{R}^2 \to \mathbf{R}^2$

2. Find the matrix that maps $\mathbf{R}^2 \to \mathbf{R}^2$ such that
$$\begin{bmatrix} 1 \\ 2 \end{bmatrix} \mapsto \begin{bmatrix} 5 \\ 1 \end{bmatrix} \text{ and } \begin{bmatrix} 3 \\ -2 \end{bmatrix} \mapsto \begin{bmatrix} -1 \\ 11 \end{bmatrix}.$$

3. Let T be a transformation between vector spaces, $\mathbf{u}$ and $\mathbf{v}$ vectors in the domain, and a and b scalars. Prove that T is linear if and only if
$$T(a\mathbf{u} + b\mathbf{v}) = aT(\mathbf{u}) + bT(\mathbf{v})$$
This can be used as an alternative definition of **linear transformation**.

4. Find a single matrix that defines a rotation of the plane through an angle of $\pi/6$ about the origin, while at the same time moves points to three times their original distance from the origin.

5. Determine the matrix that defines a reflection about the line $y = -x$.

6. Determine the matrix that defines a projection onto the line $y = -x$.

7. Find the equation of the image of the line $y = -5x + 1$ under a scaling of factor 5 in the x direction and factor 2 in the y direction.

8. Find the equation of the image of the line $y = 2x + 3$ under a shear of factor 3 in the y direction.

9. Construct a 2×2 matrix that defines a shear of factor 3 in the y direction, followed by a scaling of factor 2 in the x direction, followed by a reflection about the y axis.

10. Determine the kernel and range of the transformation defined by the matrix $\begin{bmatrix} 6 & 4 \\ 3 & 2 \end{bmatrix}$. Show that
$$\dim \ker(T) + \dim \text{range}(T) = \dim \text{domain}(T)$$

11. Find bases for the kernel and range of the transformation defined by the matrix $\begin{bmatrix} 1 & 1 & 1 \\ 0 & 1 & -1 \\ 2 & 3 & 1 \end{bmatrix}$.

12. Determine whether the linear transformations defined by the following matrices are one-to-one.

*(a) $A = \begin{bmatrix} 1 & 3 & -4 & 9 \\ 0 & 1 & 2 & 6 \\ 0 & 0 & 0 & 0 \end{bmatrix}$

(b) $B = \begin{bmatrix} -1 & 2 & 0 & 5 \\ 3 & 7 & 2 & 8 \\ -4 & 2 & 0 & 0 \\ 1 & 3 & 0 & 6 \end{bmatrix}$

13. Determine the kernel and range of the transformation

$$T(x, y, z) = (x, 2x, y - z) \text{ of } \mathbf{R}^3 \to \mathbf{R}^3$$

14. Prove that $g: P_2 \to P_3$ defined as follows is linear. Find the kernel and range of g. Give bases for these subspaces.

$$g(a_2 x^2 + a_1 x + a_0) = (a_2 - a_1)x^3 - a_1 x + 2a_0$$

15. Let D be the operation of taking the derivative. Interpret D as an operator on P_n. Prove that $D^2 - 2D + 1$ is a linear operator. Find the set of polynomials that are mapped by $D^2 - 2D + 1$ into $12x - 4$.

16. Find the pseudoinverse of the matrix $\begin{bmatrix} 1 & 2 \\ 3 & 1 \\ 4 & 2 \end{bmatrix}$.

17. Determine the least squares parabola for the data $(1, 6), (2, 2), (3, 5), (4, 9)$.

18. Let A be a matrix with pseudoinverse B. Show that

(a) $ABA = A$ (b) $BAB = B$

(c) AB is symmetric   (d) BA is symmetric

MATLAB Discussion

M.20   Transformations Defined by Matrices (Section 7.2)

Let A be an $m \times n$ matrix. Let **x** be a vector in $\mathbf{R}^n$. Write **x** as a column matrix. The transformation $T: \mathbf{R}^n \to \mathbf{R}^m$ defined by $T(\mathbf{x}) = A\mathbf{x}$ is called a matrix transformation. Such a transformation is linear.

    The function **map**(A) sketches and finds the image of the unit square, $O(0, 0)$, $P(1, 0)$, $Q(1, 1)$, $R(0, 1)$, under the transformation defined by the matrix A. The image is denoted $Q^*P^*Q^*R^*$. Let us rotate the unit square through an angle of $\pi/4$. The matrix that defines a rotation through an angle α about the origin is

$$\begin{bmatrix} \cos \alpha & -\sin \alpha \\ \sin \alpha & \cos \alpha \end{bmatrix}$$

Use this matrix with $\alpha = \pi/4$. (π is entered as pi in MATLAB.)

»**A** = [cos(pi/4)   −sin(pi/4);sin(pi/4)   cos(pi/4)];
»**map(A)**;
O* =
  0.0000   0.0000
P* =
  0.5253   0.8509
Q* =
 −0.3256   1.3762
R* =
 −0.8509   0.5253

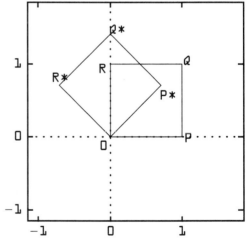

Note: Use "Enter" and "shg" (or "figure(gcf)") to alternate between command and graph windows on IBMs.

Exercises

*1.   Sketch and find the vertices of the image of the unit square under a rotation of $2\pi/3$ about the origin.

2.   Sketch and find the vertices of the image of the unit square under a rotation of $-\pi/6$.

*3. Sketch and find the vertices of the image of the unit square under a dilation of factor 3.

4. Determine the matrix that defines a reflection in the y axis. Find and sketch the image of the unit square under this transformation.

5. Consider the transformations on $\mathbf{R}^2$ defined by each of the following matrices. Find the image of the unit square under each transformation.

$$\textbf{*(a)} \begin{bmatrix} 3 & 0 \\ 1 & 4 \end{bmatrix} \qquad \textbf{(b)} \begin{bmatrix} 4 & -1 \\ 1 & 5 \end{bmatrix} \qquad \textbf{(c)} \begin{bmatrix} 0 & 3 \\ -3 & 0 \end{bmatrix}$$

6. The function **compmap**(A,B) gives the image of the unit square under the composite transformation $B \circ A$. Find the image of the unit square under each of the following transformations.

*(a) A rotation through $\pi/2$ counterclockwise, then a dilation of factor 2.

(b) A dilation of factor 4, then a reflection in the x axis.

(c) A rotation through $\pi/3$, then a reflection about the line $y = x$.

7. Construct single 2×2 matrices that define the following transformations on $\mathbf{R}^2$. Find the image of the unit square under each transformation.

*(a) A dilation of factor 3, then a shear of factor 2 in the x direction.

(b) A scaling of factor 3 in the x direction, of factor 2 in the y direction, then a reflection in the line $y = x$.

(c) A dilation of factor 2, then a shear of factor 3 in the y direction, then a rotation through $\pi/3$ counterclockwise.

8. Use **affine**(A,v) to sketch the image of the unit square under the affine transformation

$$A = \begin{bmatrix} 3 & 2 \\ 1 & 1 \end{bmatrix}, \quad \mathbf{v} = \begin{bmatrix} 3 \\ 1 \end{bmatrix}$$

9. Use **affine**(A,v) to draw the images of the unit square under the affine transformations $T_n(\mathbf{u}) = A^n\mathbf{u} + \mathbf{v}$, for $n = 1, 2, 3$, defined by the following matrix A and vector $\mathbf{v}$. What does the image approach as $n \to \infty$?

(a) $A = \begin{bmatrix} 0.5 & 0 \\ 0 & 0.5 \end{bmatrix}$, $\mathbf{v} = \begin{bmatrix} 2 \\ 2 \end{bmatrix}$

(b) $B = \begin{bmatrix} \cos(45) & -\sin(45) \\ \sin(45) & \cos(45) \end{bmatrix}$, $C = \begin{bmatrix} 0.5 & 0 \\ 0 & 0.5 \end{bmatrix}$, $A = C \circ B$, $\mathbf{v} = \begin{bmatrix} 3 \\ 2 \end{bmatrix}$

M.21   Fractals (Section 7.2)

Fractals can be generated using affine transformations on $\mathbf{R}^2$

$$T_i(u) = A_i u + v_i$$

The algorithm is as follows:

1. Let $x = 0$, $y = 0$.
2. Use random # generator to select an affine transformation T_i.
3. Let $(x', y') = T_i(x, y)$.
4. Plot (x', y').
5. Let $(x, y) = (x', y')$.
6. Repeat steps 2, 3, 4, and 5 five thousand times.

Example 1   Let us generate a Black Spleenwort fern using the following four affine transformations T_i with associated probabilities p_i.

$$T_1 = \begin{bmatrix} 0.86 & 0.03 \\ -0.03 & 0.86 \end{bmatrix} \begin{bmatrix} x \\ y \end{bmatrix} + \begin{bmatrix} 0 \\ 1.5 \end{bmatrix}, p_1 = 0.83;$$

$$T_2 = \begin{bmatrix} 0.2 & -0.25 \\ 0.21 & 0.23 \end{bmatrix} \begin{bmatrix} x \\ y \end{bmatrix} + \begin{bmatrix} 0 \\ 1.5 \end{bmatrix}, p_2 = 0.08;$$

$$T_3 = \begin{bmatrix} -0.15 & 0.27 \\ 0.25 & 0.26 \end{bmatrix} \begin{bmatrix} x \\ y \end{bmatrix} + \begin{bmatrix} 0 \\ 0.45 \end{bmatrix}, p_3 = 0.08;$$

$$T_4 = \begin{bmatrix} 0 & 0 \\ 0 & 0.17 \end{bmatrix} \begin{bmatrix} x \\ y \end{bmatrix} + \begin{bmatrix} 0 \\ 0 \end{bmatrix}, p_4 = 0.01$$

The function **fern**(n) implements the above algorithm.

»**fern(5000)**
use "command period" (Macintosh), "CTRL C" (IBM) to stop at any time
computing pts

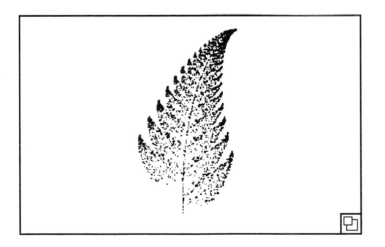

Exercise 1   Modify the function **fern** to zoom in on **(a)** the tip of the fern and **(b)** the stem of the bottom right-hand leaf. Observe the self-similar fractal characteristics

of the fern. Investigate the effects of varying the parameters of the affine transformations. Determine which parameters lead to squatter ferns and which to stunted ferns. Determine the role of each transformation in the set.

IFS Code

The affine transformations and probabilities that describe fractals can be written as matrices (IFS codes), each row corresponding to an affine transformation and probability.

$$\begin{bmatrix} x' \\ y' \end{bmatrix} = \begin{bmatrix} a & b \\ c & d \end{bmatrix} \begin{bmatrix} x \\ y \end{bmatrix} + \begin{bmatrix} e \\ f \end{bmatrix}, p$$

Exercise 2   The IFS code for a fractal called the Sierpinski triangle is

T	a	b	c	d	e	f	p
1	0.5	0	0	0.5	0.5	0.5	0.34
2	0.5	0	0	0.5	1	0	0.33
3	0.5	0	0	0.5	0	0	0.33

Run the fractal function **Sierpinski**. How many different-sized triangles have you been able to produce in your image?

Exercise 3   The IFS code for a fractal tree and "dragon" are as follows. Run the fractal function **tree**. Modify the function to get the **dragon** fractal.

Tree

T	a	b	c	d	e	f	p
1	0.42	0.42	−0.42	0.42	0	0.2	0.4
2	0.42	−0.42	0.42	0.42	0	0.2	0.4
3	0.1	0	0	0.1	0	0.2	0.15
4	0	0	0	0.5	0	0	0.05

Dragon

T	a	b	c	d	e	f	p
1	0.5	0.5	−0.5	0.5	0.125	0.625	0.5
2	0.5	0.5	−0.5	0.5	−0.125	0.375	0.5

Exercise 4   The first of the fern affine transformations can be written in the form

$$T_1 = AB \begin{bmatrix} x \\ y \end{bmatrix} + \begin{bmatrix} 0 \\ 1.5 \end{bmatrix}$$

where A is a dilation and B is a rotation.

*(a) Find A, B, and hence the dilation factor and the angle of rotation of this trans-
formation.

(b) If the angle of rotation is α, determine and use the transformation that "rotates
the fern" through an angle 1.5α, the dilation factor being unchanged.

(c) How do you make "controlled changes" in the first row of the IFS of the fern to
rotate it while preserving its size?

(d) Find an IFS that makes the fern bend to the left and that makes it become vertical.

Exercise 5   Use mathematics to find the coordinates of the tip of the fern. Use the
picture to check your answer.

M.22   Kernel and Range (Section 7.3)

Let the $m \times n$ matrix A define a linear transformation T of $\mathbf{R}^n$ to $\mathbf{R}^m$. The kernel and
range of T are subspaces found by solving systems of linear equations.

Example 1   Find the kernel and range of the linear transformation T defined by the following
matrix A.

$$A = \begin{bmatrix} 1 & 2 & 3 \\ 0 & -1 & 1 \\ 1 & 1 & 4 \end{bmatrix}$$

The kernel will consist of the elements of $\mathbf{R}^3$ such that

$$\begin{bmatrix} 1 & 2 & 3 \\ 0 & -1 & 1 \\ 1 & 1 & 4 \end{bmatrix} \begin{bmatrix} x \\ y \\ z \end{bmatrix} = \begin{bmatrix} 0 \\ 0 \\ 0 \end{bmatrix}$$

Solve this system of linear equations using MATLAB.

```
»B = [1  2  3  0;0  −1  1  0;1  1  4  0];
»rref(B)
ans =
   1   0   5   0
   0   1  −1   0
   0   0   0   0      {the reduced echelon form}
```

The system has many solutions, $x = -5r$, $y = r$, $z = r$. The kernel of T is the one-
dimensional vector space with basis $\{(-5, 1, 1)^t\}$.

The range of T will be the subspace of $\mathbf{R}^3$ spanned by the column vectors of A.
A basis for the range can be found by finding the reduced echelon form of A^t using
MATLAB. The transposes of the row vectors of this form give a basis for the range.

»**rref(A′)**
ans =
```
   1   0   1
   0   1   1
   0   0   0      {the reduced echelon form}
```

The range of T is the two-dimensional space with basis $\{(1, 0, 1)', (0, 1, 1)'\}$. Observe that the rank/nullity theorem holds:

$$\dim \text{ domain}(3) = \dim \text{ ker}(1) + \dim \text{ range}(2)$$

Exercise 1   Find bases for the kernels and ranges of the linear transformations defined by the following matrices. Show that the rank/nullity holds in each case.

*(a) $\begin{bmatrix} 1 & 1 & 5 \\ 0 & 1 & 3 \\ 2 & 1 & 7 \end{bmatrix}$

(b) $\begin{bmatrix} 1 & 2 & 3 \\ -1 & 2 & 4 \\ 1 & 6 & 10 \end{bmatrix}$

*(c) $\begin{bmatrix} 6 & 1 & 3 \\ -1 & 2 & 4 \\ 0 & 1 & 5 \end{bmatrix}$

(d) $\begin{bmatrix} -1 & 3 & 6 \\ 2 & -6 & -12 \\ 1 & -3 & -6 \end{bmatrix}$

 ## M.23   Pseudoinverse and Least Squares Curves (Section 7.5)

The pseudoinverse of a matrix A is $\text{pinv}(A) = (A'A)^{-1}A'$. It can be used to give the "best solution" possible for an overdetermined system of linear equations. It can also be used to compute the coefficients in a least squares polynomial.

Example 1   Find the pseudoinverse of $A = \begin{bmatrix} 1 & 2 \\ 2 & 4 \\ 3 & 0 \end{bmatrix}$.

»**A= [1   2;2   4;3   0]**
»**pinv(A)**   {MATLAB pseudoinverse function}
ans =
```
  -0.0000   -0.0000    0.3333
   0.1000    0.2000   -0.1667
```

Exercise 1   Verify the answer to the previous example by computing $(A'A)^{-1}A'$.

Exercise 2   Find the pseudoinverse of each of the following matrices. Check that $\text{pinv}(A)$ and $(A'A)^{-1}A'$ agree.

$$*(a) \quad \begin{bmatrix} 1 & 2 \\ 1 & 5 \\ 1 & -1 \end{bmatrix} \qquad (b) \quad \begin{bmatrix} 1 & 0 \\ 2 & 4 \\ -3 & 1 \end{bmatrix}$$

Exercise 3   Use MATLAB and various invertible matrices to investigate the pseudo-inverse of an invertible matrix. Make a conjecture and prove the conjecture.

Exercise 4   Compute, if possible, the pseudoinverse of the matrix $A = \begin{bmatrix} 1 & 2 & 3 \\ 4 & 5 & 6 \end{bmatrix}$ using $(A'A)^{-1}A'$. Investigate pseudoinverses of matrices of various sizes. Arrive at a result (with proof) about the existence of a pseudoinverse of a matrix based on the definition $(A'A)^{-1}A'$.

Least Squares Solution

If $AX = Y$ is an overdetermined system, then the least squares solution is $X = \text{pinv}(A)$. This is the "closest" we can get to a true solution.

Example 2   Find the least squares solution to the overdetermined system

$$\begin{aligned} x_1 + x_2 &= 6 \\ -x_1 + x_2 &= 3 \\ 2x_1 + 3x_2 &= 9 \end{aligned}$$

```
»A = [1   1;−1   1;2   3];   Y = [6;3;9];
»X = pinv(A)*Y
X =
   0.5000
   3.0000
```

The least squares solution is $x_1 = \frac{1}{2}$, $x_2 = 3$.

Exercise 5   Find the least squares solution to the following systems

$$*(a) \quad \begin{aligned} x_1 + x_2 &= 3 \\ x_1 - x_2 &= 1 \\ 2x_1 + 3x_2 &= 5 \end{aligned} \qquad (b) \quad \begin{aligned} 2x_1 - x_2 &= 0 \\ x_1 + 2x_2 &= 3 \\ x_1 + x_2 &= 1 \end{aligned} \qquad (c) \quad \begin{aligned} 3x_1 + 2x_2 &= 5 \\ x_1 - x_2 &= 1 \\ 2x_1 + x_2 &= 4 \end{aligned}$$

Least Squares Curve

Let $AX = Y$ be an overdetermined system of equations obtained on substituting n data points into a polynomial of degree m. The coefficients of the least squares polynomial of degree m are given by $X = \text{pinv}(A)*Y$.

Example 3   Find the least squares line for the data points (1, 1), (2, 4), (3, 2), (4, 4).

»ls   {author-defined function to compute least
                                            squares curve, based on above result}

give degree of polynomial: 1
give # of data points: 4
x values of data points: 1
: 2
: 3
: 4
y values of data points: 1
: 4
: 2
: 4

computing

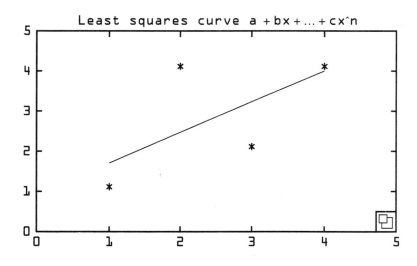

Least squares curve a + bx + ... + cx^n

coeffs of LS polynomial in ascending powers:
  1.0000   0.7000   {equation of LS curve is $y = 1 + 0.7x$}

Note: Use "Enter" and "shg" (or "figure(gcf)") to alternate between command and
graph windows on IBMs.

*Exercise 6   Find and graph the least squares line for the following data points:
(1, 1), (2, 5), (3, 9).

Exercise 7   Find and graph the least squares polynomial of degree two for the following data points: (1, 7), (2, 2), (3, 1), (4, 3).

We now give a sequence of MATLAB functions that can be used to compute least squares curves and make predictions when a graph is not needed. We introduce the functions **polyfit** and **polyval**. Polyfit returns the coefficients of a least squares polynomial in *descending order*. Polyval gives the value of a polynomial at a point.

Example 4   The following statistics give the estimated public expenditures on education, in U.S. dollars per inhabitant, for North America and Europe (*UNESCO Statistical Yearbook*, 1987). Construct least squares lines for these statistics and predict the expenditures for 1995.

	1970	1975	1980	1985
Europe	90	197	335	394
North America	317	474	816	1101

Let us use 1970 as base year, year 0, 1975 as year 5, and so forth.

```
»X= [0   5   10   15];          {x values of Europe data points}
»Y= [90   197   335   394];     {y values of data points}
»n=1;                           {fit a polynomial of degree 1}
»LScoeffs = polyfit(X,Y,n)      {give coeffs for LS polynomial of degree n}
LScoeffs =
          21.0000   96.5000     {the LS line is y = 21x + 96.5 for Europe}

»s=25;
»prediction = polyval(LScoeffs,s)   {compute value of LS polynomial at x = s}

prediction = 621.5000               {predicted expenditure for Europe for
                                       1995 is $621.50}
```

*Exercise 8   The following table gives the numbers of students, per 100,000 inhabitants, attending college in various countries. (*Statistical Yearbook of the United Nations 1983/84*, p. 296.) Find the least squares lines and use them to predict the numbers for the year 2000.

	1975	1980	1983	1984
U.K.	1308	1478	1784	1795
U.S.A.	5179	5313	5322	5185
Japan	2017	2065	2026	2006

Exercise 9   Researchers who have studied the sense of time in humans have concluded that a person's subjective estimate of time is not equal to the actual elapsed time, but is a linear function $s(t) = mt + b$ of the elapsed time, for some constants m and b. With a co-worker who has a watch, estimate intervals of 10 minutes, for a given period while you are reading. You estimate times 10 minutes, 20 minutes, . . . , and the co-worker records the actual times t_1, t_2, ... Use this data to find the least squares line that describes the behavior of your subjective time. What would your estimate be of 3 hours? Does time go slowly or quickly for you?

Exercise 10   Consider the following data. What must the y value of the last point be to ensure that the least squares line has equation $y = mx + 0.3763$ with $m > 1.4458$?

$$(1, 2.1), (2, 3.2), (4, 5.6), (6, y)$$

Check your answer.

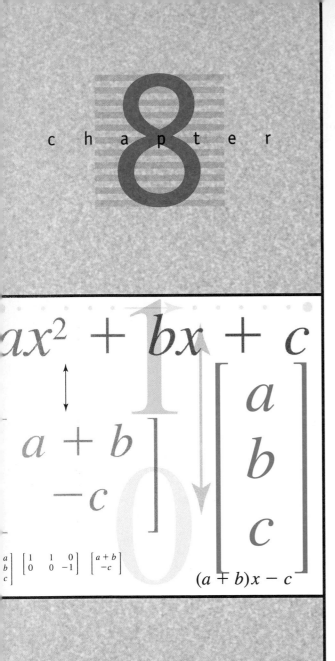

Coordinate Representations

In this chapter the reader will see how bases for a vector space lead to coordinate systems. It will be seen that every finite dimensional vector space is in a mathematical sense "the same" as $\mathbf{R}^n$, for some value of n. The mathematics will be used to analyze the Fibonacci sequence, and to understand the behavior of oscillating systems.

8.1   Coordinate Vectors

We have discussed various types of vector spaces; the vector spaces $\mathbf{R}^n$, spaces of matrices, and spaces of functions. In this section we shall find how we can use vectors in $\mathbf{R}^n$, called **coordinate vectors**, to describe vectors in any real finite dimensional vector space. This discussion leads to the conclusion that all finite dimensional vector spaces are in some mathematical sense "the same" as $\mathbf{R}^n$.

Definition   *Let U be a vector space with basis $B = \{\mathbf{u}_1, \ldots, \mathbf{u}_n\}$ and let $\mathbf{u}$ be a vector in U. We know that there exist unique scalars $a_1, \ldots, a_n$, such that*

$$\mathbf{u} = a_1\mathbf{u}_1 + \cdots + a_n\mathbf{u}_n$$

The column vector $\mathbf{u}_B = \begin{bmatrix} a_1 \\ \vdots \\ a_n \end{bmatrix}$ *is called the **coordinate vector** of $\mathbf{u}$ relative to this basis. The scalars $a_1, \ldots, a_n$ are called the **coordinates** of $\mathbf{u}$ relative to this basis. Note: We shall use a column vector form for coordinate vectors rather than row vectors. The theory develops most smoothly with this convention.*

Notation

Notation is extremely important in mathematics. Much care and attention should be paid to it—it is an intrinsic part of the mathematics. Good notation frees the mind to focus on the concepts involved, while poor notation detracts. This section and the following one challenge us to the hilt to find a suitable notation! We have attempted to select the most straightforward notation. The reader should realize that mastering the notation is an important part of the learning experience of this section. We shall be discussing different bases for the vector space U. We have selected the following notation.

We denote two different bases for a vector space U by B and B'. If $\mathbf{u}$ is a vector in U, we denote the coordinate vectors of $\mathbf{u}$ relative to these bases by $\mathbf{u}_B$ and $\mathbf{u}_{B'}$.

Example 1   Find the coordinate vectors of $\mathbf{u} = (4, 5)$ relative to the following bases B and B' of $\mathbf{R}^2$. (a) The standard basis, $B = \{(1, 0), (0, 1)\}$ and (b) $B' = \{(2, 1), (-1, 1)\}$.

Solution   (a)   By observation, we see that

$$(4, 5) = 4(1, 0) + 5(0, 1)$$

Thus $\mathbf{u}_B = \begin{bmatrix} 4 \\ 5 \end{bmatrix}$. The given representation of $\mathbf{u}$ is, in fact, relative to the standard basis.

(b)   Let us now find the coordinate vector of $\mathbf{u}$ relative to B', a basis that is not the standard basis. Let

$$(4, 5) = a_1(2, 1) + a_2(-1, 1)$$

Thus

$$(4, 5) = (2a_1, a_1) + (-a_2, a_2)$$
$$(4, 5) = (2a_1 - a_2, a_1 + a_2)$$

Comparing components leads to the following system of equations.

$$2a_1 - a_2 = 4$$
$$a_1 + a_2 = 5$$

This system has the unique solution

$$a_1 = 3, \qquad a_2 = 2$$

Thus $\mathbf{u}_{B'} = \begin{bmatrix} 3 \\ 2 \end{bmatrix}$.

These coordinate vectors have geometrical interpretation. Denote the basis vectors as follows.

$$\mathbf{u}_1 = (1, 0), \quad \mathbf{u}_2 = (0, 1) \qquad \text{and} \qquad \mathbf{u}_1' = (2, 1), \quad \mathbf{u}_2' = (-1, 1)$$

We can write

$$\mathbf{u} = 4\mathbf{u}_1 + 5\mathbf{u}_2, \quad \mathbf{u} = 3\mathbf{u}_1' + 2\mathbf{u}_2'$$

See Figure 8.1.

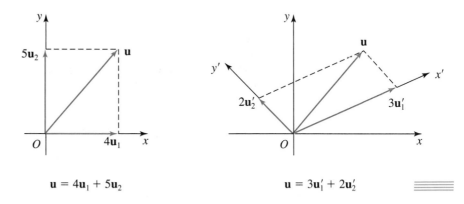

Figure 8.1 $\mathbf{u} = 4\mathbf{u}_1 + 5\mathbf{u}_2$ $\mathbf{u} = 3\mathbf{u}_1' + 2\mathbf{u}_2'$

Example 2   Find the coordinate vectors of $\mathbf{u} = 5x^2 + x - 3$ relative to the following bases B and B' of P_2.

(a)   The standard basis, $B = \{x^2, x, 1\}$

(b) $B' = \{x^2 - x + 5, 3x^2 - 1, 2x^2 + 4x - 2\}$

Solution   (a)   By observation, we see that

$$5x^2 + x - 3 = 5(x^2) + (x) - 3(1)$$

Thus $\mathbf{u}_B = \begin{bmatrix} 5 \\ 1 \\ -3 \end{bmatrix}$. The coordinates of $\mathbf{u}$ relative to the standard basis are the coefficients of $5x^2 + x - 3$.

(b)   Once again it is more laborious to find the coordinate vector of $\mathbf{u}$ relative to a basis B' that is not the standard basis. Let

$$5x^2 + x - 3 = a_1(x^2 - x + 5) + a_2(3x^2 - 1) + a_3(2x^2 + 4x - 2)$$

Thus

$$5x^2 + x - 3 = (a_1 + 3a_2 + 2a_3)x^2 + (-a_1 + 4a_3)x + (5a_1 - a_2 - 2a_3)$$

Comparing coefficients leads to the following system of equations:

$$
\begin{aligned}
a_1 + 3a_2 + 2a_3 &= 5 \\
-a_1 + 4a_3 &= 1 \\
5a_1 - a_2 - 2a_3 &= -3
\end{aligned}
$$

This system has the unique solution

$$a_1 = -\tfrac{1}{5}, \qquad a_2 = \tfrac{8}{5}, \qquad a_3 = \tfrac{1}{5}$$

Thus $\mathbf{u}_{B'} = \begin{bmatrix} -\frac{1}{5} \\ \frac{8}{5} \\ \frac{1}{5} \end{bmatrix}$

The above examples illustrate how the standard bases are the most natural bases to use to express a vector. In general a system of linear equations has to be solved to find a coordinate vector relative to another basis. However, coordinate vectors can be easily computed relative to an **orthonormal basis**. We remind the reader of the result of Theorem 5.5. If $B = \{\mathbf{u}_1, \ldots, \mathbf{u}_n\}$ is an orthonormal basis for a vector space U, then an arbitrary vector $\mathbf{v}$ in U can be expressed

$$\mathbf{v} = (\mathbf{v} \cdot \mathbf{u}_1)\mathbf{u}_1 + (\mathbf{v} \cdot \mathbf{u}_2)\mathbf{u}_2 + \cdots + (\mathbf{v} \cdot \mathbf{u}_n)\mathbf{u}_n$$

Thus the coordinate vector is

$$\mathbf{v}_B = \begin{bmatrix} \mathbf{v} \cdot \mathbf{u}_1 \\ \vdots \\ \mathbf{v} \cdot \mathbf{u}_n \end{bmatrix}$$

Example 3   Find the coordinate vector of $\mathbf{v} = (2, -5, 10)$ relative to the orthonormal basis $B\left\{(1, 0, 0), \left(0, \frac{3}{5}, \frac{4}{5}\right), \left(0, \frac{4}{5}, -\frac{3}{5}\right)\right\}$.

Solution   We get

$$(2, -5, 10) \cdot (1, 0, 0) = 2$$
$$(2, -5, 10) \cdot \left(0, \tfrac{3}{5}, \tfrac{4}{5}\right) = 5$$
$$(2, -5, 10) \cdot \left(0, \tfrac{4}{5}, -\tfrac{3}{5}\right) = -10$$

Thus $\mathbf{v}_B = \begin{bmatrix} 2 \\ 5 \\ -10 \end{bmatrix}$

There are occasions when bases other than standard or orthonormal bases best fit the situation. It becomes necessary to know how coordinate vectors relative to different bases are related. This is the topic of our next discussion.

Change of Basis

Let $B = \{\mathbf{u}_1, \ldots, \mathbf{u}_n\}$ and $B' = \{\mathbf{u}_1', \ldots, \mathbf{u}_n'\}$ be bases for a vector space U. A vector $\mathbf{u}$ in U will have coordinate vectors $\mathbf{u}_B$ and $\mathbf{u}_{B'}$ relative to these bases. We now discuss the relationship between $\mathbf{u}_B$ and $\mathbf{u}_{B'}$.

Let the coordinate vectors of $\mathbf{u}_1, \ldots, \mathbf{u}_n$ relative to the basis $B' = \{\mathbf{u}_1', \ldots, \mathbf{u}_n'\}$ be $(\mathbf{u}_1)_{B'}, \ldots, (\mathbf{u}_n)_{B'}$. The matrix P having these vectors as columns plays a central role in our discussion. It is called the **transition matrix from the basis B to the basis B'.**

Transition matrix

$$P = [(\mathbf{u}_1)_{B'} \cdots (\mathbf{u}_n)_{B'}]$$

Theorem 8.1   Let $B = \{\mathbf{u}_1, \ldots, \mathbf{u}_n\}$ and $B' = \{\mathbf{u}_1', \ldots, \mathbf{u}_n'\}$ be bases for a vector space U. If $\mathbf{u}$ is a vector in U having coordinate vectors $\mathbf{u}_B$ and $\mathbf{u}_{B'}$ relative to these bases, then

$$\mathbf{u}_{B'} = P\mathbf{u}_B$$

where P is the transition matrix from B to B'

$$P = [(\mathbf{u}_1)_{B'} \cdots (\mathbf{u}_n)_{B'}]$$

Proof   Since $\{\mathbf{u}_1', \ldots, \mathbf{u}_n'\}$ is a basis for U, each of the vectors $\mathbf{u}_1, \ldots, \mathbf{u}_n$ can be expressed as a linear combination of these vectors. Let

$$\mathbf{u}_1 = c_{11}\mathbf{u}_1' + \cdots + c_{n1}\mathbf{u}_n'$$
$$\vdots$$
$$\mathbf{u}_n = c_{1n}\mathbf{u}_1' + \cdots + c_{nn}\mathbf{u}_n'$$

If $\mathbf{u} = a_1\mathbf{u}_1 + \cdots + a_n\mathbf{u}_n$, we get

$$\mathbf{u} = a_1\mathbf{u}_1 + \cdots + a_n\mathbf{u}_n$$
$$= a_1(c_{11}\mathbf{u}_1' + \cdots + c_{n1}\mathbf{u}_n') + \cdots + a_n(c_{1n}\mathbf{u}_1' + \cdots + c_{nn}\mathbf{u}_n')$$
$$= (a_1c_{11} + \cdots + a_nc_{1n})\mathbf{u}_1' + \cdots + (a_1c_{n1} + \cdots + a_nc_{nn})\mathbf{u}_n'$$

The coordinate vector of $\mathbf{u}$ relative to B' can therefore be written

$$\mathbf{u}_{B'} = \begin{bmatrix} (a_1c_{11} + \cdots + a_nc_{1n}) \\ \vdots \\ (a_1c_{n1} + \cdots + a_nc_{nn}) \end{bmatrix} = \begin{bmatrix} c_{11} & \cdots & c_{1n} \\ \vdots & & \vdots \\ c_{n1} & \cdots & c_{nn} \end{bmatrix} \begin{bmatrix} a_1 \\ \vdots \\ a_n \end{bmatrix}$$
$$= [(\mathbf{u}_1)_{B'} \quad \cdots \quad (\mathbf{u}_n)_{B'}]\mathbf{u}_B$$

proving the theorem.

A change of basis for a vector space from a nonstandard basis to the standard basis is straightforward to accomplish. The following example illustrates the ease with which the transition matrix P is found. The columns of P are, in fact, the vectors of the first basis.

Example 4 Consider the bases $B = \{(1, 2), (3, -1)\}$ and $B' = \{(1, 0), (0, 1)\}$ of $\mathbf{R}^2$. If $\mathbf{u}$ is a vector such that $\mathbf{u}_B = \begin{bmatrix} 3 \\ 4 \end{bmatrix}$, find $\mathbf{u}_{B'}$.

Solution We express the vectors of B in terms of the vectors of B' to get the transition matrix,

$$(1, 2) = 1(1, 0) + 2(0, 1)$$
$$(3, -1) = 3(1, 0) - 1(0, 1)$$

The coordinate vectors of $(1, 2)$ and $(3, -1)$ are $\begin{bmatrix} 1 \\ 2 \end{bmatrix}$ and $\begin{bmatrix} 3 \\ -1 \end{bmatrix}$. The transition matrix P is thus

$$P = \begin{bmatrix} 1 & 3 \\ 2 & -1 \end{bmatrix}$$

(Observe that the columns of P are the vectors of the basis B.) We get

$$\mathbf{u}_{B'} = \begin{bmatrix} 1 & 3 \\ 2 & -1 \end{bmatrix} \begin{bmatrix} 3 \\ 4 \end{bmatrix} = \begin{bmatrix} 15 \\ 2 \end{bmatrix}$$

Let B and B' be bases for a vector space. We now see that the transition matrices from B to B' and B' to B are related.

Theorem 8.2   Let B and B' be bases for a vector space U, and let P be the transition matrix from B to B'. Then P is invertible and the transition matrix from B' to B is P^{-1}.

Proof   Let $\mathbf{u}$ be a vector in U having column coordinate vectors $\mathbf{u}_B$ and $\mathbf{u}_{B'}$ relative to the bases B and B'. P is given to be the transition matrix from B to B'; let Q be the transition matrix from B' to B. Then we know that

$$\mathbf{u}_{B'} = P\mathbf{u}_B \quad \text{and} \quad \mathbf{u}_B = Q\mathbf{u}_{B'}$$

We can combine these equations in two ways. Substitute for $\mathbf{u}_B$ from the second equation into the first, then substitute for $\mathbf{u}_{B'}$ from the first into the second. We get

$$\mathbf{u}_{B'} = PQ\mathbf{u}_{B'} \quad \text{and} \quad \mathbf{u}_B = QP\mathbf{u}_B$$

Since these results hold for all values of $\mathbf{u}_B$ and $\mathbf{u}_{B'}$ we have that

$$PQ = QP = I$$

Thus $Q = P^{-1}$.

The following example introduces a very useful technique for finding the transition matrix from one basis to another if neither basis is the standard basis. It makes use of the standard basis as an intermediate basis because of the ease with which one can transform to and from the standard basis.

Example 5   Consider the bases $B = \{(1, 2), (3, -1)\}$ and $B' = \{(3, 1), (5, 2)\}$ of $\mathbf{R}^2$. Find the transition matrix from B to B'. If $\mathbf{u}$ is a vector such that $\mathbf{u}_B = \begin{bmatrix} 2 \\ 1 \end{bmatrix}$, find $\mathbf{u}_{B'}$.

Solution   Let us use the standard basis $S = \{(1, 0), (0, 1)\}$ as an intermediate basis. The transition matrix P from B to S and the transition matrix P' from B' to S are

$$P = \begin{bmatrix} 1 & 3 \\ 2 & -1 \end{bmatrix}, \quad P' = \begin{bmatrix} 3 & 5 \\ 1 & 2 \end{bmatrix}$$

The transition matrix from S to B' will be $(P')^{-1}$. The transition matrix from B to B' (by way of S) is $(P')^{-1}P$. Thus the transition matrix from B to B' is

$$\begin{bmatrix} 3 & 5 \\ 1 & 2 \end{bmatrix}^{-1} \begin{bmatrix} 1 & 3 \\ 2 & -1 \end{bmatrix} = \begin{bmatrix} 2 & -5 \\ -1 & 3 \end{bmatrix} \begin{bmatrix} 1 & 3 \\ 2 & -1 \end{bmatrix} = \begin{bmatrix} -8 & 11 \\ 5 & -6 \end{bmatrix}$$

This gives

$$\mathbf{u}_{B'} = \begin{bmatrix} -8 & 11 \\ 5 & -6 \end{bmatrix} \begin{bmatrix} 2 \\ 1 \end{bmatrix} = \begin{bmatrix} -5 \\ 4 \end{bmatrix}$$

Rotation of Coordinate Axes

Let a rectangular xy coordinate system be given and a second rectangular coordinate system $x'y'$ be obtained from it through a rotation through an angle θ about the origin.

See Figure 8.2(a). An arbitrary point C in the plane has coordinates $\begin{bmatrix} x \\ y \end{bmatrix}$ relative to the xy coordinate system and coordinates $\begin{bmatrix} x' \\ y' \end{bmatrix}$ relative to the $x'y'$ coordinate system. See Figure 8.2(b). Let us determine the **coordinate transformation** that relates the two coordinate systems.

Let $\mathbf{u}$ be the vector associated with the point C. Let the unit vectors along the positive x and y axes be $\mathbf{u}_1$ and $\mathbf{u}_2$. Let the unit vectors along the positive x' and y' axes be $\mathbf{u}'_1$ and $\mathbf{u}'_2$. See Figure 8.2(c). Denote these bases $B = \{\mathbf{u}_1, \mathbf{u}_2\}$ and $B' = \{\mathbf{u}'_1, \mathbf{u}'_2\}$. We can write

$$\mathbf{u} = x\mathbf{u}_1 + y\mathbf{u}_2 \qquad \text{and} \qquad \mathbf{u} = x'\mathbf{u}'_1 + y'\mathbf{u}'_2$$

The coordinate vectors of $\mathbf{u}$ relative to these bases are thus $\begin{bmatrix} x \\ y \end{bmatrix}$ and $\begin{bmatrix} x' \\ y' \end{bmatrix}$. We see from Figure 8.2(d) that

$$\mathbf{u}_1 = \cos \theta \, \mathbf{u}'_1 - \sin \theta \, \mathbf{u}'_2$$
$$\mathbf{u}_2 = \sin \theta \, \mathbf{u}'_1 + \cos \theta \, \mathbf{u}'_2$$

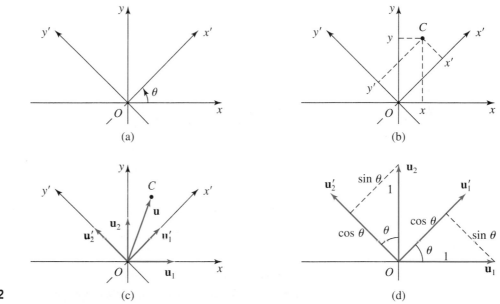

Figure 8.2

                        (c)   (d)

The transition matrix P from basis B to B' is thus

$$P = \begin{bmatrix} \cos \theta & \sin \theta \\ -\sin \theta & \cos \theta \end{bmatrix}$$

Thus the coordinate transformation for a rotation of coordinates is

$$\begin{bmatrix} x' \\ y' \end{bmatrix} = \begin{bmatrix} \cos\theta & \sin\theta \\ -\sin\theta & \cos\theta \end{bmatrix} \begin{bmatrix} x \\ y \end{bmatrix}$$

For example, consider a rotation through an angle of 45°. Suppose C is the point $\begin{bmatrix} 1 \\ 3 \end{bmatrix}$ relative to the xy coordinate system. Let us find the coordinates of C relative to the $x'y'$ coordinate system. Since $\sin 45° = \cos 45° = 1/\sqrt{2}$, we get

$$\begin{bmatrix} x' \\ y' \end{bmatrix} = \begin{bmatrix} \dfrac{1}{\sqrt{2}} & \dfrac{1}{\sqrt{2}} \\ -\dfrac{1}{\sqrt{2}} & \dfrac{1}{\sqrt{2}} \end{bmatrix} \begin{bmatrix} 1 \\ 3 \end{bmatrix} = \begin{bmatrix} 2\sqrt{2} \\ \sqrt{2} \end{bmatrix}$$

Note the similarity between this coordinate transformation and the transformation that defines a rotation of the plane through an angle θ. This arises from the fact one can interpret a transformation either as moving points in a fixed coordinate system or as changing the coordinate system around fixed points.

Isomorphisms

Let $T: U \rightarrow V$. T is said to be **one-to-one** if each element in the range of T corresponds to just one element in the domain of T. If every element of V is the image of an element of U, the transformation T is said to be **onto**.

The following definition brings the concepts of one-to-one, onto, and linearity together in one transformation.

Definition   *Let T be a one-to-one, linear transformation of U onto W. T is called an **isomorphism**. U and W are called **isomorphic** vector spaces.*

The significance of isomorphic vector spaces is that they are identical from the mathematical point of view. Since an isomorphism is one-to-one and onto, the elements of the two spaces match. Furthermore, an isomorphism is a linear transformation; thus such spaces have similar properties of vector addition and scalar multiplication. The physical appearance of the elements in the two spaces may differ, but mathematically they can be thought of as being the same space.

The act of associating a coordinate vector with every vector in a given real vector space defines a transformation of that vector space to $\mathbf{R}^n$. This transformation is linear; furthermore, we now see that it is one-to-one and onto.

Theorem 8.3　Let U be a real vector space with basis $\{\mathbf{u}_1, \ldots, \mathbf{u}_n\}$. Let $\mathbf{u}$ be an arbitrary element

of U with coordinate vector $\begin{bmatrix} a_1 \\ \vdots \\ a_n \end{bmatrix}$ relative to this basis. The following transformation

T is an isomorphism of U onto $\mathbf{R}^n$.

$$T(\mathbf{u}) = \begin{bmatrix} a_1 \\ \vdots \\ a_n \end{bmatrix}$$

Proof　A transformation T is one-to-one if $T(\mathbf{u}) = T(\mathbf{v})$ implies that $\mathbf{u} = \mathbf{v}$. Let

$$T(\mathbf{u}) = \begin{bmatrix} a_1 \\ \vdots \\ a_n \end{bmatrix} \quad \text{and} \quad T(\mathbf{v}) = \begin{bmatrix} a_1 \\ \vdots \\ a_n \end{bmatrix}$$

Thus

$$\mathbf{u} = a_1 \mathbf{u}_1 + \cdots + a_n\mathbf{u}_n \quad \text{and} \quad \mathbf{v} = a_1\mathbf{u}_1 + \cdots + a_n\mathbf{u}_n$$

This implies that $\mathbf{u} = \mathbf{v}$. Thus T is one-to-one.

We now prove that T is onto by showing that every element of $\mathbf{R}^n$ is the image

of some element of U. Let $\begin{bmatrix} b_1 \\ \vdots \\ b_n \end{bmatrix}$ be an element of $\mathbf{R}^n$. Then

$$T(b_1\mathbf{u}_1 + \cdots + b_n\mathbf{u}_n) = \begin{bmatrix} b_1 \\ \vdots \\ b_n \end{bmatrix}$$

Thus $\begin{bmatrix} b_1 \\ \vdots \\ b_n \end{bmatrix}$ is the image of the vector $b_1\mathbf{u}_1 + \cdots + b_n\mathbf{u}_n$. Therefore T is onto.

We leave it to you to show that T is linear.

An extremely important implication of the last theorem is that every real finite-dimensional vector space U is isomorphic to $\mathbf{R}^n$ for some value of n. Thus every real finite-dimensional vector space is identical, from the algebraic viewpoint, to $\mathbf{R}^n$. In developing the mathematics of $\mathbf{R}^n$, we have thus in a sense developed the mathematics of all real finite-dimensional vector spaces.

Exercise Set 8.1

In Exercises 1–4, find the coordinate vector of **u** *relative to the given basis B in* $\mathbf{R}^2$.

*1. $\mathbf{u} = (2, -3)$; $B = \{(1, 0), (0, 1)\}$

*2. $\mathbf{u} = (8, -1)$; $B = \{(3, -1), (2, 1)\}$

3. $\mathbf{u} = (5, 1)$; $B = \{(1, 0), (0, 1)\}$

4. $\mathbf{u} = (-7, -5)$; $B = \{(1, -1), (3, 1)\}$

In Exercises 5–8, find the coordinate vector of **u** *relative to the given basis B in* $\mathbf{R}^3$.

5. $\mathbf{u} = (4, 0, -2)$; $B = \{(1, 0, 0), (0, 1, 0), (0, 0, 1)\}$

*6. $\mathbf{u} = (6, -3, 1)$; $B = \{(1, -1, 0), (2, 1, -1), (2, 0, 0)\}$

*7. $\mathbf{u} = (3, -1, 7)$; $B = \{(2, 0, -1), (0, 1, 3), (1, 1, 1)\}$

8. $\mathbf{u} = (1, -6, -8)$; $B = \{(1, 2, 3), (1, -1, 0), (0, 1, -2)\}$

In Exercises 9–12, find the coordinate vector of **u** *relative to the given basis B in* P_n.

9. $\mathbf{u} = 7x - 3$; $B = \{x, 1\}$

*10. $\mathbf{u} = 2x - 6$; $B = \{3x - 5, x - 1\}$

11. $\mathbf{u} = 4x^2 - 5x + 2$; $B = \{x^2, x, 1\}$

*12. $\mathbf{u} = 3x^2 - 6x - 2$; $B = \{x^2, x - 1, 2x\}$

In Exercises 13–15, find the coordinate vector of **u** *relative to the given orthonormal basis B.*

*13. $\mathbf{u} = (5, 0, -10)$; $B = \left\{(0, -1, 0), \left(\frac{3}{5}, 0, -\frac{4}{5}\right), \left(\frac{4}{5}, 0, \frac{3}{5}\right)\right\}$

14. $\mathbf{u} = (1, 2, 3)$; $B = \left\{\left(\frac{1}{3}, \frac{2}{3}, \frac{2}{3}\right), \left(\frac{2}{3}, -\frac{2}{3}, \frac{1}{3}\right), \left(\frac{2}{3}, \frac{1}{3}, -\frac{2}{3}\right)\right\}$

*15. $\mathbf{u} = (2, 1, 4)$;

$B = \left\{(1, 0, 0), \left(0, \frac{1}{\sqrt{2}}, \frac{1}{\sqrt{2}}\right), \left(0, \frac{1}{\sqrt{2}}, -\frac{1}{\sqrt{2}}\right)\right\}$

In Exercises 16–20, find the transition matrix P from the given basis B to the standard basis B′ of $\mathbf{R}^2$. *Use this matrix to find the coordinate vectors of* **u**, **v**, *and* **w** *relative to B′.*

*16. $B = \{(2, 3), (1, 2)\}$ and $B' = \{(1, 0), (0, 1)\}$;

$$\mathbf{u}_B = \begin{bmatrix} 1 \\ 2 \end{bmatrix}, \mathbf{v}_B = \begin{bmatrix} 3 \\ -1 \end{bmatrix}, \mathbf{w}_B = \begin{bmatrix} 2 \\ 4 \end{bmatrix}$$

17. $B = \{(4, 1), (3, -1)\}$ and $B' = \{(1, 0), (0, 1)\}$;

$$\mathbf{u}_B = \begin{bmatrix} 3 \\ 0 \end{bmatrix}, \mathbf{v}_B = \begin{bmatrix} 2 \\ 1 \end{bmatrix}, \mathbf{w}_B = \begin{bmatrix} -2 \\ 1 \end{bmatrix}$$

*18. $B = \{(1, 1), (2, -3)\}$ and $B' = \{(1, 0), (0, 1)\}$;

$$\mathbf{u}_B = \begin{bmatrix} 0 \\ 1 \end{bmatrix}, \mathbf{v}_B = \begin{bmatrix} 1 \\ -2 \end{bmatrix}, \mathbf{w}_B = \begin{bmatrix} 3 \\ 1 \end{bmatrix}$$

19. $B = \{(3, 2), (2, 1)\}$ and $B' = \{(1, 0), (0, 1)\}$;

$$\mathbf{u}_B = \begin{bmatrix} 2 \\ -1 \end{bmatrix}, \mathbf{v}_B = \begin{bmatrix} 0 \\ -1 \end{bmatrix}, \mathbf{w}_B = \begin{bmatrix} 5 \\ 3 \end{bmatrix}$$

*20. $B = \{(2, -3) (-3, 4)\}$ and $B' = \{(1, 0), (0, 1)\}$;

$$\mathbf{u}_B = \begin{bmatrix} 1 \\ 1 \end{bmatrix}, \mathbf{v}_B = \begin{bmatrix} 3 \\ 0 \end{bmatrix}, \mathbf{w}_B = \begin{bmatrix} 4 \\ 2 \end{bmatrix}$$

*21. Consider the bases $B = \{(1, 0), (0, 1)\}$ and $B' = \{(5, 3), (3, 2)\}$ of $\mathbf{R}^2$. If **u** is a vector such that

$$\mathbf{u}_B = \begin{bmatrix} -2 \\ 5 \end{bmatrix}, \text{ find } \mathbf{u}_{B'}.$$ (*Hint:* The transition matrix from B to B' is the inverse of that from B' to B.)

22. Consider the bases $B = \{(1, 0), (0, 1)\}$ and $B' = \{(1, 2), (-1, -1)\}$ of $\mathbf{R}^2$. If **u** is a vector such that

$$\mathbf{u}_B = \begin{bmatrix} 8 \\ 3 \end{bmatrix}, \text{ find } \mathbf{u}_{B'}.$$

*23. Find the transition matrix P from the basis $B = \{(1, 2), (3, 0)\}$ of $\mathbf{R}^2$ to the basis $B' = \{(2, 1), (3, 2)\}$. If $\mathbf{u}_B = \begin{bmatrix} -3 \\ 2 \end{bmatrix}$, find $\mathbf{u}_{B'}$.

24. Find the transition matrix P from the basis $B = \{(3, -1), (1, 1)\}$ of $\mathbf{R}^2$ to the basis $B' = \{(2, 5), (1, 2)\}$. If $\mathbf{u}_B = \begin{bmatrix} 7 \\ -2 \end{bmatrix}$, find $\mathbf{u}_{B'}$.

*25. Consider the vector space P_2. $B = \{x^2, x, 1\}$ and $B' = \{3x^2, x - 1, 4\}$ are bases for P_2. Find the transition matrix from B to B'. Use this matrix to find the coordinate vectors of $3x^2 + 4x + 8$, $6x^2 + 4$, $8x + 12$, and $3x^2 + 4x + 4$ relative to B'. (*Hint:* The transition matrix from B to B' is the inverse of that from B' to B.)

*26. Consider the vector space P_1. $B = \{x, 1\}$ and $B' = \{x + 2, 3\}$ are bases for P_1. Find the transition matrix from B to B'. Use this matrix to find the coordinate vectors of $3x + 3$, $6x$, $6x + 9$, and $12x - 3$ relative to B'.

*27. Construct an isomorphism from the vector space of diagonal 2×2 matrices onto $\mathbf{R}^2$.

28. Construct an isomorphism from the vector space of symmetric 2×2 matrices onto $\mathbf{R}^3$.

29. Let T be a matrix transformation defined by a square matrix A. Prove that T is an isomorphism if and only if A is nonsingular.

30. Let A be a nonsingular matrix that defines an isomorphism of $\mathbf{R}^n$ onto $\mathbf{R}^n$. (See previous exercise.) Prove that A^{-1} also defines an isomorphism of $\mathbf{R}^n$ onto itself.

8.2 Matrix Representations of Linear Transformations

We have seen that a linear transformation $T: \mathbf{R}^n \rightarrow \mathbf{R}^m$ can be defined by a matrix A. In this section we introduce a way of representing a linear transformation between other vector spaces by a matrix. We lead up to this discussion by looking at the information that is necessary to define a linear transformation.

Let f be any function. We know that f is defined if its effect on *every* element of the domain is known. This is usually done by means of an equation that gives the effect of the function on an arbitrary element in the domain. For example, consider the function f defined by

$$f(x) = \sqrt{x - 3}$$

The domain of f is $x \geq 3$. The above equation gives the effect of f on every x in this interval. For example, $f(7) = 2$. Similarly, a linear transformation T is defined if its value at every vector in the domain is known. However, unlike a general function, we shall see that if we know the effect of the linear transformation on a finite subject of the domain (a basis), it will be automatically defined on all elements of the domain.

Theorem 8.4 Let $T: U \rightarrow V$ be a linear transformation. Let $\{\mathbf{u}_1, \ldots, \mathbf{u}_n\}$ be a basis for U. T is determined by its effect on the base vectors, namely by $T(\mathbf{u}_1), \ldots, T(\mathbf{u}_n)$. The range of T is spanned by $\{T(\mathbf{u}_1), \ldots, T(\mathbf{u}_n)\}$.

Thus, defining a linear transformation on a basis defines it on the whole domain.

Proof   Let $\mathbf{u}$ be an element of U. Since $\{\mathbf{u}_1, \ldots, \mathbf{u}_n\}$ is a basis for U, there exist scalars $a_1, \ldots, a_n$ such that

$$\mathbf{u} = a_1\mathbf{u}_1 + \cdots + a_n\mathbf{u}_n$$

The linearity of T gives

$$T(\mathbf{u}) = T(a_1\mathbf{u}_1 + \cdots + a_n\mathbf{u}_n)$$
$$= a_1T(\mathbf{u}_1) + \cdots + a_nT(\mathbf{u}_n)$$

Therefore $T(\mathbf{u})$ is known if $T(\mathbf{u}_1), \ldots, T(\mathbf{u}_n)$ are known.

Further, $T(\mathbf{u})$ may be interpreted to be an arbitrary element in the range of T. It can be expressed as a linear combination of $T(\mathbf{u}_1), \ldots, T(\mathbf{u}_n)$. Thus $\{T(\mathbf{u}_1), \ldots, T(\mathbf{u}_n)\}$ spans the range of T.

Example 1   Consider the linear transformation $T: \mathbf{R}^3 \rightarrow \mathbf{R}^2$, defined as follows on basis vectors of $\mathbf{R}^3$. Find $T(1, -2, 3)$.

$$T(1, 0, 0) = (3, -1), \qquad T(0, 1, 0) = (2, 1), \qquad T(0, 0, 1) = (3, 0)$$

Solution   Since T is defined on basis vectors of $\mathbf{R}^3$, it is defined on the whole space. To find $T(1, -2, 3)$, express the vector $(1, -2, 3)$ as a linear combination of the basis vectors and use the linearity of T.

$$\begin{aligned}
T(1, -2, 3) &= T(1(1, 0, 0) - 2(0, 1, 0) + 3(0, 0, 1)) \\
&= 1T(1, 0, 0) - 2T(0, 1, 0) + 3T(0, 0, 1) \\
&= 1(3, -1) - 2(2, 1) + 3(3, 0) \\
&= (8, -3)
\end{aligned}$$

We have seen that a linear transformation $T: \mathbf{R}^n \rightarrow \mathbf{R}^m$ can be defined by a matrix A:

$$T(\mathbf{u}) = A\mathbf{u}$$

where $A = [T(\mathbf{e}_1) \cdots T(\mathbf{e}_n)]$. Note that A is constructed by finding the effect of T on each of the standard basis vectors of $\mathbf{R}^n$. These ideas can be extended to a linear transformation $T: U \rightarrow V$ between general vector spaces. We shall represent the elements of U and V by coordinate vectors, and T by a matrix A that defines a transformation of coordinate vectors. As for $\mathbf{R}^n$, the matrix A is constructed by finding the effect of T on basis vectors.

Theorem 8.5   Let U and V be vector spaces with bases $B = \{\mathbf{u}_1, \ldots, \mathbf{u}_n\}$ and $B' = \{\mathbf{v}_1, \ldots, \mathbf{v}_m\}$ and $T: U \rightarrow V$ a linear transformation. If $\mathbf{u}$ is a vector in U with image $T(\mathbf{u})$ having coordinate vectors $\mathbf{a}$ and $\mathbf{b}$ relative to these bases, then

$$\mathbf{b} = A\mathbf{a}$$

where $A = [T(\mathbf{u}_1)_{B'} \quad \cdots \quad T(\mathbf{u}_n)_{B'}]$.

The matrix A thus defines a transformation of coordinate vectors of U in the "same way" as T transforms the vectors of U. See Figure 8.3. A is called the **matrix representation of T** (or **matrix of T**) with respect to the bases B and B'.

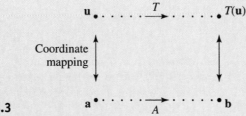

Figure 8.3

Proof Let $\mathbf{u} = a_1\mathbf{u}_1 + \cdots + a_n\mathbf{u}_n$. Using the linearity of T, we can write

$$T(\mathbf{u}) = T(a_1\mathbf{u}_1 + \cdots + a_n\mathbf{u}_n)$$
$$= a_1T(\mathbf{u}_1) + \cdots + a_nT(\mathbf{u}_n)$$

Let the effect of T on the basis vectors of U be

$$T(\mathbf{u}_1) = c_{11}\mathbf{v}_1 + \cdots + c_{1m}\mathbf{v}_m$$
$$\vdots$$
$$T(\mathbf{u}_n) = c_{n1}\mathbf{v}_1 + \cdots + c_{nm}\mathbf{v}_m$$

Thus

$$T(\mathbf{u}) = a_1(c_{11}\mathbf{v}_1 + \cdots + c_{1m}\mathbf{v}_m) + \cdots + a_n(c_{n1}\mathbf{v}_1 + \cdots + c_{nm}\mathbf{v}_m)$$
$$= (a_1c_{11} + \cdots + a_nc_{n1})\mathbf{v}_1 + \cdots + (a_1c_{1m} + \cdots + a_nc_{nm})\mathbf{v}_m$$

The coordinate vector of $T(\mathbf{u})$ is therefore

$$\mathbf{b} = \begin{bmatrix} (a_1c_{11} + \cdots + a_nc_{1n}) \\ \vdots \\ (a_1c_{1m} + \cdots + a_nc_{nm}) \end{bmatrix} = \begin{bmatrix} c_{11} & \cdots & c_{1n} \\ & \vdots & \\ c_{1m} & \cdots & c_{nm} \end{bmatrix} \begin{bmatrix} a_1 \\ \vdots \\ a_n \end{bmatrix}$$
$$= [T(\mathbf{u}_1)_{B'} \quad \cdots \quad T(\mathbf{u}_n)_{B'}]\mathbf{a}$$

proving the theorem.

Example 2 Let $T: U \rightarrow V$ be a linear transformation. T is defined relative to bases $B = \{\mathbf{u}_1, \mathbf{u}_2, \mathbf{u}_3\}$ and $B' = \{\mathbf{v}_1, \mathbf{v}_2\}$ of U and V as follows.

$$T(\mathbf{u}_1) = 2\mathbf{v}_1 - \mathbf{v}_2$$
$$T(\mathbf{u}_2) = 3\mathbf{v}_1 + 2\mathbf{v}_2$$
$$T(\mathbf{u}_3) = \mathbf{v}_1 - 4\mathbf{v}_2$$

Find the matrix representation of T with respect to these bases and use this matrix to determine the image of the vector $\mathbf{u} = 3\mathbf{u}_1 + 2\mathbf{u}_2 - \mathbf{u}_3$.

Solution The coordinate vectors of $T(\mathbf{u}_1)$, $T(\mathbf{u}_2)$, and $T(\mathbf{u}_3)$ are

$$\begin{bmatrix} 2 \\ -1 \end{bmatrix}, \begin{bmatrix} 3 \\ 2 \end{bmatrix}, \text{ and } \begin{bmatrix} 1 \\ -4 \end{bmatrix}$$

These vectors make up the columns of the matrix of T.

$$A = \begin{bmatrix} 2 & 3 & 1 \\ -1 & 2 & -4 \end{bmatrix}$$

Let us now find the image of the vector $\mathbf{u} = 3\mathbf{u}_1 + 2\mathbf{u}_2 - \mathbf{u}_3$ using this matrix. The coordinate vector of $\mathbf{u}$ is $\mathbf{a} = \begin{bmatrix} 3 \\ 2 \\ -1 \end{bmatrix}$. We get

$$A\mathbf{a} = \begin{bmatrix} 2 & 3 & 1 \\ -1 & 2 & -4 \end{bmatrix} \begin{bmatrix} 3 \\ 2 \\ -1 \end{bmatrix} = \begin{bmatrix} 11 \\ 5 \end{bmatrix}$$

$T(\mathbf{u})$ has coordinate vector $\begin{bmatrix} 11 \\ 5 \end{bmatrix}$. Thus $T(\mathbf{u}) = 11\mathbf{v}_1 + 5\mathbf{v}_2$.

Example 3   Consider the linear transformation $T: \mathbf{R}^3 \to \mathbf{R}^2$, defined by $T(x, y, z) = (x + y, 2z)$. Find the matrix of T with respect to the bases $\{\mathbf{u}_1, \mathbf{u}_2, \mathbf{u}_3\}$ and $\{\mathbf{u}_1', \mathbf{u}_2'\}$ of $\mathbf{R}^3$ and $\mathbf{R}^2$, where

$$\mathbf{u}_1 = (1, 1, 0), \quad \mathbf{u}_2 = (0, 1, 4), \quad \mathbf{u}_3 = (1, 2, 3), \quad \text{and} \quad \mathbf{u}_1' = (1, 0), \quad \mathbf{u}_2' = (0, 2)$$

Use this matrix to find the image of the vector $\mathbf{u} = (2, 3, 5)$.

Solution   We find the effect of T on the basis vectors of $\mathbf{R}^3$.

$$T(\mathbf{u}_1) = T(1, 1, 0) = (2, 0) = 2(1, 0) + 0(0, 2) = 2\mathbf{u}_1' + 0\mathbf{u}_2'$$
$$T(\mathbf{u}_2) = T(0, 1, 4) = (1, 8) = 1(1, 0) + 4(0, 2) = 1\mathbf{u}_1' + 4\mathbf{u}_2'$$
$$T(\mathbf{u}_3) = T(1, 2, 3) = (3, 6) = 3(1, 0) + 3(0, 2) = 3\mathbf{u}_1' + 3\mathbf{u}_2'$$

The coordinate vectors of $T(\mathbf{u}_1)$, $T(\mathbf{u}_2)$, and $T(\mathbf{u}_3)$ are thus $\begin{bmatrix} 2 \\ 0 \end{bmatrix}$, $\begin{bmatrix} 1 \\ 4 \end{bmatrix}$, and $\begin{bmatrix} 3 \\ 3 \end{bmatrix}$. These vectors form the columns of the matrix of T.

$$A = \begin{bmatrix} 2 & 1 & 3 \\ 0 & 4 & 3 \end{bmatrix}$$

Let us now use A to find the image of the vector $\mathbf{u} = (2, 3, 5)$. We determine the coordinate vector of $\mathbf{u}$. It can be shown that

$$\mathbf{u} = (2, 3, 5) = 3(1, 1, 0) + 2(0, 1, 4) - (1, 2, 3) = 3\mathbf{u}_1 + 2\mathbf{u}_2 + (-1)\mathbf{u}_3$$

The coordinate vector of $\mathbf{u}$ is thus $\mathbf{a} = \begin{bmatrix} 3 \\ 2 \\ -1 \end{bmatrix}$. The coordinate vector of $T(\mathbf{u})$ is

$$\mathbf{b} = A\mathbf{a} = \begin{bmatrix} 2 & 1 & 3 \\ 0 & 4 & 3 \end{bmatrix} \begin{bmatrix} 3 \\ 2 \\ -1 \end{bmatrix} = \begin{bmatrix} 5 \\ 5 \end{bmatrix}$$

Therefore $T(\mathbf{u}) = 5\mathbf{u}_1' + 5\mathbf{u}_2' = 5(1, 0) + 5(0, 2) = (5, 10)$.

We can check this result directly using the definition $T(x, y, z) = (x + y, 2z)$. For $\mathbf{u} = (2, 3, 5)$, this gives

$$T(\mathbf{u}) = T(2, 3, 5) = (2 + 3, 2 \times 5) = (5, 10)$$

While the previous example helped us better understand the idea of matrix representation, the importance of the concept of matrix representation obviously does not lie in its use to find $T(\mathbf{u})$! We now discuss its relevance.

Importance of Matrix Representations

In the previous section we saw that every real finite dimensional vector space is isomorphic to $\mathbf{R}^n$. This means that any such vector space can be discussed in terms of the appropriate $\mathbf{R}^n$. The fact that every linear transformation can now be represented by a matrix means that all the theoretical mathematics of these vector spaces and their linear transformations can be undertaken in terms of the vector spaces $\mathbf{R}^n$ and matrices.

A second reason is a computational one. The elements of $\mathbf{R}^n$ and matrices can be manipulated on computers. Thus general vector spaces and their linear transformations can be discussed on computers through these representations.

There are many ad hoc applications of matrix representations. We shall see that elementary row operations are linear transformations having matrix representations called **elementary matrices**. We shall see how these representations are used to derive a method called *LU* decomposition that is widely used for solving systems of equations on computers.

Example 4   Consider the linear transformation $T: P_2 \rightarrow P_1$ defined by $T(ax^2 + bx + c) = (a + b)x - c$. Find the matrix of T with respect to the bases $\{\mathbf{u}_1, \mathbf{u}_2, \mathbf{u}_3\}$ and $\{\mathbf{u}_1', \mathbf{u}_2'\}$ of P_2 and P_1, where

$$\mathbf{u}_1 = x^2, \quad \mathbf{u}_2 = x, \quad \mathbf{u}_3 = 1 \quad \text{and} \quad \mathbf{u}_1' = x, \quad \mathbf{u}_2' = 1$$

Use this matrix to find the image of $\mathbf{u} = 3x^2 + 2x - 1$.

Solution   Consider the effect of T on each basis vector of P_2.

$$T(\mathbf{u}_1) = T(x^2) = \quad x = \quad 1x + 0(1) = 1\mathbf{u}_1' + 0\mathbf{u}_2'$$
$$T(\mathbf{u}_2) = \quad T(x) = \quad x = \quad 1x + 0(1) = 1\mathbf{u}_1' + 0\mathbf{u}_2'$$
$$T(\mathbf{u}_3) = \quad T(1) = -1 = 0x + (-1)(1) = 0\mathbf{u}_1' + (-1)\mathbf{u}_2'$$

The coordinate vectors of $T(x^2)$, $T(x)$, and $T(1)$ are $\begin{bmatrix} 1 \\ 0 \end{bmatrix}$, $\begin{bmatrix} 1 \\ 0 \end{bmatrix}$, and $\begin{bmatrix} 0 \\ -1 \end{bmatrix}$. The matrix of T is thus

$$A = \begin{bmatrix} 1 & 1 & 0 \\ 0 & 0 & -1 \end{bmatrix}$$

Let us now use T to find the image of $\mathbf{u} = 3x^2 + 2x - 1$. The coordinate vector of $\mathbf{u}$ relative to the basis $\{x^2, x, 1\}$ is $\mathbf{a} = \begin{bmatrix} 3 \\ 2 \\ -1 \end{bmatrix}$. We get

$$\mathbf{b} = A\mathbf{a} = \begin{bmatrix} 1 & 1 & 0 \\ 0 & 0 & -1 \end{bmatrix} \begin{bmatrix} 3 \\ 2 \\ -1 \end{bmatrix} = \begin{bmatrix} 5 \\ 1 \end{bmatrix}$$

Therefore $T(\mathbf{u}) = 5\mathbf{u}_1' + 1\mathbf{u}_2' = 5x + 1$.

We visualize the way this matrix representation works in Figure 8.4. The top half of the figure shows the linear transformation T of P_2 to P_1. The bottom half is analogous to the top half, with A defining a transformation of the coordinate vectors of P_2 into the coordinate vectors of P_1 according to

$$\begin{bmatrix} 1 & 1 & 0 \\ 0 & 0 & -1 \end{bmatrix} \begin{bmatrix} a \\ b \\ c \end{bmatrix} = \begin{bmatrix} a + b \\ -c \end{bmatrix}$$

The bottom half is a **coordinate representation** of the top half.

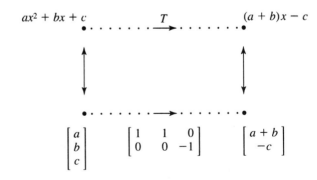

Figure 8.4

When a linear transformation is of a vector space into itself, the matrix representation is usually found relative to a single basis. The following example illustrates this.

Example 5   Let $D = \frac{d}{dx}$ be the operation of taking the derivative. D is a linear operator on P_2. Find the matrix of D with respect to the basis $\{x^2, x, 1\}$ of P_2.

Solution   We examine the effect of D on the basis vectors.

$$D(x^2) = 2x = 0x^2 + 2x + 0(1)$$
$$D(x) = 1 \;\; = 0x^2 + 0x + 1(1)$$
$$D(1) = 0 \;\; = 0x^2 + 0x + 0(1)$$

The matrix of D is thus

$$A = \begin{bmatrix} 0 & 0 & 0 \\ 2 & 0 & 0 \\ 0 & 1 & 0 \end{bmatrix}$$

The matrix A defines a linear operator on $\mathbf{R}^3$ that is analogous to D on P_2. We have

$$D(ax^2 + bx + c) = 2ax + b \quad \text{and} \quad \begin{bmatrix} 0 & 0 & 0 \\ 2 & 0 & 0 \\ 0 & 1 & 0 \end{bmatrix}\begin{bmatrix} a \\ b \\ c \end{bmatrix} = \begin{bmatrix} 0 \\ 2a \\ b \end{bmatrix}$$

See Figure 8.5.

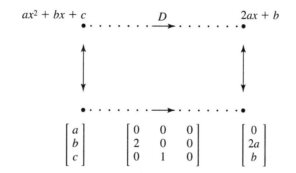

Figure 8.5

Relations between Matrix Representations

We have seen that the matrix representation of a linear transformation depends upon the bases selected. When linear transformations arise in applications, a goal is often to determine a simple matrix representation. Later in the course the reader will, for example, see how bases can be selected that will give diagonal matrix representations of certain linear transformations. At this time we discuss how matrix representations of linear operators relative to different bases are related, paving the way for this work.

Definition   *Let A and B be square matrices of the same size. B is said to be **similar** to A if there exists an invertible matrix P such that*

$$B = P^{-1}AP$$

*The transformation of the matrix A into the matrix B in this manner is called a **similarity transformation**.*

Similarity transformations are frequently used in applying linear algebra. We now find that the matrix representations of a linear operator relative to two bases are similar matrices.

Theorem 8.6   Let U be a vector space with bases B and B'. Let P be the transition matrix from B' to B. If T is a linear operator on U, having matrix A with respect to the first basis and A' with respect to the second basis, then

$$A' = P^{-1}AP$$

Proof   Consider a vector $\mathbf{u}$ in U. Let its coordinate vectors relative to B and B' be $\mathbf{a}$ and $\mathbf{a}'$. The coordinate vectors of $T(\mathbf{u})$ are $A\mathbf{a}$ and $A'\mathbf{a}'$. Since P is the transition matrix from B' to B, we know that

$$\mathbf{a} = P\mathbf{a}' \quad \text{and} \quad A\mathbf{a} = P(A'\mathbf{a}')$$

This second equation may be rewritten

$$P^{-1}A\mathbf{a} = A'\mathbf{a}'$$

Substituting $\mathbf{a} = P\mathbf{a}'$ into this equation gives

$$P^{-1}AP\mathbf{a}' = A'\mathbf{a}'$$

The effect of the matrices $P^{-1}AP$ and A' as transformations, on an arbitrary coordinate vector $\mathbf{a}'$, is the same. Thus these matrices are equal.

Example 6   Consider the linear operator $T(x, y) = (2x, x + y)$ on $\mathbf{R}^2$. Find the matrix of T with respect to the standard basis $B = \{(1, 0), (0, 1)\}$ of $\mathbf{R}^2$. Use the transformation $A' = P^{-1}AP$ to determine the matrix A' with respect to the basis $B' = \{(-2, 3), (1, -1)\}$.

Solution   The effect of T on the vectors of the standard basis is

$$T(1, 0) = (2, 1) = 2(1, 0) + 1(0, 1)$$
$$T(0, 1) = (0, 1) = 0(1, 0) + 1(0, 1)$$

The matrix of T relative to the standard basis is

$$A = \begin{bmatrix} 2 & 0 \\ 1 & 1 \end{bmatrix}$$

We now find P, the transition matrix from B' to B. Write the vectors of B' in terms of those of B.

$$(-2, 3) = -2(1, 0) + 3(0, 1)$$
$$(1, -1) = 1(1, 0) - 1(0, 1)$$

The transition matrix is

$$P = \begin{bmatrix} -2 & 1 \\ 3 & -1 \end{bmatrix}$$

Therefore

$$A' = P^{-1}AP = \begin{bmatrix} -2 & 1 \\ 3 & -1 \end{bmatrix}^{-1} \begin{bmatrix} 2 & 0 \\ 1 & 1 \end{bmatrix} \begin{bmatrix} -2 & 1 \\ 3 & -1 \end{bmatrix}$$

$$= \begin{bmatrix} 1 & 1 \\ 3 & 2 \end{bmatrix} \begin{bmatrix} 2 & 0 \\ 1 & 1 \end{bmatrix} \begin{bmatrix} -2 & 1 \\ 3 & -1 \end{bmatrix} = \begin{bmatrix} -3 & 2 \\ -10 & 6 \end{bmatrix}$$

Exercise Set 8.2

*1. Let $T: \mathbf{R}^3 \to \mathbf{R}^2$ be a linear transformation defined as follows on the standard basis of $\mathbf{R}^3$. Find $T(0, 1, -1)$.

$$T(1, 0, 0) = (2, 1), \qquad T(0, 1, 0) = (0, -2)$$
$$T(0, 0, 1) = (-1, 1)$$

*2. Let $T: \mathbf{R}^2 \to \mathbf{R}$ be a linear transformation defined as follows on the standard basis of $\mathbf{R}^2$. Find $T(3, -2)$.

$$T(1, 0) = 4, \qquad T(0, 1) = -3$$

3. Let T be a linear operator on $\mathbf{R}^2$ defined as follows on the standard basis of $\mathbf{R}^2$. Find $T(2, 1)$.

$$T(1, 0) = (2, 5), \qquad T(0, 1) = (1, -3)$$

*4. Let $T: P_2 \to P_1$ be a linear transformation defined as follows on the standard basis $\{x^2, x, 1\}$ of P_2. Find $T(3x^2 - 2x + 1)$.

$$T(x^2) = 3x + 1, \qquad T(x) = 2, \qquad T(1) = 2x - 5$$

5. Let T be a linear operator on P_2 defined as follows on the standard basis $\{x^2, x, 1\}$ of P_2. Find $T(x^2 + 3x - 2)$.

$$T(x^2) = x^2 + 3, \qquad T(x) = 2x^2 + 4x - 1,$$
$$T(1) = 3x - 1$$

*6. Let $T: U \to V$ be a linear transformation. Let T be defined relative to bases $\{\mathbf{u}_1, \mathbf{u}_2\}$ and $\{\mathbf{v}_1, \mathbf{v}_2\}$ of U and V as follows:

$$T(\mathbf{u}_1) = 2\mathbf{v}_1 + 3\mathbf{v}_2, \qquad T(\mathbf{u}_2) = 4\mathbf{v}_1 - \mathbf{v}_2$$

Find the matrix of T with respect to these bases. Use this matrix to find the image of the vector $\mathbf{u} = 2\mathbf{u}_1 + 5\mathbf{u}_2$.

7. Let $T: U \to V$ be a linear transformation. Let T be defined relative to bases $\{\mathbf{u}_1, \mathbf{u}_2\}$ and $\{\mathbf{v}_1, \mathbf{v}_2, \mathbf{v}_3\}$ of U and V as follows:

$$T(\mathbf{u}_1) = 2\mathbf{v}_1 + \mathbf{v}_2 - 3\mathbf{v}_3, \qquad T(\mathbf{u}_2) = \mathbf{v}_1 - 2\mathbf{v}_2 + \mathbf{v}_3$$

Find the matrix of T with respect to these bases. Use this matrix to find the image of the vector $\mathbf{u} = 4\mathbf{u}_1 - 7\mathbf{u}_2$.

*8. Let $T: U \to V$ be a linear transformation. Let T be defined relative to bases $\{\mathbf{u}_1, \mathbf{u}_2, \mathbf{u}_3\}$ and $\{\mathbf{v}_1, \mathbf{v}_2, \mathbf{v}_3\}$ of U and V as follows:

$$T(\mathbf{u}_1) = \mathbf{v}_1 + \mathbf{v}_2 + \mathbf{v}_3, \qquad T(\mathbf{u}_2) = 3\mathbf{v}_1 - 2\mathbf{v}_2,$$
$$T(\mathbf{u}_3) = \mathbf{v}_1 + 2\mathbf{v}_2 - \mathbf{v}_3$$

Find the matrix of T with respect to these bases. Use this matrix to find the image of the following vector:

$$\mathbf{u} = 3\mathbf{u}_1 + 2\mathbf{u}_2 - 5\mathbf{u}_3$$

9. Find the matrices of the following linear transformations of $\mathbf{R}^3 \to \mathbf{R}^2$ with respect to the standard bases of these spaces. Use these matrices to find the images of the vector $(1, 2, 3)$.

*(a) $T(x, y, z) = (x, z)$

(b) $T(x, y, z) = (3x, y + z)$

*(c) $T(x, y, z) = (x + y, 2x - y)$

10. Find the matrices of the following linear operators on $\mathbf{R}^3$ with respect to the standard basis of $\mathbf{R}^3$. Use these matrices to find the images of the vector $(-1, 5, 2)$.

*(a) $T(x, y, z) = (x, 2y, 3z)$

(b) $T(x, y, z) = (x, y, z)$

*(c) $T(x, y, z) = (x, 0, 0)$

(d) $T(x, y, z) = (x + y, 3y, x + 2y - 4z)$

*11. Consider the linear transformation $T: \mathbf{R}^3 \to \mathbf{R}^2$, defined by $T(x, y, z) = (x - y, x + z)$. Find the matrix of T with respect to the bases $\{\mathbf{u}_1, \mathbf{u}_2, \mathbf{u}_3\}$ and $\{\mathbf{u}_1', \mathbf{u}_2'\}$ of $\mathbf{R}^3$ and $\mathbf{R}^2$, where

$$\mathbf{u}_1 = (1, -1, 0), \quad \mathbf{u}_2 = (2, 0, 1)$$
$$\mathbf{u}_3 = (1, 2, 1), \quad \text{and} \quad \mathbf{u}_1' = (-1, 0), \quad \mathbf{u}_2' = (0, 1)$$

Use this matrix to find the image of the following vector:

$$\mathbf{u} = (3, -4, 0)$$

12. Consider the linear transformation $T: \mathbf{R}^2 \to \mathbf{R}^3$, defined by $T(x, y) = (x, x + y, 2y)$. Find the matrix of T with respect to the bases $\{\mathbf{u}_1, \mathbf{u}_2\}$ and $\{\mathbf{u}_1', \mathbf{u}_2', \mathbf{u}_3'\}$ of $\mathbf{R}^2$ and $\mathbf{R}^3$, where

$$\mathbf{u}_1 = (1, 3), \quad \mathbf{u}_2 = (4, -1), \quad \text{and}$$
$$\mathbf{u}_1' = (1, 0, 0), \quad \mathbf{u}_2' = (0, 2, 0), \quad \mathbf{u}_3' = (0, 0, -1)$$

Use this matrix to find the image of the vector $\mathbf{u} = (9, 1)$.

*13. Consider the linear operator $T: \mathbf{R}^2 \to \mathbf{R}^2$, defined by $T(x, y) = (2x, x + y)$. Find the matrix of T with respect to the basis $\{\mathbf{u}_1, \mathbf{u}_2\}$ of $\mathbf{R}^2$, where

$$\mathbf{u}_1 = (1, 2), \quad \mathbf{u}_2 = (0, -1)$$

Use this matrix to find the image of the vector $\mathbf{u} = (-1, 3)$.

*14. Find the matrix of the differential operator D with respect to the basis $\{2x^2, x, -1\}$ of P_2. Use this matrix to find the image of $3x^2 - 2x + 4$.

15. Let V be the vector space of functions having domain $[0, \pi]$ generated by the functions $\sin x$ and $\cos x$. Let D be the operation of taking the derivative with respect to x, and D^2 be the operation of taking the second derivative. Find the matrix of the linear operator $D^2 + 2D + 1$ of V with respect to the basis $\{\sin x, \cos x\}$ of V. Determine the image of the element $3\sin x + \cos x$ under this transformation.

16. Find the matrix of the following linear transformations with respect to the basis $\{x, 1\}$ of P_1 and $\{x^2, x, 1\}$ of P_2.

*(a) $T(ax^2 + bx + c) = (b + c)x^2 + (b - c)x$ of P_2 into itself

*(b) $T(ax + b) = bx^2 + ax + b$ of P_1 into P_2

(c) $T(ax^2 + bx + c) = 2cx + b - a$ of P_2 into P_1

*17. Find the matrix of the following linear transformation T of P_2 into P_1 with respect to the basis $\{x^2 + x, x, 1\}$ of P_2 and the basis $\{x, 1\}$ of P_1.

$$T(ax^2 + bx + c) = ax + c$$

Determine the image of $3x^2 + 2x - 1$.

18. Find the matrix of the following linear operator T on P_1 with respect to the basis $\{x + 1, 2\}$.

$$T(ax + b) = bx - a$$

Determine the image of $4x - 3$.

*19. Find the matrix of the following linear operator T on P_1 with respect to the standard basis $\{x, 1\}$ of P_1.

$$T(ax + b) = (a + b)x - b$$

Use a similarity transformation to then find the matrix of T with respect to the basis $\{x + 1, x - 1\}$ of P_1.

*20. Consider the linear operator $T(x, y) = (2x, x + y)$ on $\mathbf{R}^2$. Find the matrix of T with respect to the standard basis for $\mathbf{R}^2$. Use a similarity transformation to then find the matrix with respect to the basis $\{(1, 1), (2, 1)\}$ of $\mathbf{R}^2$.

21. Consider the linear operator $T(x, y) = (x - y, x + y)$ on $\mathbf{R}^2$. Find the matrix of T with respect to the standard basis for $\mathbf{R}^2$. Use a similarity transformation to then find the matrix with respect to the basis $\{(-2, 1), (1, 2)\}$ of $\mathbf{R}^2$.

22. *(a) Let V and W be vector spaces and let U be a subspace of V. Is it always possible to construct a linear transformation of V into W that has U as its kernel?

*(b) Construct a linear transformation of $\mathbf{R}^3$ into $\mathbf{R}^2$ that has the subspace consisting of all vectors of the form $r(1, 3, -1)$ as kernel.

(c) Construct a linear transformation of $\mathbf{R}^3$ into $\mathbf{R}^2$ that has the subspace consisting of all vectors of the form $(r + s, 2r, -s)$ as kernel.

*23. Construct a linear transformation of $\mathbf{R}^2$ into $\mathbf{R}^2$ that has the subspace consisting of all vectors of the form $r(2, -1)$ as kernel.

24. Construct a linear transformation of $\mathbf{R}^2$ into $\mathbf{R}^3$ that has the subspace consisting of all vectors of the form $r(4, 1)$ as kernel.

*25. Let $T: U \to U$ for a vector space U be defined by $T(\mathbf{u}) = \mathbf{u}$. Prove that T is linear. T is called the **identity operator** on U. Let B be a basis for U. Show that the matrix of T with respect to B is the identity matrix.

26. Let $T: U \rightarrow U$ for a vector space U be defined by $T(\mathbf{u}) = \mathbf{0}$. Prove that T is linear. T is called the **zero operator** on U. Let B be a basis for U. Show that the matrix of T with respect to B is the zero matrix.

27. Let U, V, and W be vector spaces with bases $B = \{\mathbf{u}_1, \ldots, \mathbf{u}_n\}$, $B' = \{\mathbf{v}_1, \ldots, \mathbf{v}_m\}$, and $B'' = \{\mathbf{w}_1, \ldots, \mathbf{w}_n\}$. Let $T: U \rightarrow V$ and $L: V \rightarrow W$ be linear transformations. Let P be the matrix of T with respect to B and B', and Q be the matrix representation of L with respect to B' and B''. Prove that the matrix of $L \circ T$ with respect to B and B'' is QP.

*28. Is it possible for two distinct linear transformations $T: U \rightarrow V$ and $L: U \rightarrow V$ to have the same matrix with respect to bases B and B' of U and V?

8.3 Diagonalization of Matrices

We have seen that every linear operator has a matrix representation and that this representation depends upon the basis selected. Matrix representations relative to different bases are related through similarity transformations. In this section we discuss diagonal matrix representations. The reader will see computational and geometrical applications of the mathematics.

We remind the reader that if A and B are square matrices of the same size, B is said to be **similar** to A if there exists an invertible matrix C such that $B = C^{-1}AC$. The transformation of the matrix A into the matrix B in this manner is called a **similarity transformation**.

Example 1 Consider the following matrices A and C. C is invertible. Use the similarity transformation $C^{-1}AC$ to transform A into a matrix B.

$$A = \begin{bmatrix} 7 & -10 \\ 3 & -4 \end{bmatrix}, \quad C = \begin{bmatrix} 2 & 5 \\ 1 & 3 \end{bmatrix}$$

Solution We get

$$\begin{aligned} B = C^{-1}AC &= \begin{bmatrix} 2 & 5 \\ 1 & 3 \end{bmatrix}^{-1} \begin{bmatrix} 7 & -10 \\ 3 & -4 \end{bmatrix} \begin{bmatrix} 2 & 5 \\ 1 & 3 \end{bmatrix} \\ &= \begin{bmatrix} 3 & -5 \\ -1 & 2 \end{bmatrix} \begin{bmatrix} 7 & -10 \\ 3 & -4 \end{bmatrix} \begin{bmatrix} 2 & 5 \\ 1 & 3 \end{bmatrix} \\ &= \begin{bmatrix} 6 & -10 \\ -1 & 2 \end{bmatrix} \begin{bmatrix} 2 & 5 \\ 1 & 3 \end{bmatrix} = \begin{bmatrix} 2 & 0 \\ 0 & 1 \end{bmatrix} \end{aligned}$$

In this example A is transformed into a diagonal matrix B. Not every square matrix can be "diagonalized" in this manner. In this section we shall discuss conditions under which a matrix can be diagonalized—and when it can, ways of constructing an appropriate transforming matrix C. We shall find that eigenvalues and eigenvectors play a key role in this discussion.

Theorem 8.7   Similar matrices have the same eigenvalues.

Proof   Let A and B be similar matrices. Hence there exists a matrix C such that $B = C^{-1}AC$. The characteristic polynomial of B is $|B - \lambda I|$. Substituting for B and using the multiplicative properties of determinants, we get

$$\begin{aligned}
|B - \lambda I| = |C^{-1}AC - \lambda I| &= |C^{-1}(A - \lambda I)C| \\
&= |C^{-1}||A - \lambda I||C| = |A - \lambda I||C^{-1}||C| \\
&= |A - \lambda I||C^{-1}C| = |A - \lambda I||I| \\
&= |A - \lambda I|
\end{aligned}$$

The characteristic polynomials of A and B are identical. This means that their eigenvalues are the same.

Definition   A *square matrix A is said to be **diagonalizable** if there exists a matrix C such that* $D = C^{-1}AC$ *is a diagonal matrix.*

We see in the above example that $A = \begin{bmatrix} 7 & -10 \\ 3 & -4 \end{bmatrix}$ is diagonalizable.

The following theorem tells us when it is possible to diagonalize a given matrix A—and if it is possible, how to find the similiarity transformation that diagonalizes A.

Theorem 8.8   Let A be an $n \times n$ matrix.

(a) If A has n linearly independent eigenvectors, it is diagonalizable. The matrix C whose columns consist of n linearly independent eigenvectors can be used in a similarity transformation $C^{-1}AC$ to give a diagonal matrix D. The diagonal elements of D will be the eigenvalues of A.

(b) If A is diagonalizable, then it has n linearly independent eigenvectors.

Proof

(a) Let A have eigenvalues $\lambda_1, \ldots, \lambda_n$ (which need not be distinct), with corresponding linearly independent eigenvectors $\mathbf{v}_1, \ldots, \mathbf{v}_n$. Let C be the matrix having $\mathbf{v}_1, \ldots, \mathbf{v}_n$ as column vectors.

$$C = [\mathbf{v}_1 \quad \cdots \quad \mathbf{v}_n]$$

Since $A\mathbf{v}_1 = \lambda_1\mathbf{v}_1, \ldots, A\mathbf{v}_n = \lambda_n\mathbf{v}_n$, matrix multiplication in terms of columns gives

$$AC = A[\mathbf{v}_1 \quad \cdots \quad \mathbf{v}_n]$$
$$= [A\mathbf{v}_1 \quad \cdots \quad A\mathbf{v}_n]$$
$$= [\lambda_1\mathbf{v}_1 \quad \cdots \quad \lambda_n\mathbf{v}_n]$$
$$= [\mathbf{v}_1 \quad \cdots \quad \mathbf{v}_n]\begin{bmatrix} \lambda_1 & & 0 \\ & \ddots & \\ 0 & & \lambda_n \end{bmatrix} = C\begin{bmatrix} \lambda_1 & & 0 \\ & \ddots & \\ 0 & & \lambda_n \end{bmatrix}$$

Since the columns of C are linearly independent, C is nonsingular. Thus

$$C^{-1}AC = \begin{bmatrix} \lambda_1 & & 0 \\ & \ddots & \\ 0 & & \lambda_n \end{bmatrix}$$

Therefore, if an $n \times n$ matrix A has n linearly independent eigenvectors, these eigenvectors can be used as the columns of a matrix C that diagonalizes A. The diagonal matrix has the eigenvalues of A as diagonal elements.

(b) The converse is proved by retracing the above steps. Commence with the assumption that C is a matrix $[\mathbf{v}_1 \quad \cdots \quad \mathbf{v}_n]$ that diagonalizes A. Thus there exist scalars $\gamma_1, \ldots, \gamma_n$, such that

$$C^{-1}AC = \begin{bmatrix} \gamma_1 & & 0 \\ & \ddots & \\ 0 & & \gamma_n \end{bmatrix}$$

Retracing the above steps, we arrive at the conclusion that

$$A\mathbf{v}_1 = \gamma_1\mathbf{v}_1, \ldots, A\mathbf{v}_n = \gamma_n\mathbf{v}_n$$

The $\mathbf{v}_1, \ldots, \mathbf{v}_n$ are eigenvectors of A. Since C is nonsingular, these vectors (column vectors of C) are linearly independent. Thus if an $n \times n$ matrix A is diagonalizable, it has n linearly independent eigenvectors.

Example 2

(a) Show that the following matrix A is diagonalizable.

(b) Find a diagonal matrix D that is similar to A.

(c) Determine the similarity transformation that diagonalizes A.

$$A = \begin{bmatrix} -4 & -6 \\ 3 & 5 \end{bmatrix}$$

Solution

(a) The eigenvalues and corresponding eigenvectors of this matrix were found in Example 1 of Section 4.7. They are

$$\lambda_1 = 2, \quad \mathbf{v}_1 = r\begin{bmatrix} -1 \\ 1 \end{bmatrix}; \qquad \lambda_2 = -1, \quad \mathbf{v}_2 = s\begin{bmatrix} -2 \\ 1 \end{bmatrix}$$

Since A, a 2×2 matrix, has two linearly independent eigenvectors, it is diagonalizable.

(b) *A* is similar to the diagonal matrix *D*, which has diagonal elements $\lambda_1 = 2$ and $\lambda_2 = -1$. Thus

$$A = \begin{bmatrix} -4 & -6 \\ 3 & 5 \end{bmatrix} \quad \text{is similar to} \quad D = \begin{bmatrix} 2 & 0 \\ 0 & -1 \end{bmatrix}$$

(c) It is often important to know the transformation that diagonalizes *A*. We now find the transformation and verify that it does lead to the above matrix *D*. Select two convenient linearly independent eigenvectors, say

$$\mathbf{v}_1 = \begin{bmatrix} -1 \\ 1 \end{bmatrix} \quad \text{and} \quad \mathbf{v}_2 = \begin{bmatrix} -2 \\ 1 \end{bmatrix}$$

Let these vectors be the column vectors of the diagonalizing matrix *C*.

$$C = \begin{bmatrix} -1 & -2 \\ 1 & 1 \end{bmatrix}$$

We then get

$$C^{-1}AC = \begin{bmatrix} -1 & -2 \\ 1 & 1 \end{bmatrix}^{-1} \begin{bmatrix} -4 & -6 \\ 3 & 5 \end{bmatrix} \begin{bmatrix} -1 & -2 \\ 1 & 1 \end{bmatrix}$$

$$= \begin{bmatrix} 1 & 2 \\ -1 & -1 \end{bmatrix} \begin{bmatrix} -4 & -6 \\ 3 & 5 \end{bmatrix} \begin{bmatrix} -1 & -2 \\ 1 & 1 \end{bmatrix} = \begin{bmatrix} 2 & 0 \\ 0 & -1 \end{bmatrix} = D$$

If *A* is similar to a diagonal matrix *D* under the transformation $C^{-1}AC$, then it can be shown that $A^k = CD^kC^{-1}$. This result can be used to compute A^k. Let us derive this result and then apply it.

$$D^k = (C^{-1}AC)^k = \underbrace{(C^{-1}AC) \cdots (C^{-1}AC)}_{k \text{ times}} = C^{-1}A^kC$$

This leads to

$$A^k = CD^kC^{-1}$$

Example 3   Compute A^9 for the following matrix *A*.

$$A = \begin{bmatrix} -4 & -6 \\ 3 & 5 \end{bmatrix}$$

Solution   *A* is the matrix of the previous example. Use the values of *C* and *D* from that example. We get

$$D^9 = \begin{bmatrix} 2 & 0 \\ 0 & -1 \end{bmatrix}^9 = \begin{bmatrix} 2^9 & 0 \\ 0 & -1^9 \end{bmatrix} = \begin{bmatrix} 512 & 0 \\ 0 & -1 \end{bmatrix}$$

The transformation now gives

$$A^9 = CD^9C^{-1}$$

$$= \begin{bmatrix} -1 & -2 \\ 1 & 1 \end{bmatrix} \begin{bmatrix} 512 & 0 \\ 0 & -1 \end{bmatrix} \begin{bmatrix} -1 & -2 \\ 1 & 1 \end{bmatrix}^{-1} = \begin{bmatrix} -514 & -1026 \\ 513 & 1025 \end{bmatrix}$$

In practice, powers of matrices such as A^9 can now be quickly found using computers. The importance of the result $A^k = CD^kC^{-1}$ for computing powers is primarily theoretical. It can be used to give expressions for powers of general matrices. You will appreciate the technique in the following section, when it is used in solving equations called difference equations.

The following example illustrates that not every matrix is diagonalizable.

Example 4 Show that the following matrix A is not diagonalizable.

$$A = \begin{bmatrix} 5 & -3 \\ 3 & -1 \end{bmatrix}$$

Solution Let us compute the eigenvalues and corresponding eigenvectors of A. We get

$$A - \lambda I_2 = \begin{bmatrix} 5 - \lambda & -3 \\ 3 & -1 - \lambda \end{bmatrix}$$

The characteristic equation is

$$|A - \lambda I_2| = 0$$
$$(5 - \lambda)(-1 - \lambda) + 9 = 0$$
$$\lambda^2 - 4\lambda + 4 = 0$$
$$(\lambda - 2)(\lambda - 2) = 0$$

There is a single repeated eigenvalue, $\lambda = 2$. We now find the corresponding eigenvectors. $(A - 2I_2)\mathbf{x} = \mathbf{0}$ gives

$$\begin{bmatrix} 3 & -3 \\ 3 & -3 \end{bmatrix} \begin{bmatrix} x_1 \\ x_2 \end{bmatrix} = \mathbf{0}$$

This gives $3x_1 - 3x_2 = 0$. Thus $x_1 = r$, $x_2 = r$. The eigenvectors are nonzero vectors of the form

$$r \begin{bmatrix} 1 \\ 1 \end{bmatrix}$$

The eigenspace is a one-dimensional space. A is a 2×2 matrix, but it does not have two linearly independent eigenvectors. Thus A is not diagonalizable.

Diagonalization of Symmetric Matrices

We have discussed symmetric matrices and looked at applications of these matrices. Let us now examine the diagonalization of these important matrices. The previous theorem showed us that the diagonalization of a matrix is related to its eigenvectors. The following theorem summarizes the properties of eigenvalues and eigenvectors of symmetric matrices, paving the way for results about their diagonalization.

Theorem 8.9   Let A be an $n \times n$ symmetric matrix.

(a)   All the eigenvalues of A are real numbers.

(b)   The dimension of an eigenspace of A is the multiplicity of the eigenvalue as a root of the characteristic equation.

(c)   The eigenspaces of A are orthogonal.

(d) A has n linearly independent eigenvectors.

The proof of this theorem is beyond the scope of this class. However, the results are extremely important and can be easily understood and used. We illustrate these results for the following symmetric matrix A.

$$A = \begin{bmatrix} 5 & 4 & 2 \\ 4 & 5 & 2 \\ 2 & 2 & 2 \end{bmatrix}$$

The eigenvalues and eigenvectors of this matrix were found in Section 4.7.

The characteristic equation can be written $(\lambda - 10)(\lambda - 1)^2 = 0$. The eigenvalues λ_1, λ_2 and corresponding eigenspaces V_1, V_2 are as follows:

$$\lambda_1 = 10, \quad V_1 = \left\{ r \begin{bmatrix} 2 \\ 2 \\ 1 \end{bmatrix} \right\}; \qquad \lambda_2 = 1, \quad V_2 = \left\{ s \begin{bmatrix} -1 \\ 1 \\ 0 \end{bmatrix} + t \begin{bmatrix} -1 \\ 0 \\ 2 \end{bmatrix} \right\}$$

We see that the results of the theorem are illustrated as follows:

(a)   The eigenvalues of A are real numbers; $\lambda_1 = 10$ and $\lambda_2 = 1$.

(b) $\lambda_1 = 10$ is a root of the characteristic equation of multiplicity **1**, and its eigenspace is the **1**-dimensional space having basis $\left\{ \begin{bmatrix} 2 \\ 2 \\ 1 \end{bmatrix} \right\}$.

$\lambda_2 = 1$ is a root of multiplicity **2**, and its eigenspace is the **2**-dimensional space having basis $\left\{ \begin{bmatrix} -1 \\ 1 \\ 0 \end{bmatrix}, \begin{bmatrix} -1 \\ 0 \\ 2 \end{bmatrix} \right\}$.

(c) Two vector spaces are said to be orthogonal if any vector in the one space is orthogonal to any vector in the other space. Let $\mathbf{v}_1 = r\begin{bmatrix} 2 \\ 2 \\ 1 \end{bmatrix}$ and

$\mathbf{v}_2 = s\begin{bmatrix} -1 \\ 1 \\ 0 \end{bmatrix} + t\begin{bmatrix} 1 \\ 0 \\ 2 \end{bmatrix}$ be arbitrary vectors in the two eigenspaces. It can be

seen that $\mathbf{v}_1 \cdot \mathbf{v}_2 = 0$, showing that the two eigenspaces are orthogonal.

(d) A is a 3×3 matrix. Observe that it has **3** linearly independent eigenvectors,

namely $\begin{bmatrix} 2 \\ 2 \\ 1 \end{bmatrix}$, $\begin{bmatrix} -1 \\ 1 \\ 0 \end{bmatrix}$, and $\begin{bmatrix} -1 \\ 0 \\ 2 \end{bmatrix}$.

Orthogonal Diagonalization

If a matrix C is orthogonal, then we know that $C^{-1} = C^t$. Thus if such a matrix is used in a similarity transformation, the transformation becomes $D = C^t AC$. This type of similarity transformation is, of course, easier to compute than $D = C^{-1}AC$. This is important if one is performing computations by hand, but not important when using a computer.

There is a deeper significance to "orthogonal diagonalization." Suppose one is performing the transformation to find a diagonal representation of a linear operator T. A is usually the matrix representation of T relative to an orthonormal basis B. B defines a rectangular coordinate system xy. D will be the diagonal representation of T relative to a basis B'. It is very desirable that B' be orthonormal; it will then define an orthogonal coordinate system $x'y'$. C will be the transition matrix from the basis B to B' in this context. If C is orthogonal, it preserves norms and angles, thus ensuring that B' is orthonormal, that $x'y'$ is an orthogonal coordinate system, and that the geometry is preserved (objects have the same shape in both coordinate systems). You will see a geometrical application that relates orthogonal coordinate systems in the following section.

Definition   *A square matrix A is said to be **orthogonally diagonalizable** if there exists an orthogonal matrix C such that $D = C^t AC$ is a diagonal matrix. The next theorem tells us that the set of orthogonally diagonalizable matrices is, in fact, the set of symmetric matrices.*

Theorem 8.10   Let A be a square matrix. A is orthogonally diagonalizable if and only if it is a symmetric matrix.

(continues)

Proof   Assume that A is symmetric. We give the steps that can be taken to construct an orthogonal matrix C such that $D = C'AC$ is diagonal. The previous theorem ensures us that this algorithm can be carried out.

1.   Find a basis for each eigenspace of A.

2.   Find an orthonormal basis for each eigenspace. (Use the Gram-Schmidt process if necessary.)

3.   Let C be the matrix whose columns are these orthonormal vectors.

4.   The matrix $D = C'AC$ will be a diagonal matrix.

This algorithm will be used to orthogonally diagonalize a given symmetric matrix.

Conversely, assume that A is orthogonally diagonalizable. Thus there exists an orthogonal matrix C such that $D = C'AC$. Therefore

$$A = CDC'$$

Use the properties of transpose to get

$$A' = (CDC')' = (C')'(CD)' = CDC' = A$$

Thus A is symmetric.

Example 5   Orthogonally diagonalize the following symmetric matrix A.

$$A = \begin{bmatrix} 1 & -2 \\ -2 & 1 \end{bmatrix}$$

Solution   You can verify that the eigenvalues and corresponding eigenspaces of this matrix are as follows.

$$\lambda_1 = -1, \quad V_1 = \left\{ s \begin{bmatrix} 1 \\ 1 \end{bmatrix} \right\}; \qquad \lambda_2 = 3, \quad V_2 = \left\{ r \begin{bmatrix} -1 \\ 1 \end{bmatrix} \right\}$$

Since A is symmetric, we know that it can be diagonalized to give

$$D = \begin{bmatrix} -1 & 0 \\ 0 & 3 \end{bmatrix}$$

Let us determine the transformation. The eigenspaces V_1 and V_2 are, as is to be expected, orthogonal. Use a unit vector in each eigenspace as columns of an orthogonal matrix C. We get

$$C = \begin{bmatrix} \dfrac{1}{\sqrt{2}} & -\dfrac{1}{\sqrt{2}} \\ \dfrac{1}{\sqrt{2}} & \dfrac{1}{\sqrt{2}} \end{bmatrix}$$

The orthogonal transformation that leads to D is

$$C'AC = \begin{bmatrix} \dfrac{1}{\sqrt{2}} & \dfrac{1}{\sqrt{2}} \\ -\dfrac{1}{\sqrt{2}} & \dfrac{1}{\sqrt{2}} \end{bmatrix} \begin{bmatrix} 1 & -2 \\ -2 & 1 \end{bmatrix} \begin{bmatrix} \dfrac{1}{\sqrt{2}} & -\dfrac{1}{\sqrt{2}} \\ \dfrac{1}{\sqrt{2}} & \dfrac{1}{\sqrt{2}} \end{bmatrix}$$

We now apply these tools to the study of matrix representations of linear operators.

Diagonal Matrix Representations of a Linear Operator

Let us see how to find a diagonal matrix representation of a linear operator T, if one exists. A diagonal matrix representation is usually the representation that provides most information in applications.

Let T be a linear operator on a vector space V of dimension n. Let B be a basis for V, and let A be the matrix representation of T relative to B. The matrix representation of T relative to another basis B' can be obtained using a similarity transformation $A' = P^{-1}AP$, where P is the transition matrix from the basis B' to the basis B.

Suppose A has eigenvalues $\lambda_1, \ldots, \lambda_n$, with n corresponding, linearly independent eigenvectors $\mathbf{v}_1, \ldots, \mathbf{v}_n$. If $P = [\mathbf{v}_1 \quad \cdots \quad \mathbf{v}_n]$, then we know that A' is the diagonal matrix

$$A' = \begin{bmatrix} \lambda_1 & & 0 \\ & \ddots & \\ 0 & & \lambda_n \end{bmatrix}$$

The coordinate vectors of B' relative to B are the eigenvectors $\mathbf{v}_1, \ldots, \mathbf{v}_n$. It is desirable, for the reasons stated above, to use an orthogonal transformation if possible to arrive at this basis B', which provides the diagonal matrix A'.

Example 6   Consider the linear operator $T(x, y) = (3x + y, x + 3y)$ on $\mathbf{R}^2$. Find a diagonal matrix representation of T. Determine the basis for this representation and give a geometrical interpretation of T.

Solution   Let us start by finding the matrix representation A relative to the standard basis $B = \{(1, 0), (0, 1)\}$ of $\mathbf{R}^2$. (This is the easiest representation to find.) We get

$$T(1, 0) = (3, 1) = 3(1, 0) + 1(0, 1)$$
$$T(0, 1) = (1, 3) = 1(1, 0) + 3(0, 1)$$

The coordinate vectors of $T(1, 0)$ and $T(0, 1)$ relative to B are $\begin{bmatrix} 3 \\ 1 \end{bmatrix}$ and $\begin{bmatrix} 1 \\ 3 \end{bmatrix}$. The matrix representation of T relative to the standard basis B is thus

$$A = \begin{bmatrix} 3 & 1 \\ 1 & 3 \end{bmatrix}$$

You can verify that A has the following eigenvalues and eigenvectors.

$$\lambda_1 = 4, \quad \mathbf{v}_1 = r\begin{bmatrix} 1 \\ 1 \end{bmatrix}; \qquad \lambda_2 = 2 \quad \mathbf{v}_2 = s\begin{bmatrix} -1 \\ 1 \end{bmatrix}$$

The following matrix A' is thus a diagonal matrix representation of T.

$$A' = \begin{bmatrix} \lambda_1 & 0 \\ 0 & \lambda_2 \end{bmatrix} = \begin{bmatrix} 4 & 0 \\ 0 & 2 \end{bmatrix}$$

Let us now find the basis B' that gives this representation A'. Observe that A is a symmetric matrix. Select unit orthogonal eigenvectors for the coordinate vectors of B' relative to B. The transition matrix from B to B' will then be orthogonal, and the geometry will be preserved. Let

$$B' = \left\{ \left(\frac{1}{\sqrt{2}}, \frac{1}{\sqrt{2}} \right), \left(-\frac{1}{\sqrt{2}}, \frac{1}{\sqrt{2}} \right) \right\}$$

Note that the basis B' is obtained from the basis B by rotation through $\pi/4$.

We now give a geometrical interpretation to the operator T. The standard basis B defines an xy coordinate system. Let the basis B' define an $x'y'$ coordinate system. See Figure 8.6. The matrix A' tells us that T is a scaling in the $x'y'$ coordinate system, with factor 4 in the x' direction and factor 2 in the y' direction. Thus, for example, T maps the square $PQRO$ into the rectangle $P'Q'R'O$.

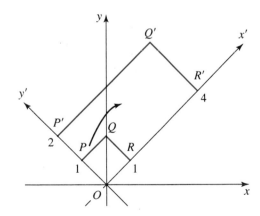

Figure 8.6

This example illustrates a situation that arises frequently in physics and engineering. One is given a mathematical description in an xy coordinate system. One then searches for a second coordinate system $x'y'$ that better fits the situation and carries out the analysis in that coordinate system. The transformation from the one system to the other is usually an orthogonal similarity transformation.

Exercise Set 8.3

1. In each of the following exercises, transform the matrix A into a matrix B using the similarity transformation $C^{-1}AC$, with the given matrix C.

   *(a) $A = \begin{bmatrix} 1 & 2 \\ -1 & 3 \end{bmatrix}$, $C = \begin{bmatrix} 2 & 5 \\ 1 & 3 \end{bmatrix}$

   (b) $A = \begin{bmatrix} -8 & 18 \\ -6 & 13 \end{bmatrix}$, $C = \begin{bmatrix} 3 & 2 \\ 2 & 1 \end{bmatrix}$

   *(c) $A = \begin{bmatrix} 0 & 4 \\ 3 & 2 \end{bmatrix}$, $C = \begin{bmatrix} 2 & 1 \\ 7 & 4 \end{bmatrix}$

2. In each of the following exercises, transform the matrix A into a matrix B using the similarity transformation $C^{-1}AC$, with the given matrix C.

   *(a) $A = \begin{bmatrix} 2 & 0 & 0 \\ -2 & 2 & 1 \\ 2 & 0 & 1 \end{bmatrix}$, $C = \begin{bmatrix} -1 & 2 & 0 \\ 2 & -3 & 1 \\ -2 & 4 & -1 \end{bmatrix}$

   (b) $A = \begin{bmatrix} 1 & 2 & 3 \\ 0 & -1 & 2 \\ 1 & 1 & 0 \end{bmatrix}$, $C = \begin{bmatrix} 3 & 5 & -1 \\ -2 & -3 & 1 \\ -1 & -2 & 1 \end{bmatrix}$

3. Diagonalize (if possible) each of the following matrices. Give the similarity transformation.

   *(a) $\begin{bmatrix} 5 & 4 \\ 1 & 2 \end{bmatrix}$ (b) $\begin{bmatrix} 2 & 1 \\ 2 & 3 \end{bmatrix}$ *(c) $\begin{bmatrix} 1 & 1 \\ 0 & 1 \end{bmatrix}$

   *(d) $\begin{bmatrix} 4 & -1 \\ 2 & 1 \end{bmatrix}$ (e) $\begin{bmatrix} 4 & 1 \\ -1 & 2 \end{bmatrix}$

4. Diagonalize (if possible) each of the following matrices. Give the similarity transformation.

   *(a) $\begin{bmatrix} -7 & 10 \\ -5 & 8 \end{bmatrix}$ (b) $\begin{bmatrix} 7 & 4 \\ -8 & 5 \end{bmatrix}$

   *(c) $\begin{bmatrix} 1 & -2 \\ 2 & -3 \end{bmatrix}$ (d) $\begin{bmatrix} 7 & -4 \\ 1 & 3 \end{bmatrix}$

   *(e) $\begin{bmatrix} a & b \\ 0 & a \end{bmatrix}$, $b \neq 0$

5. Diagonalize (if possible) each of the following matrices. Give the similarity transformation.

   *(a) $\begin{bmatrix} 15 & 7 & -7 \\ -1 & 1 & 1 \\ 13 & 7 & -5 \end{bmatrix}$ (b) $\begin{bmatrix} 5 & -2 & 2 \\ 4 & -3 & 4 \\ 4 & -6 & 7 \end{bmatrix}$

   *(c) $\begin{bmatrix} 1 & 0 & 0 \\ -2 & 1 & 2 \\ -2 & 0 & 3 \end{bmatrix}$ (d) $\begin{bmatrix} 3 & 0 & 0 \\ 1 & 2 & 0 \\ 0 & 0 & -4 \end{bmatrix}$

6. Orthogonally diagonalize each of the following symmetric matrices. Give the similarity transformation.

   *(a) $\begin{bmatrix} 1 & 2 \\ 2 & 1 \end{bmatrix}$ *(b) $\begin{bmatrix} 11 & 2 \\ 2 & 14 \end{bmatrix}$

   (c) $\begin{bmatrix} 3 & 1 \\ 1 & 3 \end{bmatrix}$ (d) $\begin{bmatrix} -1 & -8 \\ -8 & 11 \end{bmatrix}$

7. Orthogonally diagonalize each of the following symmetric matrices. Give the similarity transformation.

   *(a) $\begin{bmatrix} 1 & 5 \\ 5 & 1 \end{bmatrix}$ (b) $\begin{bmatrix} 9 & 2 \\ 2 & 6 \end{bmatrix}$

   *(c) $\begin{bmatrix} 1 & 3 \\ 3 & 9 \end{bmatrix}$ (d) $\begin{bmatrix} 1.5 & -0.5 \\ -0.5 & 1.5 \end{bmatrix}$

8. Orthogonally diagonalize each of the following symmetric matrices. Give the similarity transformation.

   *(a) $\begin{bmatrix} 0 & 2 & 0 \\ 2 & 0 & 0 \\ 0 & 0 & 1 \end{bmatrix}$ *(b) $\begin{bmatrix} 9 & -3 & 3 \\ -3 & 6 & -6 \\ 3 & -6 & 6 \end{bmatrix}$

   (c) $\begin{bmatrix} 1 & 2 & -2 \\ 2 & 4 & -4 \\ -2 & -4 & 4 \end{bmatrix}$

9. We know that if $D = C^{-1}AC$, then $A^k = CD^kC^{-1}$. Use this result to compute the following powers. (These matrices were diagonalized in Exercise 7.)

   *(a) $\begin{bmatrix} 1 & 5 \\ 5 & 1 \end{bmatrix}^8$ (b) $\begin{bmatrix} 9 & 2 \\ 2 & 6 \end{bmatrix}^5$

   (c) $\begin{bmatrix} 1 & 3 \\ 3 & 9 \end{bmatrix}^6$ *(d) $\begin{bmatrix} 1.5 & -0.5 \\ -0.5 & 1.5 \end{bmatrix}^{16}$

10. Compute the following powers. (These matrices were diagonalized in Exercise 8.)

    *(a) $\begin{bmatrix} 0 & 2 & 0 \\ 2 & 0 & 0 \\ 0 & 0 & 1 \end{bmatrix}^6$ *(b) $\begin{bmatrix} 9 & -3 & 3 \\ -3 & 6 & -6 \\ 3 & -6 & 6 \end{bmatrix}^5$

    (c) $\begin{bmatrix} 1 & 2 & -2 \\ 2 & 4 & -4 \\ -2 & -4 & 4 \end{bmatrix}^4$

11. Find a diagonal matrix representation for each of the following operators. Determine a basis for each diagonal representation (that involves an orthogonal transformation). Give a geometrical interpretation of *T*.

   *(a) $T(x, y) = (4x + 2y, 2x + 4y)$

   (b) $T(x, y) = (5x + 3y, 3x + 5y)$

   *(c) $T(x, y) = (9x + 2y, 2x + 6y)$

   (d) $T(x, y) = (14x + 2y, 2x + 11y)$

12. Find a diagonal matrix representation for each of the following operators. Determine a basis for each diagonal representation. Give a geometrical interpretation of *T*. (It is not possible to find an orthogonal transformation in these examples.)

   *(a) $T(x, y) = (8x - 6y, 9x - 7y)$

   (b) $T(x, y) = (-2x + 2y, -10x + 7y)$

   *(c) $T(x, y) = (3x - 4y, 2x - 3y)$

   (d) $T(x, y) = (7x + 5y, -10x - 8y)$

13. Prove that if *A* and *B* are similar matrices, then

   *(a) $|A| = |B|$.   (b) rank(*A*) = rank(*B*).

   *(c) Tr(*A*) = Tr(*B*).   (d) A^t and B^t are similar.

   (e) *A* is nonsingular if and only if *B* is nonsingular.

   *(f) if *A* is nonsingular, then A^{-1} and B^{-1} are similar.

14. Prove that the eigenspaces of a symmetric matrix corresponding to distinct eigenvalues are orthogonal.

15. Let *A* be diagonalizable and let *B* be similar to *A*. Prove that

   (a) *B* is diagonalizable.

   *(b) $B + kI$ is similar to $A + kI$, for any scalar *k*.

16. If *A* is similar to a diagonal matrix *D*, we say that *A* is **diagonalizable** to *D*. Let *A* be diagonalizable to *D*. Prove that A^2 is diagonalizable to D^2 and that, in general, A^n is diagonalizable to D^n.

*17. If *A* is a symmetric matrix, we know that it is similar to a diagonal matrix. Is such a diagonal matrix unique? (*Hint:* Does the order of the column vectors in the transforming matrix matter?)

18. Prove that if *A* is diagonalizable and has eigenvalues $\lambda_1, \ldots, \lambda_n$, then $|A| = \lambda_1 \cdots \lambda_n$.

*19. Matrices *A* and *B* are said to be **orthogonally similar** if there exists an orthogonal matrix *C* such that $B = C^{-1}AC$. Show that if *A* is symmetric and *A* and *B* are orthogonally similar, then *B* is also symmetric.

20. Show that if *A* and *B* are orthogonally similar and *B* and *C* are orthogonally similar, then *A* and *C* are orthogonally similar.

*8.4 Quadratic Forms, Differences, Equations, and Normal Modes

In this section we see how to use diagonalization in geometry, algebra, and engineering. The techniques used in these examples are important in the sciences and in engineering. You should master them and be able to use them whenever the need arises.

Quadratic Forms

The algebraic expression

$$ax^2 + bxy + cy^2$$

where *a*, *b*, and *c* are constants, is called a **quadratic form**. Quadratic forms play an important role in geometry. The reader can verify by multiplying out the matrices that this quadratic form can be written as follows.

$$[x \quad y] \begin{bmatrix} a & \frac{b}{2} \\ \frac{b}{2} & c \end{bmatrix} \begin{bmatrix} x \\ y \end{bmatrix}$$

Let $\mathbf{x} = \begin{bmatrix} x \\ y \end{bmatrix}$ and $A = \begin{bmatrix} a & \frac{b}{2} \\ \frac{b}{2} & c \end{bmatrix}$. We can write the quadratic form

$$\mathbf{x}'A\mathbf{x}$$

The symmetric matrix A associated with the quadratic form is called the **matrix of the quadratic form**.

Example 1   Write the following quadratic form in terms of matrices.

$$5x^2 + 6xy - 4y^2$$

Solution   On comparison with the standard form $ax^2 + bxy + cy^2$, we have

$$a = 5, \quad b = 6, \quad c = -4$$

The matrix form of the quadratic form is thus

$$[x \quad y] \begin{bmatrix} 5 & 3 \\ 3 & -4 \end{bmatrix} \begin{bmatrix} x \\ y \end{bmatrix}$$

Let us develop techniques for understanding and graphing equations of the form

$$ax^2 + bxy + cy^2 + d = 0$$

This equation includes the quadratic form $ax^2 + bxy + cy^2$. We can write the equation in matrix form

$$\mathbf{x}'A\mathbf{x} + d = 0$$

Since A is a symmetric matrix, there exists an orthogonal matrix C such that $C'AC$ is a diagonal matrix D. Further, since C is orthogonal, $C^{-1} = C'$, and thus $C'C = I$. This identity enables us to introduce $C'C$ into the equation at various locations under the guise of I, to diagonalize A, leading to a standard recognizable form of the equation. We get

$$\mathbf{x}'A\mathbf{x} + d = 0$$
$$\mathbf{x}'(CC')A(CC')\mathbf{x} + d = 0$$
$$\mathbf{x}'C(C'AC)C'\mathbf{x} + d = 0$$
$$\mathbf{x}'CDC'\mathbf{x} + d = 0$$

We have seen that $\mathbf{x}' = C'\mathbf{x}$ defines a coordinate transformation from the xy coordinate system to a coordinate system $x'y'$, where C' is the transition matrix of the transformation. Since C' is an orthogonal matrix, it preserves magnitudes and angles. Thus if the original xy coordinate system was rectangular, the new $x'y'$ system will be rectangular. In the $x'y'$ coordinate system the equation becomes

$$\mathbf{x''Dx'} + d = 0$$

Let $D = \begin{bmatrix} p & 0 \\ 0 & q \end{bmatrix}$ and $\mathbf{x'} = \begin{bmatrix} x' \\ y' \end{bmatrix}$. Thus

$$[x' \quad y'] \begin{bmatrix} p & 0 \\ 0 & q \end{bmatrix} \begin{bmatrix} x' \\ y' \end{bmatrix} + d = 0$$

$$p(x')^2 + q(y')^2 + d = 0$$

If $d \neq 0$, we get

$$\frac{(x')^2}{-d/p} + \frac{(y')^2}{-d/q} = 1$$

This is the equation of a **conic**. Its graph in the $x'y'$ coordinate system can now be sketched, leading to its graph in the original xy system. The following example illustrates the method.

Example 2   Analyze the following equation. Sketch its graph.

$$6x^2 + 4xy + 9y^2 - 20 = 0$$

Solution   This equation includes the quadratic form $6x^2 + 4xy + 9y^2$. Write the equation in matrix form as follows:

$$[x \quad y] \begin{bmatrix} 6 & 2 \\ 2 & 9 \end{bmatrix} \begin{bmatrix} x \\ y \end{bmatrix} - 20 = 0$$

The eigenvalues and corresponding eigenvectors of the matrix are found. They are

$$\lambda_1 = 10, \quad \mathbf{v_1} = r \begin{bmatrix} 1 \\ 2 \end{bmatrix}; \qquad \lambda_2 = 5, \quad \mathbf{v_2} = s \begin{bmatrix} -2 \\ 1 \end{bmatrix}$$

Normalize these eigenvectors and write them as the columns of a matrix C. We get

$$C = \begin{bmatrix} \dfrac{1}{\sqrt{5}} & \dfrac{-2}{\sqrt{5}} \\ \dfrac{2}{\sqrt{5}} & \dfrac{1}{\sqrt{5}} \end{bmatrix}$$

We now know that the equation can be transformed into the form

$$[x' \quad y'] \begin{bmatrix} 10 & 0 \\ 0 & 5 \end{bmatrix} \begin{bmatrix} x' \\ y' \end{bmatrix} - 20 = 0$$

by the coordinate transformation

$$\begin{bmatrix} x' \\ y' \end{bmatrix} = \begin{bmatrix} \dfrac{1}{\sqrt{5}} & \dfrac{2}{\sqrt{5}} \\ \dfrac{-2}{\sqrt{5}} & \dfrac{1}{\sqrt{5}} \end{bmatrix} \begin{bmatrix} x \\ y \end{bmatrix}$$

On multiplying the matrices and rearranging, the equation in the $x'y'$ coordinate system is

$$\frac{(x')^2}{2} + \frac{(y')^2}{4} = 1$$

We recognize this as being the equation of an ellipse in the $x'y'$ coordinate system. The length of the semimajor axis is 2 and the length of the semiminor axis is $\sqrt{2}$. See Figure 8.7(a).

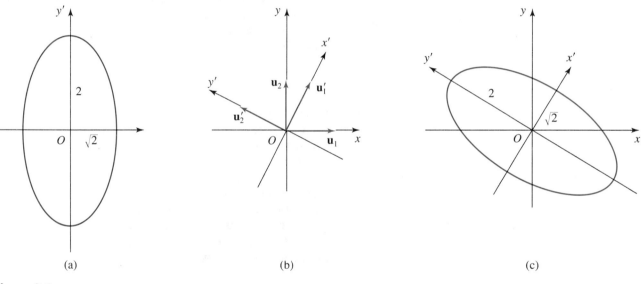

(a)   (b)   (c)

Figure 8.7

It remains to locate the $x'y'$ coordinate system relative to the xy coordinate system. Let $B = \{\mathbf{u}_1, \mathbf{u}_2\}$ and $B' = \{\mathbf{u}_1', \mathbf{u}_2'\}$ be the bases associated with the xy and $x'y'$ coordinating systems. C is the transition matrix from B' to B. Since C is orthogonal, C^t will be the transition matrix from B to B'. We get

$$\mathbf{u}_1' = \frac{1}{\sqrt{5}}\mathbf{u}_1 + \frac{2}{\sqrt{5}}\mathbf{u}_2$$

$$\mathbf{u}_2' = \frac{-2}{\sqrt{5}}\mathbf{u}_1 + \frac{1}{\sqrt{5}}\mathbf{u}_2$$

These equations lead to the locations of the x' and y' axes. See Figure 8.7(b). The graph of the equation is now known in the xy coordinate system. See Figure 8.7(c).

Difference Equations

Let a_1, a_2, a_3, ... be a sequence of real numbers. Such a sequence may be defined by giving its nth term. For example, suppose

$$a_n = n^2 + 1$$

Letting $n = 1, 2, 3, \ldots$, we get the terms of the sequence,

$$2, 5, 10, 17, \ldots$$

Furthermore, any specific term of the sequence can be found. For example, if we want a_{20}, then we let $n = 20$ in this expression for a_n and we get

$$a_{20} = (20)^2 + 1 = 401$$

When sequences arise in applications, they are often initially defined by a relationship between consecutive terms, with some initial terms known, rather than defined by the nth term. For example, a sequence might be defined by the relationship

$$a_n = 2a_{n-1} + 3a_{n-2}, \qquad n = 3, 4, 5, \ldots$$

with $a_1 = 0$ and $a_2 = 1$. Such an equation is called a **difference equation** (or **recurrence relation**), and the given terms of the sequence are called **initial conditions**. Further terms of the sequence can be found from the difference equation and initial conditions. For example, letting $n = 3, 4$, and 5, we get

$$a_3 = 2a_2 + 3a_1 = 2(1) + 3(0) = 2$$
$$a_4 = 2a_3 + 3a_2 = 2(2) + 3(1) = 7$$
$$a_5 = 2a_4 + 3a_3 = 2(7) + 3(2) = 20$$

The sequence is $0, 1, 2, 7, 20, \ldots$.

However if one wants to find a specific term such as the 20th term of this sequence, this method of using the difference equation to first find all the preceding terms is impractical. We need an expression for the nth term of the sequence. The expression for the nth term is called the **solution to the difference equation**. We now illustrate how the tools of linear algebra can be used to solve certain linear difference equations.

Consider the difference equation

$$a_n = pa_{n-1} + qa_{n-2}, \qquad n = 3, 4, 5, \ldots$$

where p and q are fixed real numbers and a_1 and a_2 are known (the initial conditions). This equation is said to be a **linear** difference equation (each a_i appears to the first power). It is of **order** 2 (a_n is expressed in terms of two preceding terms a_{n-1} and a_{n-2}). To solve this equation, introduce a second relation $b_n = a_{n-1}$. We get the system

$$a_n = pa_{n-1} + qb_{n-1}$$
$$b_n = a_{n-1}, \qquad n = 3, 4, 5, \ldots$$

These equations can be written in matrix form:

$$\begin{bmatrix} a_n \\ b_n \end{bmatrix} = \begin{bmatrix} p & q \\ 1 & 0 \end{bmatrix} \begin{bmatrix} a_{n-1} \\ b_{n-1} \end{bmatrix}, \qquad n = 3, 4, 5, \ldots$$

Let $X_n = \begin{bmatrix} a_n \\ b_n \end{bmatrix}$ and $A = \begin{bmatrix} p & q \\ 1 & 0 \end{bmatrix}$. We get

$$X_n = AX_{n-1}, \qquad n = 3, 4, 5, \ldots$$

Thus

$$X_n = AX_{n-1} = A^2X_{n-2} = A^3X_{n-3} = \cdots = A^{n-2}X_2,$$

$$\text{where } X_2 = \begin{bmatrix} a_2 \\ b_2 \end{bmatrix} = \begin{bmatrix} a_2 \\ a_1 \end{bmatrix}$$

In most applications A has distinct eigenvalues λ_1 and λ_2. It thus has two linearly independent eigenvectors, so it can be diagonalized using a similarity transformation. Let C be a matrix whose columns are linearly independent eigenvectors of A and let

$$D = C^{-1}AC = \begin{bmatrix} \lambda_1 & 0 \\ 0 & \lambda_2 \end{bmatrix}$$

Then

$$A^{n-2} = (CDC^{-1})^{n-2} = (CDC^{-1})(CDC^{-1}) \cdots (CDC^{-1}) = CD^{n-2}C^{-1}$$

$$= C\begin{bmatrix} \lambda_1 & 0 \\ 0 & \lambda_2 \end{bmatrix}^{n-2}C^{-1} = C\begin{bmatrix} (\lambda_1)^{n-2} & 0 \\ 0 & (\lambda_2)^{n-2} \end{bmatrix}C^{-1}$$

This gives

$$X_n = C\begin{bmatrix} (\lambda_1)^{n-2} & 0 \\ 0 & (\lambda_2)^{n-2} \end{bmatrix}C^{-1}X_2$$

We now illustrate the application of this result.

Example 3   Solve the difference equation

$$a_n = 2a_{n-1} + 3a_{n-2}, \quad \text{for } n = 3, 4, 5, \ldots$$

with initial conditions $a_1 = 0$, $a_2 = 1$. Use the solution to determine a_{15}.

Solution   Construct the system

$$a_n = 2a_{n-1} + 3a_{n-2}$$
$$b_n = a_{n-1}, \qquad n = 3, 4, 5, \ldots$$

We write this system in the matrix form

$$\begin{bmatrix} a_n \\ b_n \end{bmatrix} = \begin{bmatrix} 2 & 3 \\ 1 & 0 \end{bmatrix} \begin{bmatrix} a_{n-1} \\ b_{n-1} \end{bmatrix}, \qquad n = 3, 4, 5, \ldots$$

Eigenvalues and corresponding eigenvectors of the matrix $\begin{bmatrix} 2 & 3 \\ 1 & 0 \end{bmatrix}$ are

$$\lambda_1 = -1, \quad \mathbf{v}_1 = \begin{bmatrix} 1 \\ -1 \end{bmatrix}; \qquad \lambda_2 = 3, \quad \mathbf{v}_2 = \begin{bmatrix} 3 \\ 1 \end{bmatrix}$$

Let $C = \begin{bmatrix} 1 & 3 \\ -1 & 1 \end{bmatrix}$. Then $C^{-1} = \frac{1}{4}\begin{bmatrix} 1 & -3 \\ 1 & 1 \end{bmatrix}$. We get

$$\begin{bmatrix} a_n \\ b_n \end{bmatrix} = C \begin{bmatrix} (\lambda_1)^{n-2} & 0 \\ 0 & (\lambda_2)^{n-2} \end{bmatrix} C^{-1} \begin{bmatrix} a_2 \\ b_2 \end{bmatrix}$$

$$= \frac{1}{4}\begin{bmatrix} 1 & 3 \\ -1 & 1 \end{bmatrix}\begin{bmatrix} (-1)^{n-2} & 0 \\ 0 & (3)^{n-2} \end{bmatrix}\begin{bmatrix} 1 & -3 \\ 1 & 1 \end{bmatrix}\begin{bmatrix} 1 \\ 0 \end{bmatrix}, \qquad \text{since } b_2 = a_1 = 0$$

$$= \frac{1}{4}\begin{bmatrix} (-1)^{n-2} + 3(3)^{n-2} \\ (-1)(-1)^{n-2} + (3)^{n-2} \end{bmatrix} = \frac{1}{4}\begin{bmatrix} (-1)^{n-2} + (3)^{n-1} \\ (-1)^{n-1} + (3)^{n-2} \end{bmatrix}$$

Thus the solution is

$$a_n = \tfrac{1}{4}[(-1)^{n-2} + (3)^{n-1}], \qquad n = 3, 4, 5, \ldots$$

This solution can be checked. It gives $a_3 = 2$, $a_4 = 7$, $a_5 = 20$, agreeing with the values obtained by substituting $n = 3, 4, 5$ into the difference equation.

Letting $n = 15$, we get

$$a_{15} = \tfrac{1}{4}[(-1)^{15-2} + (3)^{15-1}] = \tfrac{1}{4}[-1 + 4{,}782{,}969] = 1{,}195{,}742$$

Fibonacci Sequence*

The sequence of numbers defined by the difference equation

$$a_n = a_{n-1} + a_{n-2}, \qquad n = 3, 4, 5, \ldots$$

with initial conditions $a_1 = 1$, $a_2 = 1$ is called a **Fibonacci sequence**. Each term in this sequence is the sum of the two preceding terms.

This sequence was developed by the Italian mathematician Leonardo Fibonacci in the 13th century to describe the growth in population of rabbits. Fibonacci assumed

---

* Leonardo Fibonacci (about 1170–1250) was the first great mathematician of the Christian West. He grew up and lived most of his life in Pisa, Italy. His father was secretary for commerce in Pisa, and Fibonacci represented him on business trips to Egypt, Syria, and Greece. He learned the Hindu-Arabic method of numeration and calculation and brought it back to Italy. His own writings reveal that while instruction of his own countrymen in mathematics and financial problems was his chief concern, he nevertheless also provided material on geometry for those who were interested in theoretical questions.

that a pair of rabbits produces a pair of young rabbits every month and that newborn rabbits become adults in two months. At this time, they can produce another pair. Starting with an adult pair, how many adult pairs will be in the colony after the first, second, third month, and so on? Fibonacci showed by counting that the first seven terms of the sequence are

$$1, 1, 2, 3, 5, 8, 13$$

He then made the observation that each term in the sequence was the sum of the two preceding terms, leading to the preceding difference equation. Fibonacci used the difference equation to compute the number of adult pairs in the colony at any time. We ask you to solve the Fibonacci difference equation in the exercises that follow.

Since the time of Fibonacci, scientists have used this sequence in fields such as botany, architecture, archaeology, and sociology. It is a puzzle why the sequence arises the way it does in some of these fields. For example, biologists have discovered that the arrangements of leaves around the stems of plants often follow a Fibonacci pattern. The leaves of roses and cabbages, and the scales of spruce trees follow a Fibonacci pattern. There are 5 rose leaves and 8 cabbage leaves in a full circle of the stems. There are 21 scales of spruce in a full circle of a tree.

The ratio of consecutive terms of the Fibonacci sequence is interesting. It can be shown that a_n/a_{n-1} approaches the number $(1 + \sqrt{5})/2$ as n increases. (See the following exercises.) This irrational number, $1.6180339885\ldots$, is called the **golden ratio**. Psychologists have found that rectangles whose length-to-width ratios are near the golden ratio are the most pleasing. Such artists as Leonard da Vinci, Mondrian, and Seurat have made great use of the golden ratio in their works. The dimensions of the Parthenon in Athens are based on the golden ratio.

Normal Modes of Oscillating Systems

Consider a horizontal string AB of length $4a$ and negligible mass, loaded with three particles, each of mass m. Let the masses be located at fixed distances a, $2a$, and $3a$ from A. The particles are displaced slightly from their equilibrium position and released. Let us analyze the subsequent motion, assuming that it all takes place in a vertical plane through A and B. See Figure 8.8.

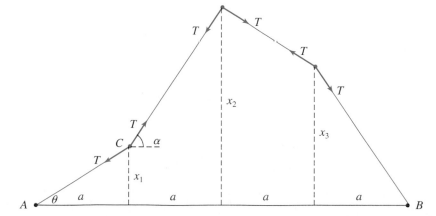

Figure 8.8

Let the vertical displacements of the particles at any instant during the subsequent motion be x_1, x_2, and x_3, as shown in Figure 8.8. Let T be the tension in the string. Consider the motion of the particle at C. The resultant force on this particle in a vertical direction is $T \sin \alpha - T \sin \theta$. If we assume that the displacements are small, the motion of each particle can be assumed to be vertical and the tension can be assumed to be unaltered throughout the motion. Since the angles are small, $\sin \alpha$ can be taken to equal $\tan \alpha$, and $\sin \theta$ to equal $\tan \theta$. Thus the resultant force at C is

$$T \tan \alpha - T \tan \theta = T\frac{(x_2 - x_1)}{a} - T\frac{x_1}{a}$$

$$= \frac{-2Tx_1}{a} + \frac{Tx_2}{a}$$

Applying **Newton's second law of motion** (force = mass × acceleration), we find that the motion of the first particle is described by the equation

$$m\ddot{x}_1 = \frac{-2Tx_1}{a} + \frac{Tx_2}{a}$$

Similarly, the motions of the other two particles are described by the equations

$$m\ddot{x}_2 = \frac{Tx_1}{a} - \frac{2Tx_2}{a} + \frac{Tx_3}{a} \quad \text{and} \quad m\ddot{x}_3 = \frac{Tx_2}{a} - \frac{2Tx_3}{a}$$

These three equations can be combined into the single matrix equation

$$\begin{bmatrix} \ddot{x}_1 \\ \ddot{x}_2 \\ \ddot{x}_3 \end{bmatrix} = \frac{T}{ma} \begin{bmatrix} -2 & 1 & 0 \\ 1 & -2 & 1 \\ 0 & 1 & -2 \end{bmatrix} \begin{bmatrix} x_1 \\ x_2 \\ x_3 \end{bmatrix}$$

Write this equation in matrix form

$$\ddot{\mathbf{x}} = \frac{T}{ma} A\mathbf{x}$$

Observe that A is a symmetric matrix. It can thus be diagonalized. Let C be a matrix consisting of eigenvectors of A as column vectors that can be used to diagonalize A. We can write the above equation

$$C^{-1}\ddot{\mathbf{x}} = C^{-1}\frac{T}{ma} ACC^{-1}\mathbf{x}$$

$$C^{-1}\ddot{\mathbf{x}} = \frac{T}{ma} C^{-1}ACC^{-1}\mathbf{x}$$

Let D be the diagonal matrix $C^{-1}AC$. Thus

$$C^{-1}\ddot{\mathbf{x}} = \frac{T}{ma} DC^{-1}\mathbf{x}$$

Introduce a new coordinate system defined by

$$\mathbf{x}' = C^{-1}\mathbf{x}$$

In this coordinate system, the equation becomes

$$\ddot{\mathbf{x}}' = \frac{T}{ma}D\mathbf{x}' \tag{1}$$

This matrix equation can be solved for $\mathbf{x}'$. The coordinate relationship

$$\mathbf{x} = C\mathbf{x}'$$

then leads to the solution in terms of the original coordinate system.

Having discussed the general method, let us now implement it for our specific equation. The eigenvalues and corresponding eigenvectors of the symmetric matrix A can be found to be

$$\lambda_1 = -2, \quad v_1 = r\begin{bmatrix} 1 \\ 0 \\ -1 \end{bmatrix}; \qquad \lambda_2 = -2 - \sqrt{2}, \quad v_2 = s\begin{bmatrix} 1 \\ -\sqrt{2} \\ 1 \end{bmatrix};$$

$$\lambda_3 = -2 + \sqrt{2}, \quad v_3 = t\begin{bmatrix} 1 \\ \sqrt{2} \\ 1 \end{bmatrix}$$

Thus let

$$C = \begin{bmatrix} 1 & 1 & 1 \\ 0 & -\sqrt{2} & \sqrt{2} \\ -1 & 1 & 1 \end{bmatrix} \quad \text{and} \quad D = \begin{bmatrix} -2 & 0 & 0 \\ 0 & -2 - \sqrt{2} & 0 \\ 0 & 0 & -2 + \sqrt{2} \end{bmatrix}$$

The equations of motion (1) are

$$\begin{bmatrix} \ddot{x}_1' \\ \ddot{x}_2' \\ \ddot{x}_3' \end{bmatrix} = \frac{T}{ma}\begin{bmatrix} -2 & 0 & 0 \\ 0 & -2 - \sqrt{2} & 0 \\ 0 & 0 & -2 + \sqrt{2} \end{bmatrix}\begin{bmatrix} x_1' \\ x_2' \\ x_3' \end{bmatrix}$$

This matrix equation represents the following system of three differential equations, each describing a simple harmonic motion.

$$\ddot{x}_1' = \frac{-2T}{ma}x_1', \qquad \ddot{x}_2' = \frac{(-2 - \sqrt{2})T}{ma}x_2', \qquad \ddot{x}_3' = \frac{(-2 + \sqrt{2})T}{ma}x_3'$$

These equations have the following solutions

$$x_1' = b_1 \cos\left(\sqrt{\frac{2T}{ma}}t + \gamma_1\right)$$

$$x_2' = b_2 \cos\left(\sqrt{\frac{(2 + \sqrt{2})T}{ma}}t + \gamma_2\right)$$

$$x_3' = b_3 \cos\left(\sqrt{\frac{(2 - \sqrt{2})T}{ma}}t + \gamma_3\right)$$

Here b_1, b_2, b_3, γ_1, γ_2, γ_3 are constants of integration that depend upon the configuration at time $t = 0$. They are not all independent. In standard interpretation of simple harmonic motion, b_1, b_2, b_3 are amplitudes and γ_1, γ_2, γ_3 are phases.

These special coordinates x_1', x_2', and x_3', in which the motion is easiest to describe, are called the **normal coordinates** of the motion. The motion in terms of the original coordinates can be obtained by using the relationship

$$\mathbf{x} + \mathbf{Cx'}$$

$$
\begin{bmatrix} x_1 \\ x_2 \\ x_3 \end{bmatrix} =
\begin{bmatrix} 1 & 1 & 1 \\ 0 & -\sqrt{2} & \sqrt{2} \\ -1 & 1 & 1 \end{bmatrix}
\begin{bmatrix} x_1' \\ x_2' \\ x_3' \end{bmatrix}
$$

$$
= x_1' \begin{bmatrix} 1 \\ 0 \\ -1 \end{bmatrix}
+ x_2' \begin{bmatrix} 1 \\ -\sqrt{2} \\ 1 \end{bmatrix}
+ x_3' \begin{bmatrix} 1 \\ -\sqrt{2} \\ 1 \end{bmatrix}
$$

Substituting for x_1', x_2', and x_3' gives

$$
\begin{bmatrix} x_1 \\ x_2 \\ x_3 \end{bmatrix} =
b_1 \cos\left(\sqrt{\frac{2T}{ma}}\, t + \gamma_1 \right) \begin{bmatrix} 1 \\ 0 \\ -1 \end{bmatrix}
+ b_2 \cos\left(\sqrt{\frac{(2 + \sqrt{2})T}{ma}}\, t + \gamma_2 \right) \begin{bmatrix} 1 \\ -\sqrt{2} \\ 1 \end{bmatrix}
$$

$$
+ b_3 \cos\left(\sqrt{\frac{(2 - \sqrt{2})T}{ma}}\, t + \gamma_3 \right) \begin{bmatrix} 1 \\ \sqrt{2} \\ 1 \end{bmatrix}
$$

The motion can thus be interpreted as a combination of the following three motions, or **modes**.

Mode 1:

$$
\begin{bmatrix} x_1 \\ x_2 \\ x_3 \end{bmatrix} =
\cos\left(\sqrt{\frac{2T}{ma}}\, t + \gamma_1 \right) \begin{bmatrix} 1 \\ 0 \\ -1 \end{bmatrix}
\qquad \text{(Note that } x_2 = 0 \text{ and } x_3 = -x_1\text{)}
$$

Mode 2:

$$
\begin{bmatrix} x_1 \\ x_2 \\ x_3 \end{bmatrix} =
\cos\left(\sqrt{\frac{(2 + \sqrt{2})T}{ma}}\, t + \gamma_2 \right) \begin{bmatrix} 1 \\ -\sqrt{2} \\ 1 \end{bmatrix}
\qquad (x_2 = -\sqrt{2}x_1 \text{ and } x_3 = x_1)
$$

Mode 3:

$$
\begin{bmatrix} x_1 \\ x_2 \\ x_3 \end{bmatrix} =
\cos\left(\sqrt{\frac{(2 - \sqrt{2})T}{ma}}\, t + \gamma_3 \right) \begin{bmatrix} 1 \\ \sqrt{2} \\ 1 \end{bmatrix}
\qquad (x_2 = \sqrt{2}x_1 \text{ and } x_3 = x_1)
$$

The actual combination will be a linear combination determined by b_1, b_2, and b_3, that will depend on the initial configuration. Each of these motions is a simple harmonic motion of a particle. These modes, called the **normal modes of oscillation**, are illustrated in Figure 8.9.

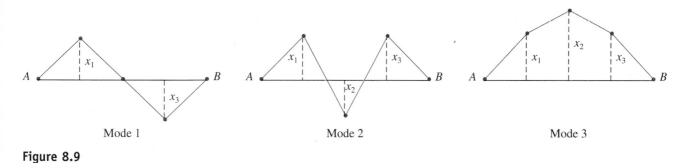

Mode 1   Mode 2   Mode 3

Figure 8.9

Exercise Set 8.4

1. Express each of the following quadratic forms in terms of matrices.

*(a) $x^2 + 4xy + 2y^2$

(b) $3x^2 + 2xy - 4y^2$

*(c) $7x^2 - 6xy - y^2$

(d) $2x^2 + 5xy + 3y^2$

*(e) $-3x^2 - 7xy + 4y^2$

(f) $5x^2 + 3xy - 2y^2$

2. Sketch the graph of each of the following equations.

*(a) $11x^2 + 4xy + 14y^2 - 60 = 0$

*(b) $3x^2 + 2xy + 3y^2 - 12 = 0$

(c) $x^2 - 6xy + y^2 - 8 = 0$

(d) $4x^2 + 4xy + 4y^2 - 5 = 0$

3. Solve the following difference equations and use the solutions to determine the given terms.

*(a) $a_n = a_{n-1} + 2a_{n-2}$, $a_1 = 1$, $a_2 = 2$. Find a_{10}.

(b) $a_n = 2a_{n-1} + 3a_{n-2}$, $a_1 = 1$, $a_2 = 3$. Find a_9.

*(c) $a_n = 3a_{n-1} + 4a_{n-2}$, $a_1 = 1$, $a_2 = -1$. Find a_{12}.

4. The Fibonacci sequence is defined by the difference equation $a_n = a_{n-1} + a_{n-2}$, with initial conditions $a_1 = 1$, $a_2 = 1$. Solve this difference equation. Show that $\dfrac{a_n}{a_{n-1}}$ approaches the number $\dfrac{1 + \sqrt{5}}{2}$ as n increases, arriving at the golden ratio.

*5. Determine the normal modes of oscillation of the system in Figure 8.10 when it is displaced slightly from equilibrium.

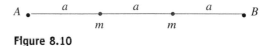

Figure 8.10

*6. The motion of a weight attached to a spring is governed by **Hooke's law:**

$$\text{tension} = k \times \text{extension},$$
where k is a constant of the spring

Consider the oscillations of the spring described in Figure 8.11. x_1 and x_2 are the displacements of weights of masses m_1 and m_2 at any instant. The extensions of the two springs at that instant are x_1 and $x_2 - x_1$. Application of Hooke's law gives the following equations of motion:

$$m_1\ddot{x}_1 = -k_1x_1 + k_2(x_2 - x_1) \quad \text{and}$$

$$m_2\ddot{x}_2 = -k_2(x_2 - x_1)$$

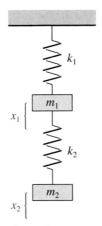

Figure 8.11

The general motion can be analyzed in terms of normal modes as in the example of this section. Analyze the motion when $m_1 = m_2 = M$, $k_1 = 3$, and $k_2 = 2$.

7. Hooke's law also applies to forces in the extended springs of the system shown in Figure 8.12. x_1 and x_2 are the displacements of balls each of mass m at any instant. Application of Hooke's law gives

$$m\ddot{x}_1 = -k_1x_1 + k_2(x_2 - x_1) \quad \text{and}$$

$$m\ddot{x}_2 = -k_2(x_2 - x_1) - k_3x_2$$

Analyze the motion in terms of normal modes if the constants of the springs are $k_1 = 1$, $k_2 = 2$, and $k_3 = 3$.

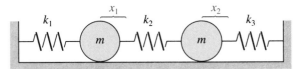

Figure 8.12

Review Exercises Chapter 8

1. Find the coordinate vector of $(-1, 18)$ relative to the basis $\{(1, 3), (-1, 4)\}$.

2. Find the coordinate vector of $3x^2 + 2x - 13$ relative to the basis $\{x^2 + 1, x + 2, x - 3\}$.

3. Find the coordinate vector of $(0, 5, -15)$ relative to the orthonormal basis $\left\{(0, 1, 0), \left(-\frac{3}{5}, 0, \frac{4}{5}\right), \left(\frac{4}{5}, 0, \frac{3}{5}\right)\right\}$.

4. Find the transition matrix P from the following basis B to the standard basis B' of $\mathbf{R}^2$. Use this matrix to find the coordinate vectors of $\mathbf{u}$, $\mathbf{v}$, and $\mathbf{w}$ relative to B'.

$$B = \{(1, 3), (5, 2)\} \quad \text{and} \quad B' = \{(1, 0), (0, 1)\};$$

$$\mathbf{u}_B = \begin{bmatrix} 3 \\ 1 \end{bmatrix}, \quad \mathbf{v}_B = \begin{bmatrix} 5 \\ -2 \end{bmatrix}, \quad \mathbf{w}_B = \begin{bmatrix} 4 \\ 1 \end{bmatrix}$$

5. Find the transition matrix P from the basis $B = \{(-1, 2), (2, 1)\}$ of $\mathbf{R}^2$ to the basis $B' = \{(4, 3), (-3, 2)\}$. If $\mathbf{u}_B = \begin{bmatrix} 4 \\ 1 \end{bmatrix}$, find $\mathbf{u}_{B'}$.

6. Let T be a linear operator on $\mathbf{R}^2$ defined as follows on the standard basis of $\mathbf{R}^2$: $T(1, 0) = (3, 2)$, $T(0, 1) = (-1, 4)$. Find $T(2, 7)$.

7. Let $T: U \to V$ be a linear transformation. Let T be defined relative to bases $\{\mathbf{u}_1, \mathbf{u}_2\}$ and $\{\mathbf{v}_1, \mathbf{v}_2, \mathbf{v}_3\}$ of U and V as follows:

$$T(\mathbf{u}_1) = \mathbf{v}_1 + 5\mathbf{v}_2 - 2\mathbf{v}_3, \qquad T(\mathbf{u}_2) = 3\mathbf{v}_1 - \mathbf{v}_2 + 2\mathbf{v}_3$$

Find the matrix of T with respect to these bases. Use this matrix to find the image of the vector $\mathbf{u} = 2\mathbf{u}_1 - 3\mathbf{u}_2$.

8. Find the matrix of $T(x, y, z) = (2x, -3y)$ relative to the standard bases. Use this matrix to find the image of $(1, 2, 3)$.

9. Find the matrix of $T(ax^2 + bx + c) = (a - b)x^2 + 2cx$ with respect to the basis $\{x^2, x, 1\}$ of P_2. Use the matrix to find the image of $2x^2 - x + 3$.

10. Consider the linear operator $T(x, y) = (3x, x - y)$ on $\mathbf{R}^2$. Find the matrix of T with respect to the standard basis for $\mathbf{R}^2$. Use a similarity transformation to then find the matrix with respect to the basis $\{(1, 2), (2, 3)\}$ of $\mathbf{R}^2$.

11. Transform the matrix $A = \begin{bmatrix} 4 & -2 \\ 1 & 1 \end{bmatrix}$ into a matrix B using the similarity transformation $C^{-1}AC$, with $C = \begin{bmatrix} 2 & 1 \\ 1 & 1 \end{bmatrix}$.

12. Diagonalize the matrix $\begin{bmatrix} 1 & 1 \\ -2 & 4 \end{bmatrix}$ using a similarity transformation. Give the transformation.

13. Diagonalize the symmetric matrix $\begin{bmatrix} 7 & -2 & 1 \\ -2 & 10 & -2 \\ 1 & -2 & 7 \end{bmatrix}$. Give the similarity transformation.

14. Let the matrix $A = \begin{bmatrix} a & b \\ b & c \end{bmatrix}$ represent an arbitrary 2×2 symmetric matrix. Prove that the characteristic equation of A has real roots for all values of a, b, and c.

15. Show that if A is a symmetric matrix having only one eigenvalue λ, then $A = \lambda I$.

16. Sketch the graph of $-x^2 - 16xy + 11y^2 - 30 = 0$.

17. Solve the difference equation $a_n = 4a_{n-1} + 5a_{n-2}$, $a_1 = 3$, $a_2 = 2$, and use the solution to find a_{12}.

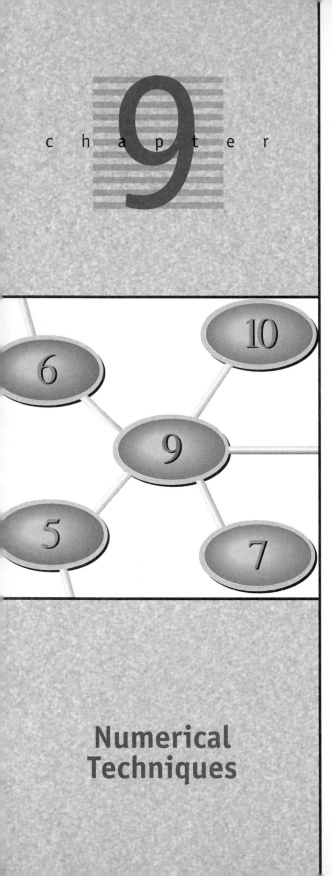

9

Numerical Techniques

In this chapter we look closely at numerical techniques for solving systems of linear equations and for computing eigenvalues and eigenvectors. We add three important methods to our library of techniques of solving systems of linear equations. We discuss the reliability of solutions of linear equations. Certain systems of equations can lead to incorrect results unless great care is taken. We discuss ways of recognizing and solving "delicate" systems.

*9.1 Gaussian Elimination

There are many elimination methods in addition to the method of Gauss-Jordan elimination for solving systems of linear equations. In this section we introduce another elimination method called **Gaussian elimination**. Different methods are suitable for different occasions. It is important to choose the best method for the purpose in mind. We shall discuss the relative merits of Gauss-Jordan elimination and Gaussian elimination. The merits and drawbacks of other methods will be discussed later.

The method of Gaussian elimination involves an *echelon form* of the augmented matrix of the system of equations. An echelon form satisfies the first three of the conditions of the reduced echelon form.

Definition   *A matrix is in **echelon form** if*

1. *Any rows consisting entirely of zeros are grouped at the bottom of the matrix.*

2. *The first nonzero element of each row is 1. This element is called a **leading 1**.*

3. *The leading 1 of each row after the first is positioned to the right of the leading 1 of the previous row.*

   (This implies that all the elements below a leading 1 are zero.)

The following matrices are all in echelon form.

$$\begin{bmatrix} 1 & -1 & 2 \\ 0 & 1 & 2 \\ 0 & 0 & 1 \end{bmatrix} \quad \begin{bmatrix} 1 & 3 & -6 & 4 \\ 0 & 0 & 1 & 3 \\ 0 & 0 & 0 & 0 \end{bmatrix} \quad \begin{bmatrix} 1 & 4 & 6 & 2 & 5 & 2 \\ 0 & 0 & 1 & 2 & 3 & 4 \\ 0 & 0 & 0 & 0 & 1 & 6 \end{bmatrix}$$

The difference between a reduced echelon form and an echelon form is that the elements above and below a leading 1 are zero in a reduced echelon form, while only the elements below the leading 1 need to be zero in an echelon form.

The Gaussian elimination algorithm is as follows.

Gaussian Elimination

1. Write down the augmented matrix of the system of linear equations.

2. Find an echelon form of the augmented matrix using elementary row operations. This is done by creating leading 1s, then zeros below each leading 1, column by column starting with the first column.

3. Write down the system of equations corresponding to the echelon form.

4. Use back substitution to arrive at the solution.

We illustrate the method with the following example.

Example 1   Solve the following system of linear equations using the method of Gaussian elimination.

$$x_1 + 2x_2 + 3x_3 + 2x_4 = -1$$
$$-x_1 - 2x_2 - 2x_3 + x_4 = 2$$
$$2x_1 + 4x_2 + 8x_3 + 12x_4 = 4$$

Solution   Starting with the augmented matrix, create zeros below the pivot in the first column.

$$\begin{bmatrix} ① & 2 & 3 & 2 & -1 \\ -1 & -2 & -2 & 1 & 2 \\ 2 & 4 & 8 & 12 & 4 \end{bmatrix} \underset{\substack{R2 + R1 \\ R3 + (-2)R1}}{\approx} \begin{bmatrix} 1 & 2 & 3 & 2 & -1 \\ 0 & 0 & ① & 3 & 1 \\ 0 & 0 & 2 & 8 & 6 \end{bmatrix}$$

At this stage, we create a zero only *below* the pivot.

$$\underset{R3 + (-2)R2}{\approx} \begin{bmatrix} 1 & 2 & 3 & 2 & -1 \\ 0 & 0 & 1 & 3 & 1 \\ 0 & 0 & 0 & ② & 4 \end{bmatrix} \underset{\frac{1}{2}R3}{\approx} \begin{bmatrix} 1 & 2 & 3 & 2 & -1 \\ 0 & 0 & 1 & 3 & 1 \\ 0 & 0 & 0 & 1 & 2 \end{bmatrix}$$

echelon form

We have arrived at the echelon form.

The corresponding system of equations is

$$x_1 + 2x_2 + 3x_3 + 2x_4 = -1$$
$$x_3 + 3x_4 = 1$$
$$x_4 = 2$$

Observe that the effect of performing the row operations in this manner to arrive at an echelon form is to eliminate variables from equations. This is called **forward elimination**. This system is now solved by **back substitution**. (The terms **forward pass** and **backward pass** are also used.) The value of x_4 is substituted into the second equation to give x_3. x_3 and x_4 are then substituted into the first equation to get x_1. We get

$$x_3 + 3(2) = 1$$
$$x_3 = -5$$

Substituting $x_4 = 2$ and $x_3 = -5$ into the first equation,

$$x_1 + 2x_2 + 3(-5) + 2(2) = -1$$
$$x_1 + 2x_2 = 10$$
$$x_1 = -2x_2 + 10$$

Let $x_2 = r$. The system has many solutions. The solutions are

$$x_1 = -2r + 10, \quad x_2 = r, \quad x_3 = -5, \quad x_4 = 2$$

The forward elimination of variables in this method was performed using matrices and elementary row operations. The back substitution can also be performed using matrices. The final matrix is then the reduced echelon form of the system. This way of performing the back substitution can be implemented on a computer. We illustrate the method for the system of equations of the previous example.

Example 2   Solve the following system of linear equations using the method of Gaussian elimination, performing back substitution using matrices.

$$\begin{aligned} x_1 + 2x_2 + 3x_3 + 2x_4 &= -1 \\ -x_1 - 2x_2 - 2x_3 + x_4 &= 2 \\ 2x_1 + 4x_2 + 8x_3 + 12x_4 &= 4 \end{aligned}$$

Solution   We arrive at the echelon form as in the previous example.

$$\begin{bmatrix} 1 & 2 & 3 & 2 & -1 \\ -1 & -2 & -2 & 1 & 2 \\ 2 & 4 & 8 & 12 & 4 \end{bmatrix} \approx \cdots \approx \begin{bmatrix} 1 & 2 & 3 & 2 & -1 \\ 0 & 0 & 1 & 3 & 1 \\ 0 & 0 & 0 & 1 & 2 \end{bmatrix}$$

echelon form

This marks the end of the forward elimination of variables from equations. We now commence the back substitution using matrices.

$$\begin{bmatrix} 1 & 2 & 3 & 2 & -1 \\ 0 & 0 & 1 & 3 & 1 \\ 0 & 0 & 0 & ① & 2 \end{bmatrix} \underset{\substack{R1 + (-2)R3 \\ R2 + (-3)R3}}{\approx} \begin{bmatrix} 1 & 2 & 3 & 0 & -5 \\ 0 & 0 & ① & 0 & -5 \\ 0 & 0 & 0 & 1 & 2 \end{bmatrix}$$

(Create zeros above the leading 1 in row 3. This is equivalent to substitution for x_4 from equation 3 into equations 1 and 2.)

$$\underset{R1 + (-3)R2}{\approx} \begin{bmatrix} 1 & 2 & 0 & 0 & 10 \\ 0 & 0 & 1 & 0 & -5 \\ 0 & 0 & 0 & 1 & 2 \end{bmatrix}$$

(Create zero above the leading 1 in row 2. This is equivalent to substituting for x_3 from equation 2 into equation 1.)

This matrix is the reduced echelon form of the original augmented matrix. The corresponding system of equations is

$$\begin{aligned} x_1 + 2x_2 \quad\quad &= 10 \\ x_3 \quad &= -5 \\ x_4 &= 2 \end{aligned}$$

Let $x_2 = r$. We get the same solution as previously,

$$x_1 = -2r + 10, \quad x_2 = r, \quad x_3 = -5, \quad x_4 = 2$$

Comparison of Gauss-Jordan and Gaussian Elimination

The method of Gaussian elimination is in general more efficient than Gauss-Jordan elimination in that it involves fewer operations of addition and multiplication. It is during the back substitution that Gaussian elimination picks up this advantage. We now illustrate how Gaussian elimination saves two operations over Gauss-Jordan elimination in the preceding example. Consider the final transformation that brings the matrix to reduced echelon form.

$$\begin{bmatrix} 1 & 2 & 3 & 0 & -5 \\ 0 & 0 & 1 & 0 & -5 \\ 0 & 0 & 0 & 1 & 2 \end{bmatrix} \underset{\text{R1} + (-3)\text{R2}}{\approx} \begin{bmatrix} 1 & 2 & 0 & 0 & 10 \\ 0 & 0 & 1 & 0 & -5 \\ 0 & 0 & 0 & 1 & 2 \end{bmatrix}$$

The aim of this transformation is to create a 0 in location (1, 3). Note that changing the 3 in location (1, 3) to 0 need not in practice involve any arithmetic operations, as one (or the computer) knows in advance that the element is to be zero. The 0 in the (1, 4) location remains unchanged; no arithmetic operations need be performed on it. The row operation R1 + (−3)R2 in fact uses only two arithmetic operations, one of multiplication, and one of addition, in changing the −5 in the (1, 5) location to 10:

$$-5 + (-3)(-5) = 10$$

addition   multiplication

On the other hand, when the zero is created in the (1, 3) location during Gauss-Jordan elimination, the (1, 4) element and the (1, 5) element are both changed, each involving two operations, one addition, the other multiplication. These two operations for the change in the (1, 4) element are two additional operations involved in Gauss-Jordan elimination.

In larger systems of equations, many more operations are saved in Gaussian elimination during back substitution. The reduction in the number of operations not only saves time on a computer but also increases the accuracy of the final answer. With each arithmetic operation, there is a possibility of round-off error on a computer. With large systems, the method of Gauss-Jordan elimination involves approximately 50% more arithmetic operations than does Gaussian elimination (see the following table).

Count of Operations for $n \times n$ System with Unique Solution

	Number of Multiplications	Number of Additions
Gauss-Jordan	$\dfrac{n^3}{2} + \dfrac{n^2}{2} \approx \dfrac{n^3}{2}$ (for large n)	$\dfrac{n^3}{2} - \dfrac{n}{2} \approx \dfrac{n^3}{2}$
Gaussian elimination	$\dfrac{n^3}{3} + n^2 - \dfrac{n}{3} \approx \dfrac{n^3}{3}$	$\dfrac{n^3}{3} + \dfrac{n^2}{2} - \dfrac{5n}{6} \approx \dfrac{n^3}{3}$

Gauss-Jordan elimination, on the other hand, has the advantages of being more straightforward for hand computations. It is superior for solving small systems, and it is the method that we shall use in this course when we solve systems of linear equations by hand.

We complete this section with a discussion of the formulas for the total number of operations involved in solving a system of n linear equations in n variables that has a unique solution, using both Gauss-Jordan elimination and Gaussian elimination. We group multiplications and divisions together as multiplications, and additions and subtractions together as additions. If there are many equations, with n large, then the term with the highest power of n dominates the other terms in the above formulas. The total number of operations with Gauss-Jordan elimination is approximately n^3 while the total in Gaussian elimination is approximately $2n^3/3$. Gaussian elimination is thus approximately 50% more efficient than Gauss-Jordan elimination.

We now derive the above formulas for Gauss-Jordan elimination, leaving it for you to arrive at the formulas for Gaussian elimination in the exercises that follow. Let us denote unknown elements in the matrices by $*$. Assume that there are no row interchanges. Note that when an element is known to become a 1 or a zero, there are no arithmetic operations involved; substitution is used. Thus, for example, there are no operations involved in the location where a leading one is created. Starting with the $n \times (n + 1)$ augmented matrix of the system, we get

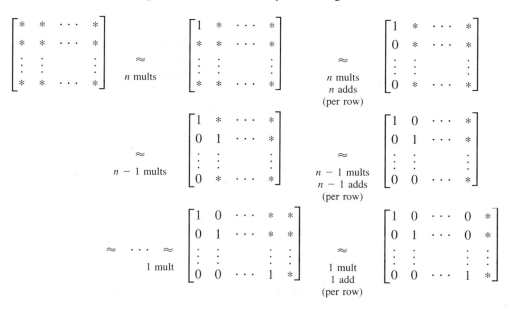

We now sum the above operations. Remember that there are $(n - 1)$ rows involved every time zeros are created "per row" in the above manner. The following formula is used for the sum of the first n integers:

$$[n + (n - 1) + \cdots + 1] = \frac{n(n + 1)}{2}$$

Total number of multiplications

$$= \underbrace{[n + (n - 1) + \cdots + 1]}_{\text{to create leading ones}} + \underbrace{(n - 1)[n + (n - 1) + \cdots + 1]}_{\text{to create zeros}}$$

$$= n[n + (n - 1) + \cdots + 1] = n\left[\frac{n(n + 1)}{2}\right] = \frac{n^3}{2} + \frac{n^2}{2}$$

Total number of additions

$$= (n - 1)[n + (n + 1) + \cdots + 1] = (n - 1)\left[\frac{n(n + 1)}{2}\right] = \frac{n^3}{2} - \frac{n}{2}$$

Exercise Set 9.1

1. Determine whether or not each of the following matrices is in echelon form.

(a) $\begin{bmatrix} 1 & 2 & 1 \\ 0 & 1 & 3 \\ 0 & 0 & 0 \end{bmatrix}$ *(b) $\begin{bmatrix} 1 & 2 & 3 & 4 \\ 0 & 0 & 1 & 0 \\ 0 & 0 & 0 & 1 \end{bmatrix}$

(c) $\begin{bmatrix} 1 & 5 & 6 & 2 \\ 0 & 1 & 0 & 4 \\ 0 & 0 & 1 & 2 \end{bmatrix}$ *(d) $\begin{bmatrix} 1 & 0 & 0 & 3 & 0 \\ 0 & 0 & 1 & 2 & 0 \\ 0 & 0 & 0 & 0 & 1 \end{bmatrix}$

2. Determine whether or not each of the following matrices is in echelon form.

*(a) $\begin{bmatrix} 1 & 2 & 4 & 3 \\ 0 & 0 & 1 & 5 \\ 0 & 0 & 0 & 7 \\ 0 & 0 & 0 & 1 \end{bmatrix}$ *(b) $\begin{bmatrix} 1 & 2 & 4 & 6 \\ 0 & 0 & 1 & 2 \\ 0 & 1 & 3 & 3 \\ 0 & 0 & 0 & 1 \end{bmatrix}$

(c) $\begin{bmatrix} 1 & 3 & 4 & 2 & 3 \\ 0 & 0 & 2 & 5 & 1 \\ 0 & 0 & 0 & 1 & 4 \\ 0 & 0 & 0 & 0 & 6 \end{bmatrix}$ (d) $\begin{bmatrix} 1 & -1 & 4 & 6 \\ 0 & 0 & 1 & 3 \\ 0 & 0 & 1 & 2 \\ 0 & 0 & 0 & 1 \end{bmatrix}$

In Exercises 3–6, solve the systems of equations using Gaussian elimination. (a) Perform the back substitution using equations. (b) Perform the back substitution using matrices.

*3. $\begin{aligned} x_1 + x_2 + x_3 &= 6 \\ x_1 - x_2 + x_3 &= 2 \\ x_1 + 2x_2 + 3x_3 &= 14 \end{aligned}$

4. $\begin{aligned} x_1 - x_2 - x_3 &= 2 \\ x_1 - x_2 + x_3 &= 2 \\ 3x_1 - 2x_2 + x_3 &= 5 \end{aligned}$

*5. $\begin{aligned} x_1 - x_2 + 2x_3 &= 3 \\ 2x_1 - 2x_2 + 5x_3 &= 4 \\ x_1 + 2x_2 - x_3 &= -3 \\ 2x_2 + 2x_3 &= 1 \end{aligned}$

*6. $\begin{aligned} x_1 - x_2 + x_3 + 2x_4 - 2x_5 &= 1 \\ 2x_1 - x_2 - x_3 + 3x_4 - x_5 &= 3 \\ -x_1 - x_2 + 5x_3 \quad\quad - 4x_5 &= -3 \end{aligned}$

*7. Consider a system of four equations in five variables. In general, how many arithmetic operations will be saved by using Gaussian elimination rather than Gauss-Jordan elimination to solve the system? Where are these operations saved?

8. Compare Gaussian elimination to Gauss-Jordan elimination for a system of four equations in six variables. Determine where operations are saved during Gaussian elimination.

*9. Can a matrix have more than one echelon form? (*Hint: Consider a 2×3 matrix and arrive at an echelon form using two distinct sequences of row operations.*)

10. Consider a system of n linear equations in n variables that has a unique solution. Show that Gaussian elimination involves $n^3/3 + n^2/2 + n/6$ multiplications and $n^3/3 - n/3$ additions to arrive at an echelon form, and then a further $n^2/2 - n/2$ multiplications and $n^2/2 - n/2$ additions to arrive at the reduced echelon form. Thus a total of $n^3/3 + n^2 - n/3$ multiplications and $n^3/3 + n^2/2 - 5n/6$ additions are involved in solving the system of equations using Gaussian elimination. (The formula for the sum of squares is $n^2 + (n-1)^2 + \cdots + 1 = [n(n+1)(2n+1)/6]$.)

11. Construct a table that gives the number of multiplications and additions in both Gauss-Jordan elimination and Gaussian elimination for systems of n equations in n variables having unique solutions, for $n = 2$ to to 10.

*9.2 The Method of *LU* Decomposition

The method of *LU* decomposition is used to solve certain systems of linear equations. The method is widely used on computers. It involves writing the matrix of coefficients as the product of a **lower triangular matrix** L and an **upper triangular matrix** U. We remind the reader that a lower triangular matrix is a square matrix with zeros in all locations above the main diagonal while an upper triangular matrix is a square matrix with zeros below the main diagonal. The following are examples of lower and upper triangular matrices.

$$\begin{bmatrix} 2 & 0 & 0 & 0 \\ 3 & -1 & 0 & 0 \\ 5 & 2 & 7 & 0 \\ 4 & 0 & -2 & 8 \end{bmatrix} \qquad \begin{bmatrix} 8 & 2 & 5 & -3 \\ 0 & 2 & 9 & 1 \\ 0 & 0 & -4 & 2 \\ 0 & 0 & 0 & 7 \end{bmatrix}$$

lower triangular matrix   upper triangular matrix

A system of equations that has a triangular matrix of coefficients is particularly straightforward to solve. If the matrix is lower triangular, a method called forward substitution is used. If the matrix is upper triangular, the method of back substitution used in Gaussian elimination is applied. The method of *LU* decomposition involves first solving a lower triangular system of equations, then an upper triangular system. The following examples illustrate the methods of forward and back substitution. We shall then bring these methods together in *LU* decomposition.

Example 1 Solve the following system of equations, which has a lower triangular matrix of coefficients.

$$\begin{aligned} 2x_1 &&&= 8 \\ x_1 + 3x_2 &&&= -2 \\ 4x_1 + 5x_2 &- x_3 &&= 3 \end{aligned}$$

Solution We solve the system by **forward substitution**. The first equation gives the value of x_1. This value is substituted into the second equation to get x_2. These values of x_1 and x_2 are then substituted into the third equation to get x_3.

1st equation gives

$$2x_1 = 8$$
$$x_1 = 4$$

2nd equation gives

$$x_1 + 3x_2 = -2$$
$$4 + 3x_2 = -2$$
$$x_2 = -2$$

3rd equation gives

$$4x_1 + 5x_2 - x_3 = 3$$
$$16 - 10 - x_3 = 3$$
$$x_3 = 3$$

The solution is $x_1 = 4$, $x_2 = -2$, $x_3 = 3$.

Example 2   Solve the following system of equations, which has an upper triangular matrix of coefficients.

$$2x_1 - x_2 + 4x_3 = \ \ 0$$
$$x_2 - \ \ x_3 = \ \ 4$$
$$3x_3 = -6$$

Solution   We use the method of **back substitution**. The third equation gives the value of x_3. This value is substituted into the second equation to get x_2. These values of x_2 and x_3 are then substituted into the first equation to get x_1.

3rd equation gives

$$3x_3 = -6$$
$$x_3 = -2$$

2nd equation gives

$$x_2 - x_3 = 4$$
$$x_2 + 2 = 4$$
$$x_2 = 2$$

1st equation gives

$$2x_1 - x_2 + 4x_3 = 0$$
$$2x_1 - 2 - \ \ 8 = 0$$
$$x_1 = 5$$

The solution is $x_1 = 5$, $x_2 = 2$, $x_3 = -2$.

Forward and back substitution can, of course, be carried out using row operations on matrices. This is the way they are implemented on a computer.

Definition   *Let A be a square matrix that can be factored into the form A = LU, where L is an upper triangular matrix and U is a lower triangular matrix. This factoring is called an **LU decomposition** of A.*

Not every matrix has an *LU* decomposition, and when it exists, it is not unique. (See the following exercises.) The method that we now introduce can be used to solve a system of linear equations if *A* has an *LU* decomposition.

Method of *LU* Decomposition

Let $AX = B$ be a system of *n* equations in *n* variables, where *A* has *LU* decomposition $A = LU$. The system can thus be written

$$LUX = B$$

The method involves writing this system as two subsystems, one of which is lower triangular and the other upper triangular.

$$UX = Y$$
$$LY = B$$

Observe that substituting for *Y* from the first equation into the second gives the original system $LUX = B$. In practice, we first solve $LY = B$ for *Y* and then solve $UX = Y$ to get the solution *X*. We now summarize these steps.

Solution of *AX= B*

1. Find the *LU* decomposition of *A*.
   (If *A* has no *LU* decomposition, the method is not applicable.)
2. Solve $LY = B$ by forward substitution.
3. Solve $UX = Y$ by back substitution.

A key to this method, of course, is being able to arrive at an *LU* decomposition of the matrix of coefficients *A*. Suppose it is possible to transform *A* into an upper triangular form *U* using the sequence of row operations that involve **adding multiples of rows to rows**. The row operation of adding *c* times row *j* to row *k* for an $n \times n$ matrix *A* can be performed by the matrix multiplication *EA*, where *E* is the matrix obtained by changing element i_{kj} of I_n to *c*. *E* is called an **elementary matrix**. Let the row operations be described by elementary matrices $E_1, \ldots, E_k$. Thus

$$E_k \quad \cdots \quad E_1 A = U$$

It is known that each elementary matrix is invertible. Multiply both sides of this equation by $E_1^{-1} \quad \cdots \quad E_k^{-1}$, in succession, to get

$$A = E_1^{-1} \quad \cdots \quad E_k^{-1} U$$

Each such elementary matrix is a lower triangular matrix. It can be shown that the inverse of a lower triangular matrix is lower triangular and that the product of lower triangular matrices is also lower triangular. (See the following exercises.) Let $L = E_1^{-1} \quad \cdots \quad E_k^{-1}$. Thus $A = LU$.

Example 3   Solve the following system of equations using *LU* decomposition.

$$2x_1 + x_2 + 3x_3 = -1$$
$$4x_1 + x_2 + 7x_3 = 5$$
$$-6x_1 - 2x_2 - 12x_3 = -2$$

Solution   Let us transform the matrix of coefficients *A* into upper triangular form *U* by creating zeros below the main diagonal as follows.

$$
\begin{array}{c}
A \\
\begin{bmatrix} 2 & 1 & 3 \\ 4 & 1 & 7 \\ -6 & -2 & -12 \end{bmatrix}
\end{array}
\underset{R2-2R1}{\approx}
\begin{bmatrix} 2 & 1 & 3 \\ 0 & -1 & 1 \\ -6 & -2 & -12 \end{bmatrix}
$$

$$
\underset{R3+3R1}{\approx}
\begin{bmatrix} 2 & 1 & 3 \\ 0 & -1 & 1 \\ 0 & 1 & -3 \end{bmatrix}
\underset{R3+R2}{\approx}
\begin{array}{c} U \\ \begin{bmatrix} 2 & 1 & 3 \\ 0 & -1 & 1 \\ 0 & 0 & -2 \end{bmatrix} \end{array}
$$

The elementary matrices that correspond to these row operations are

$$
\begin{array}{ccc}
R2-2R1 & R3+3R1 & R3+R2 \\
E_1 = \begin{bmatrix} 1 & 0 & 0 \\ -2 & 1 & 0 \\ 0 & 0 & 1 \end{bmatrix}, & E_2 = \begin{bmatrix} 1 & 0 & 0 \\ 0 & 1 & 0 \\ 3 & 0 & 1 \end{bmatrix}, & E_3 = \begin{bmatrix} 1 & 0 & 0 \\ 0 & 1 & 0 \\ 0 & 1 & 1 \end{bmatrix}
\end{array}
$$

The inverses of these matrices are

$$
E_1^{-1} = \begin{bmatrix} 1 & 0 & 0 \\ 2 & 1 & 0 \\ 0 & 0 & 1 \end{bmatrix}, \quad E_2^{-1} = \begin{bmatrix} 1 & 0 & 0 \\ 0 & 1 & 0 \\ -3 & 0 & 1 \end{bmatrix}, \quad E_3^{-1} = \begin{bmatrix} 1 & 0 & 0 \\ 0 & 1 & 0 \\ 0 & -1 & 1 \end{bmatrix}
$$

We get

$$
L = E_1^{-1}E_2^{-1}E_3^{-1} = \begin{bmatrix} 1 & 0 & 0 \\ 2 & 1 & 0 \\ -3 & -1 & 1 \end{bmatrix}
$$

Thus

$$
A = \begin{array}{c} L \\ \begin{bmatrix} 1 & 0 & 0 \\ 2 & 1 & 0 \\ -3 & -1 & 1 \end{bmatrix} \end{array}
\begin{array}{c} U \\ \begin{bmatrix} 2 & 1 & 3 \\ 0 & -1 & 1 \\ 0 & 0 & -2 \end{bmatrix} \end{array}
$$

We now solve the given system *LUX* = *B* by solving the two subsystems *LY* = *B* and *UX* = *Y.* We get

$$LY = B: \quad \begin{bmatrix} 1 & 0 & 0 \\ 2 & 1 & 0 \\ -3 & -1 & 1 \end{bmatrix} \begin{bmatrix} y_1 \\ y_2 \\ y_3 \end{bmatrix} = \begin{bmatrix} -1 \\ 5 \\ -2 \end{bmatrix}$$

This lower triangular system has solution $y_1 = -1$, $y_2 = 7$, $y_3 = 2$.

$$UX = Y: \quad \begin{bmatrix} 2 & 1 & 3 \\ 0 & -1 & 1 \\ 0 & 0 & -2 \end{bmatrix} \begin{bmatrix} x_1 \\ x_2 \\ x_3 \end{bmatrix} = \begin{bmatrix} 1 \\ 7 \\ 2 \end{bmatrix}$$

This upper triangular system has solution $x_1 = 5$, $x_2 = -8$, $x_3 = -1$.
The solution to the given system is $x_1 = 5$, $x_2 = -8$, $x_3 = -1$.

A careful analysis of the above example leads to the following method of arriving at an *LU* decomposition directly from a knowledge of the row operations that lead to an upper triangular form.

Construction of an *LU* Decomposition of *A*

1. Use row operations to arrive at *U*.
   (The operations must involve adding multiples of rows to rows. In general, if row interchanges are required to arrive at *U*, an *LU* form does not exist.)

2. The diagonal elements of *L* are 1s.
   The nonzero elements of *L* correspond to row operations.
   The row operation $R_i + cR_j$ implies that $l_{ij} = -c$.

We now give a second example to reinforce the understanding of this method of arriving at *L* and *U*.

Example 4   Solve the following system of equations using *LU* decomposition.

$$\begin{aligned} x_1 - 3x_2 + 4x_3 &= 12 \\ -x_1 + 5x_2 - 3x_3 &= -12 \\ 4x_1 - 8x_2 + 23x_3 &= 58 \end{aligned}$$

Solution   We transform the matrix of coefficients *A* into upper triangular form *U* by creating zeros below the main diagonal.

$$\begin{bmatrix} 1 & -3 & 4 \\ -1 & 5 & -3 \\ 4 & -8 & 23 \end{bmatrix} \underset{\substack{R2 + R1 \\ R3 - 4R1}}{\approx} \begin{bmatrix} 1 & -3 & 4 \\ 0 & 2 & 1 \\ 0 & 4 & 7 \end{bmatrix} \underset{R3 - 2R2}{\approx} \begin{bmatrix} 1 & -3 & 4 \\ 0 & 2 & 1 \\ 0 & 0 & 5 \end{bmatrix}$$

These row operations lead to the following *LU* decomposition of *A*.

$$A = \begin{bmatrix} 1 & 0 & 0 \\ -1 & 1 & 0 \\ 4 & 2 & 1 \end{bmatrix} \begin{bmatrix} 1 & -3 & 4 \\ 0 & 2 & 1 \\ 0 & 0 & 5 \end{bmatrix}$$

We again solve the given system $LUX = B$ by solving the two subsystems $LY = B$ and $UX = Y$.

$$LY = B: \quad \begin{bmatrix} 1 & 0 & 0 \\ -1 & 1 & 0 \\ 4 & 2 & 1 \end{bmatrix} \begin{bmatrix} y_1 \\ y_2 \\ y_3 \end{bmatrix} = \begin{bmatrix} 12 \\ -12 \\ 58 \end{bmatrix}$$

This lower triangular system has solution $y_1 = 12$, $y_2 = 0$, $y_3 = 10$.

$$UX = Y: \quad \begin{bmatrix} 1 & -3 & 4 \\ 0 & 2 & 1 \\ 0 & 0 & 5 \end{bmatrix} \begin{bmatrix} x_1 \\ x_2 \\ x_3 \end{bmatrix} = \begin{bmatrix} 12 \\ 0 \\ 10 \end{bmatrix}$$

This upper triangular system has solution $x_1 = 1$, $x_2 = -1$, $x_3 = 2$.

The solution to the given system is $x_1 = 1$, $x_2 = -1$, $x_3 = 2$.

The method of *LU* decomposition can be applied to any system of n equations in n variables, $AX = B$, that can be transformed into an upper triangular form U using row operations that involve adding multiples of rows to rows. In general, if row interchanges are required to arrive at the upper triangular form, then the matrix does not have an *LU* decomposition and the method cannot be used.

The total number of arithmetic operations needed to solve a system of equations using *LU* decomposition is exactly the same as that needed in Gaussian elimination. The method is the most efficient method available, however, for solving many systems, all having the same matrix of coefficients, when the right-hand sides are not all available in advance. In that situation, the factorization has to be done only once, using $(4n^3 - 3n^2 - n)/6$ arithmetic operations. The solutions of the two triangular systems can then be carried out in $2n^2 - n$ arithmetic operations per system, as compared to $(4n^3 + 9n^2 - 7n)/6$ arithmetic operations for full Gaussian elimination each time.

The fact that *LU* decomposition is used to solve systems of equations on computers where applicable, rather than Gaussian elimination, is a result of the usefulness of *LU* decomposition of a matrix for many types of computations. If $A = LU$, then

$$A^{-1} = (LU)^{-1} = U^{-1}L^{-1} \quad \text{and} \quad |A| = |LU| = |L||U|$$

The inverse of a triangular matrix can be computed very efficiently, and the determinant of a triangular matrix is the product of its diagonal elements. In practice, once one of these matrix functions has been computed using L and U, these matrices are then available for use in other matrix functions. For example, if $|A|$ has been computed, then L and U are known, providing a very efficient starting place for solving the system $AX = B$.

The matrix software package MATLAB (discussed in this text), for example, uses *LU* decomposition extensively. MATLAB is one of the most powerful scientific/engineering numerical linear algebra packages available.

Exercise Set 9.2

In Exercises 1–3, solve the lower triangular systems.

*1.
$$\begin{aligned} x_1 \quad &= 1 \\ 2x_1 - x_2 \quad &= -2 \\ 3x_1 + x_2 - x_3 &= 8 \end{aligned}$$

2.
$$\begin{aligned} 2x_1 \quad &= 4 \\ -x_1 + x_3 \quad &= 1 \\ x_1 + x_2 + x_3 &= 5 \end{aligned}$$

*3.
$$\begin{aligned} x_1 \quad &= -2 \\ 3x_1 + x_2 \quad &= -5 \\ x_1 + 4x_2 + 2x_3 &= 4 \end{aligned}$$

In Exercises 4–6, solve the upper triangular systems.

*4.
$$\begin{aligned} x_1 + x_2 + x_3 &= 3 \\ 2x_2 + x_3 &= 3 \\ 3x_3 &= 6 \end{aligned}$$

5.
$$\begin{aligned} 2x_1 - x_2 + x_3 &= 3 \\ x_2 - 3x_3 &= -7 \\ x_3 &= 3 \end{aligned}$$

*6.
$$\begin{aligned} 3x_1 + 2x_2 - x_3 &= -6 \\ 2x_2 + 4x_3 &= 26 \\ x_3 &= 7 \end{aligned}$$

7. Let $A = LU$. The following sequences of transformations were used to arrive at U. Find L in each case.

*(a) $R_2 - R_1,\ R_3 + R_1,\ R_3 + R_2$

(b) $R_2 + 5R_1,\ R_3 - 2R_1,\ R_3 + 7R_2$

*(c) $R_2 + \frac{1}{2}R_1,\ R_3 + \frac{1}{5}R_1,\ R_3 - R_2$

(d) $R_2 + 4R_1,\ R_3 - 2R_1,\ R_4 - 8R_1,\ R_3 - 2R_2,$
$R_4 + 2R_2,\ R_4 - 6R_3$

8. Let $A = LU$. Determine the row operations used to arrive at U for each of the following cases of L.

*(a) $\begin{bmatrix} 1 & 0 & 0 \\ 2 & 1 & 0 \\ -3 & 5 & 1 \end{bmatrix}$
(b) $\begin{bmatrix} 1 & 0 & 0 \\ -1 & 1 & 0 \\ 4 & 2 & 1 \end{bmatrix}$

*(c) $\begin{bmatrix} 1 & 0 & 0 \\ 0 & 1 & 0 \\ \frac{1}{7} & -\frac{3}{4} & 1 \end{bmatrix}$
(d) $\begin{bmatrix} 1 & 0 & 0 & 0 \\ -2 & 1 & 0 & 0 \\ 3 & 5 & 1 & 0 \\ 6 & 0 & -3 & 1 \end{bmatrix}$

In Exercises 9–16, solve the systems using the method of LU decomposition.

*9.
$$\begin{aligned} x_1 + 2x_2 - x_3 &= 2 \\ -2x_1 - x_2 + 3x_3 &= 3 \\ x_1 - x_2 - 4x_3 &= -7 \end{aligned}$$

10.
$$\begin{aligned} 2x_1 + x_2 \quad &= 9 \\ 6x_1 + 4x_2 - x_3 &= 25 \\ 2x_1 \quad + 4x_3 &= 20 \end{aligned}$$

*11.
$$\begin{aligned} 3x_1 - x_2 + x_3 &= 10 \\ -3x_1 + 2x_2 + x_3 &= -8 \\ 9x_1 + 5x_2 - 3x_3 &= 24 \end{aligned}$$

12.
$$\begin{aligned} x_1 + 2x_2 + 3x_3 &= -5 \\ 2x_1 + 8x_2 + 7x_3 &= -9 \\ -x_1 + 14x_2 - x_3 &= 15 \end{aligned}$$

13.
$$\begin{aligned} 4x_1 + x_2 + 2x_3 &= 18 \\ -12x_1 - 4x_2 - 3x_3 &= -56 \\ - 5x_2 + 16x_3 &= -10 \end{aligned}$$

*14.
$$\begin{aligned} -2x_1 \quad + 3x_3 &= 3 \\ -4x_1 + 3x_2 + 5x_3 &= 11 \\ 8x_1 + 9x_2 - 11x_3 &= 7 \end{aligned}$$

15.
$$\begin{aligned} -2x_1 \quad + 3x_3 &= 14 \\ -4x_1 + 3x_2 + 5x_3 &= 30 \\ 8x_1 + 9x_2 - 11x_3 &= -34 \end{aligned}$$

*16.
$$\begin{aligned} 2x_1 - 3x_2 + x_3 &= -5 \\ 4x_1 - 5x_2 + 6x_3 &= 2 \\ -10x_1 + 19x_2 + 9x_3 &= 55 \end{aligned}$$

In Exercises 17–19, solve the systems using the method of LU decomposition.

17.
$$\begin{aligned} 4x_1 + x_2 + 3x_3 &= 11 \\ 12x_1 + x_2 + 10x_3 &= 28 \\ -8x_1 - 16x_2 + 6x_3 &= -62 \end{aligned}$$

*18.
$$\begin{aligned} x_1 + 2x_2 - x_3 &= 2 \\ 2x_1 + 5x_2 + x_3 &= 3 \\ -x_1 - x_2 + 4x_3 &= -3 \end{aligned}$$

*19.
$$\begin{aligned} 4x_1 + x_2 - 2x_3 &= 3 \\ -4x_1 + 2x_2 + 3x_3 &= 1 \\ 8x_1 - 7x_2 - 7x_3 &= -2 \end{aligned}$$

In Exercises 20 and 21, solve the systems using the method of LU decomposition.

*20.
$$\begin{aligned} x_1 + x_2 - x_3 + 2x_4 &= 7 \\ x_1 + 3x_2 + 2x_3 + 2x_4 &= 6 \\ -x_1 - 3x_2 - 4x_3 + 6x_4 &= 12 \\ 4x_2 + 7x_3 - 2x_4 &= -7 \end{aligned}$$

21.
$$2x_1 \quad\;\; + \;\; x_3 - \;\; x_4 = 6$$
$$6x_1 + 3x_2 + 2x_3 - \;\; x_4 = 15$$
$$4x_1 + 3x_2 - 2x_3 + 3x_4 = 3$$
$$-2x_1 - 6x_2 + 2x_3 - 14x_4 = 12$$

22. Prove that the inverse (if it exists) of a lower triangular matrix is also lower triangular.

23. Prove that the product of two lower triangular matrices is lower triangular.

*24. Show that an *LU* decomposition of a matrix (if one exists) is not unique by finding two distinct decompositions of the matrix $A = \begin{bmatrix} 6 & -2 \\ 12 & 8 \end{bmatrix}$.

25. Consider the matrix $E = \begin{bmatrix} 1 & 0 & 0 \\ 0 & 0 & 1 \\ 0 & 1 & 0 \end{bmatrix}$ that defines the interchange of rows 2 and 3 of a 2 × 3 matrix. Show that E has no *LU* decomposition. This example demonstrates that not every matrix can be written in the form *LU*.

*26. Let $AX = B$ be a system where $A = LU$ and A is a 3 × 3 matrix. The formula $(4n^3 - 3n^2 - n)/6$ for the total number of arithmetic operations in arriving at L and U gives 13 operations. How many of these operations are used to compute L and how many to compute U? The formula $2n^2 - n$ for computing the total number of operations then needed to solve $AX = B$ gives 15. How many of these operations are needed to solve $LY = B$ and how many to solve $UX = Y$?

**9.3 Practical Difficulties in Solving Systems of Equations

In this course we have discussed the use of elimination methods such as Gauss-Jordan, Gauss, and *LU* decomposition for solving systems of linear equations, and we have seen applications of linear systems. In practice, the elements of the matrix of coefficients A, and the matrix of constants B, in a system of equations $AX = B$, often arise from measurements and are not known exactly. Small errors in the elements of these matrices can cause large errors in the solution, leading to very inaccurate results. In this section we look into ways of assessing such effects and ways of minimizing them.

The Condition Number of a Matrix

Nonsingular matrices have played a major role in our discussions in this course. The time has now come to look closely at their behavior, especially at the way it effects systems of equations. Like many things in life, there are good nonsingular matrices and bad ones. The following definition enables us to "sort out the good from the bad."

Definition   *Let A be a nonsingular matrix. The **condition number** of A is denoted c(A) and is defined*

$$c(A) = \|A\|\|A^{-1}\|$$

*If c(A) is small, then the matrix A is said to be **well-conditioned**. If c(A) is large, then A is **ill-conditioned**.*

Let us now see how $c(A)$ can be used to indicate the accuracy of the solution of a system of equations $AX = B$. Suppose $AX = B$ describes a given experiment where the elements of A and B arise from measurements. Such data depend on the accuracy of instruments and are rarely exact. Let small errors in A be represented by a matrix

E and the corresponding errors in X be represented by e. Thus $(A + E)(X + e) = B$. If we select appropriate norms for the vectors and matrices, then it can be shown that

$$\frac{\|e\|}{\|X + e\|} \le c(A)\frac{\|E\|}{\|A\|}$$

Thus if $c(A)$ is small, small errors in A can produce only small errors in X, and the results are reasonably accurate. The system of equations is said to be **well-behaved**. On the other hand, if $c(A)$ is large, there is the possibility that small errors in A can result in large errors in X, leading to inaccurate results. Such a system of equations is **ill-conditioned**. (Note that a large value of $c(A)$ is a warning, not a guarantee of a large error in the solution.)

If I is an identity matrix, then

$$c(I) = 1 \quad \text{and} \quad c(A) \ge 1$$

(See the following exercises.) Thus 1 is a lower bound for condition numbers. We can intuitively think of the system $IX = B$ as being a best-behaved system. The smaller $c(A)$ is, then in some sense, the closer A is to I and the better behaved the system $AX = B$ becomes. On the other end of the scale, the larger $c(A)$ is, the closer A is to being singular, and problems can occur.

The actual value of $c(A)$ will, of course, depend upon the norm used for A. A norm that is commonly used is the so-called 1-norm:

$$\|A\| = \max\{|a_{1j}| + \cdots + |a_{nj}|\} \text{ for } j = 1, \ldots, n$$

This norm is the largest number resulting from adding the absolute values of elements in each column. Some of the other norms are more reliable, but less efficient. This norm is a good compromise of reliability and efficiency.

A natural question to ask is, "How large is large for a condition number?" If $c(A)$ is written in the form $c(A) \approx 0.d \times 10^k$, then the above inequality implies that

$$\frac{\|e\|}{\|X + e\|} \le 10^k\frac{\|E\|}{\|A\|}$$

This inequality gives the following rule of thumb:

If $c(A) \approx 0.d \times 10^k$, then the components of X can usually be expected to have k fewer significant digits of accuracy than the elements of A.

Thus, for example, if $c(A) = 100 = 0.1 \times 10^3$, and the elements of A are known to five-digit precision, the components of X may have only two-digit accuracy. Much therefore depends upon the accuracy of the measurements and the desired accuracy of the solution. $c(A) = 100$ would, however, be considered large by any standard.

A similar relationship involving $c(A)$ exists between errors in the matrix of constants B and the resulting errors in the solution X. We clarify these ideas in the following example.

Example 1   The following system of equations describes a certain experiment. Let us show that this system is sensitive to changes in the coefficients and the constants on the right.

$$34.9x_1 + 23.6x_2 = 234$$
$$22.9x_1 + 15.6x_2 = 154$$

The exact solution is $x_1 = 4$, $x_2 = 4$. Let us compute the condition number. The matrix of coefficients and its inverse are

$$A = \begin{bmatrix} 34.9 & 23.6 \\ 22.9 & 15.6 \end{bmatrix}, \qquad A^{-1} = \begin{bmatrix} 3.9 & -5.9 \\ -5.725 & 8.725 \end{bmatrix}$$

We get

$$\|A\| = \max\{(34.9 + 22.9), (23.6 + 15.6)\} = \max\{57.8, 39.2\} = 57.8$$
$$\|A^{-1}\| = \max\{(3.9 + 5.725), (5.9 + 8.725)\} = \max\{9.625, 14.625\} = 14.625$$

Thus

$$c(A) = \|A\|\|A^{-1}\| = 57.8 \times 14.625 = 845.325$$

This number is very large. The system is ill-conditioned and the solution may not be reliable.

Let us change the coefficient of x_1 in the first equation from 34.9 to 34.8. The new system has solution $x_1 = 6.5574$, $x_2 = 0.2459$! The system is also sensitive to small changes in the constant terms. For example, changing the first constant term from 234 to 235 gives a new solution of $x_1 = 7.9$, $x_2 = -1.725$.

What does one do in a case like this to get meaningful results? If possible, the problem should be reformulated in terms of a well-defined system of equations. If this is not possible, then one should attempt to derive more accurate data. With more precise data, the system should be solved on a computer, using double precision arithmetic, applying techniques that minimize errors that occur during computation (called round-off errors). One would then have more faith in the results. We shall introduce scaling and pivoting techniques for reducing round-off errors later in this section.

Example 2   Find the condition numbers of the following matrices. Decide whether a system of linear equations defined by such a matrix of coefficients is well-behaved.

$$\text{(a) } A = \begin{bmatrix} 1 & 1 & -1 \\ 4 & 0 & 1 \\ 0 & 4 & 1 \end{bmatrix} \qquad \text{(b) } A = \begin{bmatrix} 1 & 1 & 1 \\ 1 & 2 & 4 \\ 1 & 3 & 9 \end{bmatrix}$$

Solution   (a)   The inverse is found to be

$$A^{-1} = \begin{bmatrix} \frac{1}{6} & \frac{5}{24} & -\frac{1}{24} \\ \frac{1}{6} & -\frac{1}{24} & \frac{5}{24} \\ -\frac{2}{3} & \frac{1}{6} & \frac{1}{6} \end{bmatrix}$$

We get

$$\|A\| = \max\{(1 + 4 + 0), (1 + 0 + 4), (1 + 1 + 1)\} = 5$$
$$\|A^{-1}\| = \max\left\{\left(\tfrac{1}{6} + \tfrac{1}{6} + \tfrac{2}{3}\right), \left(\tfrac{5}{24} + \tfrac{1}{24} + \tfrac{1}{6}\right), \left(\tfrac{1}{24} + \tfrac{5}{24} + \tfrac{1}{6}\right)\right\} = 1$$

Thus $c(A) = \|A\|\|A^{-1}\| = 5$. This value is considered small and the system thus well-behaved.

This system of equations arose in determining the currents through the electrical network in Example 2, Section 1.4. The values of the currents—namely $I_1 = 1$, $I_2 = 3$, $I_3 = 4$ amps—can be considered reliable.

(b) This example is of an ill-conditioned system, and illustrates the possibility of constructing an alternative system that is not ill-conditioned. The inverse of A is found.

$$A^{-1} = \begin{bmatrix} 3 & -3 & 1 \\ -2.5 & 4 & -1.5 \\ 0.5 & -1 & 0.5 \end{bmatrix}$$

We get

$$\|A\| = \max\{(1 + 1 + 1), (1 + 2 + 3), (1 + 4 + 9)\} = 14$$
$$\|A^{-1}\| = \max\{(3 + 2.5 + 0.5), (3 + 4 + 1), (1 + 1.5 + 0.5)\} = 8$$

Thus

$$c(A) = \|A\|\|A^{-1}\| = 112$$

This is a large number. The system is ill-conditioned.

This matrix of coefficients arose in finding the equation of a polynomial of degree two through three given points. We arrived at the matrix for a system that led to the coefficients of a polynomial of degree two through the points $(1, 6)$, $(2, 3)$, and $(3, 2)$ in Example 1, Section 1.4. The solution gave the coefficients for the polynomial $y = 11 - 6x + x^2$. How reliable is this equation? In this system the elements of the matrix of coefficients A emerge from the values of $x = 1$, $x = 2$, and $x = 3$; they are often completely accurate. Errors in the solution of $AX = B$ then arise solely from the components of B. Since $c(A) = 112 \approx 0.1 \times 10^3$, the solution could be three digits less accurate than the y values of the points. The equation is not reliable unless the y data have been measured with great accuracy.

This example is one in an important class of problems, namely that of finding polynomials through data points. The matrix of coefficients A is a **Vandermonde matrix**. Vandermonde matrices are square matrices of the following form. They are famously ill-conditioned.

$$\begin{bmatrix} 1 & a & a^2 & a^3 & \cdots \\ 1 & b & b^2 & b^3 & \cdots \\ 1 & c & c^2 & c^3 & \cdots \\ \vdots & \vdots & \vdots & \vdots & \ddots \end{bmatrix}$$

The matrix of coefficients in a curve-fitting problem such as this will be a Vandermonde matrix unless one is careful. We can prevent a Vandermonde ill-conditioned system from arising in the first place by using data that is not equispaced. The same applies to least square curve fitting. If the base points are equispaced, the resulting least squares equation is extremely sensitive to the y data. An important rule in curve fitting is to avoid using equispaced data if at all possible; otherwise very accurate data are needed to get reliable equations.

We now come to the second half of this discussion on errors in systems of linear equations. We discuss ways of reducing error during computation.

Pivoting and Scaling Techniques

When the condition number is small, Gauss-Jordan elimination yields accurate answers, even for large systems. However, with larger condition numbers, we have to take precautions. When calculations are performed, usually on a computer, only a finite number of digits can be carried; rounding numbers leads to errors that are called **round-off errors**. As computation proceeds in Gauss-Jordan elimination, these small inaccuracies produce matrices that would be encountered in using exact arithmetic on a problem with a slightly altered augmented matrix. Thus ill-conditioned systems are very sensitive to such errors. We now introduce mathematical techniques that can minimize round-off errors during computation.

The method of Gauss-Jordan elimination involves using the first equation in the system to eliminate the first variable from the other equations, and so on. In general, at the rth stage, the rth equation is used to eliminate the rth variable from all other equations. However, we are not restricted to using the rth equation to eliminate the rth variable from the other equations. Any later equation that contains the rth variable can be used to accomplish this. The coefficient of the rth variable in that equation can be taken as **pivot**.

We now introduce freedom in selection of pivots. The choice of pivots is very important in obtaining accurate numerical solutions. The following example illustrates how crucial the choice of pivots can be.

Example 3   Solve the following system of equations, working to three significant figures.

$$10^{-3}x_1 - x_2 = 1$$
$$2x_1 + x_2 = 0$$

Solution   Let us solve this system in two ways, using two selections of pivots. The augmented matrix of the system is

$$\begin{bmatrix} 10^{-3} & -1 & 1 \\ 2 & 1 & 0 \end{bmatrix}$$

Method 1 Select the (1, 1) element as pivot. (Circle the pivot at each stage.) We get

$$
\begin{bmatrix} \boxed{10^{-3}} & -1 & 1 \\ 2 & 1 & 0 \end{bmatrix}
\underset{(1/10^{-3})R1}{\approx}
\begin{bmatrix} \textcircled{1} & -1000 & 1000 \\ 2 & 1 & 0 \end{bmatrix}
$$

$$
\underset{R2 + (-2)R1}{\approx}
\begin{bmatrix} 1 & -1000 & 1000 \\ 0 & \textcircled{2000} & -2000 \end{bmatrix}
$$

$\uparrow$

This element should have been 2001, but has been rounded to three significant figures.

$$
\underset{(1/2000)R2}{\approx}
\begin{bmatrix} 1 & -1000 & 1000 \\ 0 & \textcircled{1} & -1 \end{bmatrix}
$$

$$
\underset{R1 + (1000)R2}{\approx}
\begin{bmatrix} 1 & 0 & 0 \\ 0 & 1 & -1 \end{bmatrix}
$$

The solution is $x_1 = 0$, $x_2 = -1$.

Method 2 Select the (2, 1) element as pivot. When a pivot such as this is selected, interchange rows if necessary to bring the pivot into the customary location for creating zeros in that particular column. We get

$$
\begin{bmatrix} 10^{-3} & -1 & 1 \\ \textcircled{2} & 1 & 0 \end{bmatrix}
\underset{R1 \leftrightarrow R2}{\approx}
\begin{bmatrix} \textcircled{2} & 1 & 0 \\ 10^{-3} & -1 & 1 \end{bmatrix}
$$

$$
\underset{(1/2)R1}{\approx}
\begin{bmatrix} \textcircled{1} & \frac{1}{2} & 0 \\ 10^{-3} & -1 & 1 \end{bmatrix}
$$

$$
\underset{R2 + (\ 10^{-3})R1}{\approx}
\begin{bmatrix} 1 & \frac{1}{2} & 0 \\ 0 & \boxed{-1} & 1 \end{bmatrix}
$$

$$
\underset{(-1)R2}{\approx}
\begin{bmatrix} 1 & \frac{1}{2} & 0 \\ 0 & \textcircled{1} & -1 \end{bmatrix}
$$

$$
\underset{R1 + (-1/2)R2}{\approx}
\begin{bmatrix} 1 & 0 & \frac{1}{2} \\ 0 & 1 & -1 \end{bmatrix}
$$

The solution obtained by pivoting in this manner is $x_1 = \frac{1}{2}$, $x_2 = -1$.

The exact solution to this system can be obtained by applying Gauss-Jordan elimination in the standard way, to full accuracy. It is $x_1 = \frac{1000}{2001}$, $x_2 = -\frac{2000}{2001}$. Observe that the second solution, obtained by pivoting on the (2, 1) element, is a reasonably good approximation to the exact solution; the first solution, obtained by pivoting on the (1, 1) element, is not.

Having seen that the choice of pivots is important when solving a system of linear equations, let us now use the above example to decide how pivots can be selected to improve accuracy. Why was the selection of the (2, 1) element as pivot a better choice than the selection of the (1, 1) element?

Let us return to look at the matrix prior to the reduced echelon form in Method 1. The matrix is

$$\begin{bmatrix} 1 & -1000 & 1000 \\ 0 & 1 & -1 \end{bmatrix}$$

The corresponding system of equations is

$$x_1 - 1000x_2 = 1000$$
$$x_2 = -1$$

Substituting for x_2 from the second equation into the first gives

$$x_1 - 1000(-1) = 1000$$
$$x_1 + 1000 = 1000$$

Since we are working to three significant figures, observe that $x_1 = 4$ satisfies this equation. $x_1 = -3$ also does; in fact, there are many possible values of x_1. When we work to three significant figures, the term x_1 is overpowered by the adjoining 1000 term. The introduction of the large numbers, -1000 and 1000, during the first transformation, when pivoting on the (1, 1) element, leads to this inaccuracy in x_1. Observe that in Method 2, when pivoting on the (2, 1) element, no such large numbers are introduced and the solution obtained is more accurate.

As further evidence that instability in Method 1 was caused during the first transformation, note that the condition number of the original system is 3 while the condition number of the system following the transformation is 500.75. The condition numbers of all the systems in Method 2 are small.

This discussion illustrates that the introduction of large numbers during the elimination process can cause round-off errors. The large numbers in Method 1 resulted from the choice of a small number, namely 10^{-3}, as pivot; dividing by such a small number introduces a large number. Thus the rule of thumb in selection of pivots is to choose the largest number available as pivot.

Furthermore, the above example illustrates that errors can be severe if there are large differences in magnitude between the numbers involved. The final inaccuracy in x_1 resulted from the large difference in magnitude between the coefficient of x_1, namely 1, and the coefficient of x_2, namely 1000. An initial procedure known as scaling can often be used to make all the elements of the augmented matrix of comparable magnitudes, so that one starts off on the best footing possible.

We now give the procedures that are usually used to minimize these causes of errors. It should be stressed that these procedures work most of the time and are the ones most widely adopted. There are, however, exceptions—situations where these procedures do not work. We shall not discuss them here.

Pivoting and Scaling Procedures

Scaling is an attempt to make the elements of the augmented matrix as uniform in magnitude as possible. Two operations are used in scaling a system of equations:

1.   Multiplying an equation throughout by a nonzero constant.

2.   Replacing any variable by a new variable that is a multiple of the original variable.

In practice, one attempts to scale the system so that the largest element in each row and column of the augmented matrix is of order unity. There is no automatic method of scaling.

Pivots are then selected, eliminating variables in the natural order, $x_1, x_2, \ldots, x_n$. At the rth stage the pivot is generally taken to be the coefficient of x_r in the rth or later equation that has the largest absolute value. In terms of elementary matrix transformations, we scan the relevant column from the rth location down, looking for the number that has the largest absolute value. Rows are then interchanged, if necessary, to bring the pivot into the correct location. This procedure ensures that the largest number available is selected as pivot.

We now give an example to illustrate these techniques.

Example 4   Solve the following system of equations using pivoting and scaling, working to five significant figures.

$$
\begin{aligned}
0.002x_1 + \quad\quad 4x_2 - 2x_3 &= \quad 1 \\
0.001x_1 + 2.0001x_2 + \quad x_3 &= \quad 2 \\
0.001x_1 + \quad\quad 3x_2 + 3x_3 &= -1
\end{aligned}
$$

Solution   First scale the system by letting

$$y_1 = 0.001x_1$$

The coefficients in the first column will then be of similar magnitude to the other coefficients. To make the notation uniform, let $y_2 = x_2$ and $y_3 = x_3$. We now have the properly scaled system of equations

$$
\begin{aligned}
2y_1 + \quad\quad 4y_2 - 2y_3 &= \quad 1 \\
y_1 + 2.0001y_2 + \quad y_3 &= \quad 2 \\
y_1 + \quad\quad 3y_2 + 3y_3 &= -1
\end{aligned}
$$

The condition number of the original system was 27,096. The condition number of this system is 32.6557. Things have improved!

Proceed now using pivoting, as follows:

$$
\begin{bmatrix} 2 & 4 & -2 & 1 \\ 1 & 2.0001 & 1 & 2 \\ 1 & 3 & 3 & -1 \end{bmatrix}
\quad \underset{(1/2)\text{R1}}{\approx} \quad
\begin{bmatrix} ① & 2 & -1 & 0.5 \\ 1 & 2.0001 & 1 & 2 \\ 1 & 3 & 3 & -1 \end{bmatrix}
$$

$$
\underset{\substack{\text{R2} + (-1)\text{R1} \\ \text{R3} + (-1)\text{R1}}}{\approx}
\begin{bmatrix} 1 & 2 & -1 & 0.5 \\ 0 & \boxed{0.0001} & 2 & 1.5 \\ 0 & \boxed{1} & 4 & -1.5 \end{bmatrix}
$$

Scan these elements to determine the one with largest absolute value. It is 1. This element will be pivot. Swap rows 2 and 3 to bring pivot into desired location.

$$
\underset{\text{R2} \leftrightarrow \text{R3}}{\approx}
\begin{bmatrix} 1 & 2 & -1 & 0.5 \\ 0 & ① & 4 & -1.5 \\ 0 & 0.0001 & 2 & 1.5 \end{bmatrix}
$$

$$
\underset{\substack{\text{R1} + (-2)\text{R2} \\ \text{R3} + (-0.0001)\text{R2}}}{\approx}
\begin{bmatrix} 1 & 0 & -9 & 3.5 \\ 0 & 1 & 4 & -1.5 \\ 0 & 0 & 1.9996 & 1.5002 \end{bmatrix}
$$

This element should have been 1.50015, but it has been rounded to five significant figures.

$$
\underset{(1/1.9996)\text{R3}}{\approx}
\begin{bmatrix} 1 & 0 & -9 & 3.5 \\ 0 & 1 & 4 & -1.5 \\ 0 & 0 & ① & 0.75025 \end{bmatrix}
$$

$$
\underset{\substack{\text{R1} + 9\text{R3} \\ \text{R2} + (-4)\text{R3}}}{\approx}
\begin{bmatrix} 1 & 0 & 0 & 10.252 \\ 0 & 1 & 0 & -4.501 \\ 0 & 0 & 1 & 0.75025 \end{bmatrix}
$$

The solution is $y_1 = 10.252$, $y_2 = -4.501$, $y_3 = 0.75025$.

In terms of the original variables, the solution is

$$ x_1 = 10{,}252, \qquad x_2 = -4.501, \qquad x_3 = 0.75025 $$

It is interesting to compare this solution with the solution that would have been obtained using the Gauss-Jordan method without the pivoting and scaling refinements. The solution, to five significant figures, would have been

$$ x_1 = 11{,}000, \qquad x_2 = -5, \qquad x_3 = 0.75025 $$

On substituting values into the original equations, it can be seen that the result using pivoting and scaling is very accurate, while the result obtained without these refinements is unsatisfactory. For example, let us test the first equation, $0.002x_1 + 4x_2 - 2x_3 = 1$, with each set of results. We get

With pivoting: $(0.002 \times 10{,}252) + (4 \times -4.501) - (2 \times 0.75025)$
$= 0.9995 \approx 1$

Without pivoting: $(0.002 \times 11{,}000) + (4 \times -5) - (2 \times 0.75025)$
$= 0.4995 \neq 1$

In this section we have indicated some of the precautions that should be taken when systems of linear equations are used in mathematical models. We have demonstrated some of the shortcomings of elimination methods, and some techniques that can be used to minimize errors. In the next section we introduce a different approach to solving systems of linear equations, namely iterative methods. Iterative methods give an alternative way of minimizing the effect of round-off errors in solving ill-conditioned systems.

Readers wanting more information on these topics are referred to *Applied Numerical Analysis* by Curtis F. Gerald, Addison-Wesley, 1989.

Exercise Set 9.3

1. Compute the 1-norms of the following matrices.

*(a) $\begin{bmatrix} 1 & 2 \\ 3 & 4 \end{bmatrix}$
(b) $\begin{bmatrix} 5 & 2 \\ -1 & 3 \end{bmatrix}$

*(c) $\begin{bmatrix} 1 & 2 & 0 \\ 3 & -5 & 4 \\ 6 & 1 & -2 \end{bmatrix}$
(d) $\begin{bmatrix} 1 & 24 & -3 \\ 6 & 247 & 56 \\ -5 & 7 & -219 \end{bmatrix}$

2. Find the condition numbers of the following matrices.

*(a) $\begin{bmatrix} 1 & 2 \\ 3 & 4 \end{bmatrix}$
(b) $\begin{bmatrix} 2 & -2 \\ 3 & 1 \end{bmatrix}$
*(c) $\begin{bmatrix} 5 & 2 \\ 8 & 3 \end{bmatrix}$

(d) $\begin{bmatrix} 24 & 21 \\ 5 & 5 \end{bmatrix}$
*(e) $\begin{bmatrix} 5.2 & 3.7 \\ 3.8 & 2.6 \end{bmatrix}$

3. Find the condition numbers of the following matrices. Let these be matrices of coefficients for systems of equations. How much less accurate than the elements of the matrices can the solutions be?

*(a) $\begin{bmatrix} 9 & 2 \\ 4 & 7 \end{bmatrix}$
(b) $\begin{bmatrix} 51 & 3 \\ -1 & 27 \end{bmatrix}$

*(c) $\begin{bmatrix} 300 & 1001 \\ 75 & 250 \end{bmatrix}$
(d) $\begin{bmatrix} 3 & 4 \\ 1 & 2 \end{bmatrix}$

*(e) $\begin{bmatrix} 5 & 3 \\ 4 & 2 \end{bmatrix}$

4. Find the condition numbers of the following matrices. These matrices are from applications we have discussed. Comment on the reliability of the solutions.

*(a) $\begin{bmatrix} 1 & 1 & -1 \\ 1 & 0 & 2 \\ 1 & -2 & 0 \end{bmatrix}$ (Electrical circuits application, Example 3, Section 1.4)

(b) $\begin{bmatrix} \frac{4}{5} & -\frac{1}{5} & -\frac{3}{10} \\ -\frac{1}{2} & \frac{1}{2} & 0 \\ 0 & 0 & \frac{4}{5} \end{bmatrix}$ (Leontif I/O application, Example 2, Section 2.5)

*5. Find the condition number of the matrix $\begin{bmatrix} \frac{1}{3} & \frac{1}{4} & \frac{1}{5} \\ \frac{1}{4} & \frac{1}{5} & \frac{1}{6} \\ \frac{1}{5} & \frac{1}{6} & \frac{1}{7} \end{bmatrix}$. This is an example of a **Hilbert matrix**. Hilbert matrices are defined by $a_{ij} = \dfrac{1}{i + j + 1}$. These matrices are more ill-conditioned than Vandermonde matrices!

*6. Find the condition of the matrix $A = \begin{bmatrix} 1 & k \\ 1 & 1 \end{bmatrix}$ $(k \neq 1)$. Observe how the condition number depends on the value of k. For what values of k is $c(A) > 100$?

7. In this section we have defined the condition number of matrices using 1-norms. Prove that the 1-norm satisfies the properties of norms discussed in Section 5.1, namely **(a)** $\|A\| \geq 0$, **(b)** $\|A\| = 0$ if and only if $A = 0$, **(c)** $\|cA\| = |c| \|A\|$, **(d)** $\|A + B\| \leq \|A\| + \|B\|$. These conditions are the axioms for a norm on a general vector space. Matrix norms are also required to satisfy the condition $\|AB\| \leq \|A\| \|B\|$. Show that the 1-norm also satisfies this condition.

8. (a) Let I be an identity matrix. Prove that $c(I) = 1$.

  *(b) Use the property $\|AB\| \leq \|A\|\|B\|$ of a matrix norm to show that $c(A) \geq 1$.

9. Prove that $c(A) = c(kA)$ for a nonzero constant k. What is the significance of this result in terms of systems of linear equations?

*10. Let A be a diagonal matrix. Prove that
$$c(A) = (\max\{a_{ii}\})(1/\min\{a_{ii}\}).$$

In Exercises 11–17, solve the systems of equations (if possible), using pivoting and scaling techiques when appropriate.

*11.
$$\begin{aligned} -\ x_2 + 2x_3 &= 4 \\ x_1 + 2x_2 -\ x_3 &= 1 \\ x_1 + 2x_2 + 2x_3 &= 4 \end{aligned}$$

*12.
$$\begin{aligned} -x_1 +\ x_2 + 2x_3 &= 8 \quad \text{(to 3 significant figures)} \\ 2x_1 + 4x_2 -\ x_3 &= 10 \\ x_1 + 2x_2 + 2x_3 &= 2 \end{aligned}$$

13.
$$\begin{aligned} -x_1 + 0.002x_2 &= 0 \\ x_1 \qquad\quad + 2x_3 &= -2 \\ 0.01x_2 +\ x_3 &= 2 \end{aligned}$$

14.
$$\begin{aligned} 2x_1 + \qquad\qquad\quad x_3 &= 1 \\ 0.0001x_1 + 0.0002x_2 + 0.0004x_3 &= 0.0004 \\ x_1 -\qquad 2x_2 -\qquad 3x_3 &= -3 \end{aligned}$$

*15.
$$\begin{aligned} x_2 - 0.1x_3 &= 2 \quad \text{(to 3 decimal places)} \\ 0.02x_1 + 0.01x_2 \qquad\quad &= 0.01 \\ x_1 -\quad 4x_2 + 0.1x_3 &= 2 \end{aligned}$$

16.
$$\begin{aligned} -x_1 + \qquad\quad 2x_3 &= 1 \\ 0.1x_1 + 0.001x_2 \qquad &= 0.1 \\ 0.1x_1 + 0.002x_2 \qquad &= 0.2 \end{aligned}$$

*17.
$$\begin{aligned} x_1 -\qquad 2x_2 -\qquad x_3 &= 1 \quad \text{(to 3 decimal places)} \\ x_1 - 0.001x_2 -\qquad x_3 &= 2 \\ x_1 +\qquad 3x_2 - 0.002x_3 &= -1 \end{aligned}$$

*9.4 Iterative Methods for Solving Systems of Linear Equations

We have discussed a number of elimination methods for solving systems of linear equations. We now introduce two so-called **iterative methods** for solving systems of n equations in n variables that have a unique solution, called the **Jacobi method** and the **Gauss-Seidel method**. At the close of this section we shall compare the merits of Gaussian elimination and the iterative methods.

An iterative method starts with an initial estimate of the solution, and generates a sequence of values that converges to the solution.

Jacobi Method[†]

We introduce this method with an example. Let us solve the following system of linear equations.

$$\begin{aligned} 6x + 2y -\ z &= 4 \\ x + 5y +\ z &= 3 \\ 2x +\ y + 4z &= 27 \end{aligned} \qquad (1)$$

[†] Carl Gustav Jacobi (1804–1851) was educated in Berlin and taught at Königsberg and Berlin. He created sensations with his penetrating investigations into number theory, geometry, analysis, and mechanics. Yet his tireless occupation with research did not impair his teaching. Jacobi drew much insight from algorithmic work. He was inclined to investigate mathematical problems for their intrinsic interest. He was interested in the history of mathematics.

Rewrite the equation as follows, isolating x in the first equation, y in the second equation, and z in the third equation.

$$x = \frac{4 - 2y + z}{6}$$

$$y = \frac{3 - x - z}{5} \tag{2}$$

$$z = \frac{27 - 2x - y}{4}$$

Now make an estimate of the solution, say $x = 1$, $y = 1$, $z = 1$. The accuracy of the estimate affects only the speed with which we get a good approximation to the solution. Let us label these values $x^{(0)}$, $y^{(0)}$, and $z^{(0)}$. They are called the **initial values** of the iterative process.

$$x^{(0)} = 1, \qquad y^{(0)} = 1, \qquad z^{(0)} = 1$$

Substitute these values into the right-hand side of system (2) to get the next set of values in the iterative process.

$$x^{(1)} = 0.5, \qquad y^{(1)} = 0.2, \qquad z^{(1)} = 6$$

These values of x, y, and z are now substituted into system (2) again to get

$$x^{(2)} = 1.6, \qquad y^{(2)} = -0.7, \qquad z^{(2)} = 6.45$$

This process can be repeated to get $x^{(3)}$, $y^{(3)}$, $z^{(3)}$, and so on. Repeating the iteration will give a better approximation to the exact solution at each step. For this straightforward system of equations, the solution can easily be seen to be

$$x = 2, \qquad y = -1, \qquad z = 6$$

Thus after the second iteration one has quite a long way to go.

The following theorem, which we present without proof, gives one set of conditions under which the Jacobi iterative method can be used.

Theorem 9.1   Consider a system of n linear equations in n variables having a square matrix of coefficients A. If the absolute value of the diagonal element of each row of A is greater than the sum of the absolute values of the other elements in that row, then the system has a unique solution. The Jacobi method will converge to this solution, no matter what the initial values.

Thus the Jacobi method can be used when the diagonal elements **dominate** the rows. Let us return to the previous example to see that the conditions of this theorem are indeed satisfied. The matrix of coefficients is

$$A = \begin{bmatrix} 6 & 2 & -1 \\ 1 & 5 & 1 \\ 2 & 1 & 4 \end{bmatrix}$$

Comparing the magnitude of the diagonal elements with the sums of the magnitudes of the other elements in each row, we get the results shown in Table 9.1. Observe that the diagonal elements dominate the rows; the Jacobi method can be used.

Table 9.1

Row	Absolute Value of Diagonal Element	Sum of Absolute Values of Other Elements in That Row
1	6	$\lvert 2 \rvert + \lvert -1 \rvert = 3$
2	5	$\lvert 1 \rvert + \lvert 1 \rvert = 2$
3	4	$\lvert 2 \rvert + \lvert 1 \rvert = 3$

In some systems it may be necessary to rearrange the equations before the above condition is satisfied and this method can be used.

The better the initial estimate, the quicker one gets a result to within the required degree of accuracy. Note that this method has an advantage: if an error is made at any stage, it merely amounts to a new estimate at that point.

There are many theorems similar to the one above that guarantee convergence under varying conditions. Investigations have made available swiftly converging methods that apply to various special systems of equations.

Gauss-Seidel Method*

The Gauss-Seidel method is a refinement of the Jacobi method that usually (but not always) gives more rapid convergence. The latest value of each variable is substituted at each step in the iterative process. This method also works if the diagonal elements of the matrix of coefficients dominate the rows. We illustrate the method for the previous systems of equations. As before, let us take our initial guess to be

$$x^{(0)} = 1, \qquad y^{(0)} = 1, \qquad z^{(0)} = 1$$

* Ludvig Von Seidel (1821–1896) studied at Berlin, Königsberg, and Munich and taught in Munich. He worked in probability theory and astronomy. His photometric measurements of fixed stars and planets were the first true measurements of this kind. He investigated the relationship between the frequency of certain diseases and climatic conditions in Munich. Seidel suffered from eye problems and was forced to retire young.

Substituting the latest value of each variable into system (2) gives

$$x^{(1)} = \frac{4 - 2y^{(0)} + z^{(0)}}{6} = 0.5$$

$$y^{(1)} = \frac{3 - x^{(1)} - z^{(0)}}{5} = 0.3$$

$$z^{(1)} = \frac{27 - 2x^{(1)} - y^{(1)}}{4} = 6.4250$$

Observe that we have used $x^{(1)}$, the most up-to-date value of x, to get $y^{(1)}$. We have used $x^{(1)}$ and $y^{(1)}$ to get $z^{(1)}$. Continuing, we get

$$x^{(2)} = \frac{4 - 2y^{(1)} + z^{(1)}}{6} = 1.6375$$

$$y^{(2)} = \frac{3 - x^{(2)} - z^{(1)}}{5} = -1.0125$$

$$z^{(2)} = \frac{27 - 2x^{(2)} - y^{(2)}}{4} = 6.1844$$

Tables 9.2 and 9.3 give the results obtained for this system of equations using both methods. They illustrate the more rapid convergence of the Gauss-Seidel method to the exact solution $x = 2$, $y = -1$, $z = 6$. Table 9.4 gives the differences between the solutions $x^{(6)}$, $y^{(6)}$, $z^{(6)}$, obtained in the two methods after six iterations, and the actual solution $x = 2$, $y = -1$, $z = 6$. The Gauss-Seidel method converges much more rapidly than the Jacobi method.

Table 9.2 Jacobi Method

Iteration	x	y	z
Initial Guess	1	1	1
1	0.5	0.2	6
2	1.6	−0.7	6.45
3	1.975	−1.01	6.125
4	2.024167	−1.02	6.015
5	2.009167	−1.007833	5.992917
6	2.001431	−1.000417	5.997375

Table 9.3 Gauss-Seidel Method

Iteration	x	y	z
Initial Guess	1	1	1
1	0.5	0.3	6.425
2	1.6375	-1.0125	6.184375
3	2.034896	-1.043854	5.993516
4	2.013537	-1.001411	5.993584
5	1.999401	-0.998597	5.999949
6	1.999524	-0.9998945	6.000212

Table 9.4

	$\left|x^{(6)} - 2\right|$	$\left|y^{(6)} - (-1)\right|$	$\left|z^{(6)} - 6\right|$
Jacobi Method	0.001431	0.000417	0.002625
Gauss-Seidel Method	0.000476	0.0001055	0.000212

Comparison of Gaussian Elimination and Gauss-Seidel

Limitations   Gaussian elimination is a finite method and can be used to solve any system of linear equations. The Gauss-Seidel method converges only for special systems of equations; thus it can be used only for those systems.

Efficiency   The efficiency of a method is a function of the number of arithmetic operations (addition, subtraction, multiplication, and division) involved in each method. For a system of n equations in n variables, where the solution is unique, Gaussian elimination involves $(4n^3 + 9n^2 - 7n)/6$ arithmetic operations. The Gauss-Seidel method requires $2n^2 - n$ arithmetic operations per iteration. For large values of n, Gaussian elimination requires approximately $2n^3/3$ arithmetic operations to solve the problem, while Gauss-Seidel requires approximately $2n^2$ arithmetic operations per iteration. Therefore, if the number of iterations is less than or equal to $n/3$, the iterative method requires fewer arithmetic operations.

For example, consider a system of 300 equations in 300 variables. Elimination requires 18,000,000 operations, whereas iteration requires 180,000 operations per iteration. For 100 or fewer iterations, the Gauss-Seidel method involves less arithmetic; it is more efficient. It should be pointed out that Gaussian elimination involves movement of data; for example, several rows may need to be interchanged. This is time-consuming and costly on computers. Iterative processes suffer much less from this factor. Thus, even if the number of iterations is more than $n/3$, iteration may require less computer time.

Accuracy   Round-off errors in Gaussian elimination are minimized by using pivoting. However, they can still be sizable. Errors in Gauss-Seidel are only the errors committed in the final iteration—the result of the next-to-last iteration can be interpreted as a very good initial estimate. Thus, in general, when Gauss-Seidel is applicable, it is more accurate than Gaussian elimination. This fact often justifies the use of Gauss-Seidel over Gaussian elimination even when the total amount of computation time involved is greater. For example, when a mathematical model results in an ill-conditioned system and there is no way of constructing an alternative well-behaved description, an iterative method can ensure minimizing bad effects due to computation.

Storage   Iterative methods are, in general, more economical in core-storage requirements of a computer.

For a more in-depth discussion of iterative methods, you are referred to *Numerical Analysis* by Richard L. Burden and J. Douglas Faires, Fourth Edition, Sections 7.3 and 7.4, Prindle, Webber, and Schmidt Publishers, 1989.

Exercise Set 9.4

Determine approximate solutions to the following systems of linear equations using the Gauss-Seidel iterative method. Work to two decimal places if you are performing the computation by hand. Use the given initial values. (Initial values are given solely so that answers can be compared.)

*1. $4x + y - z = 8$
$5y + 2z = 6$
$x - y + 4z = 10$
$x^{(0)} = 1,\ y^{(0)} = 2,\ z^{(0)} = 3$

2. $4x \quad\ y \quad\ = 6$
$2x + 4y - z = 4$
$x - y + 5z = 10$
$x^{(0)} = 1,\ y^{(0)} = 2,\ z^{(0)} = 3$

*3. $5x - y + z = 20$
$2x + 4y \quad\ = 30$
$x - y + 4z = 10$
$x^{(0)} = 0,\ y^{(0)} = 0,\ z^{(0)} = 0$

4. $6x + 2y - z = 30$
$-x + 8y + 2z = 20$
$2x - y + 10z = 40$
$x^{(0)} = 5,\ y^{(0)} = 6,\ z^{(0)} = 7$

*5. $5x - y + 2z = 40$
$2x + 4y - z = 10$
$-2x + 2y + 10z = 40$
$x^{(0)} = 20,\ y^{(0)} = 30,\ z^{(0)} = -40$

*6. $6x - y + z + w = 20$
$x + 8y + 2z \qquad = 30$
$-x + y + 6z + 2w = 40$
$2x \quad\ - 3z + 10w = 10$
$x^{(0)} = 0,\ y^{(0)} = 0,\ z^{(0)} = 0,\ w^{(0)} = 0$

7. Apply the Gauss-Seidel method to the following system of equations. Observe that the method does not converge. The method does not lead to a solution for this system.

$$5x + 4y - z = 8$$
$$x - y + z = 4$$
$$2x + y + 2z = 1$$
$$x^{(0)} = 0,\ y^{(0)} = 0,\ z^{(0)} = 0$$

*9.5 Eigenvalues by Iteration; Connectivity of Networks

Numerical techniques exist for evaluating certain eigenvalues and eigenvectors of various types of matrices. Here we present an iterative method called the **power method**. It can be used to determine the eigenvalue with the largest absolute value (if it exists), and a corresponding eigenvector for certain matrices. In many applications one is interested only in the **dominant** eigenvalue. Applications in geography and history that illustrate the importance of the dominant eigenvalue will be given.

Definition   *Let A be a square matrix with eigenvalues $\lambda_1, \lambda_2, \dots$. The eigenvalue λ_i is said to be a dominant eigenvalue if*

$$|\lambda_i| > |\lambda_k|, \qquad k \neq i$$

The eigenvectors corresponding to the dominant eigenvalue are called the dominant eigenvectors of A.

Example 1   Let A be a square matrix with eigenvalues -5, -2, 1, and 3. Then -5 is the dominant eigenvalue, since

$$|-5| > |-2|, \quad |-5| > |1|, \quad \text{and} \quad |-5| > |3|$$

Let B be a square matrix with eigenvalues -4, -2, 1, and 4. There is no dominant eigenvalue since $|-4| = |4|$.

The power method for finding a dominant eigenvector is based on the following theorem.

Theorem 9.2   Let A be an $n \times n$ matrix having n linearly independent eigenvectors and a dominant eigenvalue. Let $\mathbf{x}_0$ be any nonzero column vector in $\mathbf{R}^n$ having a nonzero component in the direction of a dominant eigenvector. Then the sequence of vectors

$$\mathbf{x}_1 = A\mathbf{x}_0, \ \mathbf{x}_2 = A\mathbf{x}_1, \ \dots, \ \mathbf{x}_k = A\mathbf{x}_{n-1}, \ \dots$$

will approach a dominant eigenvector of A.

Proof   Let the eigenvalues of A be $\lambda_1, \dots, \lambda_n$, with λ_1 being the dominant eigenvalue. Let $\mathbf{y}_1, \dots, \mathbf{y}_n$ be corresponding linearly independent eigenvectors.

These eigenvectors will form a basis for $\mathbf{R}^n$. Thus there exist scalars $a_1, \ldots, a_n$, such that

$$\mathbf{x}_0 = a_1\mathbf{y}_1 + \cdots + a_n\mathbf{y}_n \qquad a_1 \neq 0$$

We get

$$\begin{aligned}
\mathbf{x}_k = A\mathbf{x}_{k-1} &= A^2\mathbf{x}_{k-2} = \cdots = A^k\mathbf{x}_0 \\
&= A^k[a_1\mathbf{y}_1 + \cdots + a_n\mathbf{y}_n] \\
&= [a_1 A^k\mathbf{y}_1 + \cdots + a_n A^k\mathbf{y}_n] \\
&= [a_1(\lambda_1)^k\mathbf{y}_1 + \cdots a_n(\lambda_n)^k\mathbf{y}_n] \\
&= (\lambda_1)^k\left[a_1\mathbf{y}_1 + a_2\left(\frac{\lambda_2}{\lambda_1}\right)^k\mathbf{y}_1 + \cdots + a_n\left(\frac{\lambda_n}{\lambda_1}\right)^k\mathbf{y}_n\right]
\end{aligned}$$

Since $\left|\dfrac{\lambda_i}{\lambda_1}\right| < 1$, for $i = 2, 3, \ldots$, then $\left(\dfrac{\lambda_i}{\lambda_1}\right)^k$ will approach 0 as k increases, and the vector $\mathbf{x}_k$ will approach $(\lambda_1)^k a_1\mathbf{y}_1$, a dominant eigenvector.

The following theorem tells us how to determine the eigenvalue corresponding to a given eigenvector. Once a dominant eigenvector has been found, the dominant eigenvalue can be determined by applying this result.

Theorem 9.3 Let $\mathbf{x}$ be an eigenvector of a matrix A. The corresponding eigenvalue is given by

$$\lambda = \frac{A\mathbf{x} \cdot \mathbf{x}}{\mathbf{x} \cdot \mathbf{x}}$$

This quotient is called the **Rayleigh quotient**.

Proof Since λ is the eigenvalue corresponding to $\mathbf{x}$, $A\mathbf{x} = \lambda\mathbf{x}$. Thus

$$\frac{A\mathbf{x} \cdot \mathbf{x}}{\mathbf{x} \cdot \mathbf{x}} = \frac{\lambda\mathbf{x} \cdot \mathbf{x}}{\mathbf{x} \cdot \mathbf{x}} = \frac{\lambda(\mathbf{x} \cdot \mathbf{x})}{\mathbf{x} \cdot \mathbf{x}} = \lambda$$

If Theorem 9.2 as it now stands is used to compute a dominant eigenvector, the components of the vectors $\mathbf{x}_1, \mathbf{x}_2, \ldots$ may become very large, causing significant round-off errors to occur. This problem is overcome by dividing each component of $\mathbf{x}_i$ by the absolute value of its largest component and then using this vector (a vector in the same direction as $\mathbf{x}_i$) in the subsequent iteration. We refer to this procedure as **scaling** the vector. We now summarize the power method.

The Power Method for an $n \times n$ Matrix A

Select an arbitrary nonzero column vector $\mathbf{x}_0$ having n components.

Iteration 1: Compute $A\mathbf{x}_0$.

Scale $A\mathbf{x}_0$ to get $\mathbf{x}_1$.

Compute $\dfrac{A\mathbf{x}_1 \cdot \mathbf{x}_1}{\mathbf{x}_1 \cdot \mathbf{x}_1}$.

Iteration 2: Compute $A\mathbf{x}_1$.

Scale $A\mathbf{x}_1$ to get $\mathbf{x}_2$.

Compute $\dfrac{A\mathbf{x}_2 \cdot \mathbf{x}_2}{\mathbf{x}_2 \cdot \mathbf{x}_2}$, and so on.

Then

$$\mathbf{x}_0, \mathbf{x}_1, \mathbf{x}_2, \ldots \text{ converges to a dominant eigenvector}$$

and

$$\frac{A\mathbf{x}_1 \cdot \mathbf{x}_1}{\mathbf{x}_1 \cdot \mathbf{x}_1}, \frac{A\mathbf{x}_2 \cdot \mathbf{x}_2}{\mathbf{x}_2 \cdot \mathbf{x}_2}, \ldots \text{ converges to the}$$
$$\text{dominant eigenvalue}$$

if

A has n linearly independent eigenvectors and $\mathbf{x}_0$ has a nonzero component in the direction of a dominant eigenvector.

These conditions come from Theorem 9.2. We know in advance that some $n \times n$ matrices, such as **symmetric matrices**, have n linearly independent eigenvectors. The method can thus be applied to find the dominant eigenvalue of a symmetric matrix. One does not know in advance that the selected vector $\mathbf{x}_0$ has a nonzero component in the direction of a dominant eigenvector. There is a high probability that it will have. Further ramifications of this condition are discussed in the exercises that follow.

Example 2   Find the dominant eigenvalue and a dominant eigenvector of the following symmetric matrix.

$$\begin{bmatrix} 5 & 4 & 2 \\ 4 & 5 & 2 \\ 2 & 2 & 2 \end{bmatrix}$$

Solution   Let

$$\mathbf{x}_0 = \begin{bmatrix} 1 \\ -2 \\ 3 \end{bmatrix}$$

be an arbitrary column vector with three components. Computations were carried out to nine decimal places. We display the results, rounded to three decimal places for clarity of viewing, in Table 9.5.

Table 9.5

Iteration	$A\mathbf{x}_n$	Scaled Vector	$(\mathbf{x}_n \cdot A\mathbf{x}_n)/(\mathbf{x}_n \cdot \mathbf{x}_n)$
1	$\begin{bmatrix} 3 \\ 0 \\ 4 \end{bmatrix}$	$\mathbf{x}_1 = \begin{bmatrix} 0.75 \\ 0 \\ 1 \end{bmatrix}$	5
2	$\begin{bmatrix} 5.75 \\ 5 \\ 3.5 \end{bmatrix}$	$\mathbf{x}_2 = \begin{bmatrix} 1 \\ 0.870 \\ 0.609 \end{bmatrix}$	9.889
3	$\begin{bmatrix} 9.696 \\ 9.565 \\ 4.957 \end{bmatrix}$	$\mathbf{x}_3 = \begin{bmatrix} 1 \\ 0.987 \\ 0.511 \end{bmatrix}$	9.999
4	$\begin{bmatrix} 0.997 \\ 9.955 \\ 4.996 \end{bmatrix}$	$\mathbf{x}_4 = \begin{bmatrix} 1 \\ 0.999 \\ 0.501 \end{bmatrix}$	10
5	$\begin{bmatrix} 0.997 \\ 9.996 \\ 5 \end{bmatrix}$	$\mathbf{x}_5 = \begin{bmatrix} 1 \\ 1 \\ 0.5 \end{bmatrix}$	10

Thus after five iterations we have arrived at the dominant eigenvalue of 10, with a corresponding eigenvector of $\begin{bmatrix} 1 \\ 1 \\ 0.5 \end{bmatrix}$. These results are in agreement with the discussion of Section 4.7, where we computed the eigenvalues and eigenvectors of this matrix using determinants.

This method has the same advantage as the Gauss-Seidel iterative method discussed in the previous section in that any error in computation only means that a new arbitrary vector has been introduced at that stage. The method is very accurate; the only round-off errors that occur are those that arise in the final iteration. The method has the disadvantage that it may converge only very slowly for large matrices.

Deflation

If the matrix A is symmetric, a technique called **deflation** can be used for determining further eigenvalues and eigenvectors. The method is based on the following theorem, which we state without proof.

Theorem 9.4   Let A be a symmetric matrix with eigenvalues $\lambda_1, \ldots, \lambda_n$, labeled according to absolute value, λ_1 being the dominant eigenvalue. Let $\mathbf{x}$ be a dominant unit eigenvector. Then

1.   The matrix $B = A - \lambda_1 \mathbf{x}\mathbf{x}^t$ has eigenvalues $0, \lambda_2, \ldots, \lambda_n$. B is symmetric.

2.   If $\mathbf{y}$ is an eigenvector of B corresponding to one of the eigenvalues $\lambda_2, \ldots, \lambda_n$, it is also an eigenvector of A for the same eigenvalue.

One determines λ_1 and $\mathbf{x}$ of the matrix A using the power method. The matrix B is then constructed. Observe that λ_2 is the dominant eigenvalue of B (if one exists). Apply the power method to B to get λ_2 and a corresponding eigenvector, which is also an eigenvector of A. The process is then repeated. The disadvantage of this method is that the eigenvalues become increasingly inaccurate due to compounding of errors. Furthermore, if at any stage λ_i is not dominant, the method terminates at that point; no further eigenvalues can be computed.

Readers who are interested in further discussion of the power method and deflation are referred to *Numerical Analysis* by Richard L. Burden and J. Douglas Faires, Fourth Edition, Section 9.2, Prindle, Webber, and Schmidt Publishers, 1989.

Accessibility Index of a Network

Geographers use eigenvectors within the field of graph theory to analyze networks. The most common types of networks investigated are roads, airline routes, sea routes, navigable rivers, and canals. Here we illustrate the concepts in terms of a road network.

Consider a network of n cities linked by two-way roads. The cities can be represented by the vertices of a digraph and the roads as edges. Let us assume that it is possible to travel along roads from any city to another; the digraph is said to be **connected**. Geographers are interested in the **accessibility** of each city. The **Gould index** is used as a measure of accessibility. This index is obtained from an eigenvector corresponding to the dominant eigenvalue of an $n \times n$ matrix that describes the network.

The network consists of two-way roads. Thus, if there is an edge from vertex i to vertex j, there will also be an edge from vertex j to vertex i. The adjacency matrix A will therefore be symmetric. Construct the *augmented adjacency matrix* $B = A + I_n$. B is also symmetric. This implies that B has n linearly independent eigenvectors. There is a theorem, called the **Perron-Frobenius theorem**, which implies that the augmented adjacency matrix B of a connected digraph does have a dominant eigenvalue and that the corresponding eigenspace is one-dimensional with a basis vector having all positive components. The power method is used in practice to compute the dominant eigenvalue and a corresponding eigenvector. The components of this eigenvector give the accessibilities of the cities corresponding to the vertices of the digraph.

We now justify this use of the components of the dominant eigenvector. A highly accessible vertex should have a large number of paths to other vertices. The elements of B^m give the number of paths of length m between vertices including possible stopovers at vertices along the way. The sum of the elements of row i of B^m gives the number of m-paths out of vertex i. Let $\mathbf{e}$ be the column vector having n components, all of which are 1. Then $B^m\mathbf{e}$ will be a column vector, the ith component of which is the sum of the elements of row i of B^m. Thus the components of the vector $B^m\mathbf{e}$ will give the number of m-paths out of the various vertices. Since all the components of $\mathbf{e}$ are 1, and there is a dominant eigenvector of B having all positive components, $\mathbf{e}$ has a nonzero component in the direction of a dominant eigenvector. Theorem 9.2 tells us that as m increases, $B^m\mathbf{e}$ will approach a dominant eigenvector. As m increases, $B^m\mathbf{e}$ accounts for longer and longer paths out of each vertex. Thus the ratios of the components of the dominant eigenvector will give the relative accessibilities of the vertices.

This justification of the accessibility index is adapted from "Linear Algebra in Geography: Eigenvectors of Networks," by Philip D. Straffin, Jr., *Mathematics Magazine*, Vol. 53, No. 5, November 1980, pages 269–276. This interesting article has a discussion of the Perron-Frobenius theorem and references to applications ranging from growth of the São Paulo economy through an analysis of the U.S. interstate road system to an analysis of trade routes in medieval Russia (which we include later in this section).

We now illustrate the concepts for a simple transportation network.

Example 3   **Connectivity of a Transportation Network**
Consider the transportation network described in Figure 9.1. The adjacency matrix A and the augmented matrix B are given. The dominant eigenvalue and corresponding eigenvector of B are found to two decimal places, using the power method, to be $\lambda = 3.17$ and $\mathbf{x} = (0.46, 1, 0.85, 0.85)$. The components of the eigenvector give the relative accessibilities of the vertices. Note that vertex 2 has highest connectivity, vertex 1 has lowest connectivity, and vertices 3 and 4 have equal connectivity, as is to be expected on looking at the graph. The convention adopted by geographers is to divide the eigenvector by the sum of the components. This is a form of "normalizing"

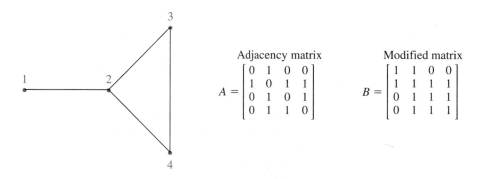

Adjacency matrix

$$A = \begin{bmatrix} 0 & 1 & 0 & 0 \\ 1 & 0 & 1 & 1 \\ 0 & 1 & 0 & 1 \\ 0 & 1 & 1 & 0 \end{bmatrix}$$

Modified matrix

$$B = \begin{bmatrix} 1 & 1 & 0 & 0 \\ 1 & 1 & 1 & 1 \\ 0 & 1 & 1 & 1 \\ 0 & 1 & 1 & 1 \end{bmatrix}$$

Figure 9.1

the vector so that the sum of the components is then 1. The components of the resulting vector are called the **Gould indices of accessibility** of the vertices. We get

$$\frac{1}{(0.46 + 1 + 0.85 + 0.85)}(0.46,\ 1,\ 0.85,\ 0.85) = (0.15,\ 0.32,\ 0.27,\ 0.27)$$

This vector gives the indices of the various vertices, as shown in Table 9.6.

Table 9.6

Vertex	Gould Index
1	0.14
2	0.32
3	0.27
4	0.27

Example 4   **Moscow as Capital of Russia**

Geographers have maintained that Moscow eventually assumed its dominant position in central Russia because of its strategic location on the medieval trade routes. Techniques of graph theory have been used to question this theory in "A Graph Theoretical Approach to Historical Geography" by F. R. Pitts, *Professional Geographer*, **17**, 1965, 15–20.

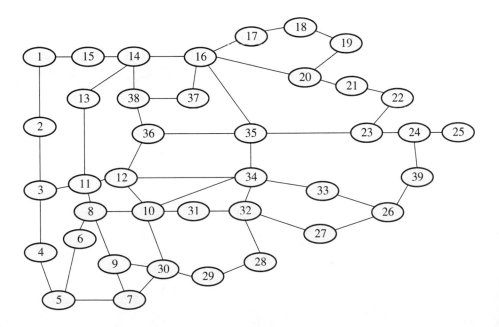

Figure 9.2   Trade routes of medieval Russia

The vertices of the graph in Figure 9.2 represent centers of population in Russia in the twelfth and thirteenth centuries, and the edges represent major trade routes. Some of the centers are Moscow (35), Kiev (4), Novgorod (1), Smolensk (3). The Gould index of accessibility of each vertex in the network was computed. This involved finding the dominant eigenvector of a 39 × 39 matrix using a computer. We list the seven highest indices in Table 9.7. Observe that Moscow ranks sixth in accessibility. The conclusion arrived at in this analysis was the sociological and political factors other than its location on the trade routes must have been important in Moscow's rise.

Table 9.7

City	Accessibility Index
Kozelsk (10)	0.0837
Kolomna (34)	0.0788
Vyazma (12)	0.0722
Bryansk (8)	0.0547
Mtsensk (30)	0.0493
Moscow (35)	0.0490
Dorogobusch (11)	0.0477

Example 5 **Comparison of Air Routes**

The components of the dominant eigenvector of the augmented adjacency matrix of a digraph are used to give internal information about a digraph, namely a comparison of the accessibility of its vertices. The dominant eigenvalues of the adjacency matrices for various digraphs, on the other hand, are used to compare the connectedness of distinct digraphs. The higher the dominant eigenvalue of a digraph, the greater its overall connectedness. Table 9.8 gives a comparison of the connectedness of internal airline networks of eight countries based on their dominant eigenvalues. It is surprising to find Sweden low on this list. Readers who are interested in reading more about this application, and others like it, are referred to Chapter 10 of *Applications of Graph Theory*, edited by Robin J. Wilson and Lowell W. Beineke, Academic Press, 1979.

Table 9.8

Country	Dominant Eigenvalue
USA	6
France	5.243
United Kingdom	4.610
India	4.590
Canada	4.511
USSR	3.855
Sweden	3.301
Turkey	2.903

Exercise Set 9.5

In Exercises 1–4, use the power method to determine the dominant eigenvalue and a dominent eigenvector of the given matrices.

*1. $\begin{bmatrix} 1 & 7 & -7 \\ -1 & 3 & -1 \\ -1 & -5 & 7 \end{bmatrix}$ **2.** $\begin{bmatrix} 9 & 4 & -4 \\ -1 & 1 & 1 \\ 7 & 4 & -2 \end{bmatrix}$

*3. $\begin{bmatrix} 13 & 7 & -7 \\ -2 & -1 & 2 \\ 12 & 7 & -6 \end{bmatrix}$ **4.** $\begin{bmatrix} 17 & 8 & -8 \\ 0 & 1 & 0 \\ 16 & 8 & -7 \end{bmatrix}$

In Exercises 5–8, use the power method to determine the dominant eigenvalue and a dominant eigenvector of the given matrices.

*5. $\begin{bmatrix} 1 & 1 & -1 & 1 \\ 1 & 1 & 1 & -1 \\ -3 & 3 & 3 & 3 \\ -3 & 3 & 1 & 5 \end{bmatrix}$

6. $\begin{bmatrix} 3 & -1 & 1 & -1 \\ -3 & 1 & -3 & 3 \\ -3 & 3 & 7 & 3 \\ -1 & 5 & 7 & 3 \end{bmatrix}$

*7. $\begin{bmatrix} 84 & 5 & -5 & 5 \\ 1 & 0 & 1 & -1 \\ -1 & 1 & 0 & 1 \\ 3 & 5 & -5 & 6 \end{bmatrix}$

8. $\begin{bmatrix} 4.5 & 5.5 & -5.5 & 5.5 \\ 1.5 & 0.5 & 1.5 & -1.5 \\ -1 & 1 & 0 & 1 \\ 3 & 5 & -6 & 7 \end{bmatrix}$

In Exercises 9–13, determine all eigenvalues and corresponding eigenvectors of the given symmetric matrices using the power method with deflation.

*9. $\begin{bmatrix} 5 & 4 & 2 \\ 4 & 5 & 2 \\ 2 & 2 & 2 \end{bmatrix}$ **10.** $\begin{bmatrix} 4 & 0 & 2 \\ 0 & 4 & 0 \\ 2 & 0 & 4 \end{bmatrix}$ ***11.** $\begin{bmatrix} 7 & 0 & 5 \\ 0 & 2 & 0 \\ 5 & 0 & 7 \end{bmatrix}$

*12. $\begin{bmatrix} 4 & 0 & 0 & 6 \\ 0 & 2 & -4 & 0 \\ 0 & -4 & 2 & 0 \\ 6 & 0 & 0 & 4 \end{bmatrix}$ **13.** $\begin{bmatrix} 5 & 4 & 1 & 1 \\ 4 & 5 & 1 & 1 \\ 1 & 1 & 4 & 2 \\ 1 & 1 & 2 & 4 \end{bmatrix}$

14. Use the power method to compute the Gould accessibility index of the vertices of the graphs in Figure 9.3. Use three interations in each case with the given initial vectors. (Initial vectors have been given so that the answers can be compared.) Comment on the choice of initial vectors!

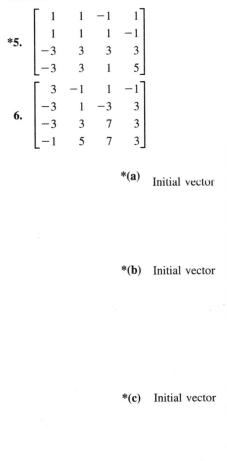

*(a)   Initial vector $\begin{bmatrix} 1 \\ 2 \\ 2 \\ 1 \end{bmatrix}$

*(b)   Initial vector $\begin{bmatrix} 1 \\ 1 \\ 3 \\ 3 \\ 1 \\ 1 \end{bmatrix}$

*(c)   Initial vector $\begin{bmatrix} 2 \\ 1 \\ 1 \\ 1 \\ 1 \end{bmatrix}$

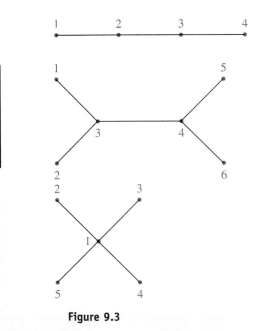

Figure 9.3

Review Exercises Chapter 9

1. Solve the following lower triangular system of equations.

$$2x_1 \qquad\qquad = 4$$
$$3x_1 - x_2 \qquad = 7$$
$$x_1 + 3x_2 + 2x_3 = 1$$

2. Let $A = LU$. The sequence of transformations $R_2 - \frac{3}{2}R_1$, $R_3 + \frac{2}{7}R_1$, $R_3 - 4R_2$ was used to arrive at U. Find L.

3. Solve the following system of equations using LU decomposition.

$$6x_1 + x_2 - x_3 = 5$$
$$-6x_1 + x_2 + x_3 = 1$$
$$12x_1 + 12x_2 + x_3 = 52$$

4. Let $A = LU$. If you are given L and U, how many arithmetic operations are needed to compute A, knowing the locations of the zeros above and below the diagonals in L and U?

5. Find the condition number of the matrix $\begin{bmatrix} 250 & 401 \\ 125 & 201 \end{bmatrix}$.

Let this matrix be the matrix of coefficients of a system of equations. How much less accurate than the elements of the matrix can the solution be?

6. Find the condition numbers of the matrices

(a) $\begin{bmatrix} 1 & 1 & -1 \\ 1 & 0 & 2 \\ 0 & 3 & 2 \end{bmatrix}$

(b) $\begin{bmatrix} 1 & -1 & -1 & 0 & 0 \\ 0 & 0 & 1 & -1 & -1 \\ 1 & 1 & 0 & 0 & 0 \\ 1 & 0 & 0 & 2 & 0 \\ 0 & 0 & 0 & 2 & 2 \end{bmatrix}$

These are the matrices of the systems of equations that describe the currents in the electrical circuits of Exercises 8 and 13 of Section 1.4. Comment on the reliability of the solutions.

7. Prove that $c(A) = c(A^{-1})$.

8. Does $c(A)$ define a linear mapping of $M_{nn} \rightarrow \mathbf{R}$?

9. Solve the following systems of equations using pivoting and scaling techniques.

(a)
$$\qquad\quad - 0.4x_2 + 0.002x_3 = 0.2$$
$$100x_1 + \quad x_2 - 0.01x_3 = 1$$
$$200x_1 + 2x_2 - 0.03x_3 = 1$$

(b)
$$\qquad\quad - x_2 + 0.001x_3 = 6$$
$$0.1x_1 + \qquad\quad 0.0002x_3 = -0.2$$
$$0.01x_1 + 0.01x_2 + 0.00003x_3 = 0.02$$

10. Solve the following system of equations using the Gauss-Seidel method with initial values $x^{(0)} = 1$, $y^{(0)} = 2$, $z^{(0)} = -3$.

$$6x + y - z = 27$$
$$2x + 7y - 3z = 9$$
$$x + y + 4z = 18$$

11. Use the power method to determine the dominant eigenvalue and corresponding eigenvector of the matrix

$$\begin{bmatrix} 1 & 3 & 4 \\ -2 & 4 & 9 \\ -5 & 8 & 7 \end{bmatrix}$$

12. Use the power method with inflation to find all eigenvalues and eigenvectors for the symmetric matrix

$$\begin{bmatrix} 3 & 2 & 1 \\ 2 & 4 & -2 \\ 1 & -2 & 3 \end{bmatrix}$$

MATLAB Discussion

M.24 *LU* Decomposition (Section 9.2)

A nonsingular matrix may be expressed as the product of lower and upper triangular matrices. Perform Gaussian elimination on the matrix to get the upper triangular form; the lower triangular form consists of the product of the inverses of the row exchange operations.

Example 1   Find the *LU* factorization of $A = \begin{bmatrix} 1 & 3 & 0 \\ 2 & 0 & 2 \\ 1 & 2 & 1 \end{bmatrix}$.

»A = [1 3 0;2 0 2;1 2 1];   {enter the matrix *A*}
»LUfact(A)   {author-defined function for *LU* factorization}

LU factorization of a matrix

All steps? y/n: n   {"all steps" option}

initial matrix A
   1   3   0
   2   0   2
   1   2   1

computing

final L matrix:
   1.0000   0   0
   2.0000   1.0000   0
   1.0000   0.1667   1.000

final U matrix:
   1.0000   3.0000   0
      0   −6.0000   2.0000
      0   0   0.6667

One of the major advantages of *LU* factorization is that, once it is found, we can solve $AX = B$ very easily. Observe that $LUX = AX = B$ so that $X = U^{-1}L^{-1}B$. Since L and U are triangular, their inverses are easy to compute.

Example 2   Find the solution of $AX = B$ using *LU* factorization, where

$$A = \begin{bmatrix} 1 & 3 & 0 \\ 2 & 0 & 2 \\ 1 & 2 & 1 \end{bmatrix}, \quad B = \begin{bmatrix} -2 & 4 \\ 2 & 0 \\ -1 & 2 \end{bmatrix}$$

»A = [1   3   0;2   0   2;1   2   1]   {enter the matrices for *A* and *B*}
»B = [−2   4;2   0;−1   2];
»lusolve(A,B)   {author-defined function to solve *AX* = *B*
                                           using the *LU* factorization}

Routine to solve the system *AX* = *B* using *LU* factorization.

All steps? y/n: n

initial matrix A
   1   3   0
   2   0   2
   1   2   1

computing

final L matrix:
   1.0000   0   0
   2.0000   1.0000   0
   1.0000   0.1667   1.0000   {display the *LU* factorization}

final U matrix:
   1.0000   3.0000   0
      0   −6.0000   2.0000
      0   0   0.6667

[press return to continue]

solving AX = LUX = B:
B:
   −2   4
    2   0
   −1   2

solution:
    1.0000   1.0000
   −1.0000   1.0000
    0.0000   −1.0000

Solution for $B = (-2, 2, -1)^t$ is $X = (1, -1, 0)^t$; for $B = (4, 0, 2)^t$ is
$X = (1, 1, -1)^t$.

Exercise 1   Find the *LU* decomposition of the following matrices.

*(a) $\begin{bmatrix} 3 & 5 \\ 1 & 2 \end{bmatrix}$ (b) $\begin{bmatrix} 4 & 1 & 3 \\ -2 & 6 & 0 \\ 1 & 5 & 1 \end{bmatrix}$ (c) $\begin{bmatrix} 2 & 0 & 2 \\ 1 & 4 & 1 \\ 2 & 5 & 5 \end{bmatrix}$

Exercise 2   Solve the following systems using *LU* decomposition.

$$*(a)\quad A = \begin{bmatrix} 4 & 1 & 3 \\ -2 & 6 & 0 \\ 1 & 5 & 1 \end{bmatrix},\quad B = \begin{bmatrix} 11 & 6 \\ 4 & -8 \\ 8 & -3 \end{bmatrix} \quad (b)\quad A = \begin{bmatrix} 3 & 5 \\ 1 & 2 \end{bmatrix},\quad B = \begin{bmatrix} 8 & 1 \\ 3 & 0 \end{bmatrix}$$

Condition Number of a Matrix (Section 9.3)

The condition number of a matrix is

$$c(A) = \|A\|\|A^{-1}\|$$

If $c(A)$ is large, the matrix is ill-conditioned. The system $AX = Y$ would then be very sensitive to small changes in A and Y and the solution could be unreliable.

We use the 1-norm for the matrix,

$$\|A\| = \max\{|a_{i1}| + \cdots + |a_{in}|\}, \text{ for } i = 1, \ldots, n$$

This norm is a good compromise in reliability and efficiency.

Example 1   Find the equation of the polynomial of degree 2 through the points $(1, 6)$, $(2, 3)$, and $(3, 2)$.

Solution   Let the polynomial be $y = a + bx + cx^2$. Thus

$$a + b + c = 6$$
$$a + 2b + 4c = 3$$
$$a + 3b + 9c = 2$$

The solution is $a = 11$, $b = -6$, $c = 1$. Thus the polynomial is $y = 11 - 6x + x^2$.
The matrix of coefficients is

$$\begin{bmatrix} 1 & 1 & 1 \\ 1 & 2 & 4 \\ 1 & 3 & 9 \end{bmatrix}$$

Let us find its condition number.

»A = [1 1 1;1 2 4;1 3 9];   {author-defined function that gives
»cn(A)   matrix inverse, determinant,
                                        1-norm, and condition number}

computing

inverse_matrix =
```
    3.0000  -3.0000   1.0000
   -2.5000   4.0000  -1.5000
    0.5000  -1.0000   0.5000
```

determinant_matrix =   2

onenorm_matrix =   14

onenorm_invmatrix =   8

condition_number =   112

The system is ill-conditioned. The y values usually correspond to measurements—they better be very accurate! A is an example of a Vandermonde matrix—such matrices are famously ill-conditioned. In a problem such as this, if data is not equispaced, a Vandermonde matrix will not arise.

$$\text{Vandermonde matrices:} \quad \begin{bmatrix} 1 & a & a^2 & a^3 & \cdots \\ 1 & b & b^2 & b^3 & \cdots \\ 1 & c & c^2 & c^3 & \cdots \\ \vdots & \vdots & \vdots & \vdots & \cdots \end{bmatrix}$$

Exercise 1   Find the condition numbers of the following matrices.

*(a) $\begin{bmatrix} 1 & 2 \\ 3 & 4 \end{bmatrix}$ (b) $\begin{bmatrix} 24 & 21 \\ 6 & 5 \end{bmatrix}$ *(c) $\begin{bmatrix} 1 & -1 & 2 \\ 4 & 3 & 0 \\ 5 & 1 & 4 \end{bmatrix}$ (d) $\begin{bmatrix} 2 & 3 & -1 \\ 4 & 5 & 3 \\ 10 & 13 & 4 \end{bmatrix}$

Exercise 2   Find the condition number of the matrix $\begin{bmatrix} \frac{1}{3} & \frac{1}{4} & \frac{1}{5} \\ \frac{1}{4} & \frac{1}{5} & \frac{1}{6} \\ \frac{1}{5} & \frac{1}{6} & \frac{1}{7} \end{bmatrix}$.

This is an example of a Hilbert matrix. Hilbert matrices are defined by $a_{ij} = \dfrac{1}{i + j + 1}$. These matrices are more ill-conditioned than Vandermonde matrices!

Exercise 3   Find a 3×3 matrix other than a Hilbert matrix that has a condition number over 900. (*Hint:* Look for a relationship between determinants and condition numbers.) Why would the determinant of a matrix be an unsatisfactory definition of condition number? Discuss.

*Exercise 4**   Find the equation of the polynomial of degree three through the points $(1, 7)$, $(2, 3)$, $(3, 9)$, and $(4, 21)$. Compute the condition number of the matrix of coefficients in the system of equations that arises.

Jacobi and Gauss-Seidel Iterative Methods (Section 9.4)

The Jacobi and Gauss-Seidel methods are iterative methods for solving certain linear systems of equations. These methods differ from the direct methods such as Gaussian elimination in that one starts with an initial guess for the solution and then computes successive approximations. Under certain conditions, these approximations get closer and closer to the true solution. The primary use of these methods is in the solution of large sparse systems of equations for which Gaussian elimination is extremely expensive. The major drawback is that the methods do not always converge to a solution. Two functions **jacobi** and **gseidel** have been written for MATLAB to implement these methods.

Example 1   Solve the following system of equations by the Jacobi method.

$$4x_1 + x_2 \qquad = 1$$
$$x_1 + 4x_2 \qquad = 1$$
$$x_2 + 4x_3 = 1$$

»A = [4   1   0   1;   {augmented matrix}
      1   4   0   1;
      0   1   4   1]
»**jacobi(A)**   {author-defined function for Jacobi iteration}
Jacobi iteration
Number of iterations: {maximum # iterations before stopping process}
   10
Tolerance: .0001   {how close to true solution before stopping process}
Initial estimate: {initial guess for the solution}
   [0   0   0]
All steps? y/n: n   {"all steps" option displays every iteration}

initial matrix
   4   1   0   1
   1   4   0   1
   0   1   4   1

computing

                              {if solution is approximated to desired tolerance within
Successful run   specified # iterations, the run is considered successful}

Number of iterations
   7
Solution
   0.2000
   0.2000
   0.2000

The Gauss-Seidel method would have given the same solution. The options are the same as for the Jacobi method. In general, the Gauss-Seidel method will converge more quickly than the Jacobi method.

As stated earlier, these iterative methods will not always converge to the solution. One condition for convergence is that the matrix of coefficients for the system of equations be strictly diagonally dominant. If it is not, then the approximations may get farther away from the true solution with each successive step. We now illustrate this situation.

Example 2   Let us apply the Gauss-Seidel method to the following system of equations.

$$x_1 + 2x_2 = 3$$
$$5x_1 + 7x_2 = 12$$

This system has the unique solution $x_1 = 1$, $x_2 = 1$. Note that the matrix of coefficients is not strictly diagonally dominant.

»**A = [1   2   3;5   7   12];**   {augmented matrix of the system}
»**gseidel(A)**
Gauss-Seidel iteration
Number of iterations: 4
Tolerance: 0.01
Initial estimate: [0   0]
All steps: y/n: y   {select the "all steps" option to examine
                                            the estimate after each iteration}

initial matrix
  1   2   3
  5   7   12

[press return at each step to continue]

iteration	...	iteration	**Desired tolerance not achieved**
1		4	**Value of X at end of run:**
X =		**X =**	6.8309
3.0000		6.8309	−3.1649
0.4286		−3.1649	

Note that, instead of getting closer to the true solution of $x_1 = 1$, $x_2 = 1$, the approximations stray.

Exercise 1   Find the solution (if possible) to the following systems using both iterative methods.

*(a) $10x_1 - x_2 + 2x_3 = 9.5$ (b) $3x_1 + 5x_3 = 10$
           $5x_1 + 10x_2 = 10$ $10x_1 + x_2 + 2x_3 = 5$
           $3x_1 + 4x_2 + 8x_3 = 5$ $5x_2 + x_3 = 0$

Exercise 2   What effect does each of the following have on the solution of systems using these iterative methods: changing the tolerance, changing the number of iterations, and changing the initial guess for the solution? Experiment with each of these inputs and draw some general conclusions.

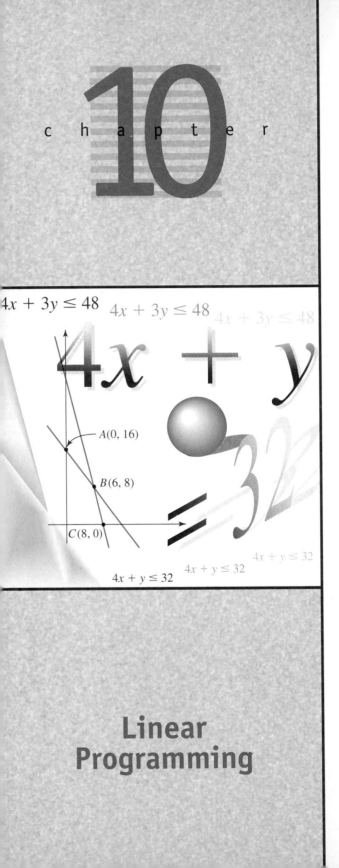

chapter

10

$4x + 3y \le 48$

$A(0, 16)$

$B(6, 8)$

$C(8, 0)$

$4x + y \le 32$

Linear Programming

We complete this course with an introduction to linear programming. Linear programming is a mathematical method for finding optimal solutions to problems. It is widely used in industry and in government.

Historically, linear programming was first developed and applied in 1947 by George Dantzig, Marshall Wood, and their associates at the U.S. Department of the Air Force; the early applications of linear programming were thus in the military field. However, the emphasis in applications has now moved to the general industrial area. Linear programming today is concerned with the efficient use or allocation of limited resources to meet desired objectives. To illustrate the current importance of linear programming, we point out that the 1975 Nobel Prize in Economic Science was awarded to two scientists, Professors Leonid Kantorovich of the former Soviet Union and Tjalling C. Koopmans of the United States, for their "contributions to the theory of optimum allocation of resources."

Kantorovich, born in 1912, was a winner of the Stalin and Lenin prizes and worked at the Mathematics Institute of the Sibirskoje Oidelenie Akademi Nauk in Novosibirsk.

(continued)

Koopmans, who was born in the Netherlands in 1913, worked with the Cowles Foundation for Research in Economics at Yale University in New Haven, Connecticut. Both economists worked independently on the problem of optimum allocation of scarce resources. Kantorovich showed how linear programming could be used to improve economic planning in the former U.S.S.R. He analyzed efficiency conditions for an economy as a whole, demonstrating the connection between the allocation of resources and the price system. According to Professor Assar Lindbeck of the International Economics Prize Committee, Kantorovich's work changed the views on economic planning in the former Soviet Union.

Koopmans developed his linear programming theory while planning the optimal movement of ships back and forth across the Atlantic in World War II. Koopmans' work led to new ways of interpreting the relationships between inputs and outputs of a production process. His methods are used to clarify the correspondence between efficiency in production and the existence of a price system.

Linear programming is now widely used in industry. Exxon Corporation, for example, uses linear programming to determine the optimal blending of gasoline. The Armour Company uses linear programming in determining specifications for a processed cheese spread. The H.J. Heinz Company uses linear programming to determine shipment schedules for its products, between its factories and warehouses.

*10.1   A Geometrical Introduction to Linear Programming

Systems of linear equations have been an important component of this course. We now turn our attention to a related topic, namely systems of linear inequalities. Linear programming involves analyzing systems of linear inequalities to arrive at optimal solutions to problems.

An expression such as $x + 2y = 6$ is called a **linear equation**. The graph of this equation is a straight line that represents all points in the plane that satisfy the equation. An expression such as $x + 2y \leq 6$ is called a **linear inequality**. It will also have a graph, namely the set of points that satisfy this condition. For example, $x = 2$, $y = 1$ satisfies the inequality while $x = 3$, $y = 4$ does not. The point $(2, 1)$ is on the graph of the inequality while the point $(3, 4)$ is not on the graph.

It can be proved that the graph of an inequality such as $x + 2y \leq 6$ will consist of all the points that lie on the line $x + 2y = 6$, together with all the points on one side of the line. A suitable test point, not on the line, can be used to determine which of the two sides is on the graph. For example, let us select $x = 0$, $y = 0$ as a test point for this inequality. It satisfies the inequality. Thus the graph consists of all points on the line $x + 2y = 6$ together with all points on the side of the line containing the origin. See Figure 10.1. The inequalities in a linear programming problem arise as mathematical descriptions of constraints. We now give examples to illustrate how limitations on resources are described by inequalities. The first example shows how a monetary constraint can be represented mathematically by an inequality. The second example shows how a time constraint can be described by an inequality.

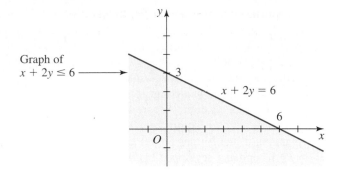

Graph of
$x + 2y \le 6$

Figure 10.1

Example 1   A company manufactures two types of cameras, the Pronto I and the Pronto II. The Pronto I costs \$10 to manufacture, while the Pronto II costs \$15. The total funds available for the production are \$23,500. We can describe this monetary constraint on the production of cameras by a linear inequality. Let the company manufacture x of Pronto I and y of Pronto II.

Total cost of producing x Pronto I cameras at \$10 per camera $= \$10x$

Total cost of producing y Pronto II cameras at \$15 per camera $= \$15y$

Thus

$$\text{Total manufacturing cost} = \$(10x + 15y)$$

Since the funds available are \$23,500, we get the constraint

$$10x + 15y \le 23{,}500$$

Example 2   A company makes two types of microcomputers, the Jupiter and the Cosmos. It takes 27 hours to assemble a Jupiter computer, and it takes 34 hours to assemble a Cosmos computer. The total labor time available for this work is 800 hours. Let us describe this time constraint by means of an inequality. Let the company assemble x Jupiter and y Cosmos computers.

Total time to assemble x Jupiters at 27 hours per micro $= 27x$ hours

Total time to assemble y Cosmos at 34 hours per micro $= 34y$ hours

Thus

$$\text{Total time to assemble computers} = (27x + 34y) \text{ hours}$$

Since the available time is 800 hours, we get the constraint

$$27x + 34y \le 800$$

Let us now look at a linear programming problem that involves determining maximum profit under both monetary and time constraints.

A Linear Programming Problem

A company manufactures two types of hand calculators, model Calc1 and model Calc2. It takes 5 hours and 2 hours to manufacture a Calc1 and a Calc2, respectively. The company has 900 hours available per week for the production of calculators. The manufacturing cost of each Calc1 is $8, and the manufacturing cost of a Calc2 is $10. The total funds available per week for production are $2800. The profit on each Calc1 is $3, and the profit on each Calc2 is $2. How many of each type of calculator should be manufactured weekly to obtain maximum profit?

In this problem there are two constraints, namely time and money. The aim of the company is to maximize profit under these constraints. We can solve the problem in three stages:

1. Construct the mathematical model—that is, the mathematical description of the problem.

2. Illustrate the mathematics by means of a graph.

3. Use the graph to determine the solution.

Step 1 **The Mathematical Model**   We first find the inequalities that describe the time and monetary constraints. Let the company manufacture x of Calc1 and y of Calc2.

Time Constraint: time to manufacture x Calc1 at 5 hours each $= 5x$ hours

time to manufacture y Calc2 at 2 hours each $= 2y$ hours

Thus   total manufacturing time $= (5x + 2y)$ hours

There are 900 hours available. Therefore

$$5x + 2y \leq 900$$

Monetary Constraint: cost of manufacturing x Calc1 at $8 each $= \$8x$

cost of manufacturing y Calc2 at $10 each $= \$10y$

Thus   total production costs $= \$(8x + 10y)$

There is $2800 available for production of calculators. Therefore

$$8x + 10y \leq 2800$$

Furthermore, x and y represent numbers of calculators manufactured. These numbers cannot be negative. We therefore get two more constraints:

$$x \geq 0, \quad y \geq 0$$

The constraints on the production process are thus described by the following **system of linear inequalities**.

$$5x + 2y \leq 900$$
$$8x + 10y \leq 2800$$
$$x \geq 0$$
$$y \geq 0$$

We next find a mathematical expression for profit.

Profit: weekly profit on *x* calculators at \$3 per calculator = \$3*x*

weekly profit on *y* calculators at \$2 per calculator = \$2*y*

Thus   total weekly profit = \$(3*x* + 2*y*)

Let us introduce a **profit function**

$$f = 3x + 2y$$

The problem thus reduces mathematically to finding values of *x* and *y* that give the maximum value of *f* under the above constraints. Such a function *f* is called the **objective function** of the linear programming problem.

Step 2 **Graphical Representation of the Constraints** We now find the points (x, y) that satisfy all the inequalities in the above system; these are called the **solutions** to the system. It can be shown that the graph of $5x + 2y \leq 900$ consists of points on and below the line $5x + 2y = 900$. Similarly, the points that satisfy the inequality $8x + 10y \leq 2800$ are on and below the line $8x + 10y = 2800$. The condition $x \geq 0$ is satisfied by all points on and to the right of the *y* axis. The condition $y \geq 0$ is satisfied by all points on and above the *x* axis. The set of solutions to the system of inequalities is the region that is common to all these regions—the shaded region in Figure 10.2. Such a region, that satisfies all the constraints, is called the **feasible region** of the linear programming problem and the points in this region are called **feasible solutions**. Any such point corresponds to a possible schedule for production. Among these points is one (or possibly more than one) that gives a maximum value to the objective functions. Such a point is called an **optimal solution** to the linear programming problem. The next step will be to find this point.

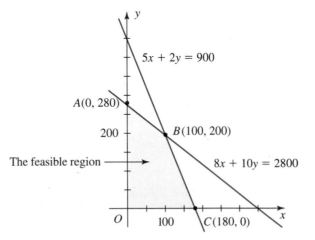

Figure 10.2

Step 3 **Determining the Optimal Solution** It would be an endless task to examine the value of the objective function $f = 3x + 2y$ at all feasible solutions. It has been proved that the maximum value of *f* will occur at a vertex of the feasible region, namely at one of the points *A*, *B*, *C*, or *O*; or, if there is more than one point at which the maximum occurs, it will be along one edge of the region, such as AB or BC. Hence we have only to examine the values of $f = 3x + 2y$ at the vertices A, B, C, and O. These vertices are found by determining the points of intersection of the lines. We get

$$A(0, 280): \quad f_A = 3(0) \quad + 2(280) = 560$$
$$B(100, 200): \quad f_B = 3(100) + 2(200) = 700$$
$$C(180, 0): \quad f_C = 3(180) + 2(0) \quad = 540$$
$$O(0, 0): \quad f_O = 3(0) \quad + 2(0) \quad = \quad 0$$

Thus the maximum value of f is 700 and it occurs at B, namely when $x = 100$ and $y = 200$. The interpretation of these results is that the maximum weekly profit is \$700 and this occurs when 100 Calc1 calculators and 200 Calc2 calculators are manufactured.

We now look at a linear programming problem that has many solutions.

Example 3 Find the maximum value of $f = 8x + 2y$ under the constraints

$$4x + \quad y \le 32$$
$$4x + 3y \le 48$$
$$x \ge \quad 0$$
$$y \ge \quad 0$$

Solution The constraints are represented graphically by the shaded region in Figure 10.3. The vertices of the feasible region are found and the value of the objective function calculated at each vertex.

$$A(0, 16): \quad f_A = 8(0) + 2(16) = 32$$
$$B(6, 8): \quad f_B = 8(6) + 2(8) \quad = 64$$
$$C(8, 0): \quad f_C = 8(8) + 2(0) \quad = 64$$
$$O(0, 0): \quad f_O = 8(0) + 2(0) \quad = \quad 0$$

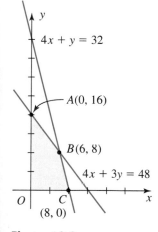

4x + y = 32

A(0, 16)

B(6, 8)

4x + 3y = 48

O

C
(8, 0)

x

Figure 10.3

The maximum value of f is 64 and this occurs at two adjacent vertices B and C. When this happens in a linear programming problem, the objective function will have the same value at every point on the edge joining the adjacent vertices. Thus f has a maximum value of 64 at all points along BC. The equation of BC is $y = -4x + 32$. Thus any point along the line $y = -4x + 32$ from (6, 8) to (8, 0) is an optimal solution. At $x = 7$, $y = 4$, for example, we have $f = 8(7) + 2(4) = 64$.

If x and y correspond to numbers of items manufactured and f to profit, as in this example, then there is a certain flexibility in the production schedule. Any schedule corresponding to a point along BC will lead to maximum profit.

Minimum Value of a Function

A problem that involves determining the minimum value of an objective function f can be solved by looking for the maximum value of $-f$, the negative of f, over the same feasible region. The general result is as follows.

> The minimum value of a function f over a region S occurs at the point(s) of maximum value of $-f$, and is the negative of that maximum value.

Let us derive this result. Let f have minimum value f_A at the point A in the region. Then if B is any other point in the region,

$$f_A \leq f_B$$

Multiply both sides of this inequality by -1 to get

$$-f_A \geq -f_B$$

This implies that A is a point of maximum value of $-f$. Furthermore, the minimum value is f_A, the negative of the maximum value $-f_A$. The above steps can be reversed, proving that the converse holds, thus verifying the result.

Example 4   Find the minimum value of $f = 2x - 3y$ under the constraints

$$
\begin{aligned}
x + 2y &\leq 10 \\
2x + y &\leq 11 \\
x &\geq 0 \\
y &\geq 0
\end{aligned}
$$

The feasible region is shown in Figure 10.4. To find the minimum value of f over the region, let us determine the maximum value of $-f$. The vertices are found and the value of $-f$ computed at each vertex.

$$
\begin{aligned}
A(0, 5): \quad &-f_A = -2(0) + 3(5) = 15 \\
B(4, 3): \quad &-f_B = -2(4) + 3(3) = 1 \\
C(\tfrac{11}{2}, 0): \quad &-f_C = -2(\tfrac{11}{2}) + 3(0) = -11 \\
O(0, 0): \quad &-f_O = -2(0) + 3(0) = 0
\end{aligned}
$$

The maximum value of $-f$ is 15 at A. Thus the minimum value of $f = 2x - 3y$ is -15 when $x = 0$, $y = 5$.

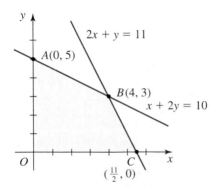

Figure 10.4

Discussion of the Method

Notice that each feasible region we have discussed is such that the whole of the segment of a straight line joining any two points within the region lies within that region. Such a region is called **convex**. See Figure 10.5. A theorem states that *the feasible region in a linear programming problem is convex.*

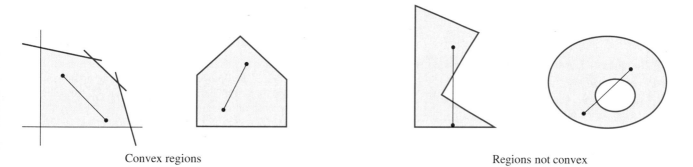

Convex regions   Regions not convex

Figure 10.5

We now give a geometrical explanation of why we can expect the maximum value of the objective function to occur at a vertex, or along one side of the feasible region. Let $S = ABCO$ in Figure 10.6(a) be a feasible region and let the objective function be $f = ax + by$. f has a value at each point within S. Let $P(x_p, y_p)$ be a point in S. Then $f_p = ax_p + by_p$. This implies that P lies on the line $ax + by = f_p$. The y intercept of this line is f_p/b. Therefore f_p will have a maximum value when the y intercept is a maximum. It can be seen from the figure that this will occur when P is the vertex B.

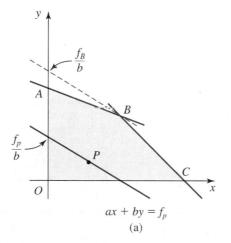
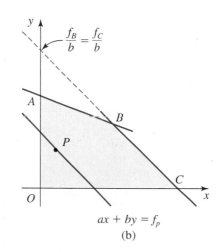

Figure 10.6   (a)   (b)

Some lines $ax + by = f_p$ will have a maximum y intercept when P is at B, as above; others will have a maximum y intercept when P is at A or when P is at C. The vertex will depend upon the slope of the line. If the line is parallel to an edge of the region S, then the y intercept will be a maximum when P is any point along that edge. See Figure 10.6(b).

Exercise Set 10.1

In Exercises 1–5, solve the linear programming problems.

***1.** Maximize $f = 2x + y$ subject to
$$4x + y \leq 36$$
$$4x + 3y \leq 60$$
$$x \geq 0, y \geq 0$$

2. Maximize $f = x - 4y$ subject to
$$x + 2y \leq 4$$
$$x + 6y \leq 8$$
$$x \geq 0, y \geq 0$$

***3.** Maximize $f = 4x + 2y$ subject to
$$x + 3y \leq 15$$
$$2x + y \leq 10$$
$$x \geq 0, y \geq 0$$

4. Maximize $f = 2x + y$ subject to
$$4x + y \leq 16$$
$$x + y \leq 7$$
$$x \geq 0, y \geq 0$$

***5.** Maximize $f = 4x + y$ subject to
$$2x + y \leq 4$$
$$6x + y \leq 8$$
$$x \geq 0, y \geq 0$$

In Exercises 6–8, solve the linear programming problems.

***6.** Minimize $f = -3x - y$ subject to
$$x + y \leq 150$$
$$4x + y \leq 450$$
$$x \geq 0, y \geq 0$$

7. Minimize $f = -2x + y$ subject to
$$2x + y \leq 440$$
$$4x + y \leq 680$$
$$x \geq 0, y \geq 0$$

***8.** Minimize $f = -x + 2y$ subject to
$$x + 2y \leq 4$$
$$x + 4y \leq 6$$
$$x \geq 0, y \geq 0$$

***9.** A company manufactures two types of hand calculators, model $C1$ and model $C2$. It takes 1 hour and 4 hours in labor time to manufacture each $C1$ and $C2$, respectively. The cost of manufacturing the $C1$ is $30 and that of manufacturing a $C2$ is $20. The company has 1600 hours of labor time available and $18,000 in running costs. The profit on each $C1$ is $10 and on each $C2$ is $8. What should the production schedule be to ensure maximum profit?

10. A company manufactures two products, X and Y. Two machines, I and II, are needed to manufacture each product. It takes 3 minutes on each machine to produce an X. To produce a Y takes 1 minute on machine I and 2 minutes on machine II. The total time available on machine I is 3000 minutes, and on machine II the time available is 4500 minutes. The company realizes a profit of $15 on each X and $7 on each Y. What should the production schedule be to make the largest profit?

***11.** A refrigerator company has two plants in towns X and Y. Its refrigerators are sold in a certain town Z. It takes 20 hours (packing, transportation, and so on) to transport a refrigerator from X to Z and 10 hours from Y to Z. It costs $60 to transport each refrigerator from X to Z, and $10 per refrigerator from Y to Z. A total of 1200 hours of labor time is available, and a total of $2400 is budgeted for transportation. The profit on each refrigerator from X is $40, and the profit on each refrigerator from Y is $20. How should the company schedule the transportation of refrigerators to maximize profit?

12. The maximum daily production of an oil refinery is 1400 barrels. The refinery can produce two types of oil: gasoline for automobiles and heating oil for domestic purposes. The production costs per barrel are $6 for gasoline and $8 for heating oil. The daily production budget is $9600. The profit per barrel is $3.50 on gasoline and $4 on heating oil. What is the maximum profit that can be realized daily, and what quantities of each type of oil are then produced?

*13. A tailor has 80 square yards of cotton material and 120 square yards of woolen material. A suit requires 2 square yards of cotton and 1 square yard of wool. A dress requires 1 square yard of cotton and 3 square yards of wool. How many of each garment should the tailor make to maximize income, if a suit and a dress each sell for $90? What is the maximum income?

14. A city has $600,000 to purchase cars. Two cars, the Arrow and the Gazelle, are under consideration, costing $4000 and $5000, respectively. The estimated annual maintenance cost on the Arrow is $400 and on the Gazelle it is $300. The city will allocate $40,000 for the total annual maintenance of these cars. The Arrow gets 24 miles per gallon and the Gazelle gets 20 miles per gallon. The city wants to maximize the "gasoline efficiency number" of this group of cars. For x Arrows and y Gazelles this number would be $24x + 20y$. How many of each model should be purchased?

*15. A car dealer imports foreign cars by way of two ports of entry, A and B. City C needs 120 cars, and city D needs 180. There are 100 cars available at A and 200 at B. It takes 2 hours to transport each car from A to C and 6 hours from B to C. It takes 4 hours and 3 hours, respectively, to transport each car from A and B to D. The dealer has 1030 hours in driver time available to move the cars. As many cars as possible should be moved from B to C, as drivers will not be available for this route in the future. What schedule will achieve this?

16. A school district is buying new buses. It has a choice of two types. The Torro costs $18,000 and holds 25 passengers. The Sprite costs $22,000 and holds 30 passengers. $572,000 has been budgeted for the new buses. A maximum of 30 drivers will be available to drive the buses. At least 17 Sprite buses must be ordered because of the desirability of having a certain number of large-capacity buses. How many of each type should be purchased to carry a maximum number of students?

*17. A manufacturer makes two types of fertilizer, X and Y, using chemicals A and B. Fertilizer X is made up of

80% chemical A and 20% chemical B. Fertilizer Y is made up of 60% chemical A and 40% chemical B. The manufacturer requires at least 30 tons of fertilizer X and at least 50 tons of fertilizer Y. He has available 100 tons of chemical A and 50 tons of B. He wants to make as much fertilizer as possible. What quantities of X and Y should be produced?

*18. A hospital wants to design a dinner menu containing two items, M and N. Each ounce of M provides 1 unit of vitamin A and 2 units of vitamin B. Each ounce of N provides 1 unit of vitamin A and 1 unit of vitamin B. The two dishes must provide at least 7 units of vitamin A and at least 10 units of vitamin B. If each ounce of M costs 8 cents and each ounce of N costs 12 cents, how many ounces of each item should the hospital serve to minimize cost?

19. A clothes manufacturer has 10 square yards of cotton material, 10 square yards of wool material, and 6 square yards of silk material. A pair of slacks requires 1 square yard of cotton, 2 square yards of wool, and 1 square yard of silk. A skirt requires 2 square yards of cotton, 1 square yard of wool, and 1 square yard of silk. The net profit on a pair of slacks is $3 and the net profit on a skirt is $4. How many skirts and how many slacks should be made to maximize profit?

20. *(a) A shipper has trucks that can carry a maximum of 12,000 pounds of cargo with a maximum volume of 9000 cubic feet. He ships for two companies: Pringle Co. has packages weighing 5 pounds each and a volume of 5 cubic feet, and Williams Co. packages weigh 6 pounds with a volume of 3 cubic feet. By contract, the shipper makes 30 cents on each package from Pringle and 40 cents on each package from Williams. How many packages from each should the shipper carry?

(b) A lawyer points out the fine print at the bottom of the Pringle contract, which says that the shipper must carry at least 240 packages from Pringle. How should the shipper now divide the work? How much did the clause cost him in profit?

*10.2   The Simplex Method

The graphical method of solving a linear programming problem introduced in the preceding section has its limitations. The method demonstrated for two variables can be extended to linear programming problems involving three variables; the feasible region then is a convex subset of three-dimensional space. However, for problems involving more than two variables, the geometrical approach becomes impractical. We now introduce the **simplex method**, an algebraic method that can be used for any number of variables. This method was developed by George B. Dantzig of the University of California, Berkeley, in 1947. The simplex method has the advantage of being readily programmable for a computer. Today, "packages" of computer programs based on the simplex method are offered commercially by software companies. Industrial customers often pay sizable monthly fees for the use of these programs. Exxon, for example, currently uses linear programming in the scheduling of its drilling operations, in the allocation of crude oil among its refineries, in the control of refinery operations, in the distribution of its products, and in the planning of business strategy. It has been estimated that linear programming accounts for between 5% and 10% of the company's total computing time. The simplex method involves reformulating constraints in terms of linear equations and then using elementary row operations in a manner very similar to that of Gauss-Jordan elimination, to arrive at the solution.

We introduce the simplex method by using it to arrive at the solution of the first linear programming problem discussed in the previous section. Let us find the maximum value of $f = 3x + 2y$ subject to the constraints

$$
\begin{aligned}
5x + 2y &\le 900 \\
8x + 10y &\le 2800 \\
x &\ge 0 \\
y &\ge 0
\end{aligned}
$$

Consider the first inequality, $5x + 2y \le 900$. For each x and y that satisfy this condition, there will exist a value for a nonnegative variable u such that

$$5x + 2y + u = 900$$

The value of u will be the number that must be added to $5x + 2y$ to bring it up to 900. Similarly, for each x and y that satisfy $8x + 10y \le 2800$, there will be a value for a nonnegative variable v such that $8x + 10y + v = 2800$. Such variables u and v are called **slack variables**—they make up the slack in the original variables.

Thus the constraints in the linear programming problem may be written

$$
\begin{aligned}
5x + 2y + u &= 900 \\
8x + 10y + v &= 2800 \\
x \ge 0,\ y \ge 0,\ u \ge 0,\ v \ge 0
\end{aligned}
$$

Finally, the objective function $f = 3x + 2y$ is rewritten in the form

$$-3x - 2y + f = 0$$

The entire problem now becomes that of determining the solution to the following system of equations:

$$5x + 2y + u \qquad\qquad = 900$$
$$8x + 10y \qquad + v \quad = 2800$$
$$-3x - 2y \qquad\qquad + f = \quad 0$$

such that f is as large as possible, under the restrictions $x \geq 0$, $y \geq 0$, $u \geq 0$, $v \geq 0$.

We have thus reformulated the problem in terms of a system of linear equations under certain constraints. The system of equations, consisting of three equations in the five variables x, y, u, v, and f will have many solutions in the region defined by $x \geq 0$, $y \geq 0$, $u \geq 0$, $v \geq 0$. Any such solution is called a **feasible solution**. A solution that maximizes f is called an **optimal solution**; this is the solution we are interested in.

We determine the optimal solution by using elementary row operations in an algorithm called the **simplex algorithm**. The method of Gauss-Jordan elimination involved selecting pivots and using them to create zeros in columns in a systematic manner. In the simplex algorithm, pivots are again selected and used to create zeros in columns, but using different criteria than were used in Gauss-Jordan elimination.

The Simplex Algorithm

1. Write the augmented matrix of the system of equations. This is called the **initial simplex tableau**.

2. Locate the negative element in the last row, other than the last element, that is largest in magnitude. (If two or more entries share this property, any one of these can be selected.) If all such entries are nonnegative, the tableau is in final form.

3. Divide each positive element in the column defined by this negative entry into the corresponding element of the last column.

4. Select the divisor that yields the smallest quotient. This element is called a **pivot element**. (If two or more elements share this property, any one of these can be selected as pivot.)

5. Use row operations to create a 1 in the pivot location and zeros elsewhere in the pivot column.

6. Repeat steps 2–5 until such negative elements have been eliminated from the last row. The final matrix is called the **final simplex tableau**. It leads to the optimal solution.

Let us apply the simplex algorithm to our linear programming problem. Starting with the initial simplex tableau, we get

$$
\begin{bmatrix}
\boxed{5} & 2 & 1 & 0 & 0 & 900 \\
8 & 10 & 0 & 1 & 0 & 2800 \\
-3 & -2 & 0 & 0 & 1 & 0
\end{bmatrix}
$$

pivot since

$$\frac{900}{5} < \frac{2800}{8}$$

$$
\approx \atop (1/5)R1
\begin{bmatrix}
\boxed{1} & \frac{2}{5} & \frac{1}{5} & 0 & 0 & 180 \\
8 & 10 & 0 & 1 & 0 & 2800 \\
-3 & -2 & 0 & 0 & 1 & 0
\end{bmatrix}
$$

$$
\approx \atop {R2 - (8)R1 \atop R3 + (3)R1}
\begin{bmatrix}
1 & \frac{2}{5} & \frac{1}{5} & 0 & 0 & 180 \\
0 & \boxed{\frac{34}{5}} & -\frac{8}{5} & 1 & 0 & 1360 \\
0 & -\frac{4}{5} & \frac{3}{5} & 0 & 1 & 540
\end{bmatrix}
$$

$$
\approx \atop (5/34)R2
\begin{bmatrix}
1 & \frac{2}{5} & \frac{1}{5} & 0 & 0 & 180 \\
0 & \boxed{1} & -\frac{4}{17} & \frac{5}{34} & 0 & 200 \\
0 & -\frac{4}{5} & \frac{3}{5} & 0 & 1 & 540
\end{bmatrix}
$$

$$
\approx \atop {R1 - (2/5)R2 \atop R3 + (4/5)R2}
\begin{bmatrix}
1 & 0 & \frac{25}{85} & -\frac{1}{17} & 0 & 100 \\
0 & 1 & -\frac{4}{17} & \frac{5}{34} & 0 & 200 \\
0 & 0 & \frac{7}{17} & \frac{2}{17} & 1 & 700
\end{bmatrix}
$$

We have arrived at the final tableau; all negative elements have been eliminated from the last row. This tableau corresponds to the following system of equations.

$$
\begin{aligned}
x + \quad \tfrac{25}{85}u - \tfrac{1}{17}v \quad\quad &= 100 \\
y - \tfrac{4}{17}u + \tfrac{5}{34}v \quad\quad &= 200 \\
\tfrac{7}{17}u + \tfrac{2}{17}v + f &= 700
\end{aligned}
$$

The solutions to this system are, of course, identical to those of the original system, since it has been derived from the original system using elementary row operations. Since $u \geq 0$, $v \geq 0$, the last equation tells us that f has a maximum value of 700 that takes place when $u = 0$ and $v = 0$. On substituting these values of u and v back into the system, we get $x = 100$, $y = 200$. Thus the maximum value of $f = 3x + 2y$ is $f = 700$, and this takes place when $x = 100$, $y = 200$. This result agrees with that obtained in the previous section, using geometry.

Note that the reasoning used in arriving at this maximum value of f implies that the element in the last row and last column of the final tableau will always correspond to the maximum value of f.

The next example illustrates the application of the simplex method for a function of three variables.

Example 1   Determine the maximum value of the function $f = 3x + 5y + 8x$ subject to the following constraints.

$$x + y + z \le 100$$
$$3x + 2y + 4z \le 200$$
$$x + 2y \le 150$$
$$x \ge 0, y \ge 0, z \ge 0$$

The corresponding system of equations with slack variables u, v, and w is

$$x + y + z + u = 100$$
$$3x + 2y + 4z + v = 200$$
$$x + 2y + w = 150$$
$$-3x - 5y - 8z + f = 0$$

with $x \ge 0, y \ge 0, z \ge 0, u \ge 0, v \ge 0, w \ge 0$

The simplex tableaux are as follows

$$\begin{bmatrix} 1 & 1 & 1 & 1 & 0 & 0 & 0 & 100 \\ 3 & 2 & ④ & 0 & 1 & 0 & 0 & 200 \\ 1 & 2 & 0 & 0 & 0 & 1 & 0 & 150 \\ -3 & -5 & -8 & 0 & 0 & 0 & 1 & 0 \end{bmatrix}$$

$$\underset{(1/4)R2}{\approx} \begin{bmatrix} 1 & 1 & 1 & 1 & 0 & 0 & 0 & 100 \\ \frac{3}{4} & \frac{1}{2} & ① & 0 & \frac{1}{4} & 0 & 0 & 50 \\ 1 & 2 & 0 & 0 & 0 & 1 & 0 & 150 \\ -3 & -5 & -8 & 0 & 0 & 0 & 1 & 0 \end{bmatrix}$$

$$\underset{\substack{R1 - R2 \\ R4 + (8)R2}}{\approx} \begin{bmatrix} \frac{1}{4} & \frac{1}{2} & 0 & 1 & -\frac{1}{4} & 0 & 0 & 50 \\ \frac{3}{4} & \frac{1}{2} & 1 & 0 & \frac{1}{4} & 0 & 0 & 50 \\ 1 & ② & 0 & 0 & 0 & 1 & 0 & 150 \\ 3 & -1 & 0 & 0 & 2 & 0 & 1 & 400 \end{bmatrix}$$

$$\underset{(1/2)R3}{\approx} \begin{bmatrix} \frac{1}{4} & \frac{1}{2} & 0 & 1 & -\frac{1}{4} & 0 & 0 & 50 \\ \frac{3}{4} & \frac{1}{2} & 1 & 0 & \frac{1}{4} & 0 & 0 & 50 \\ \frac{1}{2} & ① & 0 & 0 & 0 & \frac{1}{2} & 0 & 75 \\ 3 & -1 & 0 & 0 & 2 & 0 & 1 & 400 \end{bmatrix}$$

$$\underset{\substack{R1 - (1/2)R3 \\ R2 - (1/2)R3 \\ R4 + R3}}{\approx} \begin{bmatrix} 0 & 0 & 0 & 1 & -\frac{1}{4} & -\frac{1}{4} & 0 & \frac{25}{2} \\ \frac{1}{2} & 0 & 1 & 0 & \frac{1}{4} & -\frac{1}{4} & 0 & \frac{25}{2} \\ \frac{1}{2} & 1 & 0 & 0 & 0 & \frac{1}{2} & 0 & 75 \\ \frac{7}{2} & 0 & 0 & 0 & 2 & \frac{1}{2} & 1 & 475 \end{bmatrix}$$

maximum value of the function

This final tableau gives the following system of equations.

$$u - \tfrac{1}{4}v - \tfrac{1}{4}w = 12\tfrac{1}{2}$$
$$\tfrac{1}{2}x + z + \tfrac{1}{4}v - \tfrac{1}{4}w = 12\tfrac{1}{2}$$
$$\tfrac{1}{2}x + y + \tfrac{1}{2}w = 75$$
$$3\tfrac{1}{2}x + 2v + \tfrac{1}{2}w + f = 475$$

The constraints are

$$x \geq 0,\ y \geq 0,\ z \geq 0,\ u \geq 0,\ v \geq 0,\ w \geq 0$$

The final equation, under these constraints, implies that f has a maximum of 475 when $x = 0$, $v = 0$, $w = 0$. On substituting these values back into the equations, we get $y = 75$, $z = 12\tfrac{1}{2}$, $u = 12\tfrac{1}{2}$.

Thus $f = 3x + 5y + 8z$ has a maximum value of 475 at $x = 0$, $y = 75$, $z = 12\tfrac{1}{2}$.

The next example illustrates the application of the simplex method when there are many optimal solutions. (We use Example 3 of the previous section.)

Example 2   Find the maximum value of $f = 8x + 2y$ subject to the constraints

$$4x + y \leq 32$$
$$4x + 3y \leq 48$$
$$x \geq 0$$
$$y \geq 0$$

The simplex tableaux are as follows:

$$\begin{bmatrix} ④ & 1 & 1 & 0 & 0 & 32 \\ 4 & 3 & 0 & 1 & 0 & 48 \\ -8 & -2 & 0 & 0 & 1 & 0 \end{bmatrix} \approx \begin{bmatrix} ① & \tfrac{1}{4} & \tfrac{1}{4} & 0 & 0 & 8 \\ 4 & 3 & 0 & 1 & 0 & 48 \\ -8 & -2 & 0 & 0 & 1 & 0 \end{bmatrix}$$

$$\approx \begin{bmatrix} 1 & \tfrac{1}{4} & \tfrac{1}{4} & 0 & 0 & 8 \\ 0 & 2 & -1 & 1 & 0 & 16 \\ 0 & 0 & 2 & 0 & 1 & 64 \end{bmatrix}$$

The final tableau gives the following system of equations

$$x + \tfrac{1}{4}y + \tfrac{1}{4}u = 8$$
$$2y - u + v = 16$$
$$2u + f = 64$$

with $x \geq 0$, $y \geq 0$, $u \geq 0$, $v \geq 0$.

The last equation implies that f has a maximum value of 64 when $u = 0$. Substituting $u = 0$ into the other equations gives

$$x + \tfrac{1}{4}y \quad\;\; = \;\; 8$$
$$2y + v = 16$$

with $x \geq 0$, $y \geq 0$, $v \geq 0$.

Any point (x, y) that satisfies these conditions is an optimal solution. The second equation tells us that v can be any number in the interval $[0, 16]$ and that y can be any number in the interval $[0, 8]$. The first equation then tells us that $x = 8$ when $y = 0$, and $x = 6$ when $y = 8$.

Thus the maximum value of $f = 8x + 2y$ is 64. This is achieved at a point on the line $x + \tfrac{1}{4}y = 8$ between $(6, 8)$ and $(8, 0)$. This result agrees with that of Example 3 of the previous section when the geometric approach was used.

In this section we have introduced the simplex algorithm for maximizing a function f when all constraints are of the type $ax + by + \cdots \leq k$, where k is nonnegative, and all the variables are nonnnegative. This is called a **standard linear programming problem**. Constraints in linear programming problems can involve $=$, $\leq$, and $\geq$, and some of the variables can be negative. The way slack variables are used in setting up an initial tableau varies, depending upon the types of constraints. Once the initial tableau has been constructed, however, the above algorithm is used to arrive at a final tableau that leads to the solution. Students who are interested in reading more about linear programming are referred to *Applied Linear Programming* by Michael R. Greenberg, Academic Press, 1978. Dantzig's discovery of the simplex method ranks high among the achievements of twentieth century mathematics.

Exercise Set 10.2

In Exercises 1–9, use the simplex method to maximize the functions under the given constraints. (Exercises 1 and 2 were solved using geometry in the previous section.)

*1. Maximize $f = 2x + y$ subject to
$$4x + y \leq 36$$
$$4x + 3y \leq 60$$
$$x \geq 0, y \geq 0$$

*2. Maximize $f = x - 4y$ subject to
$$x + 2y \leq 4$$
$$x + 6y \leq 8$$
$$x \geq 0, y \geq 0$$

3. Maximize $f = 4x + 6y$ subject to
$$x + 3y \leq 6$$
$$3x + y \leq 8$$
$$x \geq 0, y \geq 0$$

*4. Maximize $f = 10x + 5y$ subject to
$$x + y \leq 180$$
$$3x + 2y \leq 480$$
$$x \geq 0, y \geq 0$$

5. Maximize $f = x + 2y + z$ subject to
$$3x + y + z \leq 3$$
$$x - 10y - 4z \leq 20$$
$$x \geq 0, y \geq 0, z \geq 0$$

*6. Maximize $f = 100x + 200y + 50z$ subject to
$$5x + 5y + 10z \leq 1000$$
$$10x + 8y + 5z \leq 2000$$
$$10x + 5y \qquad\;\; \leq 500$$
$$x \geq 0, y \geq 0, z \geq 0$$

*7. Maximize $f = 2x + 4y + z$ subject to
$$-x + 2y + 3z \le 6$$
$$-x + 4y + 5z \le 5$$
$$-x + 5y + 7z \le 7$$
$$x \ge 0,\ y \ge 0,\ z \ge 0$$

8. Maximize $f = x + 2y + 4z - w$ subject to
$$5x \qquad + 4z + 6w \le 20$$
$$4x + 2y + 2z + 8w \le 40$$
$$x \ge 0,\ y \ge 0,\ z \ge 0,\ w \ge 0$$

9. Maximize $f = x + 2y - z + 3w$ subject to
$$2x + 4y + 5z + 6w \le 24$$
$$4x + 4y + 2z + 2w \le 4$$
$$x \ge 0,\ y \ge 0,\ z \ge 0,\ w \ge 0$$

*10. A company uses three machines—I, II, and III—to produce items X, Y, and Z. The production of each item involves the use of more than one machine. It takes 2 minutes on I and 4 on II to manufacture a single X. It takes 3 minutes on I and 6 minutes on III to manufacture a Y. It takes 1 minute on I, 2 minutes on II, and 3 minutes on III to manufacture a Z. The total time available on each machine per day is 6 hours. The profits are $10, $8, and $12 on each of X, Y, and Z, respectively. How should the company allocate the production times on the machines to maximize total profit?

11. An industrial furniture company manufactures desks, cabinets, and chairs. These items involve metal, wood, and plastic. The following table gives the amounts that go into each product (in convenient units) and the profit on each item.

	Metal	Wood	Plastic	Profit
Desk	3	4	2	$16
Cabinet	6	1	1	$12
Chair	1	2	2	$ 6

If the company has 800 units of metal, 400 units of wood, and 100 units of plastic available, how should it allocate these resources to maximize total profit?

*12. A company produces washing machines at three factories, A, B, and C. The washing machines are sold in a certain city P. It costs $10, $20, and $40 to transport each washing machine from A, B, and C, respectively, to P. It involves 6, 4, and 2 hours of packing and transportation time to get a washing machine from A, B, and C, respectively, to P. There is $6000 budgeted weekly for transportation of the washing machines to P, and a total of 4000 hours of labor is available. The profit on each machine from A is $12, on each from B is $20, and on each from C is $16. How should the company schedule the transportation of washing machines from A, B, and C to P to maximize total profit?

*13. A manufacturer makes three lines of tents, all from the same material. The Aspen, Alpine, and Cub tents require 60, 30, and 15 square yards of material, respectively. The manufacturing costs of the Aspen, Alpine, and Cub are $32, $20, and $12. The material is available in amounts of 7800 square yards weekly, and the weekly working budget is $8320. The profits on the Aspen, Alpine, and Cub are $12, $8, and $4. What should the weekly production schedule of these tents be to maximize total profit?

In Exercises 14–16, use the simplex method to minimize the functions under the given constraints. (Exercise 14 was solved using geometry in the previous section.)

14. Minimize $f = -x + 2y$ subject to
$$x + 2y \le 4$$
$$x + 4y \le 6$$
$$x > 0,\ y > 0$$

*15. Minimize $f = -2x + y$ subject to
$$2x + 2y \le 8$$
$$x - y \le 6$$
$$x \ge 0,\ y \ge 0$$

*16. Minimize $f = 2x + y - z$ subject to
$$x + 2y - 2z \le 20$$
$$2x + y \qquad \le 10$$
$$x + 3y + 4z \le 15$$
$$x \ge 0,\ y \ge 0,\ z \ge 0$$

*10.3   Geometrical Explanation of the Simplex Method

We now explain, by means of an example, the sequence of transformations used in the simplex method. Let us maximize the function $f = 2x + 3y$ subject to the constraints

$$x + 2y \leq 8$$
$$3x + 2y \leq 12$$
$$x \geq 0$$
$$y \geq 0$$

(1)

The region that satisfies these constraints is $ABCO$ in Figure 10.7. Recall that any point in this region is called a **feasible solution**. A feasible solution that gives a maximum value of f is called an **optimal solution**.

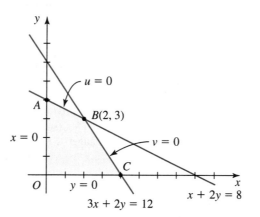

Figure 10.7

Let us reformulate the constraints using slack variables u and v:

$$x + 2y + u \qquad = 8$$
$$3x + 2y \qquad + v \quad = 12$$
$$-2x - 3y \qquad \quad + f = 0$$

(2)

with $x \geq 0$, $y \geq 0$, $u \geq 0$, $v \geq 0$.

Observe that $u = 0$ along AB since AB is a segment of the line $x + 2y = 8$. $v = 0$ along BC since BC is a segment of the line $3x + 2y = 12$. Furthermore, $x = 0$ along OA and $y = 0$ along OC. Thus the boundaries of the feasible region $ABCO$ are such that one of the variables is zero along each boundary.

Let us now look at the vertices in terms of the four variables x, y, u, and v. Vertex O is the point $x = 0$, $y = 0$, $u = 8$, $v = 12$. (One obtains the u and v values at O by letting $x = 0$ and $y = 0$ in the constraints.) Vertex A lies on OA and AB. Thus $x = 0$ and $u = 0$ at A. Substituting these values into the constraints gives $y = 4$ and $v = 4$. Thus A is the point $x = 0$, $y = 4$, $u = 0$, $v = 4$. In a similar manner we find that B is the point $x = 2$, $y = 3$, $u = 0$, $v = 0$, and C is the point $x = 4$, $y = 0$, $u = 4$, $v = 0$.

Observe that certain variables are zero while others are nonzero at each vertex. The variables that are not zero are called **basic variables**, and the remaining variables are called **nonbasic variables**. We summarize the results thus far in Table 10.1.

Table 10.1

Vertex	Coordinates	Basic Variables ($\neq 0$)	Nonbasic Variables ($= 0$)
O	$x = 0, y = 0, u = 8, v = 12$	u, v	x, y
A	$x = 0, y = 4, u = 0, v = 4$	y, v	x, u
B	$x = 2, y = 3, u = 0, v = 0$	x, y	u, v
C	$x = 4, y = 0, u = 4, v = 0$	x, u	y, v

The method that we shall now develop starts at a vertex feasible solution, O in our case, and then proceeds through a sequence of adjacent vertices, each one giving an increased value of f, until an optimal solution is reached. The initial simplex tableau corresponds (in a way to be explained later) to the situation at the initial feasible solution; further tableaux represent the pictures at other vertices. We must examine each vertex in turn to see if it is an optimal solution. If it is not an optimal solution, we have to decide which neighboring vertex to move to next. Our aims are first to develop the tools to carry out the procedure geometrically, and then to translate the geometrical concepts into analogous algebraic ones. The advantage of carrying out the procedure algebraically is that it can then be implemented on the computer. We shall find that the algebraic procedure that results is the simplex algorithm.

Let us start at O, a vertex feasible solution. The value of f at O is 0. Let us determine whether it is necessary to move from O to a neighboring vertex. There are two vertices adjacent to O, namely A and C. In moving along OA, we find that $x = 0$ and that $f(= 2x + 3y)$ increases by 3 for every unit increase in y. On the other hand, in moving along OC, $y = 0$ and f increases by 2 for every unit increase in x. Because of the larger rate of increase in f along OA, this path is selected; A becomes the next feasible solution to be examined. The value of f at A is 12. This is indeed an improvement over the value of f at O.

Let us now determine whether or not A is optimal. At the vertex O we expressed f in terms of the nonbasic variables x and y; $f = 2x + 3y$. Let us now write f in terms of the nonbasic variables of the point A, namely x and u. To do this, we substitute the value of y from the first constraint in system (2) into $f = 2x + 3y$. We get $y = 4 - (x/2) - (u/2)$. The substitution gives $f = 12 + (x/2) - (3u/2)$. If we move from A to B along AB, then $u = 0$ along this route, and f increases by $\frac{1}{2}$ for every unit increase in x. Since f increases, we move to B. The value of f at B is 13, an increase over its value at A.

We now determine whether B is an optimal solution. The nonbasic variables at B are u and v. We express f in terms of these variables. From the original restrictions (2), we get $x = 2 + (u/2) - (v/2)$ and $y = 3 - (3u/4) + (v/4)$. Thus $f = 13 - (5u/4) - (v/4)$. In moving from B to C along BC, $v = 0$ and f decreases by $\frac{5}{4}$ for every unit increase in u. Thus f has a maximum value at B. The maximum value is 13.

We have developed the geometrical ideas. Let us now translate them into algebraic form. The initial feasible solution is O. The basic variables at O are u and v, with values 8 and 12, respectively. The value of f at O is 0. The initial tableau, corresponding to system of equations (2), reflects this information.

Tableau for Vertex O

coefficients of

$$
\begin{array}{ccccc}
x & y & u & v & f \\
\end{array}
$$

$$
\left[
\begin{array}{ccccc|c}
1 & 2 & 1 & 0 & 0 & 8 \\
3 & 2 & 0 & 1 & 0 & 12 \\
-2 & -3 & 0 & 0 & 1 & 0 \\
\end{array}
\right]
\begin{array}{l}
\leftarrow u \\
\leftarrow v \\
\leftarrow f
\end{array}
\left.\begin{array}{l} \\ \\ \end{array}\right\}
\begin{array}{l}
\text{basic} \\
\text{variables}
\end{array}
$$

The last column gives the values of the basic variables and f at O. Knowing the basic variables, we also know the nonbasic variables. This is the tableau associated with vertex O.

The geometric discussion told us that O was not an optimal solution, and led us to A. The basic variables at O are u and v; at A the basic variables are y and v. In going from O to A, y replaces u as a basic variable. u is called the **departing basic variable**, and y is called the **entering basic variable**. The entering basic variable y corresponds to the largest rate of increase of f, namely 3. In terms of the above tableau, this is reflected in y being the column corresponding to the negative entry, -3, in the last row, having largest magnitude. Determining the negative entry in the last row having largest magnitude enables us to select the entering variable. If no such entry exists, then there is no appropriate entering variable and we have arrived at the final tableau.

The next step is to determine the departing variable for the tableau. In going from O to A, the departing variable is u. The first two rows of the tableau correspond to the equations

$$
\begin{aligned}
x + 2y + u \quad &= 8 \\
3x + 2y \quad + v &= 12
\end{aligned}
$$

In going along OA, $x = 0$. The equations may be written

$$
\begin{aligned}
u &= 8 - 2y \\
v &= 12 - 2y
\end{aligned}
$$

Since $u \geq 0$ and $v \geq 0$, the maximum value to which y can increase is 4, when $u = 0$ and $v = 4$ (at the point A). The u becomes a nonbasic variable. Looking at the above equations, we see that the reason u arrives at the value 0 before v is that $\frac{8}{2} < \frac{12}{2}$. In terms of the tableau this corresponds to dividing the elements in the entering variable column into the corresponding elements of the last column. The row containing the numbers that give the smallest result gives the departing variable. Thus

$$
\begin{array}{ccccc}
x & y & u & v & f \\
\end{array}
$$

$$
\left[
\begin{array}{ccccc|c}
1 & 2 & 1 & 0 & 0 & 8 \\
3 & 2 & 0 & 1 & 0 & 12 \\
-2 & -3 & 0 & 0 & 1 & 0 \\
\end{array}
\right]
\begin{array}{l}
u \quad \leftarrow \text{departing variable} \\
v \\
f
\end{array}
$$

$$
\underset{\uparrow}{\text{entering variable}}
$$

We now have a method for selecting the entering and departing variables in any tableau. Let us now transform the above tableau into the tableau having the new basic variables y and v. The new tableau should have the values of the new basic variables and the value of f at the new vertex as its last column. The initial constraints, from which the initial tableau was derived, were as follows:

$$\begin{array}{rcl} x + 2y + u & = & 8 \\ 3x + 2y \quad + v & = & 12 \\ -2x - 3y \qquad + f & = & 0 \end{array}$$

Observe that the basic variables u and v appear in a single equation each, the coefficient being 1 in each case. It is this fact that causes u and v to assume that values 8 and 12 at O when the nonbasic variables are 0. Furthermore, the fact that the last equation involves f and only nonbasic variables causes f to assume the value 0 on the right-hand side of the equation. These are the characteristics that we must attempt to obtain for the tableau that represents the vertex A, in terms of the basic variables y and v of A. This is achieved by selecting the element that lies in the entering variable column and departing variable row as pivot, creating a 1 in its location and 0s elsewhere in this column. Thus the sequence of tableaux becomes as follows:

$$\begin{array}{ccccc} x & y & u & v & f \\ \begin{bmatrix} 1 & ② & 1 & 0 & 0 \\ 3 & 2 & 0 & 1 & 0 \\ -2 & -3 & 0 & 0 & 1 \end{bmatrix} & \begin{array}{c} 8 \\ 12 \\ 0 \end{array} & \begin{array}{l} u \leftarrow \\ v \\ f \end{array} \end{array} \approx \begin{array}{ccccc} x & y & u & v & f \\ \begin{bmatrix} \frac{1}{2} & ① & \frac{1}{2} & 0 & 0 \\ 3 & 2 & 0 & 1 & 0 \\ -2 & -3 & 0 & 0 & 1 \end{bmatrix} & \begin{array}{c} 4 \\ 12 \\ 0 \end{array} & \begin{array}{l} u \leftarrow \\ v \\ f \end{array} \end{array}$$

$$\approx \begin{array}{ccccc} x & y & u & v & f \\ \begin{bmatrix} \frac{1}{2} & 1 & \frac{1}{2} & 0 & 0 \\ 2 & 0 & -1 & 1 & 0 \\ -\frac{1}{2} & 0 & \frac{3}{2} & 0 & 1 \end{bmatrix} & \begin{array}{c} 4 \\ 4 \\ 12 \end{array} & \left.\begin{array}{l} y \\ u \\ f \end{array}\right\} \begin{array}{l} \text{new basic} \\ \text{variables} \\ \text{new value of } f \end{array} \end{array}$$

Let us verify that this tableau does indeed correspond to a system of equations that gives $y = 4$, $v = 4$, and $f = 12$ in the above order, if we take y and v as basic variables. The corresponding equations are as follows:

$$\begin{array}{rcl} \frac{1}{2}x + y + \frac{1}{2}u & = & 4 \\ 2x \quad - u + v & = & 4 \\ -\frac{1}{2}x \quad + \frac{3}{2}u \qquad + f & = & 12 \end{array}$$

Taking x and u as nonbasic variables, we make them zero, and we do indeed get $y = 4$, $v = 4$, $f = 12$. This is the tableau for the vertex A.

The analysis is now repeated for this tableau. Because of the negative entry $-\frac{1}{2}$ in the last row, this value of $f = 12$ is not the optimal solution. The new entering variable is x because of the negative sign in this column. The departing variable is v since $\frac{4}{2} < 4/\frac{1}{2}$. We get

$$
\begin{array}{ccccc}
x & y & u & v & f \\
\end{array}
$$

$$
\left[
\begin{array}{ccccc|c}
\frac{1}{2} & 1 & \frac{1}{2} & 0 & 0 & 4 \\
② & 0 & -1 & 1 & 0 & 4 \\
-\frac{1}{2} & 0 & \frac{3}{2} & 0 & 1 & 12
\end{array}
\right]
\begin{array}{l}
y \\
v \leftarrow \text{departing variable} \\
f
\end{array}
$$

$$\uparrow$$
$$\text{entering variable}$$

The entering variable x will replace v in the second row. The tableaux are as follows:

$$
\approx
\left[
\begin{array}{ccccc|c}
\frac{1}{2} & 1 & \frac{1}{2} & 0 & 0 & 4 \\
① & 0 & -\frac{1}{2} & \frac{1}{2} & 0 & 2 \\
-\frac{1}{2} & 0 & \frac{3}{2} & 0 & 1 & 12
\end{array}
\right]
\begin{array}{l}
y \\
v \leftarrow \\
f
\end{array}
\approx
\left[
\begin{array}{ccccc|c}
0 & 1 & \frac{3}{4} & -\frac{1}{4} & 0 & 3 \\
1 & 0 & -\frac{1}{2} & \frac{1}{2} & 0 & 2 \\
0 & 0 & \frac{5}{4} & \frac{1}{4} & 1 & 13
\end{array}
\right]
\begin{array}{l}
y \\
x \\
f
\end{array}
$$

This is the final tableau. The basic variables are x and y, assuming the values 2 and 3, respectively. This is the tableau for the vertex B. The value of f is 13; it is the maximum value possible under the given constraints.

Let us now analyze a linear programming problem that has many solutions, to see how the final tableau is interpreted.

Example 1   Let us return to Example 2 of the previous section, a linear programming problem that has many solutions.

The objective function is $f = 8x + 2y$. The constraints are

$$
\begin{aligned}
4x + y &\le 32 \\
4x + 3y &\le 48 \\
x &\ge 0 \\
y &> 0
\end{aligned}
$$

Express the constraints using slack variables u and v.

$$
\begin{aligned}
4x + y + u \phantom{{}+v} &= 32 \\
4x + 3y \phantom{{}+u} + v &= 48 \\
-8x - 2y \phantom{{}+u+v} + f &= 0
\end{aligned}
$$

with $x \ge 0$, $y \ge 0$, $u \ge 0$, $v \ge 0$.

The simplex tableaux are

$$
\begin{array}{c}
\begin{array}{ccccc} x & y & u & v & f \end{array} \\
\begin{bmatrix}
④ & 1 & 1 & 0 & 0 & 32 \\
4 & 3 & 0 & 1 & 0 & 48 \\
-8 & -2 & 0 & 0 & 1 & 0
\end{bmatrix}
\begin{array}{l} u \leftarrow \text{departing variable} \\ v \\ f \end{array}
\end{array}
$$

$$
\uparrow
$$
$$
\text{entering variable}
$$

$$
\approx
\begin{array}{c}
\begin{array}{ccccc} x & y & u & v & f \end{array} \\
\begin{bmatrix}
① & \frac{1}{4} & \frac{1}{4} & 0 & 0 & 8 \\
4 & 3 & 0 & 1 & 0 & 48 \\
-8 & -2 & 0 & 0 & 1 & 0
\end{bmatrix}
\begin{array}{l} u \leftarrow \\ v \\ f \end{array}
\end{array}
\qquad
\approx
\begin{array}{c}
\begin{array}{ccccc} x & y & u & v & f \end{array} \\
\begin{bmatrix}
1 & \frac{1}{4} & \frac{1}{4} & 0 & 0 & 8 \\
0 & 2 & -1 & 1 & 0 & 16 \\
0 & 0 & 2 & 0 & 1 & 64
\end{bmatrix}
\begin{array}{l} x \\ v \\ f \end{array}
\end{array}
$$

This is the final tableau. It leads to a maximum value of $f = 64$. This occurs at $x = 8$, $v = 16$. Since y is not a basic variable, $y = 0$. Thus we have found the optimal solution $x = 8$, $y = 0$ (the point C in Figure 10.3). When the simplex method is applied to a problem that has many solutions, it stops as soon as it has found one optimal solution, as shown here. We now show how to extend the algorithm to find other solutions.

Observe that in the final tableau, the coefficient of y, a nonbasic variable, is 0 in the last row. Each coefficient of a nonbasic variable in this row indicates the rate at which f increases as that variable is increased. Thus, making y an entering or departing variable neither increases nor decreases f. Let us use y as an entering variable and v as a departing variable. We get the following tableaux.

$$
\begin{array}{c}
\begin{array}{ccccc} x & y & u & v & f \end{array} \\
\begin{bmatrix}
1 & \frac{1}{4} & \frac{1}{4} & 0 & 0 & 8 \\
0 & ② & -1 & 1 & 0 & 16 \\
0 & 0 & 2 & 0 & 1 & 64
\end{bmatrix}
\begin{array}{l} x \\ v \leftarrow \text{departing variable} \\ f \end{array}
\end{array}
$$

$$
\uparrow
$$
$$
\text{entering variable}
$$

$$
\approx
\begin{array}{c}
\begin{array}{ccccc} x & y & u & v & f \end{array} \\
\begin{bmatrix}
1 & \frac{1}{4} & \frac{1}{4} & 0 & 0 & 8 \\
0 & ① & -\frac{1}{2} & \frac{1}{2} & 0 & 8 \\
0 & 0 & 2 & 0 & 1 & 64
\end{bmatrix}
\begin{array}{l} x \leftarrow \\ v \\ f \end{array}
\end{array}
\qquad
\approx
\begin{array}{c}
\begin{array}{ccccc} x & y & u & v & f \end{array} \\
\begin{bmatrix}
1 & 0 & \frac{3}{8} & -\frac{1}{8} & 0 & 6 \\
0 & 1 & -\frac{1}{2} & \frac{1}{2} & 0 & 8 \\
0 & 0 & 2 & 0 & 1 & 64
\end{bmatrix}
\begin{array}{l} x \\ y \\ f \end{array}
\end{array}
$$

This tableau leads to the optimal solution $x = 6$, $y = 8$ (the point B in Figure 10.3). All points on the line between these two optimal solutions of $(8, 0)$ and $(6, 8)$, which were obtained using the simplex tableaux, are also optimal solutions.

Exercise Set 10.3

In Exercises 1–6, use the simplex method to maximize the functions under the given constraints. Determine the basic and nonbasic variables and the entering and departing variables for each tableau. Determine the optimal solution and the maximum value of the objective function directly from the final tableau. (These problems were given in the previous set of exercises. Use the tableaux that you have already derived, and check your previous answers.)

*1. Maximize $f = 2x + y$ subject to
$$4x + \ y \le 36$$
$$4x + 3y \le 60$$
$$x \ge 0, y \ge 0$$
(Exercise 1, Section 10.2)

2. Maximize $f = x - 4y$ subject to
$$x + 2y \le 4$$
$$x + 6y \le 8$$
$$x \ge 0, y \ge 0$$
(Exercise 2, Section 10.2)

*3. Maximize $f = x + 2y + z$ subject to
$$3x + \ y + \ z \le \ 3$$
$$x - 10y - 4z \le 20$$
$$x \ge 0, y \ge 0, z \ge 0$$
(Exercise 5, Section 10.2)

4. Maximize $f = 100x + 200y + 50z$ subject to
$$5x + 5y + 10z \le 1000$$
$$10x + 8y + \ 5z \le 2000$$
$$10x + 5y \ \ \ \ \ \ \le \ 500$$
$$x \ge 0, y \ge 0, z \ge 0$$
(Exercise 6, Section 10.2)

5. Maximize $f = 2x + 4y + z$ subject to
$$-x + 2y + 3z \le 6$$
$$-x + 4y + 5z \le 5$$
$$-x + 5y + 7z \le 7$$
$$x \ge 0, y \ge 0, z \ge 0$$
(Exercise 7, Section 10.2)

*6. Maximize $f = x + 2y + 4z - w$ subject to
$$5x \ \ \ \ \ \ + 4z + 6w \le 20$$
$$4x + 2y + 2z + 8w \le 40$$
$$x \ge 0, y \ge 0, z \ge 0, w \ge 0$$
(Exercise 8, Section 10.2)

Review Exercises Chapter 10

1. Maximize $f = 2x + 3y$ subject to
$$2x + 4y \le 16$$
$$3x + 2y \le 12$$
$$x \ge 0, y \ge 0$$

2. Maximize $f = 6x + 4y$ subject to
$$x + 2y \le 16$$
$$3x + 2y \le 24$$
$$x \ge 0, y \ge 0$$

3. Minimize $f = 4x + y$ subject to
$$3x + 2y \le 21$$
$$x + 5y \le 20$$
$$x \ge 0, y \ge 0$$

4. A farmer has to decide how many acres of a 40-acre plot are to be devoted to growing strawberries and how many to growing tomatoes. There will be 300 hours of labor available for the picking. It takes 8 hours to pick an acre of strawberries and 6 hours to pick an acre of tomatoes. The profit per acre is $700 on the strawberries compared to $600 on the tomatoes. How many acres of each should be grown to maximize profit?

5. A company is buying lockers. It has narrowed the choice down to two kinds, X and Y. X has a volume of 36 cubic feet, while Y has a volume of 44 cubic feet. X occupies an area of 6 square feet and costs $54, while Y occupies 8 square feet and costs $60. A total of 256 square feet of floor space is available, and $2100 in funds are available for the purchase of the lockers. At least 20 of the larger lockers are needed. How many of each type of locker should be purchased in order to maximize volume?

6. Use the simplex method to maximize $f = 2x + y + z$ under the constraints

$$x + 2y + 4z \leq 20$$
$$2x + 4y + 4z \leq 60$$
$$3x + 4y + z \leq 90$$
$$x \geq 0, y \geq 0, z \geq 0$$

7. A furniture company finishes two kinds of tables, X and Y. There are three stages in the finishing process, namely sanding, staining, and varnishing. The time in minutes involved for each of these processes are as follows:

	Sanding	Staining	Varnishing
Table X	10	8	4
Table Y	5	4	8

The three types of equipment needed for sanding, staining, and varnishing are each available for 5 hours per day. Each type of equipment can handle only one table at a time. The profit on each X table is $8 and on each Y table is $4. How many of each type of table should be finished daily to maximize total profit?

8. Minimize $f = x - 2y + 4z$ subject to

$$x - y + 3z \leq 4$$
$$2x + 2y - 3z \leq 6$$
$$-x + 2y + 3z \leq 2$$
$$x \geq 0, y \geq 0, z \geq 0$$

MATLAB Discussion

The Simplex Method in Linear Programming (Section 10.2)

Linear programming is a mathematical method for finding optimal solutions to problems. It is widely used in business, industry, and government. In the simplex method of linear programming, constraints are written in matrix form and row operations are applied to arrive at the optimal solution.

Example 1   Let us find the maximum value of $f = 3x + 2y$ subject to the constraints

$$5x + 2y \le 900$$
$$8x + 10y \le 2800$$
$$x \ge 0$$
$$y \ge 0$$

Introduce slack variables u and v, and rewrite the system of inequalities as a system of equations that includes f, in the following manner.

$$5x + 2y + u \qquad\qquad = 900$$
$$8x + 10y \quad + v \quad = 2800$$
$$-3x + 2y \qquad\qquad + f = 0$$

with $x \ge 0$, $y \ge 0$, $u \ge 0$, $v \ge 0$.

The problem becomes that of finding the maximum value of f under these constraints. The simplex method is an algorithm that uses row operations on the augmented matrix of this system to lead to the maximum value of f.

A MATLAB function called **simplex** has been written for the simplex method. Let us apply it to this problem.

```
»A = [5  2  1  0  0  900;        {enter  the augmented matrix}
      8  10  0  1  0  2800;
     -3  -2  0  0  1  0];
»simplex(A)
Rational numbers? y/n: n         {option of rational numbers format}
All steps? y/n: n                {option of displaying all the steps}
initial simplex tableau
    5    2    1    0    0    900      {displays initial tableau to check}
    8   10    0    1    0   2800
   -3   -2    0    0    1      0
—final simplex tableau—
  1.0000        0   0.2941  -0.0588        0   100.0000
       0   1.0000  -0.2353   0.1471        0   200.0000
       0        0   0.4118   0.1176   1.0000   700.0000
```

The final tableau is displayed, to be interpreted by the user. The maximum value of f is 700, and it takes place at $x = 100$, $y = 200$.

Exercises   Use the **simplex** function to maximize the following functions under the given constraints.

*1. Maximize $f = 2x + y$ subject to
$$4x + y \leq 36$$
$$4x + 3y \leq 60$$
$$x \geq 0, y \geq 0$$

2. Maximize $f = x - 4y$ subject to
$$x + 2y \leq 4$$
$$x + 6y \leq 8$$
$$x \geq 0, y \geq 0$$

*3. Maximize $f = 4x + 6y$ subject to
$$x + 3y \leq 6$$
$$3x + y \leq 8$$
$$x \geq 0, y \geq 0$$

Appendix

M.28 MATLAB Commands, Functions, and M-files

Information

help	on-screen help facility
who	lists current variables in memory
whos	lists current variables and their sizes

Interrupting and Ending

exit	exits MATLAB
quit	exits MATLAB
Ctrl.	stops execution of current command

Matrix Operators

*	multiplication
^	raise to power
'	conjugate transpose
\	left division

Matrix and Vector Functions

A(i:j,p:q)	select submatrix of A lying from row i to row j, col p to col q
A(i,:)	select row i of A
A(:,p)	select column p of A
angle(u,v)	angle between vectors
det(A)	determinant
eig(A)	eigenvalues [X,D] = eig(A) gives norm eigenvectors as columns of X and eigenvalues as diagonal elements of D
eye(m,n)	generates identity matrix
hilb(n)	generates a Hilbert matrix
inv(A)	matrix inverse
ones(m,n)	generates matrix of all ones
pinv(A)	pseudoinverse
rand(m,n)	generates a random matrix
rank(A)	rank
rref(A)	reduced echelon form
zeros(m,n)	generates matrix of all zeros

Special Functions

clock	returns vector [year month day hour minute seconds] t1 = clock records this vector
etime(t2,t1)	returns the time in seconds between t2 and t1
flops	returns cumulative number of floating point operations
flop(0)	resets the count to zero
polyfit(X,Y,n)	gives the least squares polynomial of degree *n* that fits data described by vectors X and Y
polyvalue(p,s)	evaluates the polynomial p at s

M.29 The Linear Algebra with Applications Toolbox M-files

To get information about a file:
enter help . . . e.g. help *space*

add	elementary row operation
adj	adjoint of a matrix
affine	affine transformation of unit square
circles	circles in an inner prod space
cn	condition number and matrix stats
compmap	composition of two mappings
cross	cross product of vectors
digraph	picture and info on digraph
dist	distance between two points in $\mathbf{R}^n$
div	elementary row operation
dot	dot product of two vectors in $\mathbf{R}^n$
echtest	test random 3 × 3 matrices for REF
fern	fractal fern
frac	rational form for matrix elements

galg	pattern of algorithm in Gauss elim
gangle	angle in inner product space
gdist	distance in inner product space
gelim	Gaussian elimination
ginfo	stats for Gaussian elimination
ginner	inner prod in inner product space
gjalg	pattern of alg in G/J elimination
gjdet	determinant using G/J elimination
gjelim	Gauss/Jordan elimination
gjinfo	stats for G/J elimination
gjpic	lines for G/J elimination
gjinv	matrix inv using G/J elimination
gnorm	norm in inner product space
gpic	lines for Gaussian elimination
gram	Gram/Schmidt orthogonalization
graph	graph of polynomial function
gseidel	Gauss-Seidel method
jacobi	Jacobi method

leoinv	matrix inv in Leontief model	picture	graphs of lines
leontief	predictions using Leontief	pivot	complete pivotting
ls	sketch of least squares curve	pow	powers of a matrix
lufact	*LU* decomposition of matrix	proj	vector projection in $\mathbf{R}^n$
lusolve	solution using LUfact		
		sierp	Sierpinski triangle
mag	magnitude of vector in $\mathbf{R}^n$	sim	similarity transformation
map	matrix mapping of unit square	simplex	simplex method
markov	Markov chain model	space	space/time simulation
minor	minors of a matrix	square	fractal square
mult	elementary row operation	swap	elementary row operation
ops	# ops in matrix mult	tdot	triple scalar product
		transf	elementary row operations
		tree	fractal tree
perminv	inv of a permutation		
permprod	product of permutations		
pic	function for drawing graphs		

Answers to Selected Exercises

Chapter 1

Exercise Set 1.1, page 7

1. **(a)** $x = \frac{3}{2}, y = -\frac{1}{2}$ **(c)** No solutions **(d)** General solution: $x = r, y = \frac{-r}{2} + 2$

2. **(a)** $x = \frac{6}{7}, y = \frac{1}{7}$ **(c)** $x = 4, y = 2$ **(e)** General solution: $x = 3r - 2, y = r$

3. **(a)** $c = 3$ **(c)** $c = 2$ **(e)** $c = 7$

4. **(a)** $k \neq 8; x = 5, y = 0$ **(b)** $k = 8$. General solution: $x = r, y = \frac{5 - r}{4}$ 5. $k = 6$

7. **(a)** $d \neq 6; x = 7 + \frac{6}{d - 6}, y = \frac{-2}{d - 6}$ **(b)** $d = 6$

    There are no more values of d. For each value of d, either there is a unique solution or no solution.

9. **(a)** $ad - bc \neq 0; x = \frac{ed - bf}{ad - bc}, y = \frac{af - ec}{ad - bc}$ **(b)** $ad - bc = 0$ and $af - ec \neq 0$ or $ad - bc = 0$ and $ed - bf \neq 0$

    (c) $a = kc, b = kd$, and $e = kf$ for some number k

Exercise Set C.1, page 10

1. $x_1 = 1.5, x_2 = -0.5$ 3. No solution   5. $x_1 = 11, x_2 = -4$ 6. $x_1 = 0.8571, x_2 = 0.1429$

8. $x_1 = 4, x_2 = 2$ 10. $x_1 = r, x_2 = \frac{r}{3} + \frac{2}{3}$

Exercise Set 1.2, page 18

1. **(a)** 3×3 **(c)** 2×4 **(e)** 3×5 2. $1, 4, 9, -1, 3, 8$

5. **(a)** $\begin{bmatrix} 1 & 3 \\ 2 & -5 \end{bmatrix}$ and $\begin{bmatrix} 1 & 3 & 7 \\ 2 & -5 & -3 \end{bmatrix}$ **(c)** $\begin{bmatrix} -1 & 3 & -5 \\ 2 & -2 & 4 \\ 1 & 3 & 0 \end{bmatrix}$ and $\begin{bmatrix} -1 & 3 & -5 & -3 \\ 2 & -2 & 4 & 8 \\ 1 & 3 & 0 & 6 \end{bmatrix}$

    (e) $\begin{bmatrix} 5 & 2 & -4 \\ 0 & 4 & 3 \\ 1 & 0 & -1 \end{bmatrix}$ and $\begin{bmatrix} 5 & 2 & -4 & 8 \\ 0 & 4 & 3 & 0 \\ 1 & 0 & -1 & 7 \end{bmatrix}$

6. **(b)** $7x_1 + 9x_2 = 8$ **(d)** $8x_1 + 7x_2 + 5x_3 = -1$ **(f)** $-2x_2 = 4$ **(h)** $x_1 + 2x_2 - x_3 = 6$
           $6x_1 + 4x_2 = -3$ $4x_1 + 6x_2 + 2x_3 = 4$ $5x_1 + 7x_2 = -3$ $x_2 + 4x_3 = 5$
                                 $9x_1 + 3x_2 + 7x_3 = 6$ $6x_1 = 8$ $x_3 = -2$

7. **(a)** $\begin{bmatrix} 1 & 3 & -2 & 0 \\ 1 & 2 & -3 & 6 \\ 8 & 3 & 2 & 5 \end{bmatrix}$ **(c)** $\begin{bmatrix} 1 & 2 & 3 & -1 \\ 0 & 3 & 10 & 0 \\ 0 & -8 & -1 & -1 \end{bmatrix}$ **(e)** $\begin{bmatrix} 1 & 0 & 0 & -23 \\ 0 & 1 & 0 & 17 \\ 0 & 0 & 1 & 5 \end{bmatrix}$

8. **(a)** Elements in the first column, except for the leading 1, become 0. x_1 is eliminated from all equations except the first.
    (c) The leading 1 in row 2 is moved to the left of the leading nonzero term in row 3. The second equation now contains x_2 with leading coefficient 1.

9. (a) Elements in the third column, except for the leading 1, become 0. x_3 is eliminated from all equations except the third.
  (c) The leading nonzero element in row 3 becomes 1. Equation 3 is now solved for x_3.
10. (a) $x_1 = 2, x_2 = 5$ (c) $x_1 = 10, x_2 = -9, x_3 = -7$ (e) $x_1 = \frac{4}{5}, x_2 = \frac{16}{5}, x_3 = \frac{9}{5}$
11. (a) $x_1 = 3, x_2 = 0, x_3 = 2$ (c) $x_1 = 0, x_2 = 4, x_3 = 2$ (e) $x_1 = 2, x_2 = 3, x_3 = -1$
12. (b) $x_1 = \frac{1}{2}, x_2 = \frac{3}{2}, x_3 = -\frac{1}{2}$ (d) $x_1 = -2, x_2 = -5, x_3 = -1, x_4 = 5$ (e) $x_1 = -1, x_2 = -2, x_3 = 1, x_4 = 2$
13. (a) The solutions are in turn $x_1 = 1, x_2 - 1; x_1 = -2, x_2 = 3;$ and $x_1 = -1, x_2 = 2$
  (c) The solutions are in turn $x_1 = 1, x_2 = 2, x_3 = 3; x_1 = -1, x_2 = 2, x_3 = 0;$ and $x_1 = 0, x_2 = 1, x_3 = 2$

Exercise Set C.2, page 23

3. $\begin{bmatrix} 1 & 3 & -2 & 0 \\ 1 & 2 & -3 & 6 \\ 8 & 3 & 2 & 5 \end{bmatrix}$
5. $\begin{bmatrix} 1 & 2 & 3 & -1 \\ 0 & 3 & 10 & 0 \\ 0 & -8 & -1 & -1 \end{bmatrix}$
7. $\begin{bmatrix} 1 & 0 & 0 & -23 \\ 0 & 1 & 0 & 17 \\ 0 & 0 & 1 & 5 \end{bmatrix}$
9. $x_1 = 3, x_2 = 0, x_3 = 2$

11. $x_1 = 0, x_2 = 4, x_3 = 2$
13. $x_1 = 2, x_2 = 3, x_3 = -1$
15. $x_1 = 6, x_2 = -3, x_3 = 2$

Exercise Set 1.3, page 32

1. (a) Yes (c) No (e) Yes (h) No
2. (a) No (c) Yes (e) No (g) No (i) Yes
3. (a) $x_1 = 2, x_2 = 4, x_3 = -3$ (c) $x_1 = -3r + 6, x_2 = r, x_3 = -2$
  (e) $x_1 = -5r + 3, x_2 = -6r - 2, x_3 = -2r - 4, x_4 = r$
4. (a) $x_1 = -2r - 4s + 1, x_2 = 3r - 5s - 6, x_3 = r, x_4 = s$ (c) $x_1 = 2r - 3s + 4, x_2 = r, x_3 = -2s + 9, x_4 = s,$
  $x_5 = 8$
5. (a) $x_1 = 2, x_2 = -1, x_3 = 1$ (c) $x_1 = 3 - 2r, x_2 = 4 + r, x_3 = r$ (e) $x_1 = 4 - 3r, x_2 = 1 - 2r, x_3 = r$
6. (a) No solution (c) $x_1 = 1 - 2r, x_2 = r, x_3 = -2$ (e) $x_1 = 2, x_2 = 3, x_3 = 1$
7. (a) $x_1 = 4 - 2r, x_2 = 6 + 5r, x_3 = r$ (c) $x_1 = 3 - 2r, x_2 = r, x_3 = 2,$ and $x_4 = 1$
  (e) $x_1 = -2r - 3s, x_2 = 3r - s, x_3 = r, x_4 = s$
8. (a) $x_1 = 2x_3 - 4, x_2 = x_3 + 3, x_4 = 2$ (c) $x_1 = 7 - 2r - s, x_2 = 1 + 3r - 4s, x_3 = r, x_4 = s$ (d) No solution
  (g) $x_1 = 3, x_2 = -1$
9. (a) The following system of equations has no solution, since the equations are inconsistent:

$$3x_1 + 2x_2 - x_3 + x_4 = 4$$
$$3x_1 + 2x_2 - x_3 + x_4 = 1$$

  (b) The following system of equations has the unique solution $x_1 = 1, x_2 = 2$:

$$x_1 + x_2 = 3$$
$$x_1 + 2x_2 = 5$$
$$x_1 - 2x_2 = -3$$

11. (a) $a \neq 0, ad - bc \neq 0$ (b) $a \neq 0, b = 0, ad - bc = 0$
14. (a) No (b) No
15. (b) The general solution to the first system is $x_1 = 2, x_2 = 3 - 2r, x_3 = r$, and the second system has no solution.
17. $[I_3 : X]$

Exercise Set 1.4, page 42

1. $y = 4 - 3x + x^2$ 3. $3 + 2x = y$ 4. $y = 5 + x + 2x^2$ 6. $y = -3 + x - 2x^2 + x^3$
7. $I_1 = 5, I_2 = 1, I_3 = 6$ 9. $I_1 = 2, I_2 = 1, I_3 = 3$ 11. $I_1 = 10, I_2 = 7, I_3 = 3$
13. $I_1 = \frac{7}{3}, I_2 = \frac{5}{3}, I_3 = \frac{2}{3}, I_4 = \frac{5}{6}, I_5 = \frac{1}{6}$ 15. (a) $I_1 = 3, I_2 = 1, I_3 = 2$ 17. $x_2 = 0$

Chapter 1 Review Exercises, page 45

1. (a) $x = 1 - 2r, y = r$ (b) $x = -1, y = 3$ (c) No solution   2. (a) $c = -3$ (b) $c = 4$
3. (a) $d \neq 2, x = 0, y = 2$ (b) $d = 2, x = 2 - r, y = r$
4. (a) $x_1 = 3, x_2 = -2, x_3 = 4$ (b) $x_1 = \frac{56}{9}, x_2 = \frac{17}{3}, x_3 = 2, x_4 = \frac{17}{9}$ 5. (a) Yes (b) Yes (c) No
6. (a) No solution (b) $x_1 = 2, x_2 = -1 - 2r, x_3 = r, x_4 = 3$
7. If $A \neq I_n$ there is a row with leading 1 to the right of diagonal. Let it be row i. Then rows $i + 1, \ldots, n$ all have leading 1s to the right of diagonal. But there is no column to the right of diagonal for row n. Thus row n must consist of all zeros.
8. Let E be the reduced echelon form of A. Since B is row equivalent to A, B is also row equivalent to E. But since E is in reduced echelon form, it must be the reduced echelon form of B.
9. $y = 4 - 3x + 2x^2$ 10. $I_1 = 3, I_2 = 2, I_3 = 1$

The Use of the Computer in Linear Algebra

M.2 Solving Systems of Linear Equations, page 49

1. (a) $x_1 = 2, x_2 = -1, x_3 = 1$ (b) $x_1 = 1 - 2r, x_2 = r, x_3 = -2$ 2. (a) $x_1 = 2, x_2 = 5$ (b) No solution
3. (a) $x_1 = 4, x_2 = -3, x_3 = -1$ (c) $x_1 = 2, x_2 = 3, x_3 = -1$
9. (a) 13 multiplications, 12 additions (b) 12 multiplications, 11 additions   11. (a) $y = 2x^2 + x + 5$

Chapter 2

Exercise Set 2.1, page 60

1. (a) $\begin{bmatrix} 2 & 4 \\ 3 & 9 \\ 14 & -10 \end{bmatrix}$ (c) $\begin{bmatrix} -9 & 5 \\ -3 & 0 \end{bmatrix}$ (e) Does not exist (g) $\begin{bmatrix} 8 & 4 \\ -5 & 5 \\ 4 & 4 \end{bmatrix}$

2. (a) Does not exist (d) $\begin{bmatrix} -3 & -7 & 19 \\ 27 & -35 & -7 \\ -19 & 17 & 9 \end{bmatrix}$ (f) $\begin{bmatrix} 21 \\ 6 \\ 1 \end{bmatrix}$ 3. (a) B (d) Does not exist (e) D (g) $\begin{bmatrix} 5 & 7 & 2 \\ 27 & 35 & 4 \end{bmatrix}$

4. (a) $\begin{bmatrix} 27 \\ 23 \\ 9 \end{bmatrix}$ (d) $[27]$ (f) Does not exist (h) $\begin{bmatrix} 13 & 8 & 13 \\ -5 & 76 & -33 \\ 11 & -16 & 35 \end{bmatrix}$

5. (a) $\begin{bmatrix} -12 & 2 \\ -9 & -24 \\ -20 & -24 \end{bmatrix}$ (c) $\begin{bmatrix} 8 & 2 \\ 4 & 1 \\ 0 & 6 \end{bmatrix}$ (e) Does not exist (g) $\begin{bmatrix} -14 & 0 \\ 12 & 10 \end{bmatrix}$

6. (a) 3×2 (c) Does not exist (e) 4×2 (g) Does not exist
7. (b) 2×3 (d) Does not exist (e) 3×2 (g) 2×2 9. (a) -1 (c) 9 10. (b) 9 (d) Does not exist

11. (a) -6 12. (a) -14 13. (a) $\begin{bmatrix} 2 & 3 \\ 3 & -8 \end{bmatrix} \begin{bmatrix} x_1 \\ x_2 \end{bmatrix} = \begin{bmatrix} 4 \\ -1 \end{bmatrix}$

14. **(b)** $\begin{bmatrix} 5 & 2 \\ 4 & -3 \\ 3 & 1 \end{bmatrix} \begin{bmatrix} x_1 \\ x_2 \end{bmatrix} = \begin{bmatrix} 6 \\ -2 \\ 9 \end{bmatrix}$ **(d)** $\begin{bmatrix} 2 & 5 & -3 & 4 \\ 1 & 0 & 9 & 5 \\ 3 & -3 & -8 & 5 \end{bmatrix} \begin{bmatrix} x_1 \\ x_2 \\ x_3 \\ x_4 \end{bmatrix} = \begin{bmatrix} 4 \\ 12 \\ -2 \end{bmatrix}$

15. The third row of AB is the third row of A times each of the columns of B in turn. Since the third row of A is all zeros, each of the products is zero.

19. $A_2B = [44 \quad 22 \quad 21]$

21. **(a)** $m_1 = -12$, $m_2 = 20$, $m_3 = -3$, $m_4 = 4$, $m_5 = -2$, $m_6 = -3$, $m_7 = 5$, $AB = \begin{bmatrix} 1 & 2 \\ 2 & 16 \end{bmatrix}$

(c) $m_1 = -3$, $m_2 = 12$, $m_3 = -6$, $m_4 = 1$, $m_5 = 0$, $m_6 = 0$, $m_7 = 15$, $AB = \begin{bmatrix} 8 & 1 \\ 15 & 3 \end{bmatrix}$

22. **(a)** $\begin{bmatrix} \begin{bmatrix} 1 & -1 \\ 2 & 1 \end{bmatrix} & \begin{bmatrix} 0 & 0 \\ 3 & 0 \end{bmatrix} \\ \begin{bmatrix} 0 & 4 \\ 0 & 0 \end{bmatrix} & \begin{bmatrix} 1 & 0 \\ 0 & 0 \end{bmatrix} \end{bmatrix} \begin{bmatrix} \begin{bmatrix} 1 & 1 \\ 3 & 2 \end{bmatrix} & \begin{bmatrix} 2 & 0 \\ -1 & 0 \end{bmatrix} \\ \begin{bmatrix} 0 & 1 \\ 0 & 0 \end{bmatrix} & \begin{bmatrix} 1 & 0 \\ 0 & 0 \end{bmatrix} \end{bmatrix} = \begin{bmatrix} \begin{bmatrix} -2 & -1 \\ 5 & 7 \end{bmatrix} & \begin{bmatrix} 3 & 0 \\ 6 & 0 \end{bmatrix} \\ \begin{bmatrix} 12 & 9 \\ 0 & 0 \end{bmatrix} & \begin{bmatrix} -3 & 0 \\ 0 & 0 \end{bmatrix} \end{bmatrix}$

Each of the multiplications of 2×2 matrices is executed using Strassen's method.

Exercise Set C.3, page 63

1. $\begin{bmatrix} 4 & 12 \\ -3 & 7 \end{bmatrix}$ **3.** $\begin{bmatrix} 9 & 15 \\ 0 & -6 \end{bmatrix}$ **5.** $\begin{bmatrix} 5 & 19 \\ -6 & 16 \end{bmatrix}$ **7.** $\begin{bmatrix} 9.2 & 24.4 \\ -5.1 & 10.3 \end{bmatrix}$ **9.** $\begin{bmatrix} 4 & 12 \\ -3 & 7 \end{bmatrix}$

11. $\begin{bmatrix} 0 & -5 \\ 10 & 17.5 \end{bmatrix}$ **13.** $\begin{bmatrix} 5 & 20 & -5 \\ 10 & 15 & -25 \end{bmatrix}$ **15.** Does not exist

Exercise Set C.4, page 64

1. $\begin{bmatrix} -12 & 66 \\ 6 & -18 \end{bmatrix}$ **3.** $\begin{bmatrix} -36 & 198 \\ 18 & -54 \end{bmatrix}$ **5.** $\begin{bmatrix} -15 & 75 \\ 15 & 15 \end{bmatrix}$ **7.** $\begin{bmatrix} -58800 & 65400 \\ 11400 & -10200 \end{bmatrix}$ **9.** Does not exist

10. $\begin{bmatrix} -14 & 4 & 50 \\ 69 & 26 & -219 \end{bmatrix}$ **12.** $\begin{bmatrix} -7 & -8 & 19 \\ 61 & 109 & -142 \end{bmatrix}$

Exercise Set 2.2, page 70

1. **(a)** $AB = \begin{bmatrix} 4 & 7 & 10 \\ 0 & -5 & -4 \end{bmatrix}$ and BA does not exist.   **(c)** $AD = DA = \begin{bmatrix} 4 & 4 \\ -2 & 2 \end{bmatrix}$ **2.** $ABC = \begin{bmatrix} 18 \\ -10 \\ 14 \end{bmatrix}$

4. **(a)** $\begin{bmatrix} 16 & 32 \\ 36 & 64 \end{bmatrix}$ **(c)** $\begin{bmatrix} -2 & -69 \\ 6 & 17 \end{bmatrix}$ **5.** **(a)** $\begin{bmatrix} 26 & 34 \\ 187 & 383 \end{bmatrix}$ **(c)** $\begin{bmatrix} 50 & 156 \\ 57 & 109 \end{bmatrix}$ **6.** **(a)** Does not exist   **(c)** 3×6

7. **(a)** 3×3 **(c)** Does not exist   **(e)** 2×3

8. The total number of multiplications required to compute $(AB)C$ is $mns + mrn$.   **9.** **(a)** 42   **(c)** 243

11. **(a)** 30 for $(AB)C$, 20 for $A(BC)$ **(c)** 150 for $(AB)C$, 66 for $A(BC)$ **(e)** 1400 for $(AB)C$, 2619 for $A(BC)$

13. The total number of additions required to compute AB is $mn(r - 1)$.

15. **(a)** The (i, j)th element of $A + (B + C)$ is $a_{ij} + (b_{ij} + c_{ij})$. The (i, j)th element of $(A + B) + C$ is $(a_{ij} + b_{ij}) + c_{ij} = a_{ij} + (b_{ij} + c_{ij})$. Since their elements are the same, $A + (B + C) = (A + B) + C$.

17. The (i, j)th element of cA is ca_{ij}. If $cA = O_{mn}$, then $ca_{ij} = 0$ for all i and j. So either $c = 0$ or all $a_{ij} = 0$, in which case $A = O_{mn}$.

18. **(a)** $9B^2 - AB + 2BA$ **(c)** $-2BA - 2B^2$

19. **(b)** $A^2B - AB^2 + 2BAB - 3A^2$ **(c)** $-A^3 + A^2B + BA^2 + BAB - 2ABA - 2AB^2 + B^2A$

20. **(a)** The matrices that commute with the given matrix are all matrices of the form $\begin{bmatrix} a & 0 \\ c & c + a \end{bmatrix}$, where a and c can take any real values.

21. $AX_1 = AX_2$ does not imply that $X_1 = X_2$.   **23.** $(A + B)^2 = A^2 + AB + BA + B^2$

27. **(a)** The (i, j)th element of $A + B$ is $a_{ij} + b_{ij} = 0 + 0$ if $i \neq j$, so $A + B$ is a diagonal matrix.

30. **(a)** Yes **(c)** No **(e)** Yes

31. The idempotent matrices of the given form are $\begin{bmatrix} 1 & 0 \\ 0 & 0 \end{bmatrix}$, $\begin{bmatrix} 1 & 0 \\ 0 & 1 \end{bmatrix}$, $\begin{bmatrix} 1 & b \\ 0 & 0 \end{bmatrix}$, and $\begin{bmatrix} 1 & 0 \\ c & 0 \end{bmatrix}$, where b and c are any nonzero real numbers.

33. If $AB = BA$, then $(AB)^2 = (AB)(AB) = A(BA)B = A(AB)B = A^2B^2 = AB$.

Exercise Set 2.3, page 81

1. **(a)** $A' = \begin{bmatrix} -1 & 2 \\ 2 & -3 \end{bmatrix}$; symmetric **(c)** $C' = \begin{bmatrix} 3 & 2 \\ -1 & 4 \end{bmatrix}$; not symmetric **(d)** $D' = \begin{bmatrix} 4 & -2 & 7 \\ 5 & 3 & 0 \end{bmatrix}$; not symmetric

(f) $F' = \begin{bmatrix} 1 & -1 & 3 \\ -1 & 2 & 0 \\ 3 & 0 & 4 \end{bmatrix}$; symmetric **(h)** $H' = \begin{bmatrix} 1 & 4 & -2 \\ -2 & 5 & 6 \\ 3 & 6 & 7 \end{bmatrix}$; not symmetric

2. **(a)** $\begin{bmatrix} 1 & 2 & 4 \\ 2 & 6 & 5 \\ 4 & 5 & 2 \end{bmatrix}$ **3.** **(a)** 4×2 **(c)** Does not exist **(e)** 4×3

5. **(a)** $(A + B + C)' = (A + (B + C))' = A' + (B + C)' = A' + (B' + C') = A' + B' + C'$

9. If A is symmetric, then $A = A'$. From Theorem 2.4 and Exercise 4(c) this means $A' = A = (A')'$. So A' is symmetric.

11. **(a)** $\begin{bmatrix} 0 & -1 \\ 1 & 0 \end{bmatrix}$ **(c)** If A and B are antisymmetric, then $A + B = (-A') + (-B') = -(A' + B') = -(A + B)'$.

13. $B = \frac{1}{2}(A + A')$ is symmetric and $C = \frac{1}{2}(A - A')$ is antisymmetric. $A = B + C$ **15.** **(a)** -2 **(c)** -1

17. $tr(A + B + C) = tr(A + (B + C)) = tr(A) + tr(B + C) = tr(A) + (tr(B) + tr(C)) = tr(A) + tr(B) + tr(C)$

21. $AB = \begin{bmatrix} -2 - i & 40 + 15i \\ -12 - 19i & 19 - i \end{bmatrix}$

23. $A^* = \overline{A}' = \begin{bmatrix} 2 + 3i & 2 \\ -5i & 5 + 4i \end{bmatrix} \neq A$, not hermitian; $D^* = \overline{D}' = \begin{bmatrix} -2 & 3 - 5i \\ 3 + 5i & 9 \end{bmatrix} = D$, hermitian

25. **(a)** The (i, j)th element of $A^* + B^*$ is $\overline{a_{ji}} + \overline{b_{ji}} = \overline{a_{ji} + b_{ji}}$, the (i, j)th element of $(A + B)^*$.

27. **(a)** $G' = (AA')' = (A')'A' = AA' = G$, and $P' = (A'A)' = A'(A')' = A'A = P$

28. **(a)** $g_{12} = 0$, $g_{13} = 1$, $g_{23} = 1$, so $1 \to 3 \to 2$ or $2 \to 3 \to 1$ $p_{12} = 1$, so $1 \to 2$ or $2 \to 1$, gives no information
(c) $g_{12} = 1$, $g_{13} = 2$, $g_{23} = 3$, so $1 \to 3 \to 2$ or $2 \to 3 \to 1$
   $p_{12} = 1$, $p_{13} = 2$, $p_{14} = 1$, $p_{23} = 2$, $p_{24} = 2$, $p_{34} = 2$, so $1 \to 3 \to \{2 \leftrightarrow 4\}$ or $\{2 \leftrightarrow 4\} \to 3 \to 1$
(e) $g_{12} = 1$, $g_{13} = 0$, $g_{14} = 1$, $g_{23} = 0$, $g_{24} = 0$, $g_{34} = 1$, so $2 \to 1 \to 4 \to 3$ or $3 \to 4 \to 1 \to 2$
   $p_{12} = 0$, $p_{13} = 1$, $p_{14} = 0$, $p_{23} = 1$, $p_{24} = 1$, $p_{34} = 0$, so $1 \to 3 \to 2 \to 4$ or $4 \to 2 \to 3 \to 1$

Exercise Set C.5, page 84

1. $\begin{bmatrix} 1 & 2 \\ 0 & 1 \end{bmatrix}$ **3.** $\begin{bmatrix} 2 & 5 \\ 1 & 4 \\ 0 & 3 \end{bmatrix}$ **5.** $\begin{bmatrix} 1 & 3 & 0 \\ 2 & 6 & 8 \end{bmatrix}$ **7.** $\begin{bmatrix} -3 & -19 \\ 8 & 28 \end{bmatrix}$ **8.** $\begin{bmatrix} -143 & -475 \\ 200 & 632 \end{bmatrix}$

Exercise Set 2.4, page 92

1. (a) B is the inverse of A. **(c)** B is not the inverse of A.   **2. (a)** B is the inverse of A. **(c)** B is not the inverse of A.

3. (a) $\begin{bmatrix} 1 & 0 \\ -2 & 1 \end{bmatrix}$ **(c)** $\begin{bmatrix} \frac{3}{2} & -\frac{1}{2} \\ -2 & 1 \end{bmatrix}$ **(e)** The inverse does not exist.

4. (a) $\begin{bmatrix} \frac{7}{3} & -3 & -\frac{1}{3} \\ -\frac{8}{3} & 3 & \frac{2}{3} \\ \frac{4}{3} & -1 & -\frac{1}{3} \end{bmatrix}$ **(c)** The inverse does not exist.   **5. (a)** $\begin{bmatrix} -\frac{3}{7} & \frac{5}{7} & -\frac{11}{7} \\ \frac{2}{7} & -\frac{1}{7} & -\frac{2}{7} \\ \frac{2}{7} & -\frac{1}{7} & \frac{5}{7} \end{bmatrix}$ **(c)** The inverse does not exist.

6. (a) $\begin{bmatrix} -\frac{1}{5} & -\frac{1}{5} & \frac{1}{5} & 0 \\ -\frac{1}{5} & \frac{1}{5} & 0 & \frac{1}{5} \\ \frac{1}{5} & 0 & \frac{1}{5} & -\frac{1}{5} \\ 0 & \frac{1}{5} & -\frac{1}{5} & -\frac{1}{5} \end{bmatrix}$ **(c)** $\begin{bmatrix} 1 & 0 & 1 & -1 \\ 0 & -1 & -3 & 4 \\ 1 & 0 & -1 & 2 \\ -3 & 0 & 0 & -1 \end{bmatrix}$

7. (a) $\begin{bmatrix} x_1 \\ x_2 \end{bmatrix} = \begin{bmatrix} -5 & 2 \\ 3 & -1 \end{bmatrix} \begin{bmatrix} 2 \\ 4 \end{bmatrix} = \begin{bmatrix} -2 \\ 2 \end{bmatrix}$ **(c)** $\begin{bmatrix} x_1 \\ x_2 \end{bmatrix} = \begin{bmatrix} -\frac{1}{5} & \frac{3}{5} \\ \frac{2}{5} & -\frac{1}{5} \end{bmatrix} \begin{bmatrix} 5 \\ 10 \end{bmatrix} = \begin{bmatrix} 5 \\ 0 \end{bmatrix}$ **(e)** $\begin{bmatrix} x_1 \\ x_2 \end{bmatrix} = \begin{bmatrix} 2 & -1 \\ -\frac{3}{4} & \frac{1}{2} \end{bmatrix} \begin{bmatrix} 6 \\ 1 \end{bmatrix} = \begin{bmatrix} 11 \\ -4 \end{bmatrix}$

8. (a) $\begin{bmatrix} x_1 \\ x_2 \\ x_3 \end{bmatrix} = \begin{bmatrix} \frac{1}{9} & \frac{3}{9} & \frac{5}{9} \\ \frac{3}{9} & 0 & -\frac{3}{9} \\ -\frac{2}{9} & \frac{3}{9} & -\frac{1}{9} \end{bmatrix} \begin{bmatrix} 2 \\ 0 \\ 1 \end{bmatrix} = \begin{bmatrix} \frac{7}{9} \\ \frac{3}{9} \\ -\frac{5}{9} \end{bmatrix}$ **(c)** $\begin{bmatrix} x_1 \\ x_2 \\ x_3 \end{bmatrix} = \begin{bmatrix} -40 & 16 & 9 \\ 13 & -5 & -3 \\ 5 & -2 & -1 \end{bmatrix} \begin{bmatrix} 1 \\ 3 \\ 15 \end{bmatrix} = \begin{bmatrix} 143 \\ -47 \\ -16 \end{bmatrix}$

(e) $\begin{bmatrix} x_1 \\ x_2 \\ x_3 \end{bmatrix} = \begin{bmatrix} 2 & 3 & 1 \\ 3 & 3 & 1 \\ 2 & 4 & 1 \end{bmatrix} \begin{bmatrix} 5 \\ -2 \\ 1 \end{bmatrix} = \begin{bmatrix} 5 \\ 10 \\ 3 \end{bmatrix}$

9. $\begin{bmatrix} x_1 \\ x_2 \\ x_3 \\ x_4 \end{bmatrix} = \begin{bmatrix} -\frac{14}{17} & \frac{8}{17} & \frac{4}{17} & \frac{5}{17} \\ \frac{3}{17} & -\frac{9}{17} & \frac{4}{17} & \frac{5}{17} \\ \frac{5}{17} & \frac{2}{17} & \frac{1}{17} & -\frac{3}{17} \\ \frac{18}{17} & -\frac{3}{17} & -\frac{10}{17} & -\frac{4}{17} \end{bmatrix} \begin{bmatrix} 5 \\ 6 \\ 1 \\ 7 \end{bmatrix} = \begin{bmatrix} 1 \\ 0 \\ 1 \\ 2 \end{bmatrix}$ **11.** The inverses are **(a)** $\begin{bmatrix} 3 & -8 \\ -1 & 3 \end{bmatrix}$ and **(c)** $\begin{bmatrix} 2 & -3 \\ -\frac{3}{2} & \frac{5}{2} \end{bmatrix}$

12. (a) $\frac{1}{c}A^{-1} = \frac{1}{c}A^{-1}I = \frac{1}{c}A^{-1}(cA)(cA)^{-1} = \frac{1}{c}cA^{-1}A(cA)^{-1} = I(cA)^{-1} = (cA)^{-1}$ **14.** $A = \begin{bmatrix} 6 & -2 \\ 10 & -3 \end{bmatrix}$

15. (a) $(3A)^{-1} = \begin{bmatrix} \frac{2}{3} & -\frac{1}{3} \\ -\frac{5}{3} & \frac{3}{3} \end{bmatrix}$ **(c)** $A^{-2} = (A^{-1})^2 = \begin{bmatrix} 9 & -5 \\ -25 & 14 \end{bmatrix}$

16. (a) $(2A^t)^{-1} = \frac{1}{2}\begin{bmatrix} 2 & -9 \\ -1 & 5 \end{bmatrix}$ **(c)** $(AA^t)^{-1} = \begin{bmatrix} 85 & -47 \\ -47 & 26 \end{bmatrix}$ **17.** $x = 2$ **19.** $A = \begin{bmatrix} -\frac{1}{4} & \frac{1}{4} \\ -\frac{3}{16} & \frac{1}{8} \end{bmatrix}$

21. $(A^tB^t)^{-1} = (B^t)^{-1}(A^t)^{-1} = (B^{-1})^t(A^{-1})^t = (A^{-1}B^{-1})^t$ **24. (a)** $AB = AC$ so $A^{-1}AB = A^{-1}AC$. Thus $B = C$.

27. (a) 65 multiplications and 50 additions **(b)** 150 multiplications and 100 additions   **29.** 57, 44, 26, 24, 17, 13, -1, 6

31. PEACE

Exercise Set C.6, page 95

1. $\begin{bmatrix} 1 & 0 \\ -2 & 1 \end{bmatrix}$ **3.** $\begin{bmatrix} 1.5 & -0.5 \\ 2.0 & 1.0 \end{bmatrix}$ **5.** Inverse does not exist   **7.** $\begin{bmatrix} 2.3333 & -3.0000 & -0.3333 \\ -2.6667 & 3.0000 & -0.6667 \\ 1.3333 & -1.0000 & -0.3333 \end{bmatrix}$

9. Inverse does not exist   **11.** $\begin{bmatrix} -0.4286 & 0.7143 & -1.5714 \\ 0.2857 & -0.1429 & -0.2857 \\ 0.2857 & -0.1429 & 0.7143 \end{bmatrix}$ **13.** $x_1 = -2, x_2 = 2$ **15.** $x_1 = 5, x_2 = 0$

17. $x_1 = 11, x_2 = -4$ **19.** $x_1 = -0.7778, x_2 = 0.3333, x_3 = -0.5556$ **21.** $x_1 = 143, x_2 = -47, x_3 = -16$

22. $\begin{bmatrix} 2 & 0 \\ 0 & 1 \end{bmatrix}$ **24.** $\begin{bmatrix} 6 & 14 & -7 \\ -5 & -11 & 6 \\ -3 & -6 & 5 \end{bmatrix}$

Exercise Set 2.5, page 101

1. (a) $a_{32} = .25$ **(c)** Electrical industry $(a_{43} = .30)$ **(e)** Steel industry $(a_{21} = .40)$ **2.** $\begin{bmatrix} 60 \\ 40 \end{bmatrix}, \begin{bmatrix} \frac{45}{2} \\ \frac{50}{3} \end{bmatrix}, \begin{bmatrix} 15 \\ 20 \end{bmatrix}$

4. $\begin{bmatrix} 210 \\ 175 \end{bmatrix}, \begin{bmatrix} \frac{100}{7} \\ \frac{50}{3} \end{bmatrix}, \begin{bmatrix} 40 \\ \frac{70}{3} \end{bmatrix}, \begin{bmatrix} 150 \\ 105 \end{bmatrix}$ **5.** $\begin{bmatrix} 15 \\ 24 \\ 32 \end{bmatrix}, \begin{bmatrix} 15 \\ 32 \\ 56 \end{bmatrix}, \begin{bmatrix} 30 \\ 56 \\ 48 \end{bmatrix}$ **7.** $D = \begin{bmatrix} 2.4 \\ 5 \end{bmatrix}$ **9.** $D = \begin{bmatrix} 4 \\ 0.9 \\ 0.65 \end{bmatrix}$

Exercise Set 2.6, page 108

1. (a) Stochastic   **(c)** Not stochastic   **(e)** Stochastic
3. If A and B are doubly stochastic matrices, then AB is stochastic, and A' and B' are stochastic, so $B'A' = (AB)'$ is stochastic. Thus AB is doubly stochastic.
4. (a) $p_{38} = 0.02$ **(c)** $p_{110} = 0.25$ low density residential   **5. (a)** .0780
7. 1987: city 59.37 million, suburb 124.1525 million, nonmetropolitan 56.4775 million
8. 1990: city 55,347,127, suburban 131,510,287   **11 (a)** 0.2   **(b)** White-collar 14,000, manual 16,000
13. Small 59,488, large 30,512   **15.** $AA \frac{1}{4}, Aa \frac{1}{2}, aa \frac{1}{4}$

Exercise Set 2.7, page 115

1. (a) Adjacency matrix $\begin{bmatrix} 0 & 1 & 0 \\ 1 & 0 & 1 \\ 1 & 0 & 0 \end{bmatrix}$ distance matrix $\begin{bmatrix} 0 & 1 & 2 \\ 1 & 0 & 1 \\ 1 & 2 & 0 \end{bmatrix}$

(c) Adjacency matrix $\begin{bmatrix} 0 & 1 & 1 & 1 & 1 \\ 0 & 0 & 0 & 0 & 0 \\ 0 & 0 & 0 & 0 & 0 \\ 0 & 0 & 0 & 0 & 0 \\ 0 & 0 & 0 & 0 & 0 \end{bmatrix}$ distance matrix $\begin{bmatrix} 0 & 1 & 1 & 1 & 1 \\ x & 0 & x & x & x \\ x & x & 0 & x & x \\ x & x & x & 0 & x \\ x & x & x & x & 0 \end{bmatrix}$ **2. (a)** 2   **(c)** Undefined

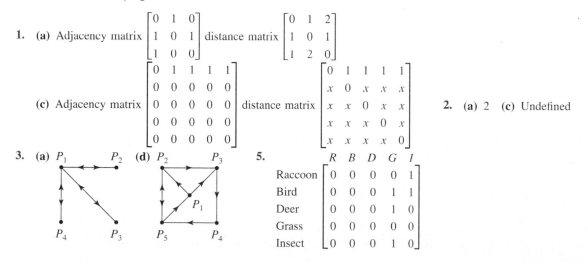

3. (a) P_1 P_2 **(d)** P_2 P_3

5.

	R	B	D	G	I
Raccoon	0	0	0	0	1
Bird	0	0	0	1	1
Deer	0	0	0	1	0
Grass	0	0	0	0	0
Insect	0	0	0	1	0

7.  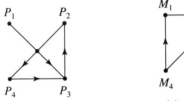 (a) $P_2 \rightarrow P_3 \rightarrow P_4 \rightarrow P_5$ length $= 3$ (b) $P_3 \rightarrow P_4 \rightarrow P_5 \rightarrow P_1 \rightarrow P_2$ length $= 4$

8. (a) $a_{22}^{(2)} = 1$ There is one 2-path from P_2 to P_2. $a_{12}^{(3)} = 1$ There is one 3-path from P_1 to P_2.
   $a_{24}^{(2)} = 0$ There is no 2-path from P_2 to P_4. $a_{24}^{(3)} = 0$ There is no 3-path from P_2 to P_4.
   $a_{31}^{(2)} = 1$ There is one 2-path from P_3 to P_1. $a_{32}^{(3)} = 1$ There is one 3-path from P_3 to P_2.
   $a_{42}^{(2)} = 1$ There is one 2-path from P_4 to P_2. $a_{41}^{(3)} = 1$ There is one 3-path from P_4 to P_1.

9. (a) There are no arcs from P_3 to any other vertex.   (c) There are three arcs from P_5.   (f) No 4-paths lead to P_3.

10. (a) There are no arcs from P_2 to any other vertex.   (c) There is an arc from P_4 to every other vertex.
    (e) There are five arcs from P_3.   (g) There are seven arcs in the digraph.   (i) Four 5-paths lead to P_3.

11. Digraph

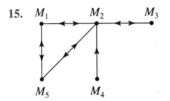

13. Digraph

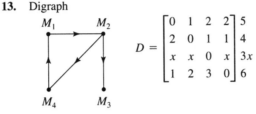

$$D = \begin{bmatrix} 0 & 1 & 2 & 2 \\ 2 & 0 & 1 & 1 \\ x & x & 0 & x \\ 1 & 2 & 3 & 0 \end{bmatrix} \begin{matrix} 5 \\ 4 \\ 3x \\ 6 \end{matrix}$$

most to least influential: M_2, M_1, M_4, M_3

15. 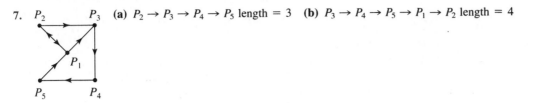   In this digraph there is one clique:

17. The digraph of A^t contains an arc from P_i to P_j, if and only if the digraph of A contains an arc from P_j to P_i.

19. If a path from P_i to P_j contains P_k twice, then there is a shorter path from P_i to P_j obtained by omitting the path from P_k to P_k.

21. $a_{ik}a_{jk} = 1$ if station k can receive messages directly from both station i and station j and $a_{ik}a_{jk} = 0$ otherwise. Thus $c_{ij} = a_{i1}a_{j1} + a_{i2}a_{j2} + \cdots + a_{in}a_{jn}$ is the number of stations that can receive messages directly from both station i and station j.

22. (a) $\begin{bmatrix} 1 & 1 & 0 & 0 \\ 1 & 1 & 0 & 0 \\ 1 & 1 & 1 & 0 \\ 1 & 1 & 1 & 1 \end{bmatrix}$ (c) $\begin{bmatrix} 1 & 1 & 1 & 1 \\ 1 & 1 & 1 & 1 \\ 1 & 1 & 1 & 1 \\ 1 & 1 & 1 & 1 \end{bmatrix}$ 23. $n-1$ 24. (b) No   28. (c) Yes

Chapter 2 Review Exercises, page 120

1. (a) $\begin{bmatrix} 28 & 0 \\ 100 & -6 \end{bmatrix}$ (b) Does not exist   (c) $\begin{bmatrix} 28 & 0 \\ 69 & -6 \end{bmatrix}$ (d) $\begin{bmatrix} -6 \\ 26 \end{bmatrix}$ (e) $\begin{bmatrix} 54 & -9 & 27 \\ 46 & -6 & 14 \end{bmatrix}$ (f) Does not exist

2. (a) 2×2 (b) 2×3 (c) 2×2 (d) Does not exist   (e) 3×2 (f) Does not exist   (g) 2×2

3. (a) 16 (b) −8   4. $m_1 = -16, m_2 = -9, m_3 = 4, m_4 = -20, m_5 = 14, m_6 = -5, m_7 = -4, AB = \begin{bmatrix} 8 & -6 \\ -9 & 23 \end{bmatrix}$

5. (a) $\begin{bmatrix} 9 & 0 \\ 5 & 4 \end{bmatrix}$ (b) $\begin{bmatrix} -4 & 1 \\ 4 & -2 \end{bmatrix}$ (c) $\begin{bmatrix} -39 & 15 \\ 24 & -12 \end{bmatrix}$ (d) $\begin{bmatrix} 4 & 2 \\ 0 & 2 \end{bmatrix}$

6. (a) 24 (b) 24 (c) 175 (d) 495   7. (a) $\begin{bmatrix} \frac{1}{9} & \frac{4}{9} \\ \frac{2}{9} & -\frac{1}{9} \end{bmatrix}$ (b) $\begin{bmatrix} 0 & 2 & -1 \\ 1 & 1 & -1 \\ -\frac{2}{3} & -1 & 1 \end{bmatrix}$ (c) $\begin{bmatrix} -40 & 16 & 9 \\ 13 & -5 & -3 \\ 5 & -2 & -1 \end{bmatrix}$

8. $\begin{bmatrix} x_1 \\ x_2 \\ x_3 \end{bmatrix} = \begin{bmatrix} 14 & -8 & -1 \\ -17 & 10 & 1 \\ -19 & 11 & 1 \end{bmatrix} \begin{bmatrix} 1 \\ 5 \\ 7 \end{bmatrix} = \begin{bmatrix} -33 \\ 40 \\ 43 \end{bmatrix}$ 9. $A = \begin{bmatrix} 3 & 6 \\ 2 & 5 \end{bmatrix}$

11. $A^2B + A^3 - BA^2 - BAB + 3ABA - 4AB^2$ 12. $(cA)^n = (cA)(cA) \cdots (cA) = c^n A^n$

13. The ith diagonal element of AA' is $(a_{i1})^2 + (a_{i2})^2 + \cdots + (a_{in})^2$, which can equal 0 only if each term is 0. Thus if $AA' = 0$, then $A = 0$.

14. If A is a symmetric matrix, then $A = A'$, so $AA' = A^2 = A'A$ and A is normal.

15. If $A = A^2$, then $A' = (A^2)' = (A')^2$, so A' is idempotent.

16. If $n < p, A^n \neq 0$, so $(A')^n = (A^n)' \neq 0$. $A^p = 0$, so $(A^p)' = (A')^p = 0$. Thus A' is nilpotent with degree of nilpotency $= p$.

17. $A = A'$, so $A^{-1} = (A')^{-1} = (A^{-1})'$, so A^{-1} is symmetric.

18. If row i of A is all zeros, then row i of AB is all zeros for any matrix B. The ith diagonal term of I_n is 1, so there is no matrix B for which $AB = I_n$.

19. $A + B = \begin{bmatrix} 5 + i & 5 - 5i \\ 6 + 10i & -3 + i \end{bmatrix}$ $AB = \begin{bmatrix} 35 + 24i & -3 + 6i \\ 7 + 6i & 12 - 6i \end{bmatrix}$ $A^* = A$, so A is hermitian.

20. If A is a real symmetric matrix, then $\overline{A} = A = A'$ so $A^* = A' = A$. Thus A is hermitian.

21. $g_{12} = 1, g_{13} = 0, g_{14} = 0, g_{15} = 0, g_{23} = 0, g_{24} = 0, g_{25} = 1, g_{34} = 1, g_{35} = 1, g_{45} = 0$, so $1 \to 2 \to 5 \to 3 \to 4$ or $4 \to 3 \to 5 \to 2 \to 1$

$p_{12} = 1, p_{13} = 0, p_{14} = 0, p_{23} = 1, p_{24} = 0, p_{34} = 1$, so $1 \to 2 \to 3 \to 4$ or $4 \to 3 \to 2 \to 1$

22. In two generations: 559,875 college educated, 490,125 noncollege-educated

23. (a) No arcs lead to vertex 4. (b) There are two arcs from vertex 3. (c) There are four 3-paths from vertex 2.
(d) No 2-paths lead to vertex 3. (e) There are two 3-paths from vertex 4 to vertex 4.
(f) There are three pairs of vertices joined by 4-paths.

MATLAB Discussion

M.3 Matrix Operations, page 122

2. (a) $\begin{bmatrix} -9 & 35 \\ 1 & 10 \end{bmatrix}$ 4. 29   5. (a) »X = A(3, :); »Y = A(:, 4); »X*A'*Y; ans = 821

M.4 Computational Considerations, page 125

3. (a) $A(BC)$, 3,864; $(AB)C$, 7,395 (b) $(A(BC))D$, 4,494; $A(B(CD))$, 65,058

M.5 Inverse of a Matrix, page 126

3. (a) $A^{-1} = \begin{bmatrix} -0.71428571428571 & 0.57142857142857 \\ 0.42857142857143 & -0.14285714285714 \end{bmatrix} = \begin{bmatrix} -5/7 & 4/7 \\ 3/7 & -1/7 \end{bmatrix}$

(b) $B^{-1} = \begin{bmatrix} 0.15116279069767 & 0.10465116279070 \\ 0.12790697674419 & -0.01162790697674 \end{bmatrix} = \begin{bmatrix} -13/86 & 9/86 \\ 11/86 & -1/86 \end{bmatrix}$

4. **(a)** $B = \begin{bmatrix} 2 & 0 \\ 0 & 1 \end{bmatrix}$ **(c)** $B = \begin{bmatrix} 6 & 14 & -7 \\ -5 & -11 & 6 \\ -3 & -6 & 5 \end{bmatrix}$

M.6 Solving Systems of Equations Using Matrix Inverse, page 127

1. $x_1 = 10,\ x_2 = -2$ 3. $x_1 = 9,\ x_2 = -14$

M.8 Leontief I/O Model, page 131

1. $X = \begin{bmatrix} 165 \\ 480 \\ 250 \end{bmatrix}$ 3. $D = \begin{bmatrix} 4 \\ .9 \\ .65 \end{bmatrix}$

M.9 Markov Chains, page 132

1. **(a)** $X_{86} = \begin{bmatrix} 184.25 \\ 55.75 \end{bmatrix},\ X_{87} = \begin{bmatrix} 183.52 \\ 56.48 \end{bmatrix},\ X_{88} = \begin{bmatrix} 182.82 \\ 57.18 \end{bmatrix},\ X_{89} = \begin{bmatrix} 182.13 \\ 57.87 \end{bmatrix},\ X_{90} = \begin{bmatrix} 181.47 \\ 58.53 \end{bmatrix} \begin{matrix} \text{City} \\ \text{Suburb} \end{matrix}$

   (b) $\begin{bmatrix} 0.99 & 0.02 \\ 0.01 & 0.98 \end{bmatrix}^4 = \begin{bmatrix} 0.9618 & 0.0765 \\ 0.0382 & 0.9235 \end{bmatrix}.\ p_{11}{}^{(4)} = 0.9618$

6. **(a)** $X_{86} = \begin{bmatrix} 59.67 \\ 124.58 \\ 55.75 \end{bmatrix},\ X_{87} = \begin{bmatrix} 59.37 \\ 124.15 \\ 56.48 \end{bmatrix},\ X_{88} = \begin{bmatrix} 59.08 \\ 123.73 \\ 57.18 \end{bmatrix},\ X_{89} = \begin{bmatrix} 58.82 \\ 123.32 \\ 57.87 \end{bmatrix},\ X_{90} = \begin{bmatrix} 58.56 \\ 122.90 \\ 58.53 \end{bmatrix} \begin{matrix} \text{City} \\ \text{Suburb} \\ \text{Nonmetro} \end{matrix}$

   (b) $\begin{bmatrix} 56 \\ 104 \\ 80 \end{bmatrix} \begin{matrix} \text{City} \\ \text{Suburb} \\ \text{Nonmetro} \end{matrix}$ **(c)** $\begin{bmatrix} 0.2333 \\ 0.4393 \\ 0.3333 \end{bmatrix} \begin{matrix} \text{City} \\ \text{Suburb} \\ \text{Nonmetro} \end{matrix}$

M.10 Digraphs, page 135

1. Distance $= 3$, Two paths $2 \to 4 \to 5 \to 1$ and $2 \to 4 \to 3 \to 1$

Chapter 3

Exercise Set 3.1, page 145

1. **(a)** 7 **(c)** 14   2. **(a)** 3 **(c)** 13   3. **(a)** $M_{11} = C_{11} = -6$ **(c)** $M_{23} = -13,\ C_{23} = 13$
4. **(a)** $M_{13} = C_{13} = 12$ **(c)** $M_{31} = C_{31} = -3$ 5. **(a)** $M_{12} = -59,\ C_{12} = 59$ **(c)** $M_{33} = -177,\ C_{33} = -177$
6. **(a)** -15 **(c)** -7 7. **(a)** 393 **(c)** 0   8. **(a)** -31 **(c)** -45 9. **(a)** -31 **(c)** 12
10. **(a)** -27 **(c)** -72 11. **(a)** -248 **(c)** 21   12. $x = 5$ or $x = -1$ 14. $x = \sqrt{3}$ or $x = -\sqrt{3}$
16. They have the same cofactor expansion using the third column.   18. **(a)** Even **(c)** Odd **(e)** Even
19. **(a)** Odd **(c)** Odd **(e)** Odd

Exercise Set C.7, page 148

1. -2 **3.** 2.13   **5.** -460 **7.** 438   **9.** 0, since row3=row1   **11.** 0, since col2=col3

Exercise Set 3.2, page 152

1. (a) -5 (c) 0   **2.** (a) 20 (c) 0   **3.** (a) -4 (c) -2 **4.** (a) -5 (c) 5
5. The second answer is correct.   **6.** (a) Row 3 is all zeros. (c) Row 3 is -3 times row 1.
7. (a) Column 3 is all zeros. (c) Row 3 is 3 times row 1.   **8.** (a) 12 (c) 9 (e) 9
9. (a) -6 (c) -6 (e) -12 **10.** (a) 3 (c) 3   **12.** Expand by row (or column) 1 at each stage.
15. Suppose the sum of the elements in each column is 0. Add each row after the first to row 1 to get a matrix B, which is row equivalent to A and has the same determinant as A. The first row of B is all zeros, so $|A| = |B| = 0$.
17. $|AB| = |A||B| = |B||A| = |BA|$

Exercise Set 3.3, page 158

1. (a) -12 (c) 27   **2.** (a) -60 (c) 0   **3.** (a) -4 (c) -20 **4.** (a) 0 (c) -6
5. (a) -6 (b) 21   **6.** (a) -9 (c) 0   **7.** (a) 0 (c) -104
12. (a) $|A^{-1}| = 1/|A|$, but if $A = A^{-1}$ then $|A| = |A^{-1}|$, so $|A| = 1/|A|$ and $|A|$ must be ± 1.
14. If $AB = I_n$, then $|A||B| = |AB| = |I_n| = 1$. If the product of two real numbers is not 0, then neither number is 0, so $|A| \neq 0$ and $|B| \neq 0$.

Exercise Set 3.4, page 166

1. (a) Invertible (c) Not invertible   **2.** (a) Invertible (c) Invertible   **3.** (a) Invertible (c) Invertible

4. (a) Not invertible (c) Invertible   **5.** (a) $\dfrac{-1}{10}\begin{bmatrix} 2 & -4 \\ -3 & 1 \end{bmatrix}$ (c) Not invertible

6. (a) $\dfrac{-1}{3}\begin{bmatrix} -7 & 9 & 1 \\ 8 & -9 & -2 \\ -4 & 3 & 1 \end{bmatrix}$ (c) $\dfrac{-1}{4}\begin{bmatrix} -6 & 2 & -2 \\ -3 & 1 & 1 \\ -8 & 4 & 0 \end{bmatrix}$ **7.** (a) $-\begin{bmatrix} -1 & 2 & 0 \\ 2 & -1 & -2 \\ 0 & -2 & 1 \end{bmatrix}$ (c) The inverse does not exist.

8. (a) $x_1 = 2$, $x_2 = 3$ (c) $x_1 = 2$, $x_2 = 3$ **9.** (a) $x_1 = -2$, $x_2 = 5$ (b) $x_1 = 1$, $x_2 = 4$
10. (a) $x_1 = 1$, $x_2 = 2$, $x_3 = -1$ (c) $x_1 = \frac{1}{2}$, $x_2 = \frac{1}{4}$, $x_3 = \frac{1}{4}$
11. (a) $x_1 = 3$, $x_2 = 0$, $x_3 = 1$ (c) $x_1 = \frac{1}{8}$, $x_2 = \frac{1}{4}$, $x_3 = \frac{1}{2}$
12. (a) $|A| = 0$, so this system of equations cannot be solved using Cramer's rule. (c) $x_1 = \frac{1}{2}$, $x_2 = \frac{3}{8}$, $x_3 = \frac{3}{4}$
13. (a) Not a unique solution (c) Not a unique solution   **14.** (a) Unique solution (c) Not a unique solution
15. $\lambda = 7$, $x_1 = x_2 = r$; $\lambda = -4$, $x_1 = \dfrac{-6r}{5}$, $x_2 = r$

17. $\lambda = 1$, $x_1 = -s - \dfrac{r}{2}$, $x_2 = s$, $x_3 = r$; $\lambda = 10$, $x_1 = x_2 = 2r$, $x_3 = r$

18. $AX = \lambda X = \lambda I_n X$, so $AX - \lambda I_n X = 0$. Thus $(A - \lambda I_n)X = 0$, and there is a nontrivial solution if and only if $|A - \lambda I_n| = 0$.

22. $A = (A^{-1})^{-1} = \dfrac{1}{|A^{-1}|}\text{adj}(A^{-1})$, so $\text{adj}(A^{-1}) = |A^{-1}|A = \dfrac{1}{|A|}A = [\text{adj}(A)]^{-1}$ from Exercise 21

25. If $|A| = \pm 1$, then $A^{-1} = \dfrac{1}{|A|}\text{adj}(A) = \pm\text{adj}(A)$, and since all elements of A are integers, all elements of $\text{adj}(A)$ are integers.

27. $AX = B_2$ has a unique solution if and only if $|A| \neq 0$ if and only if $AX = B_1$ has a unique solution.

Chapter 3 Review Exercises, page 168

1. **(a)** -7 **(b)** -18 **(c)** 29 2. **(a)** $M_{12} = -13$, $C_{12} = 13$ **(b)** $M_{31} = 1$, $C_{31} = 1$ **(c)** $M_{22} = 4$, $C_{22} = 4$

3. **(a)** 63 **(b)** 40 4. $x = 3$ or $x = 2$ 5. **(a)** -15 **(b)** 51 **(c)** -65 6. **(a)** 6 **(b)** 2 **(c)** -12

7. **(a)** 10 **(b)** -20 **(c)** 82 8. **(a)** -54 **(b)** 32 **(c)** -8 **(d)** 16 **(e)** -8 **(f)** -4

10. $|B| \neq 0$ so B^{-1} exists, and A and B^{-1} can be multiplied. Let $C = AB^{-1}$. Then $CB = AB^{-1}B = A$.

11. $|CA| = |C||A| = |A||C| = |AC|$ so that $|C^{-1}||CA| = |C^{-1}||AC|$. But $|C^{-1}||CA| = |C^{-1}CA| = |A|$ and $|C^{-1}||AC| = |C^{-1}AC|$ so that $|A| = |C^{-1}AC|$

13. If $A^2 = A$, then $|A||A| = |A|$, so $|A| = 1$ or 0. If A is also invertible, then $|A| \neq 0$, so $|A| = 1$.

14. **(a)** $\begin{bmatrix} 2 & -5 \\ -1 & 3 \end{bmatrix}$ **(b)** $\dfrac{1}{17}\begin{bmatrix} 5 & -2 \\ 1 & 3 \end{bmatrix}$ **(c)** No inverse **(d)** $\dfrac{1}{20}\begin{bmatrix} 4 & -5 & 3 \\ 36 & 10 & -18 \\ -8 & 0 & 4 \end{bmatrix}$

15. **(a)** $x_1 = 1$, $x_2 = -3$ **(b)** $x_1 = 2$, $x_2 = -1$, $x_3 = 0$

17. $|A|$ is the product of the diagonal elements, and since $|A| \neq 0$ all diagonal elements must be nonzero.

18. If $|A| = \pm 1$, then $A^{-1} = \pm\text{adj}(A)$, so $X = A^{-1}AX = A^{-1}B = \pm\text{adj}(A)B$, which has all integer components.

MATLAB Discussion

M.11 Determinants, page 170

1. **(a)** $|A| = 7$, $M(a_{22}) = -5$, $M(a_{31}) = 5$

2. All determinants are zero. **(a)** Row 3 is 3 times row 1. **(c)** Columns 2 and 3 are equal.

3. **(a)** $\begin{vmatrix} 1 & -1 & 0 & 2 \\ -1 & 1 & 0 & 0 \\ 2 & -2 & 0 & 1 \\ 3 & 1 & 5 & -1 \end{vmatrix}$ $\begin{matrix} = \\ \text{R2 + R1} \\ \text{R3 + (−2)R1} \\ \text{R4 + (−3)R1} \end{matrix}$ $\begin{vmatrix} 1 & -1 & 0 & 2 \\ 0 & 0 & 0 & 2 \\ 0 & 0 & 0 & -3 \\ 0 & 4 & 5 & -7 \end{vmatrix}$ $\begin{matrix} = \\ \text{R2} \leftrightarrow \text{R4} \end{matrix}$ $-\begin{vmatrix} 1 & -1 & 0 & 2 \\ 0 & 4 & 5 & -7 \\ 0 & 0 & 0 & -3 \\ 0 & 0 & 0 & 2 \end{vmatrix} = 0$

M.12 Matrix Inverses and Systems of Linear Equations, page 171

1. **(a)** $\dfrac{-1}{3}\begin{bmatrix} -7 & 9 & 1 \\ 8 & -9 & -2 \\ -4 & 3 & 1 \end{bmatrix}$ **(c)** $\dfrac{-1}{4}\begin{bmatrix} -6 & 2 & -2 \\ -3 & 1 & 1 \\ -8 & 4 & 0 \end{bmatrix}$

2. **(a)** $x_1 = 1$, $x_2 = 2$, $x_3 = -1$ **(c)** $x_1 = 0.5$, $x_2 = 0.25$, $x_3 = 0.25$

Chapter 4

Exercise Set 4.1, page 181

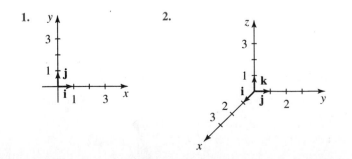

3. (a)   (c)   4. (a)

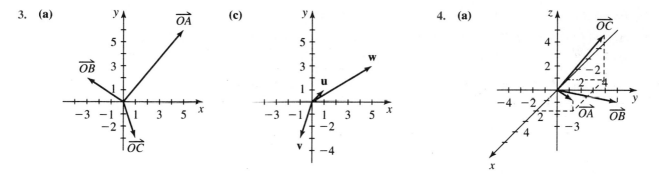

5. (a) $(3, 12)$ (c) $(1, 3)$ (d) $(-1, -2, -1)$ (g) $(-5, 20, -15, 10, -25)$
6. (b) $(13, -1)$ (d) $(17, -4)$ (e) $(19, -20)$ 7. (a) $(4, 5, 1)$ (b) $(3, 5, 8)$ (d) $(24, 23, -1)$

9. (b) $\begin{bmatrix} -14 \\ 10 \end{bmatrix}$ (d) $\begin{bmatrix} 12 \\ -25 \end{bmatrix}$ 10. (a) $\begin{bmatrix} 7 \\ 2 \\ 1 \end{bmatrix}$ (c) $\begin{bmatrix} -7 \\ 6 \\ 15 \end{bmatrix}$

11. $(16, -6, 2, 1)$, $(10, -4, 2, 0)$, $(-33, 12, -3, -3)$, $(-6, 2, 0, -1)$, $(-23, 8, -1, -3)$ 13. (a) $(8, -3)$ (c) $(6, 1)$
14. (a) $(3, 9, 7)$ 15. The resultant force is $(3, 4)$. To double the magnitude, replace $(0, 3)$ by $(3, 7)$.
16. The resultant force is $(8, 5)$. To have $(9, 7)$ resultant, replace $(3, 4)$ by $(4, 6)$.
18. (a) Scalar (b) Vector (c) Scalar (d) Scalar (e) Vector (f) Vector (g) Scalar (h) Scalar (i) Scalar

Exercise Set 4.2, page 185

1. (b) The set is the line given by the equation $x = y$. (d) This set is all of $\mathbf{R}^2$.
2. (a) The set is the xy plane. (c) The set is the plane given by the equation $y = 2x$.
3. (a) The set is the line given by $x = y = z$. (c) The set is the plane given by the equation $z = 3x$.
5. A is a subspace, B and C are not.   6. (a) Not a subspace (c) Not a subspace
7. (b) Not a subspace (d) Not a subspace   8. (a) Subspace (c) Not a subspace (e) Not a subspace
9. (a) Not a subspace (c) Not a subspace   10. (b) Subspace (c) Not a subspace (e) Not a subspace
11. (a) $\{(a, 0, 0)\}$ where a is nonnegative (b) $\{(a, b, 0)\}$ where $a \neq b$ unless $a = b = 0$
12. (a) If $(a, a + 1, b) = (0, 0, 0)$, then $a = a + 1 - b = 0$, $a = a + 1$ is impossible.
    (c) If $(a, b, a + b - 4) = (0, 0, 0)$, then $a = b = 0$ and $-4 = 0$.   13. (a) $\{(a, a^2)\}$
14. $(a, 2a, b) = (p, 2p, q)$ if and only if $a = p$ and $b = q$, so U and V consist of the same vectors. The set is a subspace of $\mathbf{R}^3$.
16. $(a, 3a, b + 2) = (p, 3p, r^3)$ if and only if $a = p$ and $b + 2 = r^3$. $U = V$ is a subspace of $\mathbf{R}^3$.

Exercise Set 4.3, page 193

1. (a) Linear combination (c) Not a linear combination
2. (a) Linear combination (d) Linear combination   3. (a) Linear combination (d) Linear combination
4. (a) Linear combination (b) Not a linear combination
5. (a) $(x_1, x_2) = \dfrac{x_1 + x_2}{2}(1, 1) + \dfrac{x_1 - x_2}{2}(1, -1)$ and $(3, 5) = 4(1, 1) + (-1)(1, -1)$
   (c) $(x_1, x_2) = (10x_1 - 3x_2)(1, 3) + (-3x_1 + x_2)(3, 10)$ and $(3, 5) = 15(1, 3) + (-4)(3, 10)$

6. **(a)** $(x_1, x_2, x_3) = a(1, 2, 3) + b(-1, -1, 0) + c(2, 5, 4)$ where $a = \frac{1}{-5}(-4x_1 + 4x_2 - 3x_3)$, $b = \frac{1}{-5}(7x_1 - 2x_2 - x_3)$,

and $c = \frac{1}{-5}(3x_1 - 3x_2 + x_3)$. $(1, 3, -2) = \frac{14}{-5}(1, 2, 3) + \frac{3}{-5}(-1, -1, 0) + \frac{8}{5}(2, 5, 4)$

(c) $(x_1, x_2, x_3) + a(5, 1, 3) + b(2, 0, 1) + c(-2, -3, -1)$ where $a = \frac{1}{-3}(3x_1 - 6x_3)$, $b = \frac{1}{-3}(-8x_1 + x_2 + 13x_3)$, and

$c = \frac{1}{-3}(x_1 + x_2 - 2x_3)$. $(1, 3, -2) = -5(5, 1, 3) + \frac{31}{3}(2, 0, 1) + \frac{8}{-3}(-2, -3, -1)$

7. **(a)** Span   **(c)** Do not span   **(d)** Span   **8.** **(a)** Span   **(c)** Do not span   **9.** $(2, 4, 3), (0, 0, 3), (2, 4, 6)$

11. $(-1, -2, -3), (2, 4, 6), \left(\frac{1}{2}, 1, \frac{3}{2}\right)$

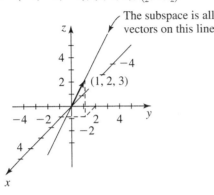

The subspace is all vectors on this line

14. $(-1, -2, 1, -3), (2, 4, -2, 6), (0.1, 0.2, -0.1, 0.3)$ **15.** $(-2, -1, 3, -4), (-1, 1, -2, 9), (20, 5, 10, 0)$

16. $a(-1, 3) = b(-2, 6)$ if and only if $2a = b$.

20. By definition, the subspace generated by $\mathbf{u}$ is the set of all vectors $a\mathbf{u}$, where a is a real number. This set is the set of all vectors on the line through the origin determined by $\mathbf{u}$.

21. If $\mathbf{v} = a\mathbf{v}_1 + b\mathbf{v}_2$, then $\mathbf{v} = \dfrac{a}{c_1}c_1\mathbf{v}_1 + \dfrac{b}{c_2}c_2\mathbf{v}_2$.

24. $\mathbf{v}_3 = b_1\mathbf{v}_1 + b_2\mathbf{v}_2$, so if $\mathbf{v} = a_1\mathbf{v}_1 + a_2\mathbf{v}_2$, then $\mathbf{v} = (a_1 - b_1)\mathbf{v}_1 + (a_2 - b_1)\mathbf{v}_2 + \mathbf{v}_3$.

Exercise Set 4.4, page 199

1. **(a)** $2(-1, 2) + (2, -4) = \mathbf{0}$ **(c)** $3(-2, 3) + (6, -9) = \mathbf{0}$

2. **(a)** If $(0, 0) = a(1, 0) + b(0, 1) = (a, b)$, then $a = b = 0$, so the vectors are linearly independent.

(c) If $(0, 0) = a(-1, 3) + b(2, 5)$, then $-a + 2b = 0$ and $3a + 5b = 0$. This system of equations has the unique solution $a = b = 0$. Thus the vectors are linearly independent.

3. **(a)** $-2(1, -2, 3) - 3(-2, 4, 1) + (-4, 8, 9) = \mathbf{0}$ **(c)** $(3, 4, 1) + 3(2, 1, 0) - (9, 7, 1) = \mathbf{0}$

4. **(a)** If $(0, 0, 0) = a(1, 2, 5) + b(1, -2, 1) + c(2, 1, 4)$, then $a + b + 2c = 0$, $2a - 2b + c = 0$, and $5a + b + 4c = 0$. This system of equations has the unique solution $a = b = c = 0$. Thus the vectors are linearly independent.

(c) If $(0, 0, 0) = a(1, 3, -4) + b(3, -1, 4) + c(1, 0, -2)$, then $a + 3b + c = 0$, $3a - b = 0$, and $-4a + 4b - 2c = 0$. This system of equations has the unique solution $a = b = c = 0$. Thus, the vectors are linearly independent.

5. **(a)** $2(2, -1, 3) + (-4, 2, -6) = \mathbf{0}$, so any set of vectors containing these vectors are linearly independent.

(c) $(3, 0, 4) + (-3, 0, -4) = \mathbf{0}$, so any set of vectors containing these vectors is linearly dependent.

6. **(a)** Dependent   **(d)** Independent   **(e)** Independent   **7.** **(a)** 2   **(c)** -7

10. Let $c_1\mathbf{v}_1 + c_2\mathbf{v}_2 = \mathbf{0}$, where c_1 and c_2 are not both zero. Then $\mathbf{0} = c_1\mathbf{v}_1 + c_2\mathbf{v}_2 = \dfrac{c_1 + c_2}{2}(\mathbf{v}_1 + \mathbf{v}_2) + \dfrac{c_1 - c_2}{2}(\mathbf{v}_1 - \mathbf{v}_2)$

and at least one of the coefficients in the last expression is not zero.

11. **(a)** Let $c_1\mathbf{v}_1 + c_2\mathbf{v}_2 + c_3\mathbf{v}_3 = \mathbf{0}$, where c_1, c_2, and c_3 are not all zero.

$\mathbf{0} = c_1\mathbf{v}_1 + c_2\mathbf{v}_2 + c_3\mathbf{v}_3 = (c_1 - c_2)\mathbf{v}_1 + c_2(\mathbf{v}_1 + \mathbf{v}_2) + c_3\mathbf{v}_3$ and at least one of the coefficients in the last expression is not zero.

12. **(a)** $a_1\mathbf{v}_1 + a_2(\mathbf{v}_1 + \mathbf{v}_2) + a_3\mathbf{v}_3 = (a_1 + a_2)\mathbf{v}_1 + a_2\mathbf{v}_2 + a_3\mathbf{v}_3 = \mathbf{0}$ only if $a_1 + a_2 = 0$, $a_2 = 0$, and $a_3 = 0$, i.e., only if $a_1 = a_2 = a_3 = 0$.

16. Independent

18. The computer is more likely to state that the vectors are linearly dependent when they are linearly independent.

Exercise Set 4.5, page 207

1. In each case below neither vector is a multiple of the other, so the vectors are linearly independent.

(a) $(x_1, x_2) = \dfrac{3x_2 - x_1}{5}(1, 2) + \dfrac{2x_1 - x_2}{5}(3, 1)$ **(c)** $(x_1, x_2) = \dfrac{x_1 + x_2}{2}(1, 1) + \dfrac{-x_1 + x_2}{2}(-1, 1)$

2. **(a)** and **(c)** In each case neither vector is a multiple of the other, so the vectors are linearly independent and therefore a basis for $\mathbf{R}^2$.   3. **(a)** Basis **(b)** Not a basis

4. **(a)** $a(1, 1, 1) + b(0, 1, 2) + c(3, 0, 1) = (0, 0, 0)$ if and only if $a + 3c = 0$, $a + b = 0$, and $a + 2b + c = 0$. The system of equations has the unique solution $a = b = c = 0$, so the three vectors are linearly independent. Thus they are a basis for $\mathbf{R}^3$.   5. **(a)** Basis **(b)** Not a basis

6. **(a)** These vectors are linearly dependent. **(b)** A basis for $\mathbf{R}^2$ can contain only two vectors.
(d) The third vector is a multiple of the second, so the set is not linearly independent. **(f)** $(1, 4)$ is not in $\mathbf{R}^3$.

7. $(1, 4, 3) = 3(-1, 2, 1) + 2(2, -1, 0)$, and $(-1, 2, 1)$ and $(2, -1, 0)$ are linearly independent, so the dimension is 2 and $(-1, 2, 1)$ and $(2, -1, 0)$ are a basis for the subspace.   8. $(1, 2, -1) = 2(1, 3, 1) - (1, 4, 3)$

10. $(-3, 3, -6) = -\frac{3}{2}(2, -2, 4)$ 12. $\{(1, 2), (0, 1)\}$ 13. $\{(1, 1, 1), (1, 0, -2), (1, 0, 0)\}$

15. **(a)** Basis $= \{(1, 1, 0), (0, 0, 1)\}$, dimension $= 2$ **(c)** Basis $= \{(1, 0, 1), (0, 1, 1)\}$, dimension $= 2$
(e) Basis $= \{(1, 0, -1), (0, 1, -1)\}$, dimension $= 2$

16. **(a)** Basis $= \{(1, 0, 1, 1), (0, 1, 1, -1)\}$, dimension $= 2$ **(c)** Basis $= \{(2, 0, 1, 0), (0, 1, 3, 0), (0, 0, 0, 1)\}$, dimension $= 3$

17. Any basis for V must contain two vectors, and the vectors $\mathbf{u}_1 = \mathbf{v}_1 + \mathbf{v}_2$ and $\mathbf{u}_2 = \mathbf{v}_1 - \mathbf{v}_2$ are linearly independent. Thus from Theorem 4.11 the set $\{\mathbf{u}_1, \mathbf{u}_2\}$ is a basis for V.

19. Any basis for V must contain n vectors. The vectors $c\mathbf{v}_1, c\mathbf{v}_2, \ldots, c\mathbf{v}_n$ are linearly independent and therefore a basis for V.

20. **(a)** False **(c)** True

22. If $m > n$, then any basis of W is linearly dependent by Theorem 4.7. But a basis is linearly independent, so $m \leq n$.

23. **(a)** False **(c)** True   24. **(b)** False **(d)** True

Exercise Set 4.6, page 215

1. **(a)** 2 **(c)** 2   2. **(a)** 2 **(c)** 1

3. **(a)** Basis $= \{(1, 0, 0), (0, 1, 0), (0, 0, 1)\}$, rank $= 3$ **(c)** Basis $\{(1, -3, 2)\}$, rank $= 1$

4. **(a)** Basis $= \{(1, 0, 0), (0, 1, 0), (0, 0, 1)\}$, rank $= 3$ **(c)** Basis $\{(1, 0, 1), (0, 1, 1)\}$, rank $= 2$

5. **(a)** Basis $\{(1, 0, 0, 3), (0, 1, 0, 2), (0, 0, 1, -1)\}$, rank $= 3$ **(b)** Basis $\{(1, 0, 3, -2), (0, 1, -2, 3)\}$, rank $= 2$

6. **(a)** $\{(1, 0, 0), (0, 1, 0), (0, 0, 1)\}$. The given vectors are linearly independent so they are a basis also.
(c) $\{(1, 0, 1), (0, 1, -2)\}$

7. **(a)** $\{(1, 3, 0, 0), (0, 0, 1, 0), (0, 0, 0, 1)\}$. The given vectors are linearly independent so they are a basis also.
(c) $\{(1, 0, 0, -0.5), (0, 1, 0, 0), (0, 0, 1, 1.5)\}$

8. $\{(1, 0, 0), (0, 1, 0), (0, 0, 1)\}$, basis for the row space of A; $\left\{ \begin{bmatrix} 1 \\ 0 \\ 0 \end{bmatrix}, \begin{bmatrix} 0 \\ 1 \\ 0 \end{bmatrix}, \begin{bmatrix} 0 \\ 0 \\ 1 \end{bmatrix} \right\}$, basis for the column space of A. Both the row

   space and the column space of A have dimension 3.

10. **(b)** The largest possible rank is the smaller of the two numbers m and n.

11. dim(column space of A) = dim(row space of A) ≤ number of rows of $A = m < n$

12. **(a)** The rank of A cannot be greater than 3, so a basis for the column space of A can contain no more than three vectors. The four column vectors in A must therefore be linearly dependent.

14. If the n columns of A are linearly independent, they are a basis for the column space of A and dim(column space of A) is n, so rank(A) = n.

Exercise Set 4.7, page 221

1. Characteristic polynomial: $\lambda^2 - 7\lambda + 6$; eigenvalues: 6, 1; corresponding eigenspaces: $r\begin{bmatrix} 4 \\ 1 \end{bmatrix}, s\begin{bmatrix} -1 \\ 1 \end{bmatrix}$

4. Characteristic polynomial: $\lambda^2 - 2\lambda + 1$; eigenvalue: 1; eigenspace: $r\begin{bmatrix} 1 \\ -2 \end{bmatrix}$

5. Characteristic polynomial: $\lambda^2 - 2\lambda - 3$; eigenvalues: 3, -1; corresponding eigenspaces: $r\begin{bmatrix} 1 \\ 1 \end{bmatrix}, s\begin{bmatrix} -1 \\ 1 \end{bmatrix}$

8. Characteristic polynomial: $\lambda^2 - 4\lambda$; eigenvalues: 0, 4; corresponding eigenspaces: $r\begin{bmatrix} 2 \\ 1 \end{bmatrix}, s\begin{bmatrix} -2 \\ 1 \end{bmatrix}$

9. Characteristic polynomial: $-\lambda^3 + 2\lambda^2 + \lambda - 2$; eigenvalues: 1, -1, 2; corresponding eigenspaces: $r\begin{bmatrix} 1 \\ 0 \\ 1 \end{bmatrix}, s\begin{bmatrix} 1 \\ -1 \\ 1 \end{bmatrix}, t\begin{bmatrix} 0 \\ 1 \\ 1 \end{bmatrix}$

10. Characteristic polynomial: $(1 - \lambda)^2(3 - \lambda)$; eigenvalues: 1, 3; corresponding eigenspaces: $r\begin{bmatrix} 1 \\ 1 \\ 1 \end{bmatrix}, s\begin{bmatrix} 0 \\ 1 \\ 1 \end{bmatrix}$

13. Characteristic polynomial: $(1 - \lambda)(2 - \lambda)(8 - \lambda)$; eigenvalues: 1, 2, 8; corresponding eigenspaces: $r\begin{bmatrix} 1 \\ -1 \\ 1 \end{bmatrix}, s\begin{bmatrix} 0 \\ 1 \\ 1 \end{bmatrix}, t\begin{bmatrix} 1 \\ 0 \\ 1 \end{bmatrix}$

15. Characteristic polynomial: $(2 - \lambda)(4 - \lambda)(2 - \lambda)(6 - \lambda)$; eigenvalues: 2, 4, 6;

   corresponding eigenspaces: $r\begin{bmatrix} 1 \\ -1 \\ 0 \\ 0 \end{bmatrix} + s\begin{bmatrix} 0 \\ 0 \\ 1 \\ 1 \end{bmatrix}, t\begin{bmatrix} 0 \\ 1 \\ 0 \\ -1 \end{bmatrix}, p\begin{bmatrix} 1 \\ 0 \\ 0 \\ 1 \end{bmatrix}$

17. Characteristic polynomial: $(1 - \lambda)^2$; eigenvalue: $\lambda = 1$; eigenvectors: $\mathbf{R}^2$; $\mathbf{v}$ in $\mathbf{R}^2$ is mapped into itself.

19. Characteristic polynomial: $(-2 - \lambda)^2$; eigenvalue: -2; eigenvectors: $\mathbf{R}^2$; $\mathbf{v}$ in $\mathbf{R}^2$ is mapped into $-2\mathbf{v}$.

20. $\lambda^2 + 1 \neq 0$ for any real value of λ, so there are no real eigenvalues.

22. $|A - \lambda I_n|$ is the product of the terms $a_{ii} - \lambda$, so the solutions of $|A - \lambda I_n| = 0$ are the values $\lambda = a_{ii}$.

24. $(A - \lambda I_n)' = A' - (\lambda I_n)' = A' - \lambda I_n$, so $|A - \lambda I_n| = |(A - \lambda I_n)'| = |A' - \lambda I_n|$, so A and A' have the same characteristic polynomial and therefore the same eigenvalues.

27. $|A - 0I_n| = |A|$, so $|A - 0I_n| = 0$ if and only if $|A| = 0$.

29. $|A - \lambda I_n| = \lambda^n + c_{n-1}\lambda^{n-1} + \cdots + c_1\lambda + c_0$. Substituting $\lambda = 0$, this becomes $|A| = c_0$.

31. (a) Characteristic polynomial: $\lambda^2 - 3\lambda + 2$; $\begin{bmatrix} 0 & 2 \\ -1 & 3 \end{bmatrix}^2 - 3\begin{bmatrix} 0 & 2 \\ -1 & 3 \end{bmatrix} + 2\begin{bmatrix} 1 & 0 \\ 0 & 1 \end{bmatrix} = \begin{bmatrix} 0 & 0 \\ 0 & 0 \end{bmatrix}$

(c) Characteristic polynomial: $\lambda^2 - 4$; $\begin{bmatrix} 6 & -8 \\ 4 & -6 \end{bmatrix}^2 - 4\begin{bmatrix} 1 & 0 \\ 0 & 1 \end{bmatrix} = \begin{bmatrix} 0 & 0 \\ 0 & 0 \end{bmatrix}$

Exercise Set 4.8, page 227

1. Metropolitan: 160 million, nonmetropolitan: 80 million.

3. $Q = \begin{bmatrix} 0.25 & 0.25 & 0.25 \\ 0.5 & 0.5 & 0.5 \\ 0.25 & 0.25 & 0.25 \end{bmatrix}$ The long-term probabilities of Types AA, Aa, and aa are 0.25, 0.5, and 0.25.

4. $P = \begin{bmatrix} 0.65 & 0.23 \\ 0.35 & 0.77 \end{bmatrix} \begin{matrix} \text{wet} \\ \text{dry} \end{matrix}$ (with column labels wet, dry)   (a) 0.67   (b) wet day: 0.4, dry day: 0.6

6. room 1 2 3 4

$P = \begin{bmatrix} 0 & \frac{1}{3} & 0 & \frac{1}{4} \\ \frac{1}{2} & 0 & \frac{1}{3} & \frac{1}{4} \\ 0 & \frac{1}{3} & 0 & \frac{1}{2} \\ \frac{1}{2} & \frac{1}{3} & \frac{2}{3} & 0 \end{bmatrix} \begin{matrix} 1 \\ 2 \\ 3 \\ 4 \end{matrix}$ The distribution of rats in rooms 1, 2, 3, and 4 is $2:3:3:4$. The long-term probabilities that a given rat will be in room 4 is $\frac{1}{3}$.

7. $P = \begin{bmatrix} 0.75 & 0.20 \\ 0.25 & 0.80 \end{bmatrix}$ Eventual distribution: 44.4%, A and 55.6%, B

9. The sum of the terms in each column of $A - I$ is 0. Thus $|A - 1I| = |A - I| = 0$ (Exercise 15, Exercise Set 3.2), and 1 is an eigenvalue of A.

Chapter 4 Review Exercises, page 229

1. (a) $(3, 0, 2)$ (b) $(11, 0, 22)$ (c) $(3, -3, 11)$ (d) $(8, -7, -3)$ (e) $(-4, -18, -22)$
2. (a) Not a subspace (b) Subspace (c) Subspace (d) Not a subspace   **3.** Only the subset (a) is a subspace.
4. (a) Linear combination (b) Linear combination   **5.** (a) Span (b) Do not span
6. (a) Linearly dependent (b) Linearly dependent (c) Linearly independent
7. (a) and (b) In each case the set consists of two linearly independent vectors. By Theorem 4.11 they are therefore a basis for $\mathbf{R}^2$.

(c) and (d) In each case the set consists of three linearly independent vectors. By Theorem 4.11 they are therefore a basis for $\mathbf{R}^3$.
8. $(10, 9, 8) = 2(-1, 3, 1) + 3(4, 1, 2)$ **9.** $\{(1, -2, 3), (4, 1, -1), (1, 0, 0)\}$
10. $\{(1, 0, 0, 1), (0, 1, 0, -2), (0, 0, 1, 3,)\}$ **11.** (a) 2 (b) 3 (c) 2   **12.** $\{(1, 0, 11, 8), (0, 1, 4, 2)\}$
13. $\mathbf{v} = a\mathbf{v}_1 + b\mathbf{v}_2 = a\mathbf{v}_1 + b\mathbf{v}_2 + 0\mathbf{v}_3$
14. If $a\mathbf{v}_1 + b\mathbf{v}_2 = \mathbf{0}$, then $a\mathbf{v}_1 + b\mathbf{v}_2 + 0\mathbf{v}_3 = \mathbf{0}$, and since the set $\{\mathbf{v}_1, \mathbf{v}_2, \mathbf{v}_3\}$ is linearly independent, this means $a = b = 0$, so $\{\mathbf{v}_1, \mathbf{v}_2\}$ must be linearly independent.
15. There are scalars c and d, not both zero, with $\mathbf{0} = c\mathbf{v}_1 + d\mathbf{v}_2 = \dfrac{3d + c}{7}(\mathbf{v}_1 + 2\mathbf{v}_2) + \dfrac{2c - d}{7}(3\mathbf{v}_1 - \mathbf{v}_2)$ and at least one of the coefficients in the last expression is not zero.
16. (a) True (b) False (c) True (d) True (e) False   **17.** (a) False (b) True (c) True (d) True (e) False

18. A is row equivalent to I_n if and only if the rows of I_n are linear combinations of the rows of A if and only if the rows of A span $\mathbf{R}^n$ if and only if $\text{rank}(A) = n$.

19. Characteristic polynomial: $(5 - \lambda)(1 - \lambda)(-2 - \lambda)$; eigenvalues: 5, 1, -2; corresponding eigenspaces: $r\begin{bmatrix} 1 \\ 1 \\ 1 \end{bmatrix}, s\begin{bmatrix} 0 \\ 1 \\ 1 \end{bmatrix}, t\begin{bmatrix} 1 \\ 0 \\ -1 \end{bmatrix}$

20. $A\mathbf{x} = \lambda\mathbf{x}$, so $\mathbf{x} = A^{-1}A\mathbf{x} = A^{-1}\lambda\mathbf{x} = \lambda A^{-1}\mathbf{x}$. A is invertible so $\lambda \neq 0$, and $\dfrac{1}{\lambda}\mathbf{x} = A^{-1}\mathbf{x}$.

21. $A\mathbf{x} = \lambda\mathbf{x}$, so $(A - kI)\mathbf{x} = A\mathbf{x} - kI\mathbf{x} = \lambda\mathbf{x} - kI\mathbf{x} = \lambda\mathbf{x} - k\mathbf{x} = (\lambda - k)\mathbf{x}$

MATLAB Discussion

M.13 Linear Combinations, Linear Dependence, Basis, page 231

1. **(a)** $(-3, 3, 7) = 2(1, -1, 2) - (2, 1, 0) + 3(-1, 2, 1)$ **(b)** Not a combination
2. **(a)** $2(-1, 3, 2) - 3(1, -1, -3) - (-5, 9, 13) = 0$ **3.** **(a)** Linearly independent, $\det(A) = 2268 \neq 0$, $\mathbf{rank}(A) = 4$
5. **(a)** $\{(1, 0, 1), (0, 1, 1)\}$

M.14 Eigenvalues and Eigenvectors and Applications, page 233

1. $\lambda_1 = 1, \mathbf{v}_1 = \begin{bmatrix} 0.4399 \\ 0.0091 \\ -0.8980 \end{bmatrix}, \lambda_2 = 1, \mathbf{v}_2 = \begin{bmatrix} -0.6017 \\ 0.7453 \\ -0.2872 \end{bmatrix}, \lambda_3 = 10, \mathbf{v}_3 = \begin{bmatrix} 0.6667 \\ 0.6667 \\ 0.3333 \end{bmatrix}.$

4. The eigenvectors of $\lambda = 1$ are vectors of the form $r\begin{bmatrix} 2 \\ 1 \end{bmatrix}$. If there is no change in total population, $2r + r = 185 + 55 = 240$, so $r = 80$. Thus the long-term prediction is that population in metropolitan areas will be 160 million and population in nonmetropolitan areas will be 80 million.

Chapter 5

Exercise Set 5.1, page 248

1. **(a)** 10 **(c)** 0 **2** **(a)** 6 **(c)** 0 **3.** **(a)** 7 **(c)** -1 **(e)** 0 **4.** **(a)** $\sqrt{5}$ **(c)** 4 **(d)** $\sqrt{10}$
5. **(a)** $\sqrt{11}$ **(c)** $3\sqrt{3}$ **(e)** $\sqrt{62}$ **6.** **(a)** $\sqrt{29}$ **(b)** $\sqrt{29}$ **(e)** $\sqrt{30}$
7. **(a)** $\left(\dfrac{1}{\sqrt{10}}, \dfrac{3}{\sqrt{10}}\right)$ **(c)** $\left(\dfrac{1}{\sqrt{14}}, \dfrac{2}{\sqrt{14}}, \dfrac{3}{\sqrt{14}}\right)$ **(d)** $\left(\dfrac{-1}{\sqrt{5}}, \dfrac{2}{\sqrt{5}}, 0\right)$
8. **(a)** $\left(\dfrac{2}{\sqrt{5}}, \dfrac{1}{\sqrt{5}}\right)$ **(c)** $\left(\dfrac{7}{3\sqrt{6}}, \dfrac{2}{3\sqrt{6}}, 0, \dfrac{1}{3\sqrt{6}}\right)$ **(e)** $(0, 0, 0, 1, 0, 0)$ **9.** **(a)** $45°$ **(b)** $60°$
10. **(a)** $\dfrac{5}{\sqrt{17}\sqrt{13}}$ **(b)** $\dfrac{13}{6\sqrt{7}}$ **(d)** $\dfrac{23}{10\sqrt{7}}$
11. **(a)** $(1, 3) \cdot (3, -1) = 1 \times 3 + 3 \times -1 = 0$, so the vectors are orthogonal.
12. **(c)** $(7, 1, 0) \cdot (2, -14, 3) = 7 \times 2 + 1 \times -14 + 0 \times 3 = 0$, so the vectors are orthogonal.
13. **(a)** Any vector of the form $(-3b, b)$ **(c)** Any vector of the form $(a, -4a)$
14. **(b)** Any vector of the form $(2b - 3c, b, c)$ **(d)** Any vector of the form $(a, b, c, -5a - c)$
  (f) Any vector of the form $(a, b, c, 2b - 3c - 5e, e)$
15. Any vector of the form $(a, -3a, -5a)$ **16.** **(a)** 5 **(c)** $5\sqrt{2}$ **17.** **(b)** $\sqrt{11}$ **(d)** $2\sqrt{5}$ **(e)** $2\sqrt{11}$

19. $\mathbf{u}$ is a scalar multiple of $\mathbf{v}$, so it has the same direction as $\mathbf{v}$. The magnitude of $\mathbf{u}$ is

$$\|\mathbf{u}\| = \frac{1}{\|\mathbf{v}\|}\sqrt{(v_1)^2 + (v_2)^2 + \cdots + (v_n)^2} = \frac{\|\mathbf{v}\|}{\|\mathbf{v}\|} = 1, \text{ so } \mathbf{u} \text{ is a unit vector.}$$

21. If $\mathbf{u} \cdot \mathbf{v} = \mathbf{u} \cdot \mathbf{w}$, then $\mathbf{u} \cdot (\mathbf{v} - \mathbf{w}) = 0$ for all vectors $\mathbf{u}$ in U. Since $\mathbf{v} - \mathbf{w}$ is a vector in U, this means that $(\mathbf{v} - \mathbf{w}) \cdot (\mathbf{v} - \mathbf{w}) = 0$. Therefore $\mathbf{v} - \mathbf{w} = \mathbf{0}$, so $\mathbf{v} = \mathbf{w}$.

23. **(a)** Vector   **(c)** Makes no sense   **(f)** Scalar   **(h)** Makes no sense   24. $c = \pm 3$

26. $(a, b) \cdot (-b, a) = a \times -b + b \times a = 0$, so $(-b, a)$ is orthogonal to (a, b).

27. (a, b) is orthogonal to (u_1, u_2) if and only if (a, b) is of the form $r(-u_2, u_1)$, i.e., (a, b) is in the vector space with basis $\{(-u_2, u_1)\}$, a one-dimensional subspace of $\mathbf{R}^2$.

30. $(\mathbf{u} + \mathbf{v}) \cdot (\mathbf{u} - \mathbf{v}) = \|\mathbf{u}\| - \|\mathbf{v}\|$, so $\|\mathbf{u}\| = \|\mathbf{v}\|$ if and only if $(\mathbf{u} + \mathbf{v}) \cdot (\mathbf{u} - \mathbf{v}) = 0$ if and only if $\mathbf{u} + \mathbf{v}$ and $\mathbf{u} - \mathbf{v}$ are orthogonal.

31. The two-dimensional subspace with basis $(1, 0, 0)$ and $(0, 1, 0)$ is orthogonal to the one-dimensional subspace with basis $(0, 0, 1)$.

34. **(b)** $\|(1, 2)\| = |2| = 2$, $\|(-3, 4)\| = |4| = 4$, $\|(1, 2, -5)\| = |-5| = 5$, and $\|(0, -2, 7)\| = |7| = 7$

35. **(c)** $d(\mathbf{x}, \mathbf{z}) = \|\mathbf{x} - \mathbf{y} + \mathbf{y} - \mathbf{z}\| \le \|\mathbf{x} - \mathbf{y}\| + \|\mathbf{y} - \mathbf{z}\| = d(\mathbf{x}, \mathbf{y}) + d(\mathbf{y}, \mathbf{z})$, from the triangle inequality

Exercise Set 5.2, page 261

1. **(a)** Orthogonal   **(b)** Not orthogonal   **(d)** Not orthogonal   2.   **(a)** Orthogonal   **(b)** Not orthogonal

3. **(a)** Orthonormal   **(c)** Orthonormal   **(e)** Not orthonormal   4. $\mathbf{v} = 3\mathbf{u}_1 + \frac{2}{5}\mathbf{u}_2 + \frac{11}{5}\mathbf{u}_3$

6. **(a)** $(3, 6)$ **(c)** $(2, 4, 6)$ **(e)** $\left(\frac{1}{3}, \frac{-1}{3}, \frac{2}{3}, 1\right)$ 7.   **(a)** $\left(\frac{24}{29}, \frac{60}{29}\right)$ **(c)** $(1, 2, 0)$ **(e)** $\left(\frac{-2}{15}, \frac{4}{15}, \frac{2}{15}, \frac{2}{5}\right)$

8. **(a)** $\left\{\left(\frac{1}{\sqrt{5}}, \frac{2}{\sqrt{5}}\right), \left(\frac{-2}{\sqrt{5}}, \frac{1}{\sqrt{5}}\right)\right\}$ **(c)** $\left\{\left(\frac{1}{\sqrt{2}}, \frac{-1}{\sqrt{2}}\right), \left(\frac{1}{\sqrt{2}}, \frac{1}{\sqrt{2}}\right)\right\}$

9. **(a)** $\left\{\left(\frac{1}{\sqrt{3}}, \frac{1}{\sqrt{3}}, \frac{1}{\sqrt{3}}\right), \left(\frac{1}{\sqrt{2}}, \frac{-1}{\sqrt{2}}, 0\right), \left(\frac{-1}{\sqrt{6}}, \frac{-1}{\sqrt{6}}, \frac{2}{\sqrt{6}}\right)\right\}$ 10.   **(a)** $\left\{\left(\frac{1}{\sqrt{5}}, 0, \frac{2}{\sqrt{5}}\right), \left(\frac{-2}{\sqrt{5}}, 0, \frac{1}{\sqrt{5}}\right)\right\}$

11. **(b)** $\left\{(1, 0, 0, 0), \left(0, \frac{1}{\sqrt{6}}, \frac{2}{\sqrt{6}}, \frac{1}{\sqrt{6}}\right), \left(0, \frac{-13}{\sqrt{210}}, \frac{4}{\sqrt{210}}, \frac{5}{\sqrt{210}}\right)\right\}$

12. Start with any vector that is not a multiple of $(1, 2, -1, -1)$ and use the Gram-Schmidt process to find a vector orthogonal to $(1, 2, -1, -1)$.

14. **(a)** $\left(\frac{5}{7}, \frac{13}{35}, \frac{86}{35}\right)$ **(c)** $\left(\frac{15}{7}, \frac{109}{35}, \frac{48}{35}\right)$ 15.   **(a)** $\left(\frac{-4}{31}, \frac{3}{31}, \frac{2}{31}, \frac{-8}{31}\right)$ **(c)** $\left(\frac{21}{62}, \frac{-4}{31}, \frac{-13}{31}, \frac{11}{62}\right)$

16. $\mathbf{v} = \mathbf{w} + \mathbf{w}_\perp = \left(\frac{3}{2}, \frac{3}{2}, -1\right) + \left(\frac{-1}{2}, \frac{1}{2}, 0\right)$ 18. $\mathbf{v} = \mathbf{w} + \mathbf{w}_\perp = \left(\frac{5}{3}, \frac{2}{3}, \frac{7}{3}\right) + \left(\frac{4}{3}, \frac{4}{3}, \frac{-4}{3}\right)$

19. $\sqrt{2}$ 21. $\dfrac{\sqrt{2730}}{14}$

22. Any set $\{(1, 2, -2), (6, 1, 4), (a, b, c)\}$ where $a + 2b - 2c = 0$ and $6a + b + 4c = 0$, e.g., $(-10, 16, 11)$

24. $\{\mathbf{u}_1, \mathbf{u}_2, \ldots, \mathbf{u}_n\}$ is a set of n linearly independent vectors in a vector space with dimension n. Therefore, by Theorem 4.11, $\{\mathbf{u}_1, \mathbf{u}_2, \ldots, \mathbf{u}_n\}$ is a basis for the vector space.

25. $(\mathbf{v} - \text{proj}_\mathbf{u}\, \mathbf{v}) \cdot \mathbf{u} = \left(\mathbf{v} - \dfrac{\mathbf{v} \cdot \mathbf{u}}{\mathbf{u} \cdot \mathbf{u}}\mathbf{u}\right) \cdot \mathbf{u} = \mathbf{v} \cdot \mathbf{u} - \dfrac{\mathbf{v} \cdot \mathbf{u}}{\mathbf{u} \cdot \mathbf{u}}\mathbf{u} \cdot \mathbf{u} = 0$

26. **(a)** $\mathbf{a}_1 = \begin{bmatrix} 1 \\ 0 \end{bmatrix}$ and $\mathbf{a}_2 = \begin{bmatrix} 0 \\ 1 \end{bmatrix}$. $\|\mathbf{a}_1\| = \|\mathbf{a}_2\| = 1$ and $\mathbf{a}_1 \cdot \mathbf{a}_2 = 0$

   (c) $\mathbf{a}_1 = \begin{bmatrix} \frac{\sqrt{3}}{2} \\ \frac{-1}{2} \end{bmatrix}$ and $\mathbf{a}_2 = \begin{bmatrix} \frac{1}{2} \\ \frac{\sqrt{3}}{2} \end{bmatrix}$ $\|\mathbf{a}_1\| = \sqrt{\frac{3}{4} + \frac{1}{4}} = 1$, $\|\mathbf{a}_2\| = \sqrt{\frac{1}{4} + \frac{3}{4}} = 1$, and $\mathbf{a}_1 \cdot \mathbf{a}_2 = \frac{\sqrt{3}}{2} \times \frac{1}{2} - \frac{1}{2} \times \frac{\sqrt{3}}{2} = 0$

30. The (i, j)th element of AA' is the dot product of the ith and jth columns of A. $AA' = I$ if and only if this dot product is 0 for $i \neq j$ and 1 for $i = j$. That is, $A' = A^{-1}$ if and only if A is an orthogonal matrix.

31. To show that A^{-1} is unitary, it is necessary to show that $(A^{-1})^{-1} = (\overline{A^{-1}})'$. Since A is unitary, $A^{-1} = \overline{A}'$. Thus $(A^{-1})' = (\overline{A}')' = \overline{A}$, so that $(\overline{A^{-1}})' = A = (A^{-1})^{-1}$.

Exercise Set 5.3, page 271

1. (a) $(8, -7, 2)$ (c) $(-8, 4, 0)$ (e) $(3, 10, 23)$ **2.** (a) $(10, 22, -6)$ (c) $(-10, -17, 6)$ (e) $(-80, 28, -54)$

3. (a) $10\mathbf{i} - 9\mathbf{j} + 7\mathbf{k}$ (c) $21\mathbf{i} - 17\mathbf{j} + 9\mathbf{k}$ (e) $-86\mathbf{i} - 93\mathbf{j} + 25\mathbf{k}$

4. (a) $(0, -14, -7)$ (c) -28 (d) $(4, -36, -26)$ (f) -196 **5.** (a) $\sqrt{329}$ (c) $3\sqrt{69}$ **6** (a) 130 (c) 30

8. $(\mathbf{i} \times \mathbf{j}) \cdot \mathbf{k} = (0 \times 0 - 0 \times 1, 0 \times 0 - 1 \times 0, 1 \times 1 - 0 \times 0) \cdot (0, 0, 1) = (0, 0, 1) \cdot (0, 0, 1) = 0 + 0 + 1 = 1$

12. $c(\mathbf{u} \times \mathbf{v}) = c(u_2v_3 - u_3v_2, u_3v_1 - u_1v_3, u_1v_2 - u_2v_1) = (cu_2v_3 - cu_3v_2, cu_3v_1 - cu_1v_3, cu_1v_2 - cu_2v_1) = c\mathbf{u} \times \mathbf{v}$

14. Hint: use the result of Exercise 11.

15. $\mathbf{u} \cdot (\mathbf{v} \times \mathbf{w}) = \begin{vmatrix} u_1 & u_2 & u_3 \\ v_1 & v_2 & v_3 \\ w_1 & w_2 & w_3 \end{vmatrix} = 0$ if and only if the row vectors $\mathbf{u}$, $\mathbf{v}$ and $\mathbf{w}$ are linearly dependent, i.e., if and only if one vector lies in the plane of the other two.

Exercise Set 5.4, page 277

1. (a) Point-normal form: $(x - 1) + (y + 2) + (z - 4) = 0$ (c) Point-normal form: $x + 2y + 3z = 0$
General form: $x + y + z - 3 = 0$ General form: $x + 2y + 3z = 0$

2. (a) $6x + y + 2z - 10 = 0$ (c) $3x - 2y + 1 = 0$

3. $(3, -2, 4)$ is normal to the plane $3x - 2y + 4z - 3 = 0$ and $(-6, 4, -8)$ is normal to the plane $-6x + 4y - 8z + 7 = 0$. $(-6, 4, -8) = -2(3, -2, 4)$, so the normals are parallel and therefore the planes are parallel.

5. Point-normal form: $2(x - 1) - 3(y - 2) + (z + 3) = 0$
General form: $2x - 3y + z + 7 = 0$

6. (a) Parametric equations: $x = 1 - t, y = 2 + 2t, z = 3 + 4t, -\infty < t < \infty$

Symmetric equations: $-x + 1 = \dfrac{y - 2}{2} = \dfrac{z - 3}{4}$

(c) Parametric equations: $x = -2t, y = -3t, z = 5t, -\infty < t < \infty$
Symmetric equations: $-x/2 = -y/3 = z/5$

7. $x = 1 + 2t, y = 2 + 3t, z = -4 + t, -\infty < t < \infty$ **9.** $x = 4 + 2t, y = -1 - t, z = 3 + 4t, -\infty < t < \infty$

11. The points P_1, P_2, and P_3 are collinear. There are many planes through the line containing these three points.

12. $-x + 2y - 4z + 26 = 0$

13. The points on the line are of the form $(1 + t, 14 - t, 2 - t)$. These points satisfy the equation of the plane, so the line lies in the plane.

16. The directions of the two lines are $(-4, 4, 8)$ and $(3, -1, 2)$. $(-4, 4, 8) \cdot (3, -1, 2) = 0$, so the lines are orthogonal. The point of intersection is $x = 1, y = 2, z = 3$.

17. The directions of the two lines are $(4, 5, 3)$ and $(1, -3, -2)$. $(4, 5, 3) \neq c(1, -3, -2)$, so the lines are not parallel. They do not intersect, so they are skew.

18. (a) A basis for the space must consist of two linearly independent vectors that are orthogonal to $(2, 3, -4)$. Two such vectors are $(0, 4, 3)$ and $(2, 0, 1)$.

(b) The point $(0, 0, 0)$ is not on the plane, so the set of all points on the plane is not a subspace of $\mathbf{R}^3$.

Chapter 5 Review Exercises, page 278

1. **(a)** -5 **(b)** -21 **(c)** 15 2. **(a)** $\sqrt{17}$ **(b)** $\sqrt{14}$ **(c)** $\sqrt{30}$ 3. **(a)** $\dfrac{1}{\sqrt{26}}$ **(b)** 0

4. One possibility: $(1, 2\ 0)$ 5. **(a)** $\sqrt{41}$ **(b)** $3\sqrt{2}$ **(c)** $5\sqrt{2}$ 6. $\pm 14\sqrt{14}$

7. **(a)** $\left(\frac{7}{5}, \frac{14}{5}\right)$ **(b)** $\left(-\frac{23}{21}, \frac{46}{21}, \frac{92}{21}\right)$

8. $\left\{\left(\dfrac{1}{\sqrt{15}}, \dfrac{2}{\sqrt{15}}, \dfrac{3}{\sqrt{15}}, \dfrac{-1}{\sqrt{15}}\right), \left(\dfrac{32}{\sqrt{1290}}, \dfrac{4}{\sqrt{1290}}, \dfrac{-9}{\sqrt{1290}}, \dfrac{13}{\sqrt{1290}}\right), \left(\dfrac{-26}{\sqrt{7310}}, \dfrac{72}{\sqrt{7310}}, \dfrac{-33}{\sqrt{7310}}, \dfrac{19}{\sqrt{7310}}\right)\right\}$

9. $\left\{\left(\dfrac{1}{\sqrt{2}}, 0, \dfrac{1}{\sqrt{2}}\right), \left(\dfrac{-1}{\sqrt{3}}, \dfrac{1}{\sqrt{3}}, \dfrac{1}{\sqrt{3}}\right)\right\}$ 10. $\left(\frac{294}{150}, \frac{345}{150}, \frac{-183}{150}\right)$ 11. $\frac{1}{10}\sqrt{10}$

12. If A is orthogonal, the rows of A form an orthogonal set. The rows of A are the columns of A', so from the definition of orthogonal matrix, A' is orthogonal. Interchange A and A' in the argument above to show that if A' is orthogonal, then A is orthogonal.

13. $\mathbf{v} = \mathbf{w} + \mathbf{w}_\perp = \left(\frac{5}{3}, \frac{5}{3}, \frac{-5}{3}\right) + \left(\frac{-2}{3}, \frac{4}{3}, \frac{2}{3}\right)$

14. Suppose $\mathbf{u}$ and $\mathbf{v}$ are orthogonal and that $a\mathbf{u} + b\mathbf{v} = \mathbf{0}$, so that $a\mathbf{u} = -b\mathbf{v}$. Then $a\mathbf{u} \cdot a\mathbf{u} = a\mathbf{u} \cdot (-b\mathbf{v}) = (-ab)\mathbf{u} \cdot \mathbf{v} = 0$, so $a\mathbf{u} = \mathbf{0}$. Thus $a = 0$ since $\mathbf{u} \neq \mathbf{0}$. In the same way $b = 0$, and so $\mathbf{u}$ and $\mathbf{v}$ are linearly independent.

15. If $\mathbf{u} \cdot \mathbf{v} = 0$ and $\mathbf{u} \cdot \mathbf{w} = 0$, then $\mathbf{u} \cdot (a\mathbf{v} + b\mathbf{w}) = \mathbf{u} \cdot (a\mathbf{v}) + \mathbf{u} \cdot (b\mathbf{w}) = a(\mathbf{u} \cdot \mathbf{v}) + b(\mathbf{u} \cdot \mathbf{w}) = 0$.

16. **(a)** $(1, -10, -3)$ **(b)** $(12, 72, 12)$ **(c)** 8 **(d)** $(-5, 18, 7)$ **(e)** 8 **(f)** -124 17. $3\sqrt{6}$

19. Each coordinate of $\mathbf{u} \times \mathbf{v}$ is the negative of the corresponding coordinate of $\mathbf{v} \times \mathbf{u}$. Thus $\mathbf{u} \times \mathbf{v} = \mathbf{v} \times \mathbf{u}$ if and only if each coordinate of $\mathbf{u} \times \mathbf{v}$ is equal to its negative, i.e., if and only if all the coordinates of $\mathbf{u} \times \mathbf{v}$ are 0.

20. $(\mathbf{u} + \mathbf{v}) \times (\mathbf{u} - \mathbf{v}) = \mathbf{u} \times (\mathbf{u} - \mathbf{v}) + \mathbf{v} \times (\mathbf{u} - \mathbf{v}) = \mathbf{u} \times \mathbf{u} - \mathbf{u} \times \mathbf{v} + \mathbf{v} \times \mathbf{u} - \mathbf{v} \times \mathbf{v}$
$= -\mathbf{u} \times \mathbf{v} + \mathbf{v} \times \mathbf{u} = \mathbf{v} \times \mathbf{u} + \mathbf{v} \times \mathbf{u} = 2(\mathbf{v} \times \mathbf{u}) = 2\mathbf{v} \times \mathbf{u}$

21. Point-normal form: $(x - 1) - (y + 3) + 3(z - 2) = 0$ 22. $-17x - 7y + 15z - 19 = 0$
General form: $x - y + 3z - 10 = 0$

23. Parametric equations: $x = 1 + 2t$, $y = -2 - t$, $z = 3 + 4t$, $-\infty < t < \infty$

Symmetric equations: $\dfrac{x - 1}{2} = -y - 2 = \dfrac{z - 3}{4}$

24. $x = -2 + t$, $y = 1 - 3t$, $z = 3 + 2t$, $-\infty < t < \infty$

25. $2(2 + t) - (4 - 3t) + (2 - 5t) = 4 + 2t - 4 + 3t + 2 - 5t = 2$, so the points on the line satisfy the equation of the plane. Therefore the line lies in the plane.

MATLAB Discussion

M.15 Dot Product, Norm, Angle, Distance, Projection, page 280

1. $\mathbf{u} \cdot \mathbf{v} = 2$, $\|\mathbf{u}\| = 6.3246$, angle between $\mathbf{u}$ and $\mathbf{v} = 87.2031°$, $\text{proj}_\mathbf{v}\ \mathbf{u} = (0.1905, -0.2381, -0.0476)$, $d(X, Y) = 5.7446$

M.16 Gram-Schmidt Orthogonalization, page 281

1. **(a)** The vectors are linearly dependent; orthonormal basis is $\{(0.4472, 0.8944, 0), (0.5963, -0.2981, 0.7454)\}$

M.17 Cross Product, page 282

1. **(a)** $(36, -27, -26)$ 2. **(a)** 108

Chapter 6

Exercise Set 6.1, page 293

1. Axiom 9 $(kl)\mathbf{u} = (kl)\begin{bmatrix} a & b \\ c & d \end{bmatrix} = \begin{bmatrix} (kl)a & (kl)b \\ (kl)c & (kl)d \end{bmatrix} = \begin{bmatrix} k(la) & k(lb) \\ k(lc) & k(ld) \end{bmatrix} = k\begin{bmatrix} la & lb \\ lc & ld \end{bmatrix} = k(l\mathbf{u})$

2. **(a)** $(f + g)(x) = x^2 + x + 1$, $(2f)(x) = 2x + 4$, and $(3g)(x) = 3x^2 - 3$

3. Axiom 7 $(c(f + g))(x) = c(f(x) + g(x)) = c(f(x)) + c(g(x)) = (cf)(x) + (cg)(x) = (cf + cg)(x)$

4. **(a)** $\mathbf{u} + \mathbf{v} = (7 - i, 4 + 7i)$ and $c\mathbf{u} = (4 - 7i, 17 + 6i)$

5. **(b)** This set is not closed under addition, so it is not a vector space.

6. **(a)** Subspace   **(c)** Not a subspace   **(f)** Not a subspace   7.   **(a)** Subspace

8. Every element of P_2 is an element of P_3. Both P_2 and P_3 are vector spaces with the same operations and the same set of scalars, so P_2 is a subspace of P_3.   **10.** No   **11.** No

12. A subspace must be closed under scalar multiplication, so for any vector $\mathbf{v}$ in the subspace, $0\mathbf{v} = \mathbf{0}$ must be in the subspace.

    (a) $\begin{bmatrix} 0 & 0 \\ 0 & 0 \end{bmatrix} \neq \begin{bmatrix} a & 1 \\ b & c \end{bmatrix}$, so the zero vector $\begin{bmatrix} 0 & 0 \\ 0 & 0 \end{bmatrix}$ is not in the subset.

13. **(a)** Subspace   **(b)** Not a subspace

14. **(a)** $c\mathbf{0} = c\mathbf{0} + \mathbf{0} = c\mathbf{0} + (c\mathbf{0} + -(c\mathbf{0})) = (c\mathbf{0} + c\mathbf{0}) + -(c\mathbf{0}) = c(\mathbf{0} + \mathbf{0}) + -(c\mathbf{0}) = c\mathbf{0} + -(c\mathbf{0}) = \mathbf{0}$
          axiom 5   axiom 6   axiom 4   axiom 7   axiom 5   axiom 6

15. **(a)** Linear combination   **(c)** Not a linear combination   **16.** **(a)** Linear combination   **(b)** Not a linear combination

17. **(a)** Linearly dependent   **(c)** Linearly independent   **(e)** Linearly independent

18. **(a)** Basis = $\{x^3, x^2, x, 1\}$, dimension = 4

    (c) Basis = $\left\{ \begin{bmatrix} 1 & 0 & 0 \\ 0 & 0 & 0 \end{bmatrix}, \begin{bmatrix} 0 & 1 & 0 \\ 0 & 0 & 0 \end{bmatrix}, \begin{bmatrix} 0 & 0 & 1 \\ 0 & 0 & 0 \end{bmatrix}, \begin{bmatrix} 0 & 0 & 0 \\ 1 & 0 & 0 \end{bmatrix}, \begin{bmatrix} 0 & 0 & 0 \\ 0 & 1 & 0 \end{bmatrix}, \begin{bmatrix} 0 & 0 & 0 \\ 0 & 0 & 1 \end{bmatrix} \right\}$, dimension = 6

    (e) Basis = $\left\{ \begin{bmatrix} 1 & 0 \\ 0 & 0 \end{bmatrix}, \begin{bmatrix} 0 & 0 \\ 0 & 1 \end{bmatrix}, \begin{bmatrix} 0 & 1 \\ 1 & 0 \end{bmatrix} \right\}$, dimension = 3

19. **(b)** No   **(d)** $\{f(x), g(x)\}$ **20.** **(a)** No   **(c)** No   **(e)** Yes

21. If $af + bg + ch = 0$, then $af' + bg' + ch' = 0$ and $af'' + bg'' + ch'' = 0$. If the Wronskian is not the zero function, this system of three homogeneous equations has the unique solution $a = b = c = 0$, so that f, g, and h are linearly independent.

22. **(c)** $\begin{vmatrix} \sin x & \cos x & x \sin x \\ \cos x & -\sin x & \sin x + x \cos x \\ -\sin x & -\cos x & -x \sin x + 2 \cos x \end{vmatrix} = \begin{vmatrix} \sin x & \cos x & x \sin x \\ \cos x & -\sin x & \sin x + x \cos x \\ 0 & 0 & 2 \cos x \end{vmatrix}$

    $= -2 \cos x \neq$ the zero function, so $\sin x$ $\cos x$, and $x \sin x$ are linearly independent.

23. If $c\mathbf{u}$ is in U, then $\dfrac{1}{c}c\mathbf{u} = \mathbf{u}$ is in U, so if $\mathbf{u}$ is not in U, then $c\mathbf{u}$ is also not in U for all $c \neq 0$.

25. **(a)** Let U be the subspace of $\mathbf{R}^2$ with basis $(1, 0)$ and let V be the subspace of $\mathbf{R}^2$ with basis $(0, 1)$. $(1, 0)$ and $(0, 1)$ are in the union of U and V but their sum $(1, 1)$ is not.

Exercise Set 6.2, page 302

3. $\langle \mathbf{u}, \mathbf{u} \rangle = 2x_1x_1 - x_2x_2 = 2x_1^2 - x_2^2 < 0$ for the vector $\mathbf{u} = (1, 2)$, so condition 4 of the definition is not satisfied.

4. $\mathbf{u} = \begin{bmatrix} a & b \\ c & d \end{bmatrix}$, $\mathbf{v} = \begin{bmatrix} e & f \\ g & h \end{bmatrix}$, and $\mathbf{w} = \begin{bmatrix} j & k \\ l & m \end{bmatrix}$

    $\langle \mathbf{u} + \mathbf{v}, \mathbf{w} \rangle = (a + e)j + (b + f)k + (c + g)l + (d + h)m = aj + ej + bk + fk + cl + gl + dm + hm$
    $= aj + bk + cl + dm + ej + fk + gl + hm = \langle \mathbf{u}, \mathbf{w} \rangle + \langle \mathbf{v}, \mathbf{w} \rangle$, so axiom 2 is satisfied.

5. (a) -16 8. (a) $-\frac{1}{2}$ (c) $\frac{1}{12}$ 9. (a) $\dfrac{2}{\sqrt{3}}$ (c) $\dfrac{7}{\sqrt{5}}$

10. $\langle f, g \rangle = \int_0^1 (x^2)(4x - 3)\, dx = \int_0^1 (4x^3 - 3x^2)\, dx = [x^4 - x^3]_0^1 = 0$, so the functions are orthogonal.   12. $\dfrac{\sqrt{15}}{4}$

13. Any function $ax + b$ with $8a + 15b = 0$ will do.   14. $\sqrt{\dfrac{76}{3}}$ 15. g is closer.   16. (a) 9

17. (a) $\sqrt{30}$ (c) $\sqrt{66}$ 18. (a) $\left\langle \begin{bmatrix} 1 & 2 \\ -1 & 1 \end{bmatrix}, \begin{bmatrix} 2 & 4 \\ 3 & -7 \end{bmatrix} \right\rangle = 2 + 8 - 3 - 7 = 0$, so the matrices are orthogonal.

19. Any matrix $\begin{bmatrix} a & b \\ c & d \end{bmatrix}$ with $a + 2b + 3c + 4d = 0$ will do.   20. (a) $3\sqrt{2}$

21. (a) $\langle \mathbf{u}, \mathbf{v} \rangle = 16 + 2i$, $\|\mathbf{u}\| = \|\mathbf{v}\| = 3\sqrt{2}$, $d(\mathbf{u}, \mathbf{v}) = 2$, not orthogonal
     (d) $\langle \mathbf{u}, \mathbf{v} \rangle = 0$, $\|\mathbf{u}\| = \sqrt{26}$, $\|\mathbf{v}\| = \sqrt{2}$, $d(\mathbf{u}, \mathbf{v}) = 2\sqrt{7}$, orthogonal
22. (a) $\langle \mathbf{u}, \mathbf{v} \rangle = 2 - 2i$, $\|\mathbf{u}\| = \|\mathbf{v}\| = \sqrt{19}$, $d(\mathbf{u}, \mathbf{v}) = \sqrt{34}$, not orthogonal
     (c) $\langle \mathbf{u}, \mathbf{v} \rangle = 10 - 10i$, $\|\mathbf{u}\| = 2\sqrt{3}$, $\|\mathbf{v}\| = \sqrt{30}$, $d(\mathbf{u}, \mathbf{v}) = \sqrt{22}$, not orthogonal
23. $\langle \mathbf{u}, \mathbf{v} \rangle = x_1\bar{y}_1 + \cdots + x_n\bar{y}_n = \bar{y}_1 x_1 + \cdots + \bar{y}_n x_n = \overline{\bar{y}_1 \bar{x}_1 + \cdots + \bar{y}_n \bar{x}_n} = \overline{\langle \mathbf{v}, \mathbf{u} \rangle} = \langle \mathbf{v}, \mathbf{u} \rangle$
24. $\langle \mathbf{u}, k\mathbf{v} \rangle = x_1 \overline{k y}_1 + \cdots + x_n \overline{k y}_n = \bar{k} x_1 \bar{y}_1 + \cdots + \bar{k} x_n \bar{y}_n = \bar{k}(x_1 \bar{y}_1 + \cdots + x_n \bar{y}_n) = \bar{k}\langle \mathbf{u}, \mathbf{v} \rangle$
26. (a) $\langle \mathbf{u}, \mathbf{v} \rangle = \mathbf{u}A\mathbf{v}^t = (\mathbf{u}A) \cdot \mathbf{v} = \mathbf{v} \cdot (\mathbf{u}A) = \mathbf{v}(\mathbf{u}A)^t = \mathbf{v}A\mathbf{u}^t = \langle \mathbf{v}, \mathbf{u} \rangle$
     (c) Using A, $\langle \mathbf{u}, \mathbf{v} \rangle = 0$, $\|\mathbf{u}\| = \sqrt{2}$, $\|\mathbf{v}\| = \sqrt{3}$, and $d(\mathbf{u}, \mathbf{v}) = \sqrt{5}$
     Using C, $\langle \mathbf{u}, \mathbf{v} \rangle = 3$, $\|\mathbf{u}\| = \sqrt{2}$, $\|\mathbf{v}\| = \sqrt{5}$, and $d(\mathbf{u}, \mathbf{v}) = 1$
     (d) $\begin{bmatrix} 1 & 0 \\ 0 & 4 \end{bmatrix}$

Exercise Set 6.3, page 310

1. $x_1^2 + 4x_2^2 = 1$ 2. (a) $\|(1, 0)\| = 2$, $\|(0, 1)\| = 3$, $\|(1, 1)\| = \sqrt{13}$, $\|(2, 3)\| = \sqrt{97}$ (c) $d((1, 0), (0, 1)) = \sqrt{13}$

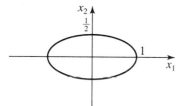

3. (c) $d((5, 0), (0, 4)) = \sqrt{281}$. In Euclidean space the distance is $\sqrt{41}$. (d) $x_1^2 + 16x_2^2 = 1$
4. $\langle (x_1, x_2), (y_1, y_2) \rangle = \frac{1}{4}x_1 y_1 + \frac{1}{25}x_2 y_2$

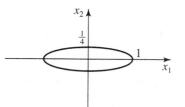

6. $d(R, Q)^2 = \|(4, 0, 0, -5)\|^2 = |-16 - 0 - 0 + 25| = 9$, so $d(R, Q) = 3$

8. $\langle(2, 0, 0, 1), (1, 0, 0, 2)\rangle = -2 - 0 - 0 + 2 = 0$, so the vectors are orthogonal.

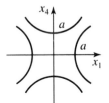

9. The equations of the circles with radii $a = 1, 2, 3$ and center at $(0, 0)$ are $|-x_1^2 + x_4^2| = a^2$.

10. 12 years, 0.8 speed of light   12. More than 8

Exercise Set 6.4, page 317

1. $g(x) = \frac{1}{3}$ 2. $\{1, 2\sqrt{3}(x - \frac{1}{2})\}$ is an orthonormal basis for $P_1[0, 1]$, $g(x) = 4e - 10 + (18 - 6e)x$.

4. $\{\pi^{-1/2}, 2\sqrt{3}\pi^{-3/2}(x - \frac{\pi}{2})\}$ is an orthonormal basis for $P_1[0, \pi]$, $g(x) = -24\pi^{-3}x + 12\pi^{-2}$.

5. (a) $\left\{\frac{1}{\sqrt{2}}, \frac{\sqrt{3}}{\sqrt{2}}x, \frac{3\sqrt{5}}{2\sqrt{2}}(x^2 - \frac{1}{3})\right\}$ is an orthonormal basis for $P_2[-1, 1]$, $g(x) = -\frac{3}{4}e + \frac{33}{4}e^{-1} + 3e^{-1}x + \frac{15}{4}(e - 7e^{-1})x^2$.

6. $\{1, 2\sqrt{3}(x - \frac{1}{2}), 6\sqrt{5}(x^2 + \frac{1}{6})\}$ is an orthonormal basis for $P_2[0, 1]$, $g(x) = -\frac{4}{7}x^2 + \frac{48}{35}x + \frac{6}{35}$.

8. $\left\{\pi^{-1/2}, 2\sqrt{3}\pi^{-3/2}(x + \frac{\pi}{2}), 6\sqrt{5}\pi^{-5/2}(x^2 - \pi x + \frac{\pi^2}{6})\right\}$ is an orthonormal basis for $P_2[0, \pi]$,

$g(x) = 180\pi^{-5}(-4 + \frac{\pi^2}{3})x^2 - 180\pi^{-4}(-4 + \frac{\pi^2}{3})x - 120\pi^{-3} + 12\pi^{-1}$.

9. The vectors that form an orthonormal basis over $[-\pi, \pi]$ also form an orthonormal basis over $[0, 2\pi]$,
$g(x) = \pi - 2(\sin x + \frac{1}{2}\sin 2x + \frac{1}{3}\sin 3x + \frac{1}{4}\sin 4x)$.

11. $g(x) = \frac{4}{3}\pi^2 + 4(\cos x + \frac{1}{4}\cos 2x + \frac{1}{9}\cos 3x + \frac{1}{16}\cos 4x) - 4\pi(\sin x + \frac{1}{2}\sin 2x + \frac{1}{3}\sin 3x + \frac{1}{4}\sin 4x)$.

13. $(1, 0, 0, 0, 0, 1, 1)$, $(0, 1, 0, 0, 1, 0, 1)$, $(0, 0, 1, 0, 1, 1, 0)$, $(0, 0, 0, 1, 1, 1, 1)$, $(0, 1, 1, 1, 1, 0, 0)$, $(1, 0, 1, 1, 0, 1, 0)$,
$(1, 1, 0, 1, 0, 0, 1)$, $(1, 1, 1, 0, 0, 0, 0)$, $(1, 1, 0, 0, 1, 1, 0)$, $(1, 0, 1, 0, 1, 0, 1)$, $(1, 0, 0, 1, 1, 0, 0)$, $(1, 1, 1, 1, 1, 1, 1)$,
$(0, 0, 1, 1, 0, 0, 1)$, $(0, 1, 0, 1, 0, 1, 0)$, $(0, 1, 1, 0, 0, 1, 1)$, $(0, 0, 0, 0, 0, 0, 0)$

14. (a) center $= (0, 0, 1, 0, 1, 1, 0)$
$(1, 0, 1, 0, 1, 1, 0)$, $(0, 1, 1, 0, 1, 1, 0)$, $(0, 0, 0, 0, 1, 1, 0)$, $(0, 0, 1, 1, 1, 1, 0)$, $(0, 0, 1, 0, 0, 1, 0)$, $(0, 0, 1, 0, 1, 0, 0)$,
$(0, 0, 1, 0, 1, 1, 1)$
(b) center $= (1, 1, 0, 1, 0, 0, 1)$
$(0, 1, 0, 1, 0, 0, 1)$, $(1, 0, 0, 1, 0, 0, 1)$, $(1, 1, 1, 1, 0, 0, 1)$, $(1, 1, 0, 0, 0, 0, 1)$, $(1, 1, 0, 1, 1, 0, 1)$, $(1, 1, 0, 1, 0, 1, 1)$,
$(1, 1, 0, 1, 0, 0, 0)$

15. (a) $(0, 0, 0, 0, 0, 0, 0)$ (c) $(0, 0, 1, 0, 1, 1, 0)$

16. (a) Since each vector in V_{23} has 23 components and there are 2 possible values (0 and 1) for each component, there are 2^{23} vectors.
(c) There are 23 ways to change exactly one component of the center vector, $23 \times \frac{22}{2} = 253$ ways to change exactly two components, and $23 \times 22 \times \frac{21}{6} = 1771$ ways to change exactly three components.
$1 + 23 + 253 + 1771 = 2048$ vectors.

Chapter 6 Review Exercises, page 318

1. $(f + g)(x) = 2x^2 + 3x + 2$, $3f(x) = 9x - 3$, and $(2f - 3g)(x) = -6x^2 + 6x - 11$ **2.** No

3. $13x^2 + 8x - 21 = 2(2x^2 + x - 3) - 3(-3x^2 - 2x + 5)$

4. $\begin{bmatrix} 1 & 0 & 0 \\ 0 & 0 & 0 \\ 0 & 0 & 0 \end{bmatrix}, \begin{bmatrix} 0 & 1 & 0 \\ 0 & 0 & 0 \\ 0 & 0 & 0 \end{bmatrix}, \begin{bmatrix} 0 & 0 & 1 \\ 0 & 0 & 0 \\ 0 & 0 & 0 \end{bmatrix}, \begin{bmatrix} 0 & 0 & 0 \\ 0 & 1 & 0 \\ 0 & 0 & 0 \end{bmatrix}, \begin{bmatrix} 0 & 0 & 0 \\ 0 & 0 & 1 \\ 0 & 0 & 0 \end{bmatrix}, \begin{bmatrix} 0 & 0 & 0 \\ 0 & 0 & 0 \\ 0 & 0 & 1 \end{bmatrix}$

5. The three given functions are linearly independent. The dimension of P_2 is 3, so the three functions are a basis.

7. $\langle f, g \rangle = 4$, $\|f\| = 1$, $\|g\| = \dfrac{\sqrt{97}}{\sqrt{3}}$, $d(f, g) = 2\dfrac{\sqrt{19}}{\sqrt{3}}$

8. $\|(1, -2)\| = \sqrt{14}$, $\|(3, 2)\| = \sqrt{30}$, $d((1, -2), (3, 2)) = 2\sqrt{14}$ **9.** The condition is necessary but not sufficient.

10. $2x_1^2 + 3x_2^2 = 1$. In the sketch below $a = \dfrac{1}{\sqrt{2}}$ and $b = \dfrac{1}{\sqrt{3}}$.   **11.** $g(x) = 2x - \frac{2}{3}$

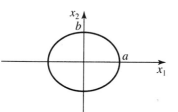

12. $g(x) = 2\pi - 1 - 4\left(\sin x + \frac{1}{2}\sin 2x + \frac{1}{3}\sin 3x + \frac{1}{4}\sin 4x\right)$

13. center = (0, 1, 1, 0, 0, 1, 1)

(1, 1, 1, 0, 0, 1, 1), (0, 0, 1, 0, 0, 1, 1), (0, 1, 0, 0, 0, 1, 1), (0, 1, 1, 1, 0, 1, 1), (0, 1, 1, 0, 1, 1, 1), (0, 1, 1, 0, 0, 0, 1),
(0, 1, 1, 0, 0, 1, 0)

Technology Explorations with MATLAB

M.18 Inner Product, Non-Euclidean Geometry, page 319

1. (a) $\|(0, 1)\| = 2$ (b) $90°$. The vectors (1, 1), and (−4, 1) are at right angles. (c) $\text{dist}((1, 0), (0, 1)) = \sqrt{5}$
(d) $\text{dis}((x, y), (0, 0)) = 1, 2, 3$; $\langle (x, y), (x, y) \rangle = 1, 4, 9$; $[x \; y]A[x \; y]^t = 1, 4, 9$; $x^2 + 4y^2 = 1, 4, 9$.

M.19 Space-Time Travel, page 322

1. Earth time 120 years, spaceship time 79.3725 years

Chapter 7

Exercise Set 7.1, page 329

1. $T(1, 2) = (2, -1)$, $T(-1, 4) = (-2, -5)$

3. $T(x_1 + x_2, y_1 + y_2, z_1 + z_2) = (0, y_1 + y_2, 0) = (0, y_1, 0) + (0, y_2, 0) = T(x_1, y_1, z_1) + T(x_2, y_2, z_2)$ and
$T(c(x, y, z)) = T(cx, cy, cz) = (0, cy, 0) = c(0, y, 0) = cT(x, y, z)$

4. (a) $T(cx, cy, cz) = (3cx, (cy)^2) = (3cx, c^2y^2) \neq c(3x, y^2) = cT(x, y, z)$ **6.** Linear   **8.** (a) Linear

10. Not linear **11.** Linear **12.** $A\mathbf{x} = \begin{bmatrix} 1 \\ 4 \\ 1 \end{bmatrix}$, $A\mathbf{y} = \begin{bmatrix} 8 \\ 7 \\ 8 \end{bmatrix}$, $A\mathbf{z} = \begin{bmatrix} 9 \\ 1 \\ 9 \end{bmatrix}$

14. $T(\mathbf{x}_1 + \mathbf{x}_2) = A(\mathbf{x}_1 + \mathbf{x}_2) + \mathbf{c} = A\mathbf{x}_1 + A\mathbf{x}_2 + \mathbf{c} \neq T(\mathbf{x}_1) + T(\mathbf{x}_2)$ if $\mathbf{c} \neq 0$, so T is not linear.

17. $T(\mathbf{x}_1 + \mathbf{x}_2) = (\mathbf{x}_1 + \mathbf{x}_2) \cdot \mathbf{y} = \mathbf{x}_1 \cdot \mathbf{y} + \mathbf{x}_2 \cdot \mathbf{y} = T(\mathbf{x}_1) + T(\mathbf{x}_2)$ and $T(c\mathbf{x}) = c\mathbf{x} \cdot \mathbf{y} = c(\mathbf{x} \cdot \mathbf{y}) = cT(\mathbf{x})$, so T is a linear transformation from $\mathbf{R}^n$ to $\mathbf{R}$.

19. **(a)** $T(\mathbf{x}) = \begin{bmatrix} -1 & -2 \\ 16 & 2 \end{bmatrix} \mathbf{x}$, $T\left(\begin{bmatrix} 5 \\ 2 \end{bmatrix}\right) = \begin{bmatrix} -9 \\ 84 \end{bmatrix}$ **(c)** $T(\mathbf{x}) = \begin{bmatrix} 6 & -2 \\ 3 & -3 \\ 0 & 4 \end{bmatrix} \mathbf{x}$, $T\left(\begin{bmatrix} -3 \\ 2 \end{bmatrix}\right) = \begin{bmatrix} -22 \\ -15 \\ 8 \end{bmatrix}$

20. $T(x, y) = (0, 2x - y)$ **22.** $T(x, y) = (x + y, 3x + y, 2x - y)$ **23.** In general, no **25.** Not linear

Exercise Set 7.2, page 347

1. **(a)** $\begin{bmatrix} 0 & -1 \\ 1 & 0 \end{bmatrix}, \begin{bmatrix} -1 \\ 2 \end{bmatrix}$ **(c)** $\begin{bmatrix} \frac{1}{\sqrt{2}} & \frac{-1}{\sqrt{2}} \\ \frac{1}{\sqrt{2}} & \frac{1}{\sqrt{2}} \end{bmatrix}, \begin{bmatrix} \frac{1}{\sqrt{2}} \\ \frac{3}{\sqrt{2}} \end{bmatrix}$ **(d)** $\begin{bmatrix} -1 & 0 \\ 0 & -1 \end{bmatrix}, \begin{bmatrix} -2 \\ -1 \end{bmatrix}$ **(f)** $\begin{bmatrix} \frac{\sqrt{3}}{2} & \frac{-1}{2} \\ \frac{1}{2} & \frac{\sqrt{3}}{2} \end{bmatrix}, \begin{bmatrix} \sqrt{3} - \frac{1}{2} \\ 1 + \frac{\sqrt{3}}{2} \end{bmatrix}$

3. $\dfrac{x^2}{9} + \dfrac{y^2}{4} = 1$ **4.** $\begin{bmatrix} 0 & -2 \\ 2 & 0 \end{bmatrix}$ **6.** $x^2 + y^2 = 9$ **7.** Radii are 1, 1.5, 2.25, 3.375, and 5.0625

8. $\begin{bmatrix} -1 & 0 \\ 0 & 1 \end{bmatrix}, \begin{bmatrix} -2 \\ 1 \end{bmatrix}$

10. Vertices of the image of the unit square are **(a)** $(0, 1), (-1, 1), (-1, 0), (0, 0)$ **(c)** $(3, 1), (3, 5), (0, 4), (0, 0)$ **(e)** $(-2, 0), (-5, 4), (-3, 4), (0, 0)$ **(g)** $(0, 2), (-2, 2), (-2, 0), (0, 0)$

11. **(a)** $\begin{bmatrix} 2 & 0 \\ 1 & -1 \end{bmatrix}$ **(c)** $\begin{bmatrix} 2 & -5 \\ 0 & 3 \end{bmatrix}$ **13.** $\begin{bmatrix} 0 & 0 \\ 0 & 1 \end{bmatrix}$ **14.** $\begin{bmatrix} \frac{1}{2} & \frac{1}{2} \\ \frac{1}{2} & \frac{1}{2} \end{bmatrix}$

15. $\begin{bmatrix} a & 0 \\ 0 & b \end{bmatrix}$ If $a = 3$ and $b = 2$, the unit square becomes a 3×2 rectangle.

16. $\begin{bmatrix} 2 & 0 \\ 0 & 3 \end{bmatrix} \begin{bmatrix} x \\ y \end{bmatrix} = \begin{bmatrix} 2x \\ 3y \end{bmatrix} = \begin{bmatrix} x' \\ y' \end{bmatrix}$. $y = 2x$, so $\dfrac{y'}{3} = 2\dfrac{x'}{2}$, and $y' = 3x'$. Thus the images of the points on the line $y = 2x$ are the points on the line $y = 3x$.

18. $\begin{bmatrix} 1 & c \\ 0 & 1 \end{bmatrix}$ **20.** $16y = 3x$

23. The columns are orthogonal and $\sin^2\theta + \cos^2\theta = 1$, so the norm of each column is 1. **26.** **(a)** $y = 3x$

27. **(a)** Square with vertices $(6, 4), (6, 6), (4, 6), (4, 4)$; circle $(x - 4)^2 + (y - 4)^2 = 4$

(c) Square with vertices $\left(\dfrac{1}{\sqrt{2}} + 3, \dfrac{1}{\sqrt{2}} + 1\right), \left(3, \dfrac{2}{\sqrt{2}} + 1\right), \left(\dfrac{-1}{\sqrt{2}} + 3, \dfrac{1}{\sqrt{2}} + 1\right), (3, 1)$; circle $(x - 3)^2 + (y - 1)^2 = 1$

29. $T^{-1}(\mathbf{u}) = A^{-1}(\mathbf{u} - \mathbf{v}) = A^{-1}\mathbf{u} + \mathbf{w}$, where $\mathbf{w} = -A^{-1}\mathbf{v}$. **30.** **(a)** $\begin{bmatrix} 0 & -2 \\ 2 & 0 \end{bmatrix}, \begin{bmatrix} -2 \\ 4 \end{bmatrix}$ **(c)** $\begin{bmatrix} 0 & -1 \\ -1 & 0 \end{bmatrix}, \begin{bmatrix} -1 \\ -2 \end{bmatrix}$

31. **(a)** $\begin{bmatrix} 3 & 6 \\ 0 & 3 \end{bmatrix}, \begin{bmatrix} 21 \\ 6 \end{bmatrix}$ **(b)** $\begin{bmatrix} 0 & 2 \\ 3 & 0 \end{bmatrix}, \begin{bmatrix} 4 \\ 9 \end{bmatrix}$ **36.** **(a)** $T = A\mathbf{u} + \mathbf{v}$ square $(2.5, 2), (2.5, 2.5), (2, 2.5), (2, 2)$
$T = A^2\mathbf{u} + \mathbf{v}$ square $(2.25, 2), (2.25, 2.25), (2, 2.25), (2, 2)$
$T = A^3\mathbf{u} + \mathbf{v}$ square $(2.125, 2), (2.125, 2.125), (2, 2.125), (2, 2)$

38. The pairs that commute are D and R, D and F, D and S, and D and H. **39.** $\begin{bmatrix} 0 & -1 & 0 \\ 1 & 0 & 0 \\ 0 & 0 & 1 \end{bmatrix}$ or $\begin{bmatrix} 0 & 1 & 0 \\ -1 & 0 & 0 \\ 0 & 0 & 1 \end{bmatrix}$

41. $\begin{bmatrix} 0 & -1 & 6 \\ 1 & 0 & -4 \\ 0 & 0 & 1 \end{bmatrix}$, square with vertices $(6, -3)$, $(5, -3)$, $(5, -4)$, $(6, -4)$ **42.** $T^{-1} = \begin{bmatrix} 1 & 0 & -h \\ 0 & 1 & -k \\ 0 & 0 & 1 \end{bmatrix}$

44. Triangle with vertices $(9, -25)$, $(-9, -35)$, and $(9, -40)$

Exercise Set 7.3, page 359

1. **(a)** Kernel $= \mathbf{0}$, range $= \mathbf{R}^2$ **(c)** Kernel $= \{(-2r, r)\}$, range $= \{(r, 2r)\}$ **(e)** Kernel $= \{(r, -2r, r)\}$, range $= \mathbf{R}^2$
  (g) Kernel $= \{(r, 0, 0)\}$, range $= \{(a, 2a, b)\}$ **(h)** Kernel $= \{(-2r, r, 0)\}$, range $= \{(a + b, -a, 2a + b)\}$
2. **(a)** Kernel $= \{(0, r, s)\}$, range $= \{(a, 0, 0)\}$ **(b)** Kernel $= \{(r, -r, 0)\}$, range $= \mathbf{R}^2$
  (e) Kernel $= \mathbf{0}$, range $= \{(3a, a - b, b)\}$ **3.** **(a)** $\mathrm{Ker}(T) = \mathbf{0}$, range$(T) = U$ **(b)** Not linear
4. **(a)** Dim range $= 2$, dim kernel $= 1$, not one-to-one   **(c)** Dim range $= 3$, dim kernel $= 1$, not one-to-one
  (f) Dim range $= 3$, dim kernel $= 0$, one-to-one   **5.** **(a)** One-to-one   **(c)** Not one-to-one   **(f)** One-to-one
6. **(a)** $\{(r, 0)\}$ **(c)** No fixed points   **(f)** $\mathbf{0}$ **9.** Dim domain$(T) = $ dim ker$(T) + $ dim range$(T) \geq 0 + $ dim range(T)
11. $\{(1, 2, r)\}$, all vectors on the line perpendicular to the xy plane and passing through the point $(1, 2, 0)$

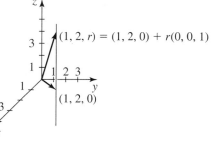

13. Not a subspace of $\mathbf{R}^2$ **15.** Not a subspace   **16.** Ker(T) is the zero polynomial, range(T) is P_2.

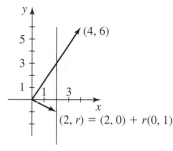

18. Ker(g) is the zero polynomial, range$(g) = \{a_3 x^3 + a_1 x + a_0\}$ with basis $\{1, x, x^3\}$.   **19.** $T(\mathbf{0}) \neq \mathbf{0}$
21. $D(x^3 - 3x^2 + 2x + 1) = 3x^2 - 6x + 2$, ker$(D) = $ the set of constant polynomials, range$(D) = P_{n-1}$
22. **(a)** $D^2(2x^3 + 3x^2 - 5x + 4) = 12x - 6$, ker$(D^2) = P_1$, range$(D^2) = P_{n-2}$ **23.** $\{x^3 - 2x^2 + 7x + r\}$

25. Ker(T) is all polynomials $a_n x^n + a_{n-1} x^{n-1} + \cdots + a_1 x + a_0$ for which $\dfrac{a_n}{n + 1} + \dfrac{a_{n-1}}{n} + \cdots + \dfrac{a_1}{2} = -a_0$,

  range$(T) = \mathbf{R}$.

26. **(b)** Not linear   **(d)** Not linear   **(f)** Ker$(T) = \left\{ \begin{bmatrix} 0 & a \\ b & c \end{bmatrix} \right\}$, range$(T) = \mathbf{R}$

  (i) Ker$(T) = \left\{ \begin{bmatrix} 0 & r \\ -r & 0 \end{bmatrix} \right\}$, range$(T)$ is the set of all symmetric matrices.   **(j)** Not linear

27. **(a)** Not linear   **(b)** Ker(T) is the set of all vectors in U which are orthogonal to **v**, range(T) = **R**.
28. T is one-to-one if and only if ker(T) is the zero vector if and only if dim ker(T) = 0 if and only if dim range(T) = dim domain(T).

Exercise Set 7.4, page 368

1. $(r, r, 1) = r(1, 1, 0) + (0, 0, 1)$ 3. $(r + 1, 2r, r) = r(1, 2, 1) + (1, 0, 0)$
5. $(-4r + 3s - 2, -5r + 5s - 6, r, s) = r(-4, -5, 1, 0) + s(3, 5, 0, 1) + (-2, -6, 0, 0)$
7. **(a)** Solutions exist   **(d)** No solutions   8.   Any particular solution of the system will do in place of $\mathbf{x}_1$.
10. $x_1 + 2x_2 + x_3 = 3$, $x_2 + 2x_3 = 7$, $x_1 + x_2 - x_3 = -4$
13. **(a)** Ker($D^2 + D - 2$) = $\{ae^{-2x} + be^x\}$ **(c)** Ker($D^2 + 2D - 8$) = $\{ae^{-4x} + be^{2x}\}$
14. **(b)** $y = re^{3x} + s - \dfrac{8x}{3}$ **(d)** $y = re^{-8x} + s + \dfrac{3e^{2x}}{20}$

Exercise Set 7.5, page 380

1. $\dfrac{1}{27}\begin{bmatrix} 11 & -1 & -16 \\ 5 & 2 & 5 \end{bmatrix}$ 3.   Does not exist   6.   Does not exist   8. $\dfrac{1}{11}\begin{bmatrix} 5 & -2 \\ 3 & 1 \end{bmatrix}$ 9. $y = -\dfrac{10}{3} + \dfrac{7x}{2}$

11. $y = -3 + 4x$ 13. $y = 10 - \dfrac{7x}{2}$ 15. $y = -\frac{3}{2} + 2x$ 17. $y = \dfrac{23}{2} - \dfrac{5x}{2}$ 19. $y = -3 + \dfrac{12x}{5}$

21. $y = 12 - 9x + 2x^2$ 23. $y = -\dfrac{29}{4} + \dfrac{223x}{20} - \dfrac{9x^2}{4}$ 25.   **(a)** $L = 3.8 + 1.06F$

26. $G = 23.545 - 0.1351S$ 28. $U = \dfrac{127{,}955{,}000 + 10{,}545D}{262{,}500}$

30. Least squares line: $\% = \dfrac{-1632.75 + 97.65Y}{4900}$
    Least squares parabola: $\% = 0.93643 - 0.05907Y + 0.00113Y^2$

32. U.K. $N = \dfrac{125{,}038 + 5{,}619Y}{98}$

    U.S.A. $N = \dfrac{511{,}049 + 623Y}{98}$

    Japan $N = \dfrac{199{,}233 - 80Y}{98}$

    The year 2000 is year 25. The least square predictions are $N = 2709$ for U.K., 5374 for U.S.A., and 2013 for Japan.

35. China $N = 555.61302 + 24.11237Y - 0.11360Y^2$
    Japan $N = 83.65595 + 1.05255Y + 0.00066Y^2$
    W. Europe $N = 121.31114 + 1.78878Y - 0.02385Y^2$
    N. America $N = 166.28642 + 3.42106Y - 0.01882Y^2$
    World $N = 2493.40194 + 48.06250Y + 0.56248Y^2$
    The year 2000 is year 50. The least square predictions are $N = 1477$ for China, 138 for Japan, 151 for W. Europe, 290 for N. America, and 6303 for World.

36. **(a)** $\dfrac{1}{3}\begin{bmatrix} 2 & 1 & 1 \\ 1 & 2 & -1 \\ 1 & -1 & 2 \end{bmatrix}$, $\left(\frac{4}{3}, \frac{5}{3}, -\frac{1}{3}\right)$ **(c)** $\dfrac{1}{6}\begin{bmatrix} 5 & 2 & -1 \\ 2 & 2 & 2 \\ -1 & 2 & 5 \end{bmatrix}$, $(1, 1, 1)$ 37.   **(a)** $\dfrac{1}{8}\begin{bmatrix} 8 & 0 & 0 \\ 0 & 4 & -4 \\ 0 & -4 & 4 \end{bmatrix}$, $(-2, -1, 1)$

39. The first step of multiplying by A^t ($A\mathbf{x} = \mathbf{y}$, so $A^tA\mathbf{x} = A^t\mathbf{y}$) is not reversible unless A^t is invertible.
41. **(a)** No   **(b)** No
42. **(b)** Pinv(A) exists if and only if $(A^tA)^{-1}$ exists if and only if A is invertible if and only if pinv(A) = A^{-1} (see Exercise 40).
    Thus pinv(pinv(A)) = pinv(A^{-1}) = $(A^{-1})^{-1} = A$.

(c) As in (b), $\text{pinv}(A) = A^{-1}$ and $\text{pinv}(A') = (A')^{-1} = (A^{-1})' = (\text{pinv}(A))'$.

44. (a) $P' = (A(A'A)^{-1}A')' = (A')'((A'A)^{-1})'A' = A((A'A)')^{-1}A' = A(A'A)^{-1}A' = P$.

Chapter 7 Review Exercises, page 382

1. (a) Linear (b) Not linear   2. $\begin{bmatrix} 1 & 2 \\ 3 & -1 \end{bmatrix}$

3. If T is linear, then $T(a\mathbf{u} + b\mathbf{v}) = T(a\mathbf{u}) + T(b\mathbf{v}) = aT(\mathbf{u}) + bT(\mathbf{v})$. If $T(a\mathbf{u} + b\mathbf{v}) = aT(\mathbf{u}) + bT(\mathbf{v})$, then $T(\mathbf{u} + \mathbf{v}) = T(\mathbf{u}) + T(\mathbf{v})$ and $T(c\mathbf{u}) = cT(\mathbf{u})$, so T is linear.

4. $\begin{bmatrix} \dfrac{3\sqrt{3}}{2} & -\dfrac{3}{2} \\ \dfrac{3}{2} & \dfrac{3\sqrt{3}}{2} \end{bmatrix}$ 5. $\begin{bmatrix} 0 & -1 \\ -1 & 0 \end{bmatrix}$ 6. $\begin{bmatrix} \dfrac{1}{2} & -\dfrac{1}{2} \\ -\dfrac{1}{2} & \dfrac{1}{2} \end{bmatrix}$ 7. $\dfrac{y}{2} = -x + 1$ 8. $y = 5x + 3$

9. $\begin{bmatrix} -2 & 0 \\ 3 & 1 \end{bmatrix}$ 10. $\text{Ker}(T) = \{(r, 3r/2)\}$, $\text{range}(T) = \{(2r, r)\}$, $\dim \ker(T) = 1$, $\dim \text{range}(T) = 1$

11. Basis for kernel $= \{(-2, 1, 1)\}$, basis for range $= \{(1, 0, 2), (1, 1, 3)\}$ 12. (a) Not one-to-one (b) One-to-one

13. $\text{Ker}(T) = \{(0, r, r)\}$, $\text{range}(T) = \{(a, 2a, b)\}$

14. $\text{Ker}(g) = \mathbf{0}$, $\text{range}(g)$ is the set of all polynomials $ax^3 + bx + c$ with basis $\{x^3, x, 1\}$

15. The only polynomial mapped into $12x - 4$ is the polynomial $12x + 20$.

16. $\dfrac{1}{65}\begin{bmatrix} -17 & 14 & 10 \\ 39 & -13 & 0 \end{bmatrix}$ 17. $y = 12.5 - 8.8x + 2x^2$

18. (a) $ABA = A(A'A)^{-1}A'A = AI_m = A$ (b) $BAB = (A'A)^{-1}A'AB = I_mB = B$ (c) See Exercise 44(a) of Section 7.5.
    (d) $BA = (A'A)^{-1}A'A = I_m$ and is therefore symmetric.

MATLAB Discussion

M.20 Transformations defined by Matrices, page 384

1. »map([cos(2*pi/3)   −sin(2*pi/3);   sin(2*pi/3)   cos(2*pi/3)])

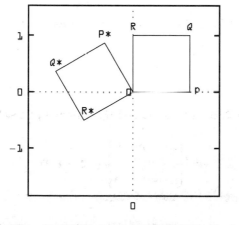

$O^* = (0, 0)$, $P^* = (-0.5000, 0.8600)$, $Q^* = (-1.3660, -1.5000)$, $R^* = (-0.8660, -1.5000)$
$A = [3 \quad 0;0 \quad 3]$, $O^* = (0, 0)$, $P^* = (3, 0)$, $Q^* = (3, 3)$, $R^* = (0, 3)$

5. **(a)** $O* = (0, 0)$, $P* = (3, 1)$, $Q* = (3, 5)$, $R* = (0, 4)$
6. **(a)** $A = [\cos(pi/2) \ -\sin(pi/2); \sin(pi/2) \ \cos(pi/2)]$, $B = [2 \ \ 0; 0 \ \ 2]$,
   $O^{\#} = (0, 0)$, $P^{\#} = (0, 2)$, $Q^{\#} = (-2, 2)$, $R^{\#} = (-2, 0)$
7. **(a)** $A = [3 \ \ 0; 0 \ \ 3]$, $B = [1 \ \ 2; 0 \ \ 1]$, $O^{\#} = (0, 0)$, $P^{\#} = (3, 0)$, $Q^{\#} = (9, 3)$, $R^{\#} = (6, 3)$

M.21 Fractals, page 385

4. **(a)** Dilation factor $= 0.8605230967$, angle of rotation $= -1.998°$

M.22 Kernel and Range, page 388

1. **(a)** Kernel $\{(-2, -3, 1)'\}$, range $\{(1, 0, 2)', (0, 1, -1)'\}$ **(c)** Kernel is zero vector, range $\{(1, 0, 0)', (0, 1, 0)', (0, 0, 1)'\}$

M.23 Pseudoinverse and Least Squares Curves, page 389

2. **(a)** $\begin{bmatrix} 0.3337 & 0 & 0.6667 \\ 0 & 0.1667 & -0.1667 \end{bmatrix}$ 5. **(a)** $x = 1.7333$, $y = 0.6$ 6. $y = 4x - 3$.

8. U.K. $N = 57.33673469Y + 1275.897959$
   U.S.A. $N = 5214.785714Y + 6.357142857$
   Japan $N = 2032.989796Y - 0.8163265306$
   Year 2000 is $y = 25$. Predictions are U.K., 2709; U.S.A., 5374; and Japan, 2013.

Chapter 8

Exercise Set 8.1, page 404

1. $\begin{bmatrix} 2 \\ -3 \end{bmatrix}$ 2. $\begin{bmatrix} 2 \\ 1 \end{bmatrix}$ 6. $\begin{bmatrix} 2 \\ -1 \\ 3 \end{bmatrix}$ 7. $\begin{bmatrix} \frac{16}{3} \\ \frac{20}{3} \\ -\frac{23}{3} \end{bmatrix}$ 10. $\begin{bmatrix} 2 \\ -4 \end{bmatrix}$ 12. $\begin{bmatrix} 3 \\ 2 \\ -4 \end{bmatrix}$

13. $\begin{bmatrix} 0 \\ 11 \\ -2 \end{bmatrix}$ 15. $\begin{bmatrix} 2 \\ \frac{5}{\sqrt{2}} \\ -\frac{3}{\sqrt{2}} \end{bmatrix}$ 16. $P = \begin{bmatrix} 2 & 1 \\ 3 & 2 \end{bmatrix}$, $\mathbf{u}_{B'} = \begin{bmatrix} 4 \\ 7 \end{bmatrix}$, $\mathbf{v}_{B'} = \begin{bmatrix} 5 \\ 7 \end{bmatrix}$, $\mathbf{w}_{B'} = \begin{bmatrix} 8 \\ 14 \end{bmatrix}$

18. $P = \begin{bmatrix} 1 & 2 \\ 1 & -3 \end{bmatrix}$, $\mathbf{u}_{B'} = \begin{bmatrix} 2 \\ -3 \end{bmatrix}$, $\mathbf{v}_{B'} = \begin{bmatrix} -3 \\ 7 \end{bmatrix}$, $\mathbf{w}_{B'} = \begin{bmatrix} 5 \\ 0 \end{bmatrix}$

20. $P = \begin{bmatrix} 2 & -3 \\ -3 & 4 \end{bmatrix}$, $\mathbf{u}_{B'} = \begin{bmatrix} -1 \\ 1 \end{bmatrix}$, $\mathbf{v}_{B'} = \begin{bmatrix} 6 \\ -9 \end{bmatrix}$, $\mathbf{w}_{B'} = \begin{bmatrix} 2 \\ -4 \end{bmatrix}$ 21. Transition matrix $\begin{bmatrix} 2 & -3 \\ -3 & 5 \end{bmatrix}$, $\mathbf{u}_{B'} = \begin{bmatrix} -19 \\ 31 \end{bmatrix}$

23. Transition matrix $\begin{bmatrix} -4 & 6 \\ 3 & -3 \end{bmatrix}$, $\mathbf{u}_{B'} = \begin{bmatrix} 24 \\ -15 \end{bmatrix}$

25. Transition matrix $\begin{bmatrix} \frac{1}{3} & 0 & 0 \\ 0 & 1 & 0 \\ 0 & \frac{1}{4} & \frac{1}{4} \end{bmatrix}$, coordinate vectors $\begin{bmatrix} 1 \\ 4 \\ 2 \end{bmatrix}$, $\begin{bmatrix} 2 \\ 0 \\ 1 \end{bmatrix}$, $\begin{bmatrix} 0 \\ 8 \\ 5 \end{bmatrix}$, $\begin{bmatrix} 1 \\ 4 \\ 2 \end{bmatrix}$

26. Transition matrix $\frac{1}{3}\begin{bmatrix} 3 & 0 \\ -2 & 1 \end{bmatrix}$, coordinate vectors $\begin{bmatrix} 3 \\ -1 \end{bmatrix}$, $\begin{bmatrix} 6 \\ -4 \end{bmatrix}$, $\begin{bmatrix} 6 \\ -1 \end{bmatrix}$, $\begin{bmatrix} 12 \\ -9 \end{bmatrix}$ 27. $T\left(\begin{bmatrix} a & 0 \\ 0 & b \end{bmatrix}\right) = (a, b)$

Exercise Set 8.2, page 413

1. $(1, -3)$ 2. 18 4. $11x - 6$ 6. $\begin{bmatrix} 2 & 4 \\ 3 & -1 \end{bmatrix}$, $24\mathbf{v}_1 + \mathbf{v}_2$ 8. $\begin{bmatrix} 1 & 3 & 1 \\ 1 & -2 & 2 \\ 1 & 0 & -1 \end{bmatrix}$, $4\mathbf{v}_1 - 11\mathbf{v}_2 + 8\mathbf{v}_3$

9. (a) $\begin{bmatrix} 1 & 0 & 0 \\ 0 & 0 & 1 \end{bmatrix}$, $(1, 3)$ (c) $\begin{bmatrix} 1 & 1 & 0 \\ 2 & -1 & 0 \end{bmatrix}$, $(3, 0)$ 10. (a) $\begin{bmatrix} 1 & 0 & 0 \\ 0 & 2 & 0 \\ 0 & 0 & 3 \end{bmatrix}$, $(-1, 10, 6)$ (c) $\begin{bmatrix} 1 & 0 & 0 \\ 0 & 0 & 0 \\ 0 & 0 & 0 \end{bmatrix}$, $(-1, 0, 0)$

11. $\begin{bmatrix} -2 & -2 & 1 \\ 1 & 3 & 2 \end{bmatrix}$, $(7, 3)$ 13. $\begin{bmatrix} 2 & 0 \\ 1 & 1 \end{bmatrix}$, $(-2, 2)$ 14. $\begin{bmatrix} 0 & 0 & 0 \\ 4 & 0 & 0 \\ 0 & -1 & 0 \end{bmatrix}$, $6x - 2$

16. (a) $\begin{bmatrix} 0 & 1 & 1 \\ 0 & 1 & -1 \\ 0 & 0 & 0 \end{bmatrix}$ (b) $\begin{bmatrix} 0 & 1 \\ 1 & 0 \\ 0 & 1 \end{bmatrix}$ 17. $\begin{bmatrix} 1 & 0 & 0 \\ 0 & 0 & 1 \end{bmatrix}$, $3x - 1$ 19. $\begin{bmatrix} 1 & 1 \\ 0 & -1 \end{bmatrix}$, $\frac{1}{2}\begin{bmatrix} 1 & 1 \\ 3 & -1 \end{bmatrix}$

20. $\begin{bmatrix} 2 & 0 \\ 1 & 1 \end{bmatrix}$, $\begin{bmatrix} 2 & 2 \\ 0 & 1 \end{bmatrix}$

22. (a) Yes, if $\dim(W) \geq \dim(V) - \dim(U)$; no, if $\dim(W) < \dim(V) - \dim(U)$
    (b) Let T be the linear transformation given by $T(1, 3, -1) = (0, 0)$, $T(1, 0, 0) = (1, 0)$ and $T(0, 1, 0) = (0, 1)$.
23. Let T be the linear transformation given by $T(2, -1) = (0, 0)$ and $T(1, 0) = (1, 0)$.
25. Let $B = \{\mathbf{u}_1, \mathbf{u}_2, \ldots, \mathbf{u}_n\}$ be a basis for U. $T(\mathbf{u}_i) = \mathbf{u}_i$, so if A is the matrix of T with respect to B, its ith column will have a 1 in the ith position and zeros elsewhere. Thus $A = I_n$.   28. No

Exercise Set 8.3, page 425

1. (a) $\begin{bmatrix} 7 & 13 \\ -2 & -3 \end{bmatrix}$ (c) $\begin{bmatrix} 92 & 53 \\ -156 & -90 \end{bmatrix}$ 2. (a) $\begin{bmatrix} 2 & 0 & 0 \\ 0 & 2 & 0 \\ 0 & 0 & 1 \end{bmatrix}$

3. (a) $\begin{bmatrix} 6 & 0 \\ 0 & 1 \end{bmatrix}$, transformation matrix: $\begin{bmatrix} 4 & -1 \\ 1 & 1 \end{bmatrix}$ (c) Cannot be diagonalized (d) $\begin{bmatrix} 2 & 0 \\ 0 & 3 \end{bmatrix}$, transformation matrix: $\begin{bmatrix} 1 & 1 \\ 2 & 1 \end{bmatrix}$

4. (a) $\begin{bmatrix} -2 & 0 \\ 0 & 3 \end{bmatrix}$, transformation matrix: $\begin{bmatrix} 2 & 1 \\ 1 & 1 \end{bmatrix}$ (c) Cannot be diagonalized (e) Cannot be diagonalized

5. (a) $\begin{bmatrix} 1 & 0 & 0 \\ 0 & 2 & 0 \\ 0 & 0 & 8 \end{bmatrix}$, transformation matrix: $\begin{bmatrix} 1 & 0 & 1 \\ -1 & 1 & 0 \\ 1 & 1 & 1 \end{bmatrix}$ (c) $\begin{bmatrix} 1 & 0 & 0 \\ 0 & 1 & 0 \\ 0 & 0 & 3 \end{bmatrix}$, transformation matrix: $\begin{bmatrix} 1 & 0 & 0 \\ 0 & 1 & 1 \\ 1 & 0 & 1 \end{bmatrix}$

6. (a) $\begin{bmatrix} 3 & 0 \\ 0 & -1 \end{bmatrix}$, transformation matrix: $\begin{bmatrix} \frac{1}{\sqrt{2}} & -\frac{1}{\sqrt{2}} \\ \frac{1}{\sqrt{2}} & \frac{1}{\sqrt{2}} \end{bmatrix}$ (b) $\begin{bmatrix} 15 & 0 \\ 0 & 10 \end{bmatrix}$, transformation matrix: $\begin{bmatrix} \frac{1}{\sqrt{5}} & \frac{-2}{\sqrt{5}} \\ \frac{2}{\sqrt{5}} & \frac{1}{\sqrt{5}} \end{bmatrix}$

7. (a) $\begin{bmatrix} 6 & 0 \\ 0 & -4 \end{bmatrix}$, transformation matrix: $\begin{bmatrix} \dfrac{1}{\sqrt{2}} & -\dfrac{1}{\sqrt{2}} \\ \dfrac{1}{\sqrt{2}} & \dfrac{1}{\sqrt{2}} \end{bmatrix}$ (c) $\begin{bmatrix} 10 & 0 \\ 0 & 0 \end{bmatrix}$, transformation matrix: $\begin{bmatrix} \dfrac{1}{\sqrt{10}} & -\dfrac{3}{\sqrt{10}} \\ \dfrac{3}{\sqrt{10}} & \dfrac{1}{\sqrt{10}} \end{bmatrix}$

8. (a) $\begin{bmatrix} 1 & 0 & 0 \\ 0 & 2 & 0 \\ 0 & 0 & -2 \end{bmatrix}$, transformation matrix: $\begin{bmatrix} 0 & \dfrac{1}{\sqrt{2}} & \dfrac{1}{\sqrt{2}} \\ 0 & \dfrac{1}{\sqrt{2}} & -\dfrac{1}{\sqrt{2}} \\ 1 & 0 & 0 \end{bmatrix}$

(b) $\begin{bmatrix} 0 & 0 & 0 \\ 0 & 15 & 0 \\ 0 & 0 & 6 \end{bmatrix}$, transformation matrix: $\begin{bmatrix} 0 & \dfrac{1}{\sqrt{3}} & \dfrac{2}{\sqrt{6}} \\ \dfrac{1}{\sqrt{2}} & -\dfrac{1}{\sqrt{3}} & \dfrac{1}{\sqrt{6}} \\ \dfrac{1}{\sqrt{2}} & \dfrac{1}{\sqrt{3}} & -\dfrac{1}{\sqrt{6}} \end{bmatrix}$

9. (a) $\begin{bmatrix} 872{,}576 & 807{,}040 \\ 807{,}040 & 872{,}576 \end{bmatrix}$ (d) $\begin{bmatrix} 32{,}768.5 & -32{,}767.5 \\ -32{,}767.5 & 32{,}768.5 \end{bmatrix}$

10. (a) $\begin{bmatrix} 64 & 0 & 0 \\ 0 & 64 & 0 \\ 0 & 0 & 1 \end{bmatrix}$ (b) $\begin{bmatrix} 258{,}309 & -250{,}533 & 250{,}533 \\ -250{,}533 & 254{,}421 & -254{,}421 \\ 250{,}533 & -254{,}421 & 254{,}421 \end{bmatrix}$

11. (a) $B' = \left\{ \left(\dfrac{1}{\sqrt{2}}, \dfrac{1}{\sqrt{2}} \right), \left(\dfrac{-1}{\sqrt{2}}, \dfrac{1}{\sqrt{2}} \right) \right\}$ The matrix representation of T relative to B' is $\begin{bmatrix} 6 & 0 \\ 0 & 2 \end{bmatrix}$. B' defines an $x'y'$ coordinate system, rotated 45° counterclockwise from the xy system. T is a scaling in the $x'y'$ system.

(c) $B' = \left\{ \left(\dfrac{2}{\sqrt{5}}, \dfrac{1}{\sqrt{5}} \right), \left(-\dfrac{1}{\sqrt{5}}, \dfrac{2}{\sqrt{5}} \right) \right\}$ The matrix representation of T relative to B' is $\begin{bmatrix} 10 & 0 \\ 0 & 5 \end{bmatrix}$. B' defines an $x'y'$ coordinate system, rotated counterclockwise through an angle θ from the xy system, where $\cos\theta = 2/\sqrt{5}$ and $\sin\theta = 1/\sqrt{5}$. T is a scaling in the $x'y'$ system.

12. (a) $B' = \{(1, 1), (2, 3)\}$ The matrix representation of T relative to B' is $\begin{bmatrix} 2 & 0 \\ 0 & -1 \end{bmatrix}$. B' defines an $x'y'$ coordinate system which is not rectangular. The transformation T is a scaling followed by a reflection about the x' axis.

(c) $B' = \{(2, 1), (1, 1)\}$ The matrix representation of T relative to B' is $\begin{bmatrix} 1 & 0 \\ 0 & -1 \end{bmatrix}$. B' defines an $x'y'$ coordinate system which is not rectangular. T is a reflection about the x' axis.

13. (a) If A and B are similar, then $B = C^{-1}AC$ so that $|B| = |C^{-1}AC| = |C^{-1}||A||C| = |A|$.
(c) If $B = C^{-1}AC$, then $\mathrm{Tr}(B) = \mathrm{Tr}(C^{-1}AC) = \mathrm{Tr}(C^{-1}(AC)) = \mathrm{Tr}((AC)C^{-1}) = \mathrm{Tr}(A(CC^{-1})) = \mathrm{Tr}(A)$.
(f) If $B = C^{-1}AC$, then $B^{-1} = (C^{-1}AC)^{-1} = C^{-1}A^{-1}C$, so B^{-1} is similar to A^{-1}.

15. (b) If $A = C^{-1}BC$, then $A + kI = C^{-1}BC + kC^{-1}IC = C^{-1}BC + C^{-1}kIC = C^{-1}(B + kI)C$.   17. Not unique

19. If $B = C^{-1}AC = C'AC$, then $B' = (C'AC)' = C'A'(C')' = C'A'C = C'AC = B$.

Exercise Set 8.4, page 437

1. (a) $[x \quad y]\begin{bmatrix} 1 & 2 \\ 2 & 2 \end{bmatrix}\begin{bmatrix} x \\ y \end{bmatrix}$ (c) $[x \quad y]\begin{bmatrix} 7 & -3 \\ -3 & -1 \end{bmatrix}\begin{bmatrix} x \\ y \end{bmatrix}$ (e) $[x \quad y]\begin{bmatrix} -3 & -\frac{7}{2} \\ -\frac{7}{2} & 4 \end{bmatrix}\begin{bmatrix} x \\ y \end{bmatrix}$

2. (a) $\dfrac{x'^2}{4} + \dfrac{y'^2}{6} = 1$. The graph is an ellipse with the lines $y = 2x$ and $x = -2y$ as axes.

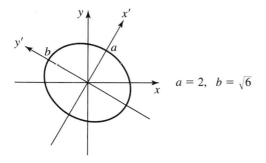

$a = 2, \quad b = \sqrt{6}$

(b) $\dfrac{x'^2}{3} + \dfrac{y'^2}{6} = 1$. The graph is an ellipse with the lines $y = x$ and $y = -x$ as axes.

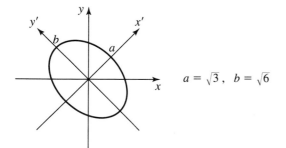

$a = \sqrt{3}, \quad b = \sqrt{6}$

3. (a) $a_n = \frac{1}{3}(2^n + 2^{n-1}) = 2^{n-1}$, $a_{10} = 2^9 = 512$ (c) $a_n = (-1)^{n-1}$, $a_{12} = (-1)^{11} = -1$

5. Mode 1: $\begin{bmatrix} x_1 \\ x_2 \end{bmatrix} = \cos(\alpha_1 t + \gamma_1)\begin{bmatrix} 1 \\ -1 \end{bmatrix}$, where $\alpha_1 = (3T/ma)^{1/2}$

   Mode 2: $\begin{bmatrix} x_1 \\ x_2 \end{bmatrix} = \cos(\alpha_2 t + \gamma_2)\begin{bmatrix} 1 \\ 1 \end{bmatrix}$, where $\alpha_2 = (T/ma)^{1/2}$

6. $\begin{bmatrix} x_1 \\ x_2 \end{bmatrix} = b_1 \cos(\alpha_1 t + \gamma_1)\begin{bmatrix} 1 \\ 2 \end{bmatrix} + b_2 \cos(\alpha_2 t + \gamma_2)\begin{bmatrix} 2 \\ -1 \end{bmatrix}$, where $\alpha_1 = (1/M)^{1/2}$, $\alpha_2 = (6/M)^{1/2}$

Chapter 8 Review Exercises, page 438

1. $\begin{bmatrix} 2 \\ 3 \end{bmatrix}$ 2. $\begin{bmatrix} 3 \\ -2 \\ 4 \end{bmatrix}$ 3. $\begin{bmatrix} 5 \\ -12 \\ -9 \end{bmatrix}$ 4. $P = \begin{bmatrix} 1 & 5 \\ 3 & 2 \end{bmatrix}$, $\mathbf{u}_{B'} = \begin{bmatrix} 8 \\ 11 \end{bmatrix}$, $\mathbf{v}_{B'} = \begin{bmatrix} -5 \\ 11 \end{bmatrix}$, $\mathbf{w}_{B'} = \begin{bmatrix} 9 \\ 14 \end{bmatrix}$

5. Transition matrix $\dfrac{1}{17}\begin{bmatrix} 4 & 7 \\ 11 & -2 \end{bmatrix}$, $\mathbf{u}_{B'} = \dfrac{1}{17}\begin{bmatrix} 23 \\ 42 \end{bmatrix}$ 6. $(-1, 32)$ 7. $\begin{bmatrix} 1 & 3 \\ 5 & -1 \\ -2 & 2 \end{bmatrix}$, $-7\mathbf{v}_1 + 13\mathbf{v}_2 - 10\mathbf{v}_3$

8. $\begin{bmatrix} 2 & 0 & 0 \\ 0 & -3 & 0 \end{bmatrix}$, $(2, -6)$ 9. $\begin{bmatrix} 1 & -1 & 0 \\ 0 & 0 & 2 \\ 0 & 0 & 0 \end{bmatrix}$, $3x^2 + 6x$ 10. $\begin{bmatrix} 3 & 0 \\ 1 & -1 \end{bmatrix}$, $\begin{bmatrix} -11 & -20 \\ 7 & 13 \end{bmatrix}$ 11. $\begin{bmatrix} 3 & 0 \\ 0 & 2 \end{bmatrix}$

12. $\begin{bmatrix} 3 & 0 \\ 0 & 2 \end{bmatrix}$, transformation matrix: $\begin{bmatrix} 1 & 1 \\ 2 & 1 \end{bmatrix}$ **13.** $\begin{bmatrix} 6 & 0 & 0 \\ 0 & 6 & 0 \\ 0 & 0 & 12 \end{bmatrix}$, transformation matrix: $\begin{bmatrix} \frac{1}{\sqrt{2}} & \frac{1}{\sqrt{3}} & \frac{1}{\sqrt{6}} \\ 0 & \frac{1}{\sqrt{3}} & -\frac{2}{\sqrt{6}} \\ -\frac{1}{\sqrt{2}} & \frac{1}{\sqrt{3}} & \frac{1}{\sqrt{6}} \end{bmatrix}$

14. The characteristic equation $\lambda^2 - (a + c)\lambda + ac - b^2 = 0$ has roots $\lambda = \frac{1}{2}(a + c \pm \sqrt{D})$ where $D = (a - c)^2 + 4b^2 \geq 0$ for all values of a, b, and c, so the roots are real.

15. $D = C^{-1}AC$, where the diagonal elements of D are all λ. Thus $D = \lambda I$ and $\lambda I = C^{-1}AC$, so that $A = CC^{-1}ACC^{-1} = C\lambda IC^{-1} = \lambda CIC^{-1} = \lambda CC^{-1} = \lambda I$.

16. $\dfrac{x'^2}{2} - \dfrac{y'^2}{6} = 1$. The graph is a hyperbola with axes $y = -2x$ and $x = 2y$.

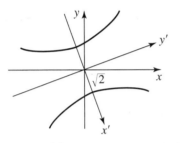

17. $a_n = \frac{1}{6}(13(-1)^{n-1} + 5^n)$, $a_{12} = 40{,}690{,}102$

Chapter 9

Exercise Set 9.1, page 446

1. **(b)** Yes **(d)** Yes   **2.** **(a)** Yes **(b)** No   **3.** $x_1 = 1$, $x_2 = 2$, $x_3 = 3$ **5.** No solution

6. $x_1 = 2 + 2r - s - t$, $x_2 = 1 + 3r + s - 3t$, $x_3 = r$, $x_4 = s$, $x_5 = t$

7. Eight operations are saved using Gaussian elimination.   **9.** Yes

Exercise Set 9.2, page 453

1. $x_1 = 1$, $x_2 = 4$, $x_3 = -1$ **3.** $x_1 = -2$, $x_2 = 1$, $x_3 = 1$ **4.** $x_1 = \frac{1}{2}$, $x_2 = \frac{1}{2}$, $x_3 = 2$

6. $x_1 = 1$, $x_2 = -1$, $x_3 = 7$ **7.** **(a)** $L = \begin{bmatrix} 1 & 0 & 0 \\ 1 & 1 & 0 \\ -1 & -1 & 1 \end{bmatrix}$ **(c)** $L = \begin{bmatrix} 1 & 0 & 0 \\ -\frac{1}{2} & 1 & 0 \\ -\frac{1}{5} & 1 & 1 \end{bmatrix}$

8. **(a)** $R2 - 2R1$, $R3 + 3R1$, $R3 - 5R2$ **(c)** $R3 - \frac{1}{7}R1$, $R3 + \frac{3}{4}R2$

9. $L = \begin{bmatrix} 1 & 0 & 0 \\ -2 & 1 & 0 \\ 1 & -1 & 1 \end{bmatrix}$, $U = \begin{bmatrix} 1 & 2 & -1 \\ 0 & 3 & 1 \\ 0 & 0 & -2 \end{bmatrix}$, $\mathbf{x} = \begin{bmatrix} -1 \\ 2 \\ 1 \end{bmatrix}$ **11.** $L = \begin{bmatrix} 1 & 0 & 0 \\ -1 & 1 & 0 \\ 3 & 8 & 1 \end{bmatrix}$, $U = \begin{bmatrix} 3 & -1 & 1 \\ 0 & 1 & 2 \\ 0 & 0 & -22 \end{bmatrix}$, $\mathbf{x} = \begin{bmatrix} 3 \\ 0 \\ 1 \end{bmatrix}$

14. $L = \begin{bmatrix} 1 & 0 & 0 \\ 2 & 1 & 0 \\ -4 & 3 & 1 \end{bmatrix}$, $U = \begin{bmatrix} -2 & 0 & 3 \\ 0 & 3 & -1 \\ 0 & 0 & 4 \end{bmatrix}$, $\mathbf{x} = \begin{bmatrix} 0 \\ 2 \\ 1 \end{bmatrix}$ **16.** $L = \begin{bmatrix} 1 & 0 & 0 \\ 2 & 1 & 0 \\ -5 & 4 & 1 \end{bmatrix}$, $U = \begin{bmatrix} 2 & -3 & 1 \\ 0 & 1 & 4 \\ 0 & 0 & -2 \end{bmatrix}$, $\mathbf{x} = \begin{bmatrix} -43 \\ -24 \\ 9 \end{bmatrix}$

18. $L = \begin{bmatrix} 1 & 0 & 0 \\ 2 & 1 & 0 \\ -1 & 1 & 1 \end{bmatrix}$, $U = \begin{bmatrix} 1 & 2 & -1 \\ 0 & 1 & 3 \\ 0 & 0 & 0 \end{bmatrix}$, $\mathbf{x} = \begin{bmatrix} 4 + 7r \\ -1 - 3r \\ r \end{bmatrix}$

19. $L = \begin{bmatrix} 1 & 0 & 0 \\ -1 & 1 & 0 \\ 2 & -3 & 1 \end{bmatrix}$, $U = \begin{bmatrix} 4 & 1 & -2 \\ 0 & 3 & 1 \\ 0 & 0 & 0 \end{bmatrix}$, no solution

20. $L = \begin{bmatrix} 1 & 0 & 0 & 0 \\ 1 & 1 & 0 & 0 \\ -1 & -1 & 1 & 0 \\ 0 & 2 & -\frac{1}{2} & 1 \end{bmatrix}$, $U = \begin{bmatrix} 1 & 1 & -1 & 2 \\ 0 & 2 & 3 & 0 \\ 0 & 0 & -2 & 8 \\ 0 & 0 & 0 & 2 \end{bmatrix}$, $\mathbf{x} = \begin{bmatrix} 1 \\ 1 \\ -1 \\ 2 \end{bmatrix}$

24. $\begin{bmatrix} 6 & -2 \\ 12 & 8 \end{bmatrix} = \begin{bmatrix} 1 & 0 \\ 2 & 1 \end{bmatrix}\begin{bmatrix} 6 & -2 \\ 0 & 12 \end{bmatrix} = \begin{bmatrix} 2 & 0 \\ 4 & 2 \end{bmatrix}\begin{bmatrix} 3 & -1 \\ 0 & 6 \end{bmatrix}$

26. 13 arithmetic operations are required to obtain U and no arithmetic operations are required for L. To solve $L\mathbf{y} = \mathbf{b}$ requires 6 operations and to solve $U\mathbf{x} = \mathbf{y}$ requires 9 operations.

Exercise Set 9.3, page 463

1. (a) 6 (c) 10   **2.** (a) 21 (c) 143 (e) $148\frac{1}{3}$

3. (a) 0.26×10^1, one significant digit less accurate (c) 0.217007×10^5, five significant digits less accurate (e) 0.36×10^2, two significant digits less accurate

4. (a) 0.3×10^1. The solution of the system of equations can have one fewer significant digits of accuracy than the elements of the matrix.

5. 4606   **6.** $k < -99$, $0.96 < k < 1$, $1 < k < 1.0417$, $k > 96.9583$ **8.** (b) $1 = \|AA^{-1}\| \le \|A\|\|A^{-1}\| = c(A)$

10. If A is a diagonal matrix with diagonal elements a_{ii}, then A^{-1} is a diagonal matrix with diagonal elements $1/a_{ii}$. $c(A) = \|A\|\|A^{-1}\| = (\max\{a_{ii}\})(1/\min\{a_{ii}\})$.

11. $x_1 = 6$, $x_2 = -2$, and $x_3 = 1$

12. $x_1 = -5.46$, $x_2 = 4.93$, and $x_3 = -1.2$ $\left(\text{exact solution: } x_1 = -\frac{82}{15}, x_2 = \frac{74}{15}, x_3 = -\frac{6}{5}\right)$

15. $x_1 = 1.000$, $x_2 = -1.000$, and $x_3 = -29.999$ (exact solution: $x_1 = 1$, $x_2 = -1$, and $x_3 = -30$)

17. $x_1 = -2.500$, $x_2 = 0.500$, and $x_3 = -4.500$

Exercise Set 9.4, page 469

1. $x = 2.40$, $y = 0.40$, and $z = 2.00$ **3.** $x = 4.51$, $y = 5.24$, and $z = 2.68$ $\left(\text{exact solution: } x = \frac{185}{41}, y = \frac{215}{41}, z = \frac{110}{41}\right)$

5. $x = 7.00$, $y = -0.43$, and $z = 2.29$ $\left(\text{exact solution: } x = 7, y = -\frac{3}{7}, z = \frac{16}{7}\right)$

6. $x = 2.29$, $y = 1.98$, $z = 5.94$, and $w = 2.33$ $\left(\text{exact solution: } x = \frac{7310}{3199}, y = \frac{6330}{3199}, z = \frac{19010}{3199}, w = \frac{7440}{3199}\right)$

Exercise Set 9.5, page 478

1. Dominant eigenvalue $= 8$, dominant eigenvector $= r\begin{bmatrix} 1 \\ 0 \\ -1 \end{bmatrix}$

3. Dominant eigenvalue $= 6$, dominant eigenvector $= r\begin{bmatrix} 1 \\ 0 \\ 1 \end{bmatrix}$ **5.** Dominant eigenvalue $= 6$, dominant eigenvector $= r\begin{bmatrix} 0 \\ 0 \\ 1 \\ 1 \end{bmatrix}$

7. Dominant eigenvalue $= 84.311$, dominant eigenvector $= r\begin{bmatrix} 1 \\ 0.011 \\ -0.011 \\ 0.040 \end{bmatrix}$

9. Eigenvalues are 10 and 1(dim 2) with corresponding eigenvectors $\begin{bmatrix} 1 \\ 1 \\ 0.5 \end{bmatrix}$ and $\begin{bmatrix} -1 \\ 0.8 \\ 0.4 \end{bmatrix}$

11. Eigenvalues are 12, 2(dim 2) with corresponding eigenvectors $\begin{bmatrix} 1 \\ 0 \\ 1 \end{bmatrix}$ and $\begin{bmatrix} 0 \\ 1 \\ 0 \end{bmatrix}$

12. Eigenvalues are 10, 6, and -2(dim 2) with corresponding eigenvectors $\begin{bmatrix} 1 \\ 0 \\ 0 \\ 1 \end{bmatrix}$, $\begin{bmatrix} 0 \\ 1 \\ -1 \\ 0 \end{bmatrix}$, and $\begin{bmatrix} -\frac{1}{3} \\ 1 \\ 1 \\ \frac{1}{3} \end{bmatrix}$

14. **(a)** 0.191, 0.309, 0.309, 0.191   **(b)** $\frac{1}{8}, \frac{1}{8}, \frac{1}{4}, \frac{1}{4}, \frac{1}{8}, \frac{1}{8}$

Chapter 9 Review Exercises, page 479

1. $x_1 = 2$, $x_2 = -1$, $x_3 = 1$ 2. $L = \begin{bmatrix} 1 & 0 & 0 \\ \frac{3}{2} & 1 & 0 \\ -\frac{2}{7} & 4 & 1 \end{bmatrix}$ 3. $x_1 = 1$, $x_2 = 3$, $x_3 = 4$

4. Number of multiplications $= \dfrac{2n^3 + 3n^2 + n}{6}$, number of additions $= \dfrac{2n^3 - 3n^2 + n}{6}$, total number of

   operations $= \dfrac{2n^3 + n}{3}$

   If the elements on the diagonal of L are all 1, number of multiplications $= \dfrac{n^3 - n}{3}$ and total number of

   operations $= \dfrac{4n^3 - 3n^2 - n}{6}$.

5. 0.3135216×10^4, four significant digits less accurate

6. **(a)** 5, one significant digit less accurate   **(b)** 12.5, two significant digits less accurate

7. $c(A) = \|A\|\|A^{-1}\| = \|A^{-1}\|\|A\| = \|A^{-1}\|\|(A^{-1})^{-1}\| = c(A^{-1})$ 8. No

9. **(a)** $x_1 = \frac{1}{50}$, $x_2 = 0$, and $x_3 = 100$ **(b)** $x_1 = -12$, $x_2 = -1$, and $x_3 = 5000$ 10. $x = 4.80$, $y = 1.20$, and $z = 3.00$

11. Dominant eigenvalue $= 11.716$, dominant eigenvector $= r\begin{bmatrix} 0.652 \\ 0.997 \\ 1 \end{bmatrix}$

12. Eigenvalues are 6, 4, 0 with corresponding eigenvectors $\begin{bmatrix} 1 \\ 2 \\ -1 \end{bmatrix}$, $\begin{bmatrix} 1 \\ 0 \\ 1 \end{bmatrix}$, $\begin{bmatrix} -1 \\ 1 \\ 1 \end{bmatrix}$

MATLAB Discussion

M.24 *LU* Decomposition, page 480

1. **(a)** $L = \begin{bmatrix} 1 & 0 \\ 0.3333 & 1 \end{bmatrix}$, $U = \begin{bmatrix} 3 & 5 \\ 0 & 0.3333 \end{bmatrix}$ 2. **(a)** $X = \begin{bmatrix} 1.0000 & 1.0000 \\ 1.0000 & -1.0000 \\ 2.0000 & 1.0000 \end{bmatrix}$

M.25 Condition Number of a Matrix, page 482

1. **(a)** 21 **(c)** 65   **4.** $y = 25 - 26\frac{1}{3}x + 9x^2 - \frac{2}{3}x^3$; $c(A) = 504$

M.26 Jacobi and Gauss-Seidel Iterative Methods, page 483

1. **(a)** Solution $x_1 = 1$, $x_2 = 0.5$, $x_3 = 0$. With a tolerance of 0.0001, the Jacobi method takes 11 iterations to converge; the Gauss-Seidel method takes 4 iterations.

Chapter 10

Exercise Set 10.1, page 494

1. Maximum is 24 at (6, 12).   **3.** Maximum is 20 at every point on the line segment joining (5, 0) and (3, 4).
5. Maximum is 6 at (1, 2).   **6.** Minimum is -350 at (100, 50).   **8.** Minimum is -4 at (4, 0).
9. Maximum profit is $6400 when the company manufactures 400 of $C1$ and 300 of $C2$.
11. Maximum profit is $2400 when the company ships x refrigerators from town X and $120 - 2x$ refrigerators from town Y, where $0 \le x \le 30$.
13. Maximum income is $5040 when the tailor makes 24 suits and 32 dresses.
15. Maximum is 54 cars moved from B to C, with 146 cars moved from B to D, 66 cars from A to C and 34 cars from A to D.
17. Maximum is 150 tons when 50 tons of X and 100 tons of Y are manufactured.
18. Minimum is 56 cents when 7 oz. of item M and no item N is served.
20. **(a)** Maximum profit is $800 when the shipper carries no packages for Pringle and 2000 packages for Williams.

Exercise Set 10.2, page 501

1. Maximum is 24 when $x = 6$ and $y = 12$.   **2.** Maximum is 4 when $x = 4$ and $y = 0$.
4. Maximum is 1600 when $x = 160$ and $y = 0$.   **6.** Maximum is 22,500 when $x = 0$, $y = 100$, and $z = 50$.
7. No maximum   **10.** Maximum is $1740 when 30 X, no Y, and 120 Z are produced.
12. Maximum profit is $7200 when 600 washing machines are transported from A to P and none are transported from B or C to P.
13. Maximum profit is $2080 if no Aspens are manufactured and the number of Alpines manufactured plus one-half the number of Cubs manufactured is 260.
15. Minimum is -5 at $x = 3$, $y = 1$.   **16.** Minimum is $-\frac{15}{4}$ at $x = 0$, $y = 0$, $z = \frac{15}{4}$.

Exercise Set 10.3, page 509

1. Optimal solution is $f = 24$ at $x = 6$, $y = 12$.
3. Optimal solution is $f = 6$ at $x = 0$, $y = 3$, $z = 0$ (x and z are nonbasic variables).
6. Optimal solution is $f = 50$ at $x = 0$, $y = 15$, $z = 5$, $w = 0$ (x and w are nonbasic variables).

Chapter 10 Review Exercises, page 509

1. Maximum value is 13 at (2, 3).   **2.** Maximum is 48 at every point on the line segment joining (8, 0) and (4, 6).
3. Minimum is 0 at (0, 0).
4. Maximum profit is $27,000 when 30 acres of strawberries and 10 acres of tomatoes are planted.
5. Maximum volume is 1456 when 16 lockers of type X and 20 of type Y are purchased.

6. Maximum is 40 when $x = 20$, $y = 0$, and $z = 0$.

7. Maximum daily profit is \$240 when the number of X tables plus one-half the number of Y tables finished is 30.

8. Minimum is -2 for all x, y, z where $z = 0$ and $y = 1 + x/2$, with $0 \leq x \leq 4/3$.

MATLAB Discussion

M.27 The Simplex Method in Linear Programming, page 511

1. $f = 24$ at $x = 6$, $y = 12$ 3. $f = 16.5$, at $x = 1.25$, $y = 2.25$

Index

Index of Applications